SOLID-PHASE ORGANIC SYNTHESIS

SOLID-PHASE ORGANIC SYNTHESIS

Concepts, Strategies, and Applications

Edited by

Patrick H. Toy

Yulin Lam

WILEY

A JOHN WILEY & SONS, INC., PUBLICATION

Published by John Wiley & Sons, Inc., Hoboken, New Jersey
Published simultaneously in Canada

Library of Congress Cataloging-in-Publication Data:

Solid-phase organic synthesis : concepts, strategies, and applications / edited by Patrick H. Toy, Yulin Lam. – 1st ed.
p. cm.
Includes bibliographical references and index.
ISBN 978-0-470-59914-3 (hardback)
1. Solid-phase synthesis. I. Toy, Patrick H. II. Lam, Yulin.
QD262S584 2012
547'.2–dc23

2011019936

Printed in the United States of America

10 9 8 7 6 5 4 3 2 1

CONTENTS

Part II Applications 205

PREFACE

Merrifield first introduced the concept of solid-phase peptide synthesis nearly half a century ago, and since then the use of heterogeneous materials to facilitate synthesis has evolved and become widespread in many contexts. For example, the automated solid-phase synthesis of oligomeric biomolecules, such as polypeptides and polynucleotides, has become the standard methodology for the production of such compounds.

The aim of this book is to highlight the state of the art regarding the use of a solid material to support and thereby facilitate organic synthesis. The book is divided into two parts: Part I introduces some general concepts and strategies, while Part II presents specific examples of the solid-phase synthesis of various classes of organic molecules. Since the field regarding solid-phase synthesis of polypeptides and polynucleotides is very mature and well understood, these topics are not included in this book. However, since the solid-phase synthesis of oligosaccharides is not yet routine and straightforward, a chapter on this subject is presented.

Part I includes chapters focusing on the linker groups used to attach the synthesis substrate to the solid support, colorimetric tests that identify the presence of functional groups, combinatorial synthesis (especially interesting due to its historical perspective), and diversity-oriented synthesis. These contributions showcase solid-phase synthesis that is currently used to facilitate the discovery of new molecular functionality. Finally, a chapter highlighting how using a support can change or increase reaction selectivity closes this part. Part II includes chapters on general asymmetric synthesis on a support, various strategies for heterocycle synthesis (including one focusing on the use of microwave heating), synthesis of radioactive organic molecules, dyes, dendrimers, and, last but not least, oligosaccharides.

It is hoped that this book will serve as an introduction and a starting point for those new to this field and interested in using concepts and techniques of solid-phase synthesis. As already mentioned, the application of this technology in the synthesis of small, nonoligomeric organic molecules is relatively underdeveloped compared to other applications, and thus new minds and different perspectives can help to advance this field.

Patrick H. Toy
Yulin Lam

ACKNOWLEDGMENTS

We would like to thank all the contributors to this book. Their time is very valuable, and thus their generosity in working on this book is priceless. We also wish to thank Tracy Yuen-Sze But, Julia Hermeke, and Jinni Lu for their editorial assistance.

CONTRIBUTORS

Prasad Appukkuttan, Laboratory for Organic & Microwave-Assisted Chemistry, Department of Chemistry, Katholieke Universiteit Leuven, Leuven, Belgium

Baburaj Baskar, Department of Chemical Biology, Max Planck Institute of Molecular Physiology, Dortmund, Germany

Young-Tae Chang, Department of Chemistry, National University of Singapore, Singapore

Wenteng Chen, College of Pharmaceutical Sciences, Zhejiang University, Hangzhou, P. R. China

Alexander Deiters, Department of Chemistry, North Carolina State University, Raleigh, NC, USA

Koichi Fukase, Department of Chemistry, Osaka University, Osaka, Japan

Warren R. J. D. Galloway, Department of Chemistry, University of Cambridge, Cambridge, United Kingdom

Marc Giulianotti, Torrey Pines Institute for Molecular Studies, San Diego, CA, USA

Young-Dae Gong, Department of Chemistry, Dongguk University, Seoul, South Korea

Kerem Goren, School of Chemistry, Tel Aviv University, Tel Aviv, Israel

Hyung-Ho Ha, Department of Chemistry, National University of Singapore, Singapore

Kirsi Harju, Division of Pharmaceutical Chemistry, University of Helsinki, Helsinki, Finland

Jan Hlaváč, Department of Organic Chemistry, Palacky University, Olomouc, Czech Republic

Raphaël Hoareau, Department of Radiology, University of Michigan, Ann Arbor, MI, USA

Richard A. Houghten, Torrey Pines Institute for Molecular Studies, San Diego, CA, USA

Brett M. Ibbeson, Department of Chemistry, University of Cambridge, Cambridge, United Kingdom

Albert Isidro-Llobet, Department of Chemistry, University of Cambridge, Cambridge, United Kingdom

Jonghoon Kim, Department of Chemistry, Seoul National University, Seoul, South Korea

Viktor Krchňák, Department of Chemistry and Biochemistry, University of Notre Dame, Notre Dame, IN, USA

Kamal Kumar, Department of Chemical Biology, Max Planck Institute of Molecular Physiology, Dortmund, Germany

Yulin Lam, Department of Chemistry, National University of Singapore, Singapore

Sung Chan Lee, Department of Chemistry, National University of Singapore, Singapore

Taeho Lee, Center for High Throughput Synthesis Platform Technology, Korea Research Institute of Chemical Technology, Daejeon, South Korea

Zhi Li, College of Pharmaceutical Sciences, Zhejiang University, Hangzhou, P. R. China

Vaibhav P. Mehta, Laboratory for Organic & Microwave-Assisted Chemistry, Department of Chemistry, Katholieke Universiteit Leuven, Leuven, Belgium

Kieron M. G. O'Connell, Department of Chemistry, University of Cambridge, Cambridge, United Kingdom

Cornelius J. O'Connor, Department of Chemistry, University of Cambridge, Cambridge, United Kingdom

Seung Bum Park, Department of Chemistry, Seoul National University, Seoul, South Korea

Moshe Portnoy, School of Chemistry, Tel Aviv University, Tel Aviv, Israel

Peter J. H. Scott, Department of Radiology, University of Michigan, Ann Arbor, MI, USA

Chai Hoon Soh, Department of Chemistry, National University of Singapore, Singapore

Miroslav Soural, Department of Organic Chemistry, Palacky University, Olomouc, Czech Republic

David R. Spring, Department of Chemistry, University of Cambridge, Cambridge, United Kingdom

Katsunori Tanaka, Department of Chemistry, Osaka University, Osaka, Japan

Yan Teng, Department of Chemistry, University of Hong Kong, Hong Kong, P. R. China

Patrick H. Toy, Department of Chemistry, University of Hong Kong, Hong Kong, P. R. China

Erik Van der Eycken, Laboratory for Organic & Microwave-Assisted Chemistry, Department of Chemistry, Katholieke Universiteit Leuven, Leuven, Belgium

Marc Vendrell, Department of Chemistry, National University of Singapore, Singapore

Jari Yli-Kauhaluoma, Division of Pharmaceutical Chemistry, University of Helsinki, Helsinki, Finland

Douglas D. Young, Department of Chemistry, North Carolina State University, Raleigh, NC, USA

Yongping Yu, College of Pharmaceutical Sciences, Zhejiang University, Hangzhou, P. R. China

Part I

CONCEPTS AND STRATEGIES

1

LINKER STRATEGIES IN MODERN SOLID-PHASE ORGANIC SYNTHESIS

Peter J. H. Scott

1.1 INTRODUCTION

The vast array of linker units available to the modern solid-phase organic chemist is impressive and allows a lot of exciting chemistry to be carried out using solid-phase techniques.[1–11] Linker units are molecules that possess a functional group that is used to attach substrates to a solid support and can release them at a later date upon treatment with the appropriate "cleavage cocktail." With this in mind, linker units have long been regarded as solid-supported protecting groups. Moreover, linker units are frequently lengthy molecules, which improve reactivity by holding substrates away from the polymer matrix to create a pseudo-solution-phase environment. Typically, linker units are conveniently categorized by the functionality left at the "cleavage site" in the target molecule (Scheme 1.1). Initially, following the late Prof. Merrifield's original investigations into preparing peptides on solid supports, solid-phase organic synthesis (SPOS) focused on strategies for preparing peptides and oligonucleotides. This focus was, in part, due to the relative simplicity of peptide chemistry that meant it could easily be adapted for use with solid-phase techniques. Moreover, the ease of automating peptide chemistry allowed straightforward preparation of multiple target peptides in parallel and signaled the beginning of combinatorial chemistry. Many of the classical linker units developed during this period (1960s–1990s) still represent some of the most widely used linker units in use today and an overview of these linker strategies is presented in Section 1.2. When employing a classical linker unit, a common (typically polar) functionality, that was the site of

Solid-Phase Organic Synthesis: Concepts, Strategies, and Applications, First Edition.
Edited by Patrick H. Toy and Yulin Lam.

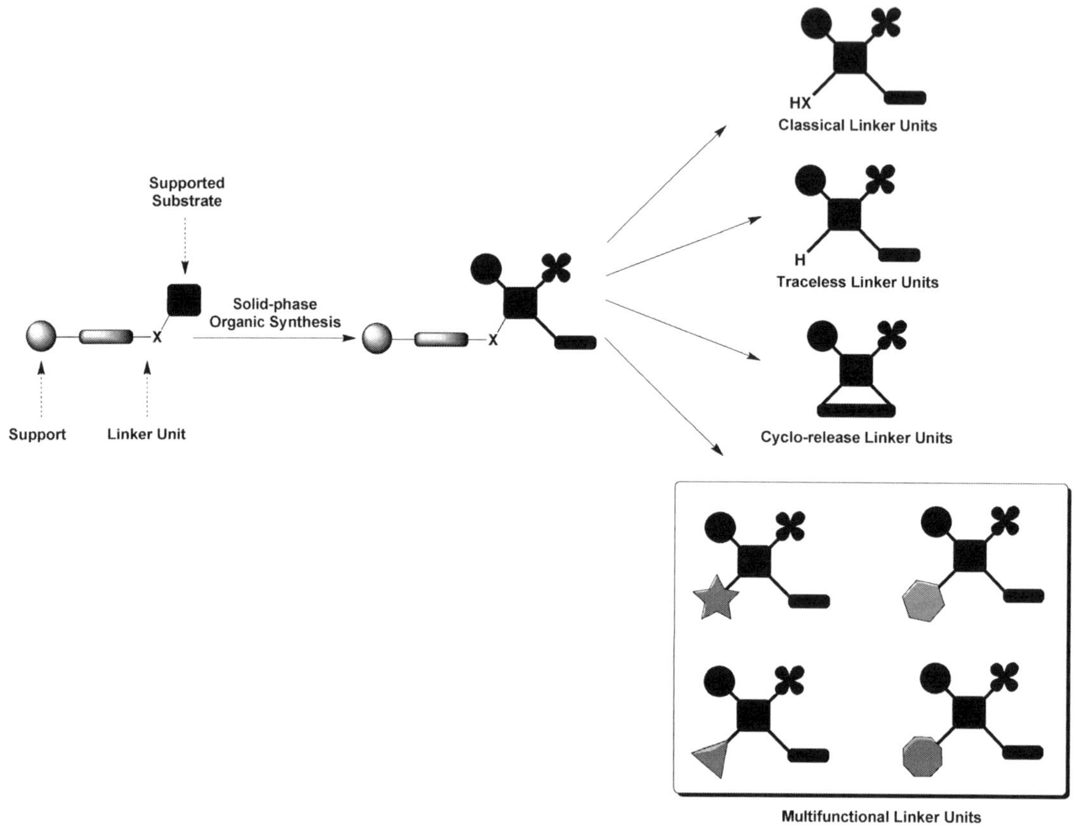

Scheme 1.1. Classification of modern linker units.

attachment of the molecule to the solid support, remains following cleavage of the target molecule.

In the 1990s, the use of solid-phase organic synthesis experienced an explosion in popularity. This was driven by the advent of combinatorial chemistry, as well as strategies such as split-and-mix, which exploited techniques for automating thousands of reactions in a parallel fashion. A combination of the ability to (i) run many solid-phase reactions in parallel using fritted tubes and commercial shakers, (ii) drive reactions to completion using excess reagents, and (iii) easily purify reactions by simple washing and filtration made SPOS particularly attractive to the combinatorial chemists.

Out of the combinatorial chemistry boom came the framework for modern solid-phase organic synthesis. While a lot of the early work with SPOS focused on reliable and relatively straightforward peptide coupling reactions, the ambitious library syntheses of the 1990s required access to a much more extensive array of solid-phase reactions. That decade saw initial strides made in adapting many well-known solution-phase reactions for use in the solid-phase arena, development that continues to the present day,[12–27] and a move beyond peptide and nucleotide chemistry toward preparation of small molecule libraries on solid phase.

In time, the vast libraries of combinatorial chemistry have given way to the smaller designed libraries of diversity-oriented synthesis (DOS). Rather than preparing multimillion compound libraries in the hope of finding new lead scaffolds, DOS concentrates on preparing smaller "focused" libraries for lead development.[28] Moreover, with the advent of chemical genetics, the interest in generating diverse compound libraries to explore chemical space has become a significant synthetic objective in its own right. These fields of research, in combination with related computational methods, are receiving much attention in the continuing quest to discover new biologically active compounds in chemical space. Reflecting these new challenges, the science of linker design in the last two decades has predominantly focused on the design and synthesis of new multifunctional linker units. Unlike the classical linker units described above that use a common cleavage cocktail for all members of a library, multifunctional linker units maximize diversity by using the cleavage step to incorporate additional structural variation into compound libraries. This final class of linker unit is discussed in Section 1.3.

1.2 CLASSICAL LINKER STRATEGIES

1.2.1 Acid and Base Cleavable Linker Units

In 1963, Merrifield reported the first example of a synthesis carried out using substrates immobilized on an insoluble polymer support.[29] In this work, the polymer Merrifield used was a chloromethylated copolymer of styrene and divinylbenzene, a polymer support that now bears his name. This polymer was functionalized with a benzyloxy group and then Merrifield was able to construct the L-Leu-L-Ala-Gly-Val tetrapeptide **1** by exploiting the Cbz protecting group strategy (Scheme 1.2). Cleavage from the ester linker unit was achieved using sodium hydroxide or a methanolic solution of sodium methoxide to generate the salt of the carboxylic acid **2** or methyl ester **3**, respectively. This work in itself represents a simple and straightforward example of multifunctional cleavage that will be discussed further later.

Reflecting this genesis in solid-phase peptide and oligonucleotide synthesis, many early linker units typically possessed a polar functional group (e.g., OH, CO_2H, NH_2, SH) that was used to attach substrates to a solid support. These linker units can be classified according to whether acidic or basic conditions are required for cleavage of target molecules, and many of them are still employed routinely in twenty-first century solid-phase organic synthesis. The main advantage is that cleavage of substrates from acid and base labile linker units can be readily achieved using mild conditions. Moreover, target molecules can frequently be isolated in sufficient purity by simple evaporation of volatile cleavage reagents.

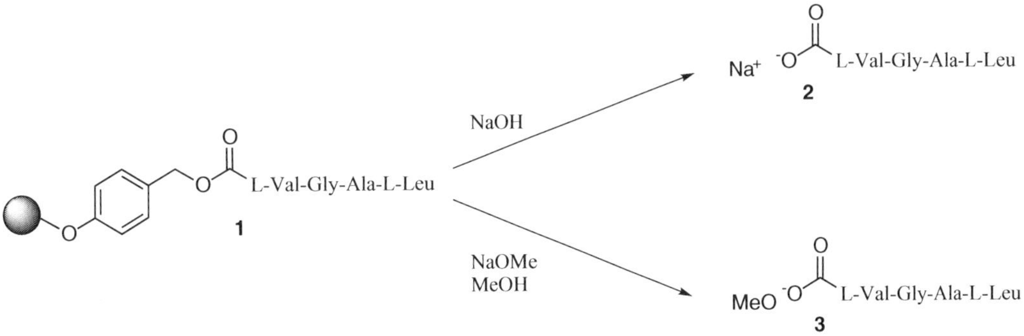

Scheme 1.2. Merrifield's original solid-phase synthesis of a tetrapeptide.

Two of the most used acid labile linker units, illustrated in Table 1.1, are the hydroxymethylphenyl linker unit reported by Wang (Table 1.1, Entry 1)[30] and the aminomethylphenyl linker (Table 1.1, Entries 2 and 3), stabilized by an additional anisole unit, developed by Rink.[31] The *para*-oxygen atom in the Wang linker has a stabilizing effect on the cation generated upon treatment with acid, allowing cleavage to be achieved using 50% trifluoroacetic acid (TFA) in dichloromethane(DCM). As a comparison, greater stabilization of the intermediate carbocation occurs in the presence of the *ortho*- and *para*-methoxy groups of the Rink linker. This enhanced stability allows cleavage to be realized under comparatively milder conditions (e.g., 0.1–50% TFA/DCM). For example, trichloroacetylurea was cleaved from the Rink linker using 5% TFA in DCM (Table 1.1, Entry 2).[32] The use of methoxy groups to afford greater stability to the intermediate carbocation has also been exploited in development of the hyperlabile SASRIN (or HMPB) linker (Table 1.1, Entry 4).[33–36] Similar to the Rink linker, cleavage of substrates from the SASRIN linker can be achieved using mild conditions such as 0.1–1% TFA.[36]

Other acid labile linker units from which substrates can be cleaved by treatment with TFA include the trityl linker units. Typically, the chlorotrityl linker unit is employed (Table 1.1, Entries 5 and 6) because it is more stable than the parent trityl linker unit, although cleavage can still be achieved using 1% TFA or acetic acid.[38,55] One advantage of using trityl linker units over, for example, the benzyl linker units discussed above is that the steric bulkiness of the trityl group makes the linkage more stable against nucleophilic bases. On the other hand, however, this steric bulkiness can cause problems if the substrate to be attached is itself a large molecule. In such situations, steric interference can reduce loading efficiency and should be taken into account before employing the trityl linker unit.

All these TFA labile linker units are well suited to SPOS using the Fmoc protective group strategy. Thus, Fmoc protecting group manipulations can be achieved using piperidine without risk of cleaving the acid labile substrate. However, if a SPOS design plans to use the Boc peptide strategy (i.e., TFA deprotection of Boc groups throughout the synthesis), then a linker unit from which substrates are cleavable with TFA is clearly not suitable. Apart from the TFA labile linkers previously discussed, a number of other acid labile linker units have been reported, allowing the ability to tailor the choice of linker unit to a given synthetic application. If it is necessary to employ the Boc protective group strategy throughout SPOS, one might select the phenylacetamide (PAM) linker (Table 1.1, Entry 7). Substrates are attached to the PAM linker through an ester linkage that is reasonably stable toward TFA. After completion of SPOS, the target molecule can then be cleaved using a stronger acid such as HF or HBr.[40]

Note that many of the linker units described above are available in multiple forms, allowing a range of substrates to be attached and cleaved. A discussion of all these related linker units is outside the scope of this chapter, but Kurosu has written a comprehensive review.[56] By way of example, multiple versions of the Rink (Table 1.1, Entries 2 and 3) and trityl linker units (Table 1.1, Entries 5 and 6)[39] are commercially available and can be selected according to the desired substrate. However, beyond these general linker units, there are also examples of substrate-specific linker units. For example, the benzhydrylamine (BHA, Table 1.1, Entry 8)[57] and Sieber (Table 1.1, Entry 9)[42–44] linkers find widespread use as acid labile carboxamide linker units, while the DHP (Table 1.1, Entry 10)[45–48] and silyl linker units (e.g., Table 1.1, Entry 11) can be used to attach alcohols to polymer supports.[58]

A number of linker units designed specifically for immobilization of amines have also been developed. One noticeable example exploits the versatility of the 9-phenylfluorenyl-9-yl group (PHFl). The PHFl group has previously been used as a protecting group for amines and was adapted into a linker unit by Bleicher (Table 1.1, Entry 12).[51] Cleavage from this

TABLE 1.1. Common Acid Cleavable Linker Units

	Linker	Cleavage Conditions	Product	References
1	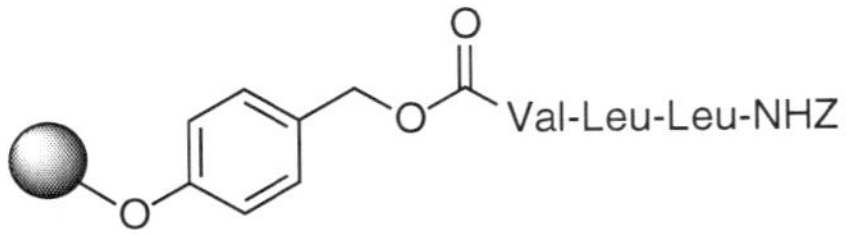	50% TFA/ DCM	HO_2C-Val-Leu-Leu-NHZ (yield: 69%)	30
2	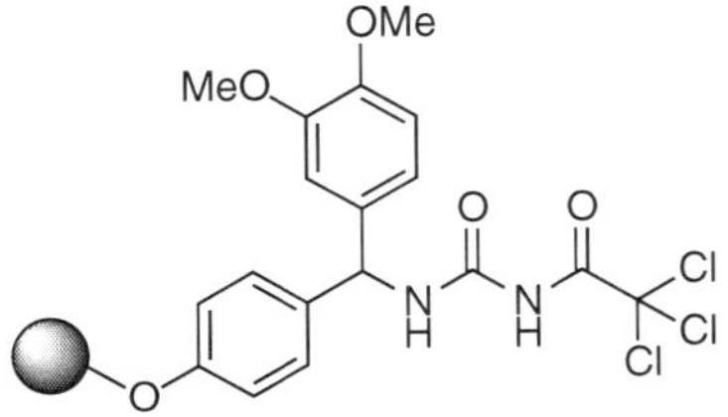	5% TFA/ DCM	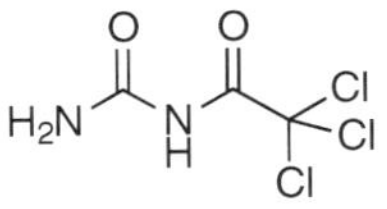 (yield: 72%)	32
3	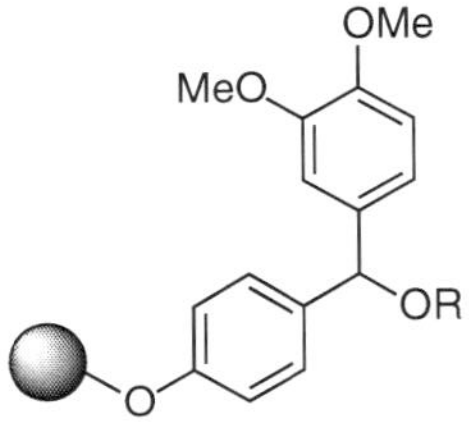	5% TFA	ROH	37
4	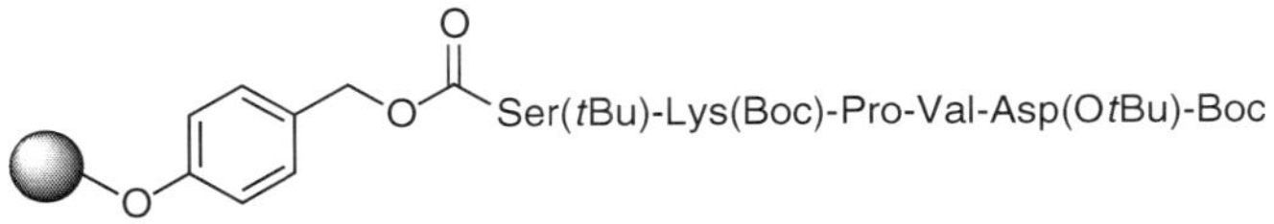	1% TFA/ DCM	Boc-Asp(O*t*Bu)-Val-Pro-Lys (Boc)-Ser(*t*Bu)-OH (crude yield: 90%, purity: 78%)	36
5	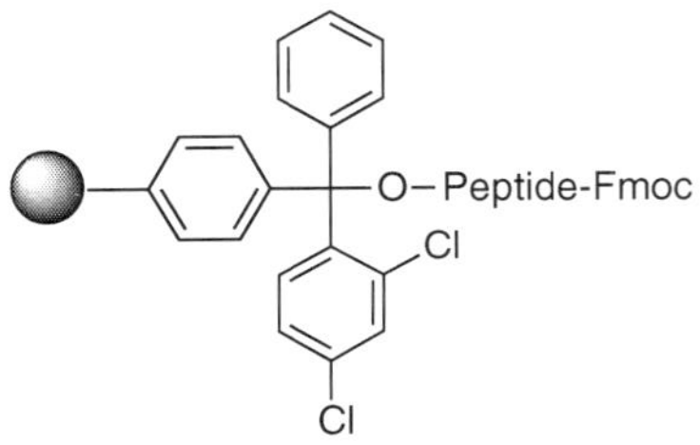	2:2:6 AcOH: TFE:DCM	Peptide (seven examples, 86–100% yield, 69–89% purity)	38

(*Continued*)

TABLE 1.1. (*Continued*)

	Linker	Cleavage Conditions	Product	References
6	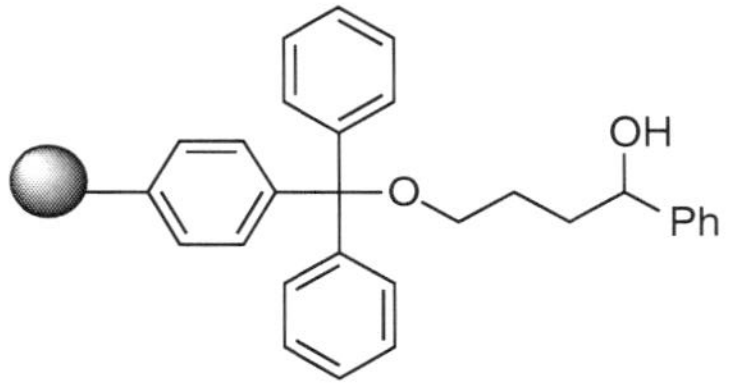	1 M HCl	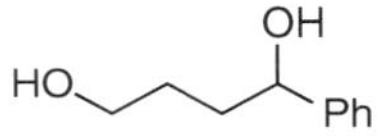 (yield: 32%)	39
7	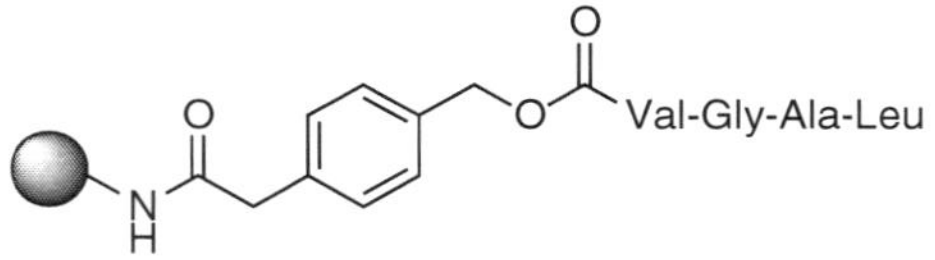	(a) 16% HBR in 1:1 AcOH: TFA; (b) 9:1 HF:anisole	Leu-Ala-Gly-Val (a: 35% yield, b: 87% yield)	40
8	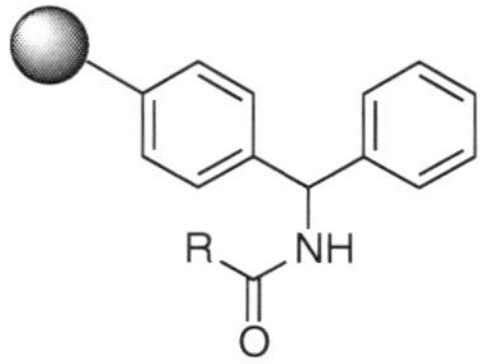	HF, 0°C	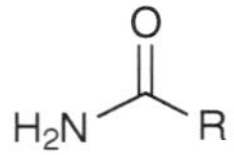	41
9	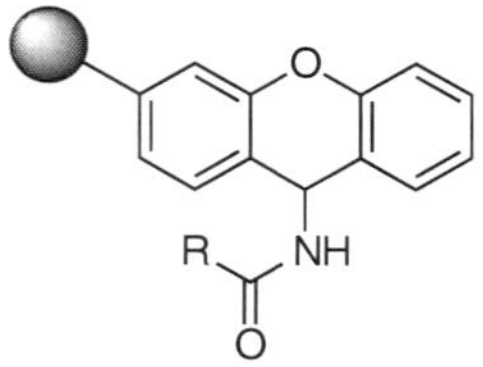	2% TFA	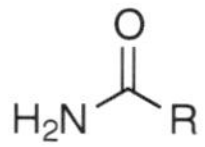	42–44
10	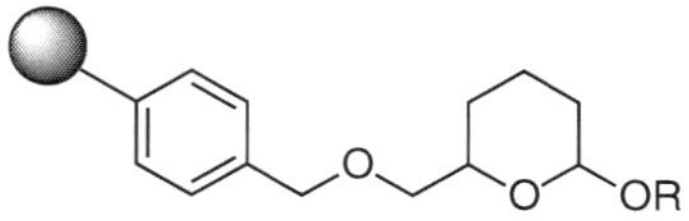	TFA–water (95:5)	ROH	45–48

11	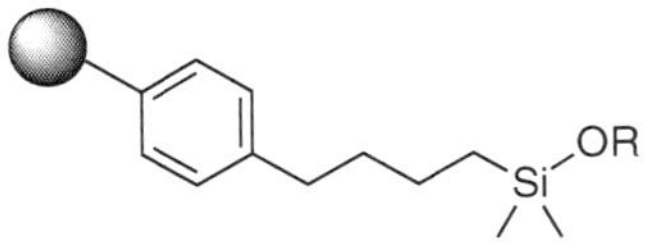	Aq HF–Pyr; TBAF, THF; AcOH, THF, H_2O	ROH	49–50
12	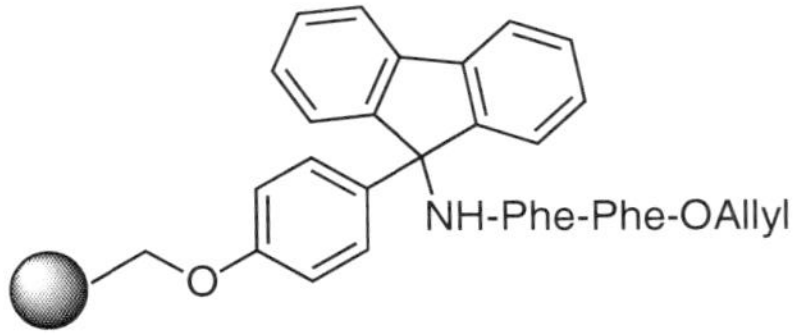	20% TFA, 2% Et_3SiH	H_2N-Phe-Phe-*O*-allyl (crude yield: 83%, purity: >95%)	51
13	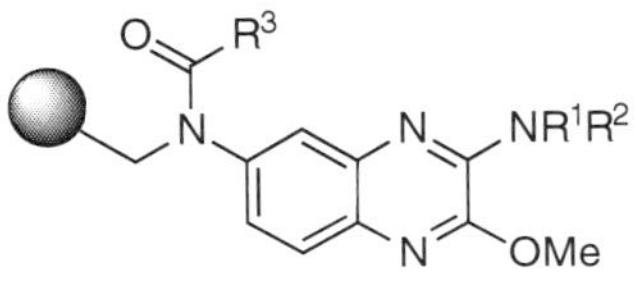	50% TFA/ DCM	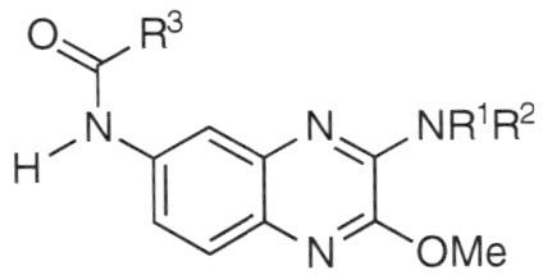 34 examples (yield: 40–89%)	52
14	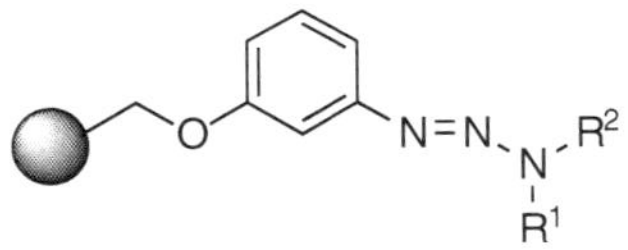	10% TFA/ DCM	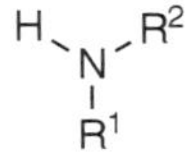	53
15	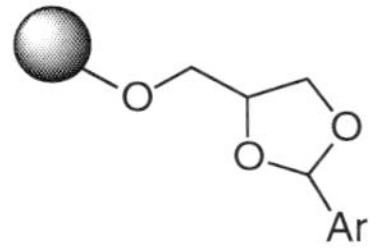	1:1 dioxane: dilute HCl	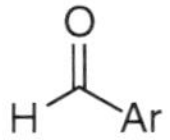	54

linker unit can be achieved by treating with 50% TFA in DCM with addition of Et_3SiH as a scavenger. Other linker units for amines have been developed based on supported aldehydes or diazonium salts. For example, amino substrates can be loaded onto aldehyde linker units (e.g., the AMEBA linker unit, Table 1.1, Entry 13) via reductive amination and subsequently cleaved upon treatment with TFA in the presence of Et_3SiH.[52,59–62] In the case of supported diazonium salts, amino substrates are loaded and form a triazene bond with the polymer support (Table 1.1, Entry 14).[53,63] The triazene linkage is stable against a range of reaction conditions but can be conveniently cleaved to release functionalized amines upon treatment with 10–50% TFA.

Finally, linker units based on common protecting groups for carbonyl groups have also been adapted for use as linker units. Acetals represent one of the most commonly employed carbonyl protecting groups. Thus, if carbonyl-containing substrates are reacted with resin-bound diols, they can be immobilized through an acetal linkage (Table 1.1, Entry 15).[54] Upon completion of SPOS, acid cleavage reforms the carbonyl group and liberates the target molecule. Note that the converse approach is also true and diols can be loaded onto resin-bound carbonyls.[64]

In the event that acid labile linker units are unacceptable for a given SPOS series because, for example, acid-sensitive substrates are being employed, alternatives are available, including mild enzyme cleavable linkers[65] or an equally extensive array of base labile linker units.[66] Merrifield employed such a base labile ester-based linker unit in his original peptide synthesis, as shown in Scheme 1.2. Thus, treating with sodium hydroxide or sodium methoxide cleaved the peptide as the carboxylic acid **2** or methyl ester **3**, respectively. Since its inception by Merrifield, saponification of substrates attached to support via ester linkages as a cleavage strategy has continued to find application in SPOS (Table 1.2). For example, saponification can be used to cleave carboxylic acids and esters (Table 1.2, Entries 1 and 2),[67,68] or alcohols, including nucleosides (Table 1.2, Entry 3)[69], by tailoring the linker and cleavage conditions accordingly.

Aminolysis, in which the nucleophile promoting cleavage is an amine, has also been widely used as a SPOS cleavage strategy. Aminolysis can be used to prepare, for example, amides using ester linkers (Table 1.2, Entry 4)[70] and sulfonamides using sulfonate ester linkers (Table 1.2, Entry 5)[70] and can be enhanced by Lewis acid catalysis (Table 1.2, Entry 6)[71]. Reflecting the importance of ureas in biologically active molecules, urea library synthesis has also been investigated using SPOS. One example of note is the preparation of tetrasubstituted ureas reported by Janda and coworkers (Table 1.2, Entry 7), in which aminolytic cleavage was used to introduce the third and fourth points of diversity.[72] Brown also developed amino cleavage for allyl phenyl ethers (Table 1.2, Entry 8).[73] This was a palladium-mediated process that Brown used to prepare a range of allylic amines. Other amines are also viable cleavage reagents for substrates attached through ester (and ester-like) linkages. For example, hydrazones (Table 1.2, Entry 9)[74] and hydroxylamines (Table 1.2, Entry 10)[75] have both been employed in nucleophilic cleavage cocktails.

Apart from the common heteroatom-derived nucleophiles described, cleavage with other nucleophiles is also possible. For example, reductive cleavage with hydride sources is possible. For ester-linked substrates, Kurth et al. reported an example in which substituted propane-1,3-diols were prepared (Table 1.2, Entry 11).[76] In related work, Chandrasekhar et al. prepared tertiary alcohols by treating an ester-linked substrate with excess Grignard reagent (Table 1.2, Entry 12).[77] If, however, it is desirable to prepare the carbonyl derivative (and not reduce all the way to the corresponding alcohol), then Weinreb-type linker units can be used (Table 1.2, Entries 13 and 14).[78] Treatment of substrates attached via such linkers

TABLE 1.2. Common Base Cleavable Linker Units

	Linker	Cleavage Conditions	Product	References
1	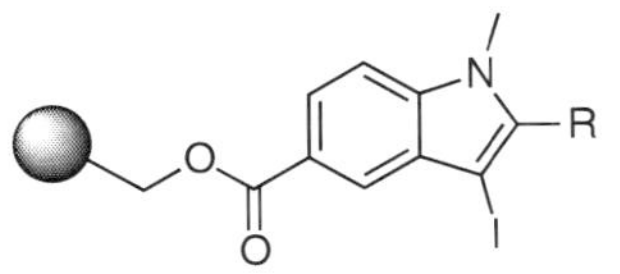	NaOMe, MeOH: THF (1:4)	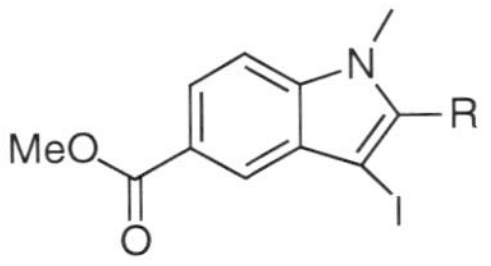 (yield: 0–99%, seven examples)	67
2	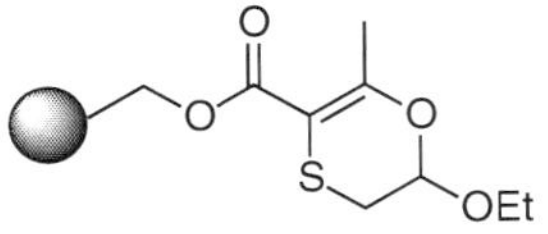	NaOMe, MeOH: THF, rt	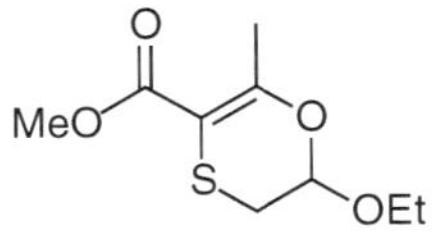 (yield: 41%)	68
3	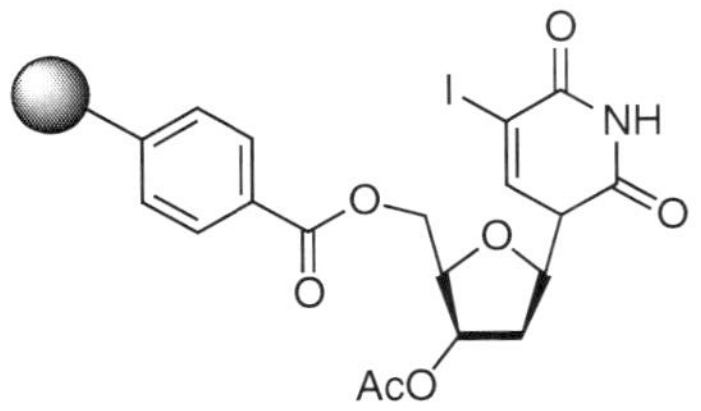	MeONa, MeOH: dioxane	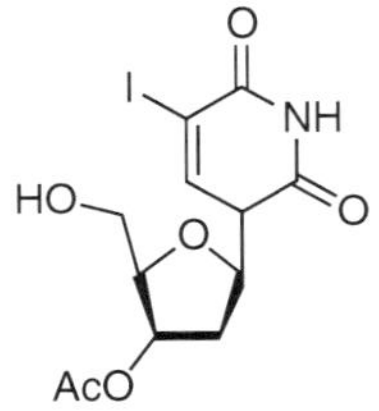 (yield: 73%)	69
4	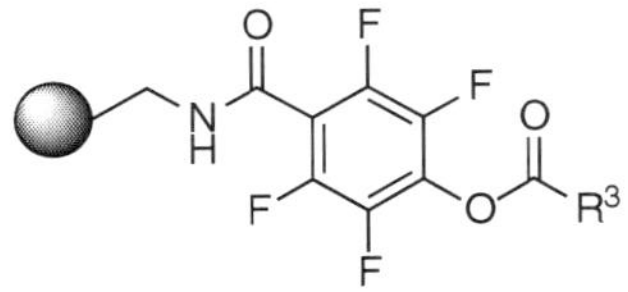	R^1R^2NH, DMF, rt	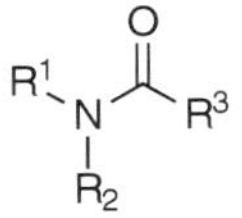 (yield: 88–100%)	70

(*Continued*)

TABLE 1.2. *(Continued)*

	Linker	Cleavage Conditions	Product	References
5	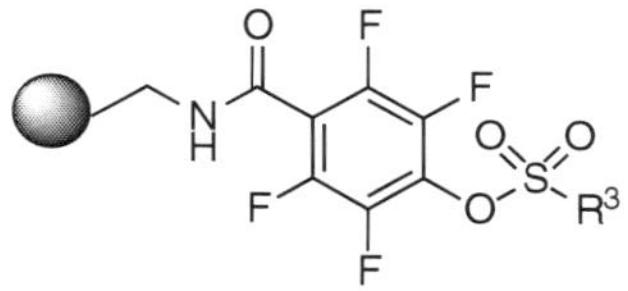	R^1R^2NH, DMF, rt	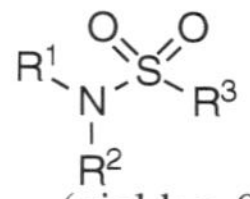 (yield: >91%)	70
6	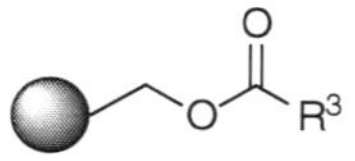	R^1R^2NH, $AlCl_3$, DCM, rt	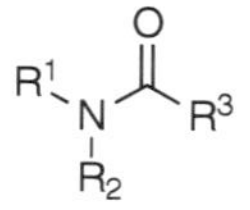 (yield: 11–74%)	71
7	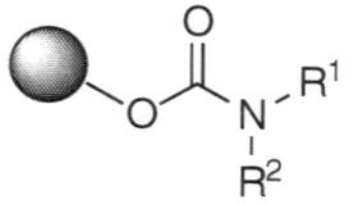	R^3R^4NH, $AlMe_3$, toluene, rt	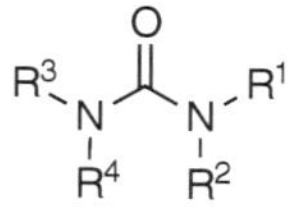 (yield: 62–100%)	72
8	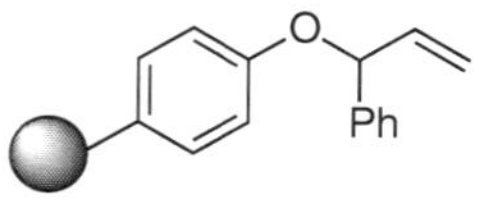	R^1R^2NH, Pd catalyst	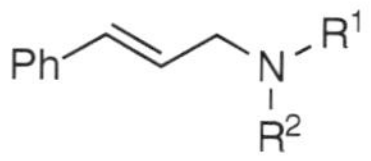 (yield: 30–77%)	73
9	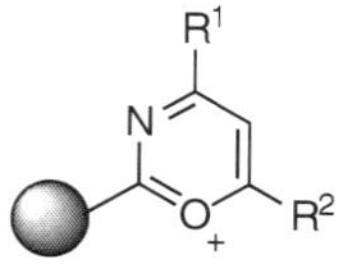	R^3-NH-NH_2	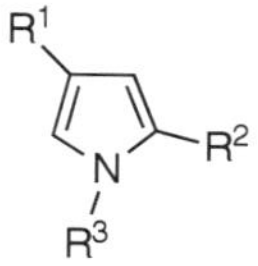 (yield: 14-25%)	74

10	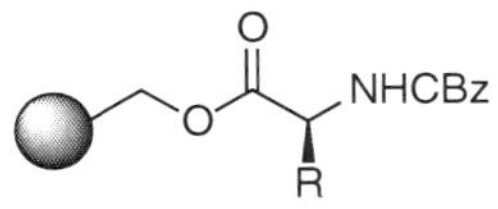	Aq NH_2OH, THF	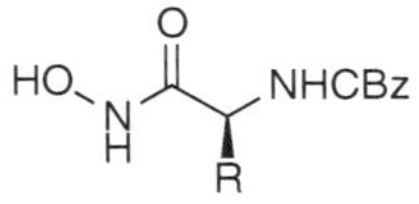	75
11	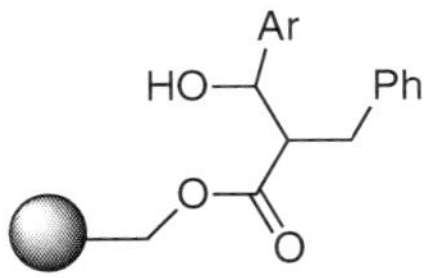	DIBAL-H	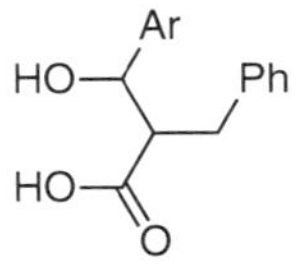 (yield: 28%)	76
12	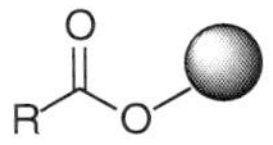	R^1-MgX, THF ether	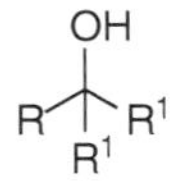	77
13	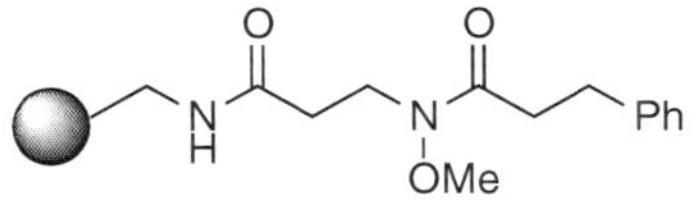	$LiAlH_4$	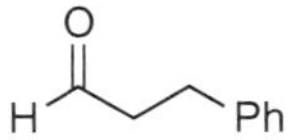	78
14		RMgCl	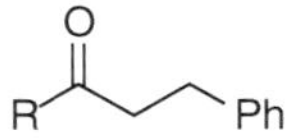 (yield: 23–77%, two examples)	78

with LAH will provide the corresponding aldehyde (Table 1.2, Entry 13), while cleavage with a Grignard reagent will give the ketone products (Table 1.2, Entry 14).

1.2.2 Cyclorelease Linker Units

As described previously, cleavage of substrates from acid and base labile linker units can be readily achieved using mild conditions. However, a significant drawback of such linker units, which has limited their application in more general organic synthesis, is that a common polar functional group is introduced into every target molecule in a compound library during cleavage. While the polar functional group might be an integral feature of the library, frequently it is not, and the presence of such functionality can greatly affect the desired (biological) activity and must be removed. The removal of such functionality can be far from straightforward, and so research aimed at developing linker units, which avoid this issue, has been extensive.

The first solution proposed to address this problem involved the use of cyclorelease linker units (Scheme 1.1).[79–81] When using such linker units to prepare cyclic species, the cyclization and cleavage steps are combined (cyclative cleavage), offering a number of benefits. First, there is no residual polar functionality left behind in the SPOS cleavage product and, second, only the final linear precursor is capable of undergoing cyclorelease. This will provide cleaved products of higher purity than other SPOS protocols because failed intermediates or other synthetic by-products generated (despite the use of excess reagents) are unable to cyclize and remain attached to the polymer support following cleavage. For example, Pavia and coworkers showed that treatment of immobilized amino acid **4** with acid did not result in cleavage of the substrate.[82] However, reaction with isocyanate provided urea **5** that on treatment with 6 M HCl cyclized to form the hydantoin **6** (Scheme 1.3). Unreacted amino acid remained bound to the polymer support providing hydantoin products in high purity.

Pavia's linker unit exploits amide or urea bond formation with concomitant displacement of the solid support, which is by far the most common approach for achieving cyclative cleavage. The first example of such an approach was Marshall's preparation of cyclic dipeptides, as shown in Table 1.3, Entry 1.[83] Besides this, such classical cyclization C–N bond forming reactions have been used to prepare ambitious synthetic targets using SPOS, including hydantoins (Table 1.3, Entry 2),[84] ureas (Table 1.3, Entry 3),[85] phthalimides

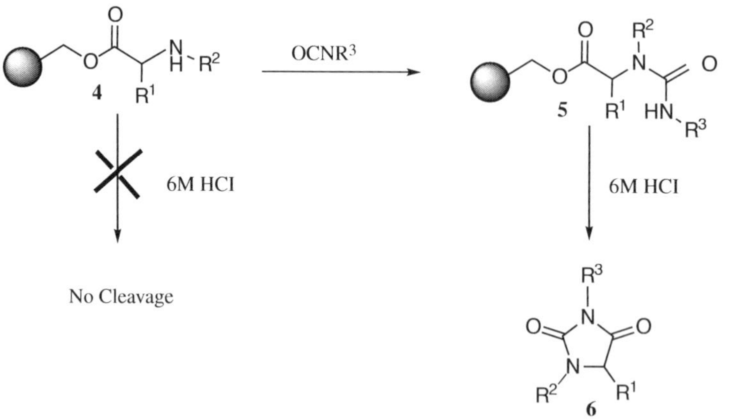

Scheme 1.3. Pavia's cyclorelease linker unit.

TABLE 1.3. Common Cyclorelease Linker Units

	Linker	Cleavage Conditions	Product	References
1	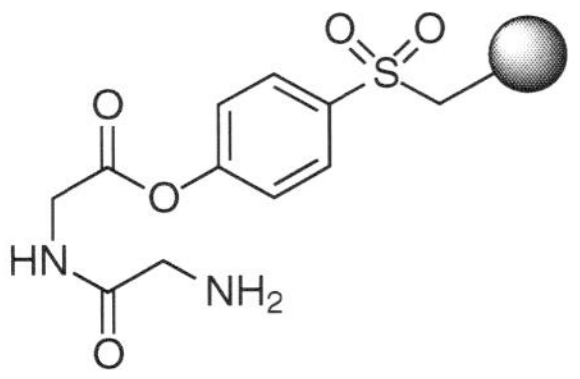	2% Et_3N/DMF	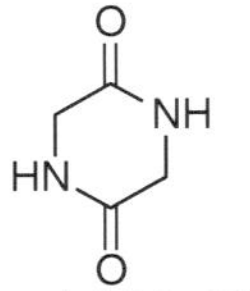 (yield: 63%)	83
2	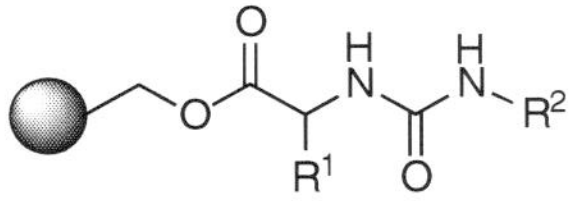	Et_3N, THF/DMF (4:1), μW	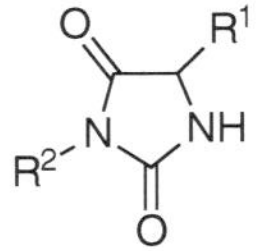	84
3	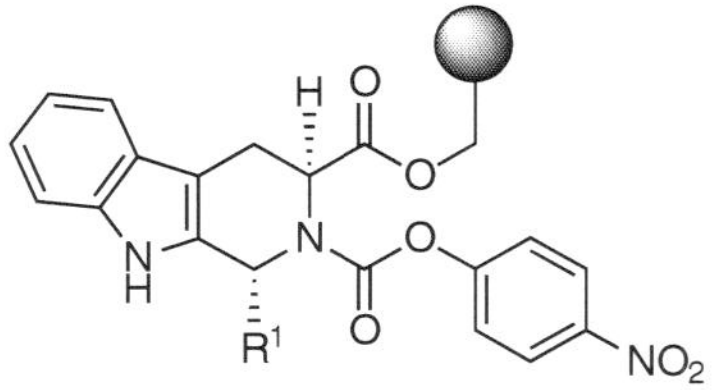	R^2-NH_2, Et_3N, DMF, 90°C	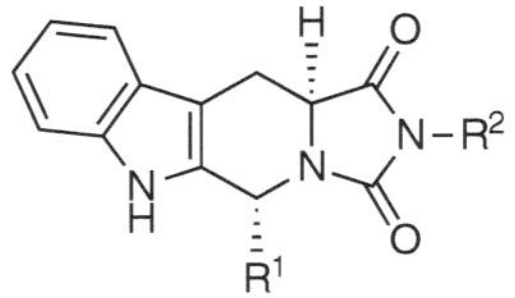 (yield: 15–44%, 12 examples)	85
4	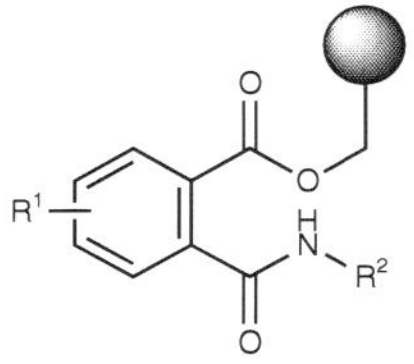	DMF, μW, 170°C	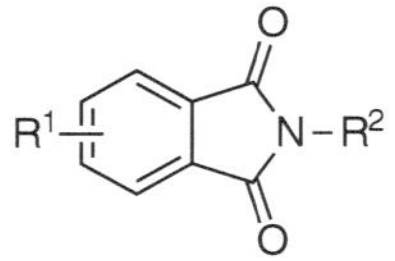 (crude yield: 51–102%, 14 examples)	86

(*Continued*)

TABLE 1.3. (*Continued*)

	Linker	Cleavage Conditions	Product	References
5	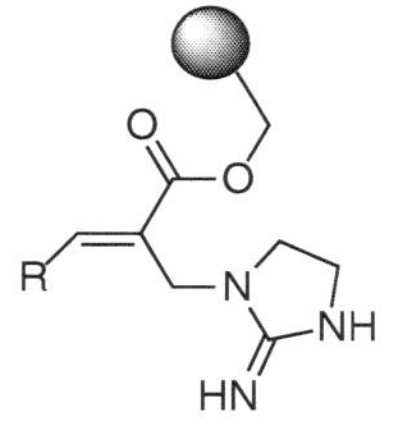	20% Et_3N, $CHCl_3$, reflux	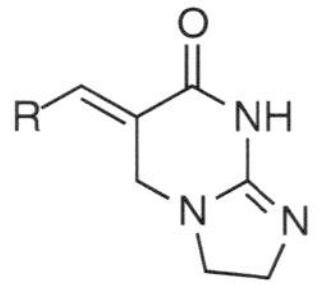 (yield: 52–94%, 15 examples)	87
6	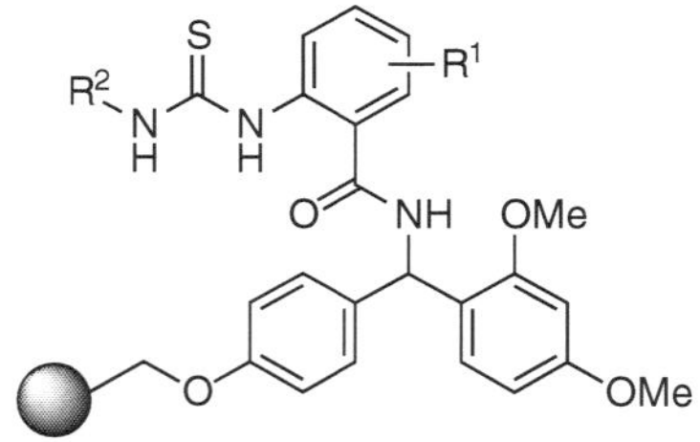	(i) R^3R^4NH, DIC; (ii) 10% AcOH, DCM	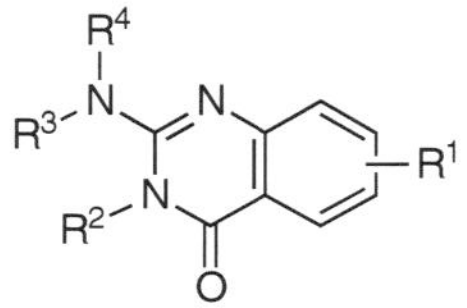 (yield: 50–67%, 12 examples)	88
7	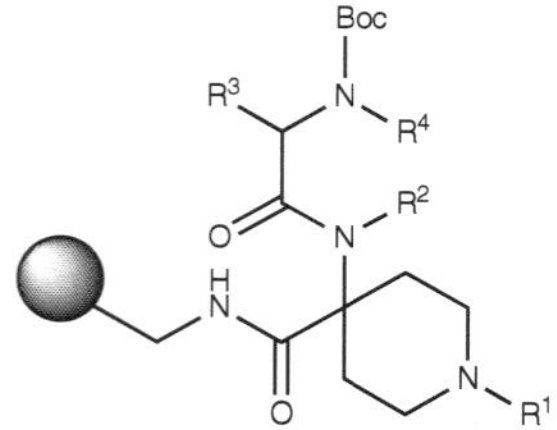	(i) 25% TFA/DCM; (ii) AcOH, toluene	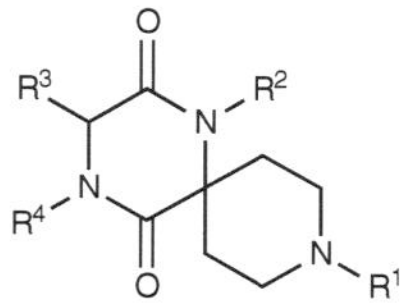	89
8	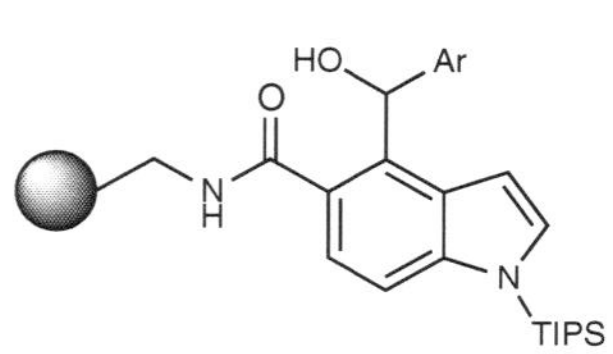	Toluene, 90°C	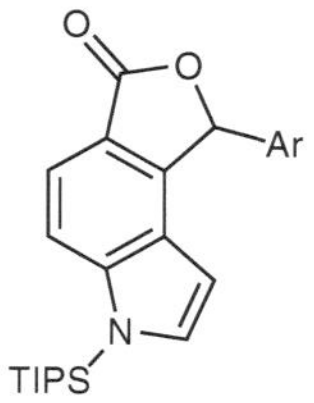	90

9	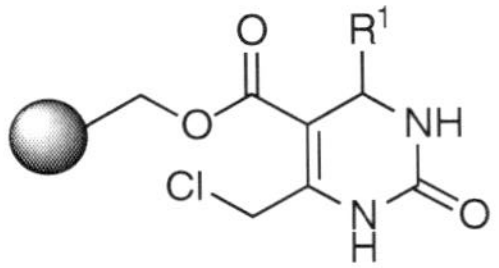	R^2NH_2, DMF, μW, 150–250°C	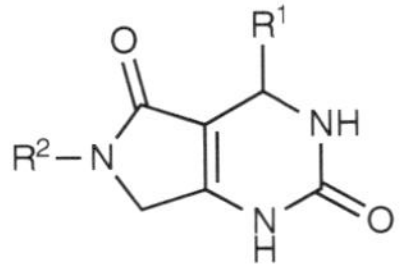 (yield: 25–55%, 22 examples)	91
10		DMF, μW, 150–250°C	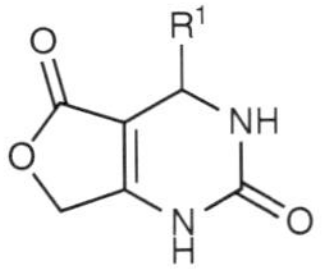 (yield: 10–77%, 12 examples)	91
11	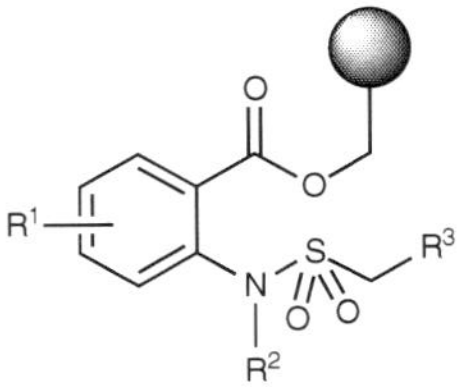	NaH, DMF	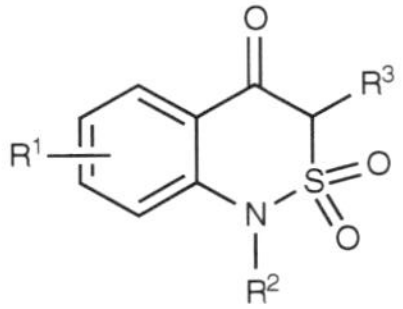 (yield: 0–52%, 28 examples)	92
12	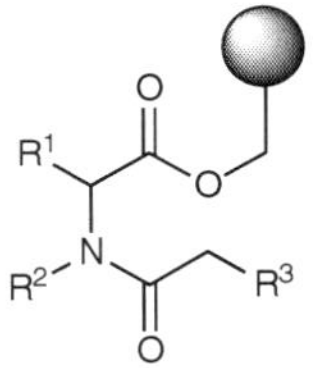	Bu_4NOH, THF/MeOH	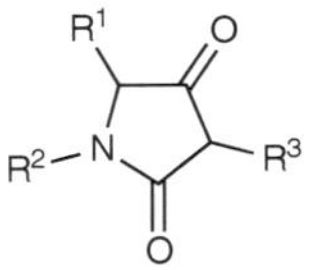 (yield: 68–91%, 11 examples)	93
13	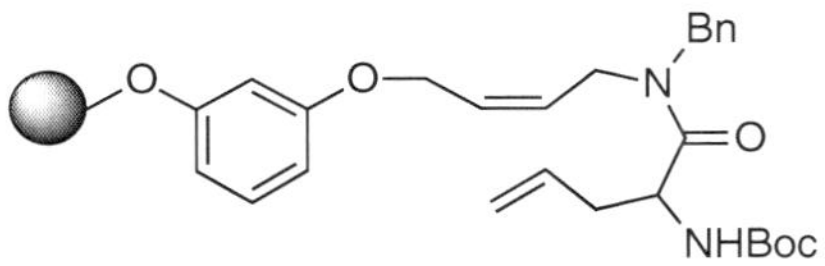	Grubbs I, 1-octene	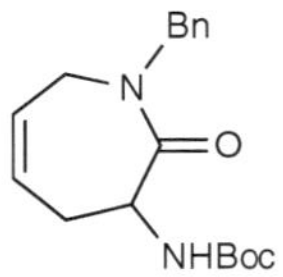 (yield: 54%)	94

(Table 1.3, Entry 4),[86] pyrimidinones (Table 1.3, Entry 5),[87] quinazolinones (Table 1.3, Entry 6),[88] and spirodiketopiperazines (Table 1.3, Entry 7).[89] Similarly, C—O bond formation is a viable cyclative cleavage strategy. Lactone formation is the most common method, such as the synthesis of phthalides reported by Tois and Koskinen (Table 1.3, Entry 8).[90] In certain cases, linker units are amenable to C—N or C—O bond forming cyclorelease, and different products can be prepared, from a common supported intermediate, by varying the cleavage conditions. This is attractive from a multifunctional cleavage viewpoint. For example, microwaving a common resin-bound intermediate in the presence and absence of an amine provided pyrrolidinones and butyrolactones, respectively (Table 1.3, Entries 9 and 10).[91]

Beyond the formation of C—N bonds and C—O bonds to achieve cyclorelease, there are also examples of C—C bond formation with concurrent cleavage. For example, Jeon prepared polymer-supported sulfonamides (Table 1.3, Entry 11).[92] Treatment with sodium hydride, exploiting the acidic proton α to the sulfone, allowed cyclization with the ester linkage and release of the cyclic sulfonamide. Alternatively, other cyclic C—C bond forming reactions have also been adapted for cyclorelease cleavage. For example, the intramolecular Claisen-like Lacey–Dieckmann reaction has been used to achieve concomitant formation and cleavage of tetramic acids (Table 1.3, Entry 12).[93]

Rhodium-mediated olefin metathesis is Nobel Prize-winning chemistry that has become increasingly powerful, and popular, since the discovery of the Grubbs I catalysts in the early 1990s. Cross-metathesis (CM) can be used to generate internal alkenes and has been exploited as a multifunctional cleavage strategy (Section 1.3.2). Likewise, the cyclic ring-closing metathesis (RCM) variant has very quickly become one of the preferred C—C bond forming reactions for routine preparation of cyclic species. Various cyclic species of differing sizes, ranging from five-membered rings to, for example, 30-membered macrocyclic species, have been generated using RCM. Such chemistry is clearly suitable for adaptation to cyclorelease SPOS and, indeed, numerous examples have been reported that have been recently reviewed.[95] For example, Table 1.3, Entry 13, illustrates van Maarseveen's preparation of seven-membered lactams, employing RCM for final cyclative cleavage.

The major advantage of using cyclorelease linker units is that the polar functional group used to attach a substrate to the polymer support remains attached to the support, rather than the target compound, upon cleavage. While this is ideal for the substrates described above, this substrate scope is limited. Noticeably, many target molecules are not cyclic or the ring size is unsuitable for cyclative cleavage. In such situations, alternative linker strategies to avoid the unwanted linking functionality are required and this initially led to development of traceless linker units and, subsequently, multifunctional linker units.

1.2.3 Traceless Linker Units

Traceless linker units are typically defined as those that leave a hydrogen residue behind upon cleavage (note that many traceless linkers can also behave as multifunctional linker units, by modifying cleavage conditions, and rather than a focus here will be discussed throughout this chapter). Traceless linkers were pioneered by Ellman and Plunkett in 1995 with the introduction of a silicon-based linker unit.[96] Ellman exploited *ipso* substitution at silicon to leave a hydrogen residue at the cleavage site of the target molecule. Proof of concept was demonstrated in the synthesis of benzodiazepines (Table 1.4, Entry 1), and this work ultimately was the catalyst for development of many traceless linker units that have

TABLE 1.4. Common Traceless Linker Units

	Linker	Cleavage Conditions	Product	References
1	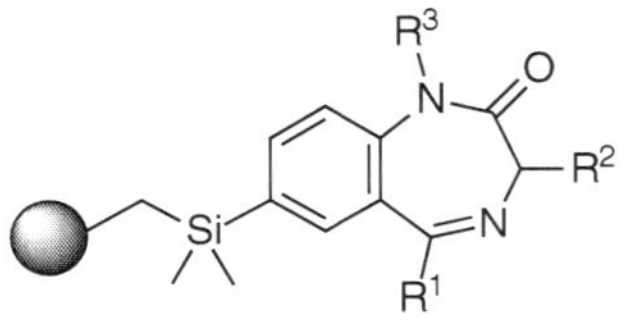	Aq HF	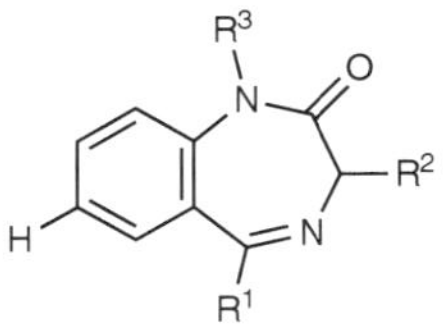 four examples (yield: 50–68%)	96
2	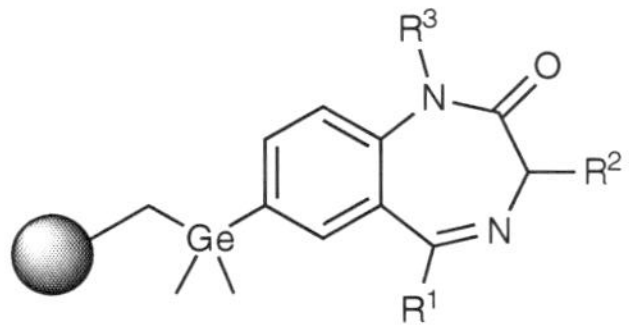	TFA	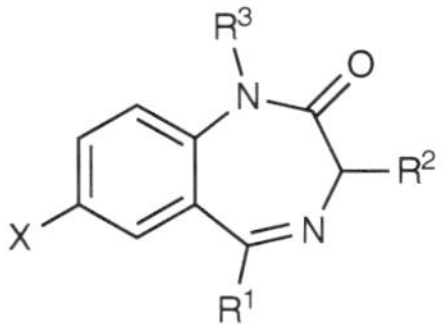 X = H (yield: 50–68%, 12 examples)	98
3	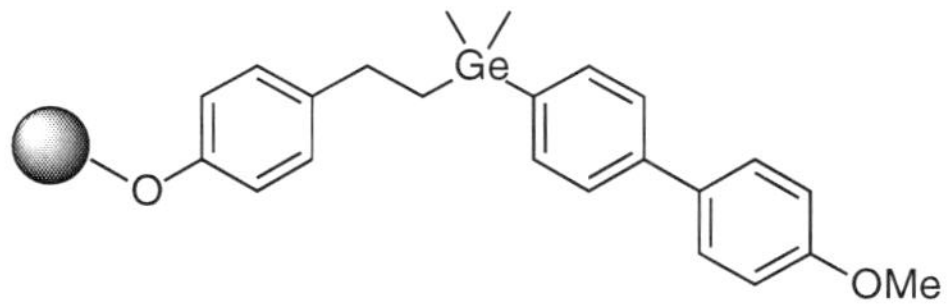	TFA, rt	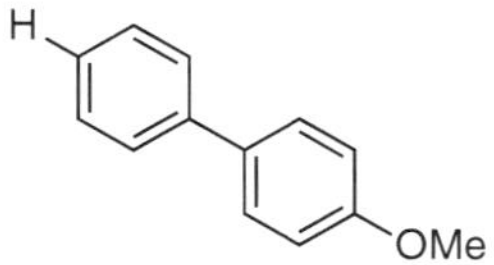	99–102
4	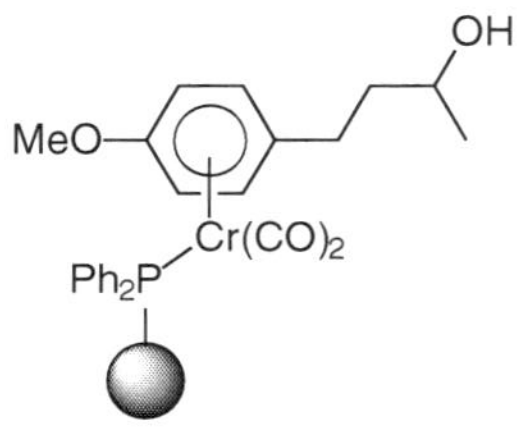	Pyridine, reflux	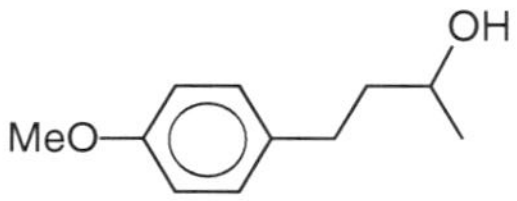 (yield: 92%)	104–105

(*Continued*)

TABLE 1.4. (*Continued*)

	Linker	Cleavage Conditions	Product	References
5	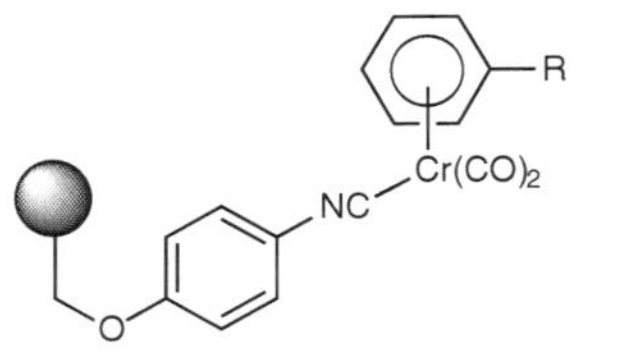	I_2 or $h\nu$/air	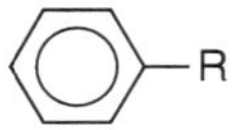 (yield: 80%)	106
6	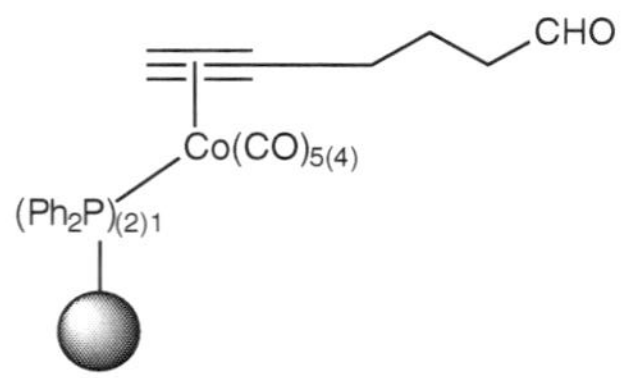	$h\nu$/air	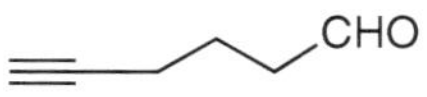	107, 108
7	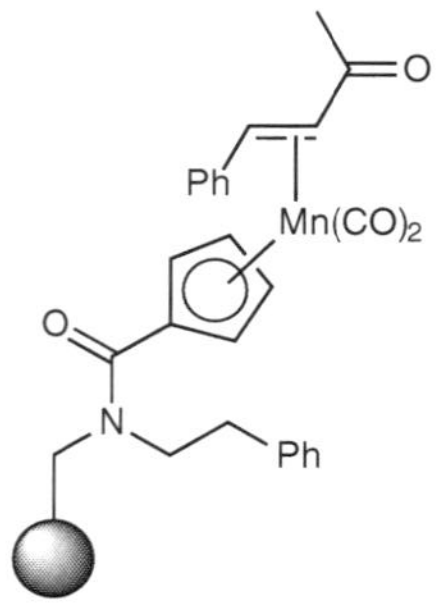	NMO	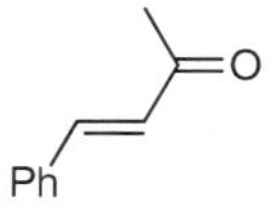	109

been reviewed.[8,10,97] Traceless cleavage using *ipso* substitution at silicon has led to the development of many silicon-based traceless linker units, which will be discussed further in Section 1.3.5. However, germanium linker units are amenable to similar chemistry. Germanium linker units were initially reported by Ellman and Plunkett (Table 1.4, Entry 2),[98] but they have been extensively developed and refined by Spivey's group (Table 1.4, Entry 3).[99–102]

An alternative traceless cleavage strategy worthy of mention is immobilization of arenes through transition metal carbonyl linker units, such as chromium (Table 1.4, Entries 4 and 5), cobalt (Table 1.4, Entry 6), and manganese (Table 1.4, Entry 7) based linker units.[103] While these linker units do not leave a hydrogen residue upon cleavage, because substrates are immobilized through the arene ring, no trace of the support remains upon cleavage, and so, for the purposes of classification, they can be considered traceless linker units in their own right. These linker units are attractive because arene rings are present in many potential substrates for SPOS. Gibson and coworkers reported the first example (Table 1.4, Entry 4) in which supported substrates were attached as $[(arene)(CO)_2(PPh_3)Cr(0)]$ complexes and then traceless cleavage could be realized simply by heating in pyridine.[104,105] Alternatively, cleavage could be achieved by treating with iodine or UV light (Table 1.4, Entry 5).[106] Other than arenes, alkynes and unsaturated carbonyl compounds are also amenable to this SPOS strategy. For example, alkyne-containing aldehydes were prepared using a cobalt linker and cleaved using UV light (Table 1.4, Entry 6),[107,108] while α,β-unsaturated ketones were immobilized on a manganese linker (Table 1.4, Entry 7) and cleaved by treatment with *N*-methylmorpholine *N*-oxide (NMO).[109]

1.2.4 Photolabile Linker Units

Photolabile linker units developed from the corresponding photolabile protecting groups are attractive linker units available to the solid-phase organic chemist because cleavage is achieved using only light.[110] Such mild cleavage conditions essentially eliminate unwanted side reactions that might otherwise occur when using, for example, strong acid or base cleavage cocktails. Early work concentrated on linker units based on the *o*-nitrobenzyloxy group, and many variants of this linker unit have since been reported. Cleavage of substrates from the *o*-nitrobenzyloxy linker can be achieved by irradiating at 350–365 nm (Table 1.5, Entry 1).[111] Related linkers based on the *o*-nitrobenzylamino (Table 1.5, Entry 2),[112–114] *o*-nitrobenzyl (Table 1.5, Entry 3),[115,116] and nitroveratryl (Table 1.5, Entry 4)[117] groups have also been reported. This allows variation in substrates that can be attached to the linker units, but cleavage is still simply a matter of irradiating with 350–366 nm light.

Photolabile linker units based on the phenacyl group have also been developed. The linker is essentially a functionalized resin since it is easily prepared by Friedel–Crafts acylation of typical polystyrene resin. Like the nitrobenzyl linkers, cleavage from the phenacyl linker units can be achieved by irradiating at 350 nm (Table 1.5, Entry 5).[118] A related linker unit is the *para*-methoxyphenacyl linker and, in this case, the *para*-methoxy group improves the efficiency of the photolysis and, thus, cleavage times are reduced.[119]

Other photolabile leaving groups including the benzoin group (Table 1.5, Entry 6),[120,121] pivaloyl group (Table 1.5, Entry 7),[122] nitroindolines (Table 1.5, Entry 8),[123] and thiohydroxamic (Table 1.5, Entry 9)[124] functionality have all been adapted as linker units for photolabile cleavage in SPOS with high degrees of success.

TABLE 1.5. Common Photolabile Linker Units

	Linker	Cleavage Conditions	Product	References
1	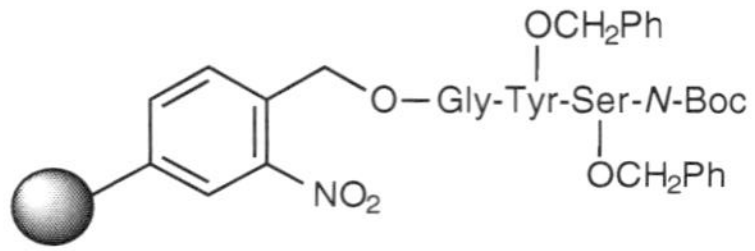	$h\nu$, $\lambda = 350$ nm	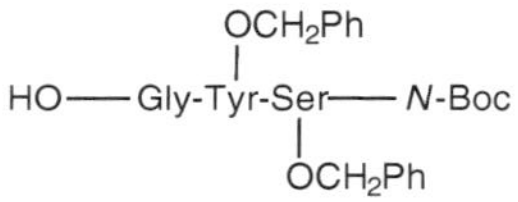 (yield: 72%)	111
2	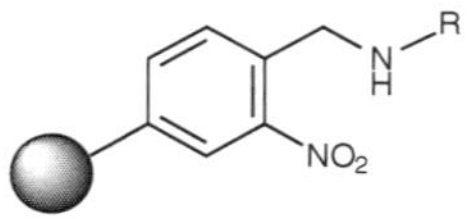	$h\nu$, $\lambda = 350$ nm	Amido peptides	112–114
3	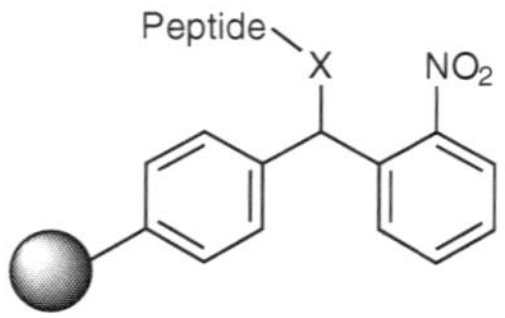	$h\nu$, $\lambda = 350$ nm	Peptides: X = O; amido peptides: X = NH	115, 116
4	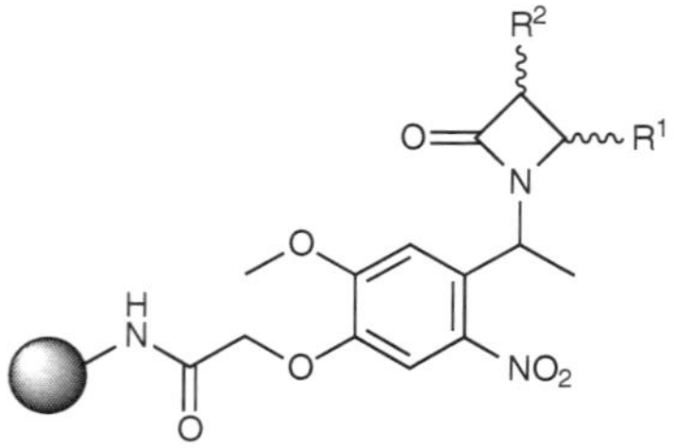	$h\nu$, $\lambda = 365$ nm	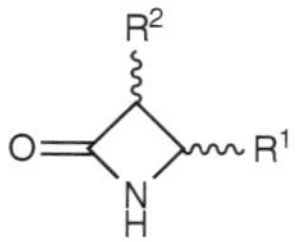 (yields: 71–90%)	117
5	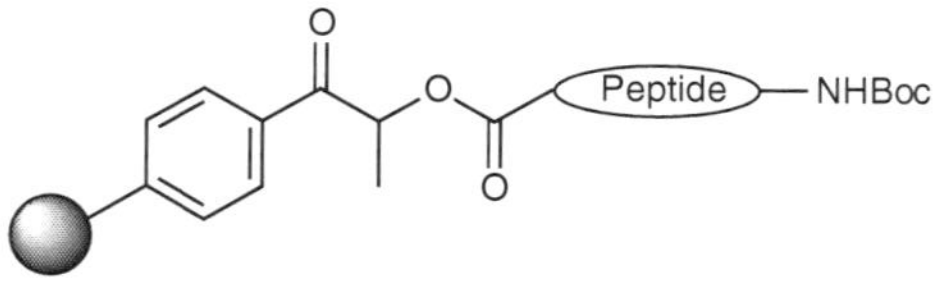	$h\nu$, $\lambda = 350$ nm	HO-peptide-NH_2	118

6	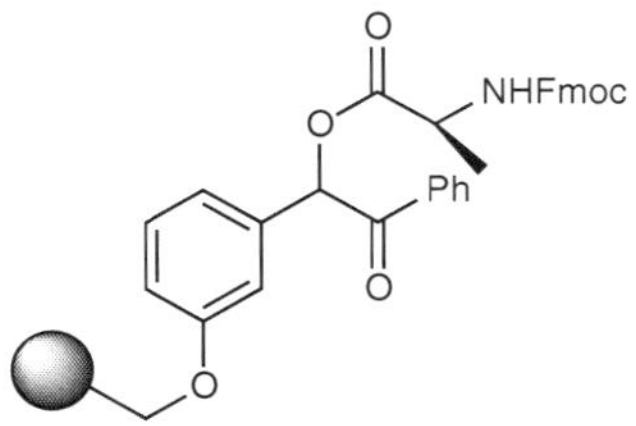	$h\nu$, $\lambda = 350$ nm	HO-Ala-Fmoc (yield: 75–97%)	120, 121
7	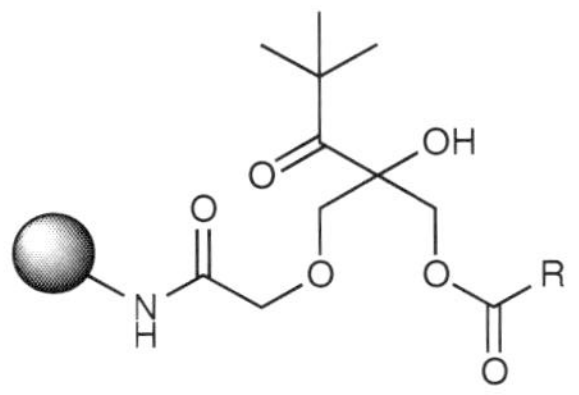	$h\nu$, $\lambda = 300$–340 nm	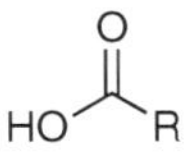 (yield: >78%)	122
8	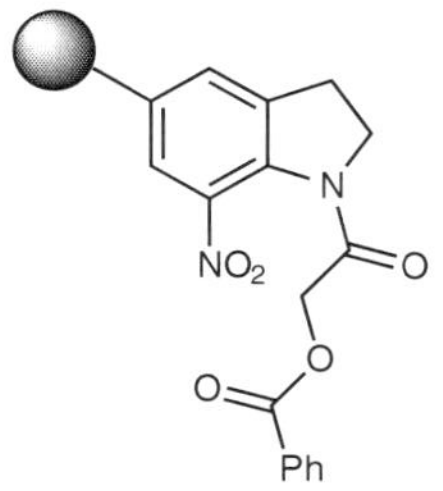	$h\nu$, $\lambda > 290$ nm, R^1R^2NH	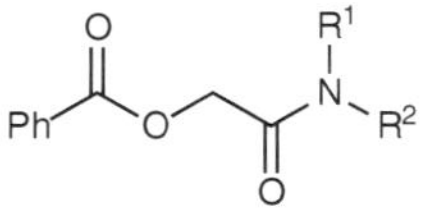 (yield: 67–95%)	123
9	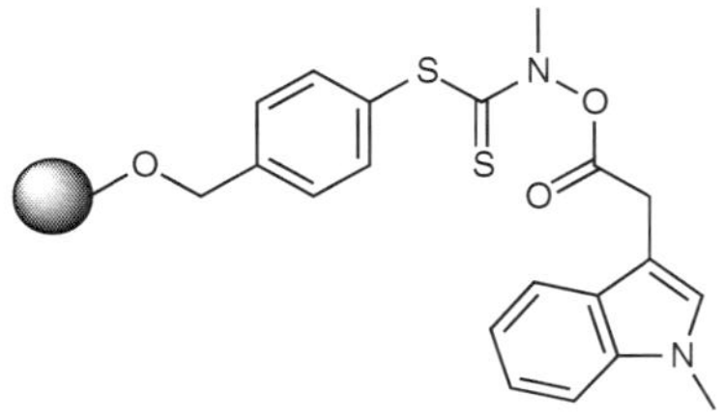	$h\nu$, $\lambda = 350$ nm, Bu_3SnH, THF	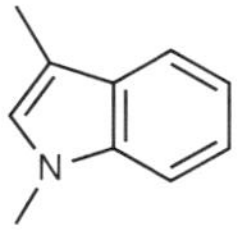 (yield: 55%)	124

1.2.5 Safety-Catch Linker Units

As outlined above, a drawback of using acid or base labile linker units is that unwanted cleavage can occur when reagents employed in the synthetic sequence resemble the cleavage conditions. One elegant solution to this problem is the safety-catch linker unit.[125,126] In such linkers, the latent bond requires activation before cleavage can occur. Many of the linker units discussed elsewhere in this chapter could be considered safety-catch linker units. For example, photolytic activation described in Section 1.2.4 and cyclorelease discussed in Section 1.2.2 are essentially safety-catch strategies. This section, however, will concentrate on synthetic activation. The first example of such an approach was a sulfonamide linker reported by Kenner et al. in 1971.[127] The sulfonamide **7** is stable to both acidic and basic conditions, making it synthetically valuable. However, alkylation of the nitrogen with, for example, diazomethane or iodoacetonitrile, gave **8**, from which substrates (e.g., carboxylic acids **9**) could be cleaved under nucleophilic conditions (Scheme 1.4).

Low loading efficiencies limited the use of Kenner's original linker, but an improved version was later reported by Ellman.[128] Kiessling and coworkers also reported an alternative palladium-catalyzed allylation strategy for activation of the linker unit for cleavage.[129] A number of other safety-catch linker units exploit the varying reactivity of sulfur in its different states. For example, a number of thioether-based linkers behave as safety-catch linkers and can be activated for cleavage by oxidation to the corresponding sulfoxides (Table 1.6, Entry 1)[130] or sulfones (Table 1.6, Entries 2 and 3).[131,132] Linkers can be activated for elimination, such as Entries 1 and 2, or nucleophilic substitution, as in the case of Entry 3. One further interesting example, reported by Li and coworkers, exploits Pummerer chemistry and has been used to prepare aldehydes and alcohols (Scheme 1.5).[133] The corresponding thioether was initially oxidized with ^{t}BuOOH/10-camphorsulfonic acid (CSA) to provide sulfoxide **10** and subsequent treatment with trifluoroacetic anhydride (TFAA) initiated the Pummerer rearrangement to give intermediate **11** and activated the linker for cleavage. Treatment with triethylamine released aldehydes (**12**), while reductive cleavage using sodium borohydride provided alcohols (**13**).

Alternatively, alkylation of the sulfur is also a viable safety-catch approach. For example, alkylation of a thioether with triethyloxonium tetrafluoroborate yielded a sulfonium ion (Table 1.6, Entry 4) that, in a report by Wagner and coworkers, activated benzyl

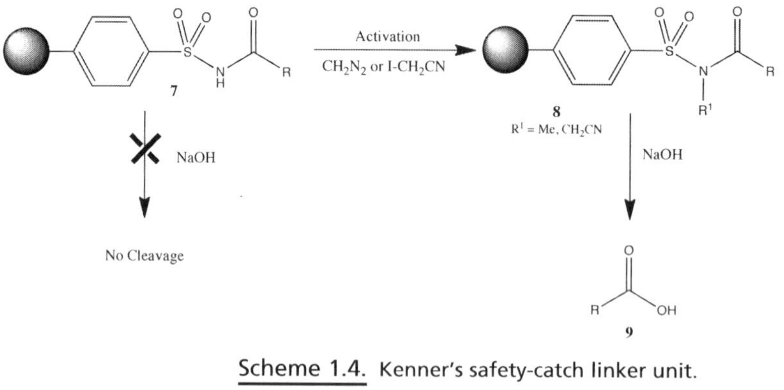

Scheme 1.4. Kenner's safety-catch linker unit.

TABLE 1.6. Common Safety-Catch Linker Units

	Linker	Cleavage Conditions	Product	References
1	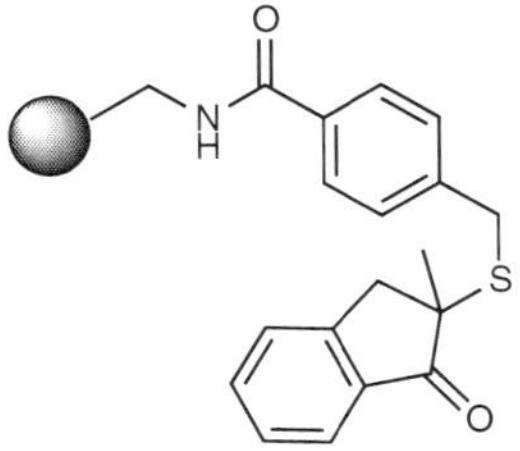	(i) 30% aq H_2O_2, HFIP, DCM; (ii) dioxane, 100°C	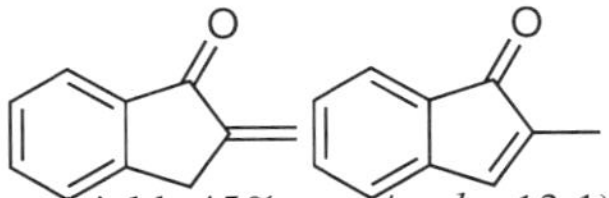 (yield: 45%; *exo*/*endo*: 13:1)	130
2	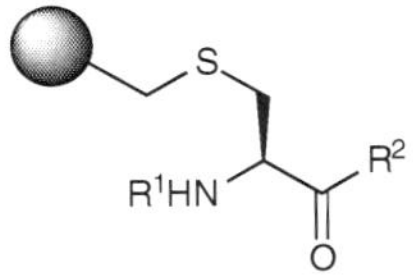	(i) *m*CPBA; (ii) DBU	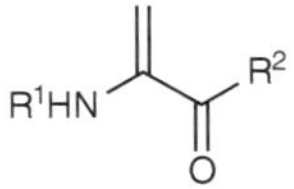 (yield: 31–86%)	131
3	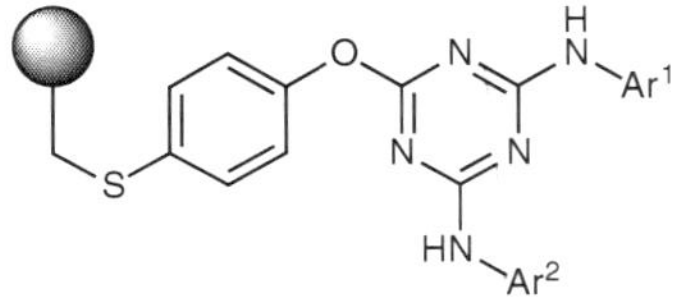	(i) CH_3CO_3H, DCM; (ii) R^1R^2NH, DMSO	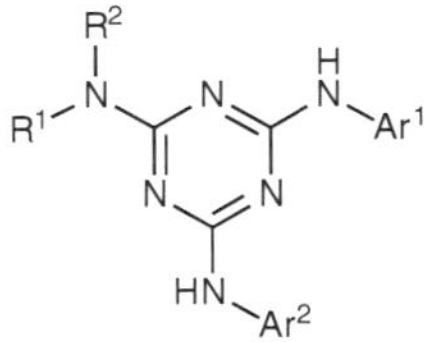	132
4	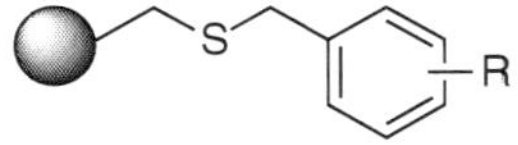	(i) EtO_3BF_4, DCM; (ii) $ArB(OH)_2$, K_2CO_3, $PdCl_2(dppf)$	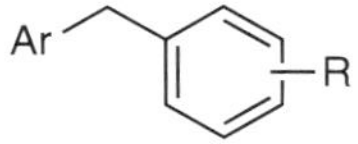 (yield: 24–99%, eight examples)	134

(*Continued*)

TABLE 1.6. (*Continued*)

	Linker	Cleavage Conditions	Product	References
5	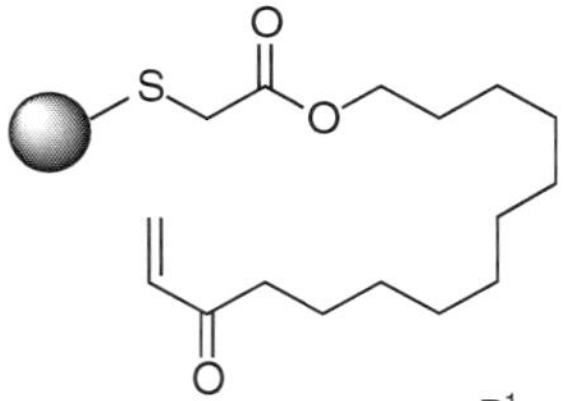	(i) MeOTf, DCM; (ii) DBU, DCM	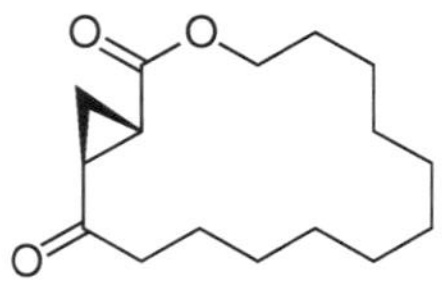	135
6	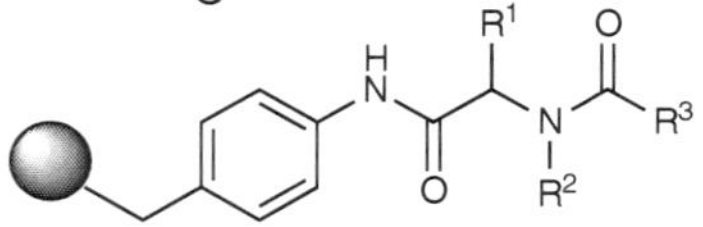	(i) Boc_2O, Et_3N, DMAP, DCM; (ii) LiOH, 5% H_2O_2/H_2O/THF	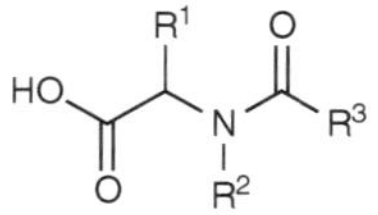	136
7	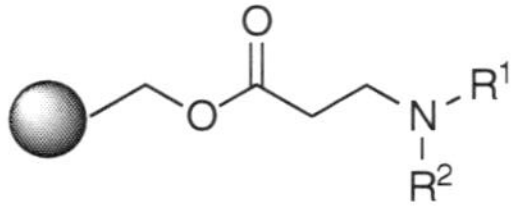	(i) R^3-X, DMF; (ii) DIPEA	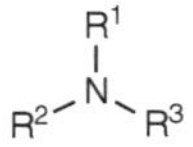	137
8	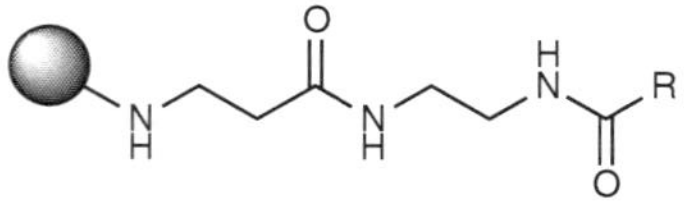	(i) MeI, 2,6-lutidine; (ii) DIPEA	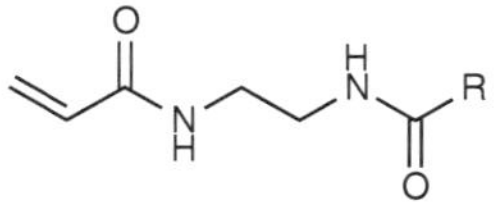	138

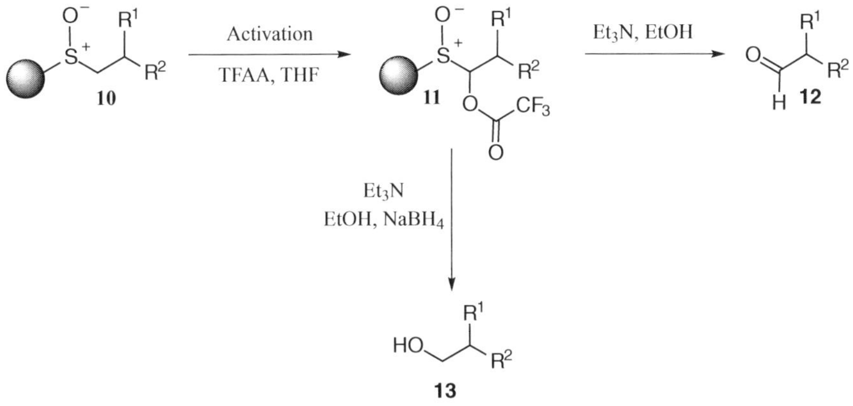

Scheme 1.5. Safety-catch linkers and the pummerer rearrangement.

groups for cleavage using Suzuki conditions to give biarylmethanes.[134] Similarly, Gennari and coworkers activated a thioether for cleavage using methyl triflate to generate the corresponding sulfur ylide.[135] The ylide then underwent an intramolecular cyclopropanation by a Michael reaction, and subsequent elimination, with concomitant cleavage of the C–S bond, to give the macrocycle exclusively as the *trans* isomer (Table 1.6, Entry 5).

A related safety-catch approach exploits activation of nitrogen-based linker units. For example, Hulme et al. reported the *N*-Boc activation strategy.[136] Supported amides could be prepared using a SPOS version of the Ugi reaction (Table 1.6, Entry 6). The amide bond was then activated for nucleophilic cleavage by introduction of the *N*-Boc group. Alternatively, Rees and colleagues developed the REM (regenerated resin after initial functionalization via Michael addition) safety-catch linker (Table 1.6, Entries 7 and 8).[137,139] After SPOS, the linker unit was activated via methylation, and subsequent β-elimination released amines (Table 1.6, Entry 7) or acrylamides (Table 1.6, Entry 8). In the case of a 1,2-dihydroquinoline linker (Scheme 1.6), substrates bound through an amide linkage (**14**) were found to be stable under acidic, basic, and reducing conditions. However, Mioskowski and coworkers were able to activate it for cleavage by oxidative aromatization to give (**15**).[140] Oxidation

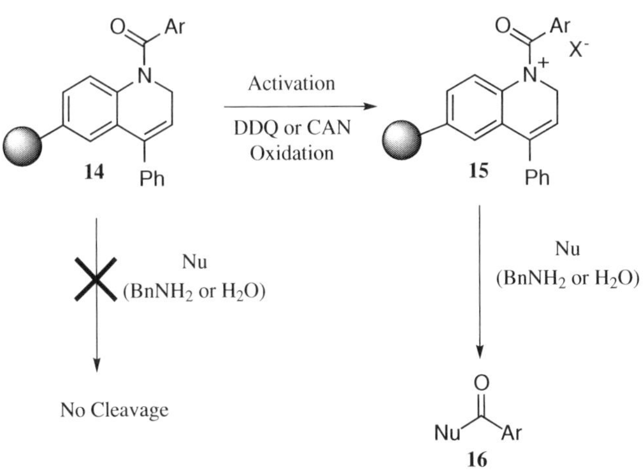

Scheme 1.6. 1,2-Dihydroquinoline linker unit.

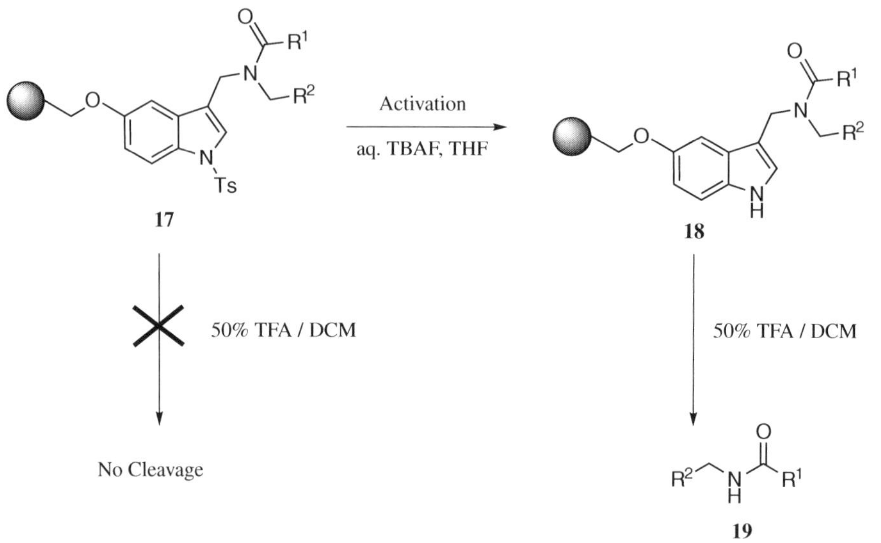

Scheme 1.7. Ley's indole safety-catch linker unit.

with DDQ or CAN resulted in concomitant aromatization, and substrates were then cleavable upon treatment with nucleophiles to give **16**.

Finally, a safety-catch linker utilizing the acidic lability of the indole core was reported by Ley and colleagues (Scheme 1.7).[141] Substrates attached to solid supports through the tosyl-protected indole (**17**) were stable in acidic conditions. However, deprotection of the tosyl group using TBAF provided activated intermediate **18**. Treatment of the activated linker with 50% TFA in DCM was then sufficient to release the target amides **19**.

1.3 MULTIFUNCTIONAL LINKER STRATEGIES

As the linker units described above have become ever more elaborate and sophisticated, they have evolved into multifunctional (or diversity) linker units. Multifunctional linker units use the cleavage step in solid-phase organic synthesis for incorporation of additional diversity into compound libraries, and the main classes of such linker units will be discussed in this section, along with representative cleavage strategies.

1.3.1 Nitrogen Linker Units

1.3.1.1 Triazene Linker Units. Owing to their multifunctionality and high stability, triazene linker units have become the most versatile diversity linker units reported to date. Initial reports of triazene linker units appeared in the mid-1990s from the groups of both Moore[142] and Tour.[143] Inspired by this work, the chemistry has been refined by Bräse, whose T1 and T2 triazene linker units have now been extensively developed for multifunctional cleavage.

The T1 linker originally found use as a traceless linker since treatment of T1 resin-bound substrates with TFA was found to release the corresponding aryl diazonium salts. Enders, in his preparation of β-lactams, was then able to show that heating the diazonium salts liberated nitrogen and a hydrogen residue was left at the cleavage site (Table 1.7, Entry 1).[144] In related

TABLE 1.7. Cleavage from the Triazene T1 Linker Units

	Linker	Cleavage Conditions	Product	References
1	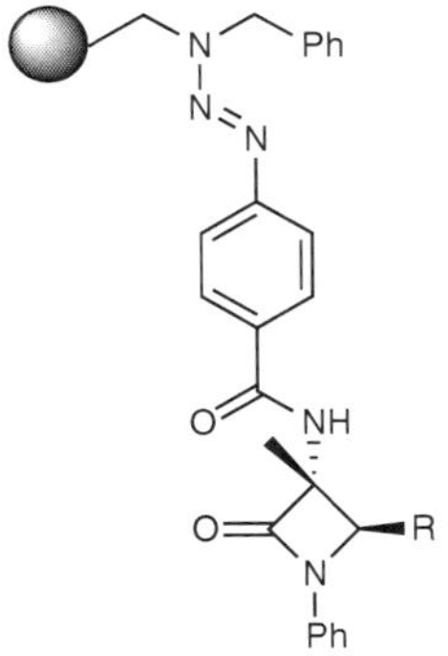	(i) 5% TFA/ DCM; (ii) DMF, 60°C, 15 min	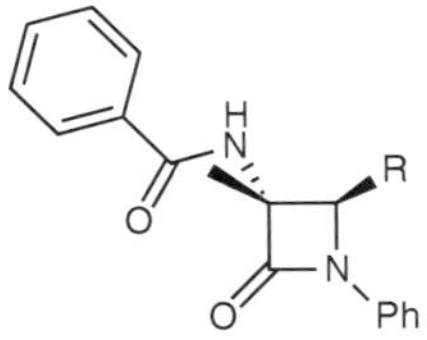 (yield: 53–71%)	144
2	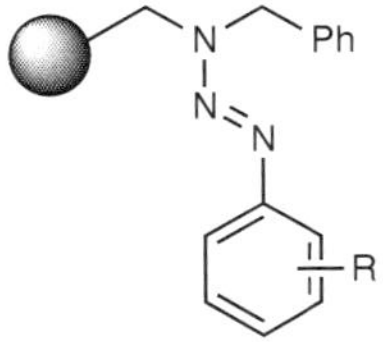	$HSiCl_3$, DCM	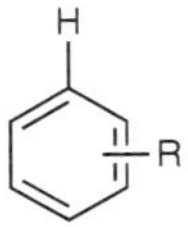 (yield: >92%, purity: >90%)	145
3	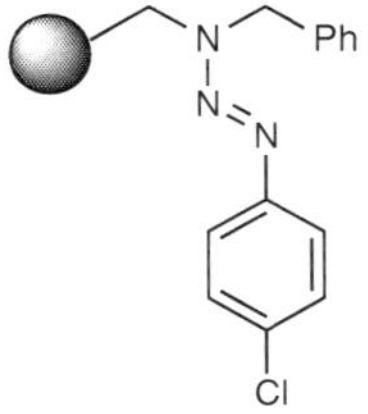	(i) BuLi, THF, −78°C; (ii) MeOH	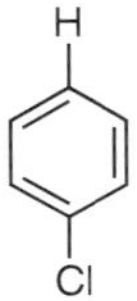	146

(*Continued*)

TABLE 1.7. (*Continued*)

	Linker	Cleavage Conditions	Product	References
4	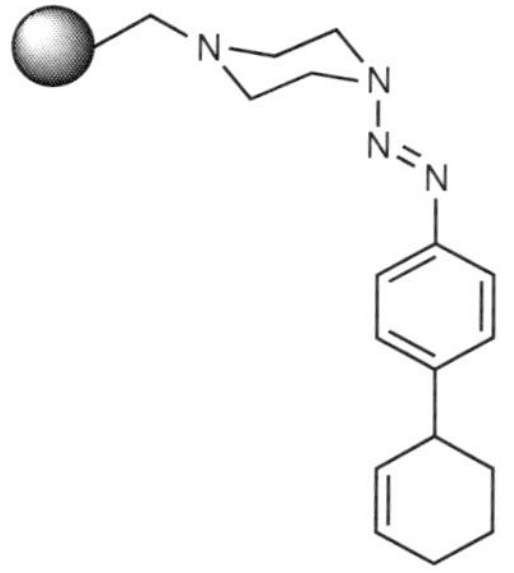	THF, Conc. HCl, 50°C, ultrasound	 (yield: 67%)	147
5	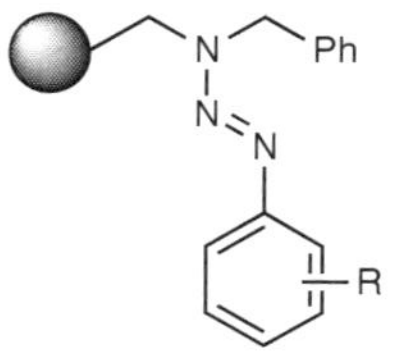	(i) TFA, MeOH; (ii) Pd cross-coupling	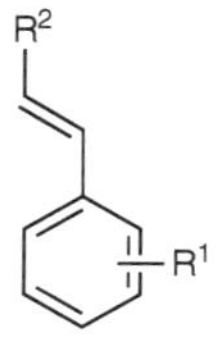	148, 149
6	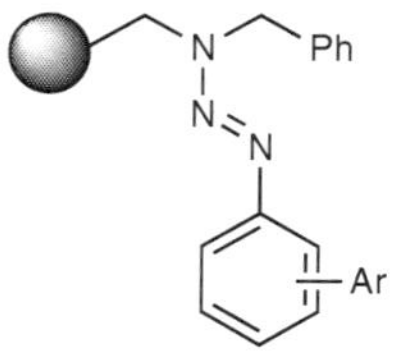	5% TFA, Me_3-SiN_3, DCM	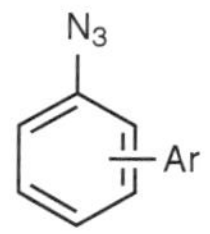 (yield: 39–73%)	152
7	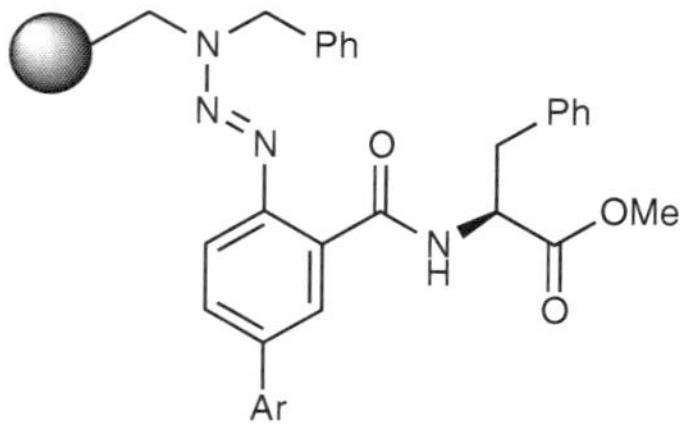	5% TFA in DCM	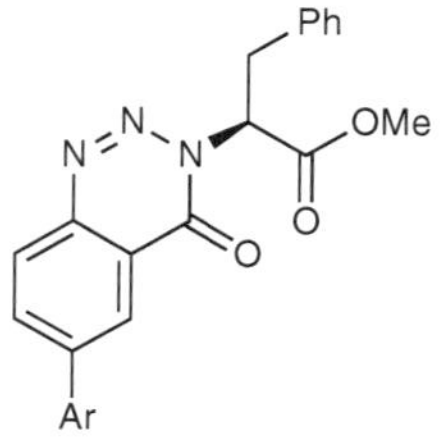 (yield: 10–29%, purity: 37–75%)	153

8	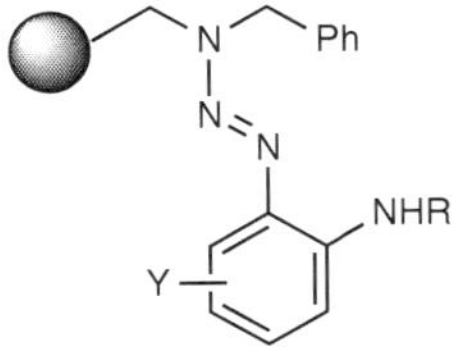	TFA/DCM	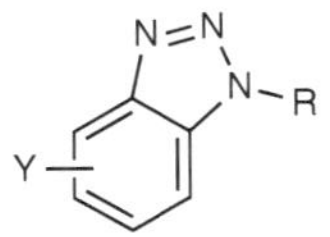 (yield: up to 83%)	154
9	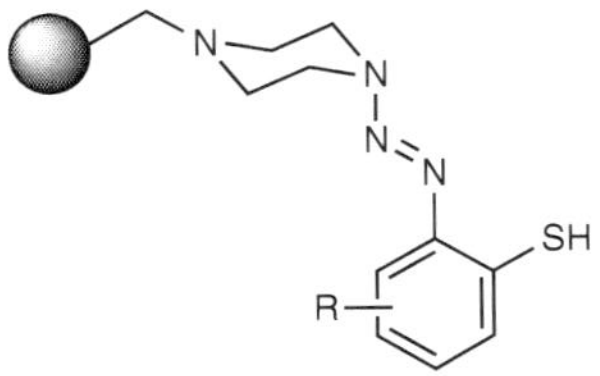	5% TFA/DCM	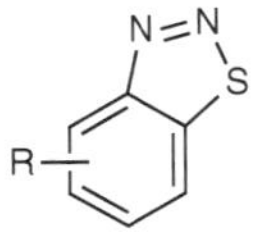 (yield: 10–63%)	155
10	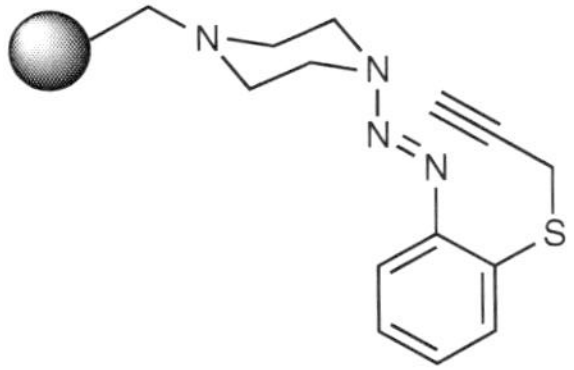	Me_3SiN_3, TFA	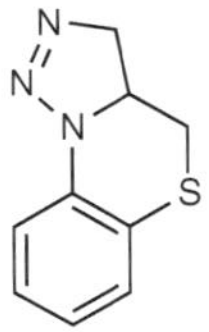 (yield: 14% over four steps)	155
11	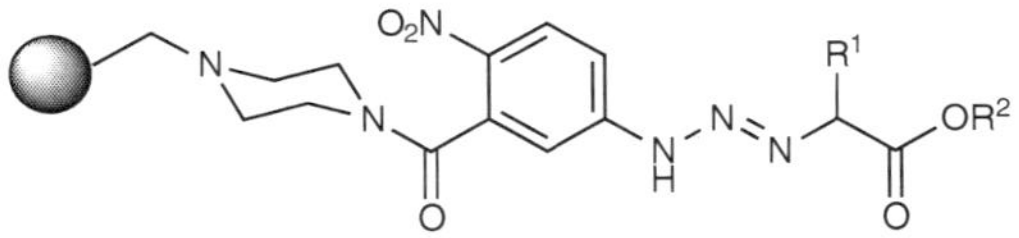	Et_3N	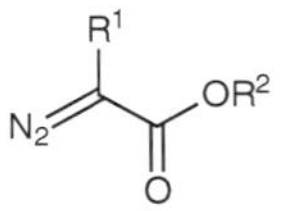 (yield: 2.5–39%)	156

work, alternative (and milder) conditions for traceless cleavage from the T1 linker were also developed by Bräse. For example, treatment of T1-bound substrates with trichlorosilane provided products in high yields and purities (Table 1.7, Entry 2).[145] Alternatively, treatment with *n*-BuLi resulted in a base-mediated fragmentation of the T1 linker and also resulted in traceless cleavage (Table 1.7, Entry 3).[146] In contrast, the related piperazinyl-type T1 linkers (Table 1.7, Entry 4) are stable to treatment with *n*-BuLi,[147] and so alternative strategies have been developed for traceless cleavage. When using these linkers, treatment with THF/conc. HCl at 50°C and concomitant application of ultrasound has proven effective in achieving traceless cleavage (Table 1.7, Entry 5).[147]

Following the discovery that aryl diazonium salts are viable electrophilic components for cross-coupling reactions, multifunctional cleavage strategies have also been worked out. For example, the diazonium salts can undergo palladium-catalyzed Heck reactions (Table 1.7, Entry 6) to introduce alkenes at the cleavage site.[148,149] Similarly, copper(I)-catalyzed cross-coupling with alkenes has also been shown.[148,149] Simple substitution with other nucleophiles is also possible. For example, treatment with trimethylsilyl azide in the presence of TFA provides the corresponding azido product (Table 1.7, Entry 7).[150–152]

Apart from the simple nucleophilic cleavage, a range of more subtle cleavage strategies have been reported, using the T1 and T1 piperazinyl-type linkers, which involve incorporating the triazene group (to varying degrees) into the final product. For example, triazinones could be prepared using a cyclorelease strategy promoted by TFA (Table 1.7, Entry 8).[153] Other heterocyclic species prepared include 1*H*-benzotriazoles (Table 1.7, Entry 9),[154] benzo[1-3]thiadiazoles (Table 1.7, Entry 10),[155] and 4*H*-[1-3]-triazolo[5,1-*c*][1-4]benzothiazines (Table 1.7, Entry 11).[155] Alternatively, treatment with triethylamine was employed to prepare diazoacetic esters (Table 1.7, Entry 12).[156]

More recently, Bräse has also introduced the T2 triazene linker unit. The T2 linkers are most commonly used for immobilization of amines (and other nitrogenous compounds). As their T1 counterparts, the T2 linkers have also proven robust linkers for SPOS. For example, amines can be cleaved by treating with TFA (Table 1.8, Entry 1),[157] while treatment with trimethylsilyl chloride is typically used when preparing (and cleaving) ureas (Table 1.8, Entry 2)[158] or amides (Table 1.8, Entry 3).[158] Alternatively, the T2 linker can also behave as a photolabile linker unit and photolytic cleavage ($\lambda = 355$ nm) by Enders et al. was used as a strategy to release amines (Table 1.8, Entry 4).[159]

Treatment of the T2 linker-bound substrates with electrophiles (e.g., Me_3SiCl; HOAc, TFA, RSO_3H) allows inclusion of an additional point of diversity upon cleavage (Table 1.8, Entry 5).[160] The mechanism proposed for such cleavage by Bräse is that the diazonium species is initially cleaved, and then displacement of nitrogen from the intermediate by the counterion (Cl^-, AcO^-, etc.) provides the products. Typically, a mixture of products is obtained using this cleavage strategy.

1.3.1.2 Hydrazone Linker Units. Hydrazones have proven versatile functional groups in organic synthesis. An extensive review of hydrazone chemistry was recently provided by Lazny and Nodzewska,[161] as well as reviews of the related hydrazone linkers.[162] The first use of a hydrazone in the capacity of a linker unit was done by Kamogawa et al. in 1983 (**20**, Scheme 1.8),[163] and it represents an early example of simple diversity cleavage. Cleavage via simple reduction ($NaBH_4$ or $LiAlH_4$) or elimination ($NaOCH_2CH_2OH$) provided alkanes (**21**) or alkenes (**22**), respectively, while treatment with potassium cyanide resulted in the corresponding nitriles (**23**).

TABLE 1.8. Cleavage from the Triazene T2 Linker Unit

	Linker	Cleavage conditions	Product	References
1	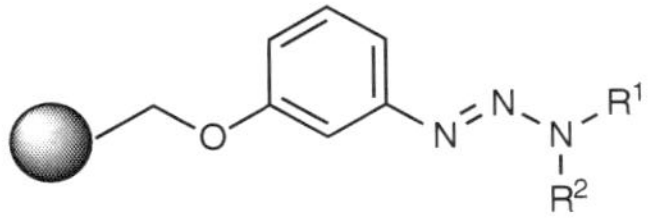	10% TFA/DCM	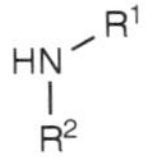 (yield: >90%)	157
2	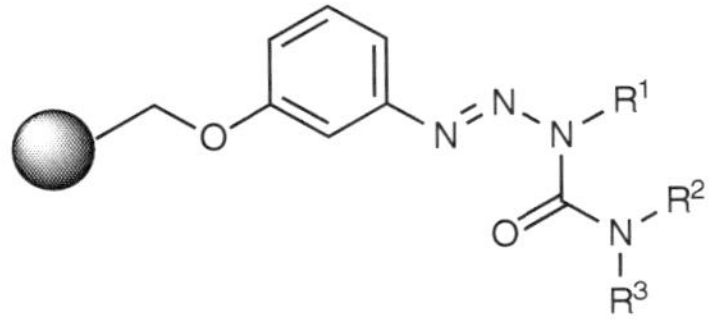	Me_3SiCl, DCM	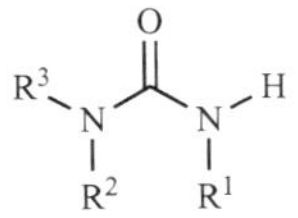 (yield: >80%)	158
3	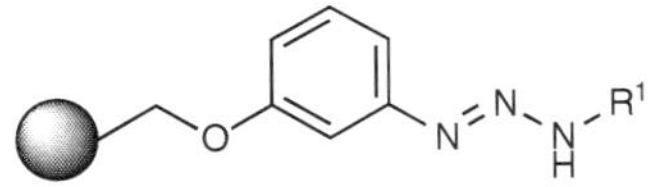	(i) R^2COCl, THF; (ii) Me_3SiCl, DCM	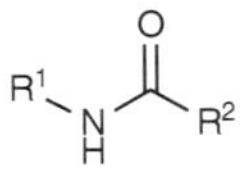 (yield: up to 75%)	158
4	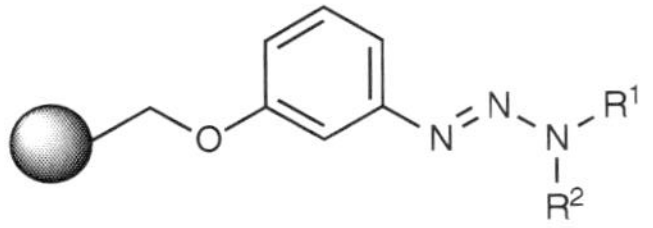	$h\nu$, $\lambda = 355$ nm	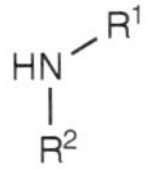 (yield: 45%)	159
5	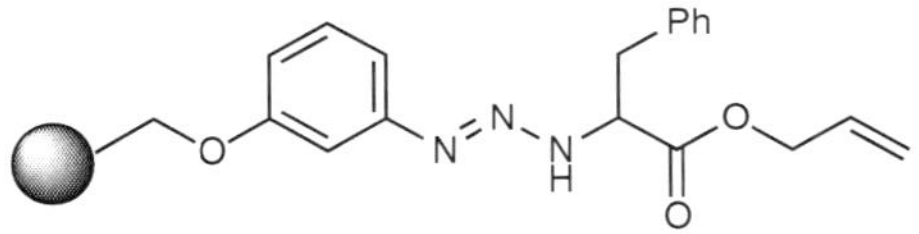	Me_3SiX (X = Cl, Br, I) or HX (X = OAc, OTfa)	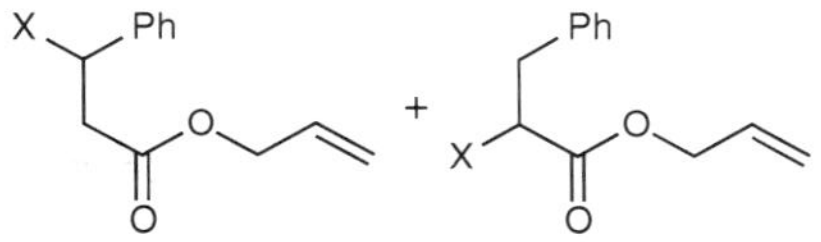 (yield: 80%, purity: 95%, ratio: 80:20–65:35)	160

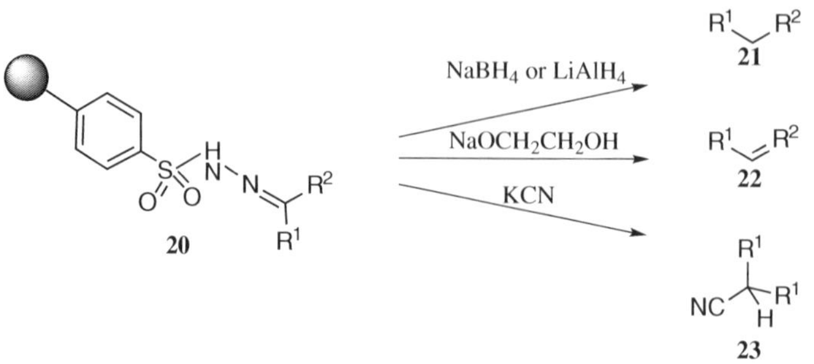

Scheme 1.8. Multifunctional cleavage from Kamogawa's hydrazone linker.

More commonly, however, and reflecting the role of hydrazones as carbonyl protecting groups in standard organic synthesis, simple acid-mediated cleavage will reform the carbonyl group (Table 1.9). For example, both Webb (Table 1.9, Entry 1)[164] and Ellman (Table 1.9, Entries 2 and 3)[165,166] have employed such a strategy to prepare peptide ketone derivatives, while addition of hydrogen peroxide to the cleavage cocktail can be used to generate carboxylic acids (Table 1.9, Entry 4).[167] Similarly, Breitinger has used a hydrazone linker in simple carbohydrate chemistry (Table 1.9, Entry 5).[168]

Beyond simple acid-mediated cleavage, a number of other cleavage strategies have been reported that show hydrazone linkers developing into quite a versatile family of multifunctional linker units. For example, in Table 1.9, Entry 6, nucleophiles react with hydrazones to introduce a second point of diversity (R^2) and then reductive cleavage was achieved by treatment with borane to provide amines. If desired, these amines can be trapped as the corresponding amides to introduce a third point of diversity (R^3), as shown in Table 1.9, Entry 7.[169] Alternatively, cleavage of substrates using *m*CPBA releases target molecules as the corresponding nitrile derivatives (Table 1.9, Entry 8).[167]

Reflecting the high impact that using hydrazones as chiral auxiliaries has had on asymmetric synthesis, recent efforts have explored the use of chiral linker units in approaches toward solid-phase asymmetric synthesis (SPAS). Efforts thus far have concentrated on supported analogues of the chiral SAMP analogues (e.g., Table 1.9, Entry 9),[170] and while the reported ee's are acceptable, they have yet to match results obtained in the analogous solution-phase reactions.

1.3.1.3 Benzotriazole Linker Units. The final class of nitrogen-based linker units is the benzotriazole linker units.[171] In the most common application of such linker units, substrates can be loaded using Mannich-type chemistry.[172] For example, treating a supported benzotriazole **24** with a mixture of amine and aldehyde provides supported amines **25** (Scheme 1.9).[173]

Cleavage can then be achieved by reduction to provide simple amines (Table 1.10, Entry 1),[174] or an additional point of diversity can be introduced by treating with an appropriate nucleophile such as a Grignard reagent (Table 1.10, Entry 2)[174] or Reformatsky reagent (Table 1.10, Entry 3).[175] Alternatively, if carbonyl compounds are loaded onto supports via a benzotriazole, then multifunctional cleavage can be achieved by treatment with nucleophiles such as enolates or amines to provide diketones (Table 1.10, Entry 4)[176] and ureas (Table 1.10, Entry 5)[177], respectively.

TABLE 1.9. Hydrazone Linker Units

	Linker	Cleavage Conditions	Product	References
1	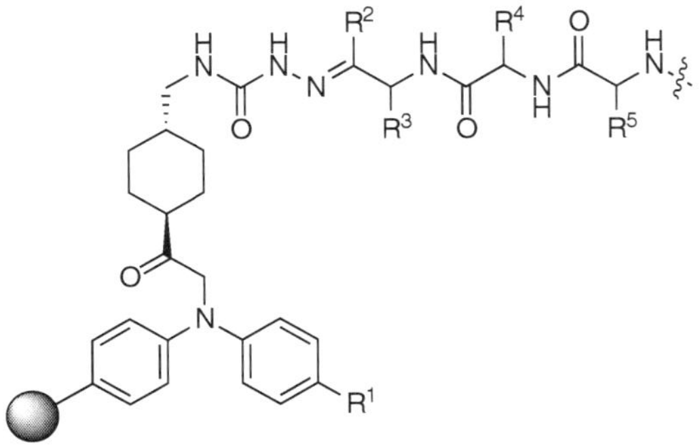	AcOH, aq HCl, THF	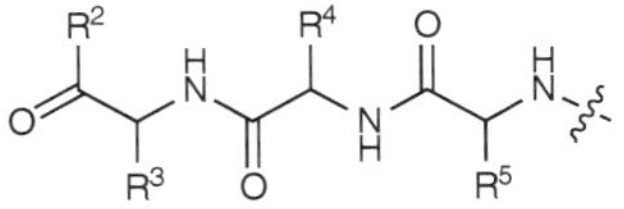	164
2	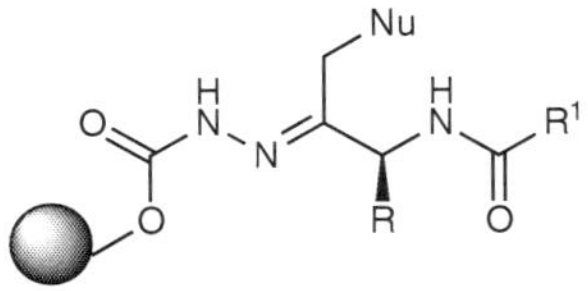	TFA/water/MeCHO/CF_3CH_2OH; Nu: SR^2, OCO-R^2, N(R)-CO-R	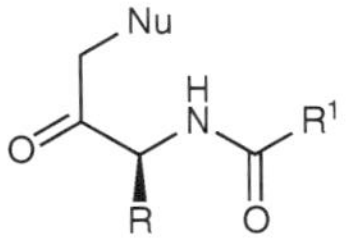	165
3	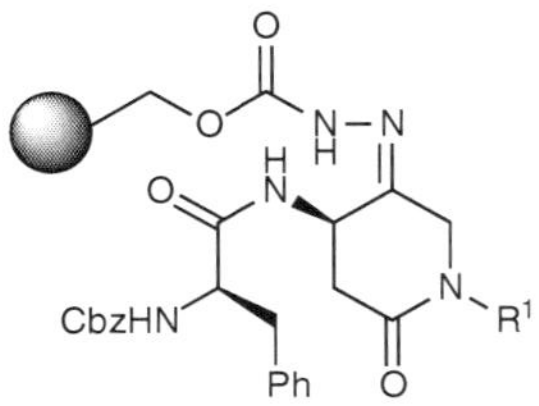	TFA/water/CF_3CH_2OH	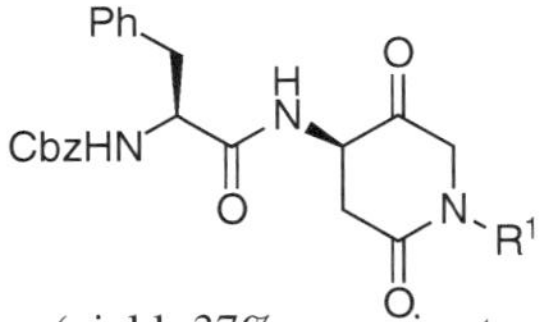 (yield: 37% over six steps)	166
4	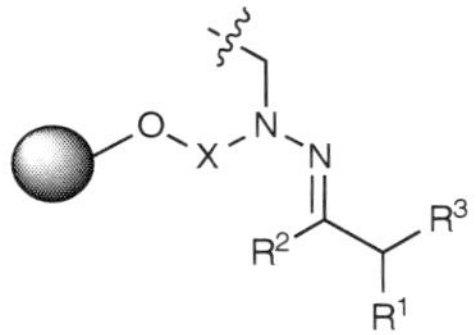	10% TFA, THF, H_2O_2	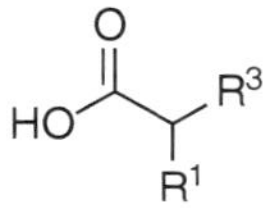 (yield: 22–40%)	167

(Continued)

TABLE 1.9. (*Continued*)

	Linker	Cleavage Conditions	Product	References
5	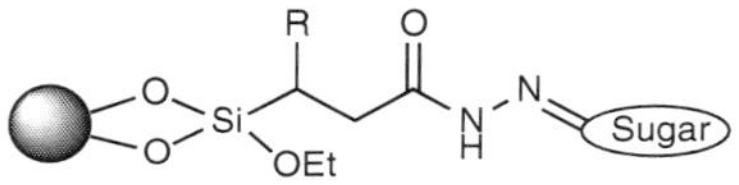	H^+	Amylose	168
6	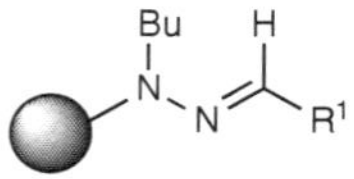	(i) R^2Li, THF; (ii) $BH_3{\cdot}THF$, THF	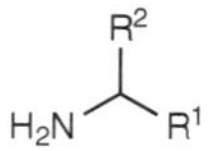	169
7	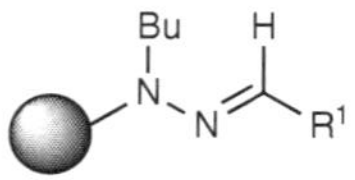	(i) R^2Li, THF; (ii) $BH_3{\cdot}THF$, THF; (iii) HCl; (iv) R^3COCl, Et_3N, DMAP	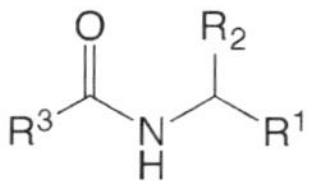	169
8	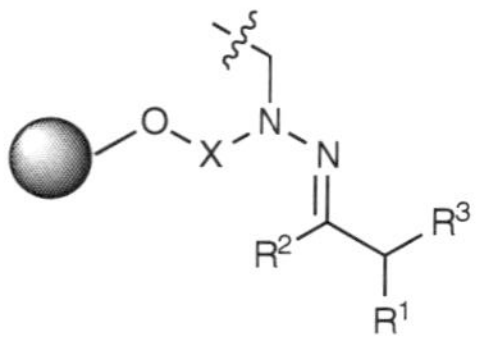	*m*CPBA	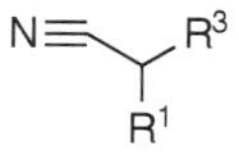 (yield: 22–90%)	167
9	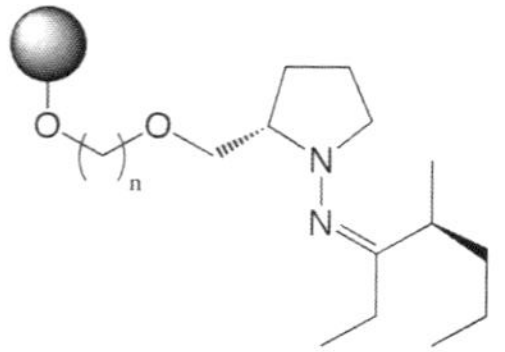	10% TFA in wet THF	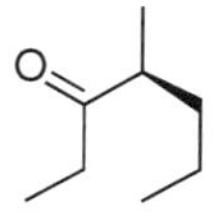 (ee: 10–73%)	170

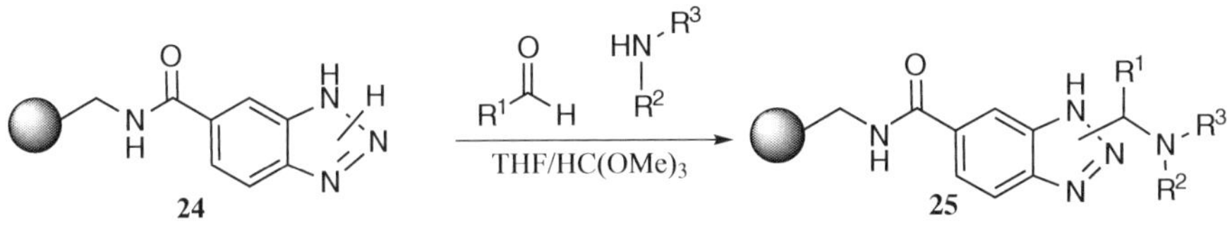

Scheme 1.9. Mannich-type chemistry with benzotriazole linker units.

1.3.2 Sulfur Linker Units

Sulfur-based linker units have been developed that utilize the reactivity of sulfur in a multitude of different forms and oxidation states.[5,178–180] The simplest linker units are the thioether-based linkers, and initially conditions for traceless cleavage of aliphatic

TABLE 1.10. Benzotriazole Linker Units

	Linker	Cleavage Conditions	Product	References
1	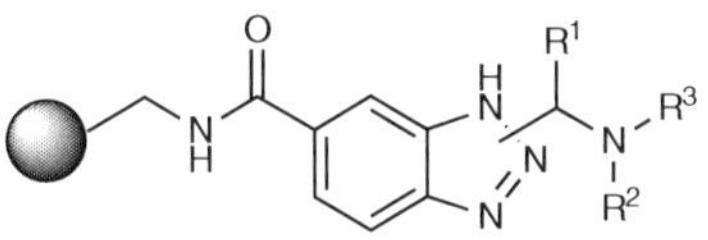	$NaBH_4$ (20 equiv), THF, 60°C	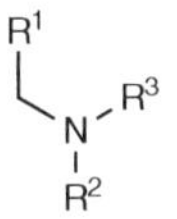	174
2	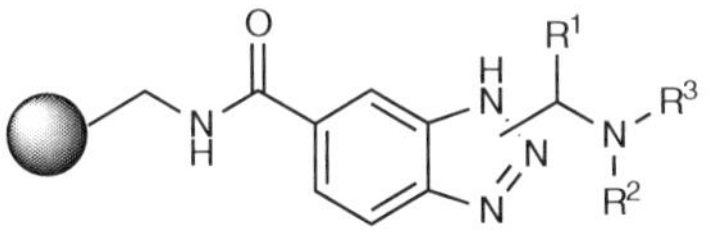	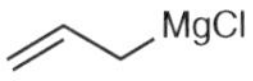 (30 equiv) HC (OMe)$_3$, 40°C	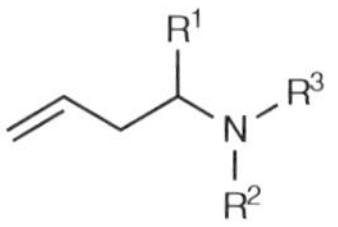	174
3	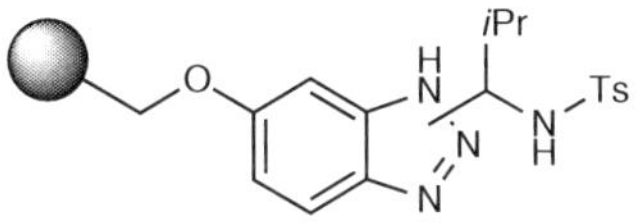	BnZnBr (4 equiv), THF, 60°C	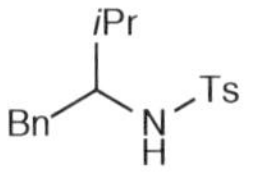 (yield: 63%, two steps)	175
4	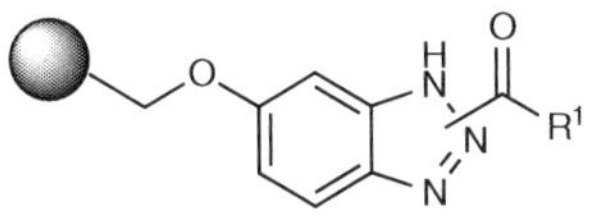	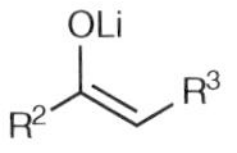 THF, −78°C–rt	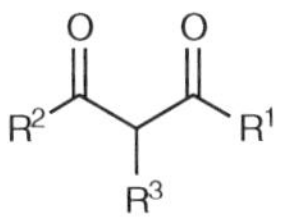 (yield: 18–41%)	176
5	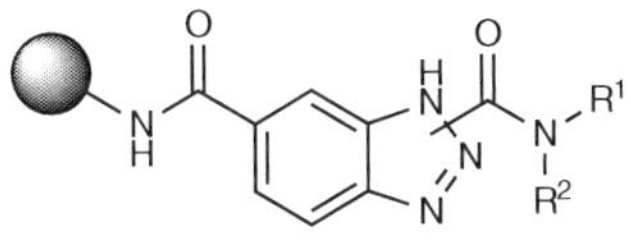	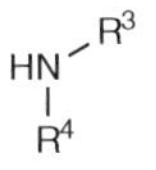 Chlorobenzene, 90°C	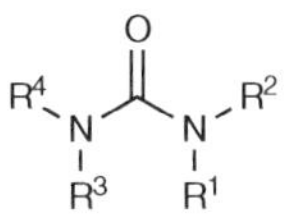 (156 examples, >80% purity)	177

substrates were reported. Such traceless cleavage could be achieved under radical conditions (Table 1.11, Entry 1).[181] However, such reactions were discovered to be sluggish and low yielding, and so a reductive desulfurization reaction using Raney Ni and hydrogen has become the preferred method for achieving such cleavage (Table 1.11, Entry 1).[181,182] Alternatively, Procter has recently shown that traceless cleavage can also be achieved using samarium(II) iodide (SmI_2), as illustrated in Table 1.11, Entry 2.[183] Simple diversity cleavage can be achieved from thioether-based linker units by treatment with a nucleophile. An early example of such an approach was demonstrated by Crosby, in 1977, who showed that treatment of supported alkylthioethers with a cocktail of sodium iodide and iodomethane released products as the corresponding alkyl iodides (Table 1.11, Entry 3).[184] Such an approach can also be used to generate bromides and has found application in carbohydrate chemistry (Table 1.11, Entries 4 and 5), as reported by Schmidt[185,186] and Kunz.[187–190] In the case of Schmidt's work (Table 1.11, Entry 4), the sugar could be isolated as the bromide or additional diversity could be incorporated by addition of methanol in a Lemieux-type glycosylation reaction at the anomeric center.[185] Beyond halogens, other nucleophiles can also be used during cleavage. For example, Hennequin treated resin-bound quinazolines with oxindoles to prepare a library of oxindole quinazolines (Table 1.11, Entry 6).[191] Alternatively, generation of disulfides inter- (Table 1.11, Entry 7) or intramolecularly (Table 1.11, Entry 8) is also possible.[192,193] In contrast to nucleophilic cleavage, treatment with a base will promote eliminative cleavage and this was demonstrated, by Baer and Masquelin, during preparation of a library of 2,4-diaminothiazoles (Table 1.11, Entry 9).[194] A related linker unit is the 1,3-propanedithiol linker unit.[195–198] Like the analogous acetal linker units previously described, this linker can be used as a linker for carbonyl compounds and cleavage can be achieved by treating with [bis(trifluoroacetoxy)iodo]benzene[195] or anhydrous periodic acid (Table 1.11, Entry 10).[196,198]

Cleavage of substrates from sulfur resins continues to be reported, and it has been shown that such cleavage strategies can be enhanced by prior activation of the sulfide by alkylation to generate sulfonium ions, or by oxidation to the sulfoxide or sulfone. This activation strategy is briefly discussed in Section 1.2.5 as it has been exploited for safety-catch linker strategies. For example, alkylation of thioethers to provide sulfonium ions was discussed as a safety-catch strategy for preparing macrocycles (Table 1.6, Entry 5) [135] and biarylmethanes (Table 1.6, Entry 4) [134]. However, such an approach has also been used in the context of a multifunctional linker unit. Thus, polymer-supported thioether **26** was methylated with methyl triflate to provide the sulfonoium intermediate **27**. Treatment with DBU then generated an ylide, which could be reacted with a range of aldehydes to generate a small family of epoxides (**28–30**, Scheme 1.10).[135]

Oxidation to the sulfoxide or sulfone can also be used as a method to activate sulfur linker units. Typically, it is easier to oxidize all the way to the sulfone, but specialized strategies have been developed that allow intermediate oxidation to the sulfoxide. Moreover, sulfoxides can be loaded onto resins directly[199,214], but it is far more common to oxidize the corresponding supported thioether.[130,133,215] For example, Bradley prepared a sulfoxide linker unit (Table 1.11, Entry 11) by treating the corresponding supported thioether with a mixture of hydrogen peroxide and hexafluoroisopropanol.[130] Heating at 100°C in dioxane released the product (as a mixture of *exo* and *endo*). Related cleavage by refluxing in benzene was also reported by Toru (Table 1.11, Entry 12).[199] Alternative cleavage from Toru's linker could also be achieved by treating with TBAF to effect desilylsulfination (Table 1.11, Entry 13). Alternatively, as described in Section 1.2.5,

TABLE 1.11. Sulfur-Based Linker Units

	Linker	Cleavage Conditions	Product	References
1	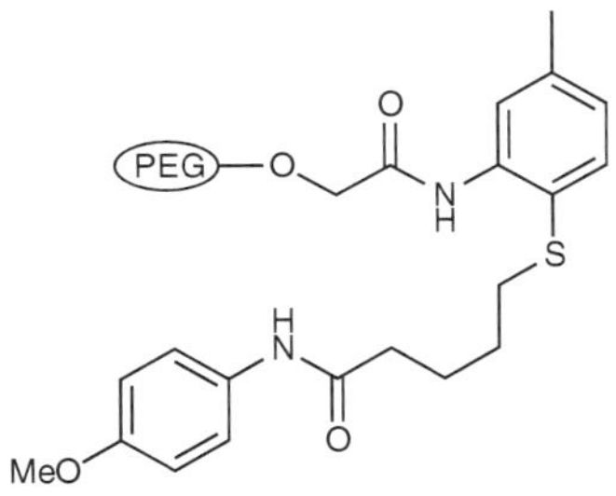	A: Bu_3SnH, PhH, reflux; B: H_2, Raney Ni, MeOH:EtOH 20°C	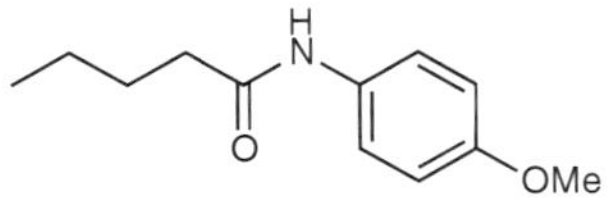 (yield: A 40%; B 94%)	181, 182
2	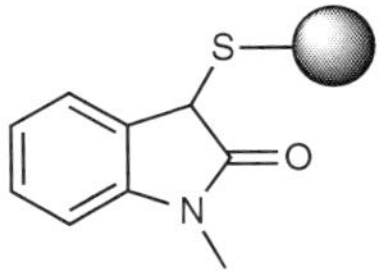	SmI_2, DMPU, THF, rt	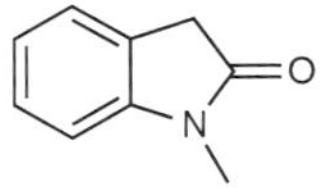 (yield: 47% over four steps)	183
3	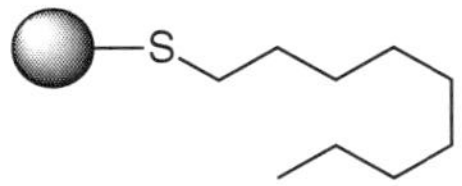	NaI, MeI, DMF	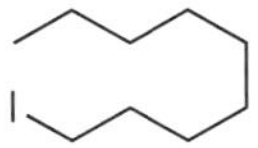	184
4	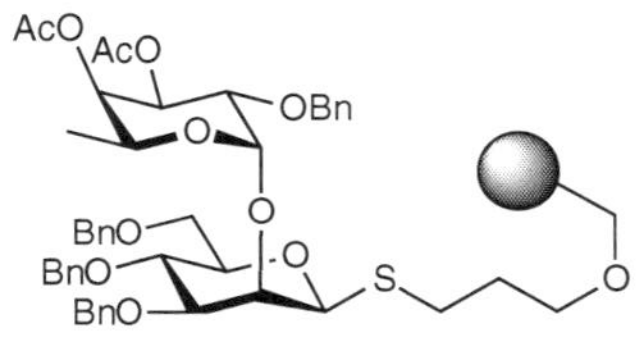	NBS, DTPB, THF:MeOH	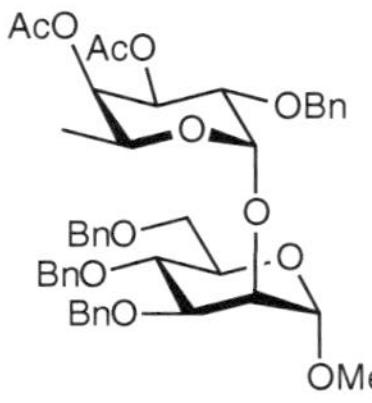 (yield: 54% over two steps)	185, 186

(*Continued*)

TABLE 1.11. (*Continued*)

	Linker	Cleavage Conditions	Product	References
5	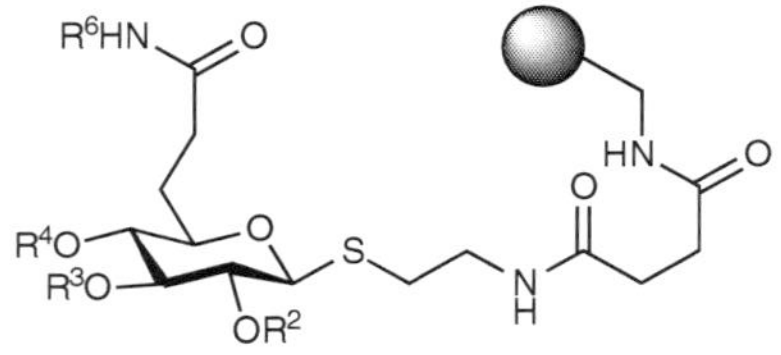	NBS, DTBP, EtOH, DCM; Br_2, DTBP, DCM	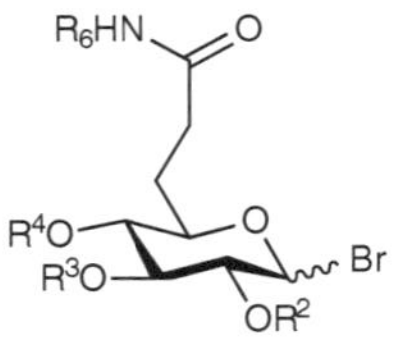	187–190
6	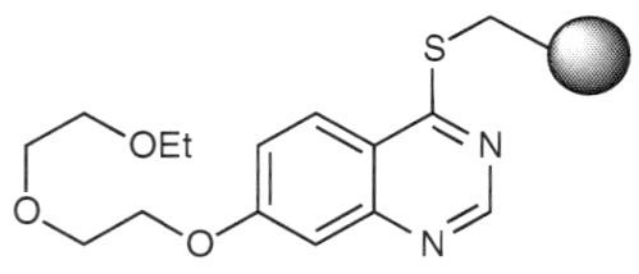	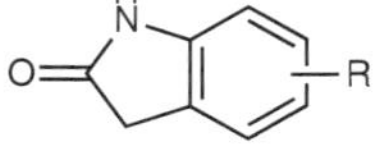 (i) NaH, DMSO, 100°C; (ii) SCX Silica; (iii) 2% NH_3 in DCM/MEOH	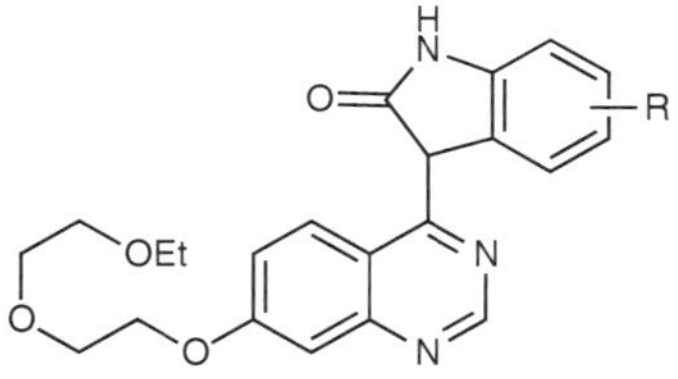 (seven examples; yield: 35–72%)	191
7	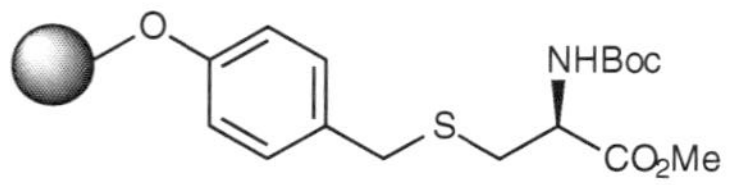	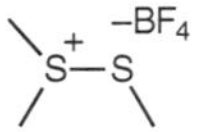 DMF, rt	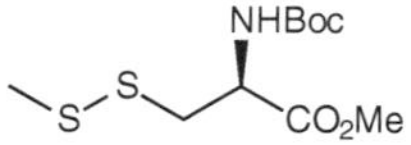 (yield: 93%)	192
8	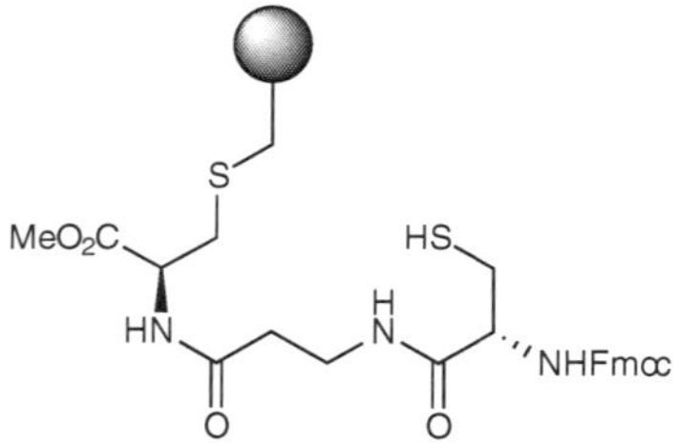	NCS, DMS, DCM, 0°C	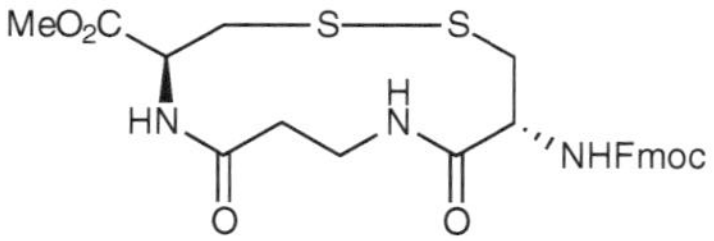 (yield: 13% over nine steps)	193

9	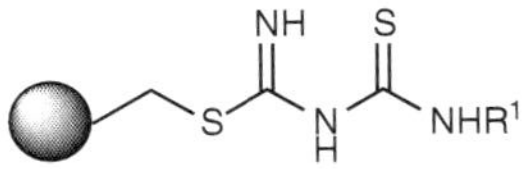	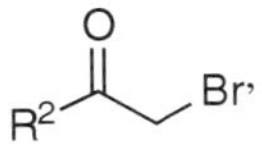 DMF 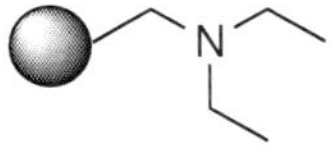	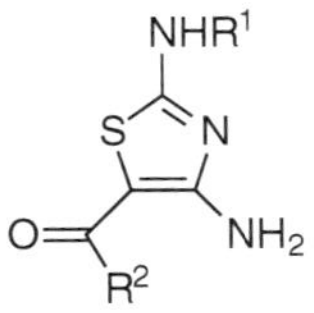	194
10	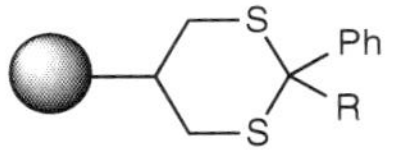	H_5IO_6, 0°C–rt	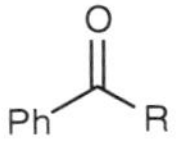	198
11	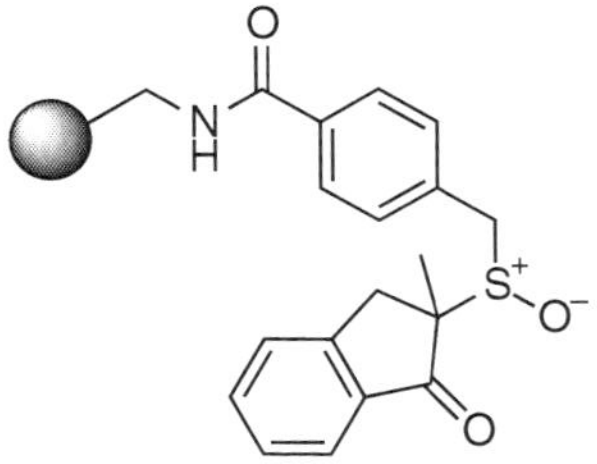	Dioxane, 100°C	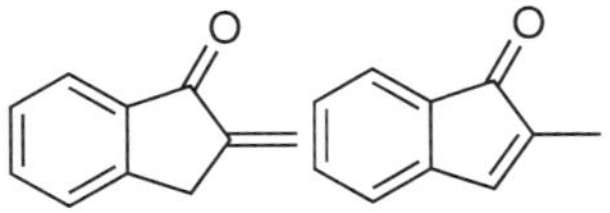 (yield: 45%; *exo*/*endo*: 13:1)	130
12	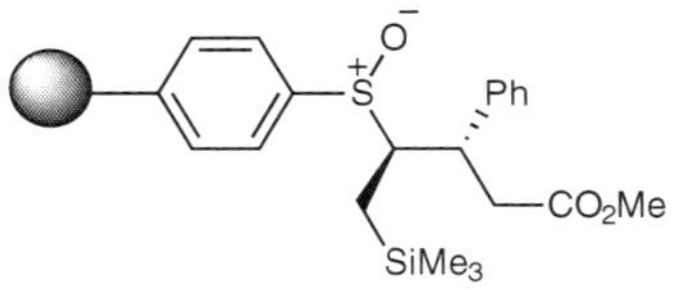	Benzene reflux	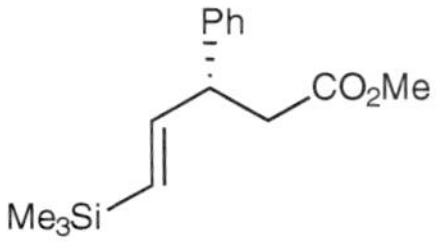 (yield: 51% over three steps; 90% ee)	199

(*Continued*)

TABLE 1.11. (*Continued*)

	Linker	Cleavage Conditions	Product	References
13	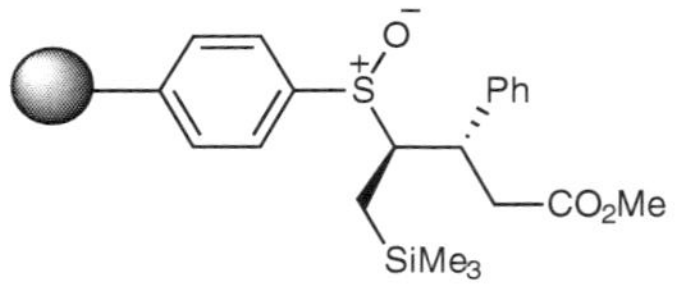	TBAF, THF, 0°C	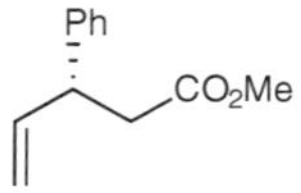 (yield: 56% over three steps; 90% ee)	199
14	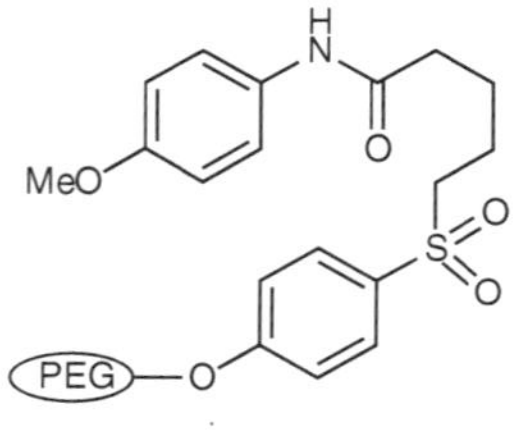	5% Na/Hg, NaH_2PO_4, MeOH/ DMF (1:8), rt	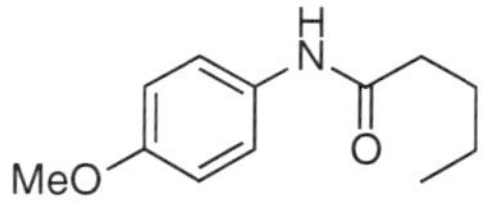 (yield: 97%)	200
15	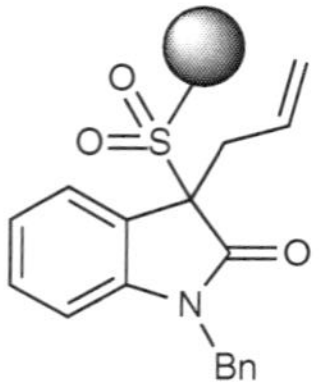	SmI_2, DMPU, THF, rt	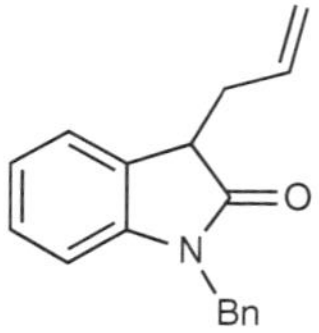 (yield: 30% over six steps)	183
16	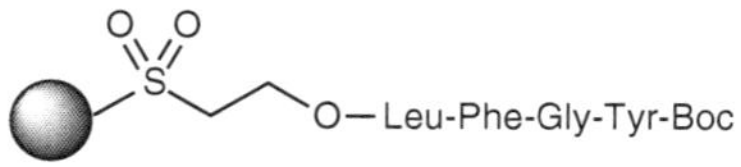	NaOH	HO-Leu-Phe-Gly-Tyr-Boc (yield: 60%)	201

17	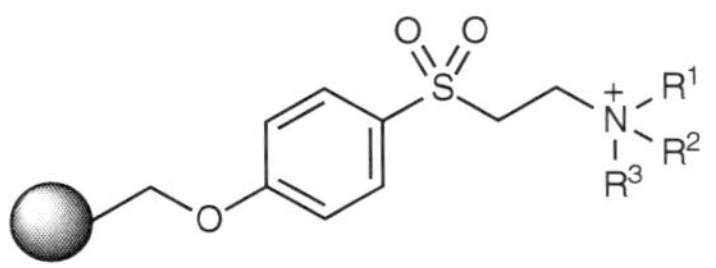	DIPEA (5 equiv)	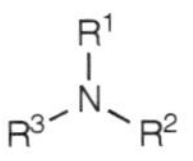 (yield: 65–83%)	202
18	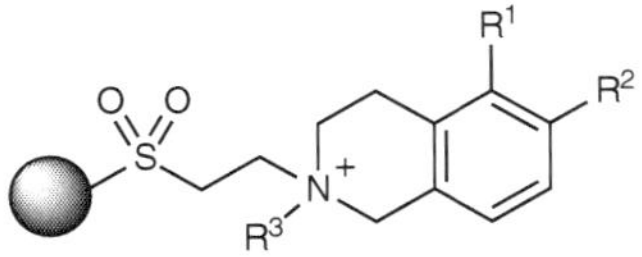	DIPEA	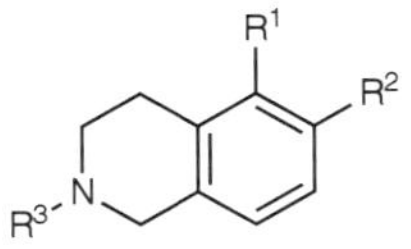 (yield: 25–100%)	203
19	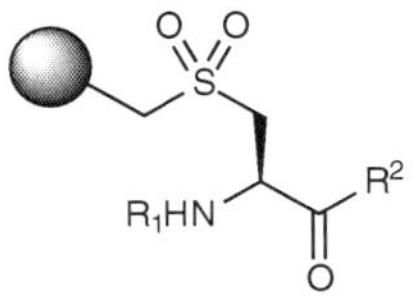	DBU, DCM, rt	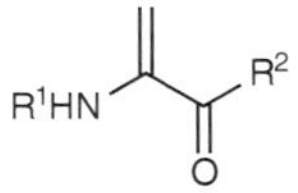 (yield: 31–86%)	131
20	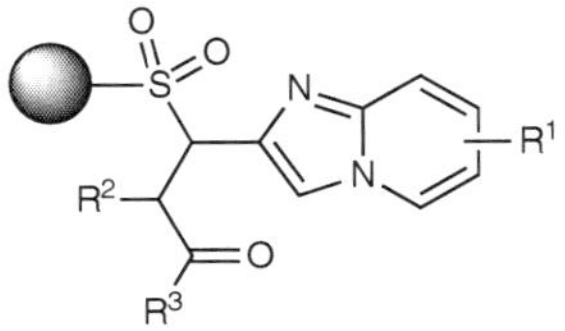	10% NaOH, DCM, rt	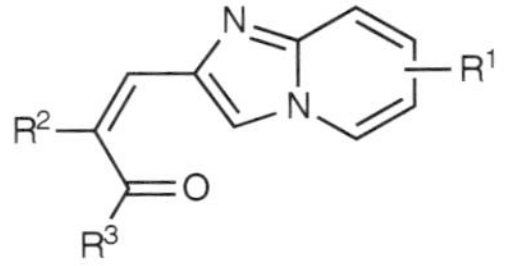 (yield: 10–26% over five steps)	204
21	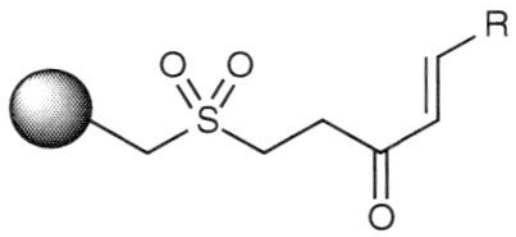	$BnNH_2$, THF, rt	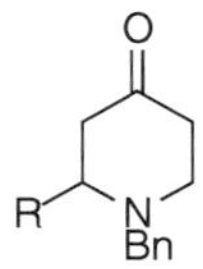 (yield: 50–75%)	205

(*Continued*)

TABLE 1.11. (*Continued*)

	Linker	Cleavage Conditions	Product	References
22	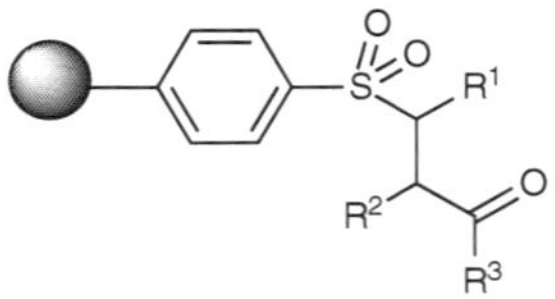	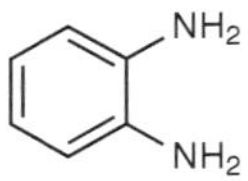	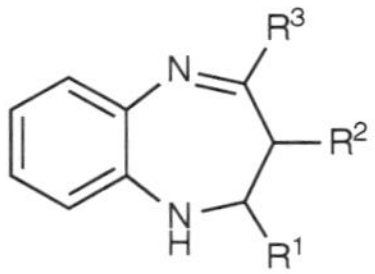 (yield: 35%)	206
23	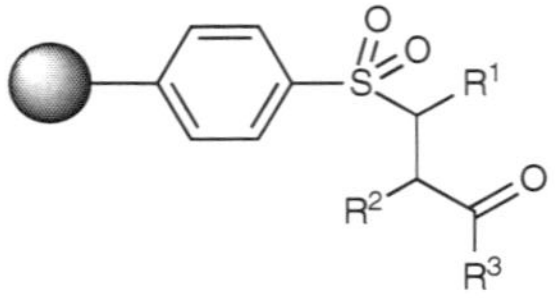	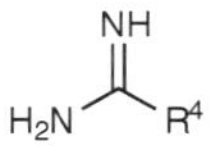	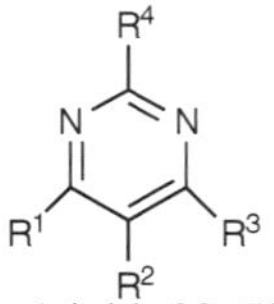 (yield: 20–53%)	206
24	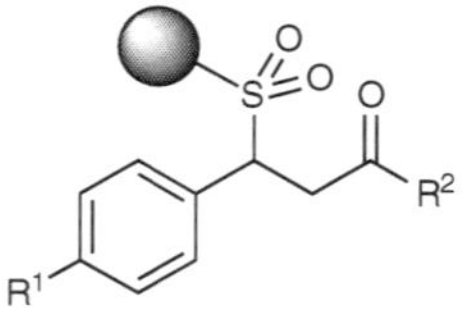	Swern oxidation	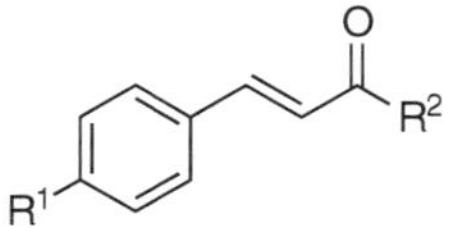 (yield: 82–90%)	207
25	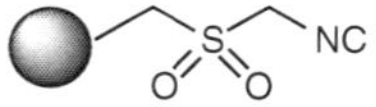	ArCHO, Bu_4NOH	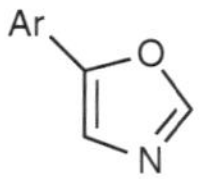 (yield: 25–50%)	208
26	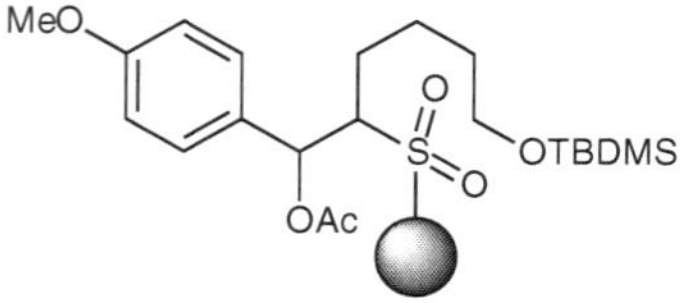	SmI_2, DMPU	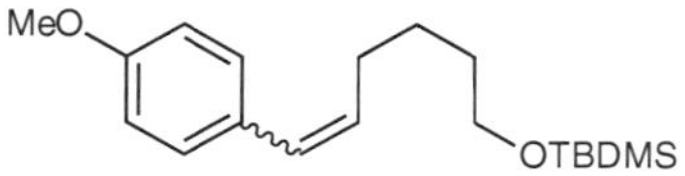 (yield: 27% over five steps)	209

27	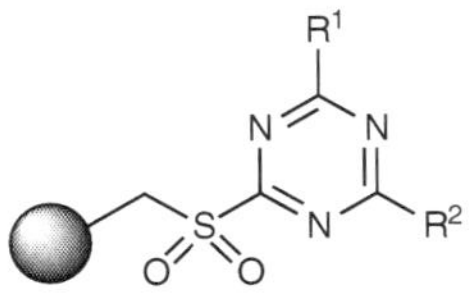	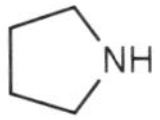, dioxane	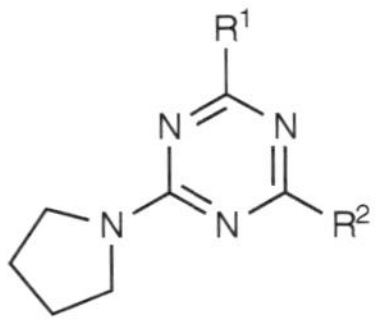 (yield: 46–65%)	210
28	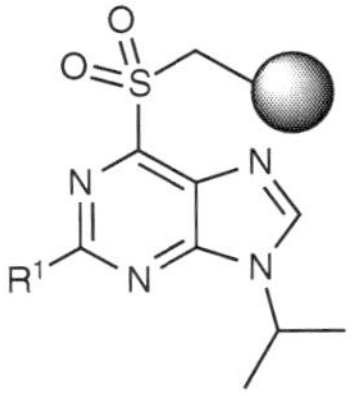	R^2NH_2	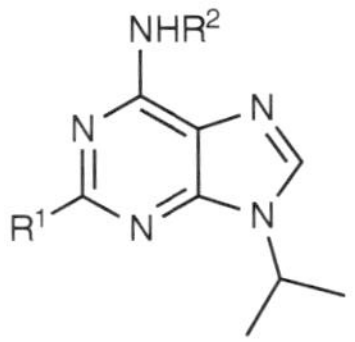 (yield: 10–25%)	211
29	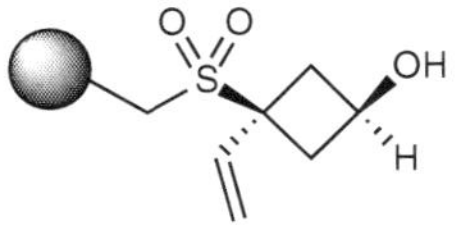	*i*PrMgCl, CuI, THF	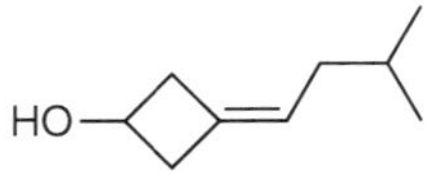 (yield: 10%)	212
30	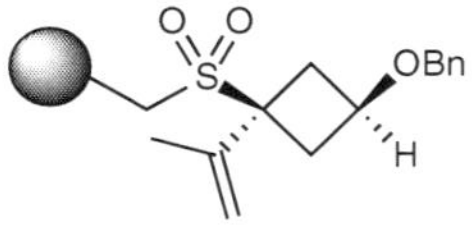	$Pd(PPh_3)_4$, THF	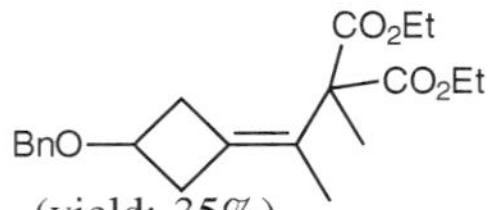 (yield: 35%)	213

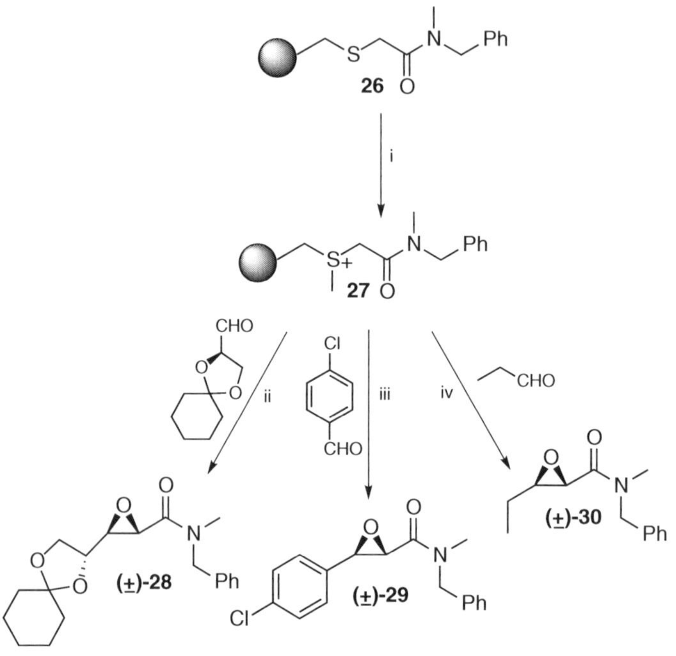

Scheme 1.10. Sulfonium-based multifunctional linker unit. (i) MeOTf, DCM, rt, 1 h; (ii) DBU, MeCN, rt, 1.5 h; (iii) DBU, DCM, rt, 3 h; (iv) DBU, DCM, rt, 1.5 h.

diversity (and safety-catch) cleavage can be achieved using Pummerer chemistry (Scheme 1.5).[133]

Sulfones can be prepared on-resin (as lithium phenyl sulfinate) by bubbling sulfur dioxide through a suspension of lithiated polystyrene resin.[212] However, analogous to sulfoxides, it is far more common to simply oxidize the corresponding thioethers with, for example, *m*CPBA[216], sodium periodate[217], or Oxone® ($KHSO_5$).[183] Traceless cleavage can be achieved from sulfone linkers using a dissolving metal reduction (Table 1.11, Entry 14)[200], or using Procter's attractive samarium chemistry (Table 1.11, Entry 15).[183] Alternatively, eliminative cleavage is possible (Scheme 1.11), via either type-1 that eliminates the product while generating resin-bound vinyl sulfones (Table 1.11, Entries 16–18)[202,203,217], or type-2 cleavage that eliminates olefinic products (Table 1.11, Entries 19–21).[131,204,205]

By varying the cleavage cocktail, it is also possible to generate very diverse libraries of heterocyclic species upon cleavage from sulfone linkers (Table 1.11, Entries 22–25). Such work has been extensively developed by Lam[206,218–220], Kurth[207,221,222], and Ganesan[208], among others, while De Clereq adapted the Julia–Lythgoe olefination into a cleavage approach (Table 1.11, Entry 26).[209] Alternatively, nucleophilic cleavage from sulfone linkers is also possible including cleavage using, for example, amines (Table 1.11, Entries

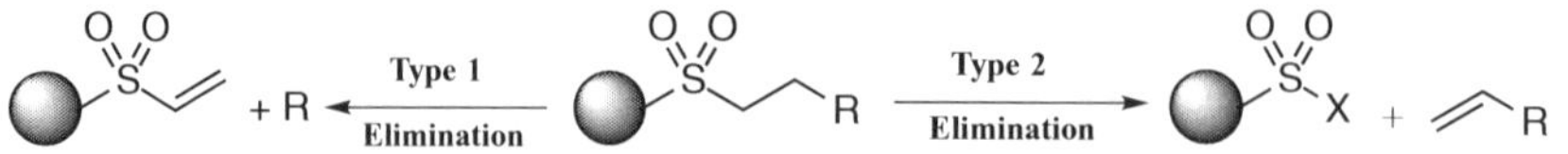

Scheme 1.11. Eliminative cleavage strategies.

27 and 28).[210] Other examples include cleavage from vinyl sulfones using organometallic approaches, as reported by Kurth (Table 1.11, Entries 29 and 30).[212,213] Similar techniques have also been reported by Blechert[223] and Brown[224] using ester-linked substrates.

Alkanesulfonate esters, such as mesylates and tosylates, and their more reactive perfluoroalkanesulfonyl counterparts, such as trifaltes and nonaflates, represent some of the best leaving groups available in organic synthesis. Reflecting this, both scaffolds have been developed into linker units for SPOS. Alkanesulfonate esters are widely used in nucleophilic substitution reactions and extensive examples of analogous multifunctional cleavage have been reported (Table 1.12, Entries 1–5).[225–230] For example, Roush was able to cleave trisaccharides using iodide, sodium acetate, or sodium azide to provide sugars ready for additional substitution if required (Table 1.12, Entry 1).[227] Related cleavage using Multipin™ systems was also reported by Takahashi.[228] The true extent of diversity that can be introduced into target libraries using this approach has been explored by Nicolaou, who prepared macrocyclic α-sulfonated ketones and then achieved multifunctional cleavage using many different nucleophiles (Table 1.12, Entries 2–5).[225,226] While nucleophilic cleavage of aliphatic sulfonate esters is quite common, analogous cleavage of the corresponding aryl sulfonate esters is comparatively rare. However, the discovery that they are viable substrates for cross-coupling reactions has been exploited in multifunctional cleavage approaches (Table 1.12, Entries 6–8).[231–233] Similarly, aryl perfluoroalkane sulfonate (PFS) esters are widely used as substrates for cross-coupling reactions, and PFS linker units, which exploit this, have also been developed by Pan and Holmes (Table 1.12, Entries 9 and 10).[234,235] Such cleavage can be traceless by using Pd-mediated transfer hydrogenation (Table 1.12, Entries 6 and 9)[233,234], or multifunctional by employing, for example, Suzuki conditions (Table 1.12, Entries 7 and 10)[232] or Grignard reagents (Table 1.12, Entry 8).[231] However, due to the complex synthetic sequences involved in preparing PFS linkers, their use has been limited. To address this issue, fluoroarylsulfonate linkers were reported, independently, by both Cammidge[236] and Ganesan[237] in 2004 (Table 1.12, Entries 11–13). Preparation of fluoroarylsulfonate linkers is more straightforward than their PFS counterparts, and analogous cleavage using cross-coupling conditions (Table 1.12, Entries 11 and 12) or transfer hydrogenation (Table 1.12, Entry 13) is viable.

Finally, thioesters are carboxylic acid derivatives that are known precursors to a wide range of compounds including alcohols and ketones. Thus, thioesters have been developed into linker units[238–241], although perhaps not to the extent expected due to difficulties involved in preparing resin-bound analogues. Kobayashi showed that reductive cleavage with lithium borohydride provided alcohols (Table 1.12, Entry 14)[238,239], a technique also employed by Bradley (Table 1.12, Entry 15).[240] However, Bradley extended the cleavage chemistry further, preparing tertiary alcohols using Grignard cleavage (Table 1.12, Entry 16) or ketones using softer organocuprate cleavage (Table 1.12, Entry 17).

1.3.3 Phosphorus Linker Units

Phosphorus reagents find widespread application in organic synthesis and, reflecting this, are playing increasingly important roles in modern SPOS. Beyond the many examples of immobilized phosphorus reagents as heterogeneous ligands for metal-catalyzed reactions, linker units based on phosphorus chemistry have also been developed.[242] These linker units are advantageous because phosphine oxide, a common by-product of many organophosphorus reactions, remains bound to the support, allowing facilitated purification strategies.

TABLE 1.12. Further Examples of Common Sulfur-Based Linker Strategies

	Linker	Cleavage Conditions	Product	References
1	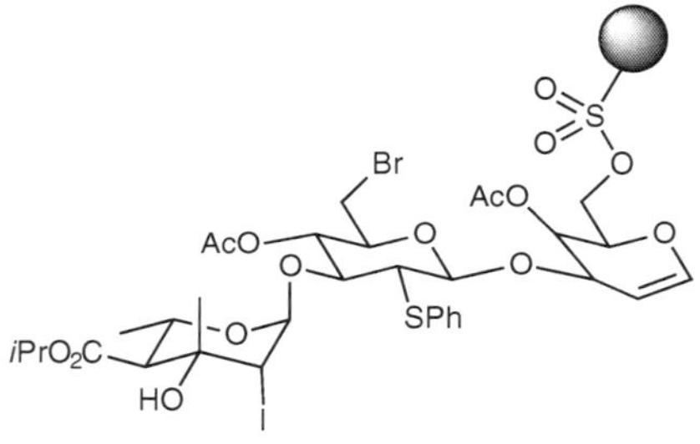	NaNu (Nu = I, OAc, N_3)	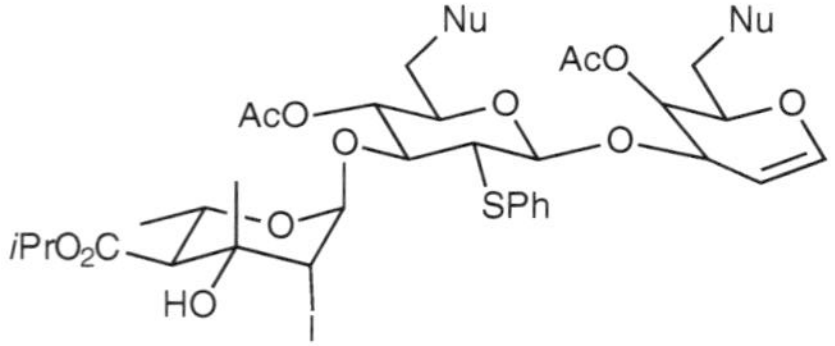	227
2		RXH (PhSH or MeOH)	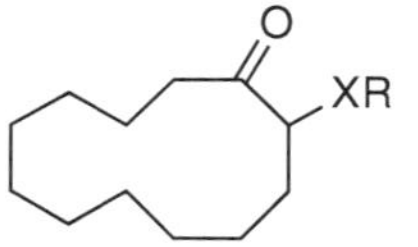 (yield: 95%; X = S and X = O)	225, 226
3	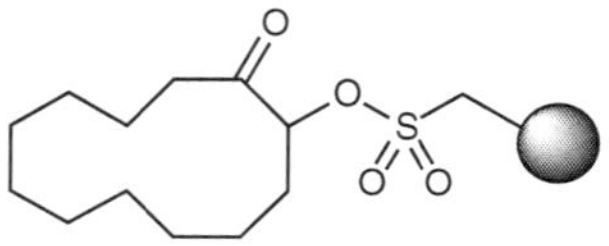	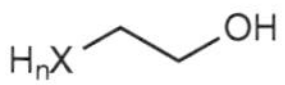	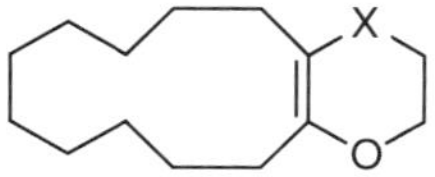 (yield: 60% (X = NH); 88% (X = S); 63% (X = O))	
4		PPTS, 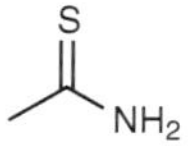	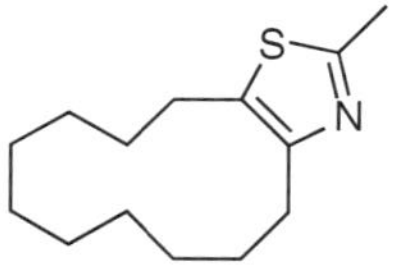 (yield: 83%)	
5		*hν*	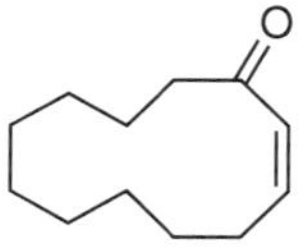 (yield: 84%)	

6	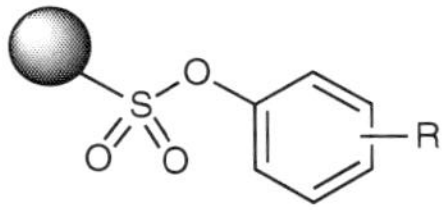	Et_3N, HCO_2H, $Pd(OAc)_2$, dppp, DMF	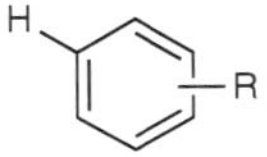	233
7	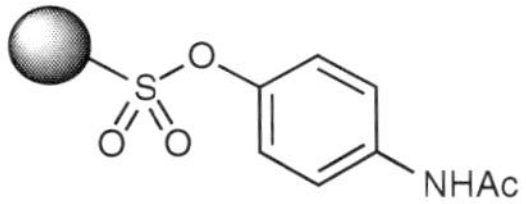	R-$B(OH)_2$, K_3PO_4, PCy_3, $NiCl_2(PCy_3)_2$, dioxane, 130°C; R-$B(OH)_2$, K_3PO_4, XPHOS, Pd $(OAc)_2$, 120°C	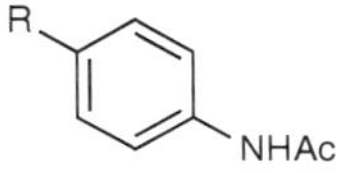 (yields: 60–65%)	232
8	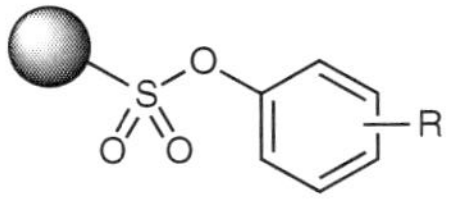	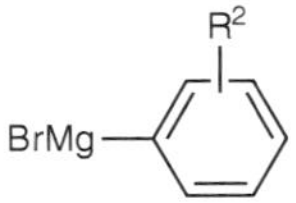 Et_3N, DCM	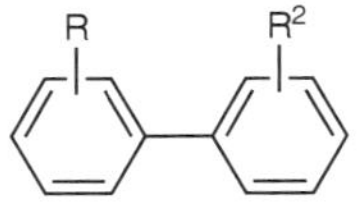 (yields: 64–81%)	231
9	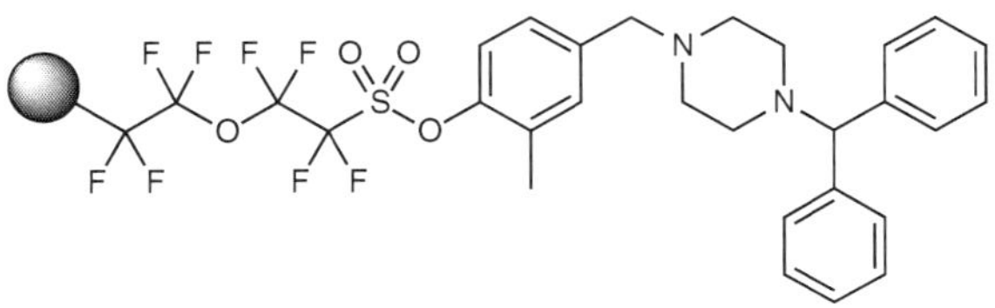	$Pd(OAc)_2$, dppp, DMF, Et_3N, HCO_2H	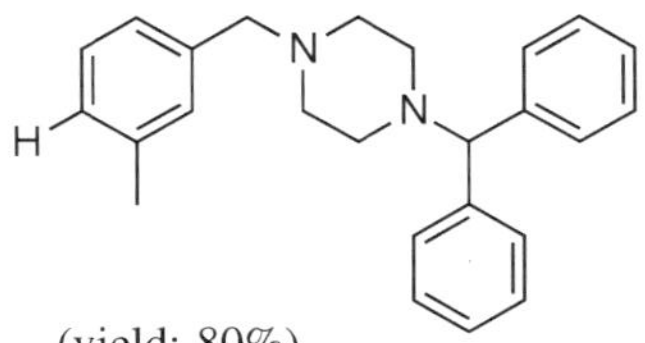 (yield: 80%)	234
10	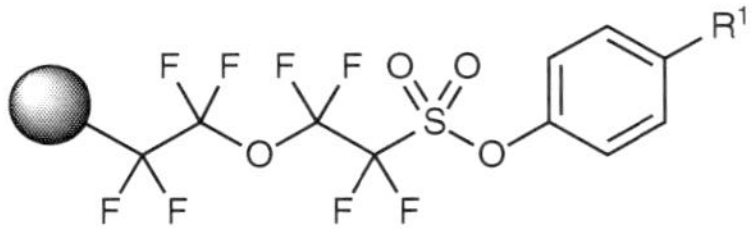	$PdCl_2$(dppf), EtN, DMF	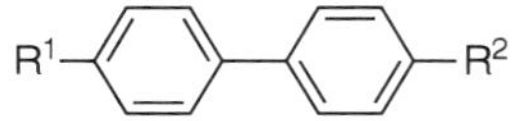 (yield: 62–88%, 10 examples)	235

(*Continued*)

TABLE 1.12. (*Continued*)

	Linker	Cleavage Conditions	Product	References
11		Ar-B(OH)$_2$, $PdCl_2$(dppf), K_2CO_3, THF/H_2O	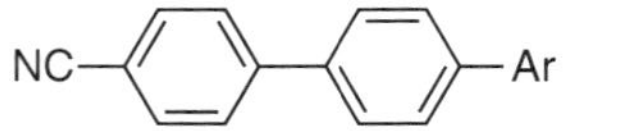	236
12	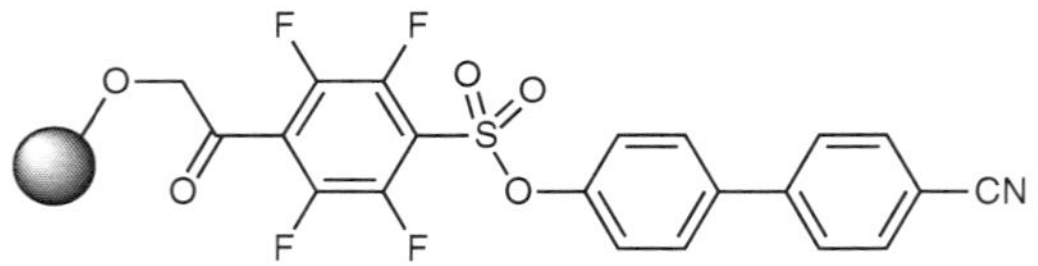	C_6H_{13}-ZnI, Ni$(PPh_3)_2Cl_2$, PPh_3, LiCL, THF, reflux	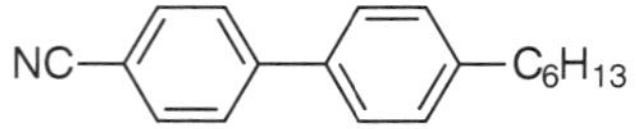 (yield: 75%)	236
13		$Pd(OAc)_2$, dppf, HCO_2H, Et_3N, 100°C	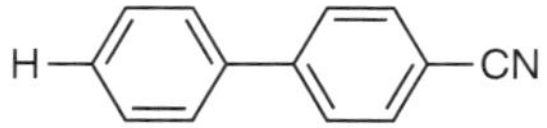 (yield: 52–75%)	236
14	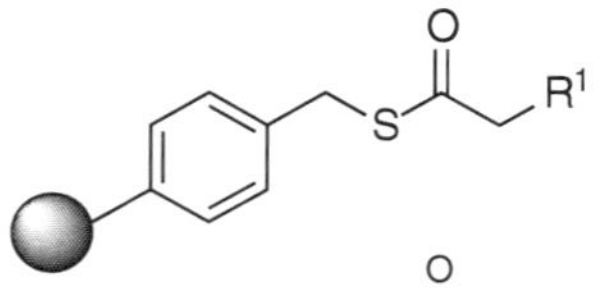	$LiBH_4$, Et_2O, rt	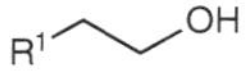	238, 239
15	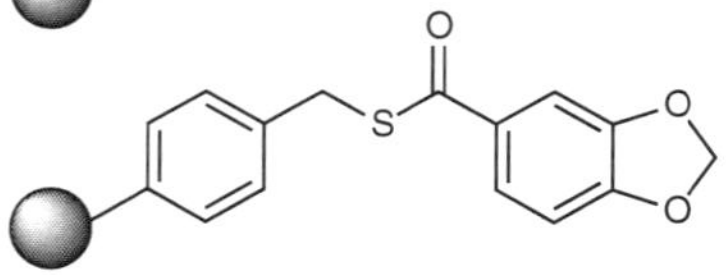	$LiBH_4$, THF, rt	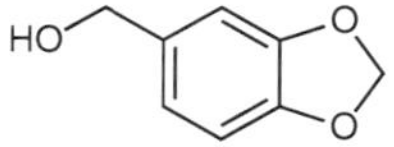 (yield: 83%)	240
16		PhMgBr, THF, 0°C	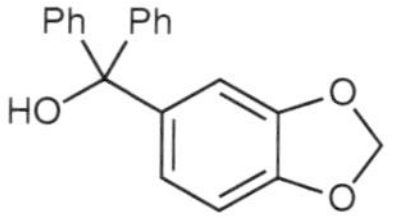 (yield: 45%)	240
17		Bu_2CuLi, THF, −78°C	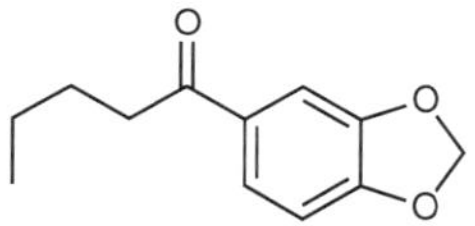 (yield: 53%)	240

The triaryl (or trialkyl) phosphine-mediated Wittig reaction is one of the most important olefin forming reactions available to the organic chemist. Supported ylides have been known for a considerable time and, indeed, a range of solid-phase Wittig reactions, in which diversity has been introduced by varying the aldehyde, have been reported since the first examples by Camps in 1971[243] and McKinley in 1972[244]. However, it is only more recently that the solid-phase Wittig reaction has truly begun to be exploited as a multifunctional linker strategy.[245–250] For example, Hughes showed that inter- and intramolecular cleavage was possible from supported ylides (Table 1.13, Entries 1 and 2, respectively).[245] Moreover, as for many of the linker units discussed herein, phosphorus-based linker units can function as traceless or multifunctional linker units with careful selection of an appropriate cleavage cocktail. Thus, Hughes also demonstrated that treatment with sodium methoxide and methanol allowed traceless cleavage of the corresponding alkane (Table 1.13, Entry 3).[245]

Beyond the original Wittig reaction, the Horner–Wittig and Horner–Wadsworth–Emmons (HWE) variants have also proven invaluable reactions for generating olefins. In the case of the HWE reaction, olefination of carbonyls can be achieved using phosphonate esters containing electron-withdrawing groups alpha to the nucleophilic carbanion. SPOS variants of the HWE have been reported (Table 1.13, Entry 4),[251–253] including an intramolecular variant employed to prepare macrolactones (Table 1.13, Entry 5).[254]

While the most common examples of diversity cleavage using phosphorus linkers have focused on this powerful olefination chemistry, other pertinent examples should be mentioned. Noticeably, cyanophosphoranes can be oxidatively cleaved (ozone or dimethyldioxirane) in the presence of a nucleophile (alcohol or amine) to provide α-keto esters and α-keto amides (Table 1.13, Entry 6).[255]

Finally, the palladium-catalyzed cross-coupling reactions with supported enol phosphonates were reported by Steel and coworkers (Table 1.13, Entry 7).[256] Polymer-supported lactam enol phosphonates were prepared and multifunctional cleavage was demonstrated, using Suzuki conditions, to provide aryl enamines in good yields.

1.3.4 Selenium and Tellurium Linker Units

Building on the many examples of thioether linker units, larger numbers of linker units have been reported that utilize the related reactivity profiles of selenium and tellurium components to achieve multifunctional cleavage.[5,179,180,257,258] Such linkers tend to be straightforward and can actually be considered functionalized resins. For example, selenyl chloride (**31**) and selenyl bromide (**32**) resins are electrophilic in nature and can be used to load nucleophilic species (Scheme 1.12) to give, for example, **33**. Alternatively, reactivity can be reversed by treating the selenyl halide resin with lithium (or sodium) borohydride to provide the corresponding supported lithium selenide (**34**)—a nucleophilic source of selenium onto which electrophilic substrates can be loaded to give species such as **35**. Radical loading strategies have also been reported, but they are much less common.

By far the most common cleavage strategy for releasing substrates from selenium linker units is oxidative cleavage. For example, many groups have reported cleavage using hydrogen peroxide (Table 1.14, Entries 1–5)[259–263], *tert*-butylhydroperoxide (Table 1.14, Entries 6 and 7)[264,265], or *meta*-chloroperbenzoic acid (*m*CPBA) (Table 1.14, Entries 8–10).[266–268] The mechanism proceeds via oxidative cleavage, with initial oxidation of the selenium to the corresponding selenoxide. Elimination then provides alkenes (Table 1.14, Entries 1–4) or, in certain cases, alkynes (Table 1.14, Entry 5). In an interesting example reported by Nicolaou, it was shown that cleavage of a pyran bearing a free hydroxyl group proceeded as expected to

TABLE 1.13. Phosphorus Linker Units

	Linker	Cleavage Conditions	Product	References
1		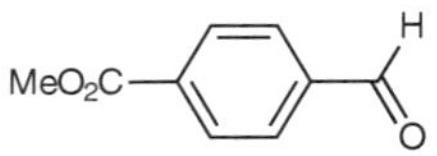 NaOMe, MeOH, reflux	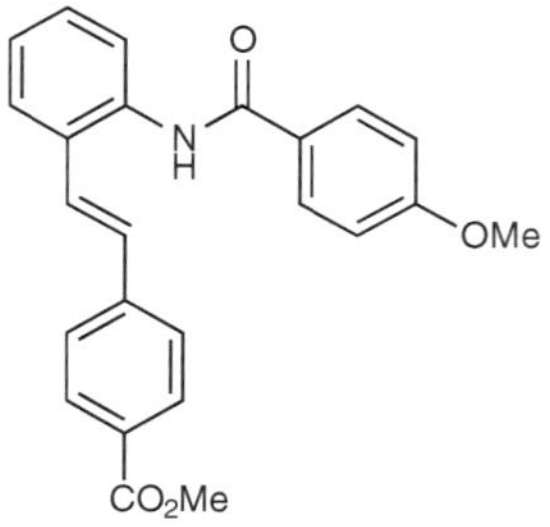 (yield: 82%, *E*/*Z*: 3:1)	245
2	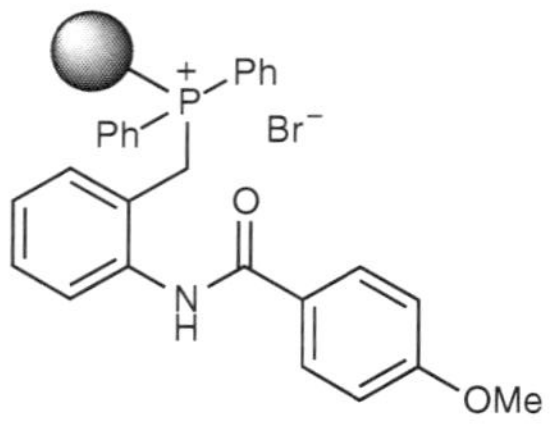	(i) Toluene, DMF, distill; (ii) KO^tBu, reflux	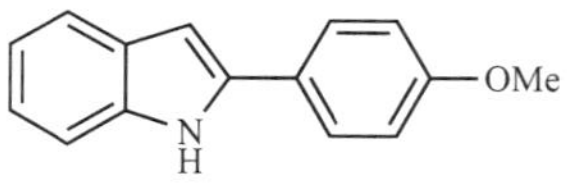	245
3		NaOMe, MeOH, reflux	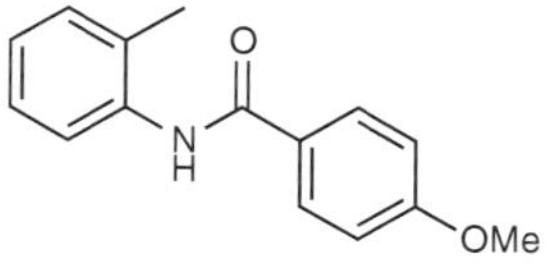 (yield: 81%)	245

4	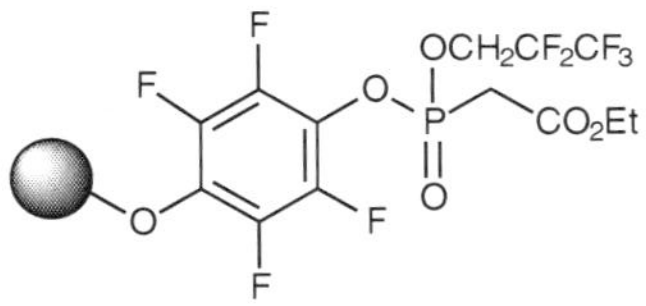	RCHO, NaH, 25°C	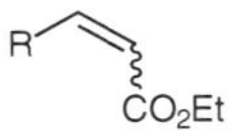 (yield: 46–96%)	251
5	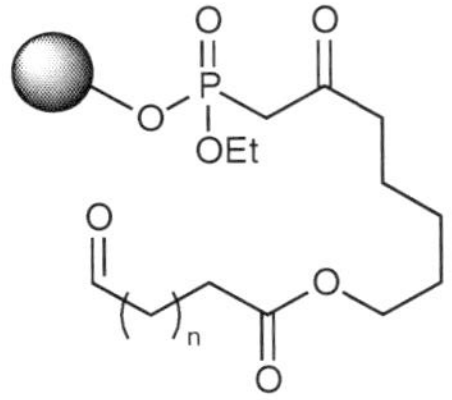	K_2CO_3, 18-crown-6, 65°C	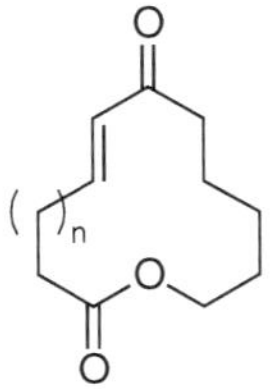 (yield: 58% ($n = 7$), 62% ($n = 9$); *E*/*Z*: >9:1)	254
6	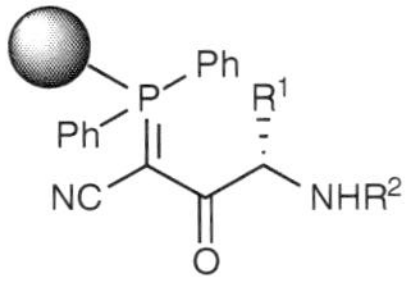	Nu (R^3OH or R^3NH_2), DMSO, DCM, rt	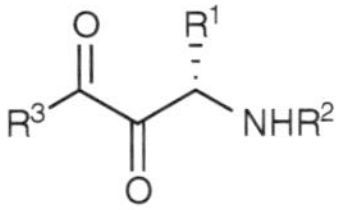 (yield: 30–65% over 4 steps, 11 examples)	255
7	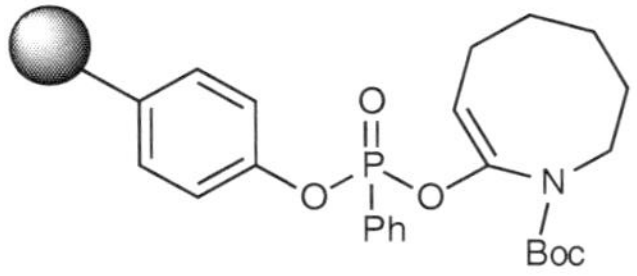	$Pd(PPh_3)_4$, $ArB(OH)_2$, Na_2CO_3, DME/H_2O/EtOH, 80°C	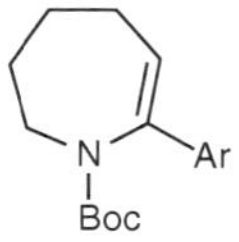 (yield: 21–72% over two steps)	256

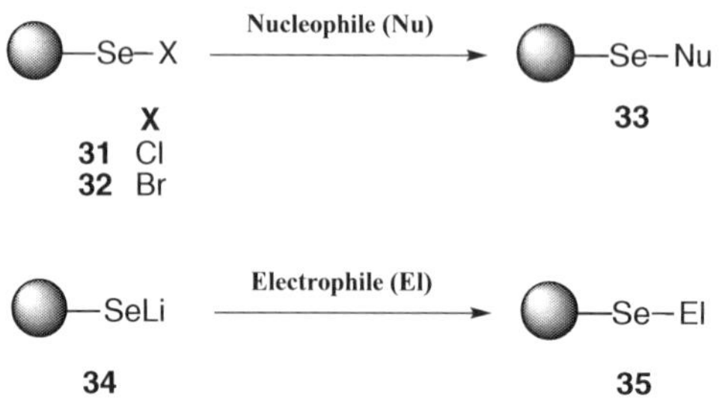

Scheme 1.12. Common selenium linker units.

yield the dihydropyran (Table 1.14, Entry 9). However, if the free hydroxyl group was protected with a TBS group (Table 1.14, Entry 10), then analogous cleavage released the corresponding tetrahydropyran.

Beyond common oxidative cleavage, nucleophilic cleavage from selenium and tellurium linkers can also occur. The nucleophilic substitution can be halogenation (Table 1.14, Entries 11 and 12),[269,270] or an organometallic such as copper acetylide (Table 1.14, Entry 13).[271] Finally, homolytic cleavage via a radical mechanism has also proven a powerful cleavage technique. Such cleavage is traceless and can be mediated by AIBN and tributyltin hydride (Table 1.14, Entry 14)[272] or AIBN/tris(trimethylsilyl)silane (Table 1.14, Entry 15).[273] Tellurium linker units are cleavable via the same mechanisms (Table 1.14, Entry 16)[274,275], although there does not appear to be any significant advantage to using them over selenium linker units.

1.3.5 Silyl and Germyl Linker Units

The use of silyl ethers as protecting groups for the hydroxyl functionality is well known, and their adaption into linker units was a welcome addition to the SPOS literature. Thus, many silyl linker units have been reported for alcohols, and a selection is illustrated in Table 1.15. Owing to the large number of reported examples, a complete discussion of each is beyond the scope of this chapter, but Spivey has written a complete review.[58] Much like deprotection of their solution-phase counterparts, cleavage from silyl linkers can be achieved using, for example, HF (Table 1.15, Entries 1–3 and 6)[276], TBAF (Table 1.15, Entries 2–5)[49,276,277], AcOH (Table 1.15, Entries 1 and 4)[50], or TFA (Table 1.15, Entry 5). Beyond alcohols, other traditional silyl linker units are useful for SPOS with other substrates such as amines (Table 1.15, Entry 6)[278].

Beyond their use as standard linker units, silicon-based linker units have found extensive use as traceless linker units for aromatics by exploiting *ipso*-substitution, under acidic conditions, to leave a hydrogen residue at the cleavage site (see also Section 1.2.3). Such cleavage is also achieved using, for example, HF (Table 1.16, Entries 1–3)[96], TFA (Table 1.16, Entries 2 and 3)[279], or TBAF (Table 1.16, Entry 4).[280] Beyond the traditional traceless silyl linker units, reactivity toward acidic cleavage can be increased by incorporation of a β-amide into the linker unit (Table 1.16, Entry 2).[279] One interesting example was the silyl linker reported by Showalter (**36**, Scheme 1.13).[281] Treating substrates attached to this linker with TBAF at 45°C resulted in traceless cleavage (**37**), while analogous treatment with TBAF at rt cleaved the dialkylarylsilanol (**38**).

By varying the electrophile, this class of linker can also be utilized in a multifunctional approach. In its simplest form, this has involved halogenation. For example, cleavage strategies for leaving bromine (Table 1.16, Entry 3)[282,283] or iodine (Table 1.16,

TABLE 1.14. Selenium and Tellurium Linker Units

	Linker	Cleavage Conditions	Product	References
1	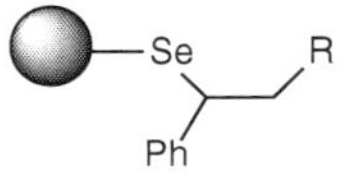	HOOH	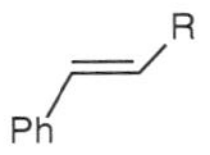 (yields: 71–91%, seven examples)	259
2	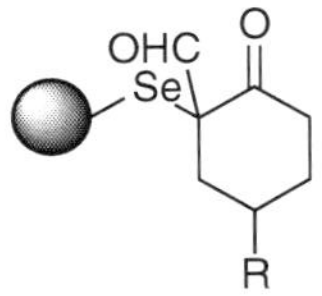	HOOH	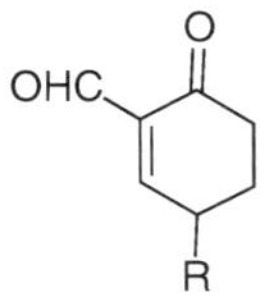 (yield: 85–90%, four examples)	260
3	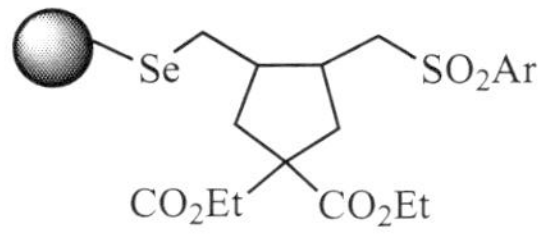	HOOH	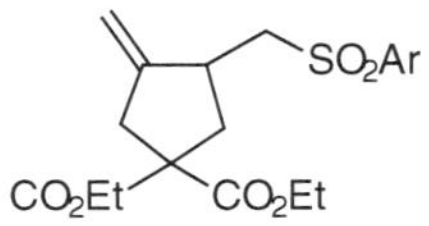 (yield: 44–58%, eight examples)	263
4	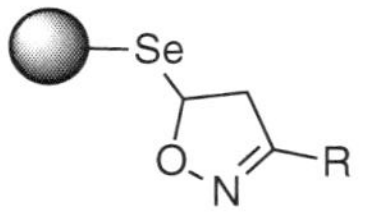	HOOH	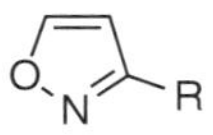 (yield: 78–88%, 32 examples)	262
5	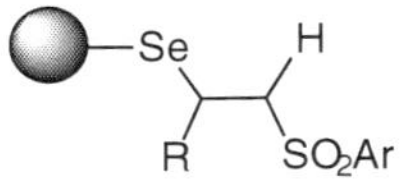	HOOH	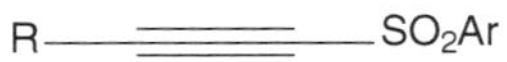 (yield: 69–90%, 13 examples)	261

(Continued)

TABLE 1.14. (*Continued*)

	Linker	Cleavage Conditions	Product	References
6	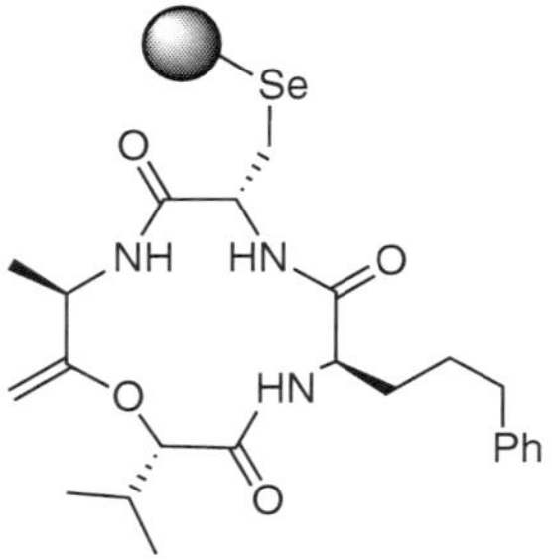	*t*BuOOH	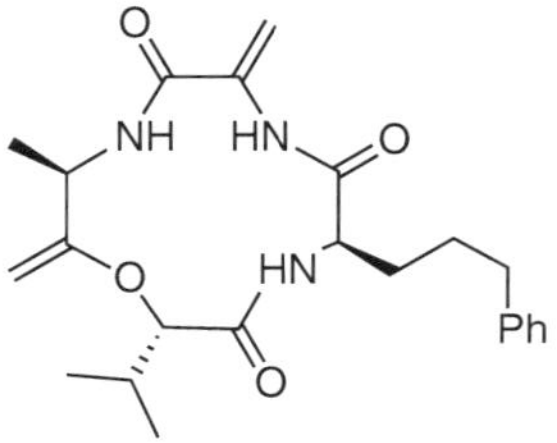 (yield: 92%)	264
7	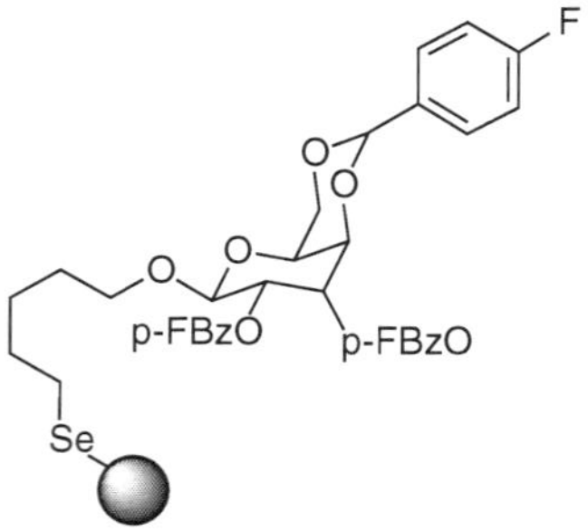	*t*BuOOH	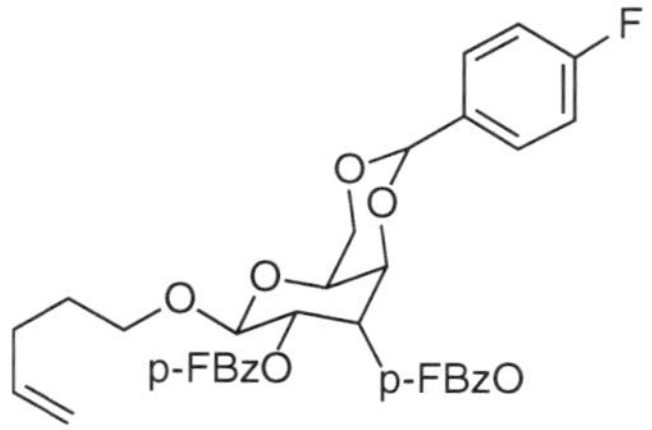 (yield: 92%)	265
8	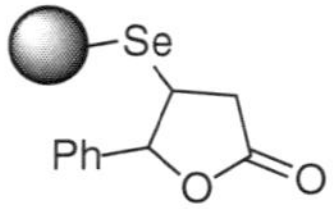	*m*CPBA	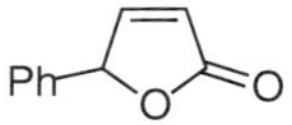 (yield: 46%)	267

9	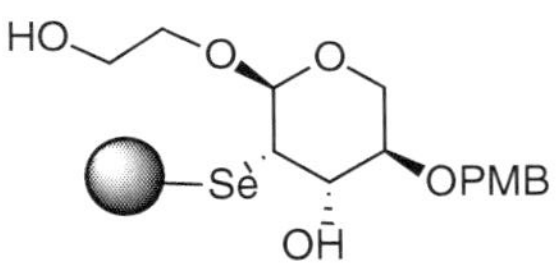	*m*CPBA	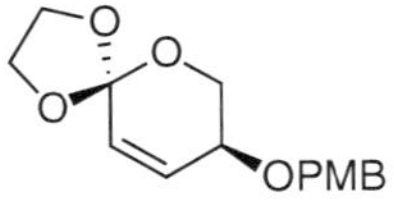 (yield: 81%)	266, 268
10	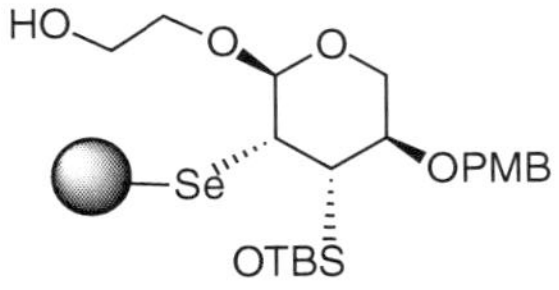	*m*CPBA	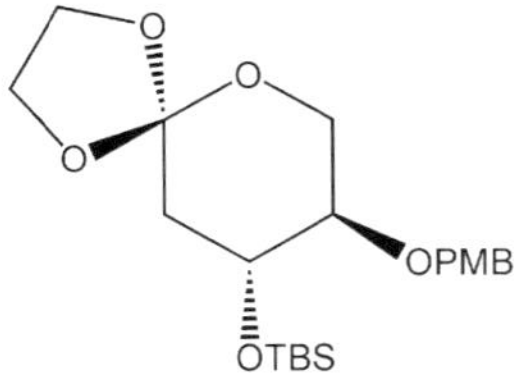 (yield: 85%)	266, 268
11	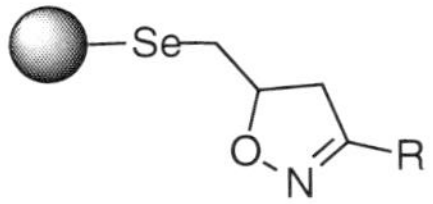	MeI	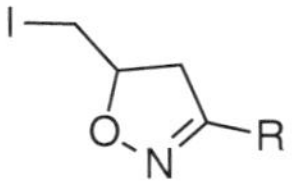 (yield: 71–87%, seven examples)	269
12	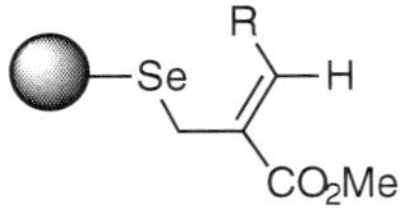	MeI	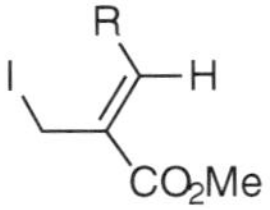 (yield: 78–95%, 10 examples)	270
13	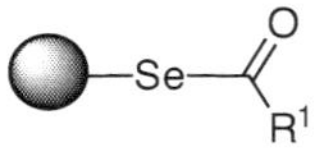	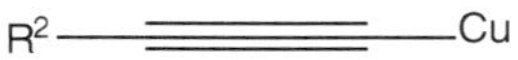	 (yield: 73–92%, nine examples)	271

(*Continued*)

TABLE 1.14. (*Continued*)

	Linker	Cleavage Conditions	Product	References
14	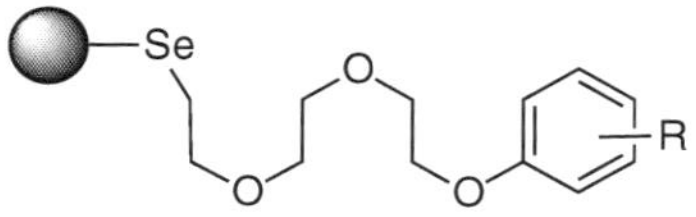	AIBN, Bu_3SnH	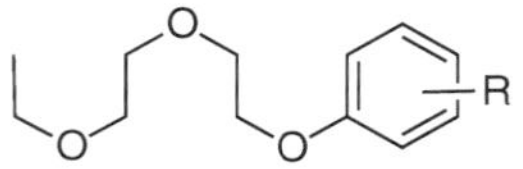 (yield: 78–88%, three examples)	272
15	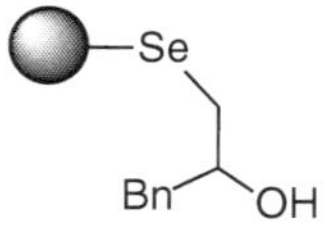	AIBN, $(Me_3Si)_3SiH$	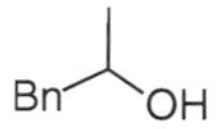 (yield: 97%)	273
16	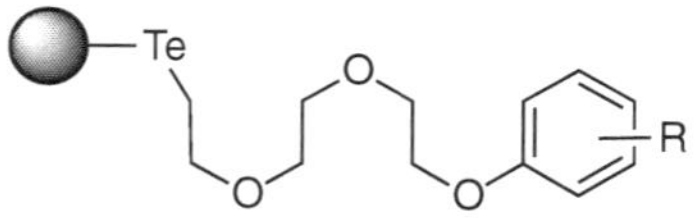	AIBN, Bu_3SnH	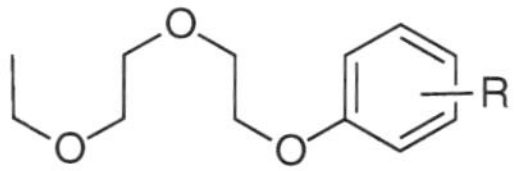 (yield: 73–83%, three examples)	274, 275

TABLE 1.15. Silyl Linker Units for Alcohols

	Linker	Cleavage Conditions	Product	References
1	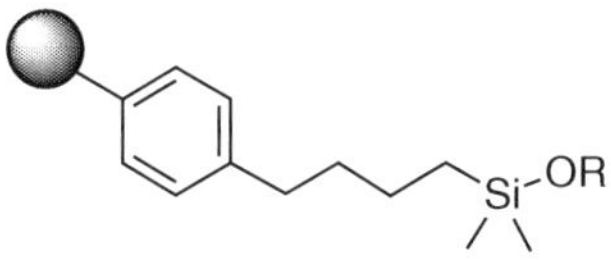	Aq HF–Pyr; TBAF, THF; AcOH, THF, H_2O	ROH	49, 50
2	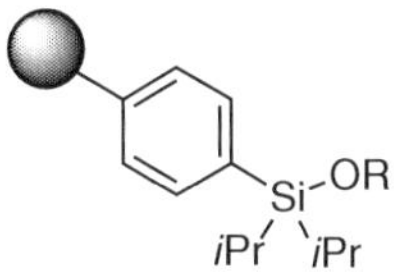	Aq HF–Pyr; TBAF, THF	ROH	
3	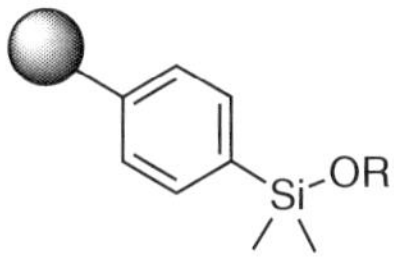	HCl (aq), dioxane; HF (aq); TBAF	ROH	276
4	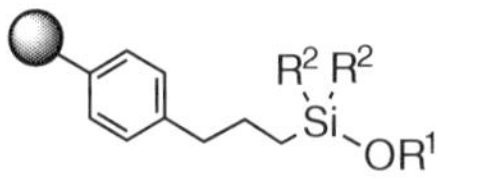	Aq HF–Pyr; TBAF, THF; TFA, MeOH; ACOH, MeOH, rt	R^1-OH	
5	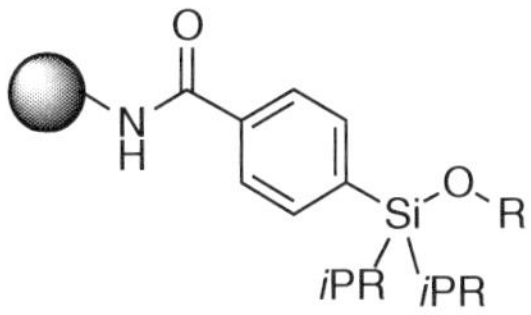	1 M TBAF, THF, rt	ROH	277
6	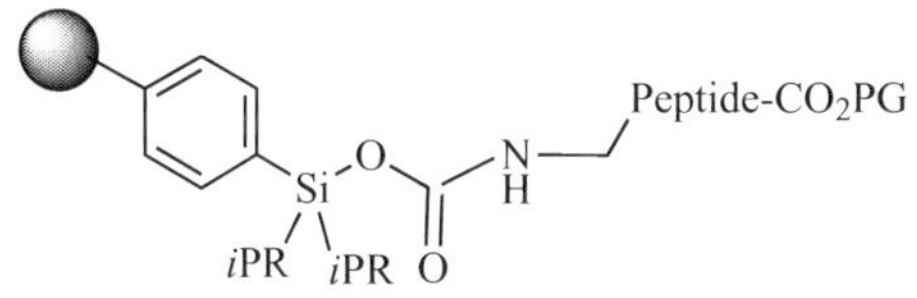	HF, MeCN	H_2N-peptide-CO_2PG	278

Entries 3 and 5)[282–284] residues at the cleavage site have been reported. Similar results have been obtained using germyl linkers (Table 1.16, Entries 6 and 7). Traceless cleavage can be achieved by treating with TFA (Table 1.16, Entries 6 and 7)[98], while Spivey demonstrated diversity cleavage to leave halogens at the cleavage site (Table 1.16, Entry 7).[99–102]

Beyond simple halogenation of aromatics, other substrates supported through silyl linker units can be cleaved in a diversity fashion. For example, allylsilanes can be combined

TABLE 1.16. Multifunctional Silyl and Germyl Linkers

	Linker	Cleavage Conditions	Product	References
1	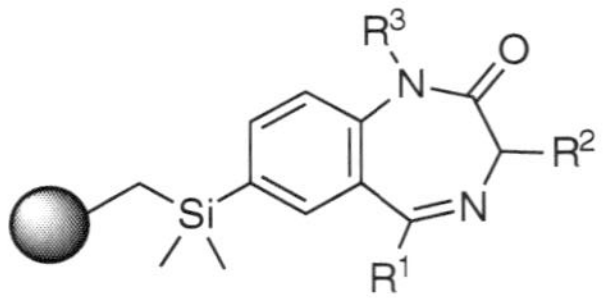	Aq HF	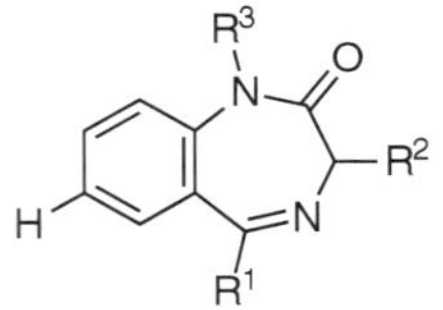 four examples (yield: 50–68%)	96
2	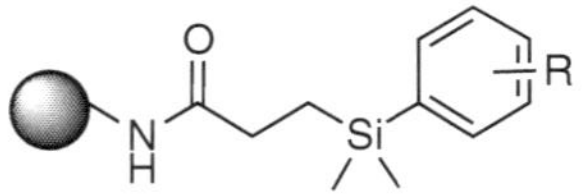	TFA vapor	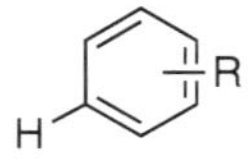	279
3	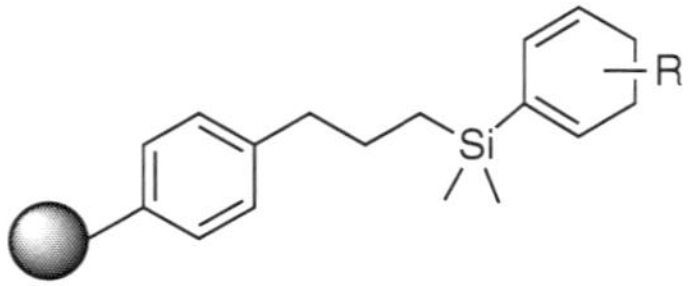	X = H: TFA; X = Br: Br_2, DCM; X = I: ICl, DCM, rt	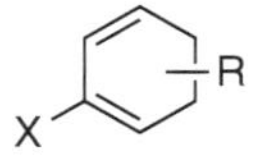 X = H, Br, I	282, 283
4	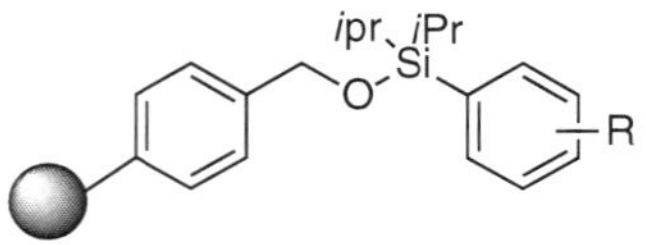	TBAF, THF	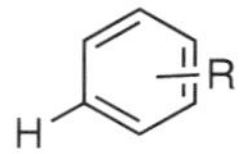	280
5	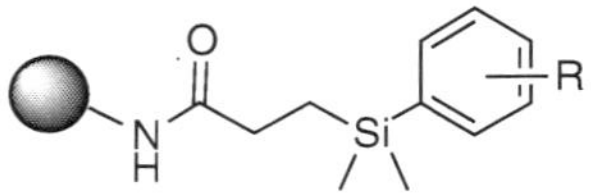	ICl, DCM, rt	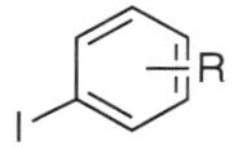	284

6	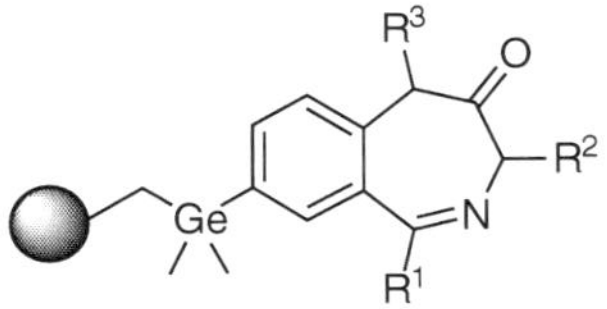	X = H: TFA; X = Br: Br_2	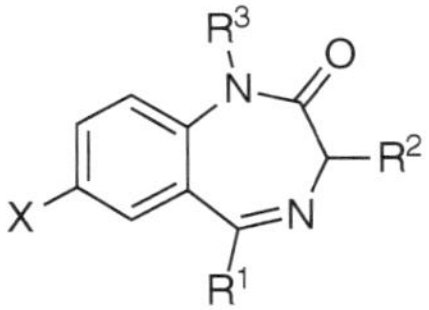 X = H, Br (yield: 50–68%, 12 examples)	98
7	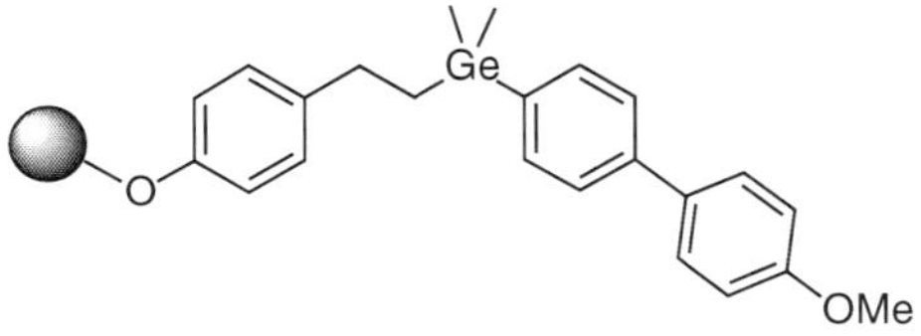	X = H: TFA, rt; X = Cl: NCS, TFA, reflux; X = Br: Br_2, rt; X = I: ICl, rt	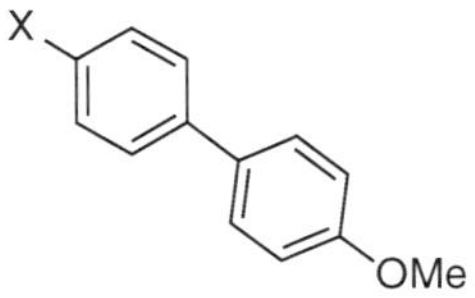 X = H, Cl, Br, I	99–102
8	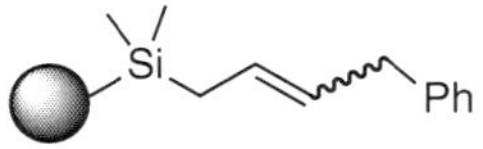	3% TFA/DCM	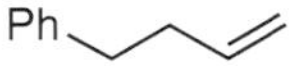	285
9		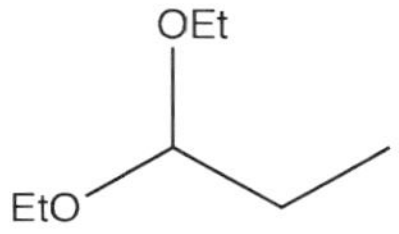	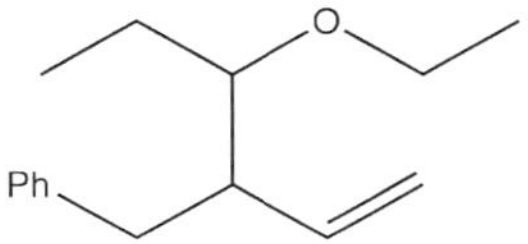	285
10	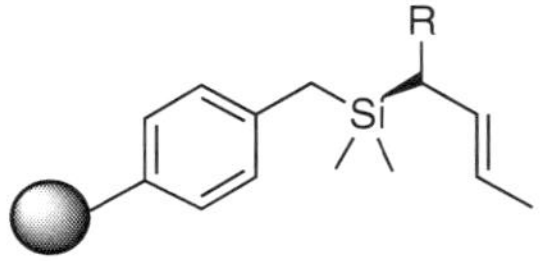	R^1CHO, $TiCl_4$, DCM, −78°C	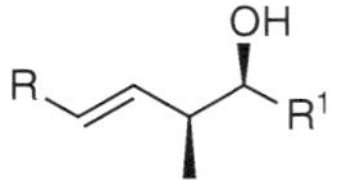 (yield: 34–54%, ee: 93–99% *syn*/*anti*: >93:7, four examples)	286

(*Continued*)

TABLE 1.16. (*Continued*)

	Linker	Cleavage Conditions	Product	References
11	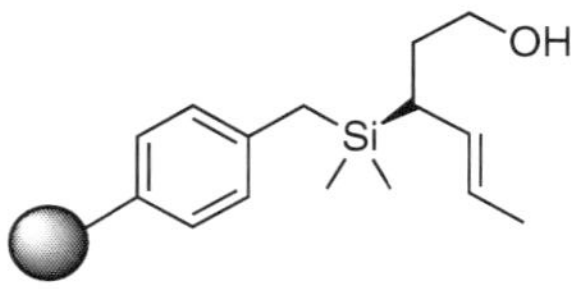	R^1CHO, TMSOTf, DCM, −78°C	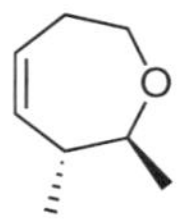 (yield: 59–70%, ee: 97–98%, *syn*/*anti*: >91:9, five examples)	286
12		TFA	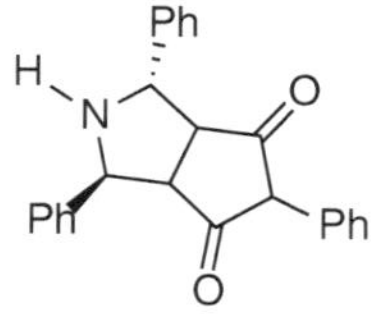 (yield: 90%)	287
13	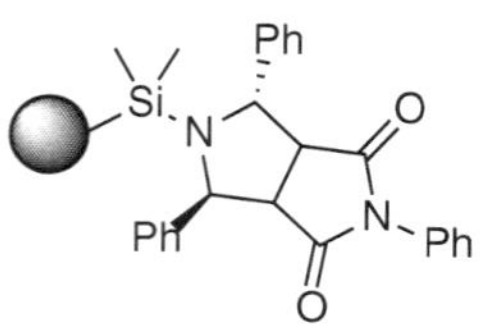	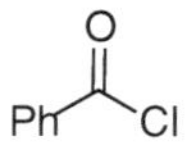	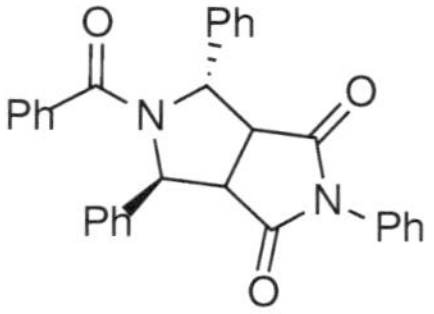 (yield: 70%)	287
14		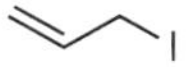	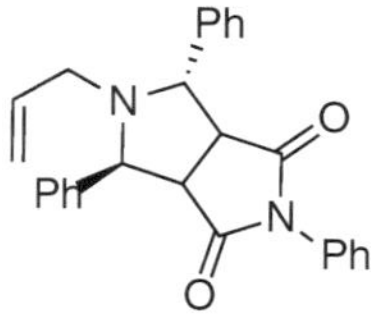 (yield: 45%)	287

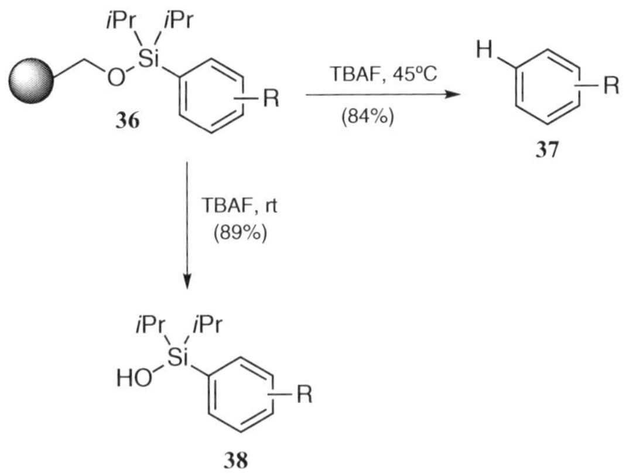

Scheme 1.13. Showalter's silyl linker unit.

with a range of electrophiles to generate carbocations, which fragment to liberate alkenes (Table 1.16, Entries 8–11). The first example was reported by Blechert, who showed that treating immobilized allylsilanes with TFA released alkenes (Table 1.16, Entry 8), while treatment with acetals in the presence of $TiCl_4$ cleaved more diverse products (Table 1.16, Entry 9).[285] The concept was extended by Suginome and Ito to include ciral allylsilane scaffold. Treatment with aldehydes, in the presence of a Lewis acid, cleaved chiral alcohols (Table 1.16, Entry 10) and cycloheptenes (Table 1.16, Entry 11), with good conservation of stereochemistry.[286]

Finally, silicon–nitrogen bonds are also labile in the presence of electrophiles, and this concept has been used by Komatsu in development of a silylimine linker (Table 1.16, Entries 12–14).[287] Multifunctional cleavage was achieved with a range of electrophiles. For example, treatment with TFA left a hydrogen residual at the point of attachment (Table 1.16, Entry 12), while benzoyl (Table 1.16, Entry 13) and allyl (Table 1.16, Entry 14) groups were introduced by cleavage with benzoyl chloride and allyl iodide, respectively.

1.3.6 Boron and Stannane Linker Units

The palladium-catalyzed cross-coupling reactions revolutionized carbon–carbon bond formation following their introduction from the late 1970s to early 1980s. Two of the earliest reported examples were the Stille[288] and Suzuki–Miyaura[289] reactions in which, for example, aryl halides can be cross-coupled with organostannanes or organoborons, respectively, in the presence of a palladium catalyst. Both these pronucleophilic components have been adapted into linker strategies so that the vast array of commercially available aryl (and vinyl) halides can be exploited in palladium-mediated multifunctional cleavage.[290]

The Stille reaction has become an extremely powerful C–C bond forming reaction in organic synthesis.[291–294] The first example of a stannane linker unit was reported by Kuhn in 1994.[295] A significant advantage of using such polymer-supported tin species is that following cleavage using, for example, Stille conditions, the trialkyltin by-products remain attached to the polymer. This allows their straightforward removal by standard SPOS filtration techniques, overcoming purification problems that have consistently hampered the use of organostannanes in traditional organic synthesis. For example, Kuhn loaded a range of alkenes and alkynes onto a stannane linker unit and multifunctional cleavage was demonstrated through palladium-catalyzed cross-coupling with a range of electrophiles (Table 1.17, Entries 1–3).[295] So attractive is the idea of leaving trialkyltin by-products attached to the resin that Nicolaou exploited the technique in his synthesis of (*S*)-zearalenone on a solid support.[296] In this natural product synthesis, the Stille reaction was intramolecular and so the bifunctional linker unit behaved concomitantly as a multifunctional and cyclorelease linker. The supported iodide precursor was treated with $Pd(PPh_3)_4$ and cleavage provided (*S*)-zearalenone (Table 1.17, Entry 4).

Beyond the Stille reaction, halogen displacement of trialkyltin groups is a widely used reaction in organic synthesis. Thus, halogenation of substrates attached to supports via stannane linker units has also been reported. This reaction is particularly attractive for radiochemists and has been used to prepare radiopharmaceuticals labeled with iodine-123 or iodine-131, such as [^{131}I]MIBG (Table 1.17, Entry 5), and [^{123}I]Congo red (Table 1.17, Entry 6).[297–299]

Despite the advantages of using the polymer-supported trialkyltin species outlined above, the toxicity of tin compounds still favors use of less toxic alternatives when available. Most noticeably, the ready availability of less toxic, water-soluble organoboronates has led to the Suzuki–Miyaura reaction becoming the preferred palladium-catalyzed cross-coupling reaction. The reaction was first reported in 1981[289], and many protocols have since been reported and reviewed.[291,292,300–305] A number of solid-phase variants have been developed. For example, Burgess reported a boron linker unit (Table 1.18, Entry 1) using Rink amide resin.[306] A drawback of this approach was that diversity resulting from the boronic acid had to be introduced early in solution phase. This was overcome by loading boronic acids directly onto resin-bound diols, as reported by both Hall[307] and Carboni[308] groups. Cleavage from each of these linker units has been demonstrated using Suzuki chemistry to provide biphenyls (Table 1.18, Entries 1 and 2).[306,309] The pseudo-high-dilution conditions of solid phase minimize production of homo-coupled boronic acids, which is a commonly observed side reaction of solution-phase Suzuki reactions.

Beyond Suzuki chemistry, and reflecting the increasingly important role of boronic acids in organic synthesis, alternative cleavage strategies are also possible using the organoboronate linker units. For example, Carreaux and Carboni demonstrated that traceless cleavage is possible if resin-bound organoboronates are treated with Tollen's reagent (Table 1.18, Entry 3).[310] An alternative (and also traceless) mode of cleavage is to treat with water (Table 1.18, Entry 4) or methanol (Table 1.18, Entry 5) to cleave the boronic acids and methyl boronate esters, respectively.[307,308,311,312]

Finally, additional multifunctional cleavage strategies are possible. Oxidative cleavage with hydrogen peroxide and sodium hydroxide can be used to generate phenols (Table 1.18, Entry 6)[308], while rhodium(I)-mediated 1,2-addition to aldehydes provides secondary alcohols (Table 1.18, Entry 7) and 1,4-addition to α,β-unsaturated ketones yields ketones (Table 1.18, Entry 8).[309]

1.3.7 Bismuth Linker Units

The use of triarylbismuth(III) species, particularly in arylation reactions, is a growing area of research in organic synthesis that has been recently reviewed.[313] Unlike the related palladium-catalyzed cross-coupling reactions, which often require development of substrate-specific reaction conditions, the organobismuth reactions appear more versatile. This simple chemistry, in addition to the broad range of applications, makes

TABLE 1.17. Organostannane Linker Units

	Linker	Cleavage Conditions	Product	References
1	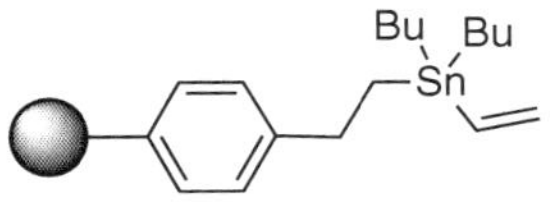	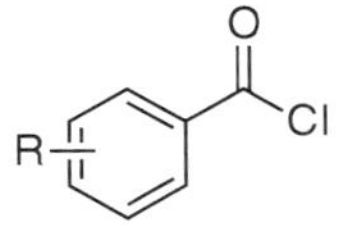 $Pd(PPh_3)_4$, toluene	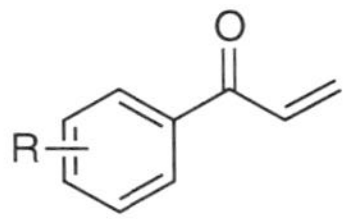 (yield: 95%)	295
2	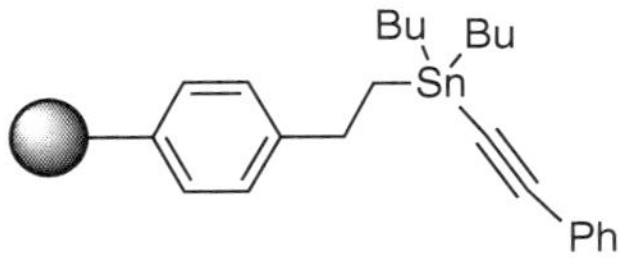	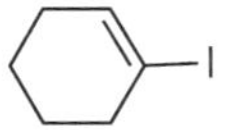 $Pd(PPh_3)_4$, toluene	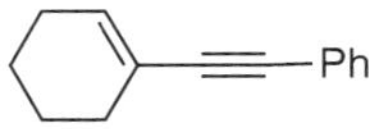 (yield: 51%)	295
3	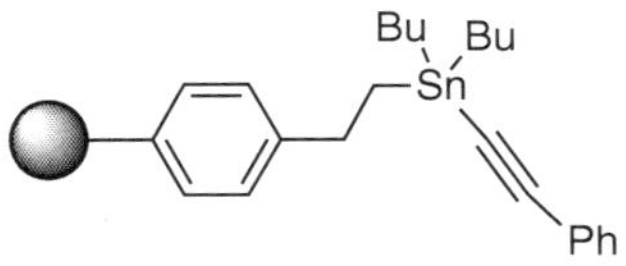	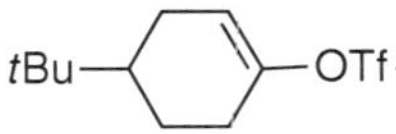 $Pd(PPh_3)_4$, toluene	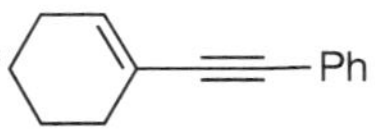 (yield: 53%)	295
4	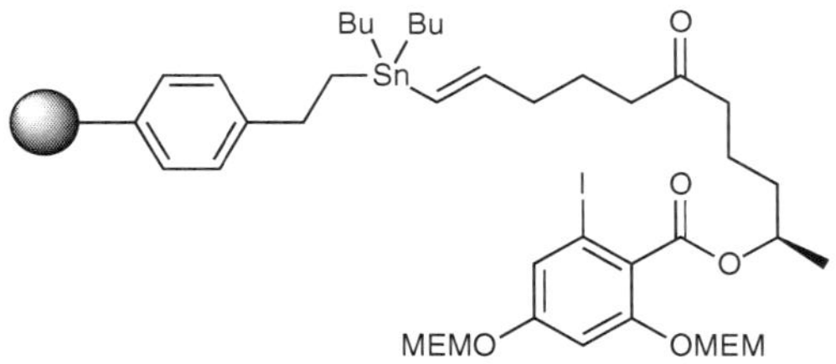	$Pd(PPh_3)_4$, toluene	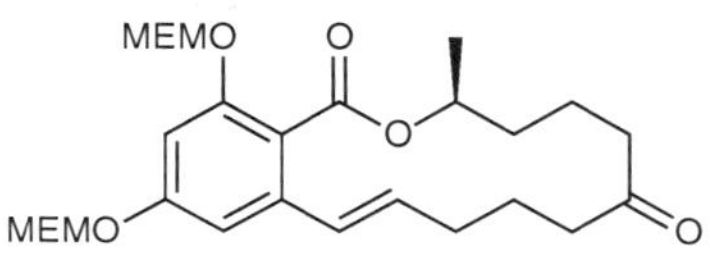 (yield: 54%)	296

(*Continued*)

TABLE 1.17. (*Continued*)

	Linker	Cleavage Conditions	Product	References
5	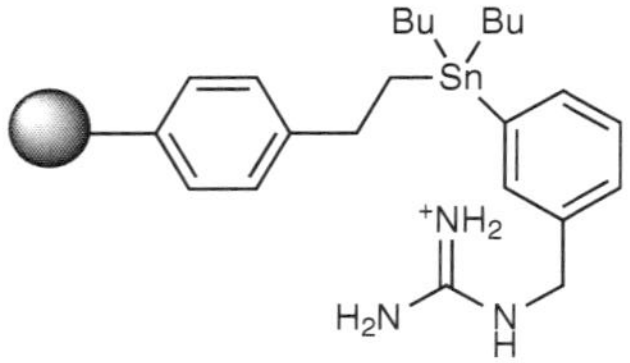	$Na^{131}I$, NaH_2PO_4, H_2O_2, HOAc, MeOH	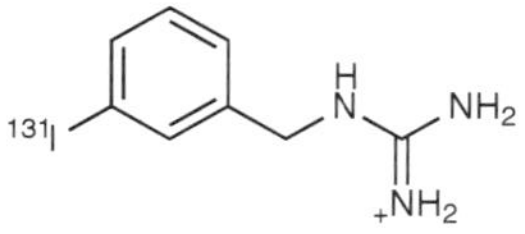	297
6	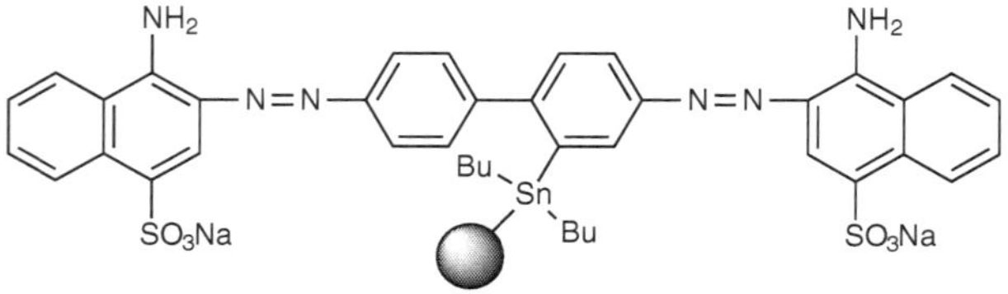	$Na^{123}I$, AcOH	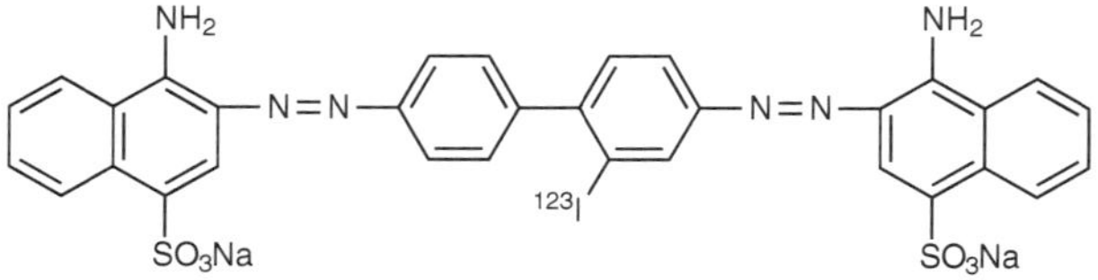	298

TABLE 1.18. Organoboronate Linker Units

	Linker	Cleavage Conditions	Product	References
1	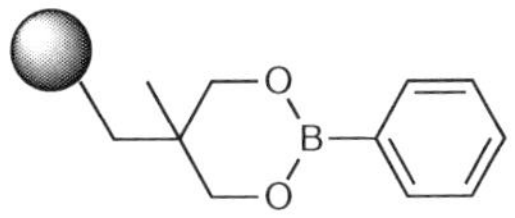	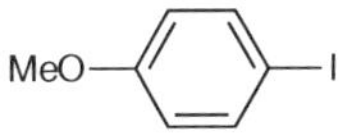 $PdCl_2$(binap), K_3PO_4	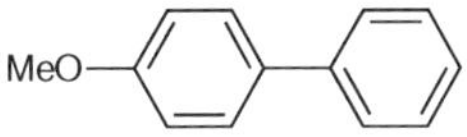 (yield: 85%)	306
2	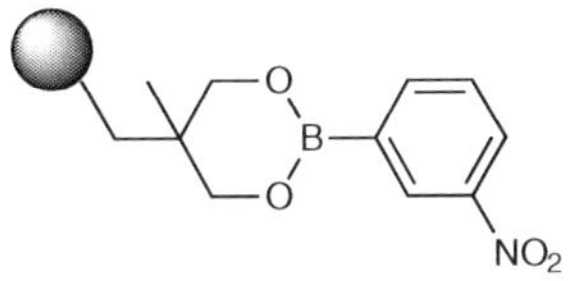	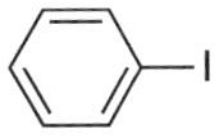 $PdCl_2$(dppf), K_3PO_4	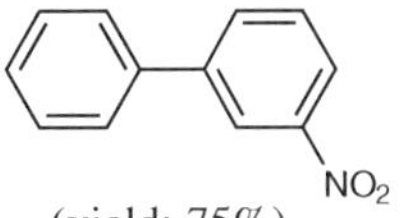 (yield: 75%)	309
3	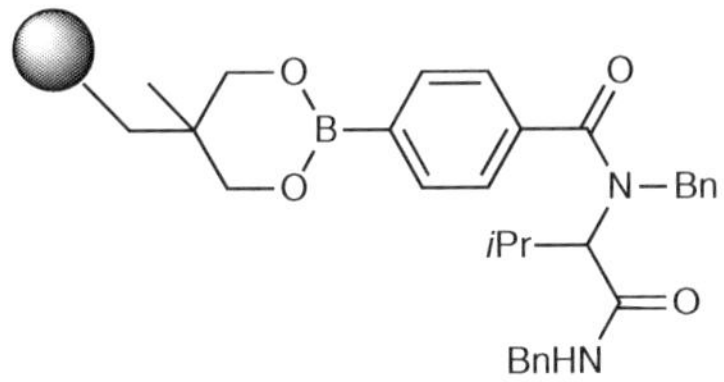	$Ag(NH_3)_2NO_3$ (0.5 M in water, 10 equiv)	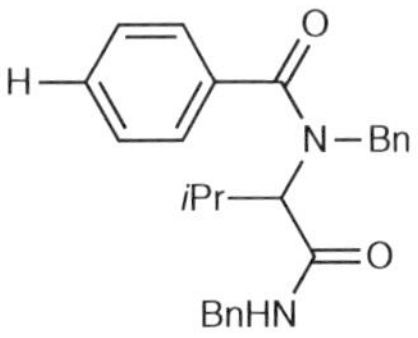 (yield: 57%)	310
4	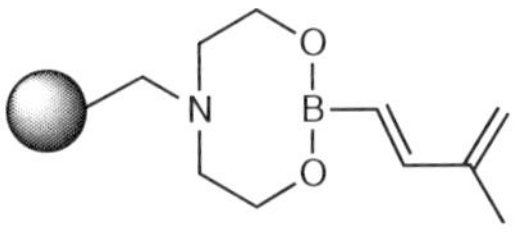	THF, H_2O (9:1)	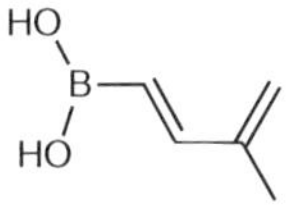 (yield: 95%)	307
5	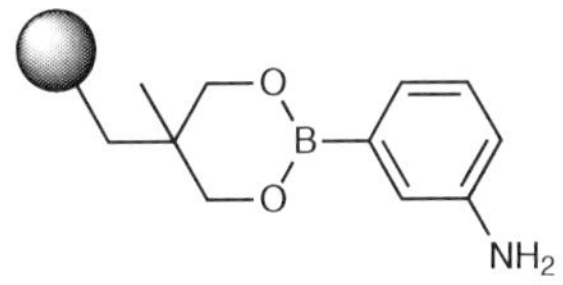	THF, MeOH, DCM (5:5:2)	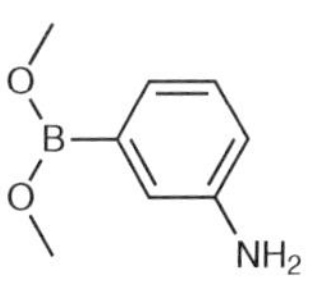	
6	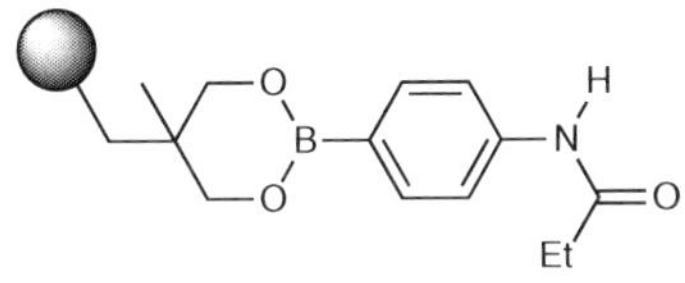	(i) H_2O_2, NaOH; (ii) HCl to pH 4	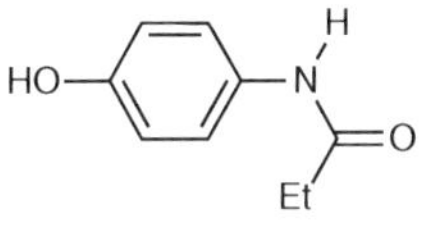 (yield: 74%)	308
7	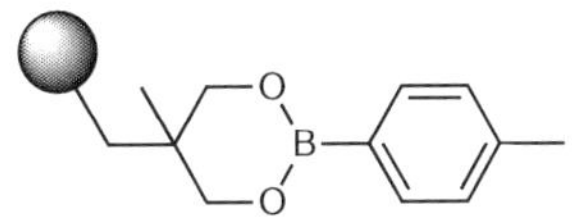	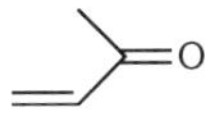 Rh(acac)CO_2, PPh_3, THF: MeOH:H_2O (7:6:1)	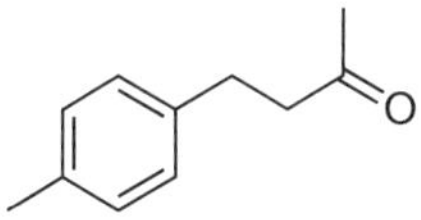	309
8	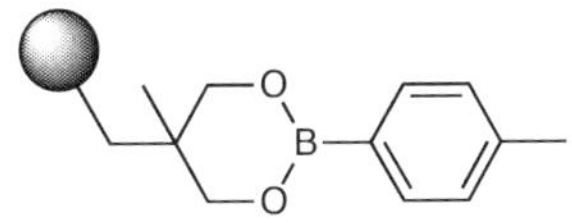	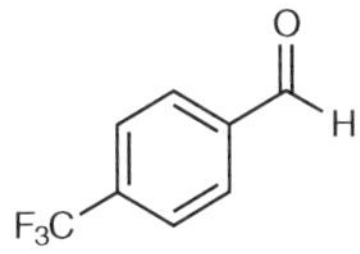 Rh(acac)CO_2, dppf, DME/H_2O	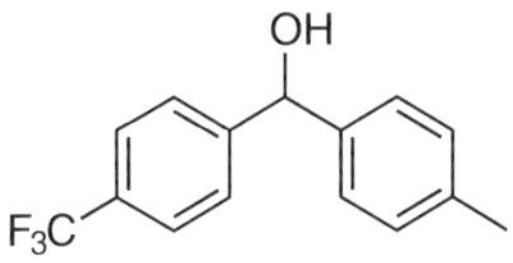	309

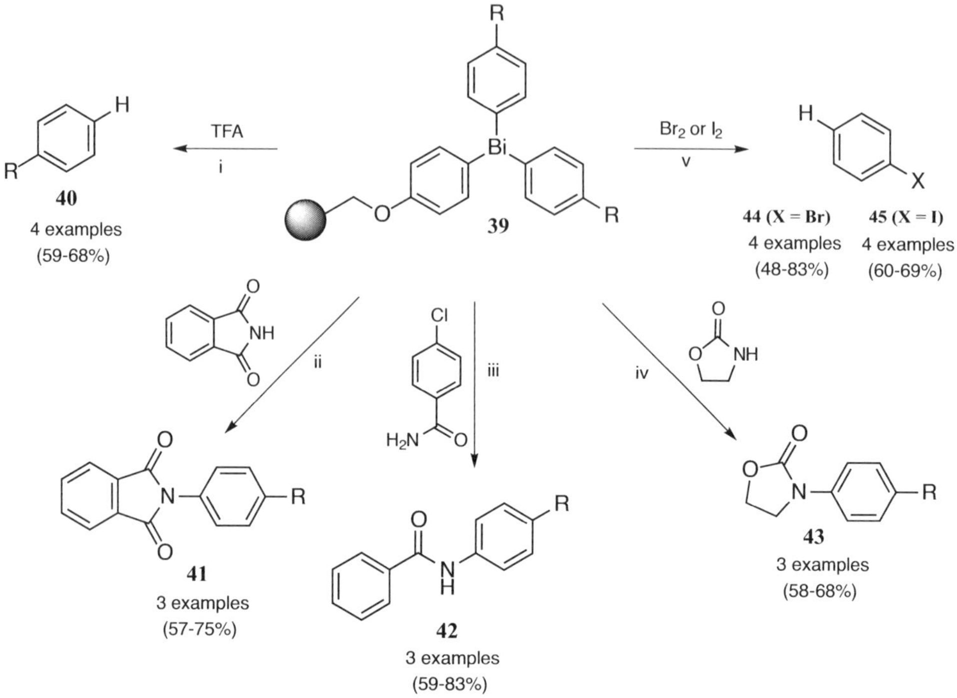

Scheme 1.14. Cleavage from triarylbismuth linkers. (i) TFA/DCM (1:10), rt, 1 h; (ii) phthalimide, $Cu(OAc)_2$ (1.5 equiv), pyridine, DCM, 40°C; (iii) 4-chlorobenzamide, $Cu(OAc)_2$ (1.5 equiv), pyridine, DCM, 40°C; (iv) 2-oxazolidone, $Cu(OAc)_2$ (1.5 equiv), Et_3N, DCM, 40°C; (v) Br_2 (2 equiv), DCE, 60°C; or I_2 (2 equiv), THF, 60°C.

triaryl bismuth species attractive candidates for linker scaffolds, and, indeed, linker units based around triaryl bismuth and triarylbismuth diacetate cores have been reported by Ruhland[314,315] and reviewed.[316] Triaryl bismuth linker unit **39** can be prepared by reacting a resin-bound Grignard reagent with the corresponding diarylbismuth triflate. It is important to recognize that the resin-bound triarylbismuth species are unsymmetrical. Reports have shown that the most electron-deficient aryl groups are transferred during an arylation reaction, and so the phenoxy spacer unit is essential to ensure reaction of a single arylbismuth group. A range of cleavage strategies has been reported, including traceless cleavage using TFA to leave a hydrogen residue at the cleavage site (**40**). Alternatively, multifunctional cleavage can occur when the bismuth linker is treated with imides, amides, or carbamates, in the presence of copper(II) acetate (1.5 equiv), to provide small libraries of substituted *N*-phenylphthalimides (**41**), *N*-phenylbenzamides (**42**) or *N*-phenyloxazolidin-2-ones (**43**) (Scheme 1.14). Finally, halogenation is also possible, and treatment with bromine or iodine gave the corresponding bromides (**44**) and iodides (**45**), respectively.

A drawback associated with the triarylbismuth linker was the relatively high levels of copper required to achieve cleavage, making purification of cleaved products challenging. To address this problem, reaction of triarylbismuth linker **39** with diacetoxy iodobenzene gave the more reactive triarylbismuth(V) diacetate linker **46**. Cleavage from the diacetate linker could then be achieved using 10% copper(II), simplifying purification, and Ruhland demonstrated diversity cleavage to provide libraries of various aromatic species (**47–51**) from a common intermediate (Scheme 1.15).

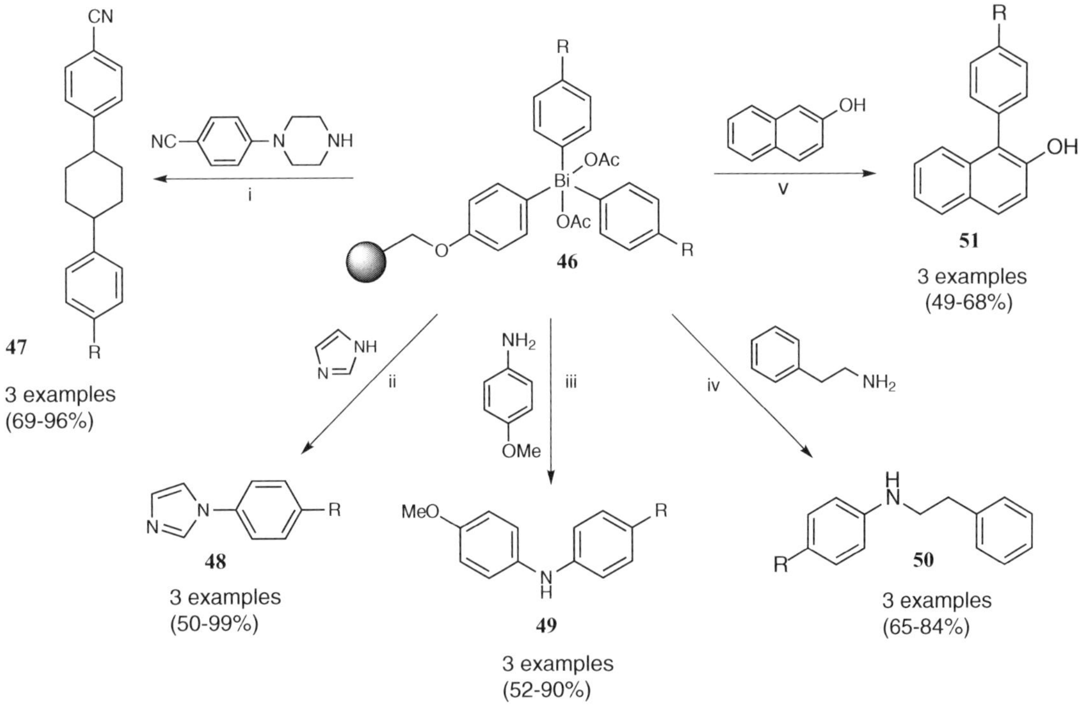

Scheme 1.15. Cleavage from triarylbismuth diacetate linkers. (i) 2 equiv 4-(piperazin-1-yl) benzonitrile, 10% $Cu(OAc)_2$, THF/Et_3N (25% v/v), rt; (ii) 2 equiv imidazole, 10% Cu $(OAc)_2$, THF, 50°C; (iii) 1.5 equiv anisole, 10% $Cu(OPiv)_2$, THF, rt; (iv) 1.5 equiv 2-phenylethanamine, 10% $Cu(OAc)_2$, THF, rt; (v) 1.5 equiv 2-naphthol, 1.2 equiv TMG, THF, rt.

1.3.8 Alkene Linker Units

Primarily, linker units releasing alkenes have focused on using rhodium-mediated RCM and CM[317], and examples are illustrated in Table 1.18. The former cleavage strategy can be considered both multifunctional and cyclorelease (Section 1.2.2), while the latter allows multifunctional cleavage. The first approach to using RCM as a cleavage strategy provided cycloolefins. For example, seven-membered lactams were cleaved by RCM in the first report of such cleavage by van Maarseveen in 1996[94], who later also used RCM cleavage to prepare additional azacycles (Table 1.19, Entry 1)[318], and even macrocyclic species such as epothilone and related analogues (Table 1.19, Entry 2)[319], 12-membered lactams (Table 1.19, Entry 3)[320], and cyclic tetrapeptides (Table 1.19, Entry 4).[321] Bicyclic products are also within the scope of this cleavage strategy, such as the bisfunctionalized *cis*-fused pyranofurans prepared by the Overkleef group (Table 1.19, Entry 5).[322] Alternatively, if the cycloolefin is left on the support following cleavage, then the RCM strategy can be used to prepare terminal olefins. Such an approach was used by Peters and Blechert in 1997 to prepare functionalized styrenes (Table 1.19, Entry 6)[323], by Waldmann during the synthesis of 6-*epi*-dysidiolde (Table 1.19, Entry 7)[324,325], and by Knerr and Schmidt to prepare substituted oligosaccharides (Table 1.19, Entry 8).[326,327] Finally, Seeberger has extended the concept to multifunctional cleavage using cross-metathesis (Table 1.19, Entry 9)[328,329]. The supported sugars were cleaved using olefin cross-metathesis to leave an unsaturated side chain at the cleavage site ready for oligosaccharide synthesis.

TABLE 1.19. Multifunctional Cleavage Using RCM and CM Reactions

	Linker	Cleavage Conditions	Product	References
1	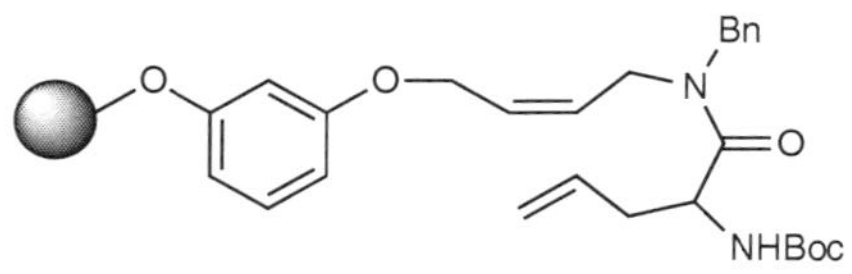	Grubbs I, 1-octene	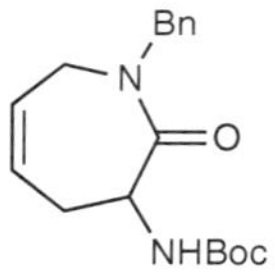 (yield: 54%)	94
2	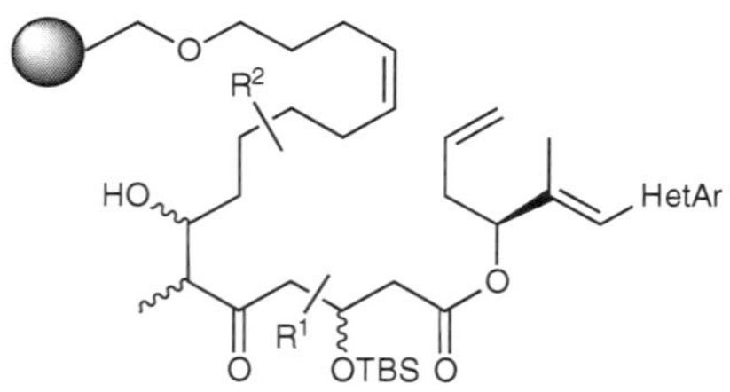	(i) Grubbs I (75%); (ii) 20% TFA/ DCM; (iii) 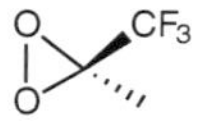	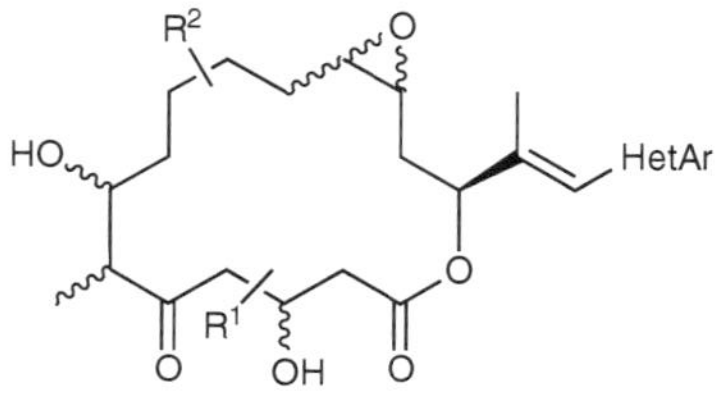	319
3	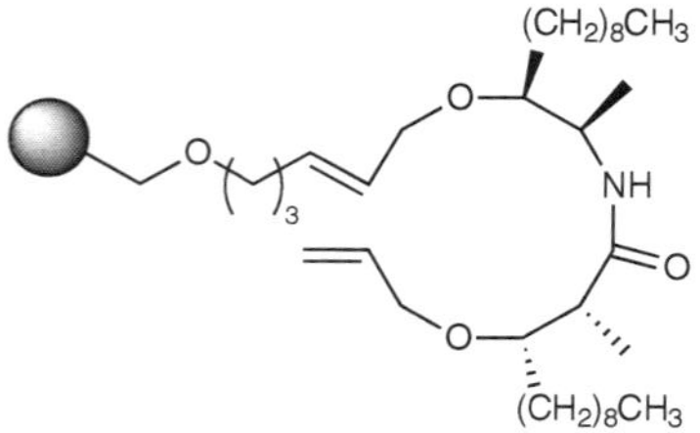	Grubbs I, DCM	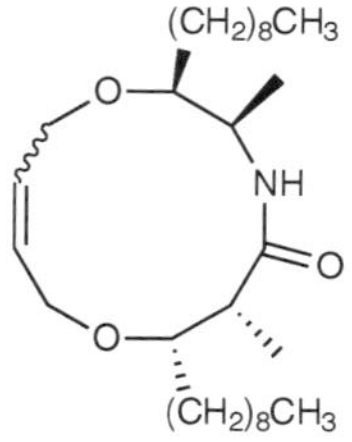 (yield: 30%, *E*/*Z*: 3:1)	320

4	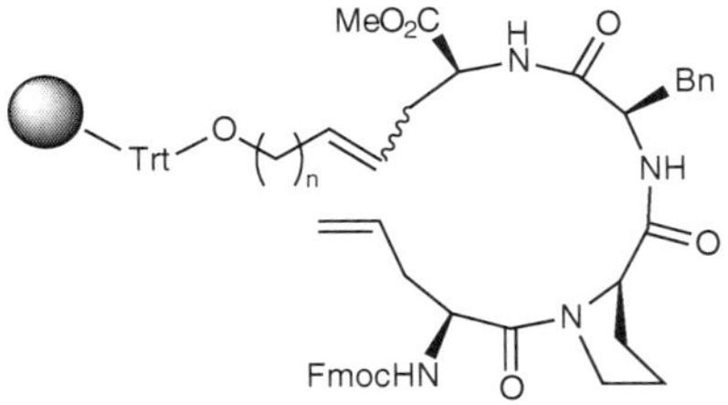	Grubbs I, DCM	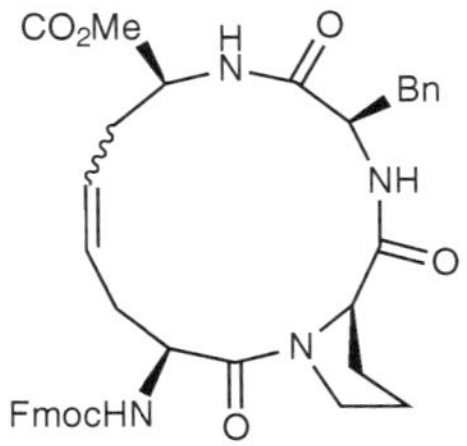 (yield: 30% ($n = 1$); 70% ($n = 8$))	321
5	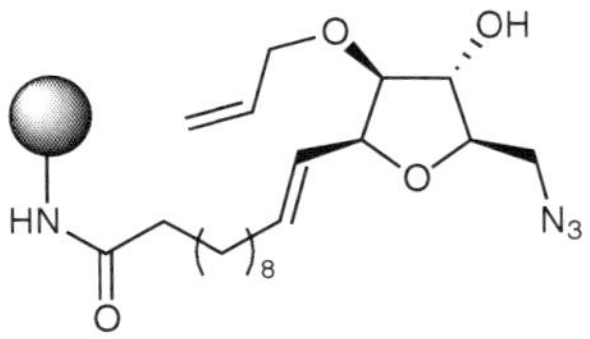	(i) R^1-NCO, Et_3N; (ii) Me_3P, H_2O, dioxane, followed by R^2-COCl, DIPEA; (iii) Grubbs II	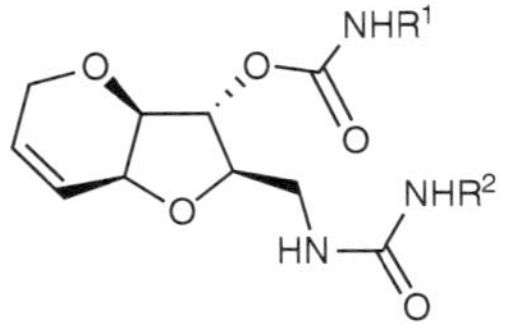	322
6	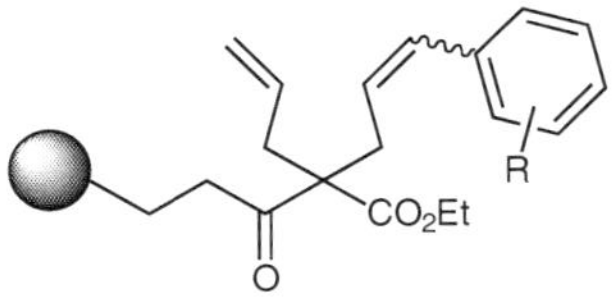	$(Cy_3P)_2Cl_2Ru{=}CHPh$, DCM	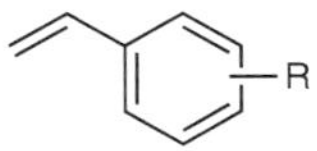 (five examples, yield: 26–55%)	323

(Continued)

TABLE 1.19. (*Continued*)

	Linker	Cleavage Conditions	Product	References
7	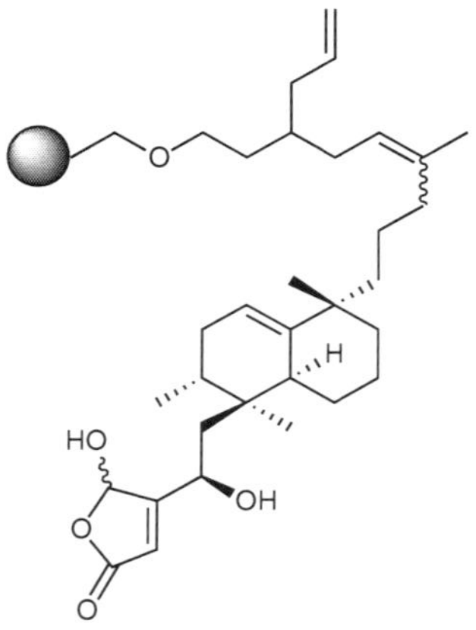	Grubbs I, DCM	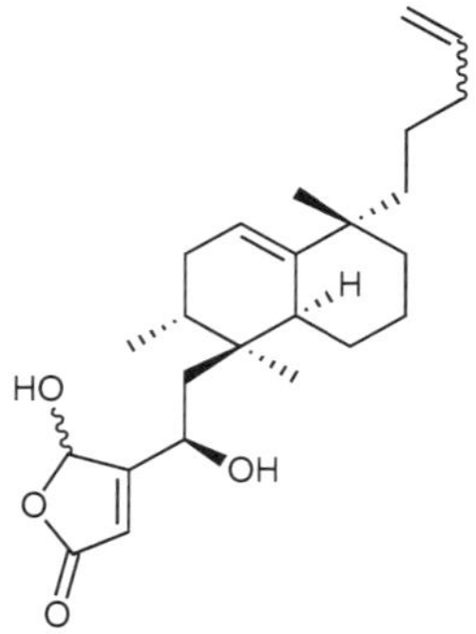 (yield: 14% over 11 steps)	324, 325
8	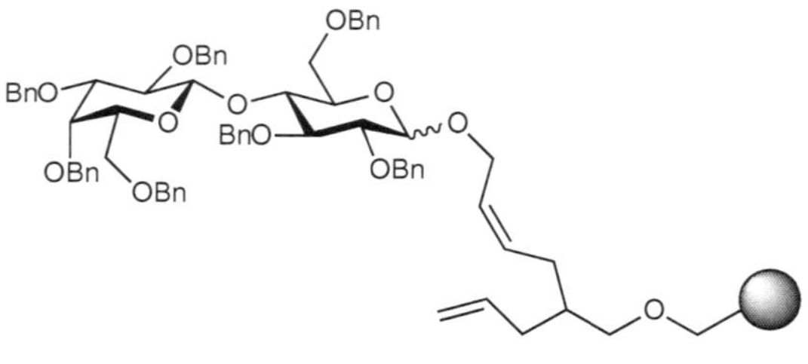	(Cy$_3$-P)$_2$Cl$_2$Ru=CHPh, DCM	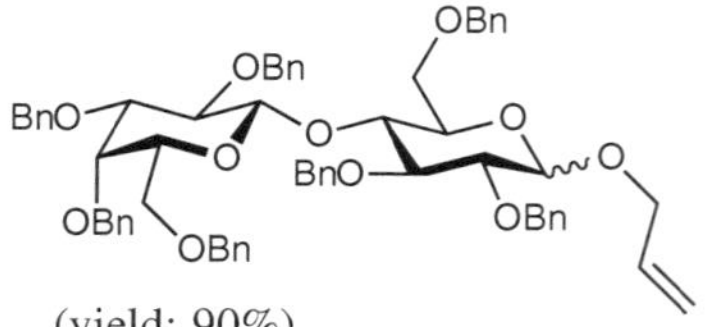 (yield: 90%)	326, 327
9	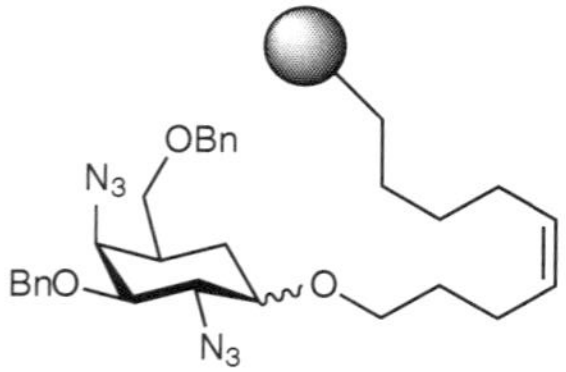	(H$_2$-Imes)(3-Br-py)$_2$Cl$_2$-Ru=CHPh, 1-pentene, DCM	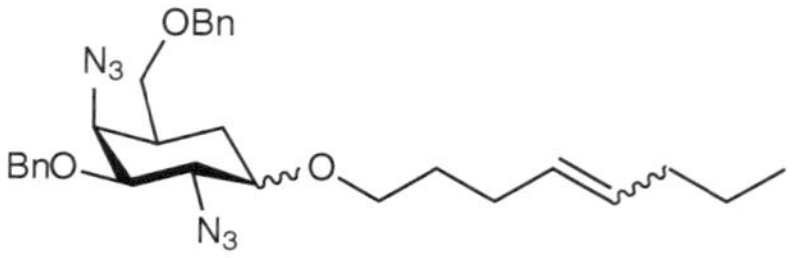 (yield: 85% over two steps)	328, 329

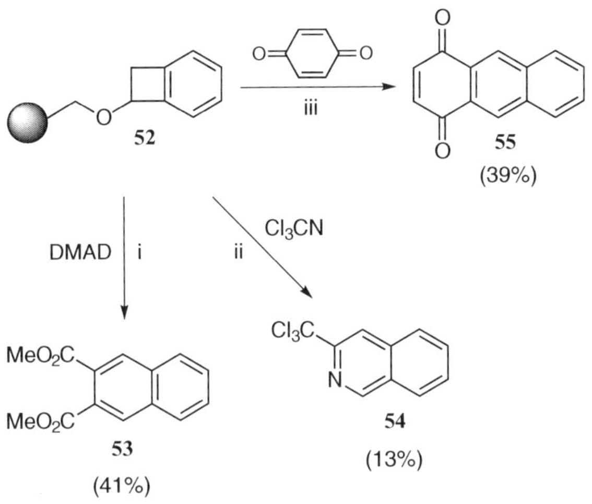

Scheme 1.16. Cleavage using Diels–Alder reactions. (i) DMAD, toluene, 105°C; (ii) Cl_3CN, toluene, 105°C; (iii) benzoquinone, toluene, 105°C.

Beyond RCM and CM strategies, Craig has reported cleavage using Diels–Alder reactions (Scheme 1.16).[330] [4 + 2] Cycloaddition (with concomitant aromatization) of the *o*-quinodimethane precursor (**52**) with dimethylacetylene dicarboxylate (DMAD), trichloroacetonitrile, and benzoquinone provided dimethyl naphthalene-2,3-dicarboxylate (**53**), 3-(trichloromethyl)isoquinoline (**54**), and 2,3-naphthoquinone (**55**), respectively. The diverse products from a single polymer-supported intermediate, such as the bismuth linkers discussed previously, make Craig's multifunctional linker unit attractive for approaches toward diversity-oriented synthesis.

1.4 CONCLUSIONS

This chapter showcases the decades of ingenuity and inventiveness that have gone into developing and designing the many linker units available today. Many of these linker units have enhanced diverse library synthesis and enable rapid and efficient exploration of chemical space. Thus, such multifunctional linker units are already impacting the science of diversity-oriented synthesis and should continue to play an important role in combinatorial library synthesis. Nevertheless, there is scope for further developmental efforts in the art and science of linker design. For example, many of the linker units described exhibit considerable substrate dependence, creating a continuing need for more versatile linker units. The recent developments in transition metal-mediated cleavage and cross-metathesis, in addition to the more versatile linkers such as the triazene, sulfur, and bismuth linkers, will play a prominent part in the continuing role of future SPOS.

REFERENCES

1. P. J. H. Scott, (Ed.), *Linker Strategies in Solid-phase Organic Synthesis*, Wiley, Chichester, UK, **2009**. 677 pp.
2. I. W. James, *Tetrahedron* **1999**, *55*, 4855.

3. F. Guillier, D. Orain, M. Bradley, *Chem. Rev.* **2000**, *100*, 2091.
4. P. Blaney, R. Grigg, V. Sridharan, *Chem. Rev.* **2002**, *102*, 2607.
5. L. A. McAllister, R. A. McCormick, D. J. Procter, *Tetrahedron* **2005**, *61*, 11527.
6. K. Gordon, S. Balasubramanian, *J. Chem. Technol. Biotechnol.* **1999**, *74*, 835.
7. F. Z. Dörwald, *Organic Synthesis on Solid-phase*. 2nd ed., Wiley-VCH, Weinheim, **1999**. 530. pp.
8. A. C. Comely, S. E. Gibson, *Angew. Chem., Int. Ed.* **2001**, *40*, 1012.
9. P. J. H. Scott, P. G. Steel, *Eur. J. Org. Chem.* **2006**, 2251.
10. S. Bräse, S. Dahmen, *Chem. Eur. J.* **2000**, *6*, 1899.
11. N. Jung, M. Wiehn, S. Bräse, *Top. Curr. Chem.* **2007**, *278*, 1.
12. P. H. H. Hermkens, H. C. J. Ottenheijm, D. C. Rees, *Tetrahedron* **1996**, *52*, 4527.
13. P. H. H. Hermkens, H. C. J. Ottenheijm, D. C. Rees, *Tetrahedron* **1997**, *53*, 5643.
14. S. C. Booth, P. H. H. Hermkens, H. C. J. Ottenheijm, D. C. Rees, *Tetrahedron* **1998**, *54*, 15385.
15. R. E. Dolle, *Mol. Diversity* **1998**, *3*, 199.
16. R. E. Dolle, *Mol. Diversity* **1998**, *4*, 233.
17. R. E. Dolle, K. Nelson, *J. Comb. Chem.* **1999**, *1*, 235.
18. R. E. Dolle, *J. Comb. Chem.* **2000**, *2*, 383.
19. R. E. Dolle, *J. Comb. Chem.* **2001**, *3*, 477.
20. R. E. Dolle, *J. Comb. Chem.* **2002**, *4*, 369.
21. R. E. Dolle, *J. Comb. Chem.* **2003**, *5*, 693.
22. R. E. Dolle, *J. Comb. Chem.* **2004**, *6*, 623.
23. R. E. Dolle, *J. Comb. Chem.* **2005**, *7*, 739.
24. R. E. Dolle, B. L. Bourdonnec, G. A. Morales, K. J. Moriarty, J. M. Salvino, *J. Comb. Chem.* **2006**, *8*, 597.
25. R. E. Dolle, B. L. Bourdonnec, A. J. Goodman, G. A. Morales, K. J. Moriarty, J. M. Salvino, W. Zhang, *J. Comb. Chem.* **2007**, *9*, 855.
26. R. E. Dolle, B. L. Bourdonnec, A. J. Goodman, G. A. Morales, C. J. Thomas, J. M. Salvino, *J. Comb. Chem.* **2008**, *10*, 753.
27. R. E. Dolle, B. L. Bourdonnec, A. J. Goodman, G. A. Morales, C. J. Thomas, W. Zhang, *J. Comb. Chem.* **2009**, *11*, 739.
28. D. S. Tan, *Nat. Chem. Bio.* **2005**, *1*, 74.
29. R. B. Merrifield, *J. Am. Chem. Soc.* **1963**, *85*, 2149.
30. S. S. Wang, *J. Am. Chem. Soc.* **1973**, *95*, 1328.
31. H. Rink, *Tetrahedron Lett.* **1987**, *28*, 3787.
32. M. Meusel, M. Gütschow, *Tetrahedron Lett.* **2005**, *46*, 2231.
33. R. C. Sheppard, B. J. Williams, *Int. J. Pept. Protein Res.* **1982**, *20*, 451.
34. A. R. Katritzky, D. Toader, K. Watson, J. S. Kiely, *Tetrahedron Lett.* **1997**, *38*, 7849.
35. M. Mergler, J. Gosteli, P. N. R. Grogg, R. Tanner, *Chimia* **1999**, *53*, 29.
36. M. Mergler, R. Nyfeler, R. Tanner, J. Gosteli, P. Grogg, *Tetrahedron Lett.* **1988**, *29*, 4009.
37. R. S. Garigipati, *Tetrahedron Lett.* **1997**, *38*, 6807.
38. K. Barlos, D. Gatos, J. Kallitsis, G. Papaphotiu, P. Sotiriu, Y. Wenqing, W. Schäfer, *Tetrahedron Lett.* **1989**, *30*, 3943.
39. B. Borhan, J. A. Wilson, M. J. Gasch, Y. Ko, D. M. Kurth, M. J. Kurth, *J. Org. Chem.* **1995**, *60*, 7375.
40. A. R. Mitchell, B. W. Erickson, M. N. Ryabstev, R. S. Hodges, R. B. Merrifield, *J. Am. Chem. Soc.* **1976**, *98*, 7357.
41. P. G. Pietta, G. R. Marshall, *J. Chem. Soc. D* **1971**, 650.

42. P. Sieber, *Tetrahedron Lett.* **1987**, *28*, 2107.
43. J. P. Malkinson, R. A. Falconer, *Tetrahedron Lett.* **2002**, *43*, 9549.
44. M. Meisenbach, H. Echner, W. Voelter, *Chem. Commun.* **1997**, 849.
45. L. A. Thompson, J. A. Ellman, *Tetrahedron Lett.* **1994**, *35*, 9333.
46. S. Ma, D. Duan, Y. Wang, *J. Comb. Chem.* **2002**, *4*, 239.
47. X. Wang, Y. Choe, C. S. Craik, J. A. Ellman, *Bioorg. Med. Chem. Lett.* **2002**, *12*, 2201.
48. M. Ramaseshan, J. W. Ellingboe, Y. L. Dory, P. Deslongchamps, *Tetrahedron Lett.* **2000**, *41*, 4743.
49. M. R. Tremblay, P. Wentworth, G. E. Lee, K. D. Janda, *J. Comb. Chem.* **2000**, *2*, 698.
50. Y. Hu, J. A. Porco, Jr, J. W. Labadie, O. W. Gooding, *J. Org. Chem.* **1998**, *63*, 4518.
51. K. H. Bleicher, C. Lutz, Y. Wüthrich, *Tetrahedron Lett.* **2000**, *41*, 9037.
52. M.-K. Jeon, D.-S. Kim, H. J. La, Y.-D. Gong, *Tetrahedron Lett.* **2005**, *46*, 4979.
53. S. Bräse, J. Köbberling, D. Enders, R. Lazny, M. Wang, *Tetrahedron Lett.* **1999**, *40*, 2105.
54. C. C. Leznoff, J. Y. Wong, *Can. J. Chem.* **1973**, *51*, 3756.
55. M. K. W. Choi, P. H. Toy, *Tetrahedron* **2004**, *60*, **2903.**
56. M. Kurosu. in Scott, P. J. H. (Ed.), *Linker Strategies in Solid-Phase Organic Synthesis*, Wiley, Chichester, UK, **2009**. 27.
57. P. G. Pietta, G. R. Marshall, *J. Chem. Soc. D.* **1970**, 650.
58. A. C. Spivey, C. M. Diaper. in Scott, P. J. H. (Ed.), *Linker Strategies in Solid-Phase Organic Synthesis*, Wiley, Chichester, UK, **2009**. 467.
59. J. L. Castro, V. G. Matassa, *J. Org. Chem.* **1994**, *59*, 2289.
60. A. M. Fivush, T. M. Wilson, *Tetrahedron Lett.* **1997**, *38*, 7151.
61. X. Ouyang, N. Tamayo, A. S. Kiselyov, *Tetrahedron Lett.* **1999**, *55*, 2827.
62. Y. Aoki, S. Kobayashi, *J. Comb. Chem.* **1999**, *1*, 371.
63. S. Areniyadis, A. Wagner, C. Mioskowski, *Tetrahedron Lett.* **2002**, *43*, 9717.
64. J. M. J. Fréchert, G. Pellé, *Chem. Commun.* **1975**, 225.
65. M. Bejugam, S. L. Flitsch. in Scott, P. J. H. (Ed.), *Linker Strategies in Solid-Phase Organic Synthesis*, Wiley, Chichester, UK, **2009**. 221.
66. A. Porcheddu, G. Giacomelli, in Scott, P. J. H. (Ed.), *Linker Strategies in Solid-Phase Organic Synthesis*, Wiley, Chichester, UK, **2009**. 77.
67. T. Yao, D. Yue, R. C. Larock, *J. Comb. Chem.* **2005**, *7*, 809.
68. G. A. Strohmeier, O. C. Kappe, *J. Comb. Chem.* **2002**, *4*, 154.
69. L. A. Agrofoglio, S. R. Challand, *Acyclic, Carbocyclic and L-Nucleosides*, Kluwer Academic Publishers, Dordrecht, The Netherlands, **1998**.
70. J. M. Salvino, N. V. Kumar, J. Airey, T. Kiesow, K. Crawford, R. Mathew, K. Paul, M. D. Drew, D. Engers, E. Orton, D. Krolikolski, T. Herpin, M. Gardyan, G. McGeehan, R. Labaudiniere, *J. Comb. Chem.* **2000**, *2*, 691.
71. D. R. Barn, J. R. Morphy, D. C. Rees, *Tetrahedron Lett.* **1996**, *37*, 3213.
72. S.-H. Lee, H. Matsushita, G. Koch, J. Zimmermann, B. Clapham, K. D. Janda, *J. Comb. Chem.* **2004**, *6*, 822.
73. M. Fisher, R. C. D. Brown, *Tetrahedron Lett.* **2001**, *42*, 8227.
74. C. Vanier, A. Wagner, C. Mioskowski, *J. Comb. Chem.* **2004**, *6*, 846.
75. S. M. Dankwardt, *Synlett* **1998**, 761.
76. M. J. Kurth, L. A. A. Randall, C. Chen, C. Melander, R. B. Miller, K. McAlister, G. Reitz, R. Kang, T. Nakatsu, C. Green, *J. Org. Chem.* **1994**, *59*, 5862.
77. S. Chandrasekhar, M. B. Padmaja, A. Raza, *J. Comb. Chem.* **2000**, *2*, 246.
78. T. Q. Dinh, R. W. Armstrong, *Tetrahedron Lett.* **1996**, *37*, 1161.

79. A. Ganesan, in Scott, P. J. H. (Ed.), *Linker Strategies in Solid-Phase Organic Synthesis*, Wiley, Chichester, UK, **2009**. 135.
80. K.-H. Park, M. J. Kurth, *Drugs Fut.* **2000**, *25*, 1265.
81. O. Seitz, *Nachr. Chem.* **2001**, *49*, 912.
82. S. H. DeWitt, J. S. Kiely, C. J. Stankovic, M. C. Schroeder, D. M. R. Cody, M. R. Pavia, *Proc. Natl. Acad. Sci. USA* **1993**, *90*, 6909.
83. E. Flanigan, G. R. Marshall, *Tetrahedron Lett.* **1970**, *11*, 2403.
84. E. Colacino, F. Lamaty, J. Martinez, I. Parrot, *Tetrahedron Lett.* **2007**, *48*, 5317.
85. D. Bonnet, A. Ganesan, *J. Comb. Chem.* **2002**, *4*, 546.
86. B. Martin, H. Sekljic, C. Chassaing, *Org. Lett.* **2003**, *5*, 1851.
87. R. Pathak, A. K. Roy, S. Kanojia, S. Batra, *Tetrahedron Lett.* **2005**, *46*, 5289.
88. A. P. Kesarwani, G. K. Srivastava, S. K. Rastogi, B. Kundu, *Tetrahedron Lett.* **2002**, *43*, 5579.
89. H. Habashita, M. Kokubo, S. Hamano, N. Hamanaka, M. Toda, S. Shibayama, H. Tada, K. Sagawa, D. Fukushima, K. Maeda, H. Mitsuya, *J. Med. Chem.* **2006**, *49*, 4140.
90. J. Tois, A. Koskinen, *Tetrahedron Lett.* **2003**, *44*, 2093.
91. R. Prez, T. Beryozkina, O. I. Zbruyev, W. Haas, C. O. Kappe, *J. Comb. Chem.* **2002**, *4*, 501.
92. M.-K. Jeon, M.-S. Kim, J.-J. Kwon, Y.-D. Gong, D.-H. Lee, *Tetrahedron* **2008**, *64*, 9060.
93. B. A. Kulkarni, A. Ganesan, *Tetrahedron Lett.* **1998**, *39*, 4369.
94. J. H. van Maarseveen, J. A. J. den Hartog, V. Engelen, E. Finner, G. Visser, C. G. Kruse, *Tetrahedron Lett.* **1996**, *37*, 8249.
95. A. D. Piscopio, J. E. Robinsion, *Curr. Opin. Chem. Biol.* **2004**, *8*, 245.
96. M. J. Plunkett, J. A. Ellmann, *J. Org. Chem.* **1995**, *60*, 6006.
97. C. Gil, S. Bräse, *Curr. Opin. Chem. Biol.* **2004**, *8*, 230.
98. M. J. Plunkett, J. A. Ellman, *J. Org. Chem.* **1997**, *62*, 2885.
99. A. C. Spivey, C. M. Diaper, H. Adams, A. J. Rudge, *Chem. Commun.* **1999**, 835.
100. A. C. Spivey, C. M. Diaper, H. Adams, A. J. Rudge, *J. Org. Chem.* **2000**, *65*, 5253.
101. A. C. Spivey, R. Srikaran, C. M. Diaper, D. J. Turner, *Org. Biomol. Chem.* **2003**, *1*, 1638.
102. A. C. Spivey, D. J. Turner, M. L. Turner, S. Yeates, *Synlett* **2004**, 111.
103. S. E. Gibson, A. A. Walke, in Scott, P. J. H. (Ed.), *Linker Strategies in Solid-Phase Organic Synthesis*, Wiley, Chichester, UK, **2009**, 525.
104. A. C. Comely, S. E. Gibson, N. J. Hales, M. A. Peplow, *J. Chem. Soc., Perkin Trans. 1*, **2001**, 2526.
105. S. E. Gibson, N. J. Hales, M. A. Peplow, *Tetrahedron Lett.* **1999**, *40*, 1417.
106. S. Maiorana, C. Baldoli, E. Licandro, L. Casiraghi, E. De Magistris, A. Paio, S. Provera, P. Seneci, *Tetrahedron Lett.* **2000**, *41*, 7271.
107. A. C. Comely, S. E. Gibson, N. J. Hales, *Chem. Commun.* **1999**, 2075.
108. A. C. Comely, S. E. Gibson, N. J. Hales, C. Johnstone, A. Stevenazzi, *Org. Biomol. Chem.* **2003**, *1*, 1959.
109. Z. Zhang, S. D. Lepore, *Tetrahedron Lett.* **2002**, *43*, 7357.
110. C. Bochet, S. Mercier,in Scott, P. J. H. (Ed.), *Linker Strategies in Solid-Phase Organic Synthesis*, Wiley, Chichester, UK, **2009**, 151.
111. D. H. Rich, S. K. Gurwara, *J. Chem. Soc., Chem. Commun.* **1973**, 610.
112. A. Ajayaghosh, V. N. R. Pillai, *J. Org. Chem.* **1990**, *55*, 2826.
113. M. Renil, V. N. R. Pillai, *Tetrahedron Lett.* **1994**, *35*, 3809.
114. K. S. Kumar, V. N. R. Pillai, *Tetrahedron* **1999**, *55*, 10437.
115. A. Ajayaghosh, V. N. R. Pillai, *J. Org. Chem.* **1987**, *52*, 5714.

116. A. Ajayaghosh, V. N. R. Pillai, *Tetrahedron Lett.* **1995**, *36*, 777.
117. B. Ruhalnd, A. Bhandari, E. M. Gordon, M. A. Gallop, *J. Am. Chem. Soc.* **1996**, *118*, 253.
118. S. S. Wang, *J. Org. Chem.* **1976**, *41*, 3258.
119. D. Bellof, M. Mutter, *Chimia* **1985**, *39*, 317.
120. A. Routledge, C. Abell, S. Balasubramanian, *Tetrahedron Lett.* **1997**, *38*, 1227.
121. H. B. Lee, S. Balasubramanian, *J. Org. Chem.* **1999**, *64*, 3454.
122. S. Peukert, B. Giese, *J. Org. Chem.* **1998**, *63*, 9045.
123. K. C. Nicolaou, B. S. Safina, N. Winssinger, *Synlett* **2001**, 900.
124. J. R. Horton, L. M. Stamp, A. Routledge, *Tetrahedron Lett.* **2000**, *41*, 9181.
125. S. Lebreton, M. Pátek,in Scott, P. J. H. (Ed.), *Linker Strategies in Solid-Phase Organic Synthesis* Wiley, Chichester, UK, **2009**, 195.
126. M. Pátek, M. Lebl, *Biopolymers (Pept. Sci.)* **1998**, *47*, 353.
127. G. W. Kenner, J. R. McDermott, R. C. J. Sheppard, *J. Chem. Soc. D* **1971**, 636.
128. B. J. Backes, A. A. Virgilio, J. A. Ellman, *J. Am. Chem. Soc.* **1996**, *118*, 3055.
129. Y. He, J. P. Wilkins, L. L. Kiessling, *Org. Lett.* **2006**, *8*, 2483.
130. H. E. Russell, R. W. A. Luke, M. Bradley, *Tetrahedron Lett.* **2000**, *41*, 5287.
131. M. Yamada, T. Miyajima, H. Horikawa, *Tetrahedron Lett.* **1998**, *39*, 289.
132. S. Lebreton, N. Newcombe, M. Bradley, *Tetrahedron* **2003**, *59*, 10213.
133. C.-H. Tai, H.-C. Wu, W.-R. Li, *Org. Lett.* **2004**, *6*, 2905.
134. C. Vanier, F. Lorgé, A. Wagner, C. Mioskowski, *Angew. Chem., Int. Ed.* **2000**, *39*, 1679.
135. E. L. Porta, U. Piarulli, F. Cardullo, A. Paio, S. Provera, P. Seneci, C. Gennari, *Tetrahedron Lett.* **2002**, *43*, 761.
136. C. Hulme, J. Peng, G. Morton, J. M. Salvino, T. Herpin, R. Labaudiniere, *Tetrahedron Lett.* **1998**, *39*, 7227.
137. J. R. Morphy, Z. Rankovic, D. C. Rees, *Tetrahedron Lett.* **1996**, *37*, 3209.
138. C. J. Ciolli, S. Kalagher, P. J. Belshaw, *Org. Lett.* **2004**, *6*, 1891.
139. A. R. Brown, D. C. Rees, Z. Rankovic, J. R. Morphy, *J. Am. Chem. Soc.* **1997**, *119*, 3288.
140. S. Arseniyadis, A. Wagner, C. Mioskowski, *Tetrahedron Lett.* **2004**, *45*, 2251.
141. J. J. Scicinski, M. S. Congreve, S. V. Ley, *J. Comb. Chem.* **2004**, *6*, 375.
142. J. K. Young, J. C. Nelson, J. S. Moore, *J. Am. Chem. Soc.* **1994**, *116*, 10841.
143. L. Jones, J. S. Schumm, J. M. Tour, *J. Org. Chem.* **1997**, *62*, 1388.
144. S. Schunk, D. Enders, *Org. Lett.* **2000**, *2*, 907.
145. M. E. P. Lormann, S. Dahmen, S. Bräse, *Tetrahedron Lett.* **2000**, *41*, 3813.
146. M. E. P. Lormann, S. Dahmen, F. Avemaria, F. Lauterwasser, S. Bräse, *Synlett* **2002**, 915.
147. S. Bräse, D. Enders, J. Köbberling, F. Avemaria, *Angew. Chem., Int. Ed.* **1998**, *37*, 3413.
148. S. Bräse, S. Dahmen, M. E. P. Lormann, *Methods Enzymol.* **2003**, *369*, 127.
149. S. Bräse, *Acc. Chem. Res.* **2004**, *37*, 805.
150. M. E. P. Lormann, C. H. Walker, M. Es-Sayed, S. Bräse, *Chem. Commun.* **2002**, 1296.
151. F. Avemaria, V. Zimmermann, S. Bräse, *Synlett* **2004**, 1163.
152. K. Knepper, S. Vanderheiden, S. Bräse, *Eur. J. Org. Chem.* **2006**, 1886.
153. C. Gil, A. Schwögler, S. Bräse, *J. Comb. Chem.* **2004**, *6*, 38.
154. V. Zimmermann, S. Bräse, *J. Comb. Chem.* **2007**, *9*, 1114.
155. M. Kreis, C. F. Nising, M. Schroen, K. Knepper, S. Bräse, *Org. Biomol. Chem.* **2005**, *3*, 1835.
156. M. Schroen, S. Bräse, *Tetrahedron* **2005**, *61*, 12186.
157. S. Bräse, J. Köbberling, D. Enders, M. Wang, R. Lazny, S. Brandtner, *Tetrahedron Lett.* **1999**, *40*, 2105.

158. S. Bräse, S. Dahmen, M. Pfefferkorn, *J. Comb. Chem.* **2000**, *2*, 710.
159. D. Enders, C. Rijksen, E. Bremus-Köbberling, A. Gillner, J. Köbberling, *Tetrahedron Lett.* **2004**, *45*, 2839.
160. C. Pilot, S. Dahmen, F. Lauterwasser, S. Bräse, *Tetrahedron Lett.* **2001**, *42*, 9179.
161. R. Lazny, A. Nodzewska, *Chem. Rev.* **2010**, *110*, 1386.
162. R. Lazny,in Scott, P. J. H. (Ed.), *Linker Strategies in Solid-Phase Organic Synthesis*, Wiley, Chichester, UK, **2009**. 303.
163. H. Kamogawa, A. Kanzawa, M. Kadoya, T. Naito, M. Nanasawa, *Bull. Chem. Soc. Jpn.* **1983**, *56*, 762.
164. A. M. Murphy, R. Dagnino, P.L. Vallar, Jr, A. J. Trippe, S. L. Sherman, R. H. Lumpkin, S. Y. Tamura, T. R. Webb, *J. Am. Chem. Soc.* **1992**, *114*, 3156.
165. A. Lee, L. Huang, J. A. Ellman, *J. Am. Chem. Soc.* **1999**, *121*, 9907.
166. L. Huang, J. A. Ellman, *Bioorg. Med. Chem. Lett.* **2002**, *12*, 2993.
167. R. Lazny, A. Nodzewska, M. Sienkiewicz, K. Wolosewicz, *J. Comb. Chem.* **2005**, *7*, 109.
168. H.-G. Breitinger, *Tetrahedron Lett.* **2002**, *43*, 6127.
169. J. H. Kirchhoff, S. Bräse, D. Enders, *J. Comb. Chem.* **2001**, *3*, 71.
170. R. Lazny, A. Nodzewska, B. Zabicka, *J. Comb. Chem.* **2008**, *10*, 986.
171. D. K. Whelligan, in Scott, P. J. H. (Ed.), *Linker Strategies in Solid-Phase Organic Synthesis*, Wiley, Chichester, UK, **2009**. 317.
172. A. R. Katritzky, X. Lan, J. Z. Yang, O. V. Denisko, *Chem. Rev.* **1998**, *98*, 409.
173. K. Schiemann, H. D. H. Showalter, *J. Org. Chem.* **1999**, *64*, 4972.
174. A. Paio, A. Zaramella, R. Ferritto, N. Conti, C. Marchioro, P. Seneci, *J. Comb. Chem.* **1999**, *1*, 317.
175. A. R. Katritzky, S. A. Belyakov, D. O. Tymoshenko, *J. Comb. Chem.* **1999**, *1*, 173.
176. A. R. Katritzky, A. Pastor, M. Voronkov, D. O. Tymoshenko, *J. Comb. Chem.* **2001**, *3*, 167.
177. A. Paio, R. F. Crespo, P. Seneci, M. Ciraco, *J. Comb. Chem.* **2001**, *3*, 354.
178. P. J. H. Scott, in Scott, P. J. H. (Ed.), *Linker Strategies in Solid-Phase Organic Synthesis*, Wiley, Chichester, UK, **2009**. 443.
179. J. T. W. Kan, P. H. Toy, *J. Sulfur Chem.* **2005**, *26*, 509.
180. F. Zaragoza, *Angew. Chem., Int. Ed.* **2000**, *39*, 2077.
181. K. W. Jung, X.-Y. Zhao, K. D. Janda, *Tetrahedron Lett.* **1996**, *37*, 6491.
182. K. W. Jung, X.-Y. Zhao, K. D. Janda, *Tetrahedron* **1997**, *53*, 6645.
183. L. A. McAllister, S. Brand, R. de Gentile, D. J. Procter, *Chem. Commun.* **2003**, 2380.
184. G. A. Crosby, M. Kato, *J. Am. Chem. Soc.* **1977**, *99*, 278.
185. J. Rademann, R. Schmidt, *J. Org. Chem.* **1997**, *62*, 3650.
186. J. Rademann, R. Schmidt, *Tetrahedron Lett.* **1996**, *37*, 3989.
187. T. Opatz, C. Kallus, T. Wunberg, H. Kunz, *Tetrahedron* **2004**, *60*, 8613.
188. C. Kallus, T. Opatz, T. Wunberg, W. Schmidt, S. Henke, H. Kunz, *Tetrahedron Lett.* **1999**, *40*, 7783.
189. T. Opatz, C. Kallus, T. Wunberg, W. Schmidt, S. Henke, H. Kunz, *Eur. J. Org. Chem.* **2003**, 1527.
190. T. Wunberg, C. Kallus, T. Opatz, S. Henke, W. Schmidt, H. Kunz, *Angew. Chem., Int. Ed.* **1998**, *37*, 2503.
191. L. F. Hennequin, S. P. L. Blanc, *Tetrahedron Lett.* **1999**, *40*, 3881.
192. T. Zoller, J.-B. Ducep, C. Tahtaoui, M. Hibert, *Tetrahedron Lett.* **2000**, *41*, 9989.
193. P. Bishop, C. Jones, J. Chmielewski, *Tetrahedron Lett.* **1993**, *34*, 4469.
194. R. Baer, T. Masquuelin, *J. Comb. Chem.* **2001**, *3*, 16.

195. C. M. Huwe, H. Künzer, *Tetrahedron Lett.* **1999**, *40*, 683.
196. V. Bertini, F. Lucchesini, M. Pocci, A. D. Munno, *Tetrahedron Lett.* **1998**, *39*, 9263.
197. V. Bertini, F. Lucchesini, M. Pocci, S. Alfei, A. De Munno, *Synlett* **2003**, 1201.
198. V. Bertini, M. Pocci, F. Lucchesini, A. Silvana, A. De Munno, *Synlett* **2003**, 864.
199. S. Nakamura, Y. Uchiyama, S. Ishikawa, R. Fukinbara, Y. Watanabe, T. Toru, *Tetrahedron Lett.* **2002**, *43*, 2381.
200. X.-Y. Zhao, K. D. Janda, *Bioorg. Med. Chem. Lett.* **1998**, *8*, 2439.
201. S. B. Katti, P. K. Misra, W. Haq, K. B. Mathur, *Chem. Commun.* **1992**, 843.
202. F. E. K. Kroll, R. Morphy, D. Rees, D. Gani, *Tetrahedron Lett.* **1997**, *38*, 8573.
203. P. Heinonen, H. Lönnberg, *Tetrahedron Lett.* **1997**, *38*, 8569.
204. Y. Chen, Y. Lam, Y.-H. Lai, *Org. Lett.* **2002**, *4*, 393.
205. A. Barco, S. Benetti, C. De Risi, P. Marchetti, G. P. Pollini, V. Zanirato, *Tetrahedron Lett.* **1998**, *39*, 1973.
206. K.-H. Kong, Y. Chen, X. Ma, W. K. Chui, Y. Lam, *J. Comb. Chem.* **2004**, *6*, 928.
207. W.-C. Cheng, C.-C. Lin, M. J. Kurth, *Tetrahedron Lett.* **2002**, *43*, 2967.
208. B. A. Kulkarni, A. Ganesan, *Tetrahedron Lett.* **1999**, *40*, 5633.
209. J. N. P. D'herde, P. J. De Clereq, *Tetrahedron Lett.* **2003**, *44*, 6657.
210. A. Chucholowski, T. Masquuelin, D. Obrecht, J. Stadelwieser, J. M. Villalgordo, *Chimia* **1996**, *50*, 525.
211. V. Brun, M. Legraverend, D. S. Grierson, *Tetrahedron* **2002**, *58*, 7911.
212. C. Halm, J. Evarts, M. J. Kurth, *Tetrahedron Lett.* **1997**, *38*, 7709.
213. W.-C. Cheng, C. Halm, J. B. Evarts, M. M. Olmstead, M. J. Kurth, *J. Org. Chem.* **1999**, *64*, 8557.
214. D. R. Dragoli, M. T. Burdett, J. A. Ellman, *J. Am. Chem. Soc.* **2001**, *123*, 10127.
215. C. Rolland, G. Hanquet, J.-B. Ducep, G. Solladié, *Tetrahedron Lett.* **2001**, *42*, 7563.
216. L. M. Gayo, M. J. Suto, *Tetrahedron Lett.* **1997**, *38*, 211.
217. R. Schwyzer, E. Felder, P. Failli, *Helv. Chim. Acta.* **1984**, *67*, 1316.
218. Y. Chen, Y. Lam, Y.-H. Lai, *Org. Lett.* **2003**, *5*, 1067.
219. W. Li, Y. Chen, Y. Lam, *Tetrahedron Lett.* **2004**, *45*, 6545.
220. W. Li, Y. Lam, *J. Comb. Chem.* **2005**, *7*, 644.
221. W.-C. Cheng, M. J. Kurth, *J. Org. Chem.* **2002**, *67*, 4387.
222. W.-C. Cheng, M. Wong, M. M. Olmstead, M. J. Kurth, *Org. Lett.* **2002**, *4*, 741.
223. S. C. Schürer, S. Blechert, *Synlett* **1998**, 166.
224. R. C. D. Brown, M. Fisher, *Chem. Commun.* **1999**, 1547.
225. K. C. Nicolaou, T. Montagnon, T. Ulven, P. S. Baran, Y.-L. Zhong, F. Sarabia, *J. Am. Chem. Soc.* **2002**, *124*, 5718.
226. K. C. Nicolaou, P. S. Baran, Y.-L. Zhong, *J. Am. Chem. Soc.* **2000**, *122*, 10246.
227. J. A. Hunt, W. R. Roush, *J. Am. Chem. Soc.* **1996**, *118*, 9998.
228. T. Takahashi, S. Tomida, H. Inoue, T. Doi, *Synlett* **1998**, 1261.
229. E. W. Baxter, J. K. Reuter, S. O. Nortey, A. B. Reitz, *Tetrahedron Lett.* **1998**, *39*, 979.
230. J. K. Reuter, S. O. Nortey, E. W. Baxter, G. C. Leo, A. B. Reitz, *Tetrahedron Lett.* **1998**, *39*, 975.
231. C.-H. Cho, H. Park, Park M-A, T. Y. Ryoo, Y.-S. Lee, K. Park, *Eur. J. Org. Chem.* **2005**, 3177.
232. H. Tsukamoto, R. Suzuki, Y. Kondo, *J. Comb. Chem.* **2006**, *8*, 289.
233. S. Jin, D. P. Holub, D. J. Wustrow, *Tetrahedron Lett.* **1998**, *39*, 3651.
234. Y. Pan, C. P. Holmes, *Org. Lett.* **2001**, *3*, 2769.
235. Y. Pan, B. Ruhland, C. P. Holmes, *Angew. Chem., Int. Ed.* **2001**, *40*, 4488.

236. A. N. Cammidge, Z. Ngaini, *Chem. Commun.* **2004**, 1914.
237. J. D. Revell, A. Ganesan, *Chem. Commun.* **2004**, 1916.
238. S. Kobayashi, I. Hachiya, S. Suzuki, M. Moriwaki, *Tetrahedron Lett.* **1996**, *37*, 2809.
239. S. Kobayashi, T. Wakabayashi, M. Yasuda, *J. Org. Chem.* **1998**, *63*, 4868.
240. P. J. May, M. Bradley, D. C. Harrowven, D. Pallin, *Tetrahedron Lett.* **2000**, *41*, 1627.
241. Y. Bolshan, M. J. Tomaszewski, V. Santhakumaar, *Tetrahedron Lett.* **2007**, *48*, 4925.
242. P. G. Steel, T. M. Woods,in Scott, P. J. H. (Ed.), *Linker Strategies in Solid-Phase Organic Synthesis*, Wiley, Chichester, UK, **2009**. 331.
243. F. Camps, J. Castells, J. Font, F. Vela, *Tetrahedron Lett.* **1971**, *12*, 1715.
244. S. V. McKinley, J. W. Rakshys, *J. Chem. Soc., Chem. Commun.* **1972**, 134.
245. I. Hughes, *Tetrahedron Lett.* **1996**, *37*, 7595.
246. W. Heitz, R. Michels, *Liebigs Ann. Chem.* **1973**, 227.
247. M. Bernard, W. T. Ford, *J. Org. Chem.* **1983**, *48*, 326.
248. W. Heitz, R. Michels, *Angew. Chem., Int. Ed. Engl.* **1972**, *11*, 298.
249. S. D. Clarke, C. R. Harrison, P. Hodge, *Tetrahedron Lett.* **1980**, *21*, 1375.
250. M. H. Bolli, S. V. Ley, *J. Chem. Soc., Perkin Trans. 1*, **1998**, 2243.
251. G. Cainelli, M. Contento, F. Manescalchi, R. Regnoli, *J. Chem. Soc., Perkin Trans. 1* **1980**, 2516.
252. A. G. M. Barrett, S. M. Cramp, R. S. Roberts, F. Zecri, *Org. Lett.* **1999**, *1*, 579.
253. S. L. X. Martina, R. J. K. Taylor, *Tetrahedron Lett.* **2004**, *45*, 3279.
254. K. C. Nicolaou, J. Pastor, N. Winssinger, F. Murphy, *J. Am. Chem. Soc.* **1998**, *120*, 5132.
255. J. Rademann, S. Weik, *Angew. Chem., Int. Ed.* **2003**, *42*, 2491.
256. I. B. Campbell, J. Guo, E. Jones, P. G. Steel, *Org. Biomol. Chem.* **2004**, *2*, 2725.
257. T. Y. S. But, P. H. Toy,in Scott, P. J. H. (Ed.), *Linker Strategies in Solid-Phase Organic Synthesis*, Wiley, Chichester, UK, **2009**, 391.
258. G. Guazzelli, M. Miller, D. J. Procter, in Scott, P. J. H. (Ed.), *Linker Strategies in Solid-Phase Organic Synthesis*, Wiley, Chichester, UK, **2009**. 419.
259. X. Huang, W. Xu, *Tetrahedron Lett.* **2002**, *43*, 5495.
260. X. L. Liu, X. C. Wang, S. R. Sheng, X. Huang, *Chin. Chem. Lett.* **2004**, *15*, 1009.
261. H. Qian, X. Huang, *Tetrahedron Lett.* **2002**, *43*, 1059.
262. S. R. Sheng, Q. Xin, X. L. Liu, W. K. Sun, R. Guo, X. Huang, *Synthesis* **2006**, 2293.
263. H. Qian, X. Huang, *J. Comb. Chem.* **2003**, *5*, 569.
264. E. Horikawa, M. Kodaka, Y. Nakahara, H. Okuno, K. Nakamura, *Tetrahedron Lett.* **2001**, *42*, 8337.
265. M. Mogemark, L. Gustafsson, C. Bengtsson, M. Elofsson, J. Kihlberg, *Org. Lett.* **2004**, *6*, 4885.
266. K. C. Nicolaou, K. C. Fylaktakidou, H. J. Mitchell, F.L. van Delft, R.M. Rodriguez, S.R. Conley, Z. Jin, *Chem. Eur. J.* **2000**, *6*, 3166.
267. K. Fujita, K. Watanabe, A. Oishi, Y. Ikeda, Y. Taguchi, *Synlett* **1999**, 1760.
268. K. C. Nicolaou, H. J. Mitchell, K. C. Fylaktakidou, H. Suzuki, R. M. Rodríguez, *Angew. Chem., Int. Ed.* **2000**, *39*, 1089.
269. W.-M. Xu, Y.-G. Wang, M.-Z. Miao, X. Huang, *Synthesis* **2005**, 2143.
270. S. R. Sheng, M. G. Hu, Q. Xin, *Chin. Chem. Lett.* **2007**, *18*, 377.
271. H. Qian, L.-X. Shao, X. Huang, *Synlett* **2001**, 1571.
272. T. Ruhland, K. Andersen, H. Pedersen, *J. Org. Chem.* **1998**, *63*, 9204.
273. S. Berlin, C. Ericsson, L. Engman, *J. Org. Chem.* **2003**, *68*, 8386.
274. T. Ruhland, J. Torang, H. Pedersen, J. C. Madsen, K. S. Bang, *Synthesis* **2004**, 2323.
275. T. Ruhland, J. Torang, H. Pedersen, J. C. Madsen, K. S. Bang, *Synthesis* **2005**, 1635.

276. B. R. Stranix, H. Q. Liu, G. D. Darling, *J. Org. Chem.* **1997**, *62*, 6183.
277. P. H. Seeberger, X. Beebe, G. D. Sukenick, S. Pochapsky, S. J. Danishefsky, *Angew. Chem., Int. Ed. Engl.* **1997**, *36*, 491.
278. B. H. Lipshutz, Y. J. Shin, *Tetrahedron Lett.* **2001**, *42*, 5629.
279. K. A. Newlander, B. Chenera, D. F. Veber, N. C. F. Yim, M. L. Moore, *J. Org. Chem.* **1997**, *62*, 6726.
280. C. A. Briehn, T. Kirschbaum, P. Bauerle, *J. Org. Chem.* **2000**, *65*, 352.
281. L. S. Harikrishnan, H. D. H. Showalter, *Tetrahedron* **2000**, *56*, 515.
282. Y.-S. Lee, R. B. Silverman, *J. Am. Chem. Soc.* **1999**, *121*, 8407.
283. Y. Lee, R. B. Silverman, *Org. Lett.* **2000**, *2*, 303.
284. S. D. Brown, R. W. Armstrong, *J. Org. Chem.* **1997**, *62*, 7076.
285. M. Schuster, N. Lucas, S. Blechert, *Chem. Commun.* **1997**, 823.
286. M. Suginome, T. Iwanami, Y. Ito, *J. Am. Chem. Soc.* **2001**, *123*, 4356.
287. M. Komatsu, H. Okada, T. Akaki, Y. Oderaotoshi, S. Minakata, *Org. Lett.* **2002**, *4*, 3505.
288. D. Milstein, J. K. Stille, *J. Am. Chem. Soc.* **1978**, *100*, 3637.
289. N. Miyauri, T. Yanagi, A. Suzuki, *Synth. Commun.* **1981**, *11*, 513.
290. P. J. H. Scott,in: Scott, P. J. H. (Ed.), *Linker Strategies in Solid-Phase Organic Synthesis*, Wiley, Chichester, UK, **2009**, 505.
291. Y. Lunxiang, J. Liebscher, *Chem. Rev.* **2007**, *107*, 133.
292. S. Stanforth, *Tetrahedron* **1998**, *54*, 263.
293. M. V. N. De Souza, *Curr. Org. Synth.* **2006**, *3*, 313.
294. P. Espinet, A. M. Echavarren, *Angew. Chem., Int. Ed.* **2004**, *43*, 4704.
295. H. Kuhn, W. P. Neumann, *Synlett* **1994**, 123.
296. K. C. Nicolaou, N. Winssinger, J. Pastor, F. Murphy, *Angew. Chem., Int. Ed.* **1998**, *37*, 2534.
297. D. H. Hunter, X. Zhu, *J. Labelled Compd. Radiopharm.* **1999**, *42*, 653.
298. G. W. Kabalka, V. Namboodiri, M. R. Akula, *J. Labelled Compd. Radiopharm.* **2001**, *44*, 921.
299. P. A. Culbert, D. H. Hunter, *React. Polym.* **1993**, *19*, 247.
300. S. R. Chemler, D. Trauner, S. J. Danishefsky, *Angew. Chem., Int. Ed.* **2001**, *40*, 4544.
301. A. Suzuki, *J. Organomet. Chem.* **1999**, *576*, 147.
302. N. Miyauri, A. Suzuki, *Chem. Rev.* **1995**, *95*, 2457.
303. H. Douchet, *Eur. J. Org. Chem.* **2008**, 2013.
304. N. T. S. Phan, Van Der Sluys M, C. W. Jones, *Adv. Synth. Cat.* **2006**, *348*, 609.
305. K. C. Nicolaou, P. G. Bulger, D. Sarlah, *Angew. Chem., Int. Ed.* **2005**, *44*, 4442.
306. W. Li, K. Burgess, *Tetrahedron Lett.* **1999**, *40*, 6527.
307. D. G. Hall, J. Tailor, M. Gravel, *Angew. Chem., Int. Ed.* **1999**, *38*, 3064.
308. B. Carboni, C. Pourbaix, F. Carreaux, H. Deleuze, B. Maillard, *Tetrahedron Lett.* **1999**, *40*, 7979.
309. C. Pourbaix, F. Carreaux, B. Carboni, *Org. Lett.* **2001**, *3*, 803.
310. C. Pourbaix, F. Carreaux, B. Carboni, H. Deleuze, *Chem. Commun.* **2000**, 1275.
311. M. Gravel, D. G. Hall, *Spec. Chem. Mag.* **2003**, *23*, 31.
312. M. Gravel, K. A. Thompson, M. Zak, C. Berube, D. G. Hall, *J. Org. Chem.* **2002**, *67*, 3.
313. G. I. Elliott, J. P. Konopelski, *Tetrahedron* **2001**, *57*, 5683.
314. L. K. Rasmussen, M. Begtrup, T. Ruhland, *J. Org. Chem.* **2006**, *71*, 1230.
315. L. K. Rasmussen, M. Begtrup, T. Ruhland, *J. Org. Chem.* **2004**, *69*, 6890.
316. P. J. H. Scott, in Scott, P. J. H. (Ed.), *Linker Strategies in Solid-Phase Organic Synthesis*, Wiley, Chichester, UK, **2009**, 519.

317. J. H. van Maarseveen,in Scott, P. J. H. (Ed.), *Linker Strategies in Solid-Phase Organic Synthesis*, Wiley, Chichester, UK, **2009**. 537.
318. J. J. N. Veerman, J. H. van Maarseveen, G. M. Visser, C. G. Kruse, H. E. Schoemaker, H. Hiemstra, F. P. J.T. Rutjes, *Eur. J. Org. Chem.* **1998**, 2583.
319. K. C. Nicolaou, N. Winssinger, J. Pastor, F. Murphy, *Angew. Chem., Int. Ed.* **1998**, *37*, 2534.
320. S. Sasmal, A. Geyer, M. E. Maier, *J. Org. Chem.* **2002**, *67*, 6260.
321. J. Pernerstorfer, M. Schuster, S. Blechert, *Chem. Commun.* **1997**, 1949.
322. M. S. M. Timmer, M. Verdoes, L. A. J.M Sliedregt, G. A. van der Marel, J. H. van Boom, H. S. Overkleeft, *J. Org. Chem.* **2003**, *68*, 9406.
323. J. U. Peters, S. Blechert, *Synlett* **1997**, 348.
324. D. Brohm, N. Philippe, S. Metzger, A. Bhargava, O. Müller, F. Lieb, H. Waldmann, *J. Am. Chem. Soc.* **2002**, *124*, 13171.
325. D. Brohm, S. Metzger, A. Bhargava, O. Müller, F. Lieb, H. Waldmann, *Angew. Chem., Int. Ed.* **2001**, *41*, 307.
326. L. Knerr, R. R. Schmidt, *Synlett* **1999**, 1802.
327. L. Knerr, R. R. Schmidt, *Eur. J. Org. Chem.* **2000**, 2803.
328. T. Kanemitsu, P. H. Seeberger, *Org. Lett.* **2003**, *5*, 4541.
329. R. B. Andrade, O. J. Plante, L. G. Melean, P. H. Seeberger, *Org. Lett.* **1999**, *1*, 1811.
330. D. Craig, M. J. Robson, S. J. Shaw, *Synlett* **1998**, 1381.

2

COLORIMETRIC TESTS FOR SOLID-PHASE ORGANIC SYNTHESIS

Yan Teng and Patrick H. Toy

2.1 INTRODUCTION

While the use of a solid support in organic synthesis imparts advantages with regard to performing and working up reactions, monitoring of the reactions becomes more complicated than when performing a reaction using traditional solution-phase methodology. For example, thin layer chromatography cannot be used to monitor a solid-phase reaction unless a cleavage reaction is performed to remove some of the synthesis substrate/product mixture from the support. Therefore, a popular methodology for "on-bead" reaction monitoring of the appearance or disappearance of a functional group from a solid-support attached synthesis substrate is to use a colorimetric test.[1,2] In such tests, the solid support is treated with reagents that react with the functional group to be monitored, imparting a visible color to the material when a reaction occurs. Such tests are generally reliable, fast, and easy to perform. Perhaps their main drawback is that they are generally destructive and sometimes require the consumption of a significant amount of the synthesis product. This can be critical if the synthesis is performed on a small scale and only a few support beads are used per synthetic target.

In this chapter, the reported colorimetric tests for solid-phase organic synthesis are presented and the examples are organized based on the functional group they were originally reported to detect. Clearly, since the reactivities of many functional groups such as alcohols and thiols are similar, some of the tests presented are useful in multiple different applications.

Solid-Phase Organic Synthesis: Concepts, Strategies, and Applications, First Edition.
Edited by Patrick H. Toy and Yulin Lam.

2.2 FUNCTIONAL GROUP TESTS

2.2.1 Amine Groups

It is perhaps not surprising, given the history of solid-phase synthesis, that the most widely tested functional groups attached to a solid support are amines.

2.2.1.1 Ninhydrin (Kaiser) Test[3]

Basic Reagent:

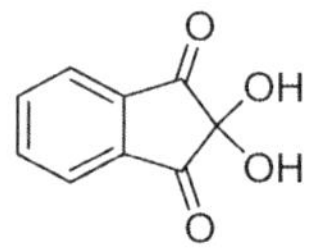

Ninhydrin

Other Reagents: Phenol and potassium cyanide.

Comments: Commercially available. Detects primary amines to a limit of 5 μmol/g. Provides an intense blue color or in some cases a brown/reddish brown color. First used to detect incomplete coupling reactions in peptide synthesis.

2.2.1.2 TNBSA Test[4,5]

Basic Reagent:

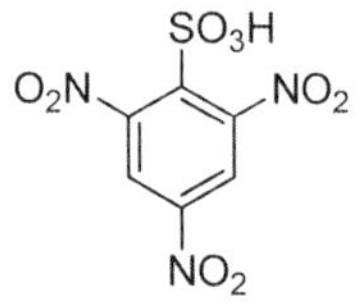

2,4,6-Trinitrobenzenesulfonic acid

Other Reagent: *N,N*-Diisopropylethylamine (used sometimes).[5]

Comments: Commercially available. Detects primary and secondary amines to a limit of 3 mmol/g. Provides an orange or red color.

2.2.1.3 Bromophenol Blue Test[6–8]

Basic Reagent:

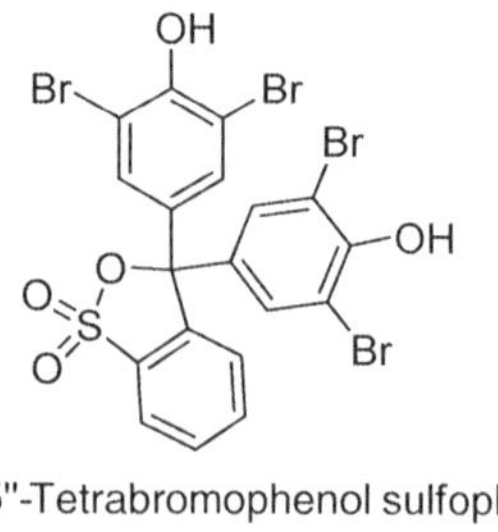

3',3",5',5"-Tetrabromophenol sulfophthalein

Comments: Commercially available. A noninvasive qualitative and quantitative test. Provides a deep blue color.

2.2.1.4 Chloranil Test[9–11]

Basic Reagent:

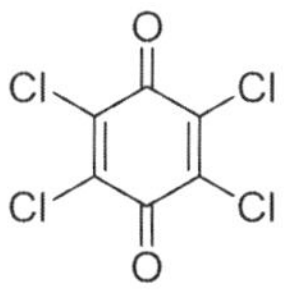

Tetrachloro-*p*-benzoquinone
(chloranil)

Other Reagents: Acetaldehyde (for primary and secondary amines)[9,10] and acetone (for secondary amines).[9]

Comments: Commercially available. Detects primary[9] and secondary[10] aliphatic amines as well as primary aromatic amines[11] to a limit of 5 μmol/g for aromatic amines. Provides a green/blue color[9,10] or a red color.[11]

2.2.1.5 DABITC Test[12]

Basic Reagent:

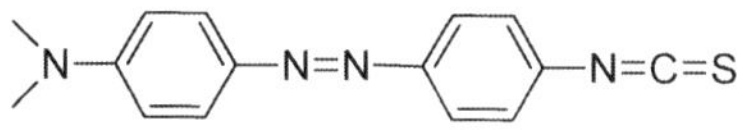

4-*N,N'*-Dimethylaminoazobenzene-
4'-isothiocyanate

Comments: Commercially available. Detects primary and secondary amines. Provides a yellow or orange color.

2.2.1.6 MGI Test[12]

Basic Reagent:

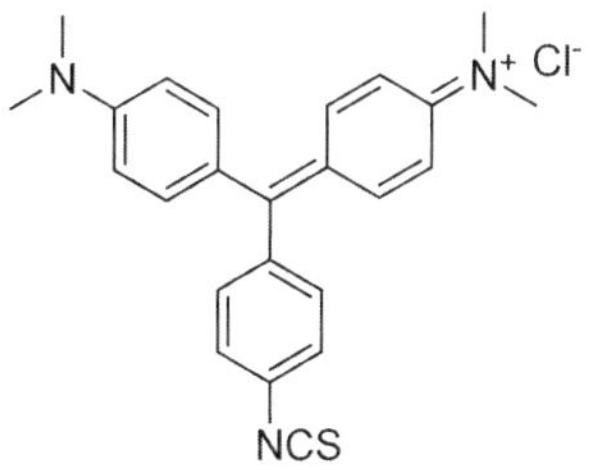

Malachite green isothiocyanate
(MGI)

Comments: Commercially available. Detects primary and secondary amines. Provides a green or deep blue color.

2.2.1.7 Isatin Test[12]

Basic Reagent:

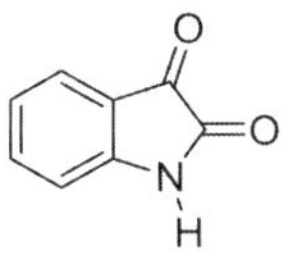

2,3-Indolinedione (isatin)

Other Reagent: Boc-phenylalanine.

Comments: Commercially available. Used specifically for the analysis of proline. Provides a blue color.

2.2.1.8 DESC Test[13]

Basic Reagent:

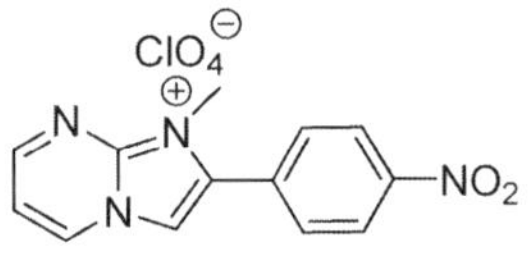

1-Methyl-2-(4'-nitrophenyl)imidazo[1,2-α]pyrimidinium perchlorate salts (DESC)

Other Reagent: *N,N*-Diisopropylethylamine (for secondary amines).

Comments: Not commercially available. Detects primary and secondary amines to a limit of 9 μmol/g for primary amines. Can also be used to detect thiols. Provides a red to dark orange color.

2.2.1.9 NPIT Test[14]

Basic Reagent:

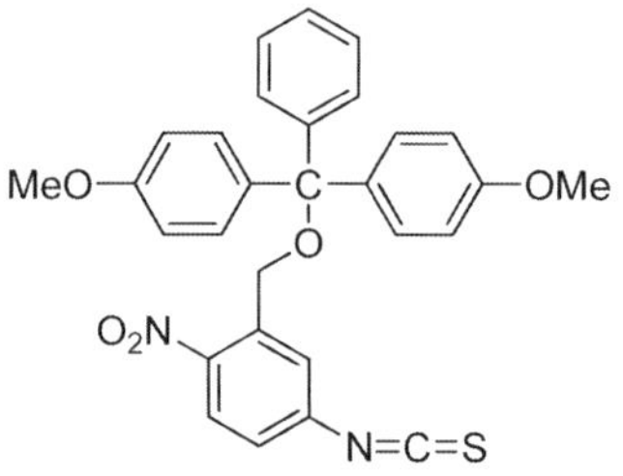

Nitro-phenylisothiocyanate-*O*-trityl (NPIT)

Other Reagent: Trifluoroacetic acid.

Comments: Not commercially available. Detects less reactive primary and secondary amines. Provides a red color.

2.2.1.10 NF31 Test[15–18]

Basic Reagent:

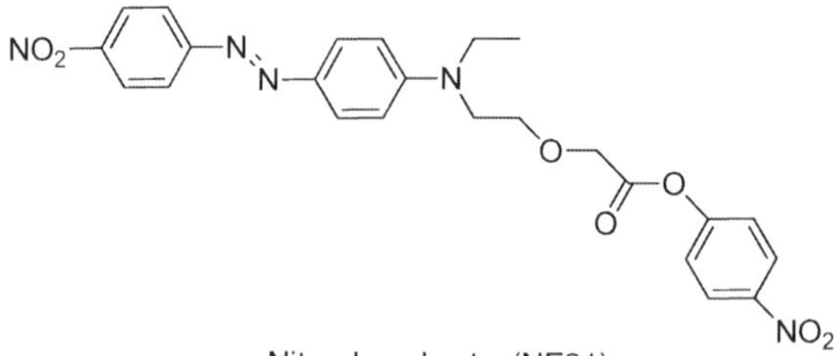

p-Nitrophenyl ester (NF31)

Other Reagent: 4-Dimethylaminopyridine (for detecting alcohols, phenols, and thiols—NF31/DMAP test).[17]

Comments: Commercially available. Detects sterically hindered primary amines, secondary amines,[15] and aromatic amines.[16] In the presence of DMAP, alcohols, phenols, and primary thiols can also be detected.[17] Detection limit of 3.4 μmol/g for aniline groups[16] and 3–5 μmol/g for NF31/DMAP test.[17] Provides a red color. May give positive results for free hydroxyl groups even without DMAP.[18]

2.2.1.11 Nondestructive NF31 Test[19]

Basic Reagent:

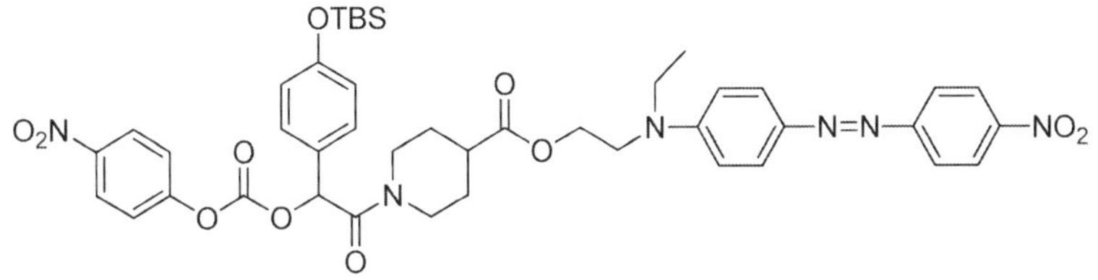

Other Reagents: 1,8-Diazabicyclo[5.4.0]undec-7-ene and 4-dimethylaminopyridine.

Comments: Not commercially available. Detects primary and secondary amines as well as thiols. Tested resin can be regenerated and used for ensuing reaction. Provides a red color.

2.2.1.12 Naphthol Test[20]

Basic Reagent:

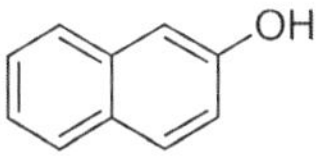

β-Naphthol

Other Reagent: $NaNO_2$.

Comments: Commercially available. Detects aromatic amines. Provides a red color.

2.2.1.13 2-Amino-3-chloro-1,4-naphthoquinone Test[21]

Basic Reagent:

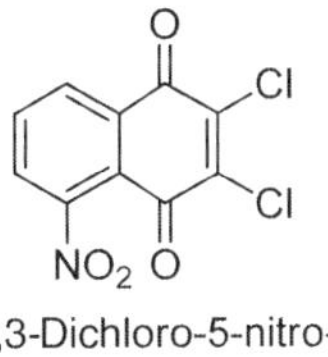

2,3-Dichloro-5-nitro-1,4-naphthoquinone

Other Reagents: 2,6-Di-*tert*-butylpyridine and trifluoroacetic acid.

Comments: Commercially available. Detects primary and secondary alkyl amines, and aromatic amines to a limit of 10 μmol/g. Quantitative analysis based on the chromogens cleaved from the resin. Provides a red color.

2.2.2 Alcohols

There are some significant applications regarding the detection of hydroxyl groups on solid support. Unsurprisingly, some of the above-mentioned reagents, such as NF31, can been used to detect hydroxyl groups in addition to amine groups.

2.2.2.1 PNBP Test[22–24]

Basic Reagent:

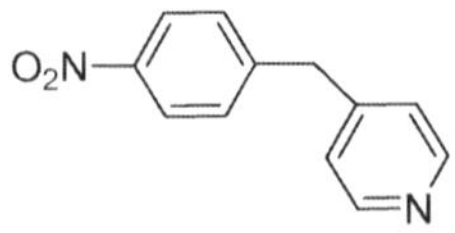

p-Nitrobenzylpyridine
(PNBP)

Other Reagents: 4-Toluenesulfonyl chloride, piperidine (for detecting alcohols),[22] and triethylamine (for chlorine groups).[24]

Comments: Commercially available. Detects primary, secondary, and tertiary alcohols as well as phenols[22] (negative result was obtained for some tertiary alcohols[25]) to a limit of 16 μmol/g.[22] Provides a violet to pink color for alcohol and chlorine detection[22,24] and a red color for chloroacetyl detection.[23]

2.2.2.2 TCT–AliR and TCT–Fluorescein Test[26,27]

Basic Reagents:

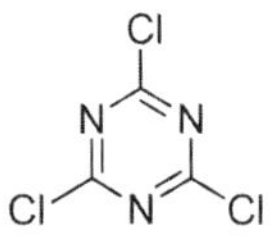

2,4,6-Trichloro[1,3,5]triazine
(TCT)

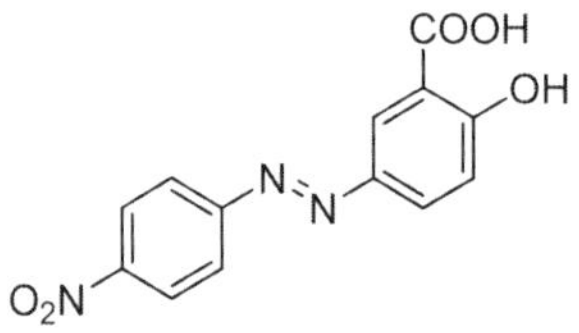

Alizarin R

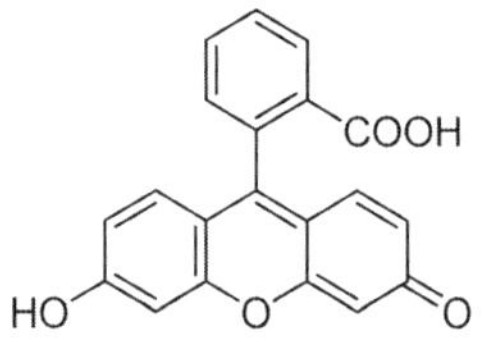

Fluorescein

Other Reagents: *N*-Methylmorpholine, dimethoxytrityl chloride, and trichloroacetic acid (for quantification).

Comments: Commercially available. Detects primary, secondary, and tertiary alcohols as well as phenols. Other nucleophilic groups, such as primary and secondary amines, and thiols can also be detected. Quantification analysis of primary and secondary alcohol groups can be performed with the TCT–AliR test. Provides a red color in the TCT–AliR test and a yellow-green color in the TCT–fluorescein test.

2.2.2.3 Diphenyldichlorosilane–Methyl Red Test[25]

Basic Reagents:

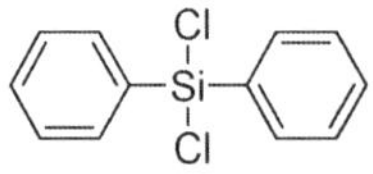

Diphenyldichlorosilane

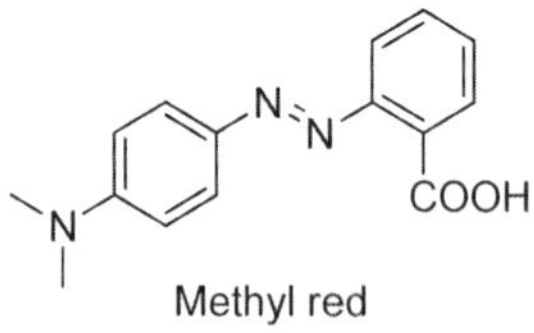

Methyl red

Other Reagents: Triethylamine and formic acid (for further confirmation).

Comments: Commercially available. Reportedly reliable result for tertiary alcohols and other alcohols and phenols to a detection limit of 0.07 mmol/g. Provides an orange to red color, and a purple color with formic acid rinsing.

2.2.2.4 9-Anthronylnitrile Test[28]

Basic Reagent:

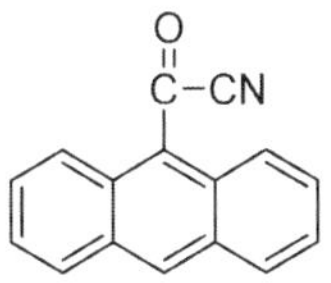

9-Anthronylnitrile

Other Reagent: Quinuclidine.

Comments: Commercially available. Provides direct quantification based on the consumption of 9-anthronylnitrile using UV–visible spectroscopic measurements to a detection limit of 0.05 mmol/g and fluorescent bead under 350–360 nm UV illumination. Also works for unhindered primary amines.

2.2.2.5 NMA Test[29]

Basic Reagent:

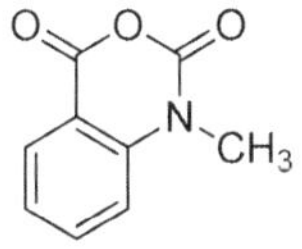

N-Methylisatoic anhydride (NMA)

Comments: Commercially available. Detects primary and secondary alcohols to a detection limit of 3 μmol/g. Also works for primary amines. Fluorescent beads using a 365 nm UV lamp.

2.2.2.6 Protecting Group NPB Test[30]

Basic Reagent:

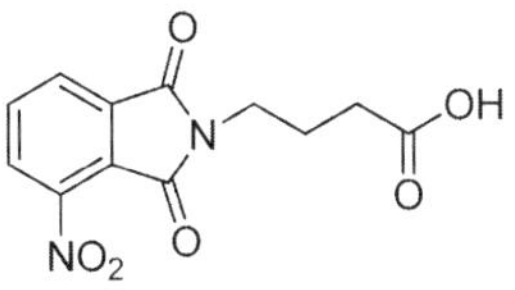

Nitrophthalimidobutyric (NPB) acid

Other Reagents: *N,N'*-Dicyclohexylcarbodiimide, 4-dimethylaminopyridine, and hydrazine acetate.

Comments: Not commercially available. After cleavage, nitrophthalhydrazide results in an orange-colored solution.

2.2.2.7 Methyl Red/DIC Test[31]

Basic Reagent:

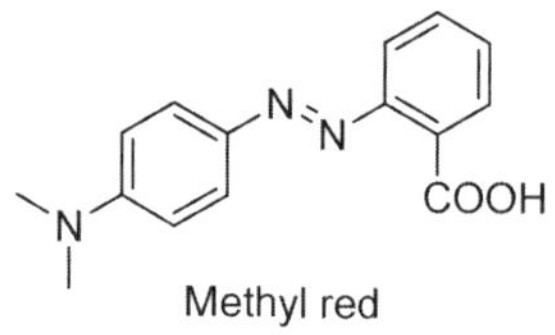

Methyl red

Other Reagents: 1,3-Diisopropylcarbodiimide (DIC) and 4-dimethylaminopyridine.

Comments: Commercially available. Methyl red reacts directly with hydroxyl groups without intermediates to a detection limit of 0.042 mmol/g. Provides an orange-red color.

2.2.2.8 Other Methods

The NF31/DMAP test described above also detects alcohols and phenols.[17]

2.2.3 Thiol Groups

2.2.3.1 Ellman's Test[32–34]

Basic Reagent:

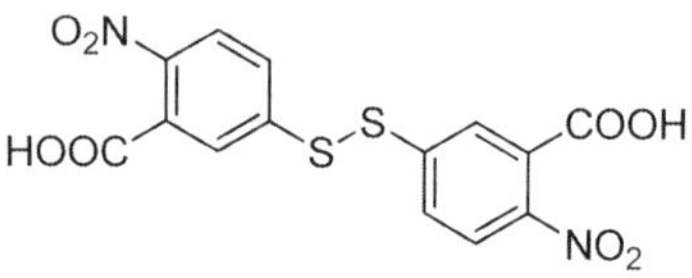

5,5'-Dithio-bis(2-nitrobenzoic acid)
(DNTB or Ellman's reagent)

Comments: Commercially available. Provides quantitative analysis and a yellow color.

2.2.3.2 Other Methods

The DESC,[13] NF31/DMAP,[17] nondestructive NF31,[19] TCT–AliR, and TCT–fluorescein[26,27] tests described above also detect thiol groups.

2.2.4 Halogen Groups

2.2.4.1 Fluorescein Test[2]

Basic Reagent:

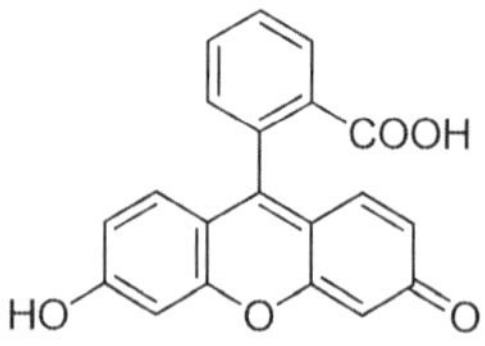

Fluorescein

Other Reagent: Triethylamine.

Comments: Commercially available. Provides a fluorescent green color. May not be suitable for hindered or secondary halogens due to possible elimination reactions.

2.2.4.2 Other Methods

The PNBP test described above can also be used to detect chloroacetyl[23] and chloromethyl[24] groups.

2.2.5 Carboxylic Acid Groups

2.2.5.1 Malachite Green Test[35]

Basic Reagent:

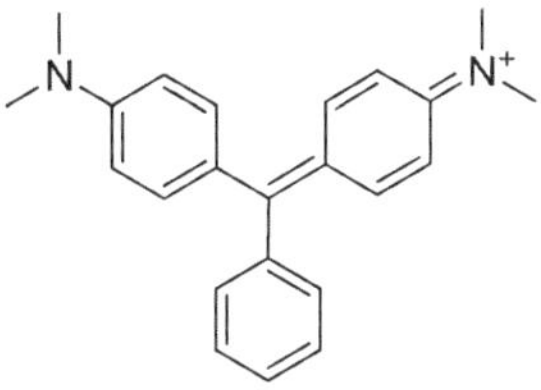

Malachite green
(MG)

Other Reagent: Triethylamine.

Comments: Commercially available. Provides indirect quantitative analysis and a green color.

2.2.5.2 PDAM Test[28]

Basic Reagent:

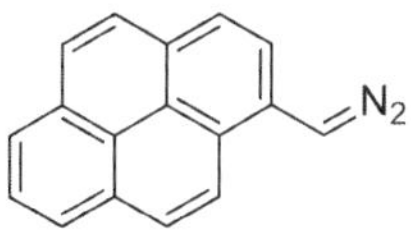

1-Pyrenyldiazomethane
(PDAM)

Comments: Commercially available. Provides direct quantification based on the consumption of PDAM using UV–visible spectroscopic measurements and fluorescent beads under 350–360 nm UV illumination.

2.2.6 Aldehyde and Ketone Groups

2.2.6.1 Fluorescent Dansylhydrazine Test[36]

Basic Reagent:

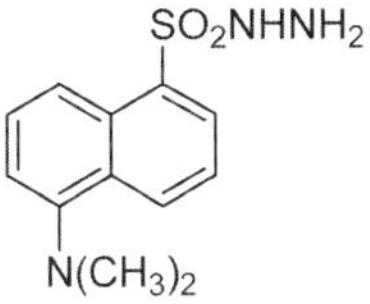

Dansylhydrazine

Comments: Commercially available. Provides direct quantification of aldehyde and ketone groups based on the consumption of dansylhydrazine using fluorescence spectroscopy and fluorescent bead under 350–360 nm UV illumination.

2.2.6.2 p-Anisaldehyde Test[37]

Basic Reagent:

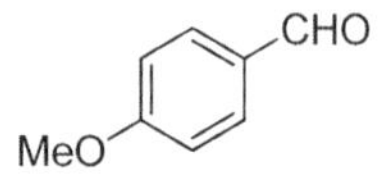

p-Anisaldehyde

Other Reagents: Sulfuric acid and acetic acid.

Comments: Commercially available. Detects aromatic and aliphatic aldehyde groups. Provides an orange to red color.

2.2.6.3 Purpald Test[38]

Basic Reagent:

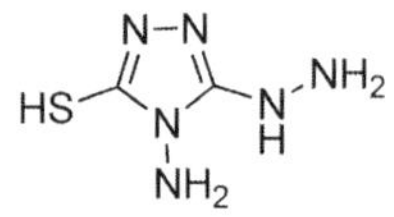

4-Amino-3-hydrazino-
5-mercapto-1,2,4-triazole
(purpald)

Other Reagent: Sodium hydroxide.

Comments: Commercially available. Test is specific for aldehyde groups to a detection limit of 20 μmol/g. Provides a dark brown to purple color.

2.3 CONCLUSIONS

After the introduction of Kaiser's ninhydrin test for the on-bead detection of primary amines, many other such colorimetric tests have been reported in the literature, not only for amines but also for other functional groups. Many of these tests are simple and quick to perform and use readily available reagents. Furthermore, some even allow quantitative determination of functional group loading levels. Given the current high level of interest in the development of fluorescent dyes, it is easy to imagine that such tools will be widely useful in monitoring solid-phase organic synthesis reactions.

REFERENCES

1. C. Kay, O. E. Lorthioir, N. J. Parr, M. Congreve, S. C. McKeown, J. J. Scicinski, S. V. Ley, *Biotechnol. Bioeng.* **2000**, *71*, 110–118.
2. F. Gaggini, A. Porcheddu, G. Reginato, M. Rodriquez, M. Taddei, *J. Comb. Chem.* **2004**, *6*, 805–810.
3. E. Kaiser, R. L. Colescott, C. D. Bossinger, P. I. Cook, *Anal. Biochem.* **1970**, *34*, 595–598.
4. W. S. Hancock, J. E. Battersby, *Anal. Biochem.* **1976**, *71*, 260–264.

5. J. Vazquez, G. Qushair, F. Albericio, *Methods Enzymol.* **2003**, *369*, 21–35.
6. V. Krchňák, J. Vágner, M. Lebl, *Int. J. Pept. Protein Res.* **1988**, *32*, 415–416.
7. V. Krchňák, J. Vágner, P. Safar, M. Lebl, *Collect. Czech. Chem. Commun.* **1988**, *53*, 2542–2548.
8. M. Flegel, R. C. Sheppard, *J. Chem. Soc. Chem. Commun.* **1990**, 536–538.
9. T. Christensen, *Acta Chem. Scand. B* **1979**, *33*, 763–766.
10. T. Vojkovsky, *Peptide Res.* **1995**, *8*, 236–237.
11. J. Mařík, A. Song, K. S. Lam, *Tetrahedron Lett.* **2003**, *44*, 4319–4320.
12. A. Shah, S. S. Rahman, V. de Biasi, P. Camilleri, *Anal. Commun.* **1997**, *34*, 325–328.
13. S. Claerhout, D. S. Ermolat'ev, E. V. Van der Eycken, *J. Comb. Chem.* **2008**, *10*, 580–585.
14. S. S. Chu, S. H. Reich, *Bioorg. Med. Chem. Lett.* **1995**, *5*, 1053–1058.
15. A. Madder, N. Farcy, N. G. C. Hosten, H. De Muynck, P. J. De Clercq, J. Barry, A. P. Davis, *Eur. J. Org. Chem.* **1999**, 2787–2791.
16. S. E. Van der Plas, P. J. De Clercq, A. Madder, *Tetrahedron Lett.* **2007**, *48*, 2587–2589.
17. J. Caroen, J. Van der Eycken, *Tetrahedron Lett.* **2009**, *50*, 41–44.
18. L. L. G. Carrette, D. Verzele, A. Madder, *Tetrahedron Lett.* **2010**, *51*, 2106–2108.
19. S.-J. Yang, X. Z. Tian, I. Shin, *Org. Lett.* **2009**, *11*, 3438–3441.
20. L. De Luca, G. Giacomelli, A. Porcheddu, M. Salaris, M. Taddei, *J. Comb. Chem.* **2003**, *5*, 465–471.
21. C. Blackburn, *Tetrahedron Lett.* **2005**, *46*, 1405–1409.
22. O. Kuisle, M. Lolo, E. Quiñoá, R. Riguera, *Tetrahedron* **1999**, *55*, 14807–14812.
23. S. Manabe, Y. Ito, *J. Am. Chem. Soc.* **2002**, *124*, 12638–12639.
24. F. Galindo, B. Altava, M. I. Burguete, R. Gavara, S. V. Luis, *J. Comb. Chem.* **2004**, *6*, 859–861.
25. B. A. Burkett, R. C. D. Brown, M. M. Meloni, *Tetrahedron Lett.* **2001**, *42*, 5773–5775.
26. M. E. Attardi, A. Falchi, M. Taddei, *Tetrahedron Lett.* **2000**, *41*, 7395–7399.
27. M. E. Attardi, A. Falchi, M. Taddei, *Tetrahedron Lett.* **2001**, *42*, 2927.
28. B. Yan, L. Liu, C. A. Astor, Q. Tang, *Anal. Chem.* **1999**, *71*, 4564–4571.
29. R. E. Fake, A. Routledge, *Tetrahedron Lett.* **2004**, *45*, 8925–8926.
30. K.-S. Ko, G. Park, Y. Yu, N. L. Pohl, *Org. Lett.* **2008**, *10*, 5381–5384.
31. S. Komba, S. Sasaki, S. Machida, *Tetrahedron Lett.* **2007**, *48*, 2075–2078.
32. G. L. Ellman, *Arch. Biochem. Biophys.* **1959,** *82*, 70–77.
33. T. J. Novak, S. G. Pleva, J. Epstein, *Anal. Chem.* **1980**, *52*, 1851–1855.
34. J. P. Badyal, A. M. Cameron, N. R. Cameron, D. M. Coe, R. Cox, B. G. Davis, L. J. Oates, G. Oye, P. G. Steel, *Tetrahedron Lett.* **2001**, *42*, 8531–8533.
35. M. E. Attardi, G. P. Porcu, M. Taddei, *Tetrahedron Lett.* **2000**, *41*, 7391–7394.
36. B. Yan, W. Li, *J. Org. Chem.* **1997**, *62*, 9354–9357.
37. J. Vazquez, F. Albericio, *Tetrahedron Lett.* **2001**, *42*, 6691–6693.
38. J. J. Cournoyer, T. Kshirsagar, P. P. Fantauzzi, G. M. Figliozzi, T. Makdessian, B. Yan, *J. Comb. Chem.* **2002**, *4*, 120–124.

3

PRACTICAL ASPECTS OF COMBINATORIAL SOLID-PHASE SYNTHESIS

Jan Hlaváč, Miroslav Soural, and Viktor Krchňák

3.1 INTRODUCTION

In the middle of the nineteenth century, gold was discovered at Coloma in California by the American pioneer James Wilson Marshall. The discovery started the famous "Forty-Niners" gold rush in the West that brought great expectations and excitement, as well as disappointment. At the end of the last millennium, the West was in a fever again, but this time the vision of wealth was to be attained using a new technology that promised to enormously accelerate the drug discovery process. The technique, known as combinatorial chemistry, once again caused great excitement and substantial disappointment. Since the early times of combinatorial chemistry ("combi-chem") in Tucson[1] and San Diego[2] (both papers were published side by side in *Nature* in 1991), drug discovery companies have been driven to find the modern equivalent of a "Mother Lode."[3] In this case, the treasure is represented by a compound collection from which biologically active and unique compounds would be mined and subsequently developed into drugs. Thus, chemical combinatorial methods became a field of great interest during the last decade of the twentieth century. The numerous attributes, advances, and large amount of interest in combinatorial chemistry are demonstrated by the number of new journals dedicated to the subject, the number of new combinatorial chemistry companies, and the amount of money invested in combinatorial chemistry and related technologies. There are several excellent books and review articles covering combinatorial technologies, and the interested reader is referred to Refs [11] and [12] for a summary of synthetic procedures and also to recent general reviews of the subject.[13–18]

Solid-Phase Organic Synthesis: Concepts, Strategies, and Applications, First Edition.
Edited by Patrick H. Toy and Yulin Lam.

3.1.1 What Is Combinatorial Chemistry

Combinatorial chemistry is a technique for producing a collection of compounds (a library) from a given number of starting materials in a combinatorial fashion.[19] Several definitions of combinatorial chemistry have been offered, such as "combinatorial chemistry is a new subfield of chemistry with the goal of synthesizing very large numbers of chemical entities by condensing a small number of reagents together in all possible combinations" by Czarnik[21] or "combinatorial synthesis is the intentional construction of a collection of molecules based on logical design and involving the selective combination of building blocks by means of simultaneous chemical reactions. The collection of molecules resulting from a combinatorial synthesis is a combinatorial library" by Curran and Wipf.[22]

The starting materials used to prepare or "build" the required products are generally referred to as building blocks. While the number of building blocks increases in an arithmetic progression, the number of compounds prepared increases in a geometric progression. This is achieved by combining each of the building blocks from the first combinatorial step with all the building blocks used in the subsequent combinatorial steps. The technique is best illustrated with an example. Suppose a series of benzimidazoles are to be synthesized according to the chemistry illustrated in Scheme 3.1 using 10 primary amines, 10 *o*-fluoronitrobenzenes, and 10 carboxylic acids.[23] Each amine is immobilized via a reductive amination procedure on a support functionalized with an aldehyde moiety. When each resin is combined (reacted) with the 10 *o*-fluoronitrobenzenes, 100 ($10 \times 10 = 100$) *o*-nitroanilines are formed. After reduction of the nitro group, each of the *o*-phenylenediamine intermediates is derivatized with each of the 10 acids, producing 1000 ($10 \times 10 \times 10 = 1000$) benzimidazole precursors, which upon cleavage and cyclization afford 1000 benzimidazoles. In summary, using 30 building blocks ($10 + 10 + 10$), 1000 compounds ($10 \times 10 \times 10$) are prepared.

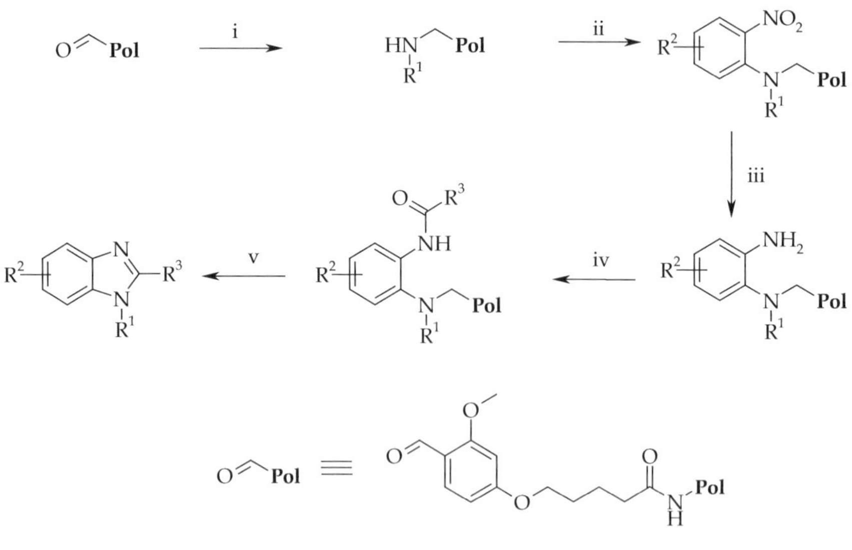

Scheme 3.1. Combinatorial synthesis of benzimidazoles: using 30 ($10 + 10 + 10$) building blocks to prepare 1000 ($10 \times 10 \times 10$) products in a combinatorial fashion. Reagents: (i) amine/NaBH(AcO)$_3$ in DMF/AcOH; (ii) o-fluoronitrobenzene, DMSO, rt, overnight; (iii) SnCl$_2$·2H$_2$O in NMP, rt, overnight; (iv) acid chloride/DIEA in DCM, rt, overnight; (v) AcOH, 80°C, overnight.

Each chemical transformation that uses a series of building blocks is referred to as a combinatorial step; the above example represents a synthesis with three combinatorial steps. A compound array prepared using only one set of building blocks is considered to be a parallel synthesis rather than a combinatorial synthesis.

Combinatorial syntheses can be performed in two ways: (i) split-and-pool (also referred to as the split-and-mix or the split-and-recombine method)[1,2,24] and its variant directed sorting[4,25–31] and (ii) parallel synthesis.[5,6,32,33] A hybrid approach is the split-and-split technique.[34,35]

Although combinatorial chemistry may be applied to a variety of areas of chemistry, to date, the majority of practical applications have been dedicated to drug discovery projects. Combinatorial chemistry is undoubtedly an efficient method for the preparation of compounds in a variety of library sizes. This approach, however, also makes combinatorial chemistry an efficient method for the production of useless or "inactive" compounds. For this reason, the technique has sometimes been referred to as an "irrational design" process, in contrast to the accepted "rational" drug design process traditionally practiced by medicinal chemists.

3.1.2 What Is Not Combinatorial Chemistry

Although solid-phase synthesis is frequently linked to combinatorial chemistry, this is not a requirement. Other synthetic methodologies, such as solution-phase synthesis[36–38] and soluble polymer-supported synthesis,[39–42] have also been used to effect the combinatorial synthesis process. However, solid-phase synthesis allows the most efficient combinatorial synthesis. The advantages and problems with solid-supported synthesis are described in later chapters. Thus, combinatorial chemistry is not solid-phase chemistry, albeit combinatorial chemistry can be advantageously performed on the solid phase.

Combinatorial chemistry encompasses a wide range of techniques and synthetic goals, and library sizes may range from millions of compounds to a few tens of compounds. Combinatorial chemistry also involves the synthesis of small collections of compounds that may be prepared in a completely manual manner or by the use of simple tools. The equipment and human resources required to meet the goals of a particular combinatorial project vary considerably, and hence the equipment required to successfully pursue combinatorial chemistry ranges from cheap, simple, manually operated apparatus to multimillion dollar automated robots. Although nobody today would prepare chemical libraries without efficient tools, their presence in combinatorial synthesis is not required. Thus, combinatorial chemistry is not the synthesis of chemical libraries using a million dollar robot or any other instrumentation, albeit combinatorial chemistry can be performed using such a robot or instrumentation.

Solid-phase combinatorial synthesis can be performed using the split-and-pool technique based on the combination of variously substituted compounds together for the same reaction in an appropriate reaction step, as well as by parallel synthesis, in which all compounds are segregated during all the reaction steps (see next chapters). Although parallel synthesis is an efficient way to prepare arrays of structurally unrelated compounds, it is not necessarily a combinatorial approach conventionally based on substituent modifications of one structural motif. Thus, combinatorial chemistry is not parallel synthesis, albeit combinatorial chemistry can be performed in parallel fashion.

The obvious requirement for a synthesis to be regarded as combinatorial is that it is undertaken in a combinatorial manner, independent of the synthetic approach, size of a library, and degree of automation.

3.1.3 History of Combinatorial Chemistry: Breakthrough Discoveries That Shaped the Future of the Combinatorial Chemistry Field

Although no single invention triggered scientific and research activities in combinatorial chemistry, there were several milestone discoveries that helped reveal the concepts of combinatorial technologies and prepare the general chemical community for the "combinatorial way of thinking."

3.1.3.1 Solid-Phase Synthesis. The first, and possibly the most crucial, was solid-phase peptide synthesis by Bruce Merrifield[43] published in 1963. The solid-phase synthesis of peptides by Bruce Merrifield is undoubtedly considered to be a landmark in chemistry as a whole and had a critical impact on combinatorial chemistry. Although the concept of solid-phase peptide synthesis is, similar to many other ingenious ideas, very simple and straightforward, Bruce Merrifield was the first to recognize its potential and develop a workable method.

The repetitive and tedious nature of solution-phase peptide synthesis prompted Merrifield to develop the solid-phase peptide synthesis. Contemporary solid-phase synthesis follows the original Merrifield concept (Scheme 3.2), but differs only in the types of amine protecting groups, carboxyl group activating agents, and the variety of linkers that are now available. In a typical Merrifield solid-phase peptide synthesis procedure, an *N*-protected (*t*-butyloxycarbonyl group (Boc)) amino acid is attached to an insoluble polymer support, chloromethyl-copoly(styrene-1% divinylbenzene) resin, via an ester bond. The protecting group of the resulting resin-bound ester is cleaved, most often by TFA diluted with DCM, the TFA salt of the liberated amino group is neutralized, and the amino groups thus produced are acylated with the next amino acid. The most common method of activation of the carboxyl group of the amino acid uses a carbodiimide, often in the presence of HOBt. This two-step cycle is repeated until the target peptide is assembled. At the end of the synthesis, the side-chain protecting groups are removed and the peptide is cleaved from the resin; in this example, liquid HF is used.

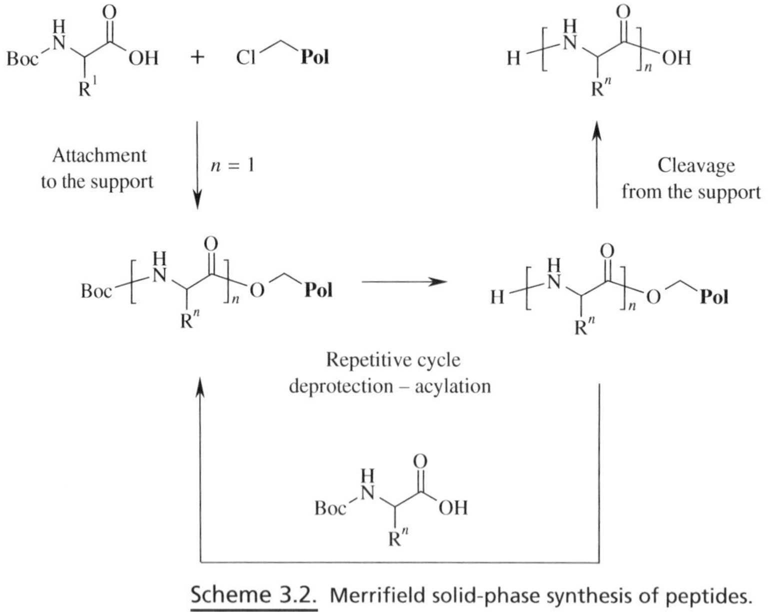

Scheme 3.2. Merrifield solid-phase synthesis of peptides.

Solid-phase synthesis in general has several inherent characteristics. As it is not possible to purify the growing substrate during solid-phase synthesis, the reaction has to be driven as close to completion as possible. Therefore, a large excess of reagents is often used. Suppose each reaction proceeded with 95% conversion, a very acceptable yield for solution-phase synthesis. However, for a 10-step synthesis, the purity of the final product would be 63% ($0.95^9 \times 100$). Therefore, an efficient synthesis must be developed before library synthesis.

The three most attractive aspects of solid-phase organic synthesis include the following: (i) a very simple separation of synthetic intermediates bound to the solid support from soluble components of a reaction mixture by simple filtration and washing of the resin. A consequence of this is the ability to use a high boiling reaction solvent, such as DMF, DMSO, NMP, and so on, without the need to evaporate the solvent. (ii) A high concentration of reactants in solution that facilitates reaction completion. (iii) A simple repetitive process (adding reagents, mixing, washing) allows integration and automation of solid-phase synthesis.

Other aspects of solid-phase synthesis that need to be considered are as follows: (i) solid-phase synthesis starts with attachment of the first building block and ends with release of the target compound from the insoluble support, thus adding two chemical transformations to the total number of synthetic steps. (ii) In solution-phase synthesis, the isolation of products may often be difficult or time-consuming. In some cases, chemical transformations may take only minutes, and thus three or more consecutive reactions on solid phase may be more time efficient when compared to one chemical reaction in solution, followed by a tedious isolation procedure to provide the intermediate for the next reaction step. (iii) The kinetics of solid-phase reactions tend to be slower than those of the equivalent solution-phase reaction. This is generally because the kinetics of a reaction on the solid phase are controlled by diffusion/accessibility of the reagents into the bead. (iv) Although the ability of a large excess of reagents to drive reactions to completion is an advantage, this requires the use of greater quantities of reactants that may be expensive or difficult to obtain. This may play a critical role in designing, or choosing between, solid- versus solution-phase synthetic routes.

Although it took many years for solid-phase synthesis to become an established and recognized tool for the synthesis of peptides, this technology started a new era in organic synthesis. It should also be noted that a number of groups recognized from an early stage that "solid-phase" synthesis could also be applied to the preparation of small organic molecules. Most notable among these are the Frechet,[44] Leznoff,[45] and Rapoport[46] groups.

3.1.3.2 Pooling Strategy. The potential of combining a number of different solid-phase-bound substrates for a reaction with a single reagent was recognized for the first time by Ronald Frank in 1983, who used cellulose-based paper disks as solid-phase supports for the synthesis of oligonucleotides.[4] Each disk contained a different substrate (oligonucleotide), and only four reaction vessels and four reagents (appropriate nucleotides) were required for the synthesis of any number of oligonucleotides. After each synthetic cycle, the disks were reshuffled in such a way that each of four flow reactors contained disks with different immobilized oligonucleotides that received the same type of reagent (a nucleobase). "Whenever growing chains on different entities have to be elongated with the same building block, these entities are gathered in the same reaction vessel." This was actually the introduction and the first application of a directed sorting method, a method now commonly used in a variety of different ways with different solid supports.

3.1.3.3 Parallel Synthesis. Frank's group later turned its attention to peptide synthesis and applied the paper disk method to the synthesis of peptides.[5] Methods of peptide synthesis have been continuously developed and improved, and during the 1980s, several methods for multiple solid-phase synthesis appeared, starting the efforts directed toward the concurrent synthesis of arrays of compounds.[33,47] Aside from Frank's synthesis on cellulose disks, Geysen's pin peptide synthesis[7,8] in 1983 and Furka's split-and-pool method[9,10,25] in 1988 represent the most important milestones in multiple synthesis.

Paper as a solid support fascinated Frank, and he later developed another unique and ingenious technique, the spot synthesis (SPOT).[6,32] In SPOT synthesis, peptides are synthesized on functionalized membrane sheets as spots that could be as large as 10 mm or as small as 1 mm. The spot synthesis technique became a popular tool for peptide synthesis[6,32,33] and was soon automated.[48] The first semiautomated SPOT synthesizer, the ASP222, was launched by ABIMED Analysen-Technik of Germany in 1993.[49] The automated method provided an economical way to synthesize very large numbers of peptides. The Intavis synthesizer Auto-Spot (www.intavis.com) allowed the simultaneous synthesis of 1600 peptides. The chemical and physical problems associated with the use of cellulose sheets as synthesis support were, for the most part, overcome by developing synthetic membranes from polypropylene and Teflon.[50–53]

The SPOT technique was brought to the next level by Affymax scientists. Leighton and Pirrung formulated the "peptides-on-chip" concept in 1989 (quoted in Ref. [54]) in analogy to computer chips. The method, referred to as "very large-scale immobilized polymer synthesis" (VLSIPS), enabled the parallel synthesis of peptide, and later oligonucleotide, arrays immobilized on a two-dimensional surface. The synthesis of peptides is carried out using photolabile protecting groups, and the identity of each peptide is determined by its spatial location on the surface. The light-directed, spatially addressable parallel chemical synthesis was published in 1991.[55] For the synthesis of peptides, amino acids having the amino group protected with photolabile groups have been used for derivatization of a glass surface (Figure 3.1).

The photolabile nitroveratryloxycarbonyl (Nvoc) group was removed by light from selected areas using a mask (Figure 3.2). The pattern of the mask allowed light exposure, and thus removal of the protecting groups only from selected areas. For the acylation, the entire surface was exposed to the amino acid; however, acylation occurred only at spots where the amino protecting group had been removed. The whole process was repeated depending on the number of acylations by different amino acids that were required.

The first synthesized array contained 1024 peptides in a 32 × 32 grid, with each peptide residing in a 400 μm × 400 μm area. In addition to peptides and oligonucleotides, light-directed synthesis has also been applied to prepare arrays of oligocarbamates.[56]

These discoveries introduced the combinatorial concept to the chemical society;[57] however, the scientific community in general had begun to accept the new combinatorial

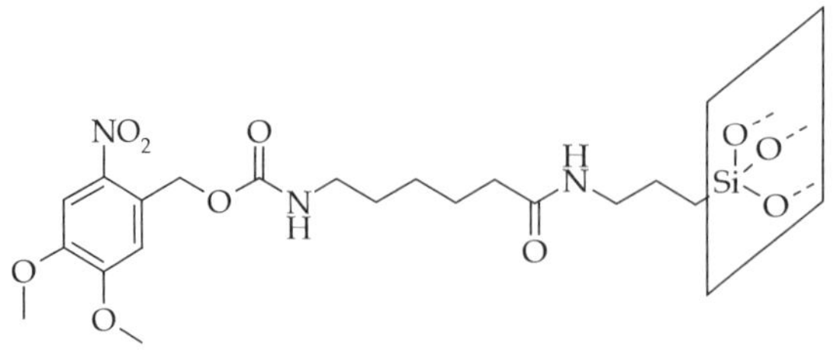

Figure 3.1. Derivatization of a glass surface for light-directed synthesis.

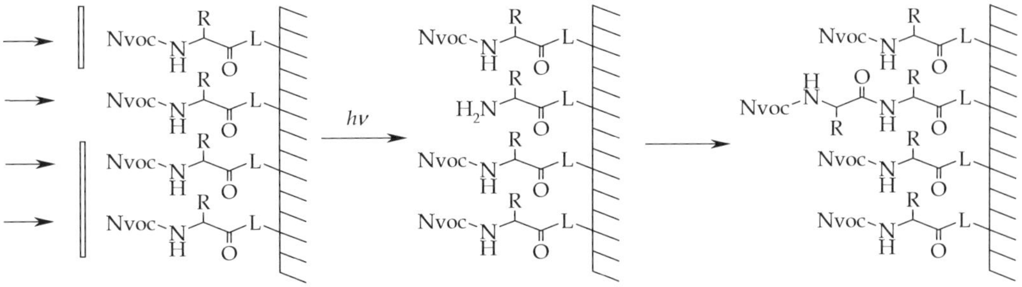

L: linker for immobilization of the C-terminal amino acid

Figure 3.2. Principle of light-directed synthesis.

approach by the time two papers describing peptide libraries appeared side by side in the same issue of *Nature* in 1991, the one-bead–one-compound (OBOC) technique of Kit Lam[1] and the organized mixtures of Richard Houghten.[2]

3.2 STRATEGIES IN COMBINATORIAL SOLID-PHASE SYNTHESIS

Combinatorial syntheses can be performed in three basic ways that differ by several critical features:

1. Parallel synthesis
   - The number of reaction vessels equals the number of synthesized compounds
   - A large number of reaction vessels potentially need to be handled at a time
   - Quantity per compound is not limited
2. Split-and-pool (also referred to as split-and-mix, portioning-mixing, or split-and-recombine)
   - The number of reaction vessels equals the number of building blocks: polymer-supported intermediates receiving the same building block are combined into one reaction vessel
   - Quantity depends on loading per solid-phase particle/container
   - Tracking of the chemical history is required
3. Split–split technique (hybrid approach)
   - The number of reaction vessels equals the number of intermediates
   - Quantity per compound is not limited
   - It is used for synthesis of relatively smaller libraries of compounds (tens to hundreds)

There exist two modifications of the split-and-pool concept: the directed and random split-and-pool methods. The key differences are as follows:

In the directed split-and-pool method:

- The chemist controls the distribution of compounds in a library. Any combination of building blocks can be excluded from the synthesis.

- The chemical history of the particles is recorded (e.g., each resin formulation is labeled, such as with an alphanumeric code).

In the random split-and-pool method:

- The distribution of compounds in a library is driven by statistical probabilities due to the random split process. Each compound is synthesized numerous times when the number of beads exceeds several times the number of compounds, or only a subset of compounds is produced when the number of beads is lower than the number of possible combinations of building blocks.
- The chemical history of the beads is lost. After each combinatorial step, the resin beads from all reaction vessels are pooled and randomly split into reaction vessels for the next combinatorial step.

The quantity criteria for the random and directed split-and-pool techniques are also different. The random split-and-pool method is suited for the synthesis of smaller quantities (nanomolar range) of large-sized libraries, such as millions of compounds, whereas the directed split-and-pool technique is suited for the synthesis of larger quantities (micromolar range) of smaller compound collections (i.e., several hundred to several thousand compounds).

3.2.1 Random Split-and-Pool Method

The split-and-pool method was presented for the first time in 1988 by Arpad Furka at two international meetings[9,10] and was published in 1991.[25] The split-and-pool concept represents the most efficient method for the preparation of large arrays of compounds in a truly combinatorial manner. This technique is illustrated by the following example.

Suppose we intend to synthesize all possible tripeptides made by using the 20 common amino acids in each position. The total number of tripeptides is $20 \times 20 \times 20 = 8000$. However, one does not need to perform the synthesis in 8000 reaction vessels. Only 20 reaction vessels and 20 activated amino acids are needed. The starting resin beads are split into 20 reaction vessels (only five are shown in Figure 3.3), and each portion is derivatized with one amino acid, A1 to A20. In this first combinatorial step, 20 different resin-bound

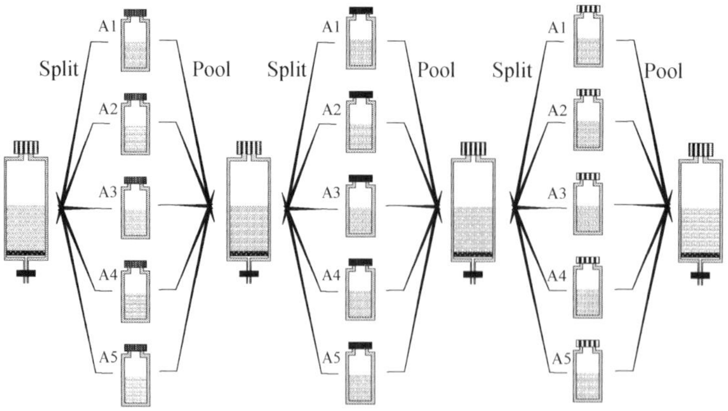

Figure 3.3. Principle of the split-and-pool method.

amino acids are prepared. All 20 resins are then combined, mixed thoroughly, and redistributed in equal portions into the same 20 reaction vessels. At this stage, each reaction vessel contains the same mixture of 20 resin-bound amino acids. In the second combinatorial step, reaction vessels are charged with amino acids from A1 to A20. As a result of the acylation, each reaction vessel contains a mixture of 20 dipeptides, having all amino acids at the carboxy terminus and one particular amino acid at the amino-terminal end. The resin-bound dipeptides from all vessels are then combined (mixed) together again. The resulting slurry is composed of all possible 400 dipeptides. The next step is obvious. The resin is again distributed into the 20 reaction vessels and reacted with 20 activated amino acids, one amino acid per reaction vessel. Each of the 20 reaction vessels contains 400 tripeptides, and all the 20 vessels together contain 8000 tripeptides.

At that time, the method was designed to synthesize equimolar mixtures of peptides. However, the consequences of the split-and-pool method were more far-reaching. The same concept was later independently used by Kit Lam for his "OBOC" concept and Richard Houghten for the synthesis of organized mixtures of peptides.

3.2.1.1 One-Bead–One-Compound Concept. The split-and-pool synthetic strategy is undoubtedly the most efficient method to prepare large numbers (millions) of compounds. A critical feature of the split-and-pool method is the fact that any bead at any time can be present only in one reaction vessel and therefore reacting with only one amino acid (in more general terms, with one building block). This means that there is only one chemical entity on each bead (excluding side products). The consequence of this is that any bead picked from a mixture of millions of beads contains only one compound. This is the basic premise of the OBOC concept. The distribution of beads is driven by statistics and multiple beads can contain the same compound if the number of compounds is substantially lower than the number of beads.

Kit Lam was the first to realize this consequence and apply the technique to lead discovery.[1] The OBOC technique includes three basic steps: library synthesis, library screening, and structure determination. As the split-and-pool concept provided a straightforward route to sizable libraries (millions of compounds), the critical steps of the OBOC technique became the identification of "positive" compounds within a library and subsequent determination of their structure. Two approaches for screening the OBOC libraries were developed: (i) the library compounds were screened while attached to the resin beads using the so-called "on-bead screening" and (ii) the compounds were cleaved from beads and screened in solution. Various techniques have been developed to overcome the loss of information that follows complete cleavage of the compounds from the beads. These techniques involve the partial cleavage of compounds from beads using multiple cleavable linkers,[59–62] gaseous cleavage[63] or photocleavable linkers,[64–71] and coding (see below).

3.2.1.2 Encoding Methods for the OBOC Technique. When screening is completed, one or several beads are identified that carry a compound of interest. Owing to the "mix" phenomenon, the chemical history (i.e., individual building blocks reacted with this particular bead) and the structure of the compound on the bead are not known. Consequently, the last step in the process is structure determination.

SEQUENCING. In the initial experiments of the Lam group, peptide libraries were screened and Edman peptide degradation[72] was used to determine the sequence of amino acids on any particular bead. There was more than enough compound on each bead for complete sequencing of the peptide (approximately 5 pmol are needed), and commercially available fully automated sequencers were used. However, the decoding of peptide structures with the use of Edman degradation is slow and expensive. Moreover, difficulties arise with the OBOC technique when non-alpha amino acids are used in a peptide sequence or other "unusual" (nonsequenceable) structural elements are included in library compounds. For such libraries, sequencing of peptides by mass spectrometry (MS) has been developed[73–76] and databases for MS sequencing of peptide libraries are commercially available. Because simultaneous cleavage and ionization occur under laser irradiation, peptides covalently attached to a single polymeric bead by a photosensitive linker can be directly sequenced by matrix-assisted laser desorption ionization (MALDI) MS.[75,76] Another approach to peptide sequence determination with the use of MS was described by Youngquist et al.[77] It is based on the partial (10%) capping of a growing peptide chain in each step of the synthesis. Each bead thus contains all partial sequences, and the synthetic history can easily be interpreted from mass differences between these capped peptides.

Chemical Encoding. Chemical coding tags have been added to the bead during the synthetic steps so that the synthetic history of each compound bead in the chemical library could be recorded. For any combinatorial step, each particular chemical transformation (e. g., attachment of a building block) was followed by a coding step. As a result of this procedure, two independent structures were synthesized on each bead. The "screening" arm contained the target structure, while the "coding" arm contained decodable information of the chemical history of the bead. The initial idea of coding was suggested by Brenner and Lerner,[72] who contemplated the use of polynucleotides for coding peptide libraries. Nucleic acid coding has been applied in several cases;[78,79] however, its application for coding organic libraries is limited due to the incompatibility of nucleic acid chemistry with other organic reactions. The obvious coding arm is rather a peptide that can be sequenced, as in the case of peptide libraries.[80,81] Later, various tags that give a unique identifier to an individual bead, allowing deconvolution of the compounds structure, were developed and reported.[26,28,82–88] The tags are generally cleaved from the resin using an orthogonal cleavage method, similar to that used to detach the screening compound from the bead. Coding subunits are those that can be easily identified by well-established analytical microtechniques: amino acids (Edman degradation and HPLC),[89,90] electrophoretic tags (halocarbon molecules determined after silylation by gas chromatography),[91,92] or amines.[93] Probably the most efficient decoding method of specific tags consists of the use of mass spectrometry,[94,95] which is simple, fast, and has great potential for automation.

Bilayer Beads. If both the library compounds and their corresponding coding tags coexist at the bead surface where the interaction between target proteins and library compounds occurs, the coding tags may interfere with the screening. To eliminate such interference, "bilayer beads" have been developed.[96] Such bilayer beads allow the preparation of library compounds on the outer layer of each bead and the coding tags in the bead interior. Thus, only the testing compounds are exposed to the target proteins. In addition, these bilayer beads can be used to prepare OBOC combinatorial libraries that are down-substituted on the bead surface but fully substituted in the bead interior. This configuration enables one to screen at a much higher stringency and yet have enough peptides or coding tags retained in the bead interior for structure determination.

Rainbow Beads. Recently, color tagging of resin beads has also been developed.[97] Polymer beads displaying chemical compounds or families of compounds were stained with oil-based organic dyes (red, blue, green, black, yellow) that were used as coding tags. The

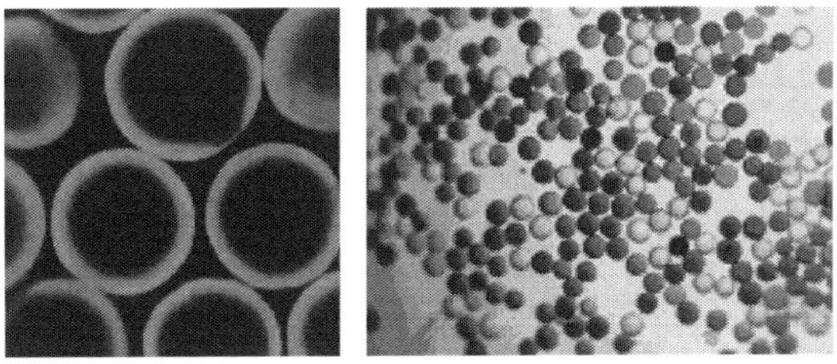

Figure 3.4. Confocal photomicrograph of bilayer beads (the outer layer was labeled with FITC) and photomicrograph of rainbow beads. Left: Adopted with permission from Ref. [96], Copyright Wiley-VCH Verlag GmbH & Co. KGaA. Right: Adopted with permission from Ref. [97], Copyright 2010 American Chemical Society.

color dyes do not affect cell binding to the compounds displayed on the surface of the beads. The colored particles (so-called rainbow beads) were applied in a multiplex manner to discover and profile ligands against cell surface receptors. A major advantage of this straightforward and powerful method is that only an ordinary inverted microscope is needed for the analysis, instead of sophisticated (and expensive) fluorescent microscopes or flow cytometers (Figure 3.4).

3.2.1.3 Organized Mixtures. In 1991 and then in 1992, Richard Houghten presented two novel concepts of synthesizing and screening "organized mixtures" of peptides.[2,98] These methods enable step-by-step identification of key structural elements of active compounds. Two approaches were developed to allow the identification of active compound(s) in the library: iterative deconvolution[2] and positional scanning.[98] Both methods share the same idea for determination of the active compounds: screening sets of peptide mixtures (sublibraries) where each mixture (sublibrary) contains a unique structural element, that is, one amino acid at a particular position in a peptide. Identification of the most active mixture reveals the type of amino acid(s) required for a given biological activity at that position.

Iterative Deconvolution. The first iterative library was composed of 400 separate hexapeptide mixtures, in which the first two positions from the amino terminus contained single individual amino acids. Out of the 20 proteinogenic amino acids, 400 dipeptides were formed (AA, AC, and so on through YW, YY). The remaining four positions were mixtures of 19 amino acids (cysteine was excluded), so that each mixture was composed of 19^4 or 130,321 individual peptides. Thus, a total of $400 \times 130{,}321$ (52,128,400) hexapeptides were synthesized. This library was prepared using the T-bag method[33] in combination with split-and-pool synthesis.[99]

The iterative deconvolution process is explained here using a simple example of a tetrapeptide library made of four different amino acids (R1 to R4, Figure 3.5), where R1 to R4 represent individual amino acids and X represents a mixture of those four amino acids. In the initial screening of this library (Step 1 of Figure 3.6) in a bioassay, the most active

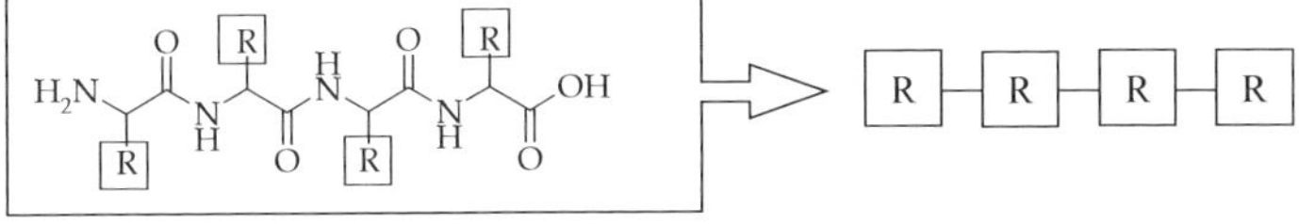

Figure 3.5. A tetrapeptide iterative deconvolution procedure.

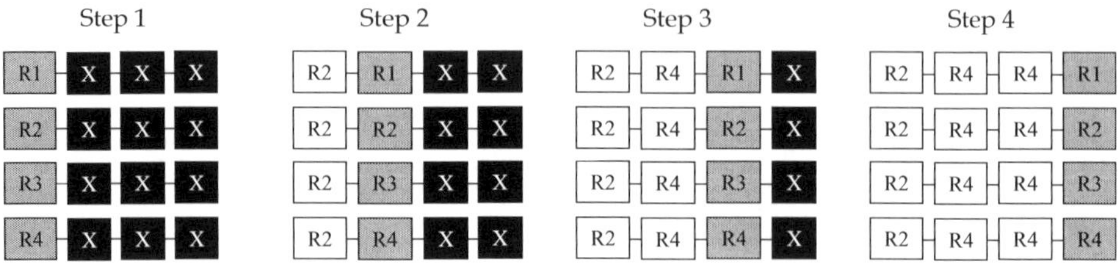

Figure 3.6. Iterative deconvolution libraries where R1 to R4 represent single amino acids and X represents mixtures of those amino acids.

peptide mixture(s) are identified. Suppose the second mixture (sublibrary) was the most active one. In the next step, four sublibraries are prepared, each containing the amino acid R2 in the first position and four individual amino acids in position 2. The screening of these four mixtures showed mixture #4 to be the most active. Accordingly, a new set of four mixtures is made with amino acids R2 and R4 present at positions #1 and #2, respectively, and one amino acid *per* mixture present at position #3. The fourth position contains a mixture of four amino acids. This process is repeated for the remaining undefined position by four single peptides being synthesized and the most active peptide being identified.

Positional Scanning Libraries. The positional scanning library format enables the identification of active compounds directly from the initial library screening data, thus avoiding the iterative synthesis and screening process associated with the above library format.[98] A typical positional scanning library is composed of n (n = the number of diversity positions) sublibraries. Accordingly, a tetrapeptide library is composed of four independent sublibraries (Figure 3.7), in which one position (R1 to R4) is individually defined, and the other three positions include a mixture of amino acids (X).

Screening of all sublibraries in a given bioassay provides information about the most effective amino acids at each position for the biological effect of interest, as well as about the relative specificity of each position (i.e., the fewer the amino acids found to be active at a position, the more specific that position is). The synthesis of all possible combinations of the most active amino acids at each position yields a range of individual peptides that are then tested to determine their individual activities. If, for example, two amino acids were found to be highly effective at each position of the above hexapeptide library, 2^6 or 64 individual peptides would be synthesized and tested based on those screening data.

Alternatively, each of the sublibraries can serve as the starting point for the iterative synthesis and screening process described above. It has to be kept in mind that amino acids identified at different positions need not be part of the same motif or may represent only the most significant amino acid(s) of the same motif, which can be placed anywhere in the

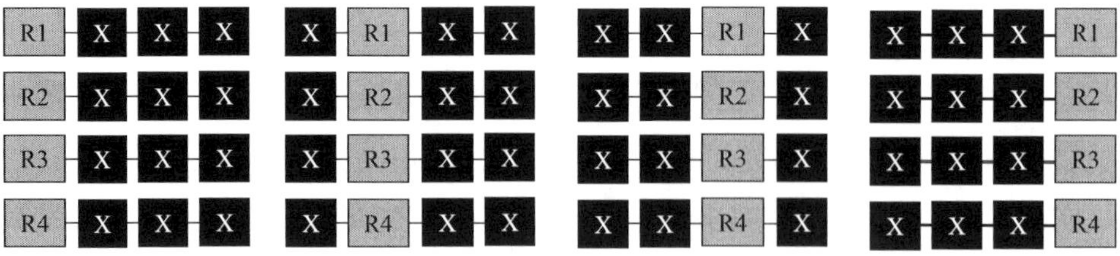

Figure 3.7. Positional scanning libraries where R1 to R4 represent single amino acids and X represents a mixture of those amino acids.

peptide sequence. For example, if arginine is the key amino acid in the potential enzyme ligand (substrate, inhibitor), and the minimal length of the ligand is a dipeptide, arginine can be identified as the key residue in almost any position of a hexapeptide library.

3.2.2 Directed Split-and-Pool Method in Practice

The split-and-pool synthesis not only simplifies the complexity of the combinatorial synthetic process but also offers additional important benefits. To undertake a full range of solid-phase chemical reactions, elaborate reaction conditions are needed for some chemical transformations. These include, but are not limited to, low temperature and inert atmosphere conditions. Parallel synthesis of 1000 compounds requires the handling of 1000 reaction vessels. To add sensitive reagents (e.g., butyl lithium) in timely manner at low temperature (−78°C) under an inert atmosphere during parallel synthesis is not a trivial task. It can be done if sophisticated automated synthesizer equipment is used. Such syntheses can be performed easily in a manual fashion using the split-and-pool method, which requires only a limited number of reaction vessels. The split-and-pool methodology has been used for the synthesis of complex and diversity-oriented combinatorial libraries.[100,101]

To make the directed split-and-pool methodology practical, two issues need to be addressed: (i) formulation of the solid support and (ii) tracking the chemical history of individual units of the solid support.

3.2.2.1 Formulation of Solid-Phase Supports for the Directed Split-and-Pool Technique. Resin beads are the most frequently used formulation of insoluble support for solid-phase organic synthesis. However, the yield from one typical bead (100 μm) is only about 100 pmol. The need for substantially higher yield from one unit of solid phase triggered the development of different formulations of solid support. Two different scenarios were followed: compartmentalization of resin beads (T-bags, wafers, Kans, capsules, and resin plugs) and development of modular solid support (SynPhase Lanterns, rods).

T-BAGS. To take advantage of the pooling strategy with resin beads, Richard A. Houghten invented polypropylene meshed packets that are similar in appearance to tea bags.[33] (Figure 3.8). The use of T-bags substantially increased the productivity of manual solid-phase synthesis. The most time-consuming operation of solid-phase synthesis is washing the resin beads. As the solid-phase chemistry practitioner has to do this repeatedly, integrating this and the other common operations required during solid-phase synthesis allows hundreds of individual syntheses to be carried out simultaneously. Since in the T-bag

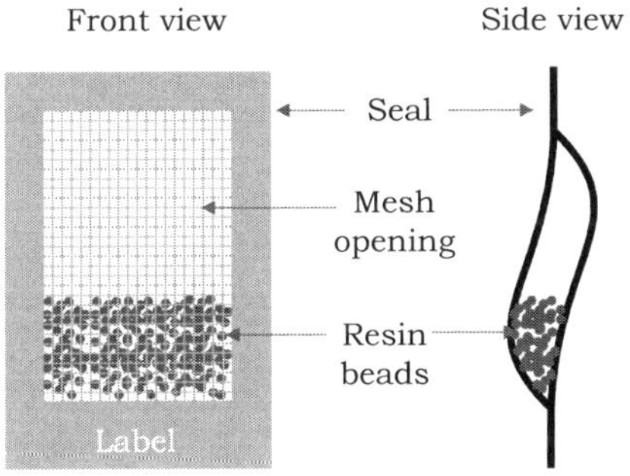

Figure 3.8. T-bag for solid-phase synthesis.

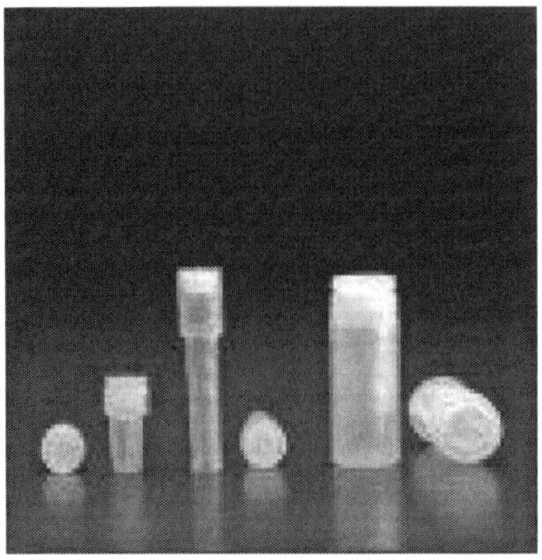

Figure 3.9. IRORI MicroKan, MiniKan, and MacroKan reactors. From web pages IRORI.

approach the resin beads containing the same intermediate are physically separated (e.g., enclosed in a bag) and distinguished in some way (e.g., with an alphanumeric code), all the bags may be combined in one washing vessel and washed at the same time. The advantage of T-bags is the easy handling of large numbers of reaction vessels; the disadvantages are a larger volume of solvent/solution and the difficulty of automation. The typical size of a T-bag is $30 \times 50\,mm^2$, and it provides 100–500 μmol of product. The application of T-bags is exemplified by the synthesis of small molecule libraries[102] and peptides.[103] The Houghten pilot work[33] has been cited more than 1300 times.

KANS. T-bag compartmentalization of resin beads inspired the design of canister-like containers (MacroKan for 300 mg, MiniKan for 100 mg, and MicroKan for 30 mg of resin) for automated handling by IRORI (Figure 3.9).[104,105] The MicroKan reactor was fabricated with high-purity polypropylene mesh, and resin beads were placed inside it. The pore size of the mesh (75 μm) was such that properly sized resin beads (>75 μm) could not cross through the mesh, while reagents and solvents could freely flow in and out of the reactor.

WAFERS AND CAPSULES. Another solution of compartmentalization of resin beads into permeable containers was reported by Beattie and Frost, who invented porous wafers that housed insoluble supports for the multiple solid-phase synthesis of oligonucleotides and peptides.[106–108] The porous wafer was made from a Teflon ring covered on both sides by a porous Teflon membrane to form a cylindrical permeable container. The use of wafers was reduced to practice in a specialized column-based oligonucleotide synthesizer.[109]

Resin capsules represent an alternative solution to Teflon wafers designed for multiple/combinatorial solid-phase organic synthesis.[110] The capsules consist of a polyethylene ring sealed with peek mesh on both sides (Figure 3.10). The reaction vessel consists of a sealed Teflon tube, in which washing resin beads can be carried out in a continuous flow manner.

RESIN PLUGS. Resin plugs represent a different resin bead formulation applicable to the pooling strategy (Figure 3.11).[111] In this process, resin beads are mixed with finely powdered ultrahigh molecular mass polyethylene and the 1:1 mixture is heated to soften the polyolefin matrix. After cooling, resin plugs are formed that are used as a modular solid-phase support. The typical plug size is $6 \times 12\,mm^2$ and contains resin with a loading of 50–200 μmol per plug. Resin plugs were commercially available from Polymer Laboratories; however, the product has been discontinued. Mechanical instability caused disintegration of the plugs and the loss of resin beads during synthesis.

Figure 3.10. Resin capsules and Teflon tube serving as a reaction vessel.

SynPhase Lanterns. An alternative modular support, the SynPhase Crown and later SynPhase Lantern, was invented by scientists at Mimotopes.[112] A rigid polypropylene mold of a shape resembling a Chinese Lantern was grafted with a layer of polystyrene (Figure 3.12). The solid-phase synthesis took place in the derivatized layer of the polystyrene graft. Three sizes of polystyrene Lanterns are available: the A-series with a loading of 75 μmol, the D-series with a 35 μmol loading, and the L-series with a 15 μmol loading. The SynPhase Crown differed in shape. SynPhase Lanterns appeared to be the support of choice for diversity-oriented synthesis (DOS) of drug-like molecules, and they were reportedly used for syntheses of over 300,000 DOS compounds.[113]

Other Formulations. Another form of solid support is monolithic disks. Monolithic polymer rods or cylinders (diameter 8–10 mm, length 50 mm) were prepared by copolymerization of styrene with divinylbenzene or poly(ethylene glycol) diacrylate as the cross-linker.[114] The rods were then cut to produce disks (thickness 1–2 mm) for solid-phase synthesis. Polymer disks were also used as scavengers[115] and reagents[116] for solution-phase synthesis. Disks of various contours were applied in shape-encoded combinatorial chemical libraries.[117]

3.2.2.2 Chemical History of the Resin Formulations. The most straightforward solution to track the chemical history during split-and-pool synthesis is labeling paper disks,[4] T-bags,[33] wafers,[107] or capsules[110] with an alphanumeric code

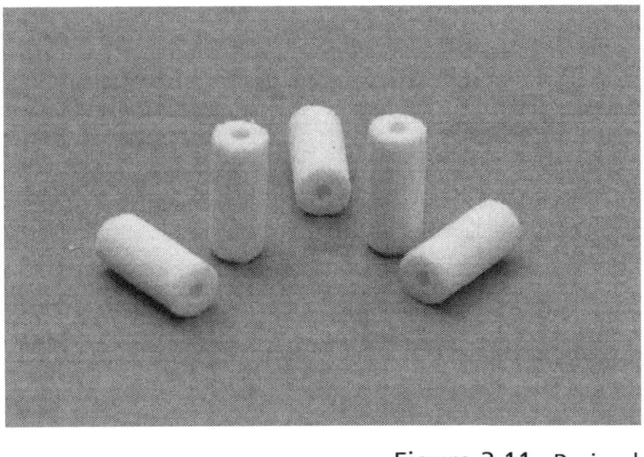

Figure 3.11. Resin plugs.

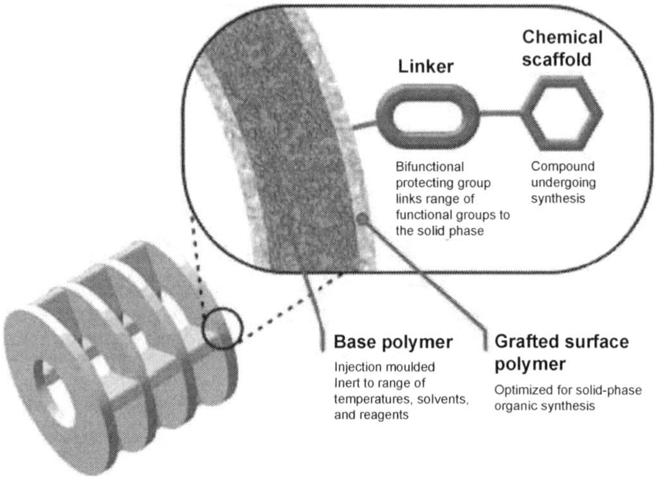

Figure 3.12. Construction of SynPhase Lanterns. Adopted from www.mimotopes.com.

readable by a chemist, who manually redistributes them for the subsequent combinatorial step. Necklace coding represents an alternative solution for manual synthesis, and radiofrequency (Rf) tagging and optical encoding were developed for automated sorting.

NECKLACE CODING. Necklace coding is a spatially addressable directed split-and-pool procedure for tracking the chemical history of a synthesis carried out on SynPhase Crowns and Lanterns.[118] Individual solid-phase particles (SynPhase Crowns, Lanterns) are manually strung on a Teflon thread and the position of a particle on the thread (necklace) encodes the previous chemical history. Figure 3.13 shows the synthesis of five nitroanilines using five Lanterns, each with a different immobilized amine. Five reaction vessels were charged with Lanterns containing the immobilized backbone amide linker (BAL) and reductively aminated with five different amines, one amine per reaction vessel. One Lantern from each vessel was then manually strung on a Teflon thread.

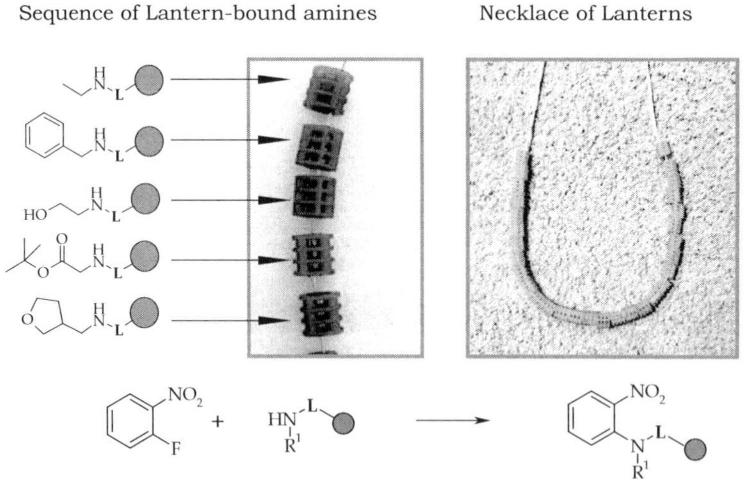

Figure 3.13. Tracking the chemical history by necklace coding.

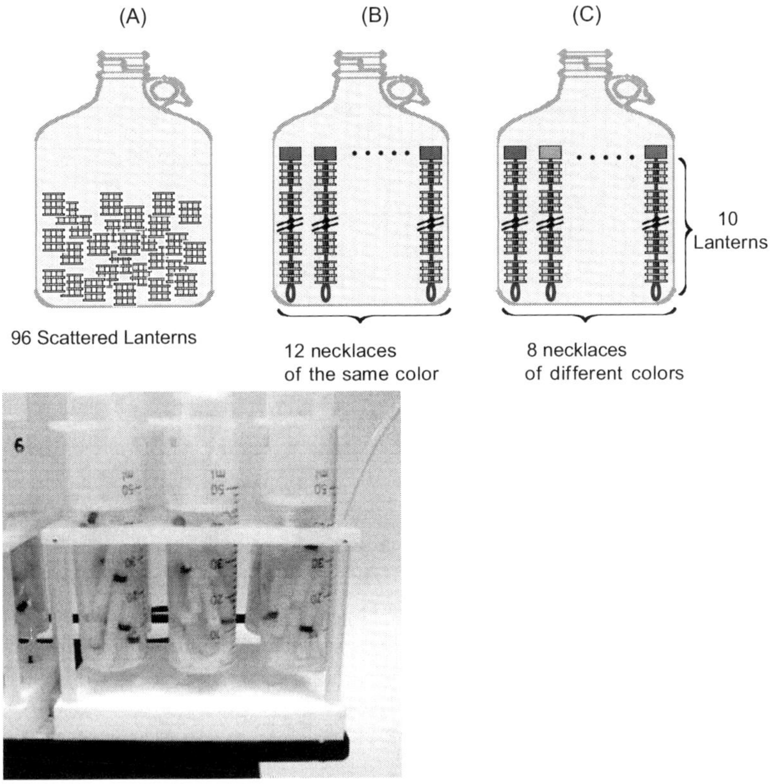

Figure 3.14. The Encore technique: principle and reality.

The sequential position of a Lantern identified the amine. Subsequently, reaction with 2-fluoronitrobenzene was carried out in one reaction vessel with a necklace of Lanterns. This simplified example illustrates a concept applied to the manual synthesis of sizable combinatorial libraries.[119] A similar concept was later reported by Furka.[26,120]

The concept of necklace coding was implemented in the Encore technique, which combines three different coding methods: sequential position on a necklace for the first combinatorial step, color coding of individual necklaces for the second combinatorial step, and reaction vessel coding as indication of the identity of the last building block.[119,121] Accordingly, the technique was termed Encore (encoding by a **n**ecklace, **co**lor, and reaction vessel) (Figure 3.14). A three-combinatorial step library of $10 \times 12 \times 8 = 960$ compounds was synthesized in the following way. The first combinatorial step was performed in 10 reaction vessels, each charged with 96 scattered Lanterns (panel A). Then the Lanterns were stringed to form 96 identical necklaces, each necklace including 10 Lanterns. Twelve necklaces were placed into each of eight reaction vessels to perform the second reaction step (panel B). The third combinatorial step is performed in 12 reaction vessels with 8 necklaces per vessel, each tagged with a different color (panel C). The logistics of the synthesis is based on the Encore technique developed for directed solid-and-pool combinatorial synthesis and has been used for the synthesis of combinatorial libraries.[119,121,122]

Spindles and Cogs Coding. An alternative to necklace coding is the tagging of SynPhase Lanterns with colored plastic tags. These plastic tags included colored "Cogs" and "Spindles" attached to Lanterns. "Cogs" and "Spindles" are available in eight different

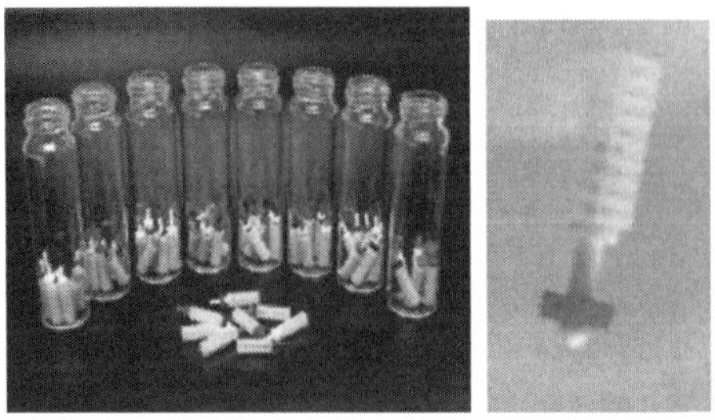

Figure 3.15. SynPhase Lanterns encoded by spindles and cogs. Left: Adopted with permission from Ref. [125], Copyright 2010 American Chemical Society. Right: Adopted with permission from Ref. [124], Copyright Wiley-VCH Verlag GmbH & Co. KGaA.

colors. Although the colored SynPhase Lanterns were originally developed for synthesis optimization, they have frequently been used for combinatorial SPOS of smaller molecules, as well as peptide libraries (Figure 3.15).[123–125]

Radiofrequency Tagging and Optical Encoding. To automate the process of MicroKan redistribution, individual containers were tagged. Radiofrequency tagging[27,85,104] and optical encoding[28,105] of containers enabled computer-assisted reading of the tag and automation of the directed split-and-pool process (Figure 3.16). A Rf tag consists of a microcircuit and an antenna encapsulated in glass. The microcircuit includes a rectifier, a transmitter/receiver, a logic control, and a unique nonvolatile 40-bit ID code (a total of over one trillion available ID codes). It is a passive device, meaning that it does not have an internal power source and is in a nonoperative mode most of the time. The use of Rf tagging can be illustrated by the synthesis of 72 analogues of distamycin using SynPhase Lanterns.[126]

3.3 EQUIPMENT AND INSTRUMENTATION

Combinatorial synthesis on solid phase can be performed in any organic laboratory without any need for dedicated instrumentation.

3.3.1 Manual Solid-Phase Synthesis

The original Merrifield reaction vessel was made from a glass tube and equipped with a side opening and a porous glass frit to facilitate the separation of solid and liquid phases

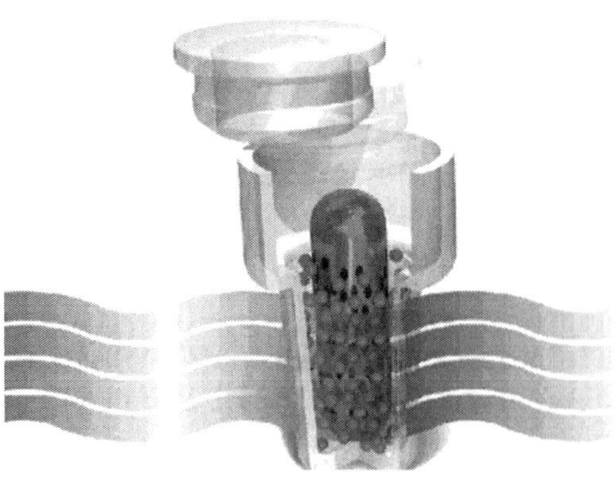

Figure 3.16. Rf-tagged IRORI MicroKan. From web pages IRORI.

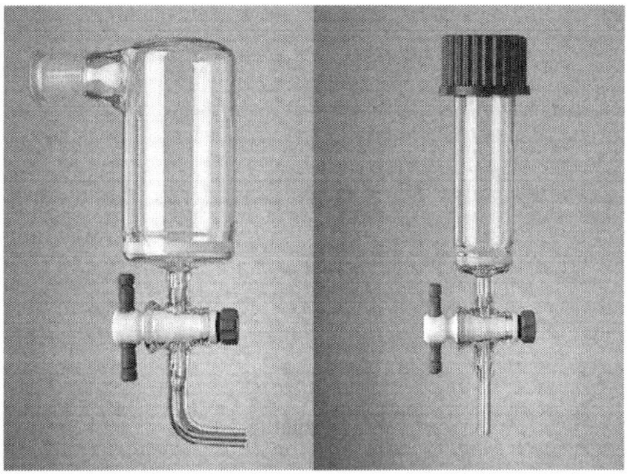

Figure 3.17. Traditional reaction vessels for manual solid-phase synthesis. (From http://www.pepnet.com/).

(Figure 3.17, panel A). The same kind of reaction vessel has remained, with the only difference being the position of the opening (panel B), which simplifies manufacturing.

The price and fragility of a glass reaction vessel prompted a search for cheaper and potentially disposable replacement reaction vessels. Standard medical syringes, made of polyethylene, are rather chemically resistant. A disk made from a porous polypropylene sheet inserted into a syringe converts a medical syringe into a simple, inexpensive, and disposable reaction vessel for solid-phase synthesis (Figure 3.18).[127] The plunger can be pulled and pushed along inside the barrel, allowing the syringe to take in and expel liquid while the resin is trapped inside.

Syringes are typically kept on a tumbler or a shaker during the reaction, can be used at temperatures up to 100°C, and can be sonicated or used in a microwave oven. However, polypropylene syringes do not tolerate prolonged (overnight) exposure to certain solvents, including methylene chloride, tetrahydrofuran, and dioxane. The syringe barrel tends to swell, resulting in the plunger not providing a leakproof seal. This problem can be solved by transferring the resin slurry into a new syringe or by using "locking sleeves." A locking sleeve is a polypropylene ring that can be moved up and down the syringe barrel to "lock" the required volume. The locking sleeve contracts the swollen barrel and the plunger seals the syringe (www.torviq.com). Leaching plasticizers from the syringe does not represent a risk since they are removed with the wash solvent. Five different sizes of reaction vessels are

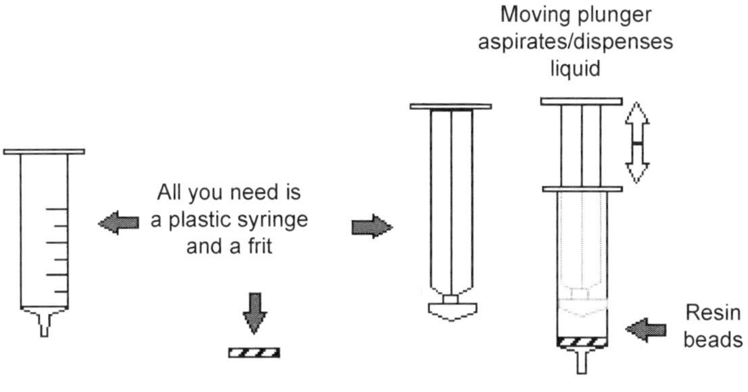

Figure 3.18. Plastic reaction vessels for solid-phase synthesis.

TABLE 3.1. Recommended Syringe Load

Syringe volume (mL)	Resin quantity
3	100 mg
5	300 mg
10	500 mg
20	1 g
50	3 g

available (www.torviq.com). The selection of the proper size depends on the amount of resin. Table 3.1 shows the maximum recommended load for typical polystyrene resin.

3.3.2 Integrated Semiautomated Synthesis

The repetitive nature of solid-phase synthesis—washing, reaction, and washing—inspired integration of common tasks and assembling reaction vessels into blocks that allowed the washing of resin beads in numerous reaction vessels at the same time. Several variations of blocks were designed, such as MultiBlock, Domino Block, Bill-Board, and MiniBlock.

MULTIBLOCK. In the late eighties, Krchňák and Vágner designed the MultiBlock, the first reaction block for parallel solid-phase synthesis (Figure 3.19). The MultiBlock consists of five parts: (i) a Teflon block that holds 42 reactors, polypropylene syringes equipped with a plastic frit[127]; (ii) a vacuum adapter that connects each reactor to a vacuum line and allows the rapid washing of resin under continuous flow; (iii) two Teflon plates with 42 stoppers to which the Teflon block is attached during reactions; (iv) a glass cover to allow mixing of resin for library synthesis using the split-and-pool approach. The materials used for its construction include Teflon, polypropylene, glass, and stainless steel.

DOMINO BLOCK. To further simplify and accelerate the manual parallel synthesis on the solid phase, a new generation of a synthetic block, the Domino Block, was designed (Figure 3.20).[128] The Domino Block consists of enclosed reaction vessels, polypropylene syringes, attached to a Teflon manifold that clamps the syringes and connects them to a selection valve. Liquid is removed from the closed reaction vessels by connecting the Domino Block to an evacuated waste container by turning the selection valve to vacuum. The selection valve then connects the Domino Block to a reservoir with washing solvent and the evacuated reaction vessels are filled with solvent. The Domino Block has a footprint of a

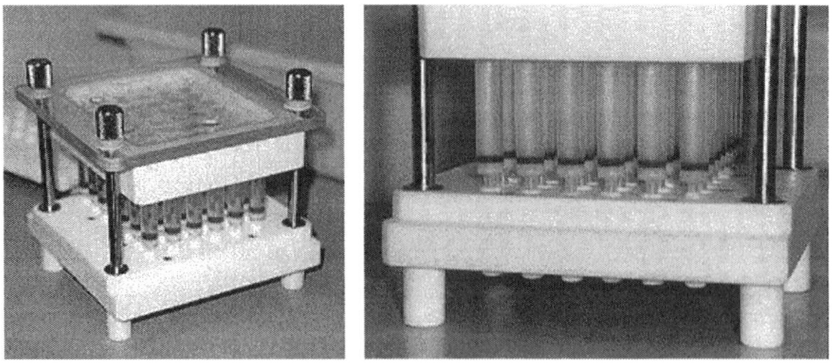

Figure 3.19. MultiBlock for manual parallel solid-phase synthesis.

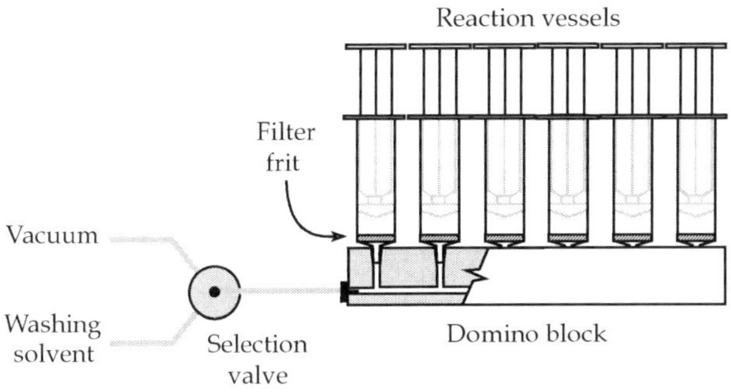

Figure 3.20. Principle of the Domino Block.

standard 96-well plate, and during the washing procedure, the Domino Block is placed on a plate shaker.

The commercial version of the Domino Block synthesizer contains four reservoirs for different solvents and accommodates four Domino Blocks (Figure 3.21) (www.torviq.com). Combinatorial solid-phase synthesis in Domino Blocks is carried out in a manner analogous to synthesis in T-bags or MicroKans. Syringes with different resin-bound intermediates that receive identical reagents are placed on the same Domino Block. After finishing one chemical reaction, the syringes are redistributed, and one Domino Block contains syringes that receive the same reagents.

The explosion of combinatorial chemistry in the nineties triggered the construction of a variety of reaction blocks for multiple parallel synthesis. Some of them are still in production, including the MiniBlock developed by Bohdan, Inc. (now part of Mettler-Toledo, www.mt.com/autochem) and Solid-Phase Synthesis Reaction blocks by J-KEM (www.jkem.com). The Bohdan MiniBlock reactor can hold up to 48 disposable polypropylene-fritted tubes in a 48-well format (6×8 array) and accommodates IRORI's MicroKans. The MiniBlock can be equipped with accessories for synthesis under inert atmosphere or at elevated temperatures.

Bill-Board. William L. Scott designed the Bill-Board, a simple and very useful apparatus for parallel solid-phase synthesis[129] that uses 3.5 mL fritted glass reaction vessels

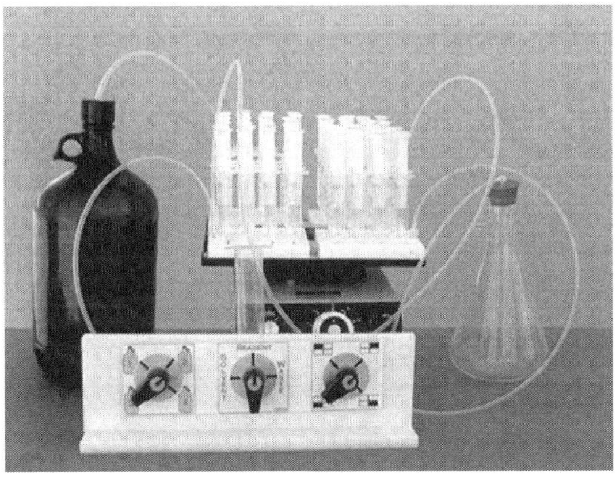

Figure 3.21. Domino Block synthesizer.

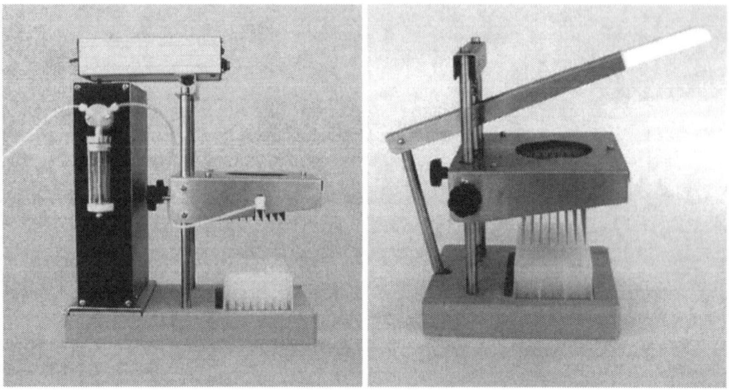

Figure 3.22. Aspirator and dispenser for solid-phase synthesis in plates.

equipped with polypropylene screw caps with Teflon faced silicon septa. The Bill-Board was used in the synthesis of unnatural amino acids on the solid phase[130] and for the project of Distributed Drug Discovery.[131–133] The apparatus is available from Leads Metal Products, Indianapolis, IN (larry@leadsmetal.com).

Synthesis in Plates. The synthesis of compounds in single milligram quantities can be performed in 96-well plates. Typically, resin beads are separated from solvent using filtration through a porous material. In an alternative procedure, the resin beads are allowed to settle and the solvent above the resin bed is aspirated using an aspirator.[134] Solvent delivery into 96 wells is carried out using a 96-well dispenser (www.torviq.com).

In most solvents used in solid-phase synthesis, the resin beads settle. Settling is relatively fast (tens of seconds). After the resin beads have settled, stainless steel needles connected to an evacuated waste container are slowly immersed into the wells of a plate. The needle removes the liquid from the surface without disturbing the resin bed. For washing the resin beads in 96-well microplates, two manually operated dedicated workstations are used, the dispenser for liquid delivery and the aspirator for removing the liquid (Figure 3.22).

3.3.3 Fully Automated Synthesizers (Gone with the Wind)

Because of the inherent ability of solid-phase synthesis to be integrated and automated, numerous instruments were built from the onset of solid-phase chemistry, and this development culminated after the introduction of combinatorial chemistry methods. Operational simplicity of solid-phase synthesis contributed to the development of multiple solid-phase synthesis, where numerous reaction vessels are handled at the same time. In 1989, Schnorrenberg and Gerhardt[47] introduced the automated multiple synthesis of peptides in parallel fashion. Multiple synthesis in a continuous flow manner was also later reported.[134–137]

For continuous peptide synthesis, a multiple automated robotic synthesizer (MARS) was designed and constructed by Selectide Corporation.[138] Its main feature was a timing procedure for handling multiple synthetic tasks that eliminated any unnecessary respite time and kept the robot arm continuously in operation. Polypropylene syringes equipped at the bottom with polypropylene frits served as physically independent reaction vessels. All operations were performed by the robotic arm, which was equipped with a specially designed gripper to hold a syringe and to aspirate and dispense liquid. Typically, the MARS concurrently synthesized 5–15 peptides of different length, and once one peptide was finished, it automatically started the synthesis of the next peptide in the queue, ensuring a continuous flow of peptides.

The boom of automation in the nineties has ceased, and the past leaders in the production of automatic synthesizers have also either ceased to exist (Argonaut, Charybdis) or stopped manufacturing SPOS synthesizers (Advanced ChemTech). Those fully automated computer-assisted synthesizers contained a multiplicity of reaction vessels and each vessel was dedicated to the synthesis of one compound, and thus the synthesis was carried out in parallel fashion rather that taking advantage of the split-and-pool method. Argonaut designed Nautilous, characterized by inert, temperature controlled, and dedicated reaction vessels for solid-phase synthesis. Charbydis Technologies designed the Calypso System, composed from modular reaction block systems for parallel synthesis, and also the Iliad PS2 family of HTOS Synthesizers. Although automatic SPOS are not being manufactured, some of the synthesizers occasionally appear on eBay. The automated solid-phase synthesis of low-weight molecule libraries was reviewed in 2006.[139]

Unlike solid-phase organic synthesizers, automated synthesizers for peptide synthesis are commercially available (Advanced ChemTech, Protein Technologies, to name just two old players in this field). The automated solid-phase synthesis of peptides was reviewed in 2002,[140] and then again with aspects of peptidomimetic development.[141] Since the landscape changes frequently, the latest information can always be obtained on the Internet.

Peptide synthesizer 433A of Applied Biosystems, Inc. was also used for the first automated synthesis of oligosaccharides using both glycosyl phosphates and trichloroacetimidates,[142] an example of meaningful strategy focused on the development of new chemistry applicable to existing instruments rather than the development of a new instrument for existing chemistry. The state-of-the-art automated synthesis of oligosaccharides from the perspective of diagnostic and therapeutic agent development has been recently reviewed.[143–145]

An important application of solid-phase automated synthesis is the production of oligonucleotides via phosphoramidite solid-phase chemistry. Modern instruments work under cost-effective and high-throughput synthesis with 96-well, 384-well, and even 1536-well microplates.[146] The use of automated synthesis of oligonucleotides was reviewed in the nineties.[147,148]

3.3.4 Instruments for Sorting

Sorting solid-phase particles/containers can be done in manual fashion without any requirements for particular instrumentation. However, this tedious and sometimes error-prone operation has also been integrated and automated.

The Encore synthesizer was developed to integrate directed sorting using the Encore technique (Figure 3.23). The Encore synthesizer integrated the process of assembling Lanterns into sequences (necklace coding) after the first combinatorial step. After completion of the synthesis, the synthesizer facilitated integrated disassembly of the necklaces and distribution of the individual Lanterns into the wells of 96-well plates for final release of the compounds from the Lanterns. Thus, there was no need to handle the Lanterns individually, the process was integrated, and assembly of the Lanterns into linear sequences (necklaces) required only a few minutes. A simple computer program kept track of individual compounds (location in 96-well plates) and calculated their molecular weights. Individual tools for the Encore synthesizer have been described in detail.[121] The Encore synthesizer was used to prepare a 480-member library of potential synthetic PS scramblases.[122]

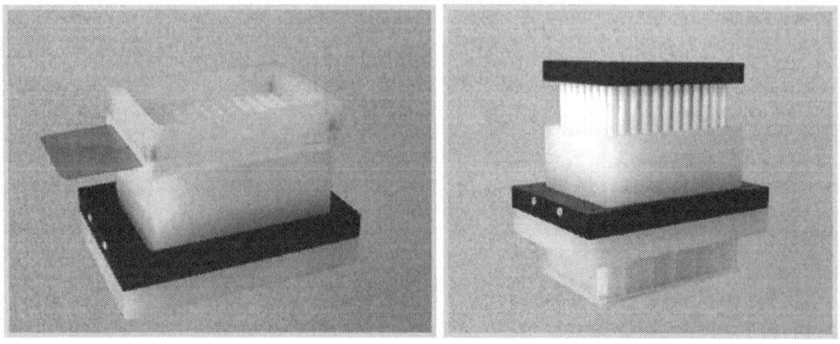

Figure 3.23. Encore synthesizer.

The IRORI platform used the directed sorting technique with Rf tagging (or 2D-coded caps) and Kan reactors as permeable containers for resin beads.[104,105] In the manual version of the sorter, the chemist placed Kans one by one on an Rf reading station and the computer program directed the Kan to a particular reaction vessel (Figure 3.24a). The light-emitting diode placed on the vessel became illuminated and the chemist placed the Kan into the vessel. The automatic sorter distributed Kans into reaction vessels and was designed for the synthesis of very large libraries (up to 100,000 compounds). The system suffered from two shortcomings (not including the steep price). This technology required dedicated larger resin beads and washing of the resin beads in Kans was problematic;[149] efficient washing of MiniKans required high solvent volumes.[150] IRORI was acquired by Nexus Biosystems and the automatic sorter is no longer commercially available.

Alternatives. A sophisticated robot for 2D and 3D spatially addressed arrays for the high-throughput automated synthesis of combinatorial libraries was designed by Selectide scientists.[151] A directed sort and combine synthesis with spatially arranged arrays used macroscopic supports (Mimotopes Lanterns).

3.4 CHARACTERIZATION AND PURIFICATION

The identity of compounds is an important factor in SPOS to reliably find and characterize structure of hit and lead compounds. The purity of compounds is also an important factor for receiving the correct response from high-throughput screening

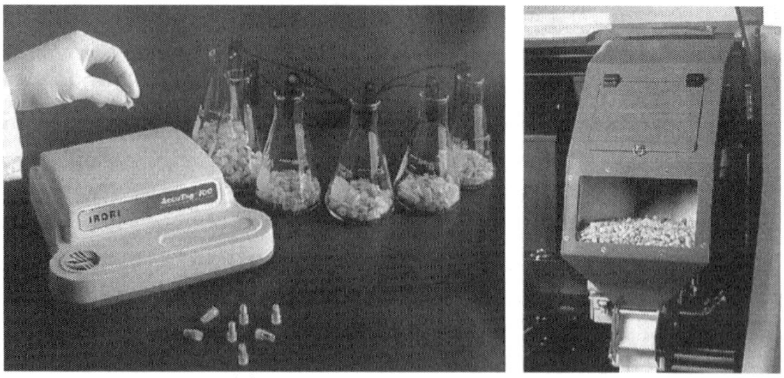

(a) (b)

Figure 3.24. Manual (a) and automatic (b) IRORI sorter. From webpage.

(HTS). Impurities, which are responsible for screening noise in HTS (reactives, fluorescents, quenchers, aggregators, chelators, cytotoxics, and reducing compounds), negatively affect the results of hit identification as well as the lead optimization process. Currently, the trend in chemical library synthesis is focused on the synthesis of smaller targeted libraries, including fully characterized and purified compounds on the scale of tens of milligrams. Larger compound quantities enable, among others, building archives of compounds for future use in various types of biological assays. As an example, the NIH has sponsored pilot-scale chemical diversity libraries for use in high-throughput biological screening by the Molecular Libraries Probe Production Centers Network (MLPCN), which requires 20 mg of chromatography-purified compounds in purities of >90% (LC at 215 nm).

Identification and purification of chemical compounds coming from the library are obviously a crucial step before their testing. Conventional methods had to be modified to efficiently apply to a large number of compounds. The important factors for identification, quantification, and purification include not only the separation power, but also a short amount of time and economy of the process. These demands lead to the development of high-throughput identification and high-throughput purification systems.

Expeditious characterization and purity determination of individual members of sizable chemical libraries are usually performed with the use of LC/UV/MS systems. Current instrumentation allows 24 h/day and 7 days/week unattended operation. An example of an instrument with a capacity of up to several thousand compounds/day can be demonstrated by the system developed by Liling.[152]

Since the UV response as a measure of purity and quantity determination is questionable due to the different extinction coefficients of individual components of a crude preparation, alternative detection systems were advocated, such as the evaporative light scattering detector (ELSD)[153] and the chemiluminescent nitrogen detector (CLND).[154] In addition, NMR techniques were adopted that allowed structural confirmation, purity characterization, compound solubility, solution stability, and biological testing from a single DMSO stock solution.[155] The quality control process for descriptions of chemical library members was discussed by Yan in 2004.[156]

Because during solid-phase synthesis impurities originate from incomplete reactions, side reactions, rearrangements, and so on and accumulate on the resin after each synthetic step, the crude released target products require purification. Although some synthetic routes enabled obtaining target compounds with a high purity by simple treatment of the crude preparation with ether,[157] most syntheses are followed by liquid chromatography (LC) purification to obtain compounds of adequate quality for structure determination/verification and subsequent use, which is mostly screening for biological activities.

Purification of drug discovery library members are usually made with the use of semipreparative HPLC systems, typically providing 10–100 mg quantities of purified material. The modern instruments are fully automated and configured to automatically inject the sample, run the separation, analyze fractions, flush and equilibrate column, and send a notification message to the user when all samples have been processed. The system can also be configured with a mass spectrometer for direct mass detection and a collection trigger for a diagnostic ion. Such automated systems are manufactured by several companies, including Waters Corporation, Agilent Technologies, Inc., and Shimadzu. Problems with poor ionization and material recovery were addressed in an automated semipreparative LC/MS chromatograph capable of monitoring the presence of target compounds in waste solvents.[158]

HPLC separations are almost exclusively performed in reversed phase (RP-HPLC). Certain limitations include the limited solubility of compounds in water/organic solvent mixture (typically water/acetonitrile). An alternative, normal-phase LC has also been automated, and a cyano-silica column was successfully applied for a library of 23 compounds.[159]

Supercritical fluid chromatography (SFC) was successfully employed in automated purification instrumentation and was used for the purification of chemical libraries of up to several thousand compounds. The advantage in comparison to traditional HPLC is solvent evaporation (primary mobile phase is carbon dioxide) and the dramatic reduction in expensive and toxic organic solvents. It is estimated that the solvent requirement for SFC is almost 50× less than that for HPLC, and the time necessary for separation is shorter by one-thirds.[160] The use of SFC in the drug discovery process was evaluated by a Sanofi-Aventis research group and exemplified by the purification of chiral compounds.[161] The SFC semipreparative chromatograph for the high-throughput purification of combi-chem libraries was described in detail by Berger in 2000.[162] An example of a commercially available automated system is the SF3™ supercritical fluid chromatography system manufactured by Gilson. This system is suitable for processing up to 108 samples for analytical runs and up to 60 samples for semipreparative runs.

The purification of chemical libraries has also been performed with the use of flash chromatography systems. Current instrumentation allows the simultaneous semiautomated purification of several compounds. An example system is the Quad3 (Biotage, Inc.), which allows simultaneous purification on up to 12 columns prepacked with sorbent for normal or reverse-phase separation. This system was successfully used for peptide library purification,[163] as well as for low-weight molecule libraries.[164,165] A natural product library including several hundred compounds was effectively purified with the use of an updated FlashMaster II,[166] capable of 10 fully automated independent separations in one run (Figure 3.25).

Different criteria apply to the identification and purification of compounds in OBOC libraries designed for hit discovery. This method typically provides a large number of compounds in very small quantities (100 pmol per one standard size bead). Thus, the use of conventional methods for identification (MS and NMR) and purification (LC) is limited but possible.[167,168] The screening of OBOC libraries is typically carried out without the identification and purification of individual compounds, and only the active hits are subjected to postscreening identification (see Section 3.2.1.1). Thus, the step subsequent

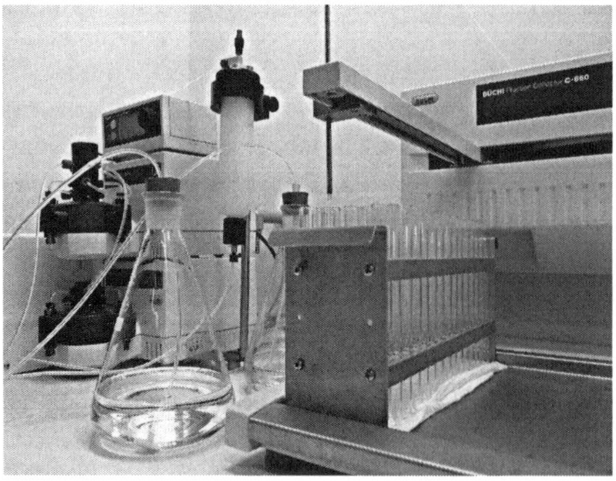

Figure 3.25. Flash chromatograph Sepacore® produced by Büchi Labor technik AG.

to hit identification includes preparation of the primarily identified compounds in milligram quantities and subsequent testing of a purified sample to confirm the activity.

3.5 CONCLUSIONS

Combinatorial solid-phase synthesis has become an integrated component of diverse scientific research projects, with drug discovery being the most prominent. The time spent on chemistry development and optimization needs to be rewarded by economical (i.e., fast with an acceptable price tag) access to diverse novel compounds. We strive to portray combinatorial solid-phase synthesis methods as an efficient route that can be performed in any organic chemistry laboratory with no prior history in solid-phase synthesis using simple or nondedicated instrumentation without compromising the number or quality of synthesized compounds. Solid-phase and library synthesis and the logistics of synthesis are not tailored to any specific instrumentation. The more time a laboratory spends on solid-phase syntheses, the more instrumentation can be dedicated to speed up and simplify the synthesis through the integration of common tasks. All chemical protocols are directly transferable to more sophisticated automated or semiautomated synthesizers if the synthetic throughput should warrant investment in such instruments.

Combinatorial solid-phase synthesis is not limited to any number or quantity of compounds. A thoughtful approach to the design of chemical libraries, selection of the most suitable instrumentation, and adequate identification and purification of the target compounds must be applied to achieve rewarding results. In summary, combinatorial chemistry is a tool that can accelerate the pace of a project and substantially contribute to its success.

ACKNOWLEDGMENTS

The work was supported by the Department of Chemistry and Biochemistry, University of Notre Dame, Department of Organic Chemistry, Palacký University, and the Ministry of Education, Youth and Sport of the Czech Republic (ME09057). The authors thank to Naděžda Cankařová and Soňa Křupková for critical review of the chapter.

REFERENCES

1. K. S. Lam, S. E. Salmon, E. M. Hersh, V. J. Hruby, W. M. Kazmierski, R. J. Knapp, A new type of synthetic peptide library for identifying ligand-binding activity. *Nature* **1991**, *354*, 82–84.
2. R. A. Houghten, C. Pinilla, S. E. Blondelle, J. R. Appel, C. T. Dooley, J. H. Cuervo, Generation and use of synthetic peptide combinatorial libraries for basic research and drug discovery. *Nature* **1991**, *354*, 84–86.
3. Although papers by Frank[4–6] Geysen,[7,8] and Furka[9,10] had previously introduced the concept of combinatorial chemistry, the papers did not trigger the expected interest.
4. R. Frank, W. Heikens, G. Heisterberg-Moutsis, H. Blocker, A new general approach for the simultaneous chemical synthesis of large number of oligonucleotides: segmental solid support. *Nucleic Acids Res.* **1983**, *11*, 4365–4377.
5. R. Frank, R. Doring, Simultaneous multiple peptide synthesis under continuous flow conditions on cellulose paper discs as segmental solid supports. *Tetrahedron* **1988**, *44*, 6031–6040.

6. R. Frank, S. Guler, S. Krause, W. Lindenmaier, Facile and rapid 'spot-synthesis' of large numbers of peptides on membrane sheets, in E. Giralt, D. Andreu, (Eds), *Peptides*, ESCOM, Leiden, 1990, pp. 151–152.
7. M. H. Geysen, R. H. Meloen, S. J. Barteling, Use of peptide synthesis to probe viral antigens for epitopes to a resolution of a single amino acid. *Proc. Natl. Acad. Sci. USA* **1984**, *81*, 3998–4002.
8. M. H. Geysen, S. J. Rodda, T. J. Mason, *A priori* delineation of a peptide which mimics a discontinuous antigenic determinant. *Mol. Immunol.* **1986**, *23*, 709–715.
9. A. Furka, F. Sebestyen, M. Asgedom, G. Dibo, More peptides by less labour. (Poster Presentation). *Proceedings of the 10th International Symposium of Medicinal Chemistry (Budapest)*, 1988, p. 288.
10. A. Furka, F. Sebestyen, M. Asgedom, G. Dibo, Cornucopia of peptides by synthesis. (Poster Presentation). *Proceedings of the 10th International Symposium of Medicinal Chemistry (Budapest)*, 1988, p. 47.
11. B. A. Bunin, *The Combinatorial Index*, Academic Press, San Diego, 1998.
12. F. Zaragoza, *Organic synthesis on Solid Phase*, Wiley-VCH, Weinheim, 2000.
13. S. E. Hall, Recent advances in solid-phase synthesis. *Mol. Diversity* **1999**, *4*, 131–142.
14. S. Booth, P. H. H. Hermkens, H. C. J. Ottenheijm, D. C. Rees, Solid-phase organic reactions III: a review of the literature Nov 96–Dec 97. *Tetrahedron* **1998**, *54*, 15385–15443.
15. A. R. Brown, P. H. H. Hermkens, H. C. J. Ottenheijm, D. C. Rees, Solid-phase synthesis. *Synlett* **1998**, 817–827.
16. R. E. Dolle, Comprehensive survey of combinatorial libraries with undisclosed biological activity: *1992–1997. Mol. Diversity* **1998**, *4*, 233–256.
17. P. H. H. Hermkens, H. C. J. Ottenheijm, D. Rees, Solid-phase organic reactions II: a review of the literature Nov 95–Nov 96. *Tetrahedron* **1997**, *53*, 5643–5678.
18. K. S. Lam, M. Lebl, V. Krchnak, The "one-bead–one-compound" combinatorial library method. *Chem. Rev.* **1997**, *97*, 411–448.
19. In fact, the technique should have been referred to as permutational chemistry, since permutation of units A and B provides AB and BA as distinctly different entities, whereas the combination of A and B is AB only, that is, arranged without regard of order. Also, the terminology "library" has been debated as being "as far from the library as one can get in this world."[20]
20. R. Hoffmann, Not a library. *Angew. Chem., Int. Ed.* **2001**, *40*, 3337–3340.
21. A. W. Czarnik, Combinatorial definitions. *Chem. Eng. News* **1997**, *75*, 7.
22. D. P. Curran, P. Wipf, Combinatorial definitions. *Chem. Eng. News* **1997**, *75*, 6–7.
23. J. Smith, V. Krchnak, A solid-phase traceless synthesis of benzimidazoles with three combinatorial steps. *Tetrahedron Lett.* **1999**, *40*, 7633–7636.
24. A Directed Sorting Method intentionally prepares each selected compound from the combinatorial array only once, and the abundance of compounds in the split and pool method is driven by statistics due to the random distribution of pooled solid-phase particles.
25. A. Furka, F. Sebestyen, M. Asgedom, G. Dibo, General method for rapid synthesis of multicomponent peptide mixtures. *Int. J. Pept. Protein Res.* **1991**, *37*, 487–493.
26. A. Furka, Redistribution in combinatorial synthesis. A theoretical approach. *Comb. Chem. High Throughput Screening* **2000**, *3*, 197–210.
27. K. C. Nicolaou, X. Y. Xiao, Z. Parandoosh, A. Senyei, M. P. Nova, Radiofrequency encoded combinatorial chemistry. *Angew. Chem., Int. Ed.* **1995**, *34*, 2289–2291.
28. C. Y. Xiao, C. F. Zhao, H. Potash, M. P. Nova, Combinatorial chemistry with laser optical encoding. *Angew. Chem., Int. Ed.* **1997**, *36*, 780–782.
29. J. W. Guiles, C. L. Lanter, R. A. Rivero, A visual tagging process for mix and sort combinatorial chemistry. *Angew. Chem., Int. Ed.* **1998**, *37*, 926–928.

30. C. Zhao, S. Shi, D. Mir, D. Hurst, R. Li, X. Y. Xiao, J. Lillig, A. W. Czarnik, Polystyrene grafted fluoropolymer microtubes: new supports for solid-phase organic synthesis with useful performance at high temperature. *J. Comb. Chem.* **1999**, *1*, 91–95.
31. X. Y. Xiao, Z. Parandoosh, M. P. Nova, Design and synthesis of a taxoid library using radiofrequency encoded combinatorial chemistry. *J. Org. Chem.* **1997**, *62*, 6029–6033.
32. R. Frank, SPOT synthesis: an easy technique for positionally addressable, parallel chemical synthesis on a membrane support. *Tetrahedron* **1992**, *48*, 9217–9232.
33. R. A. Houghten, General method for the rapid solid-phase synthesis of large numbers of peptides: specificity of antigen–antibody interaction at the level of individual amino acids. *Proc. Natl. Acad. Sci. USA* **1985**, *82*, 5131–5135.
34. V. Krchnak, Semi-automated high throughput combinatorial solid-phase organic synthesis. *Biotechnol. Bioeng. (Comb. Chem.)* **1999**, *61*, 135–141.
35. P. Brooking, A. Doran, P. Grimsey, N. W. Hird, W. S. MacLachlan, M. Vimal, Split–Split. A multiple synthesiser approach to efficient automated parallel synthesis. *Tetrahedron Lett.* **1999**, *40*, 1405–1408.
36. R. J. Booth, J. C. Hodges, Solid-supported reagents strategies for rapid purification of combinatorial synthesis product. *Acc. Chem. Res.* **1999**, *32*, 18–26.
37. D. M. Coe, R. Storer, Solution-phase combinatorial chemistry. *Mol. Diversity* **1999**, *4*, 31–38.
38. C. Blackburn, F. Albericio, S. A. Kates, Functionalized resins and linkers for solid-phase synthesis of small molecules. *Drugs Future* **1998**, *22*, 1007–1025.
39. M. Hori, K. D. Janda, A soluble polymer approach to the “Fishing Out” principle: synthesis and purification of β-amino alcohols. *J. Org. Chem.* **1998**, *63*, 889–894.
40. X. Y. Zhao, W. A. Metz, F. Sieber, K. D. Janda, Expanding on the purification methodology of polyethylene glycol (PEG) bound molecules: the synthesis of 3,5-pyrazolidinediones. *Tetrahedron Lett.* **1998**, *39*, 8433–8436.
41. H. S. Han, K. D. Janda, A soluble polymer-bound approach to the sharpless catalytic asymmetric dihydroxylation (AD) reaction: preparation and application of a [$(DHQD)_2$PHAL-PEG-OMe] ligand. *Tetrahedron Lett.* **1997**, *38*, 1527–1530.
42. X. Y. Zhao, K. W. Jung, K. D. Janda, Soluble polymer synthesis: an improved traceless linker methodology for aliphatic CH bond formation. *Tetrahedron Lett.* **1997**, *38*, 977–980.
43. R. B. Merrifield, Solid-phase peptide synthesis. I. The synthesis of a tetrapeptide. *J. Am. Chem. Soc.* **1963**, *85*, 2149–2154.
44. J. M. J. Frechet, Synthesis and applications of organic polymers as supports and protecting groups. *Tetrahedron* **1981**, *37*, 663–683.
45. C. C. Leznoff, The use of insoluble polymer supports in general organic synthesis. *Acc. Chem. Res.* **1978**, *11*, 327–333.
46. J. I. Crowley, H. Rapoport, Solid-phase organic synthesis: novelty of fundamental concept? *Acc. Chem. Res.* **1976**, *9*, 135–144.
47. G. Schnorrenberg, H. Gerhardt, Fully automatic simultaneous multiple peptide synthesis in micromolar scale: rapid synthesis of series of peptides for screening in biological assays. *Tetrahedron* **1989**, *45*, 7759–7764.
48. H. Gausepohl, C. Boulin, M. Kraft, R. W. Frank, Automated multiple peptide synthesis. *Peptide Res.* **1992**, *5*, 315–320.
49. R. Frank, The SPOT-synthesis technique: synthetic peptide arrays on membrane supports—principles and applications. *J. Immunol. Methods* **2002**, *267*, 13–26.
50. Z. Wang, R. A. Laursen, Multiple peptide synthesis on polypropylene membranes for rapid screening of bioactive peptides. *Peptide Res.* **1992**, *5*, 275–280.
51. R. Frank, Strategies and techniques in simultaneous solid-phase synthesis based on the segmentation of membrane type supports. *Bioorg. Med. Chem. Lett.* **1993**, *3*, 425–430.

52. B. Gao, M. P. Esnouf, Multiple interactive residues of recognition: elucidation of discontinuous epitopes with linear peptides. *J. Immunol.* **1996**, *157*, 183–188.
53. B. Gao, M. P. Esnouf, Elucidation of the core residues of an epitope using membrane-based combinatorial peptide libraries. *J. Biol. Chem.* **1996**, *271*, 24634–24638.
54. M. C. Pirrung, Spatially addressable combinatorial libraries. *Chem. Rev.* **1997**, *97*, 473–488.
55. S. P. A. Fodor, R. J. Leighton, M. C. Pirrung, L. Stryer, L. A. Tsai, D. Solas, Light-directed, spatially addressable parallel chemical synthesis. *Science* **1991**, *251*, 767–773.
56. C. Y. Cho, E. J. Moran, S. R. Cherry, J. C. Stephans, S. P. A. Fodor, C. L. Adams, A. Sundaram, J. W. Jacobs, P. G. Schultz, An unnatural biopolymer. *Science* **1993**, *261*, 1303–1305.
57. Nature uses combinatorial biology for antibody selection, and this general concept led scientists to introduce phage display in a combinatorial fashion to make peptides.[58]
58. J. K. Scott, G. P. Smith, Searching for peptide ligands with an epitope library. *Science* **1990**, *249*, 386–390.
59. P. Kocis, V. Krchnak, M. Lebl, Symmetrical structure allowing the selective multiple release of a defined quantity of peptide from a single bead of polymeric support. *Tetrahedron Lett.* **1993**, *34*, 7251–7252.
60. M. Lebl, M. Patek, P. Kocis, V. Krchnak, V. J. Hruby, S. E. Salmon, K. S. Lam, Multiple release of equimolar amounts of linker free peptides from the polymeric carrier. *Int. J. Pept. Protein Res.* **1993**, *41*, 201–203.
61. S. E. Salmon, K. S. Lam, M. Lebl, A. Kandola, P. S. Khattri, S. Wade, M. Patek, P. Kocis, V. Krchnak, D. Thorpe, S. Felder, Discovery of biologically active peptides in random libraries: solution-phase testing after staged orthogonal release from resin beads. *Proc. Natl. Acad. Sci. USA* **1993**, *90*, 11708–11712.
62. M. Cardno, M. Bradley, A simple multiple release system for combinatorial library and peptide analysis. *Tetrahedron Lett.* **1996**, *37*, 135–138.
63. C. K. Jayawickreme, G. F. Graminski, J. M. Quillan, M. R. Lerner, Creation and functional screening of a multi-use peptide library. *Proc. Natl. Acad. Sci. USA* **1994**, *91*, 1614–1618.
64. E. B. Akerblom, Six new photolabile linkers for solid-phase synthesis. 2. Coupling of various building blocks and photolytic cleavage. *Mol. Diversity* **1998**, *4*, 53–69.
65. E. B. Akerblom, A. S. Nygren, K. H. Agback, Six new photolabile linkers for solid-phase synthesis. 1. Methods of preparation. *Mol. Diversity* **1998**, *3*, 137–148.
66. S. Peukert, B. Giese, The pivaloylglycol anchor group: a new platform for a photolabile linker in solid-phase synthesis. *J. Org. Chem.* **1998**, *63*, 9045–9051.
67. C. Dell'Aquila, J. L. Imbach, B. Rayner, Photolabile linker for the solid-phase synthesis of base-sensitive oligonucleotides. *Tetrahedron Lett.* **1997**, *38*, 5289–5292.
68. D. L. Whitehouse, S. N. Savinov, D. J. Austin, An improved synthesis and selective coupling of a hydroxy based photolabile linker for solid-phase organic synthesis. *Tetrahedron Lett.* **1997**, *38*, 7851–7852.
69. S. J. Teague, Facile synthesis of a *o*-nitrobenzyl photolabile linker for combinatorial chemistry. *Tetrahedron Lett.* **1996**, *37*, 5751–5754.
70. B. B. Brown, D. S. Wagner, H. M. Geysen, A single-bead decode strategy using electrospray ionization mass spectrometry and a new photolabile linker: 3-Amino-3-(2-nitrophenyl)propionic acid. *Mol. Diversity* **1995**, *1*, 4–12.
71. C. P. Holmes, D. G. Jones, Reagents for combinatorial organic synthesis: development of a new *o*-nitrobenzyl photolabile linker for solid-phase synthesis. *J. Org. Chem.* **1995**, *60*, 2318–2319.
72. S. Brenner, R. A. Lerner, Encoded combinatorial chemistry. *Proc. Natl. Acad. Sci. USA* **1992**, *89*, 5381–5383.
73. K. Biemann, S. A. Martin, Mass spectrometric determination of the amino acid sequence of peptides and proteins. *Mass Spectrom. Rev.* **1987**, *6*, 1–76.

74. B. J. Egner, G. J. Langley, M. Bradley, Solid-phase chemistry: direct monitoring by matrix-assisted laser desorption/ionization time of flight mass spectrometry. A tool for combinatorial chemistry. *J. Org. Chem.* **1995**, *60*, 2652–2653.

75. G. Siuzdak, J. Lewis, Applications of mass spectrometry in combinatorial chemistry. *Biotechnol. Bioeng.* **1998**, *61*, 127–134.

76. M. C. Fitzgerald, K. Harris, Shevlin, Ch. G. G. Siuzdak, Direct characterization of solid-phase resin-bound molecules by mass spectrometry. *Bioorg. Med. Chem. Lett.* **1996**, *6*, 979–982.

77. R. S. Youngquist, G. R. Fuentes, M. P. Lacey, T. Keough, Generation and screening of combinatorial peptide libraries designed for rapid sequencing by mass spectrometry. *J. Am. Chem. Soc.* **1995**, *117*, 3900–3906.

78. M. C. Needels, D. G. Jones, E. H. Tate, G. L. Heinkel, L. M. Kochersperger, W. J. Dower, R. W. Barrett, M. A. Gallop, Generation and screening of an oligonucleotide-encoded synthetic peptide library. *Proc. Natl. Acad. Sci. USA* **1993**, *90*, 10700–10704.

79. J. Nielsen, S. Brenner, K. D. Janda, Synthetic methods for the implementation of encoded combinatorial chemistry. *J. Am. Chem. Soc.* **1993**, *115*, 9812–9813.

80. J. M. Kerr, S. C. Banville, R. N. Zuckermann, Encoded combinatorial peptide libraries containing non-natural amino acids. *J. Am. Chem. Soc.* **1993**, *115*, 2529–2531.

81. V. Nikolaev, A. Stierandova, V. Krchnak, B. Seligmann, K. S. Lam, S. E. Salmon, M. Lebl, Peptide-encoding for structure determination of nonsequenceable polymers within libraries synthesized and tested on solid-phase supports. *Peptide Res.* **1993**, *6*, 161–170.

82. H. P. Nestler, P. A. Bartlett, W. C. Still, A general method for molecular tagging of encoded combinatorial chemistry libraries. *J. Org. Chem.* **1994**, *59*, 4723–4724.

83. J. J. Baldwin, J. J. Burbaum, I. Henderson, M. H. J. Ohlmeyer, Synthesis of a small molecule combinatorial library encoded with molecular tags. *J. Am. Chem. Soc.* **1995**, *117*, 5588–5589.

84. J. J. Baldwin, J. J. Burbaum, D. Chelsky, L. W. Dillard, I. Henderson, G. Li, M. H. J. Ohlmeyer, T. L. Randle, J. C. Reader, Combinatorial libraries encoded with electrophoric tags. *Eur. J. Med. Chem.* **1995**, *30*, S349–S358.

85. E. J. Moran, S. Sarshar, J. F. Cargill, M. M. Shahbaz, A. Lio, A. M. M. Mjalli, R. W. Armstrong, Radio frequency tag encoded combinatorial library method for the discovery of tripeptide-substituted cinnamic acid inhibitors of the protein tyrosine phosphatase PTP1B. *J. Am. Chem. Soc.* **1995**, *117*, 10787–10788.

86. J. Vagner, G. Barany, K. S. Lam, V. Krchnak, N. F. Sepetov, J. A. Ostrem, P. Strop, M. Lebl, Enzyme-mediated spatial segregation on individual polymeric support beads: application to generation and screening of encoded combinatorial libraries. *Proc. Natl. Acad. Sci. USA* **1996**, *93*, 8194–8199.

87. Z. J. Ni, D. Maclean, C. P. Holmes, M. A. Gallop, Encoded combinatorial chemistry: binary coding using chemically robust secondary amine tags. *Methods Enzymol.* **1996**, *267*, 261–272.

88. H. M. Geysen, C. D. Wagner, W. M. Bodnar, C. J. Markworth, G. J. Parke, F. J. Schoenen, D. S. Wagner, D. S. Kinder, Isotope or mass encoding of combinatorial libraries. *Chem. Biol.* **1996**, *3*, 679–688.

89. V. Nikolaiev, A. Stierandova, V. Krchnak, B. Seligmann, K. S. Lam, S. E. Salmon, M. Lebl, Peptide-encoding for structure determination of nonsequenceable polymers within libraries synthesized and tested on solid-phase supports. *Peptide Res.* **1993**, *6*, 161–170.

90. J. M. Kerr, S. C. Banville, R. N. Zuckermann, Encoded combinatorial peptide libraries containing non-natural amino acids. *J. Am. Chem. Soc.* **1993**, *115*, 2529–2531.

91. M. H. J. Ohlmeyer, R. N. Swanson, L. Dillard, J. C. Reader, G. Asouline, R. Kobayashi, M. Wigler, W. C. Still, Complex synthetic chemical libraries indexed with molecular tags. *Proc. Natl. Acad. Sci. USA* **1993**, *90*, 10922–10926.

92. H. P. Nestler, P. A. Bartlett, W. C. Still, A general method for molecular tagging of encoded combinatorial chemistry libraries. *J. Org. Chem.* **1994**, *59*, 4723–4724.
93. Z. J. Ni, D. Maclean, C. P. Holmes, M. M. Murphy, B. Ruhland, J. W. Jacobs, E. M. Gordon, M. A. Gallop, Versatile approach to encoding combinatorial organic syntheses using chemically robust secondary amine tags. *J. Med. Chem.* **1996**, *39*, 1601–1608.
94. A. Song, J. Zhang, C. B. Lebrilla, K. S. Lam, A novel and rapid encoding method based on mass spectrometry for "one-bead-one-compound" small molecule combinatorial libraries. *J. Am. Chem. Soc.* **2003**, *126*, 6180–6188.
95. S. H. Hwang, A. Lehman, X. Cong, M. M. Olmstead, K. S. Lam, C. B. Lebrilla, M. J. Kurth, OBOC small-molecule combinatorial library encoded by halogenated mass-tags. *Org. Lett.* **2004**, *6*, 3829–3832.
96. R. Liu, X. Wang, A. Song, T. Bao, K. S. Lam, Development and applications of topologically segregated bilayer beads in one-bead one-compound combinatorial libraries. *QSAR Comb. Sci.* **2005**, *24*, 1127–1140.
97. J. Luo, H. Zhang, W. Xiao, P. R. Kumaresan, C. Shi, C. X. Pan, O. H. Aina, K. S. Lam, Rainbow beads: a color coding method to facilitate high-throughput screening and optimization of one-bead one-compound combinatorial libraries. *J. Comb. Chem.* **2008**, *10*, 599–604.
98. C. Pinilla, J. R. Appel, P. Blanc, R. A. Houghten, Rapid identification of high affinity peptide ligands using positional scanning synthetic peptide combinatorial libraries. *Biotechniques* **1992**, *13*, 901–905.
99. M. A. Bednarek, J. P. Springer, B. R. Cunningham, A. M. Bernick, M. Bodanszky, Reactions of C-terminal omega-amino acid residues in liquid hydrogen fluoride. *Int. J. Pept. Protein Res.* **1993**, *42*, 10–13.
100. D. Lee, J. K. Sello, S. L. Schreiber, Pairwise use of complexity-generating reactions in diversity-oriented organic synthesis. *Org. Lett.* **2000**, *2*, 709–712.
101. K. C. Nicolaou, D. Vourloumis, T. Li, J. Pastor, N. Winssinger, Y. He, S. Ninkovic, F. Sarabia, H. Vallberg, F. Roschangar, N. P. King, M. Ray, V. Finlay, P. Giannakakou, P. Verdier-Pinard, E. Hamel, Designed epothilones: combinatorial synthesis, tubulin assembly properties, and cytotoxic action against taxol-resistant tumor cells. *Angew. Chem., Int. Ed.* **1997**, *36*, 2097–2101.
102. S. Arutyunyan, A. Nefzi, Synthesis of chiral polyaminothiazoles. *J. Comb. Chem.* **2010**, *12*, 315–317.
103. C. Vizcaino, D. Restrepo-Montoya, D. Rodriguez, L. F. Nino, M. Ocampo, M. Vanegas, M. T. Reguero, N. L. Martinez, M. E. Patarroyo, M. A. Patarroyo, Computational prediction and experimental assessment of secreted/surface proteins from mycobacterium tuberculosis H37Rv. *PLoS Comput. Biol.* **2010**, *6*, 1–19.
104. X. Y. Xiao, R. Li, H. Zhuang, B. Ewing, K. Karunaratne, J. Lillig, R. Brown, K. C. Nicolaou, Solid-phase combinatorial synthesis using MicroKan reactors, Rf tagging, and directed sorting. *Biotechnol. Bioeng.* **2000**, *71*, 44–50.
105. K. C. Nicolaou, J. A. Pfefferkorn, H. J. Mitchell, A. J. Roecker, S. Barluenga, G. Q. Cao, R. L. Affleck, J. E. Lillig, Natural product-like combinatorial libraries based on privileged structures. 2. Construction of a 10,000-membered benzopyran library by directed split-and-pool chemistry using NanoKans and optical encoding. *J. Am. Chem. Soc.* **2000**, *122*, 9954–9967.
106. K. L. Beattie, J. D. I. Frost,Porous wafer for segmented synthesis of biopolymers. US Patent 5,175,209, 1992.
107. K. L. Beattie, R. F. Fowler, Solid-phase gene assembly. *Nature* **1991**, *352*, 548–549.
108. K. L. Beattie, G. D. Hurst, Synthesis and use of oligonucleotide libraries, in R. Epton, (Ed.), *Innovation Perspective in Solid-phase Synthesis, 3rd International Symposium*, Mayflower Worldwide Limited, Birmingham, 1994, pp. 69–76.

109. K. L. Beattie, N. J. Logsdon, R. S. Anderson, J. M. Espinosa-Lara, R. Maldonado-Rodriguez, J. D. I. Frost, Gene synthesis technology: recent development and future prospects. *Biotechnol. Appl. Biochem.* **1988**, *10*, 510–521.

110. I. Bouillon, M. Soural, V. Krchnak, Resin capsules: permeable containers for parallel/combinatorial solid-phase organic synthesis. *J. Comb. Chem.* **2008**, *10*, 714–720.

111. B. Atrash, M. Bradley, R. Kobylecki, D. Cowell, J. Reader, Revolutionizing resin handling for combinatorial synthesis. *Angew. Chem., Int. Ed.* **2001**, *40*, 938–941.

112. F. Rasoul, F. Ercole, Y. Pham, C. T. Bui, Z. Wu, S. N. James, R. W. Trainor, G. Wickham, N. J. Maeji, Grafted supports in solid-phase synthesis. *Biopolymers* **2000**, *55*, 207–216.

113. T. D. Ryba, K. M. Depew, L. A. Marcaurelle, Large scale preparation of silicon-functionalized SynPhase polystyrene lanterns for solid-phase synthesis. *J. Comb. Chem.* **2008**, *11*, 110–116.

114. N. Hird, I. Hughes, D. Hunter, M. G. J. T. Morrison, D. C. Sherrington, L. Stevenson, Polymer discs: an alternative support format for solid-phase synthesis. *Tetrahedron* **1999**, *55*, 9575–9584.

115. J. A. Tripp, F. Svec, J. M. J. Frechet, Grafted macroporous polymer monolithic disks: a new format of scavengers for solution-phase combinatorial chemistry. *J. Comb. Chem.* **2001**, *3*, 216–223.

116. J. A. Tripp, F. Svec, M. J. Fréchet, Solid-phase acylating reagents in new format: macroporous polymer disks. *J. Comb. Chem.* **2001**, *3*, 604–611.

117. A. R. Vaino, K. D. Janda, Euclidean shape-encoded combinatorial chemical libraries. *Proc. Natl. Acad. Sci. USA* **2000**, *97*, 7692–7696.

118. J. Smith, J. Gard, W. Cummings, A. Kaniszai, V. Krchnak, Necklace-coded polymer-supported combinatorial synthesis of 2-arylaminobenzimidazoles. *J. Comb. Chem.* **1999**, *1*, 368–370.

119. V. Krchnak, J. Smith, J. Vagner, Solid-phase traceless synthesis of selected nitrogen-containing heterocyclic compounds. The Encore technique for directed sorting of modular solid support. *Collect. Czech. Chem. Commun.* **2001**, *66*, 1078–1106.

120. A. Furka, J. W. Christensen, E. Healy, H. R. Tanner, H. Saneii, String synthesis. A spatially addressable split procedure. *J. Comb. Chem.* **2000**, *2*, 220–223.

121. V. Krchnak, V. Padera, The encore technique: a novel approach to directed split-and-pool combinatorial synthesis, in B. A. Bunin, G. A. Morales, (Eds), *Methods in Enzymology*, Vol. 369, Elsevier, 2003, pp. 112–124.

122. R. Shukla, Y. Sasaki, V. Krchnak, B. D. Smith, Identification of synthetic phosphatidylserine (PS) translocases from a combinatorial library prepared by directed split-and-pool synthesis. *J. Comb. Chem.* **2004**, *6*, 703–709.

123. L. Feliu, G. Subra, J. Martinez, M. Amblard, Spiroimidazolidinone library derivatives on SynPhase lanterns. *J. Comb. Chem.* **2003**, *5*, 356–361.

124. G. Subra, L. Soulere, P. Hoffman, C. Enjalbal, J. Martinez, Parallel and mixture combined approach: rapid cheap synthesis and characterization of a 4096-tripeptides library. *QSAR Comb. Sci.* **2003**, *22*, 646–651.

125. Z. Wu, N. J. Ede, Novel solid-phase synthesis of 1,2-dialkoxyindoles. *Org. Lett.* **2003**, *5*, 2935–2938.

126. F. Brucoli, P. W. Howard, D. E. Thurston, Efficient solid-phase synthesis of a library of distamycin analogs containing novel biaryl motifs on SynPhase lanterns. *J. Comb. Chem.* **2009**, *11*, 576–586.

127. V. Krchňák, J. Vágner, Color-monitored solid-phase multiple peptide synthesis under low-pressure continuous flow conditions. *Peptide Res.* **1990**, *3*, 182–193.

128. V. Krchnak, V. Padera, The domino blocks: a simple solution for parallel solid-phase organic synthesis. *Bioorg. Med. Chem. Lett.* **1998**, *22*, 3261–3264.

129. W. L. Scott, R. A. Schonegg, C. L. Cwi,Vessel handling system useful for combinatorial chemistry. US Patent 5,785,927, 1998.

130. W. L. Scott, F. Delgado, K. Lobb, R. Pottorf, M. J. O'Donnell, Solid-phase synthesis of amino amides and peptide amides with unnatural side chains. *Tetrahedron Lett.* **2001**, *42*, 2073–2076.
131. W. L. Scott, M. J. O'Donnell, Distributed drug discovery, Part 1: linking academia and combinatorial chemistry to find drug leads for developing world diseases. *J. Comb. Chem.* **2008**, *11*, 3–13.
132. W. L. Scott, J. Alsina, C. O. Audu, E. Babaev, L. Cook, J. L. Dage, L. A. Goodwin, J. G. Martynow, D. Matosiuk, M. Royo, J. G. Smith, A. T. Strong, K. Wickizer, E. M. Woerly, Z. Zhou, M. J. O'Donnell, Distributed drug discovery, Part 2: global rehearsal of alkylating agents for the synthesis of resin-bound unnatural amino acids and virtual D3 catalog construction. *J. Comb. Chem.* **2008**, *11*, 14–33.
133. W. L. Scott, C. O. Audu, J. L. Dage, L. A. Goodwin, J. G. Martynow, L. K. Platt, J. G. Smith, A. T. Strong, K. Wickizer, E. M. Woerly, M. J. O'Donnell, Distributed drug discovery, Part 3: using D3 methodology to synthesize analogs of an anti-melanoma compound. *J. Comb. Chem.* **2008**, *11*, 34–43.
134. V. Krchnak, A. S. Weichsel, M. Lebl, S. Felder, Automated solid-phase organic synthesis in micro-plate wells. Synthesis of *N*-(alkoxy-acyl)amino alcohols. *Bioorg. Med. Chem. Lett.* **1997**, *7*, 1013–1016.
135. A. Holm, M. Meldal, Multiple column peptide synthesis, in E. Bayer, G. Jung, (Eds), *Peptides*, Walter de Gruyter, Berlin, 1989, pp. 208–2210.
136. V. Krchnak, J. Vagner, I. Hirsch, Simultaneous synthesis of sequence unrelated peptides derived from proteins of human papilomaviruses. *Collect. Czech. Chem. Commun.* **1988**, *53*, 2645–2653.
137. V. Krchnak, J. Vagner, O. Mach, Multiple continuous-flow solid-phase peptide synthesis. Synthesis of an HIV antigenic peptide and its omission analogues. *Int. J. Pept. Protein Res.* **1989**, *33*, 209–213.
138. V. Krchnak, D. Cabel, M. Lebl, MARS: multiple automatic robot synthesizer for continuous flow of peptides. *Peptide Res.* **1996**, *9*, 45–49.
139. N. Jung, A. Encinas, S. Braese, Automated synthesis of heterocycles on solid supports. *Curr. Opin. Drug Discovery Dev.* **2006**, *9*, 713–728.
140. H. Gausepohl, C. Behn, Automated synthesis of solid-phase bound peptides, in J. Koch, M. Mahler, (Eds), *Peptide Arrays on Membrane Supports Synthesis and Application*, Springer-Verlag, Berlin, **2002**, pp. 55–68.
141. M. P. Brennan, D. Cox, A. J. Chubb, Peptide diversity in drug discovery. *Front. Drug Des. Discovery* **2007**, *3*, 395–432.
142. O. J. Plante, E. R. Palmacci, P. H. Seeberger, Automated solid-phase synthesis of oligosaccharides. *Science* **2001**, *291*, 1523–1527.
143. B. Castagner, P. Seeberger, Automated solid-phase oligosaccharide synthesis. *Top. Curr. Chem.* **2007**, 289–309.
144. L. Krock, P. H. Seeberger, Automated synthesis of oligosaccharides provides access to chemical tools and vaccine candidates. *Chimia* **2008**, *62*, 8–12.
145. P. H. Seeberger, Automated oligosaccharide synthesis. *Chem. Soc. Rev.* **2008**, *37*, 19–28.
146. J. Y. Cheng, H. H. Chen, Y. S. Kao, W. C. Kao, K. Peck, High throughput parallel synthesis of oligonucleotides with 1536 channel synthesizer. *Nucleic Acids Res.* **2002**, *30*, e93.
147. A. Frauendorf, J. W. Engels, Automated synthesis, structure, and biological activity of backbone-modified oligonucleotides. The antisense approach. *Stud. Nat. Prod. Chem.* **1993**, *13* (Bioactive Natural Products, Pt. A), 257–294.
148. R. Vinayak, Chemical synthesis, analysis, and purification of oligoribonucleotides. *Methods Comp. Methods Enzymol.* **1993**, *5*, 7–18.
149. A. Ortega, M. Erra, E. Navarro, R. S. Roberts, D. Fernandez-Forner, Perfluorous solid-phase organic synthesis (PF-SPOS) and IRORI technology for combinatorial chemistry. *QSAR Comb. Sci.* **2006**, *25*, 598–604.

150. A. R. Poreddy, O. F. Schall, T. A. Osiek, J. R. Wheatley, D. D. Beusen, G. R. Marshall, U. Slomczynska, Hydroxamate-based iron chelators: Combinatorial syntheses of desferrioxamine B analogues and evaluation of binding affinities. *J. Comb. Chem.* **2004**, *6*, 239–254.

151. M. Patek, P. Safar, M. Smrcina, E. Wegrzyniak, K. Bjergarde, A. Weichsel, P. Strop, 2D and 3D spatially addressed arrays for high-throughput automated synthesis of combinatorial libraries. *J. Comb. Chem.* **2004**, *6*, 43–49.

152. L. Fang, J. Cournoyer, M. Demee, J. Zhao, D. Tokushige, B. Yan, High-throughput liquid chromatography ultraviolet/mass spectrometric analysis of combinatorial libraries using an eight-channel multiplexed electrospray time-of-flight mass spectrometer. *Rapid Commun. Mass Spectrom.* **2002**, *16*, 1440–1447.

153. L. Fang, M. Wan, M. Pennacchio, J. Pan, Evaluation of evaporative light-scattering detector for combinatorial library quantitation by reversed phase HPLC. *J. Comb. Chem.* **2000**, *2*, 254–257.

154. E. W. Taylor, M. G. Qian, G. D. Dollinger, Simultaneous on-line characterization of small organic molecules derived from combinatorial libraries for identity, quantity, and purity by reversed-phase HPLC with chemiluminescent nitrogen, UV, and mass spectrometric detection. *Anal. Chem.* **1998**, *70*, 3339–3347.

155. V. Pinciroli, R. Biancardi, N. Colombo, M. Colombo, V. Rizzo, Characterization of small combinatorial chemistry libraries by 1H NMR. Quantitation with a convenient and novel internal standard. *J. Comb. Chem.* **2001**, *3*, 434–440.

156. B. Yan, New combichem QC standards guard the supply of compounds and ensure HTS results. *Mod. Drug Discovery* **2004**, February, 30–34.

157. M. Soural, V. Krchnak, Efficient solid-phase synthesis of 2-substituted-3-hydroxy-4(1*H*)-quinolinone-7-carboxamides with two diversity positions. *J. Comb. Chem.* **2007**, *9*, 793–796.

158. J. FitzGibbons, S. Op, A. Hobson, L. Schaffter, Novel approach to optimization of a high-throughput semipreparative LC/MS system. *J. Comb. Chem.* **2009**, *11*, 592–597.

159. M. Kagan, M. Chlenov, S. Melnikov, O. McConnell, A. C. Bach, G. Carter, A. Failli, T. J. Caggiano, J. S. Shumsky, D. Lubda, Normal-phase automated mass-directed HPLC purification of a pyrrolobenzodiazepine library with vasopressin agonist activity. *J. Comb. Chem.* **2009**, *11*, 704–719.

160. W. C. Ripka, G. Barker, J. Krakover, High-throughput purification of compound libraries. *Drug Discovery Today* **2001**, *6*, 471–477.

161. M. Maftouh, C. Granier-Loyaux, E. Chavana, J. Marini, A. Pradines, Y. V. Heyden, C. Picard, Screening approach for chiral separation of pharmaceuticals: Part III. Supercritical fluid chromatography for analysis and purification in drug discovery. *J. Chromatogr. A* **2005**, *1088*, 67–81.

162. T. A. Berger, K. Fogleman, T. Staats, P. Bente, I. Crocket, W. Farrell, M. Osonubi, The development of a semi-preparatory scale supercritical-fluid chromatograph for high-throughput purification of "combi-chem" libraries. *J. Biochem. Biophys. Methods* **2000**, *43*, 87–111.

163. J. Y. Lee, I. Im, T. R. Webb, D. McGrath, M. R. Song, Y. C. Kim, Combinatorial synthesis and biological evaluation of peptide-binding GPCR-targeted library. *Bioorg. Chem.* **2009**, *37*, 90–95.

164. I. Im, T. R. Webb, Y. D. Gong, J. I. Kim, Y. C. Kim, Solid-phase synthesis of tetrahydro-1,4-benzodiazepine-2-one derivatives as a β-turn peptidomimetic library. *J. Comb. Chem.* **2004**, *6*, 207–213.

165. M. W. Nowicki, L. B. Tulloch, L. Worralll, I. W. McNae, V. Hannaert, P. A. M. Michels, L. A. Fothergill-Gilmore, M. D. Walkinshaw, N. J. Turner, Design, synthesis and trypanocidal activity of lead compounds based on inhibitors of parasite glycolysis. *Bioorg. Med. Chem.* **2008**, *16*, 5050–5061.

166. G. R. Eldridge, H. C. Vervoort, C. M. Lee, P. A. Cremin, C. T. Williams, S. M. Hart, M. G. Goering, M. O'Neil-Johnson, L. Zeng, High-throughput method for the production and analysis of large natural product libraries for drug discovery. *Anal. Chem.* **2002**, *74*, 3963–3971.

167. S. K. Sarkar, R. S. Garigipati, J. L. Adams, P. A. Keifer, An NMR method to identify nondestructively chemical compounds bound to a single solid-phase-synthesis bead for combinatorial chemistry applications. *J. Am. Chem. Soc.* **1996**, *118*, 2305–2306.

168. R. Jelinek, A. P. Valente, K. G. Valentine, S. J. Opella, Two-dimensional NMR spectroscopy of peptides on beads. *J. Magn. Reson.* **1997**, *125*, 185–187.

4

DIVERSITY-ORIENTED SYNTHESIS

Kieron M. G. O'Connell, Warren R. J. D Galloway, Brett M. Ibbeson, Albert Isidro-Llobet, Cornelius J. O' Connor, and David R. Spring

4.1 INTRODUCTION

The deliberate synthesis of small molecule collections (libraries) that cover large areas of chemical space is described as diversity-oriented synthesis (DOS). These libraries are usually created with a view to discovering some novel biological activity but without a particular target in mind. Instead, it is hoped that the range of molecular architectures and potential bonding interactions present in the library can provide interesting and specific biological activity across a range of targets. This chapter will discuss the range of synthetic strategies applied to DOS, with particular reference to those carried out on solid phase.

4.2 SMALL MOLECULES AND BIOLOGY

The ability of small molecules to interact with biological macromolecules such as proteins in a selective, often reversible, and dose-dependent manner, and to exert specific effects, has led to them being regarded as powerful tools for the study of biological systems.[1,2] The use of small molecules in this manner to selectively perturb biological function underpins the whole of medicinal chemistry as well as forming the basis for the field of chemical genetics.[3]

Solid-Phase Organic Synthesis: Concepts, Strategies, and Applications, First Edition.
Edited by Patrick H. Toy and Yulin Lam.

The discovery of novel molecular entities or structural classes capable of these selective interactions represents a significant challenge. In cases where the biological target is well defined, rational design of potential ligands is often possible, particularly when the structure of a natural substrate is known. However, for other recondite disease states or if a novel mode of binding is sought, this is not possible, and the discovery of novel therapeutic agents largely relies on the screening of large collections of small molecules. Clearly, the composition, in terms of the structures of the molecules included in these collections, is an extremely important consideration.[4] The greater the structural variation between compounds, the higher the likelihood of achieving broad-ranging and distinct biological activity across the library.[5,6] The presence of multiple structural classes or scaffolds within a library being tested against a single target also increases the likelihood of discovering a molecule capable of binding in a novel manner.[7]

The molecules that comprise these libraries, or indeed any compound collections, may be obtained from natural (natural products) or nonnatural (chemical synthesis) sources. If all known natural products are taken into account, they represent an extremely diverse and complex collection of molecules. They also show a wide range of biological activities and have been used medicinally for millennia. However, despite this, it is not realistic or even desirable to produce large libraries consisting solely of natural products, predominately due to difficulties in sourcing, isolating, and identifying the bioactive components, as well as purifying and chemically derivatizing these often extremely complex molecular structures. Therefore, in terms of producing large numbers of compounds for biological screening, deliberate chemical synthesis is considered the most efficient approach.

With the advent of combinatorial chemistry in the 1990s, their came the use of split–pool techniques that, when combined with advances in automation, made possible the efficient synthesis of literally millions of compounds.[8] These libraries, however, were generally made up of broadly similar structures, with any variation between molecules resulting from appendage alteration of building blocks assembled around a common scaffold. Libraries of this sort have had limited success in the discovery of novel biologically active agents, a fact that is mainly attributed to the relative lack of diversity within the libraries.[9] Therefore, it is believed that the quality, in terms of structural complexity and diversity, of these libraries is as important, if not more so, than the number of the compounds synthesized.[4]

Many proprietary and commercially available libraries are synthesized in this combinatorial fashion and so suffer from these limitations. Another criticism of these libraries is that they may be too heavily biased toward certain predefined criteria such as the Lipinski rule of 5.[10] These rules for bioavailability have proved very useful in drug discovery; however, they may be unnecessarily limiting when it comes to the discovery of ligands for novel biological targets such as protein–protein interactions. Despite these limitations, many of the principles of combinatorial chemistry are shared by DOS.

A review from 1997 by Spaller et al. suggested that combinatorial libraries may be considered to fall into two categories: "focused" libraries where a number of closely related analogues based on a privileged structure are synthesized with a known target in mind, and "prospecting" libraries where an entirely new lead compound is sought and so the objective is to screen a large number of structurally varied molecules in the hope of finding a lead with a novel mode of action.[11] These so-called prospecting libraries can probably be considered to be the first examples of diversity-oriented syntheses as they were produced with similar aims of achieving high levels of structural variety and bioactive chemical space coverage. The term "diversity-oriented synthesis" was coined by Schreiber and coworkers and first appeared in the literature in 2000.[12,13] It was from this point that the ideas behind diversity-oriented synthesis began to become formalized in terms of forward synthetic analysis and strategies.

4.3 DIVERSITY-ORIENTED SYNTHESIS, TARGET-ORIENTED SYNTHESIS, AND COMBINATORIAL CHEMISTRY

The aim of efficiently synthesizing large numbers of structurally diverse compounds is not easy to realize for a number of reasons. Principal among these is the fact that the synthetic challenge of producing a successful DOS must be approached from the opposite direction to traditional chemical synthesis.[4,12] In both target-oriented synthesis (TOS) (of natural or nonnatural products) and focused library synthesis, a target structure is in mind at the beginning of the campaign. This structure is then broken down rationally to simpler starting materials through the well-established process of retrosynthetic analysis. In DOS, this is not possible and any synthetic analysis that can be carried out must be done so in the forward direction; that is, once starting materials are chosen, the reactions that can be carried out on them to generate diverse structures must be carefully considered. Generally, DOS strategies make use of complexity generating reactions to quickly build up molecular scaffolds and product–substrate relationships where the product of a given reaction then provides a suitable substrate for the next.[12] Figure 4.1 shows the synthetic strategies involved in TOS, focused library synthesis, and DOS.

Another obvious difficulty when carrying out diversity-driven synthesis is the fact that any given reaction will always provide the same structural feature, at least when carried out in an intermolecular fashion. For example, the coupling of an amine with a carboxylic acid will always form a peptide bond and a classic Diels–Alder reaction will always produce a

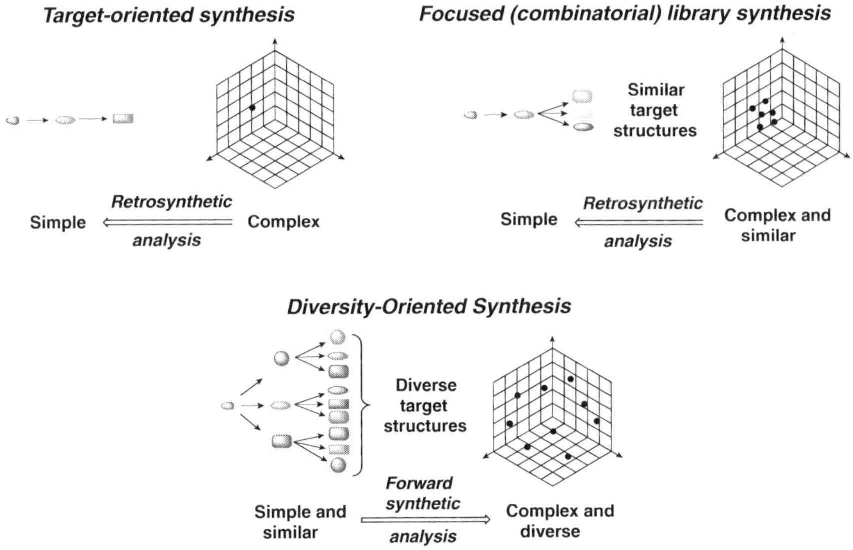

Figure 4.1. Planning strategies and end goals involved in target-oriented synthesis, focused library synthesis (combinatorial synthesis), and diversity-oriented synthesis. The first two approaches use retrosynthetic analysis to design the synthesis of target compounds. Diversity-oriented synthesis uses forward synthetic analysis to produce libraries that occupy diffuse regions of chemical space.

cyclohexene or cyclohexadiene. This means that from a diversity generating point of view if the same reactions are to be carried out on a number of substrates, these substrates must be chosen very carefully. The next thing to consider is exactly what constitutes diversity.

4.4 MOLECULAR DIVERSITY

The terms "diversity" and DOS are used widely and with some legitimacy in the literature to describe the synthesis of compound collections. While it is true that any collection of more than one compound must contain some degree of diversity as the molecules are not identical, this is not really in the spirit of DOS, where the goal should be the incorporation of the maximum possible structural diversity within a given compound collection. To provide a visually gratifying and conceptually simple comparison of the degrees of molecular diversity incorporated into compound collections, Spandl et al. proposed the consideration of the molecular diversity spectrum (Figure 4.2), where one extreme is a single-molecule target-oriented synthesis and the other extreme is complete chemical space coverage.[14] In this context, the aim of diversity-oriented synthesis is to produce a compound collection that could qualitatively be considered to be toward the right-hand side of the spectrum. The consideration of molecular diversity on a sliding scale or spectrum such as this is useful in as much as it shows that when planning a DOS, one should aim to achieve the highest possible level of structural diversity; however, it does not really do very much toward defining molecular diversity. It is not possible to define molecular diversity absolutely and analysis performed using any degree of human intuition will by necessity involve some degree of subjectivity. However, there are two main ways in which molecular diversity is usually considered: first, the synthetic strategy can be analyzed as there are only a finite number of ways that diversity can be incorporated synthetically, and second diversity can be analyzed through the computational analysis of library compounds and their relative positions in chemical space (see below).

From a synthetic point of view, when planning a diversity-oriented synthesis, four potential types of molecular diversity to be incorporated are consistently highlighted in the literature:[1,4,14,15]

1. *Appendage or Building Block Diversity*: Variation resulting from the choice of starting materials or "building blocks" used, usually resulting in the variation of R-groups around a single scaffold. This is the approach most frequently used, almost by definition, in combinatorial libraries.
2. *Functional Group Diversity*: Variation of the functional groups present in a molecule generally but also at specific sites within the gross structure. This gives

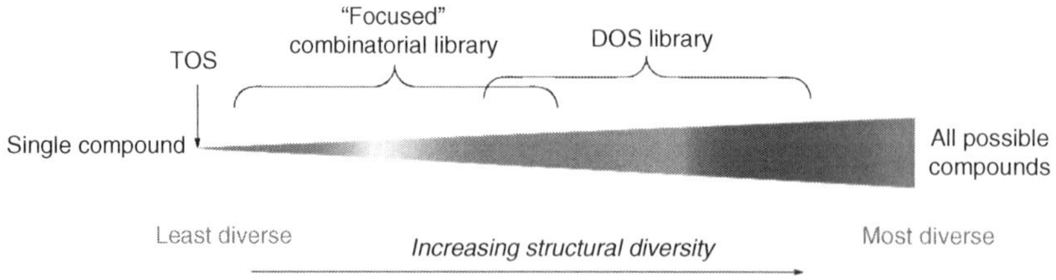

Figure 4.2. Molecular diversity spectrum: a representation of the relative degrees of molecular diversity achieved using TOS, focused library synthesis, and DOS.

the potential for interactions with different polar, apolar, or charged groups present in macromolecules.

3. *Stereochemical Diversity*: Variation in the orientation of functional groups and potential macromolecule-interacting elements. Clearly, this is very important as nature is a three-dimensional (3D) environment.
4. *Scaffold or Skeletal Diversity*: Variation in the overall molecular framework, typically considered to be variation in ring structures and other rigidifying elements, resulting in molecules with distinct scaffolds and consequently molecular shapes.

There is a widespread consensus that in terms of producing functionally (biologically) diverse molecules, the most important type of diversity that can be incorporated into a library is scaffold diversity.[1,16,17] The rationale behind this is that biomacromolecules are (on a molecular scale) large 3D environments with certain defined potential binding pockets and surfaces, and as such they will interact only with small molecules that have complementary 3D structure. Libraries that contain compounds based around distinct molecular scaffolds and so distinct 3D shape and structure should therefore cover a large range of potential binding partners.

4.4.1 Molecular Diversity and Chemical Space

Chemical space encompasses all theoretically possible compounds and is therefore to all intents and purposes infinite, limited only by the imagination of chemists and current synthetic methodology.[7,18,19] Molecules occupy discrete points and regions within chemical space based on their comparable physical and topological properties. The regions of chemical space that are populated by biologically active molecules are considered to comprise bioactive chemical space. Known biologically active molecules are usually aggregated into relatively small areas of chemical space and tend to be found in clusters of molecules that interact with particular targets. In the context of chemical space, it can be said that one of the aims of DOS is to discover novel regions of bioactive chemical space.

As chemical space theoretically comprises all possible compounds, it is clearly not possible to definitively measure it and the position of molecules within it. However, a representation of chemical space can be generated using large numbers of chemical descriptors. Such a representation is more correctly called multidimensional chemical descriptor space.[20] Chemical descriptors are usually based on the physical properties of a given molecule such as molecular weight and log *P* and also on topological features such as geometrical positions and orientations of bonds and groups within the molecule. The use of an algorithm based on a large number of chemical descriptors allows a molecule's position in that particular multidimensional chemical descriptor space to be calculated and this then provides an estimate of the position of a compound in the theoretical chemical space.

To give a visually accessible representation of this multidimensional descriptor space, it is necessary to use principal component analysis (PCA) to condense the multidimensional information into two- and three-dimensional scatter plots.[21] These plots provide the means to easily interpret the comparison of molecular diversity/chemical space coverage between given sets of compounds. They are not infallible, however, and it is conceivable that molecules' absolute and relative positions within the plot could vary depending on the chemical descriptors used. The use of large numbers of chemical descriptors though should minimize this. Figure 4.3 is an example of a chemical space analysis produced using chemical descriptors and PCA.

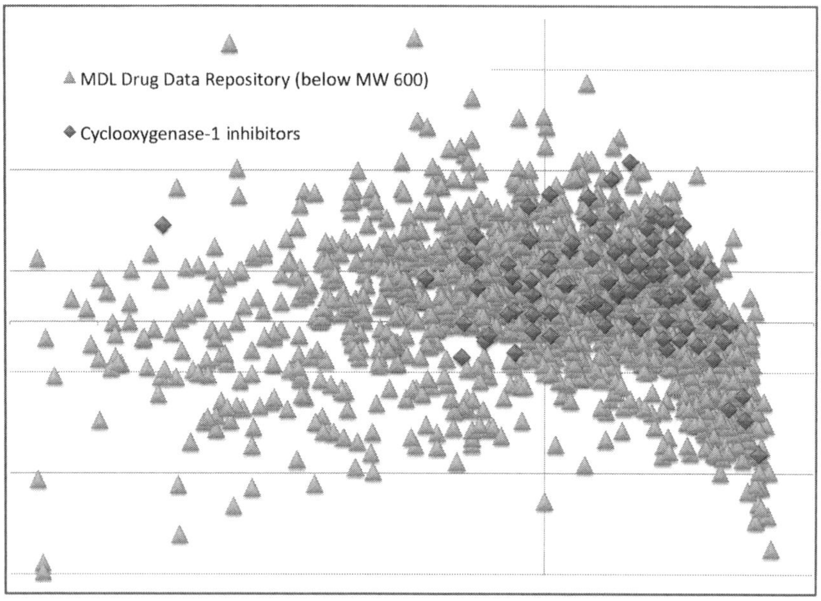

Figure 4.3. Chemical space analysis plot of cyclooxygenase-1 inhibitors (diamonds) and MDL Drug Repository compounds (triangles), created using chemical descriptors and principal component analysis. Plot shows that the cyclooxygenase-1 inhibitors occupy a fairly broad area of chemical space.

4.4.2 Synthetic Strategies for Creating Molecular Diversity

As mentioned previously, the challenge of creating molecular diversity is a considerable one, requiring strategies that differ from those used in the majority of traditional chemical synthesis. Since the conception of DOS, two distinct strategies for the generation of molecular and particularly skeletal diversity have been identified in the literature.[1] They are

1. The *reagent-based* approach, where subjecting a given molecule to a range of conditions allows the synthesis of a number of distinct compounds.
2. The *substrate-based* approach, where a number of starting materials containing "pre-encoded skeletal information" are transformed under the same reaction conditions into a range of different structures (Figure 4.4).

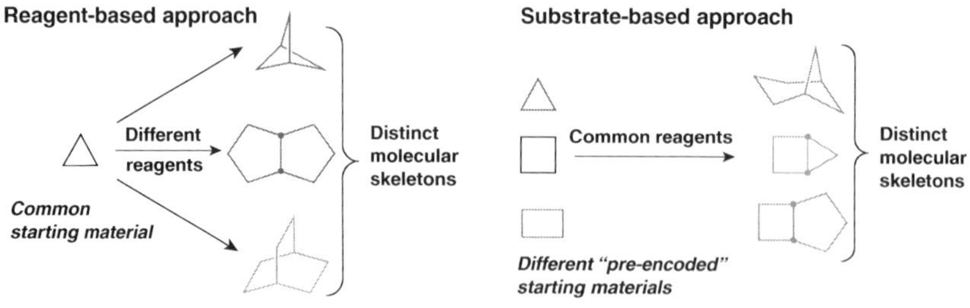

Figure 4.4. Two general strategies used to access chemical and skeletal diversity.

These strategies are not orthogonal to each other and many diversity-oriented syntheses will contain aspects of both. Reagent-based diversification, also known as a "*branching*" reaction pathway, can be used at any stage of a divergent synthesis, either during the early stages to produce diverse functionality or at the later stages to transform prefunctionalized molecules into distinct molecular scaffolds. There are generally considered to be two approaches to reagent-based diversification: (i) the use of the so-called pluripotent functionality where a single functional group can be transformed into distinct functional groups for further diversification or into distinct molecular scaffolds and (ii) the use of a densely functionalized molecule where different functional groups can be transformed under distinct reaction conditions, and this approach is usually used to pair functional groups and so create skeletal diversity. Substrate-based diversification is generally used in the later stages of a DOS to produce distinct molecular scaffolds by intramolecular reaction, and for this reason it is often referred to as a "folding" reaction pathway. While a DOS may contain aspects of both these approaches, generally the strategy used to produce skeletal diversity within the library is considered the most important.

Some of these ideas were further refined by Schreiber when he suggested the use of the *build/couple/pair* strategy for the production of optimal small molecule collections for biological screening.[22] In the build stage, the required, ideally chiral, starting materials are synthesized or obtained from the chiral pool. These starting materials are then coupled to produce densely functionalized molecules; multicomponent reactions are often used at this stage to couple three or more building blocks together. The pair stage then involves the intramolecular reaction of attached functional groups, usually under reagent control, to generate diverse molecular skeletons. This approach has been widely adopted in the literature.[23–25]

4.5 DIVERSITY-ORIENTED SYNTHESIS ON SOLID PHASE

Solid-phase technology has been a consistent feature in the synthesis of DOS libraries over the past 10 years and it is likely that this trend will continue for years to come. It is worth noting though that many elegant examples of diversity-oriented syntheses have been carried out using fluorous technology as a means for generic purification[26,27] and many more carried out purely in solution with purification achieved by traditional methods.[24,28–30] The remainder of this chapter, however, will focus for the most part on those examples that have employed solid-phase technology to their advantage.

4.5.1 Reagent-Based Strategies

In an early example of a DOS, Kwon et al. took advantage of solid-phase synthesis to achieve the synthesis of a library consisting of up to 29,400 discrete compounds comprising 10 distinct polycyclic skeletons (Scheme 4.1).[31] This strategy relied on the use of consecutive Diels–Alder reactions performed on polymer-bound trienes. The polymer-bound trienes were synthesized in three steps: the reaction of hydroxyl-substituted aryl aldehydes (**1**) with a macrobead attached alkyl silyl triflate (**2**), the reaction of the aldehydes with indium dust and 5-bromo-1,3-pentadiene, and finally mesylation and elimination of the resulting benzylic alcohols to furnish the desired trienes (**3**). These trienes were then reacted with a number of dienophiles displaying a range of creativities to give highly complex polycyclic structures.

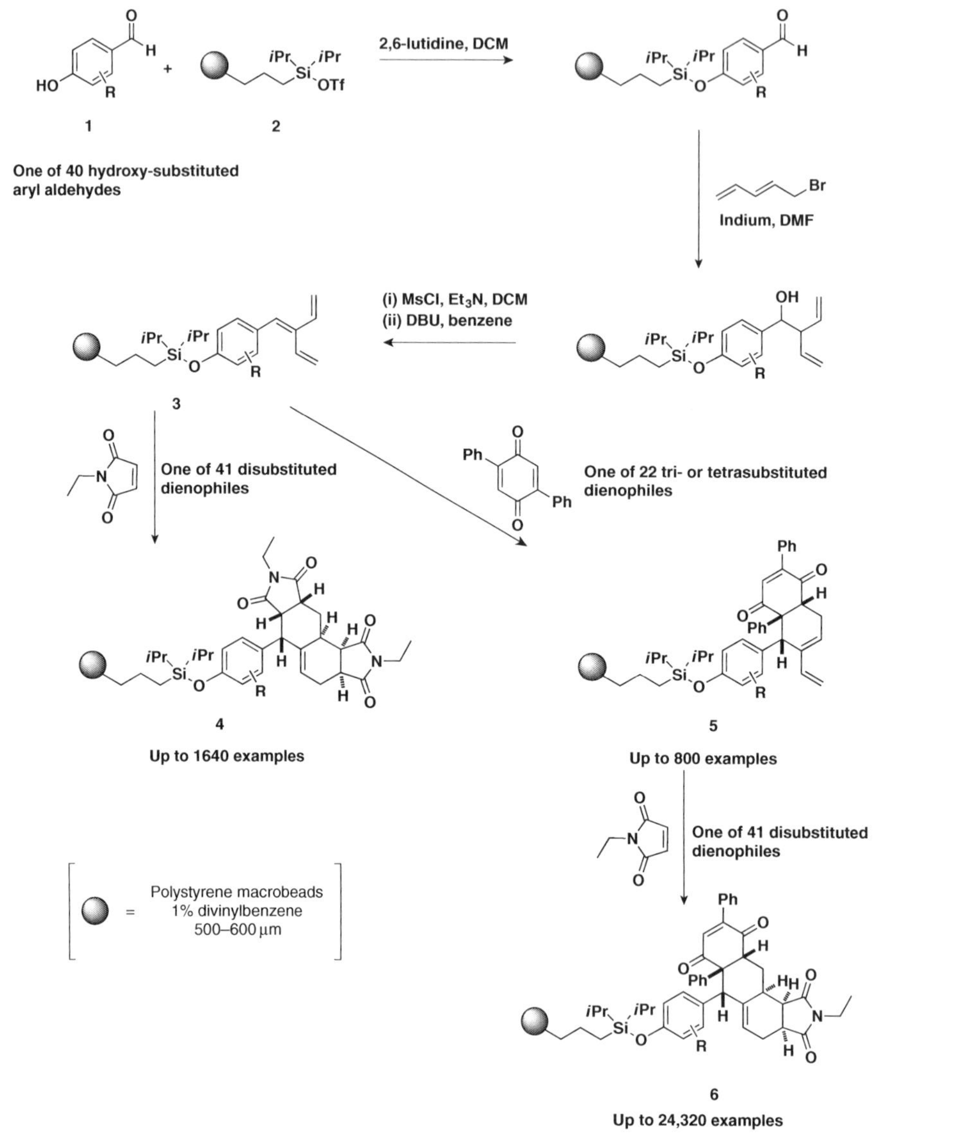

Scheme 4.1. Consecutive Diels–Alder approach to DOS employed by Kwon et al.

The authors found that by varying the level of steric hindrance associated with the dienophiles, they were able to modify the outcome of the potential double Diels–Alder reaction. When disubstituted dienophiles such as maleimides were used, the double Diels–Alder reaction occurred smoothly to yield a tetracyclic skeleton (**4**). If tri- or tetrasubstituted dienophiles were used, only one Diels–Alder reaction occurred to give a bicyclic diene-containing skeleton (**5**) that could either be treated as a discrete product or reacted further with a more reactive dienophile to again give a tetracyclic structure (**6**).

By combinatorial variation of the dienophiles and polymer-bound aldehydes used, the synthesis of 29,400 compounds was possible, broken down into 800 bicyclic dienes and 28,600 potential tetracycles. Further skeletal variation was achieved using different heterocyclic dienophiles for some examples. This work effectively illustrates the scale of synthetic campaigns that can be attempted on solid phase; however, by current standards, it shows a relatively low level of skeletal diversity for such a large number of compounds.[27,32]

Work by Taylor et al. from 2004 neatly illustrates another potential advantage of solid-phase synthesis; this work was based on the generation of reactive dihydroisoquinoline and dihydropyridine intermediates to be used in further reactions.[33] These species are known to be unstable when stored for any period of time, even under "seemingly inert" conditions; however, once attached to a solid support, they proved immune to degradation even after storage for 30 days at room temperature.

These intermediates, which were bound to a polymer support *via* a silicon-based linker similar to the one used by Kwon et al., were generated in a three-step reaction sequence. Both the dihydroisoquinoline and the dihydropyridine were synthesized following the same sequence and were subjected to identical subsequent reactions; however, for clarity only the dihydropyridine route will be discussed here. First, 3-(4-pyridyl)propan-1-ol was attached covalently to the macrobead using a standard procedure, the pyridine was then alkylated with 2-bromobenzyl bromide, and finally the resulting iminium salt was trapped with vinylmagnesium bromide. The resulting dihydropyridine (**7**) was then able to partake in a number of reactions, including a 1,3-dipolar cycloaddition with an electron-deficient azide followed by extrusion of molecular nitrogen to give **8**, a [2 + 2] cycloaddition with dimethyl acetylenedicarboxylate (DMDA) followed by ring expansion to give eight-membered ring **9,** and a Diels–Alder reaction with *N*-benzylmaleimide to generate bridged bicycle **10** (Scheme 4.2).

This methodology produced 12 distinct molecular skeletons, all of which possessed a pendant hydroxyl group resulting from cleavage of the solid support, which allowed the use of microarray technology for protein binding assays.

Work from 2008 by Thomas et al. used an imidazolidone linker to immobilize their enone substrate onto a silyl polystyrene support (Scheme 4.3).[34] Besides simplifying purification, this linker allowed two-point binding of a chiral catalyst for a subsequent Diels–Alder reaction and also gave the opportunity for divergent cleavage to increase the appendage diversity of the library. The exocyclic acyl group could be cleaved by hydrolysis, forming the acid, reduced to the aldehyde, esterified, or reacted with an amine to form an amide. The initial reagent-based strategy for this DOS was to treat the alkene as a pluripotent functional group that was able to partake in three initial branching reactions: [3 + 2] cycloaddition, dihydroxylation, and [4 + 2] cycloaddition. These initial products were then transformed into 242 discrete compounds consisting of 18 distinct molecular scaffolds, including a novel *cis–trans*-fused 7-5-7 tricycle (**11**) generated by ring-opening–ring-closing metathesis of a decorated norbornene (**12**).

The compounds produced in this library were screened for their effects on three strains of UK epidemic *Staphylococcus aureus*: methicillin-susceptible *S. aureus* (MSSA) and two strains of methicillin-resistant *S. aureus* (EMRSA 15 and EMRSA 16). Three compounds from the library were found to restrict the growth of these strains; the most potent compound "gemmacin" (Table 4.1) showed considerably higher activity against the strains of MRSA than the widely used antibiotics erythromycin and oxacillin. In the original DOS, gemmacin was made racemically, but the enantiomerically pure compounds were subsequently synthesized and showed comparable activity, with (−)-gemmacin being slightly more potent. Studies were carried out to discover the mode of action of (−)-gemmacin and the results of an assay to test the generation of reactive oxygen species proved positive, suggesting that gemmacin may act as a selective cell membrane disruptor. The discovery of gemmacin is a good example of how the DOS approach can be used to discover new structural classes of biologically active molecules.[34,35]

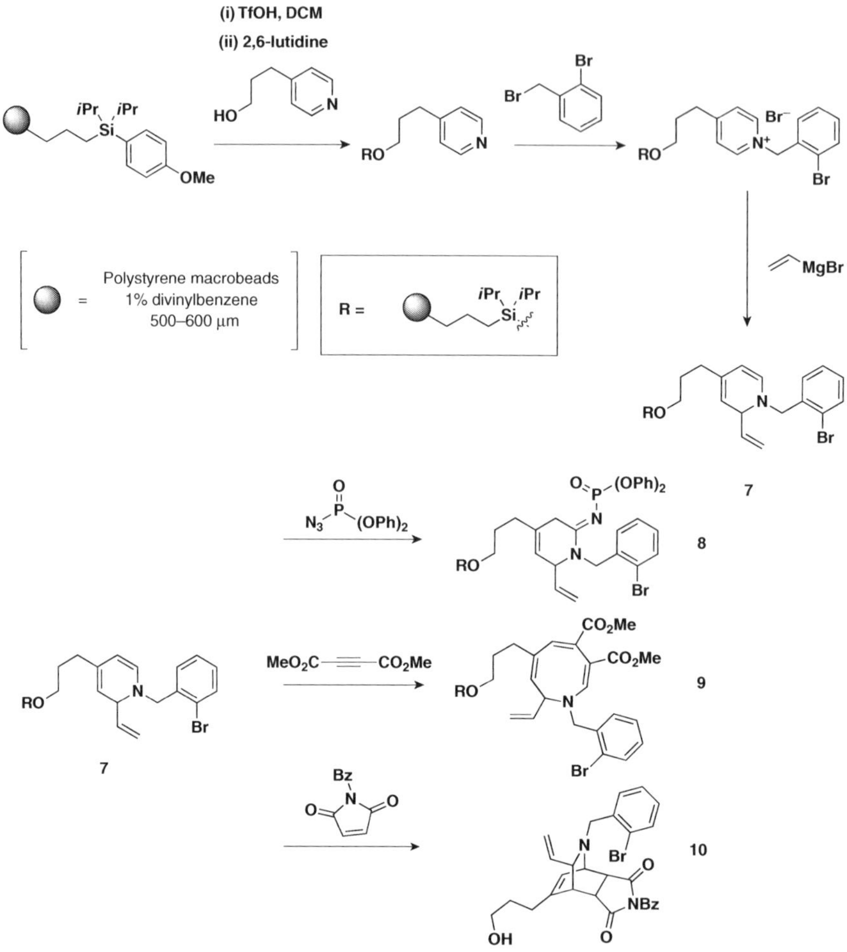

Scheme 4.2. Taylor et al.'s use of supported dihydropyridines in DOS.

4.5.2 Substrate-Based Strategies

In the first formal example of a substrate-based diversification strategy, Burke et al. made use of the Achmatowicz reaction to generate a range of molecular skeletons in a combinatorial fashion.[36] Their strategy was based on the use of macrobead-bound 2,5-disubstituted furan species; the functionalization of the attached alkyl chains then provided the necessary pre-encoded skeletal information for the "folding" of these furan species into vastly different structures. Under oxidative conditions, the relatively stable furans can be transformed into highly electrophilic enedione species. The pre-encoded skeletal information was manifested by the presence of one, two, or zero nucleophilic hydroxyl groups that could then trap the reactive intermediate intramolecularly to generate monocyclic, bridged bicyclic, and linear species, respectively (Scheme 4.4).

The enedione intermediate was generated under mild oxidative conditions using *N*-bromosuccinimide (NBS) and pyridinium *para*-toluenesulfonic acid (PPTS) and, once formed, the intramolecular reactions occurred spontaneously. The diol substrates underwent the expected NBS-mediated ring expansion followed by bicycloketalization to give [3.2.1]bicycle **13**, the single hydroxyl substrates underwent the same ring expansion

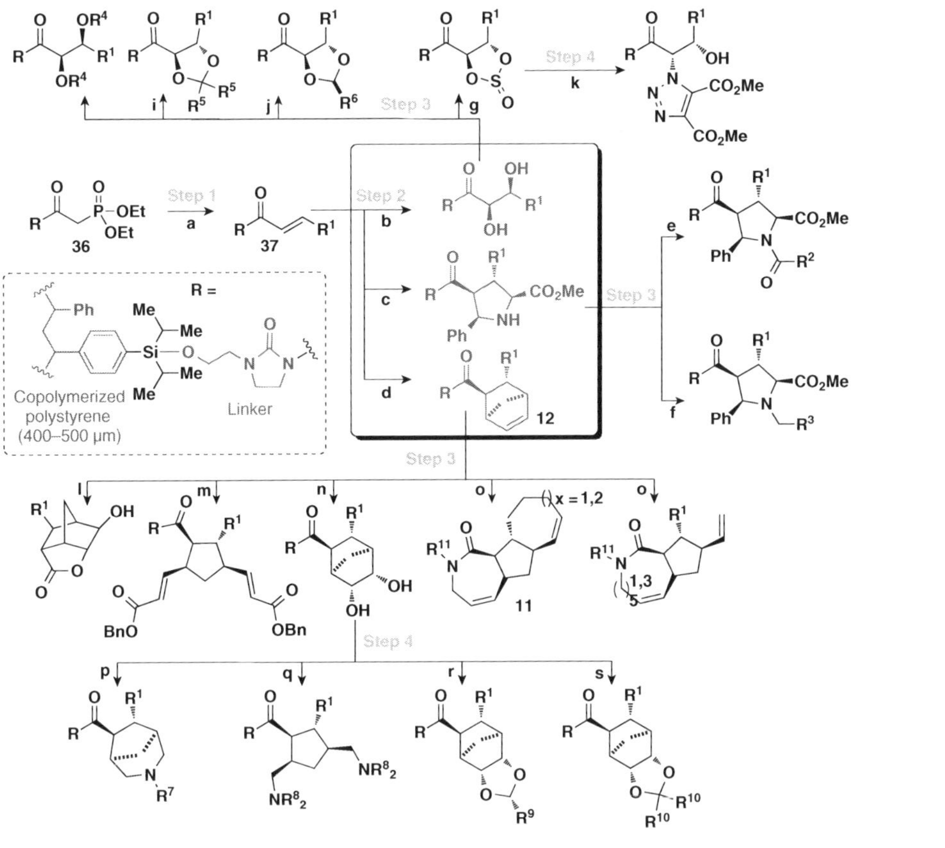

Scheme 4.3. Diversity-oriented synthesis of 242 compounds based of 18 discrete molecular frameworks by Thomas et al. Conditions: (a) LiBr, 1,8-diazabicyclo[5.4.0]undec-7-ene, R^1CHO, MeCN; (b) AD-mix, (DHQD)PHAL, THF:H_2O (1:1); (c) (R)-QUINAP, AgOAc, i-Pr_2NEt, THF, −78°C → 25°C; (d) chiral bis(oxazoline), $Cu(OTf)_2$, 3 MS, CH_2Cl_2, C_5H_6; (e) R^2COCl, DMAP, pyridine, CH_2Cl_2; (f) R^3CHO, BH_3 · pyridine, MeOH; (g) $SOCl_2$, pyridine, CH_2Cl_2, 40°C; (h) R^4Br, Ag_2O, CH_2Cl_2, 40°C; (i) $R^5C(O)R^5$, TsOH, DMF, 65°C; (j) R^6CHO, TsOH, DMF, 65°C; (k) NaN_3, DMF, 100°C, and then dimethyl acetylenedicarboxylate, PhMe, 65°C; (l) mCPBA, CH_2Cl_2 then MeOH, 65°C; (m) $CH_2{=}CHCO_2Bn$, PhMe, 120°C, Grubbs II, $CH_2{=}CH_2$; (n) OsO_4, NMO, $CH_3C(O)CH_3$:H_2O (10:1); (o) RNH_2, Me_2AlCl, PhMe, 120°C, then NaH, $R^{11}X$, DMF, THF, and finally PhMe, 120°C, Grubbs II, $CH_2{=}CH_2$; (p) $NaIO_4$, THF:H_2O (1:1) and then R^7NH_2, $NaB(OAc)_3H$, CH_2Cl_2; (q) $NaIO_4$, THF:H_2O (1:1) and then R^8NHR^8, $NaB(OAc)_3H$, CH_2Cl_2; (r) R^9CHO, DMF, TsOH, 60°C; (s) $R^{10}C(O)R^{10}$, DMF, TsOH, 60°C. DMF = *N,N*-dimethylformamide; THF = tetrahydrofuran; DMAP = *N,N*-dimethylaminopyridine; (DHQD)PHAL = hydroquinidine 1,4-phthalazinediyl diether; mCPBA = *meta*-chloroperbenzoic acid; Ts = *para*-toluenesulfonyl; Grubbs II = 1,3-(bis(mesityl)-2-imidazolidinylidene)dichloro(phenylmethylene)(tricyclohexylphosphine)ruthenium; NMO = 4-methylmorpholine *N*-oxide.

followed by PPTS-mediated dehydration to give alkylidene pyran-3-one **14**, and finally the acetylated substrates gave the corresponding *trans*-enediones (**15**) resulting from furan ring opening followed by olefin isomerization. Following this strategy and using combinatorial variation of building blocks, they were able to produce a library of ~1260 compounds.

TABLE 4.1. Gemmacin and the Comparable effects of the Enantiomers of Gemmacin, Erythromycin, and Oxacillin on Three Strains of *S. aureus*

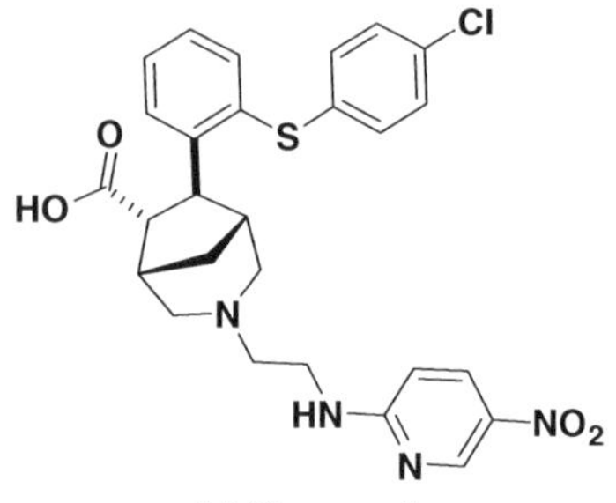

(–)-Gemmacin

	MIC (μg/mL)		
	MSSA	EMRSA 15	EMRSA 16
(±)-Gemmacin	2	16	32
(−)-Gemmacin	Not determined	8	16
(+)-Gemmacin	Not determined	16	32
Erythromycin	0.5	>64	>64
Oxacillin	0.5	>32	>32

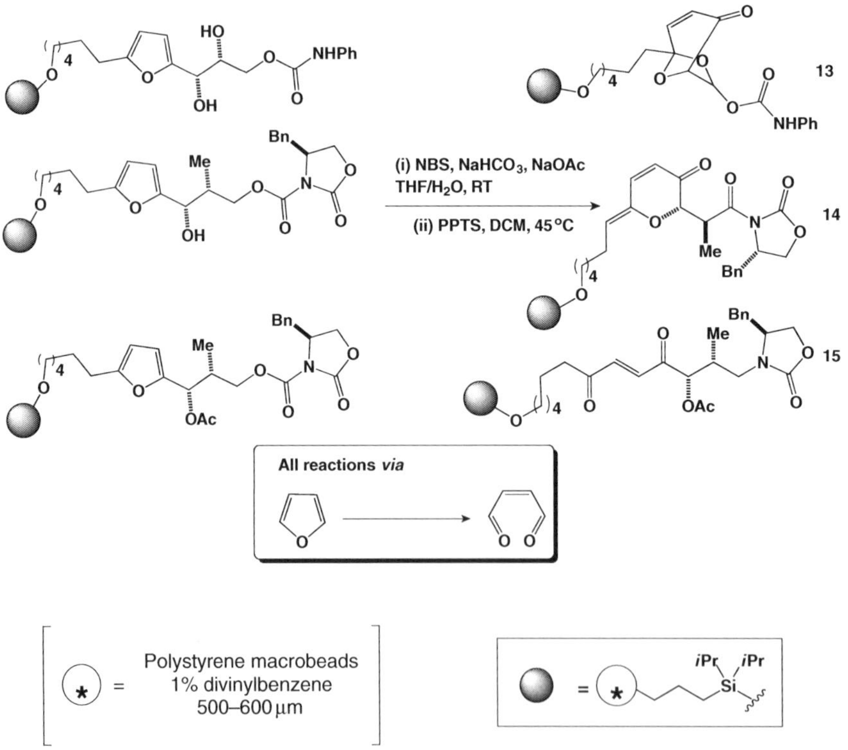

Scheme 4.4. Use of the Achmatowicz reaction in a folding pathway for DOS by Burke et al.

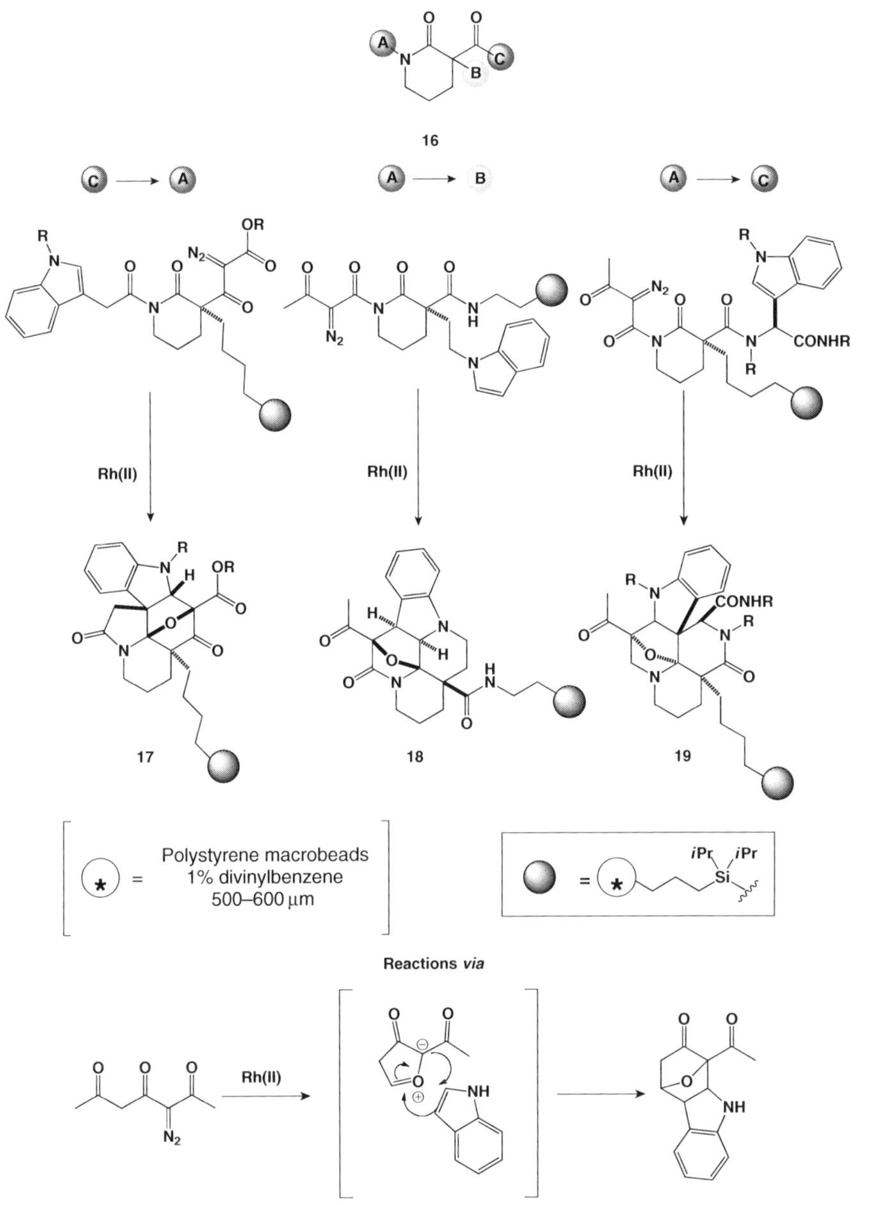

Scheme 4.5. Oguri and Schreiber's DOS of polycyclic alkaloid scaffolds using a folding pathway based on an intramolecular 1,3-dipolar cycloaddition.

Another outstanding example of a folding reaction pathway carried out on solid phase comes from Oguri and Schreiber's work on the DOS of polycyclic indole alkaloid-like scaffolds (Scheme 4.5).[37] By attaching different reactive groups, an α-diazocarbonyl, an indole, and a solid support, at three different points (A, B, and C) around a common piperidinone scaffold (**16**), they were able to create a range of pentacyclic scaffolds (**17–19**) by pairing the reactive groups attached to different positions. The pairing reaction used in this case was an extremely effective complexity generating reaction, based on methodology developed by Padwa and Mejia-Oneto.[38] The reaction involves the initial generation of a rhodium carbenoid from the α-diazocarbonyl species, and this carbenoid then cyclizes with an adjacent carbonyl group to give a cyclic carbonyl ylid that is able to undergo 1,3-dipolar cycloaddition with the electron-rich 2,3-double bond of the attached indole moieties.

4.5.3 Build/Couple/Pair Strategies

The above work by Oguri and Schreiber can be considered to be an example of the build/couple/pair strategy as the initial building blocks A, B, and C must first be built, then coupled to the piperidinone template, and finally the reactive functionality paired intramolecularly to yield the products. Another example of the build/couple/pair strategy incorporating a folding pathway can be found in the work of Mitchell and Shaw (Scheme 4.6).[39]

In their work, the build phase involved the synthesis of a solid-supported methoxy oxazole (**20**) and various azide-decorated aromatic aldehydes and benzyl bromides. The methoxy oxazole was then able to take part in an enantioselective Suga–Ibata reaction with

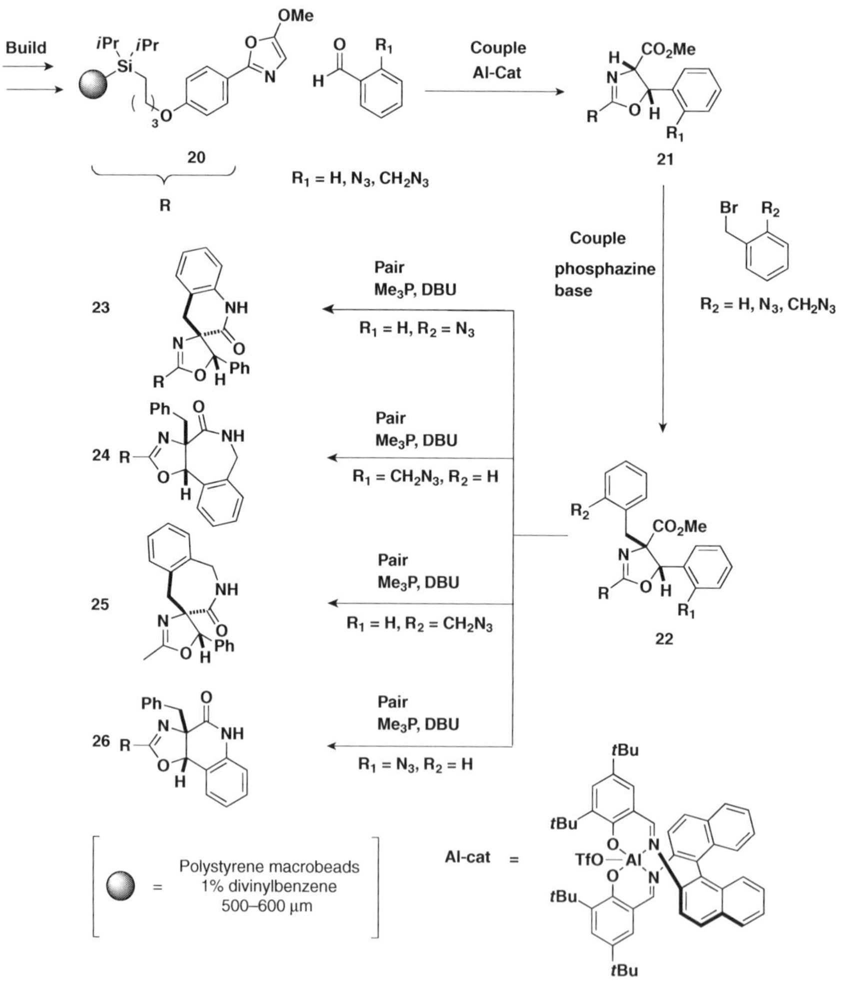

Scheme 4.6. Mitchell and Shaw's folding pathway to produce fused and spirotricyclic compounds.

the aryl aldehydes to give **21**, the enolate of which was then alkylated with the benzyl bromides to give **22**. These substrates were then subjected to Staudinger-type reductive cyclization conditions to give tricycles **23–26**. This folding pathway is appealing from a skeletal diversity generating standpoint because depending on how the azide functionality is attached, the process can generate fused and spirocyclic compounds that have very different 3D structures and shapes. The authors also draw attention to the fact that the relative configurations of six- and seven-membered ring lactams give rise to significantly different dihedral angles associated with the two substituents attached to the oxazoline core of the final structures, resulting in further 3D chemical space coverage within the compound collection. These two examples are illustrations of the use of the build/couple/pair strategy in combination with a substrate-based approach to skeletal diversity construction.

A particularly elegant example of the build/couple/pair strategy combined with reagent-based skeletal diversity construction can be found in the solution-phase work carried out by Comer et al. (Scheme 4.7).[28] Their strategy involved the synthesis of a number of substituted β-nitrostyrenes and alkylated 1,3-dicarbonyls in the build phase that were then coupled by enantioselective Michael addition of the dicarbonyls to the β-nitrostyrenes using a cinchona alkaloid-derived organocatalyst to give densely functionalized molecules such as **27** and **28**.

They were then able to generate skeletal diversity by two different reagent-based strategies, by pairing different functionality orthogonally and pairing the same functional groups using different reagents. The functional groups available for pairing were (depending on the particular substrate) alkene, alkyne, and nitro and ester groups. The alkene and alkyne groups could be paired by enyne metathesis to give conjugated dienes such as

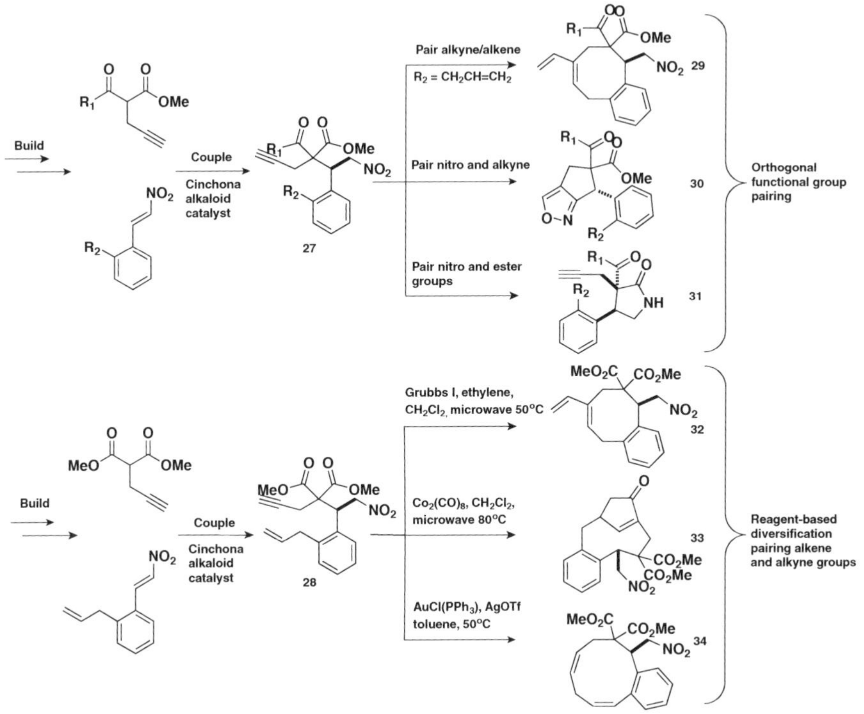

Scheme 4.7. Functional group pairing strategy used by Comer et al.

29, the nitro and alkyne groups could be paired by cycloaddition to give isoxazole-containing bicycles such as **30**, and the nitro and ester groups could be paired to give lactams such as **31**.

In addition to the already mentioned enyne metathesis to give **32**, the enyne functionality in **28** could also be paired using a cobalt-mediated Pauson–Khand reaction to give **33** and gold(I)-catalyzed cycloisomerization to give **34**. These reactions constitute an attractive branching pathway, and enyne cyclizations such as these are widely used in DOS. Another point to note about this particular pathway is that each of these products can be used in further reactions: **32** can undergo Diels–Alder reaction, the α,β-unsaturated ester **33** can undergo conjugate addition, and the 10-membered ring of cycloisomerization product **34** can be fused to a five-membered lactam by the subsequent pairing of the ester and nitro groups.

4.6 DIVERSITY-ORIENTED SYNTHESIS AROUND PRIVILEGED SCAFFOLDS

The above examples cover the range of planning strategies that are generally applied to "pure" diversity-oriented syntheses both in solid phase and in solution. Another planning strategy widely used is DOS around privileged scaffolds.[40–42] In this approach, divergent synthesis is applied around a core structure or scaffold with well-established biological activity. In many ways, DOS around a privileged scaffold is somewhat of a contradiction in terms as clearly the overall diversity of any library produced will be limited by the presence of a single core structure in all the compounds synthesized. Therefore, it is fair to say that rather than being purely diversity driven, these libraries can be considered to be more focused, both in their synthetic planning and end goals, than those produced from more authoritarian DOS. Despite this, DOS around a privileged scaffold is an effective way of exploring areas of prevalidated chemical space, with the aim of achieving increased potency and selectivity for a given target. If one were to consider the molecular diversity spectrum, then the libraries produced from this approach would be expected to lie somewhere between traditional combinatorial and DOS libraries. However, there are many situations where the DOS around privileged scaffold approach is especially applicable.

A recent example of DOS around a privileged scaffold can be found in the work of Park and coworkers (Scheme 4.8).[43] In a follow-up to previous work on DOS around the benzopyran scaffold,[44] they synthesized a library of 434 compounds on a solid support, the average purity of which, with no further purification, was 85%. Their synthetic effort began with the synthesis of eight distinct solid-supported chromanones (**35**), and these compounds differed in the substitution around the aromatic ring, and consequently the position of the linker unit, and also in the substituent R_1 that could comprise geminal dimethyl groups or a spirocyclopentane moiety. These compounds then underwent triflation to generate vinyl triflates (**36**), which were then subjected to several palladium-catalyzed transformations: Suzuki coupling (path A), Stille coupling with tributyl(vinyl)tin followed by a subsequent Diels–Alder reaction with substituted maleimides (path B), and Negishi coupling with an alkynyl zinc species followed by copper-catalyzed azide–alkyne cycloaddition with a range of azides (path C). The Diels–Alder products (**37**) were then able to undergo substrate-controlled asymmetric hydrogenation (path D) and aromatization (path E). Following this scheme, they generated five core molecular scaffolds based on the benzopyran motif; further appendage diversity around these scaffolds was then achieved through the choice of coupling partner.

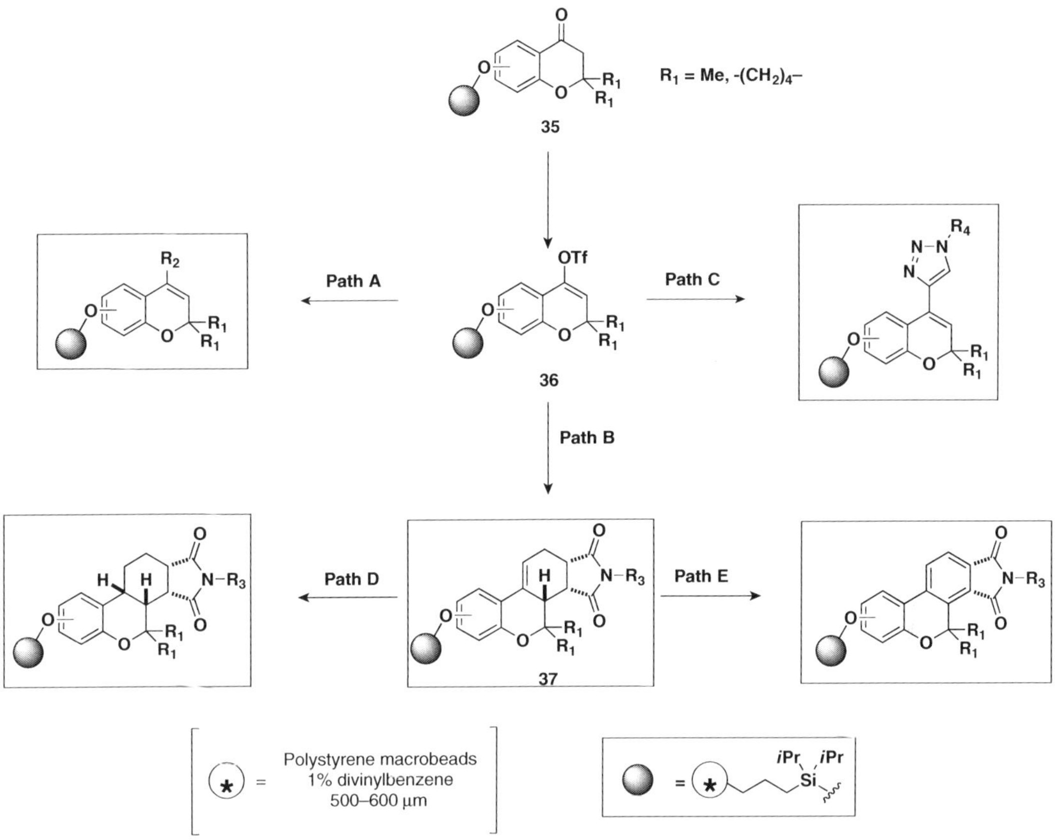

Scheme 4.8. DOS around a privileged benzopyran scaffold.

4.7 DIVERSITY LINKER UNITS IN SOLID-PHASE ORGANIC SYNTHESIS

The majority of linker units used in the examples discussed in this chapter can be cleaved, under a single set of conditions, to leave behind a single functional group, usually a hydroxyl group. However, in the work of Thomas et al., the linker could be cleaved in four different ways to incorporate further appendage diversity into the library; therefore, it can be considered to be an example of a *diversity linker unit*.[45] The use of such linkers could be of great value to the field of DOS as introducing further molecular diversity and potential complexity at the cleavage stage of a solid-phase library synthesis would add considerably to the overall step efficiency and economy of the process.

The linker unit used by Thomas et al. was fairly limited in the degree of diversity that it could introduce; however, it is possible to generate a high degree of diversity, even including scaffold diversity, during the cleavage of certain linker units. This is well illustrated by the triazine-based linker **38** developed by Brase and coworkers. Treatment of this linker with TFA generates a highly reactive electrophilic aryl diazonium salt that can be intercepted by a range of nucleophiles to generate a diverse array of structures including aryl cycloalkanes and benzotriazoles (Scheme 4.9).[46,47]

Another class of linkers or more specifically cleavage strategies with potential application to DOS are those strategies that involve cleavage with concomitant ring closure

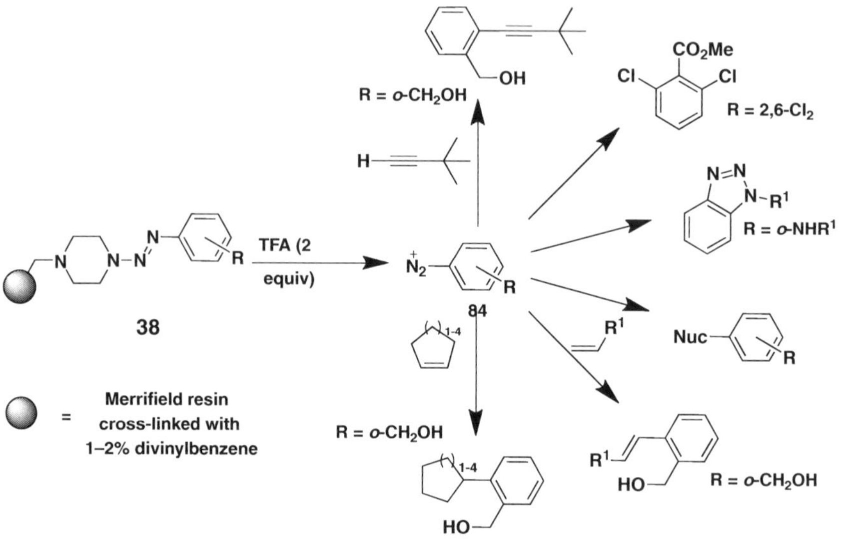

Scheme 4.9. Diversity linker unit; cleavage produces a highly electrophilic diazonium species that can be intercepted by a range of nucleophiles.

to give cyclic structures. An example of such a process can be found in the work of Veerman et al., who used ring-closing metathesis (RCM) to release linear amides from solid support, while simultaneously synthesizing six- and seven-membered ring lactams (Scheme 4.10).[48] This strategy is valuable for DOS as it allows the incorporation of further skeletal diversity into a library of compounds.

4.8 CONCLUSIONS

In the 10 years since the DOS concept was first formally described, a number of imaginative strategies for the creation of libraries of structurally complex and diverse molecules have been developed. The synthesis of libraries that cover large areas of chemical space has proved a challenging yet rewarding endeavor, with extremely diverse compound collections being successfully created. Over the years, solid-phase synthesis has proved to be a useful enabling technology for DOS, allowing multistep syntheses to be performed efficiently, and any purification necessary is performed simply and generically. Indeed, all the largest DOS libraries produced have taken advantage of solid-phase techniques. The creative use of

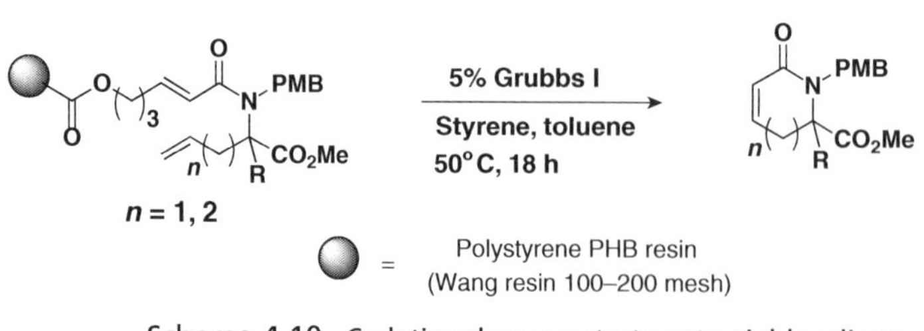

Scheme 4.10. Cyclative cleavage strategy to yield cyclic products by RCM.

solid-phase synthesis and linker units has allowed syntheses that would otherwise not be possible and also provided additional handles for reactivity on immobilized substrates.

In the future, the ideas that underpin DOS, such as maximal chemical space coverage and efficient complex structure generation, will remain. However, we may also see DOS being used in a more focused way, directed toward the synthesis of novel or unusual chemical structures and architectures, and used more in the field of fragment-based drug discovery. Whatever directions DOS takes in the future, it is likely that solid-phase synthesis will remain integral to the field.

REFERENCES

1. M. D. Burke, S. L. Schreiber, *Angew. Chem. Int. Ed.* **2004**, *43*, 46–58.
2. D. S. Tan, *Nat. Chem. Biol.* **2005**, *1*, 74–84.
3. D. R. Spring, *Chem. Soc. Rev.* **2005**, *34*, 472–482.
4. D. R. Spring, *Org. Biomol. Chem.* **2003**, *1*, 3867–3870.
5. W. H. Sauer, M. K. Schwarz, *J. Chem. Inf. Comput. Sci.* **2003**, *43*, 987–1003.
6. W. R. J. D. Galloway, D. R. Spring, *Expert Opin. Drug Discov.* **2009**, *4*, 467–472.
7. C. Lipinski, A. Hopkins, *Nature* **2004**, *432*, 855–861.
8. D. S. Tan, M. A. Foley, M. D. Shair, S. L. Schreiber, *J. Am. Chem. Soc.* **1998**, *120*, 8565–8566.
9. S. Borman, *Chem. Eng. News* **2004**, *82*, 32–40.
10. C. A. Lipinski, F. Lombardo, B. W. Dominy, P. J. Feeney, *Adv. Drug Delivery Rev.* **1997**, *23*, 3–25.
11. M. R. Spaller, M. T. Burger, M. Fardis, P. A. Bartlett, *Curr. Opin. Chem. Biol.* **1997**, *1*, 47–53.
12. S. L. Schreiber, *Science* **2000**, *287*, 1964–1969.
13. D. Lee, J. K. Sello, S. L. Schreiber, *Org. Lett.* **2000**, *2*, 709–712.
14. R. J. Spandl, A. Bender, D. R. Spring, *Org. Biomol. Chem.* **2008**, *6*, 1149–1158.
15. R. J. Spandl, M. Diaz-Gavilan, K. M. O'Connell, G. L. Thomas, D. R. Spring, *Chem. Rec.* **2008**, *8*, 129–142.
16. J. P. Kennedy, L. Williams, T. M. Bridges, R. N. Daniels, D. Weaver, C. W. Lindsley, *J. Comb. Chem.* **2008**, *10*, 345–354.
17. A. A. Shelat, R. K. Guy, *Nat. Chem. Biol.* **2007**, *3*, 442–446.
18. T. I. Oprea, J. Gottfries, *J. Comb. Chem.* **2001**, *3*, 157–166.
19. T. I. Oprea, *Curr. Opin. Chem. Biol.* **2002**, *6*, 384–389.
20. S. Fergus, A. Bender, D. R. Spring, *Curr. Opin. Chem. Biol.* **2005**, *9*, 304–309.
21. S. J. Haggarty, *Curr. Opin. Chem. Biol.* **2005**, *9*, 296–303.
22. T. E. Neilsen, S. L. Schreiber, *Angew. Chem., Int. Ed.* **2008**, *47*, 48–56.
23. T. Luo, S. L. Schreiber, *J. Am. Chem. Soc.* **2009**, *131*, 5667–5674.
24. T. Uchida, M. Rodriquez, S. L. Schreiber, *Org. Lett.* **2009**, *11*, 1559–1562.
25. A. Zhou, D. Rayabarapu, P. R. Hanson, *Org. Lett.* **2009**, *11*, 531–534.
26. E. E. Wyatt, S. Fergus, W. R. Galloway, A. Bender, D. J. Fox, A. T. Plowright, A. S. Jessiman, M. Welch, D. R. Spring, *Chem. Commun.* **2006**, 3296–3298.
27. D. Morton, S. Leach, C. Cordier, S. Warriner, A. Nelson, *Angew. Chem., Int. Ed.* **2009**, *48*, 104–109.
28. E. Comer, E. Rohan, L. Deng, J. A. J. Porco, *Org. Lett.* **2007**, *9*, 2123–2126.
29. M. Diaz-Gavilan, W. R. Galloway, K. M. O'Connell, J. T. Hodkingson, D. R. Spring, *Chem. Commun.* **2010**, *46*, 776–778.

30. D. Pizzirani, T. Kaya, P. A. Clemons, S. L. Schreiber, *Org. Lett.* **2010**, *12*, 2822–2825.
31. O. Kwon, S. B. Park, S. L. Schreiber, *J. Am. Chem. Soc.* **2002**, *124*, 13402–13404.
32. W. R. J. D. Galloway, M. Diaz-Gavilan, A. Isidro-Llobet, D. R. Spring, *Angew. Chem. Int. Ed.* **2009**, *48*, 1194–1196.
33. S. J. Taylor, A. M. Taylor, S. L. Schreiber, *Angew. Chem. Int. Ed.* **2004**, *43*, 1681–1685.
34. G. L. Thomas, R. J. Spandl, F. G. Glansdorp, M. Welch, A. Bender, J. Cockfield, J. A. Lindsay, C. Bryant, D. F. Brown, O. Loiseleur, H. Rudyk, M. Ladlow, D. R. Spring, *Angew. Chem. Int. Ed.* **2008**, *47*, 2808–2812.
35. A. Robinson, G. L. Thomas, R. J. Spandl, M. Welch, D. R. Spring, *Org. Biomol. Chem.* **2008**, *6*, 2978–2981.
36. M. D. Burke, E. M. Berger, S. L. Schreiber, *Science* **2003**, *302*, 613–618.
37. H. Oguri, S. L. Schreiber, *Org. Lett.* **2005**, *7*, 47–50.
38. J. M. Mejia-Oneto, A. Padwa, *Org. Lett.* **2004**, *6*, 3241–3244.
39. J. M. Mitchell, J. T. Shaw, *Angew. Chem. Int. Ed.* **2006**, *45*, 1722–1726.
40. A. Reayi, P. Arya, *Curr. Opin. Chem. Biol.* **2005**, *9*, 240–247.
41. C. Zhou, A. V. Dubrovsky, R. C. Larock, *J. Org. Chem.* **2006**, *71*, 1626–1632.
42. M. E. Welsch, S. A. Snyder, B. R. Stockwell, *Curr. Opin. Chem. Biol.* **2010**, *14*, 347–361.
43. S. Oh, H. J. Jang, S. K. Ko, Y. Ko, S. B. Park, *J Comb Chem*, *12*, 548–558.
44. S. K. Ko, H. J. Jang, E. Kim, S. B. Park, *Chem. Commun.* **2006**, 2962–2964.
45. P. J. H. Scott, P. G. Steel, *Eur. J. Org. Chem.* **2006**, 2251–2268.
46. S. Brase, *Acc. Chem. Res.* **2004**, *37*, 805–816.
47. S. Brase, M. Schroen, *Angew. Chem. Int. Ed.* **1999**, *38*, 1071–1073.
48. J. J. N. Veerman, J. H. van Maarseveen, G. M. Visser, C. G. Kruse, H. E. Schoemaker, H. Hiemstra, F. P. J. T. Rutjes, *Eur. J. Org. Chem.* **1998**, 2583–2589.

5

DIVERSITY-ORIENTED SYNTHESIS OF PRIVILEGED HETEROCYCLES USING DIVERGENT STRATEGY

Seung Bum Park and Jonghoon Kim

5.1 INTRODUCTION

Since the completion of the Human Genome Project in 2003, biomedical communities have been focusing on the elucidation of gene functions and the associated control of gene products by small molecule modulators,[1] which led to the birth of a new interdisciplinary research field the chemical biology.[2] The core research areas in chemical biology are the systematic identification of small molecule modulators that act as perturbing agents in biological systems and the use of these agents for the control of specific gene products.[3] Therefore, there is a great demand for the development of drug-like compound library that covers a wide range of molecular diversity for the identification of specific small molecule modulators. Natural products and their analogues have been extensively used as therapeutic agents or functional modulators of specific biological processes; approximately 60% of anticancer agents and 75% of antibiotics in the pharmaceutical market are derived from natural products. In fact, the rich bioactivity of natural products must be caused by the selection process.

Natural products have been created by various organisms as secondary metabolites and the organisms with these cytotoxic chemical weapons can survive in the biochemical warfare over billion years. The organisms we observe today are the survival of the fittest, and natural products isolated from these organisms can have such a diverse biological activity, especially excellent cytotoxicity: that is the reason why most of anticancer agents and antibiotics have originated from natural products. However, natural products were not as effective as in other therapeutic areas such as type 2 diabetes, depression, osteoporosis:

Solid-Phase Organic Synthesis: Concepts, Strategies, and Applications, First Edition.
Edited by Patrick H. Toy and Yulin Lam.

that is one of main reasons for a high demand for the high-throughput synthesis of natural product-like small molecules and subsequent high-throughput evaluation on their biological activities as drug candidates. Another reason is that the demand for the small molecule collection grew faster than the collection of natural compounds isolated from various organisms, leading to the construction of an efficient platform for the new chemical entities to overcome the limited availability and accessibility of small molecule collections.

To address these demands, combinatorial techniques were introduced in 1990s and utilized for the synthesis of a large number of compounds. Combinatorial chemistry contributed to the efficiency of drug discovery in the late stage via a fast construction of focused library for structure–activity relationship study. However, the increase in the size of compound collections did not improve the increase in the overall number of newly approved new chemical entities: A huge size of small molecule library was constructed by routine synthetic exercise with a limited number of core skeleton, but these collections of small molecule failed miserably in populating chemical space or more precisely the "multidimensional descriptor space," defined as the total descriptor space that encompasses all possible small molecules. In fact, the estimated small molecule space is greater than 10^{63} distinct molecules, a number that is not accessible by synthesis. More important, only a small portion of chemical space is relevant to the biological activity.[4]

To overcome these drawbacks, the organic chemistry community came up with a new synthetic strategy—diversity-oriented synthesis (DOS)—for efficiently populating the chemical space with skeletally and stereochemically diverse small molecules synthesized by complexity-generating reactions starting from simple building blocks.[5] The chemical synthesis endeavors for efficient access to new core skeletons are designed to explore the biologically relevant areas in chemical space. The DOS strategy for constructing natural product-like and drug-like small molecules has been proven to be effective for the discovery of bioactive small molecule modulators and therapeutic agents.[6] In addition to DOS concept, Waldmann and coworkers also claimed that a natural product (a biologically validated chemical structure) is a good starting point and stated the importance of the combinatorial construction of new chemical entities around the privileged substructure derived from natural products: they designated this approach as biology-oriented synthesis (BIOS).[7]

Many research groups including mine have been focused on populating the chemical space through the construction of novel chemical structures embedded with privileged substructures that are commonly found structural motifs in a wide variety of bioactive natural products and therapeutic agents.[8] Incorporation of privileged substructures into these core skeletons can result in the enhanced possibility of potential interaction between small molecules and various biopolymer molecules because the privileged substructures might have some preference to protein of interests due to the evolutionary similarity in their 3D structural modules. Moreover, we focused on the construction of rigid polyheterocyclic skeletons embedded with privileged substructures, with the aim of producing compounds that are selective to specific gene products or biological pathways; we designated this approach as a privileged substructure-based DOS (pDOS) strategy. We envision this selectivity due to prepaid entropic penalty imposed on the rigid core skeletons having limited conformational flexibility. On the basis of this background, we introduce some examples in this chapter for the construction of various polyheterocycles through the incorporation of privileged substructures such as diazabicycles, β-carbolines, oxopiperazines, and benzodiazepines through a divergent approach from a cyclic iminium as a single key intermediate along with pDOS strategy.

5.2 DIVERGENT SYNTHESIS OF NATURAL PRODUCT-LIKE POLYHETEROCYCLES USING A CYCLIC IMINIUM AS A SINGLE KEY INTERMEDIATE

Our research group developed practical solid-phase strategies for the divergent synthesis of various heterocycles such as diazabicycle,[9] tetrahydro-β-carboline,[10] Δ^5-2-oxopiperazine,[11] and tetrahydro-1,4-benzodiazepine[12] through cyclic iminium as a common intermediate. As shown in Figure 5.1, four novel core skeletons synthesized by this divergent pDOS approach were derived from the structures of bioactive natural products or therapeutic agents. (1) Diaza-bridged bicyclic structure is the simplified structural motif from naturally occurring bioactive polycyclic products such as yondelis, saframycin A, and cribrostatin V (see Figure 5.1a). The recombination of diaza-bridged bicyclic motif with other privileged substructures, L-DOPA (L-3,4-dihydroxyphenylalanine) and indole, allows the construction of novel polyheterocyclic core skeletons. (2) Tetrahydro-β-carboline is also

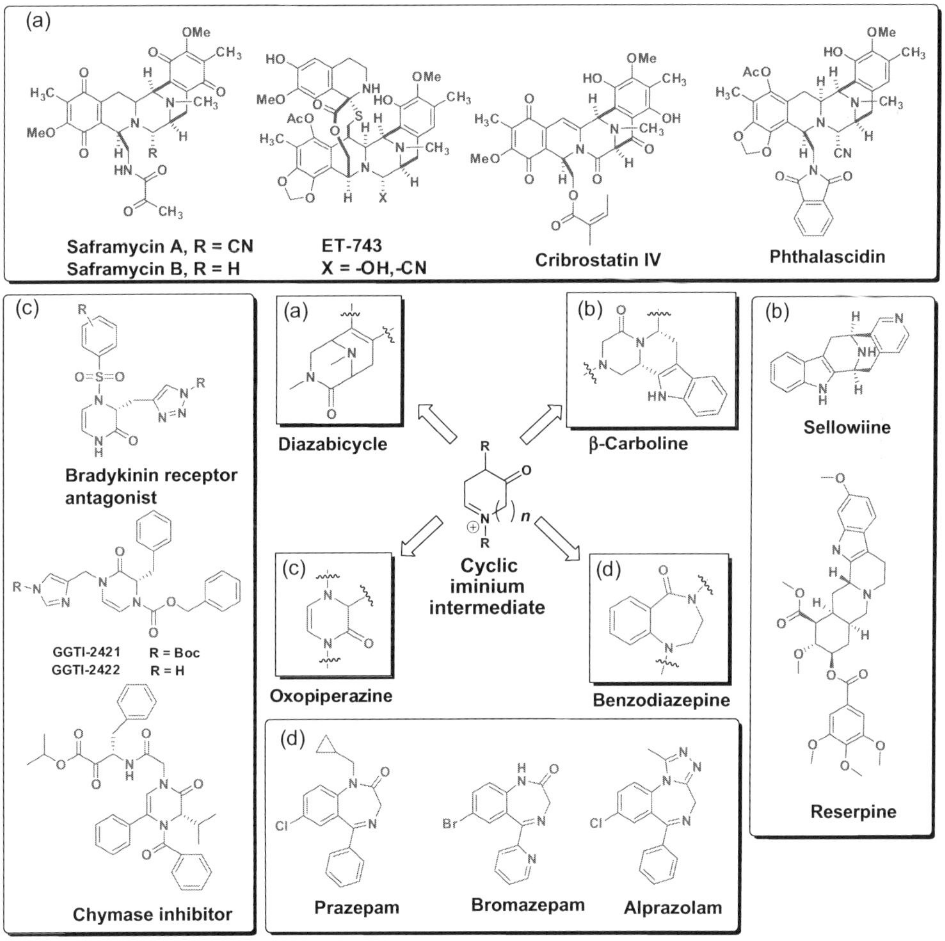

Figure 5.1. Four novel core skeletons extracted or designed structure from natural products and peptide mimic motif: (a) diaza-bridged heterocycles, (b) tetrahydro-β-carboline alkaloid, (c) Δ^5-2-oxopiperazines, and (d) tetrahydro-1,4-benzodiazepine.

a prominent heterocycle, found in complex natural products, having various biological activities such as anticancer, anti-inflammatory, anti-HIV, and antidepressant (see Figure 5.1b). (3) Oxopiperazine structure is mimicking cyclic dipeptide motif and commonly used as a privileged substructure for the construction of drug-like small molecule library through the bioisostere strategy of normal dipeptide bond with nonpeptide ligands (see Figure 5.1c). (4) The benzodiazepines, a family of well-known privileged structures, are nonpeptidic β-turn mimetics and commonly classified as central nervous system (CNS) suppressants due to their anxiolytic, anticonvulsant, sedative, and muscle relaxant activities. Benzodiazepine has been extensively used in various marketed drugs, such as alprazolam, bromazepam, chlorazepate, and valium (see Figure 5.1d).

The syntheses of these four novel core skeletons were carried out in a parallel solid-phase fashion using acid labile resins such as SASRIN™ or bromoacetal resin. Under the treatment of neat formic acid, our substrates were designed to be cleaved from solid supports and synchronized with *in situ* intramolecular formation of cyclic *N*-iminium intermediates followed by intramolecular nucleophilic addition (Type I, II, and III), olefin migration (Type IV), and hydride addition (the Leukart-Wallach reduction, Type V), with high yields and excellent control of stereochemistry and regiochemistry (see Figure 5.2). The robustness and practicality of our novel synthetic methods were validated by the successful construction of pilot library embedded with natural product-like polyheterocycles with excellent overall yields and purities. The synthetic exercise will be described in detail in the following sections.

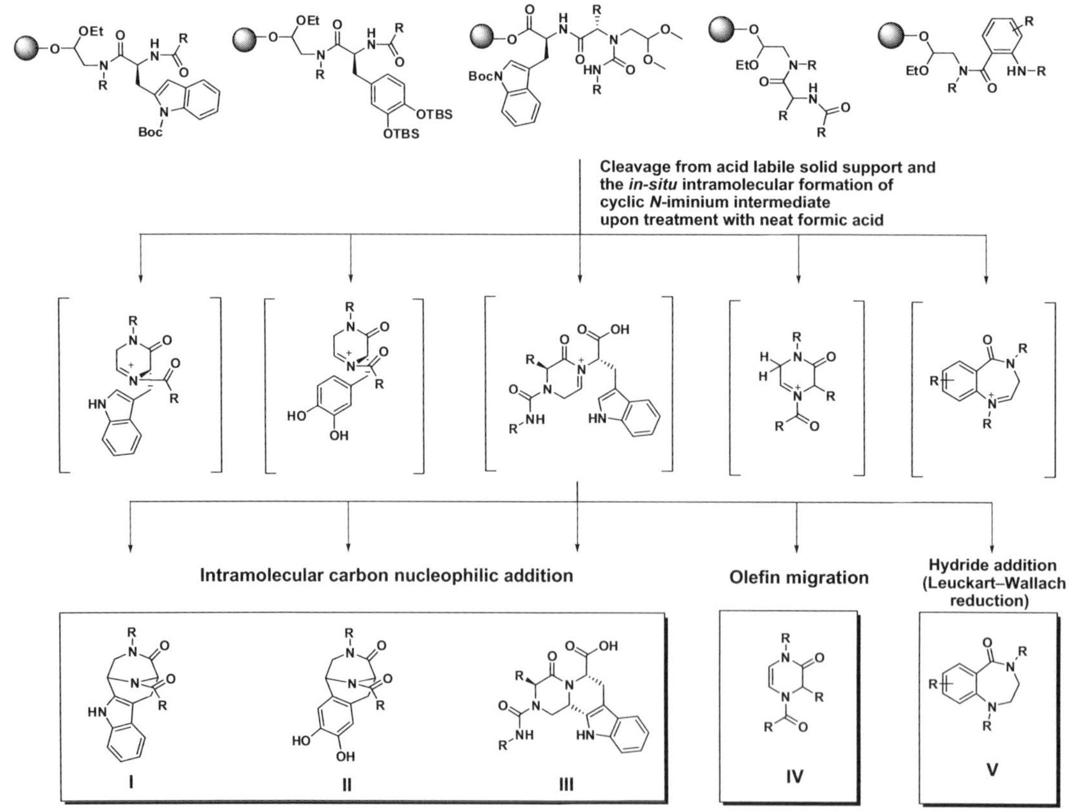

Figure 5.2. Divergent syntheses of unique polyheterocyclic scaffolds from cyclic iminium intermediates.

5.2.1 Practical Solid-Phase Synthesis of Diaza-briged Heterocycle and Tetrahydro-beta-carboline Through Intramolecular Pictet–Spengler Cyclization (Type I, II, and III)

5.2.1.1 Synthesis of Diaza-Bridged Heterocycles (Type I and II). In this section, we introduce the facile and practical solid-phase synthetic method for the construction of small molecule library with novel polyheterocyclic core skeletons embedded with 3,9-diazabicyclo[3.3.1]non-6-en-2-one (diaza-bridged) core skeleton that was recombined with privileged indole and L-DOPA moieties. The generation of 3,9-diazabicyclo[3.3.1]non-6-en-2-one was successfully achieved by the application of Pictet–Spengler intramolecular cyclization in the solid-phase parallel synthesis. In fact, the Pictet–Spengler reaction of electron-rich arene with iminium intermediate via electrophilic aromatic substitution was proposed as the key chemical transformation in biosynthetic pathway of bioactive polycyclic natural products such as yondelis, saframycin A, and cribrostatin V and utilized for the total synthesis of these natural products. However, these synthetic methods using Pictet–Spengler intermolecular reaction were not appropriate for library construction because of the poor diastereoselectivity.[13]

To address this issue, we developed a novel regio- and diastereoselective solid-phase synthetic route for the construction of small molecule library containing new core skeletons (Scheme 5.1). Newly released aldehydes from solid supports were subjected to *in situ* generation of *N*-acyliminium intermediate in solution, which can react with electron-rich arene such as indole and dimethoxyphenyl, the side chain of tryptophan, and L-DOPA. For this synthetic endeavor, we first selected bromoacetal resin **1** as an acid labile solid support because we envisioned the intramolecular Pictet–Spengler cyclization with cyclic iminium intermediates. This *N*-acyl iminium can be generated via the condensation of amide nitrogen with *in situ* unmasked aldehydes from the acetal linker under the acidolytic cleavage condition. We actually prepared bromoacetal resin in-house from Wang resin (loading level: 1.6 mmol/g) because of low loading levels of commercial bromoacetal resin. The synthesis was initiated by the incorporation of benzyl amine into bromoacetal resin **1** through simple nucleophilic substitution in a DMSO solution at 60°C. For the diversification of diaza-bridged core skeletons through the recombination with privileged indole and DOPA moieties by Pictet–Spengler reaction, secondary amines **2** on solid supports were coupled with Fmoc-Trp(Boc)-OH or Fmoc-L-DOPA(DiTBS)-OH. After the completion of HATU-assisted amidation and subsequent Fmoc deprotection step, the resulting primary amines were subjected to further modification via amide or urea formation as another diversity element for the library realization. In fact, there are many other amine modification methods such as alkylation and sulfonylation, but we only focused on amidation and urea formation in this pilot library. The final step was performed under neat formic acid to synchronize three consecutive chemical transformations in a single step: masked aldehydes of solid-bound substrates were liberated from solid-bound acetal moiety under the acidolytic cleavage condition and the resulting aldehyde reacts with amide nitrogen to generate cyclic *N*-acyliminium intermediates that undergo the Pictet–Spengler intramolecular cyclization with electron-rich carbon nucleophiles, indole, or DOPA moiety.

However, the final Pictet–Spengler intramolecular cyclization required the further optimization because there are two modes of nucleophilic attack by indole or DOPA moiety. When indoles underwent Pictet–Spengler-type cyclization with cyclic *N*-acyliminium moieties that were generated *in situ* from masked aldehyde and amide nitrogen under acidic conditions, two modes of nucleophilic attack became possible—at carbon positions C2 and C3 in indole (Scheme 5.2a). C2 attack can afford the formation of a 3,9-diazabicyclo

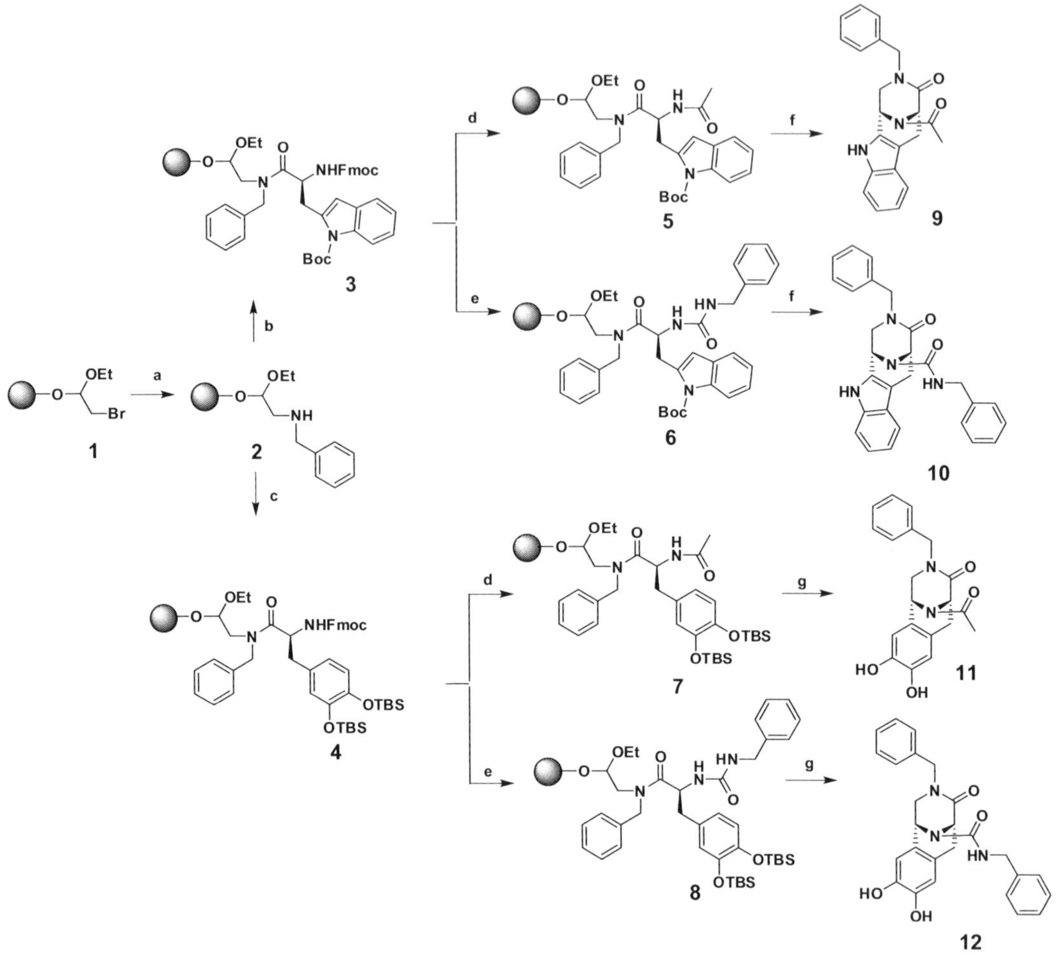

Scheme 5.1. Solid-phase synthesis of 3,9-diazabicyclo[3.3.1]non-6-en-2-one core skeleton recombined with privileged indole and DOPA moieties through Pictet–Spengler reaction. (a) Benzylamine, DMSO, 60°C; (b) Fmoc(*N*-Boc)TrpOH, HATU, DIPEA, DMF, room temperature (rt); (c) Fmoc(*O*-DiTBS)DOPA, HATU, DIPEA, DMF, rt; (d) (i) 25% piperidine, rt; (ii) acetic acid, DIC, HOBt, DIPEA, rt; (e) benzyl isocyanate, DIPEA, DCE, rt; (f) neat HCO_2H, rt; (g) neat HCO_2H, 60°C.

[3.3.1]non-6-en-2-one derivative **9** as a single diastereomer. When the C3 carbon nucleophile of indole attacks cyclic *N*-acyliminiums, it would be expected to form a five-membered spiro intermediate that readily undergoes the cationic migration followed by hydride elimination to yield 3,9-diazabicyclo[3.3.1]non-6-en-2-one derivative **9** under this reaction condition. Therefore, the final treatment of neat formic acid accomplished the formation of a series of diaza-bridged polyheterocyclic **9** in nearly quantitative yields with excellent regio- and diastereoselectivity.

To diversify the diaza-bridged core skeletons, the Pictet–Spengler-type intramolecular cyclization was applied with a DOPA moiety, which is also an attractive privileged scaffold and bears nucleophilic carbon centers on an aryl group. The simplified system using the dihydroxyphenyl group and aminoacetal, as shown in Scheme 5.2b, was utilized to identify the optimum reaction condition and validate the regio- and diastereoselectivity of the final intramolecular cyclization step. Unlike the indole ring, there are two possible regioisomers,

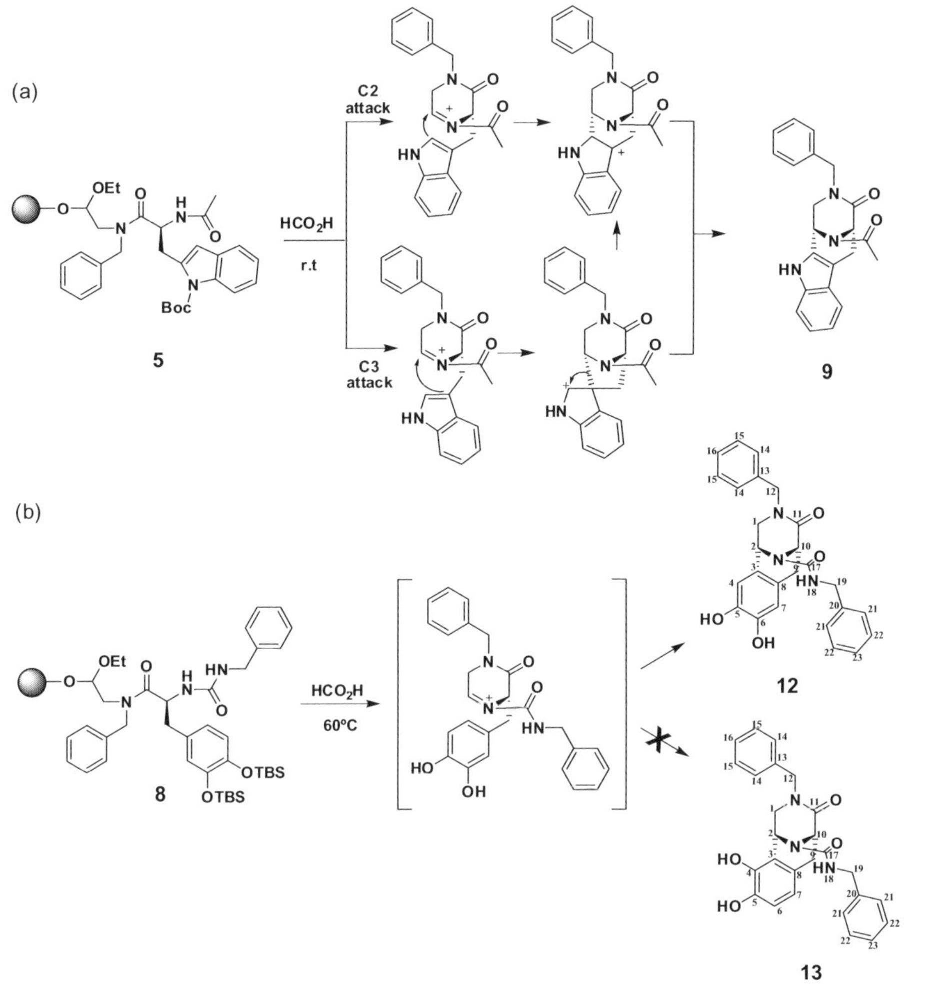

Scheme 5.2. Regio- and diastereoselective Pictet–Spengler cyclization of cyclic iminium with (a) indole and (b) dihydroxyphenyl nucleophile.

12 and **13,** under the optimized condition using L-DOPA moiety. To confirm the regioselectivity of this chemical transformation for the synthesis of *N*-substituted DOPA-derived diaza-bridged heterocycles, we pursued extensive NMR studies, including ^{1}H COSY, NOESY, HSQC, and HSBC experiments. We confirmed that there were no vicinal aryl protons of the other diastereomer **13** (a proton at the C6 position (6.47 ppm) and a proton at the C7 position (6.52 ppm)) through the no correlation between these down-fielded aryl protons in the COSY spectrum. The formation of **12** was confirmed by the HMBC experiment with the clear correlation between a carbon at the C2 position and a proton at the C4 position (6.47 ppm). The following two critical supporting NMR data clearly demonstrated that the optimized condition yielded **12** as a single diastereoisomer through the intramolecular Pictet–Spangler cyclization. The exceptional diastereoselectivity on the newly generated chiral center through Pictet–Spengler cyclization was rationalized by the chemoinformatics calculation with "Insight2000" of Acceryls®. There are two possible modes of nucleophilic attack on *in situ* generated acyliminium ion: *si* face and *re* face.

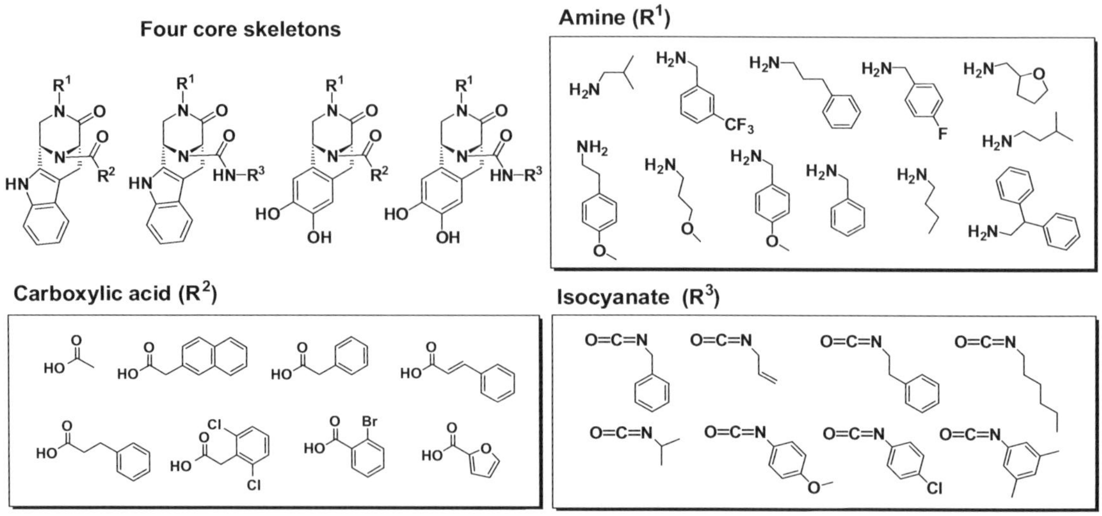

Figure 5.3. Set of building blocks for the library construction of a 3,9-diazabicyclo[3.3.1]non-6-en-2-one skeleton.

However, the predefined stereochemistry of the α-carbon at the C10 position from L-amino acids (DOPA or tryptophan) significantly influenced the stereochemical outcome of diaza-bridged heterocycles, which favors *re* face attack because of the distance between nucleophilic carbon and acyliminium ion.

After the confirmation of regio- and diastereoselective cyclization of indole and dihydroxyphenyl moieties, this solid-phase synthetic approach was exemplified by the preparation of a 384-member pilot library of a 3,9-diazabicyclo[3.3.1]non-6-en-2-one skeletons and diversified at two bridging nitrogen atoms using various building blocks, primary amine (R^1-NH_2), carboxylic acid (R^2-COOH), and isocyanate (R^3-NCO), without further purification (Figure 5.3). The average purity of this library was 90% and more than 95% of the crude library members exhibited an exceptional purity (greater than 85%), which was determined by LC/MS analysis with a PDA detector.

5.2.1.2 Synthesis of Tetrahydro-beta-carbolines (Type III). Tetrahydro-β-carboline is often found in complex natural products and is frequently associated with biological activity, such as anticancer, anti-inflammatory, anti-HIV, antidepressant, and erectile dysfunction.[14] The biological importance and therapeutic potential of tetrahydro-β-carboline alkaloids have stimulated interests in the development of efficient methodology not only in conventional organic synthesis[15] but also in solid-phase combinatorial synthesis.[16] Based on previous reports on syntheses of tetrahydro-β-carboline ring systems via Pictet–Spengler reaction[16], L-tryptophan, L-tryptophan methyl ester, or its *N*-substituted derivatives, such as L-abrine methyl ester and *N*-benzyltryptophan methyl ester, were most frequently utilized as a key element with achiral aldehyde in asymmetric intermolecular fashion.[17] There are also a few examples of asymmetric Pictet–Spengler reactions using chiral aldehydes and similar reactions using azalactones as equivalents of arylacetaldehyde.[18] Therefore, we wished to delineate the efficient synthesis of tetrahydro-β-carboline alkaloids through solid-phase methodology.

On the basis of the successful application of intramolecular tandem Pictet–Spengler cyclization of active *N*-acyliminium intermediates to construct small molecules library embedded with diaza-bridged heterocycles in Section 5.2.1.1, we also developed a solid-phase strategy for the synthesis of tetrahydro-β-carboline alkaloid core skeleton, of which the biological importance and therapeutic potentials have stimulated interests in the development of an efficient methodology not only in conventional organic synthesis but also in solid-phase combinatorial synthesis. The key transformation is intramolecular tandem Pictet–Spengler cyclization of active *N*-acyliminium intermediates with indole moiety from L-tryptophan in an asymmetric fashion similar to the case of diaza-bridged heterocycle synthesis. Prior to formation of tetrahydro-β-carboline alkaloid core skeleton in solid phase, we synthesized the representative compound **18** in solution phase to validate the synthetic strategy, as illustrated in Scheme 5.3. In brief, the synthesis of representative tetrahydro-β-carboline **18** was launched by esterification of FmocTrpOH **14**. The resulting compound **15** was Fmoc deprotected and converted to secondary amine by reductive amination with dimethoxyacetaldehyde and $NaBH_3CN$. The resulting secondary amine was converted to compound **17** by acetylation using general condition, acetic anhydride, and DIPEA in dichloromethane. The final step was the key transformation from linear peptide to cyclic tetrahydro-β-carboline alkaloid. The general condition of a Pictet–Spengler cyclization was under strong acids, such as TFA, toluenesulfonic acid, HCl, and sulfuric acid. As shown in Section 5.2.1.1, we successfully demonstrated an efficient library construction using Pictet–Spengler tandem cyclization with neat formic acid. Therefore, we applied our condition into this synthetic route and obtained the desired compound **18** in high yield (85%). The X-ray crystal structure of compound **18** proved the molecular connectivity as well as the new chiral center (*S* diastereomer) generated by diastereochemically enriched intramolecular Pictet–Spengler cyclization.

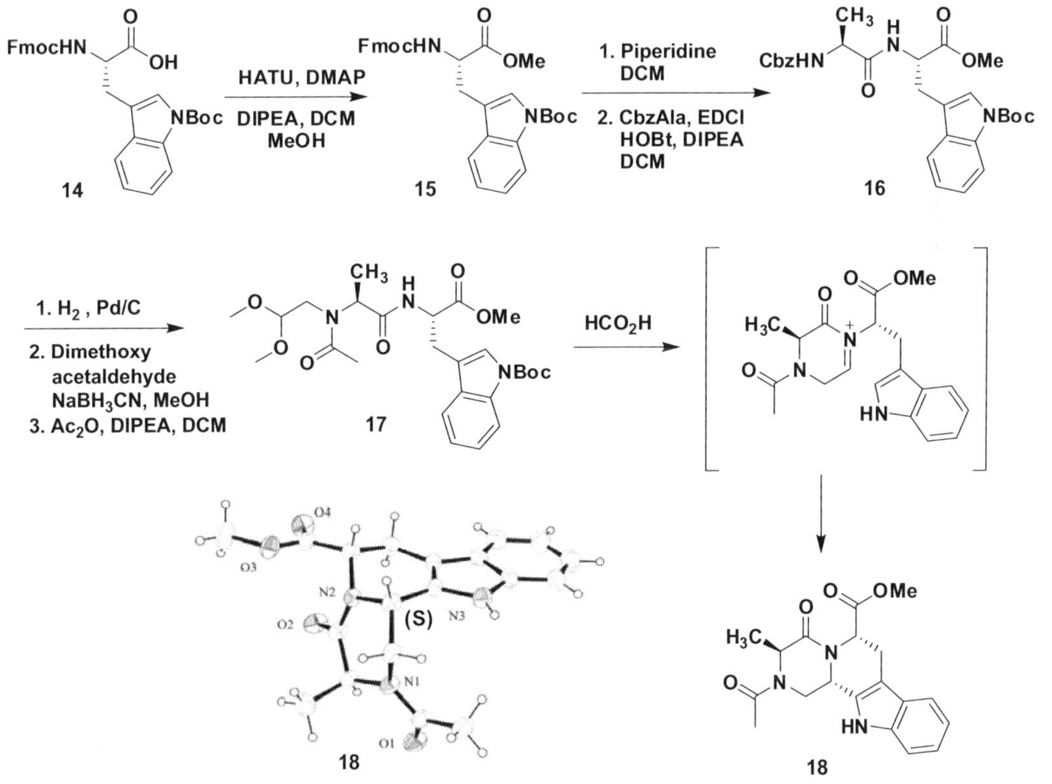

Scheme 5.3. Solution-phase validation of designed template and X-ray crystal structure of compound **18**.

After the successful validation in the solution phase, we tried to apply this synthetic methodology to solid-phase parallel synthesis. First, we considered optimum linker system on the solid support among acid labile polymer resins, and SASRIN resin was eventually selected because of the efficient cleavage upon treatment with neat formic acid. To test SASRIN resin for this synthetic pathway, we attached FmocTrpOH to benzyl alcohol moiety on polymer supports through the ester linkage and then unreacted alcohol moieties on the solid support were capped with acetic anhydride/pyridine to ensure the purity of final products. After the completion of esterification and subsequent deprotection of Fmoc group, the resulting amine was coupled with Fmoc-L-Ala-OH to form a linear dipeptide **20**. To introduce masked aldehyde functional group, the reductive amination using dimethoxyacetaldehyde was performed with a mixture of methanol and DMF (1:1, v/v) because of swelling property of solid support after second Fmoc deprotection of dipeptides **20** on solid supports. The resulting secondary amines were treated with benzyl isocyanates in the presence of DIPEA. The final step for the chemical transformation from linear peptide **21** to cyclic tetrahydro-β-carboline alkaloid **22** was successfully proceeded with the treatment of neat formic acid at 50°C through a series of reactions: cleavage of compounds from solid support, acid-catalyzed aldehyde formation and subsequent acyl-iminium formation, and Pictet–Spengler intramolecular cyclization. The resulting solution was collected and excess formic acids were evaporated, and subsequent lyophilization with 50% water/acetonitrile yielded the desired product as yellow solid. Tetrahydro-β-carboline alkaloids **22** with urea group were also synthesized in superior yields and purities. After the feasibility test, we started the construction of a pilot library with tetrahydro-β-carboline alkaloid core skeleton.

For the construction of a pilot library, we chose R^1 and R^2 groups as the diversity points on tetrahydro-β-carboline alkaloid core skeleton **22**. The R^1 position was diversified by urea formation reaction using eight commercially available isocyanates. The R^2 position was diversified through the amide coupling of 12 Fmoc-protected natural and unnatural amino acids for the synthesis of pilot library (Scheme 5.4). With these building blocks, an array of 96 compounds was synthesized in excellent purities through solid-phase parallel synthesis platform.

5.2.2 Practical Solid-Phase Synthesis of Δ^5-2-Oxopiperazines via *N*-Acyliminium Ion Cyclization (Type IV)

In this section, a practical solid-phase strategy for the synthesis of Δ^5-2-oxopiperazines via *N*-acyliminium ion cyclization will be introduced. In fact, piperazines, diketopiperazines, and piperazinones are important pharmacophores used as peptidomimetic moieties for the discovery of novel bioactive small molecules (Figure 5.4).[19] The transformation of peptides into conformationally restricted peptide analogues is a generally accepted approach to design nonpeptide ligands that target protein receptors. Peptidyl-prolyl isomerase inhibitor,[20] chymase inhibitor,[21] substance P antagonist,[22] bradykinin receptor antagonist,[23] and geranylgeranyltransferase I (GGTase-I) and farnesyltransferase (FTase) inhibitors[24] are excellent examples of peptidomimetic drugs discovered using oxopiperazine as a simple, yet critical, core skeleton that conformationally mimics the dipeptide moiety. Some of these targets are typical G-protein-coupled receptors with seven putative membrane spanning regions that are used for the perturbation of their functions with peptidomimetic ligands.[25] FTase and GGTase-I are also promising molecular targets, and the small molecule peptoid inhibitors as substrate mimics of these enzymes can specifically block the malignant transformation caused by mutated Ras proteins. In particular, GGTase-I has attracted

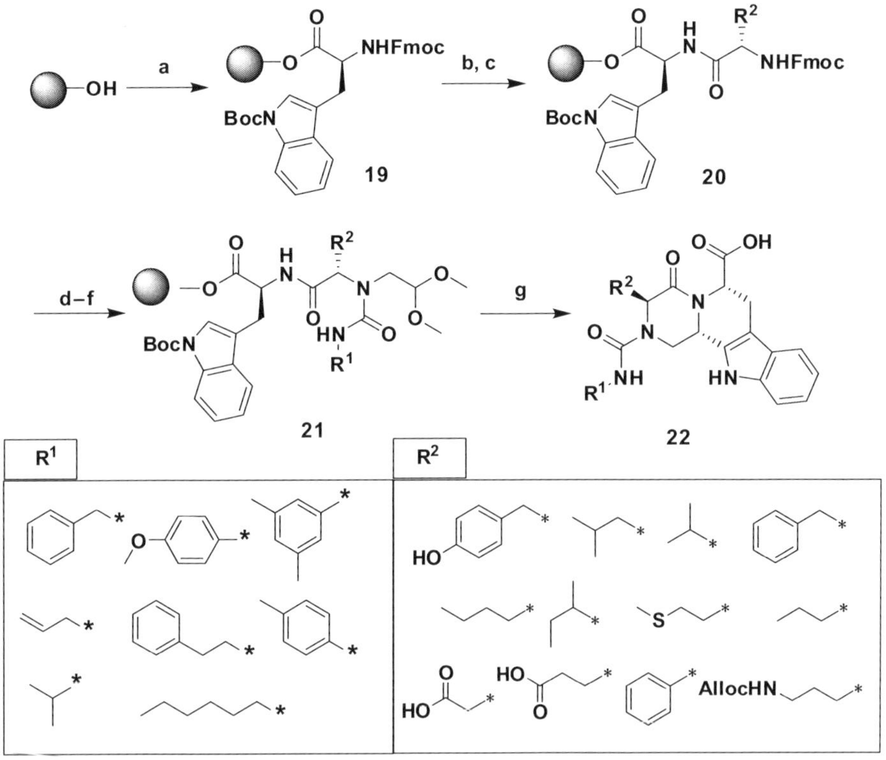

Scheme 5.4. Solid-phase parallel synthesis of tetrahydro-β-carboline alkaloid library. (a) FmocTrpOH, HATU, DMAP, DIPEA, DMF; (b) 25% piperidine in DMF; (c) Fmoc-protected amino acid, DIC, HOBt, DIPEA, DMF; (d) 25% piperidine in DMF; (e) dimethoxyacetaldehyde, $NaBH_3CN$, MeOH/DMF (1:1); (f) R^1NCO, DIPEA, DCE; (g) neat formic acid, 50°C, 3 h.

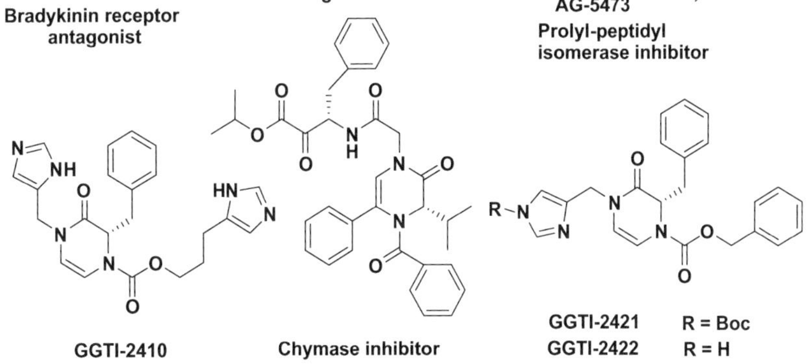

Figure 5.4. Bioactive Δ^5-2-oxopiperazines currently under the drug development.

attention due to its critical roles in promoting tumorigenesis and metastasis through a complex signaling pathway with various substrates such as RhoC, RhoA, Rac-1, Cdc42, R-Ras, and TC-21.[24,26]

Although Δ^5-2-oxopiperazine core skeleton is well known as a remarkable pharmacophore, there are few examples of reported synthetic methods aiming at practical construction of a Δ^5-2-oxopiperazine library in the solid phase.[27] Therefore, to overcome limitation of the library construction, we developed a practical procedure for efficient synthesis of a pilot library embedded with privileged Δ^5-2-oxopiperazine core skeleton through solid-phase parallel synthesis.

The *N*-acyliminium-type nucleophilic tandem cyclization has been recognized as a powerful reaction for the construction of heterocyclic and bicyclic ring systems in solid- and solution-phase syntheses.[28] The key *N*-acyliminium intermediates, formed by tandem acidolytic cleavage with subsequent *in situ* cyclization, are excellent electrophiles that react with various nucleophiles such as oxygen, nitrogen, sulfur, and even electron-rich aromatic carbon. Actually, we proved usefulness of this important intermediate by practical applications of the Pictet–Spengler intramolecular cyclization with carbon nucleophiles such as indole and DOPA moieties through that shown in Figure 5.5, Path A. However, if there are no efficient intra- and intermolecular nucleophiles after formation of *N*-acyliminium ions, the elimination of beta-hydrogen will be possible by counter base even in weakly acidic condition because of the distinct difference of stability between iminium ion and enamide (Figure 5.5, Path B). Therefore, we focused on the selective enamide generation without any nucleophilic addition in an intra- and intermolecular fashion to synthesize stable Δ^5-2-oxopiperazines from intracyclic *N*-acyliminum ions. The major aspects of this transformation are the electronic effect of R^3 substituents on *in situ* generated unstable *N*-acyliminium and the development of an appropriate acidic condition for the generation of the Δ^5-2-oxopiperazine moiety. The efficient rearrangement of *N*-acyliminium into enamide is influenced by the introduction of electron-withdrawing *N*-substituents (R^3 in Figure 5.5) such as carbonyl or amide. In addition to the effects of the substituents, the generation of the enamide moiety depends on the proton source for cyclization.

The solid-phase synthetic method for obtaining the unsaturated Δ^5-2-oxopiperazine moiety, enamide form, was developed through the optimization process using representative

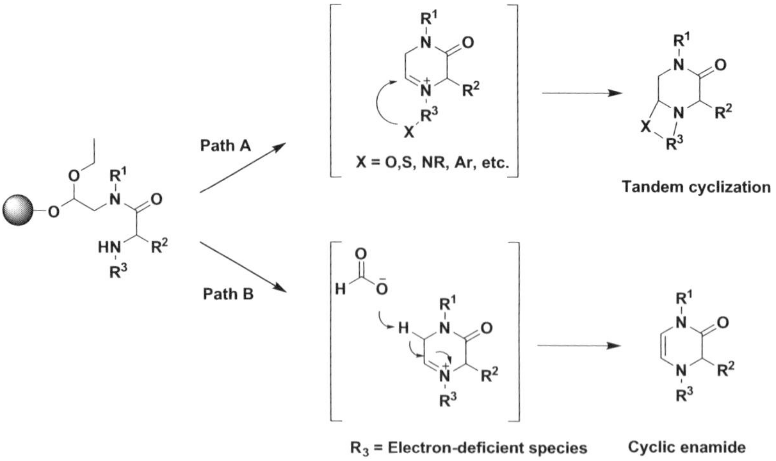

Figure 5.5. Two general pathways through the key acyliminium intermediate.

compounds with various R^3 substituents such as acetyl, benzyloxy carbonyl, and benzyl amido groups for the introduction of electron-withdrawing substituents, amide (**26a**), carbamate (**26b**), and urea (**26c**), respectively. Other substituents, namely, R^1 and R^2, were fixed with benzyl and hydroxybenzyl groups for easy comparison. We also screened various acid conditions for efficient hydrogen transformation and finally optimized the acid-catalyzed enamide cyclization by using neat formic acid, which is a relatively weak acid. As shown in Scheme 5.5, the representative compounds **27a**, **27b**, and **27c** were successfully synthesized in high yields and exceptional purities (> 90%) using the optimized reaction condition on a solid support.

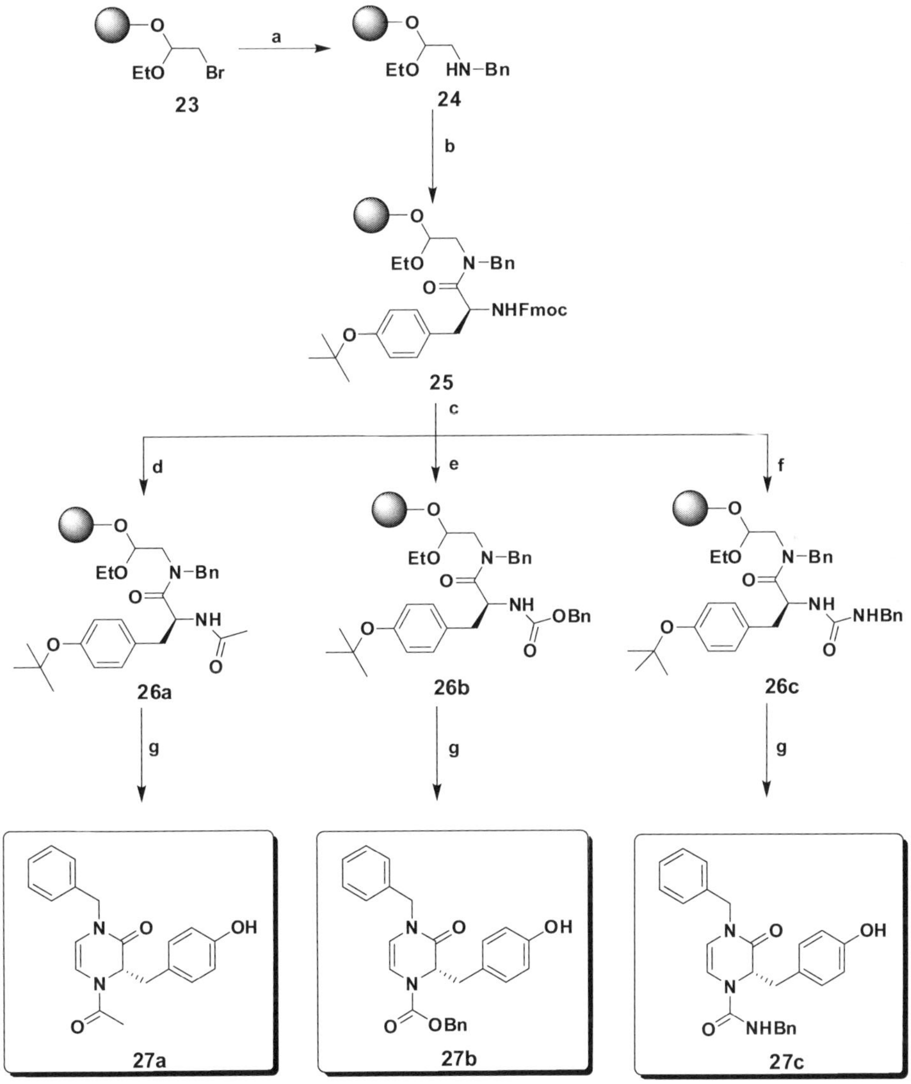

Scheme 5.5. Feasibility test of unsaturated oxopiperazine scaffold. (a) $BnNH_2$, DMSO, 60°C; (b) Fmoc(*O-t*-Bu)TyrOH, HATU, DIPEA, DMF, rt; (c) 25% piperidine, rt; (d) Ac_2O, Py, DCE, rt; (e) BnOCOCl, DIPEA, DCE; (f) BnNCO, DIPEA, DCE, rt; (g) neat HCO_2H, 60°C, 3 h.

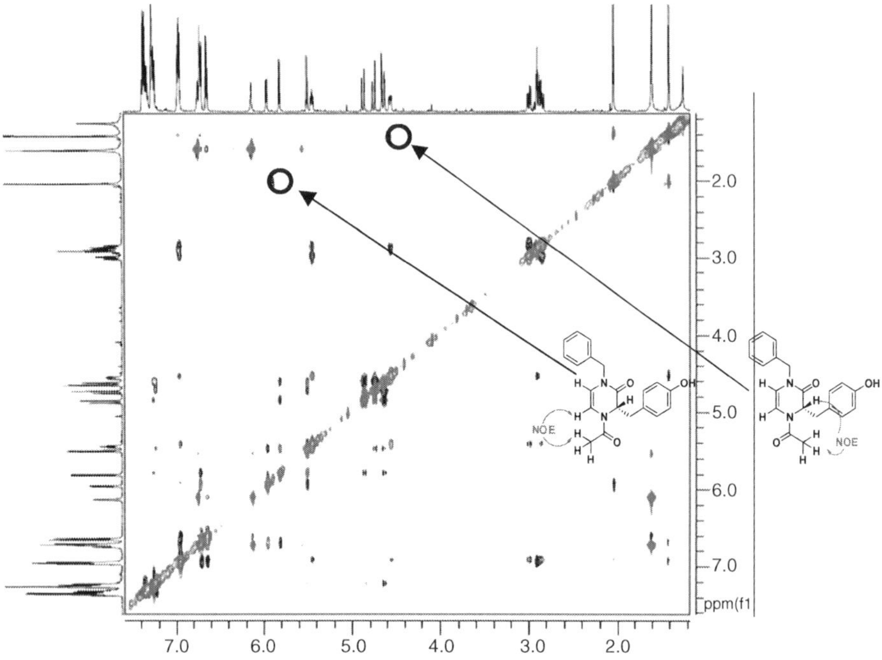

Figure 5.6. NOE correlation between two rotamers of the representative compound **27a**.

However, significant amounts of by-products were produced during the synthesis of the desired enamides even under the optimized reaction condition in the case of electrically neutral proton or electron-donating alkyl groups, which were introduced by the reductive amination of aldehydes, at the R^3 position (data not shown). Therefore, we concluded that the *in situ* generated *N*-acyliminium ion should be activated with an electron-withdrawing group for efficient rearrangement initiated by deprotonation for the synthesis of cyclic enamides (unsaturated Δ^5-2-oxopiperazines). Interestingly, **27a** and **27b** were identified as mixtures of two isomers by NMR analysis; however, they were inseparable in the reverse HPLC condition. By performing extensive NMR analysis, we could confirm that the two isomers were rotational isomers due to the rigid structure of the unsaturated oxopiperazine. As shown in Figure 5.6, the rotamers of **27a** were evidently obtained by the nuclear Overhauser effect (NOE) between acetyl and vinyl protons along with a tyrosine alpha proton (Figure 5.6).

After the confirmation of the selective and efficient monocyclization to unsaturated oxopiperazine in the solid phase, the pilot library with embedded Δ^5-2-oxopiperazine core skeleton was successfully constructed. The molecular diversity of the core skeleton was expanded by the introduction of various R^1 and R^2 groups using commercially available primary amine and a series of commercially available carboxylic acids, chloroformates, and isocyanate as shown in Figure 5.7. The average purity of this library was 90% and the purities of more than 90% of the crude library members exceeded 85%.

5.2.3 Novel Application of the Leuckart–Wallach Reaction for the Synthesis of a Tetrahydro-1,4-benzodiazepin-5-one Library (Type V)

The series of compounds containing the tetrahydro-1,4-benzodiazepine scaffold are an important class of prototypical "privileged" structures associated with various biological

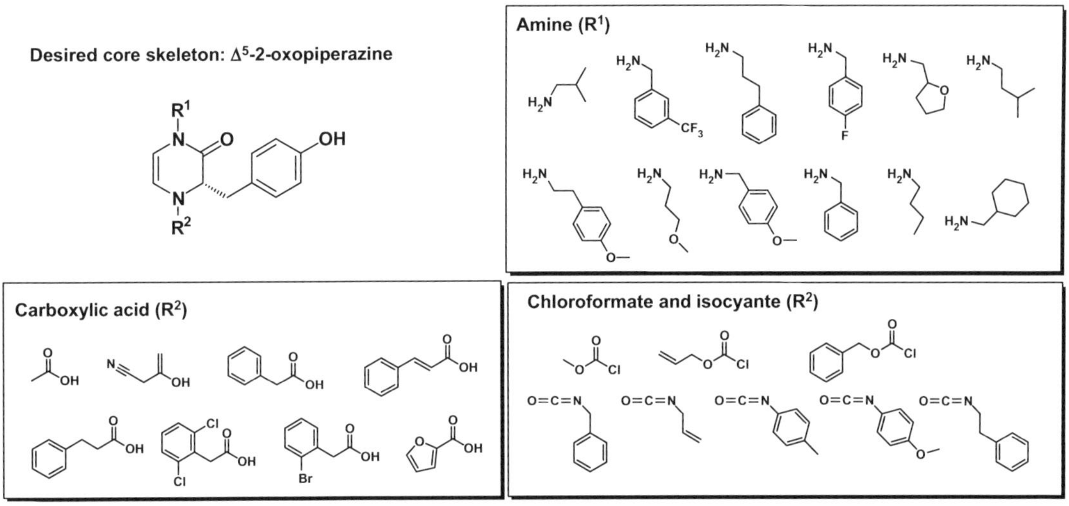

Figure 5.7. Set of building blocks for the library construction of Δ^5-2-oxopiperazine skeletons.

activities and therapeutic uses. The benzodiazepine family is commonly classified as a CNS suppressant due to its anxiolytic, anticonvulsant, sedative, and muscle relaxant activities. It is used in various marketed drugs such as Alprazolam, Bromazepam, Chlorazepate, and Valium.[29] 1,4-Benzodiazepines also demonstrate therapeutic activities and are used as antibiotics,[30] antiulcers,[31] and anti-HIV agents;[32] they are also used as Ras farnesyltransferase inhibitors.[33] Interestingly, 1,4-benzodiazepin-5-ones have not been studied extensively compared to other 1,4-benzodiazepine series such as 1,4-benzodiazepin-2-one and 1,4-benzodiazepin-2,5-dione in terms of their synthesis or biological activities. There is a list of reports on the synthetic methodology of 1,4-benzodiazepin-5-ones that involves aromatic substitution and Schmidt rearrangement (Figure 5.8).[34] However, these methods involve multiple steps with harsh reaction conditions, and usually limited derivatization is possible through modification only after the formation of the core skeleton.

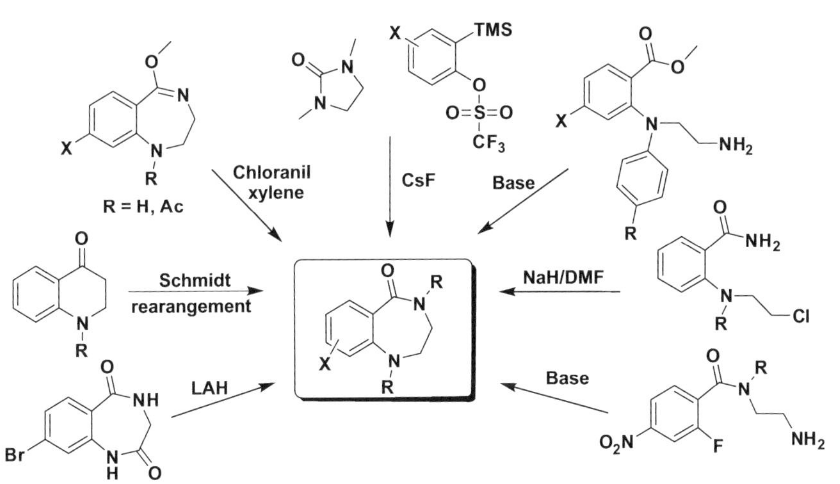

Figure 5.8. Reported methods and our synthetic strategy for the construction of tetrahydro-1,4-benzodiazepin-5-ones.

With expectations of interesting biological activities, we initiated the development of a practical synthetic pathway for library construction. Wang and Claudsdale reported a two-step method for synthesizing dihydro-1,4-benzodiazepin-5-one, which can be reduced to tetrahydro-1,4-benzodiazepin-5-one.[35] However, we aimed to identify a single-step transformation instead of a transformation with three individual steps. As described in Section 5.2.2, the reactive *N*-acyliminium ions were successfully transformed into more stable enamide form in neat formic acid without any nucleophilic addition. However, the saturated cyclized product can be synthesized with neat formic acid as the reducing agent through reductive amination, known as the Leuckart–Wallach (LW) reaction, as well as unsaturated enamide product. Fortunately, reduced amine products were not observed through LW reaction in a previous six-membered ring system, even after a systematic screening of various acidic conditions and, in fact, it has been reported that the elimination of proton for formation of enamide is the more favorable pathway in six-membered oxopiperazine system.[36]

Therefore, we designed a practical and novel synthetic pathway for the synthesis of tetrahydro-1,4-benzodiazepin-5-one through LW reaction by application of seven-membered benzodiazepine system instead of six-membered ring for the development of a new privileged core structure other than already synthesized skeletons using cyclic iminium ion intermediate by nucleophilic addition or hydrogen transformation (Scheme 5.6). From the extensive screening of the reaction conditions for the successful application of LW reaction, we recognized that iminium formation and hydrogen transfer did not proceed under any precedent reaction conditions; complex mixtures with varying quantities of unsaturated dihydro-1,4-benzodiazepin-5-one were identified. However, under the modified LW reaction condition using neat formic acid at 60°C, we obtained the desired products with excellent yields and purities (Figure 5.9). Subsequently, we tried to introduce not alkyl group but also other functional groups into R^2 to examine the scope of this reaction pathway. However, when R^2 is an electron-withdrawing group, such as an acetyl, carbamate, or urea group, a messy reaction pattern was observed under the same reaction conditions, with uncyclized compounds as the major by-products; this pattern is probably caused by the inefficient cyclic imine formation of acyl aniline moieties along with *in situ* generated aldehydes (data not shown). Thus, the aniline amine was modified only by reductive amination with aldehyde or ketone to introduce alkyl group.

The robustness and practicality of our novel synthetic pathway was validated by the successful construction of a 96-member pilot library with excellent overall yields and purities (Scheme 5.7). The molecular diversity of the core skeleton was expanded by the introduction of various R^1 elements using commercially available primary amines. The diversity scope of the R^2 position was extended by reductive amination with various

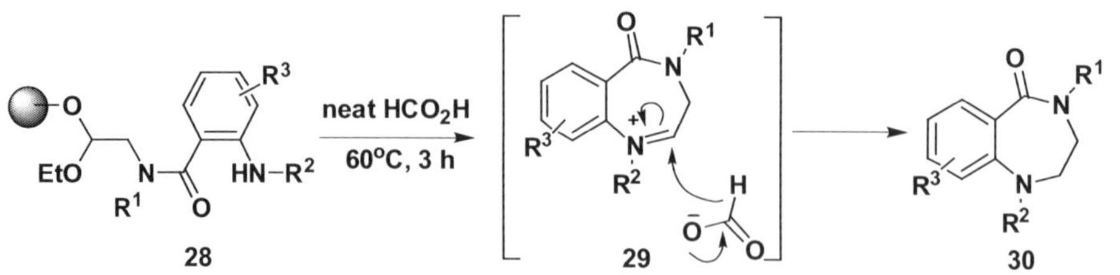

28 29 30

Scheme 5.6. Key transformation toward tetrahydro-1,4-benzodiazepin-5-one via intramolecular Leuckart–Wallach reduction.

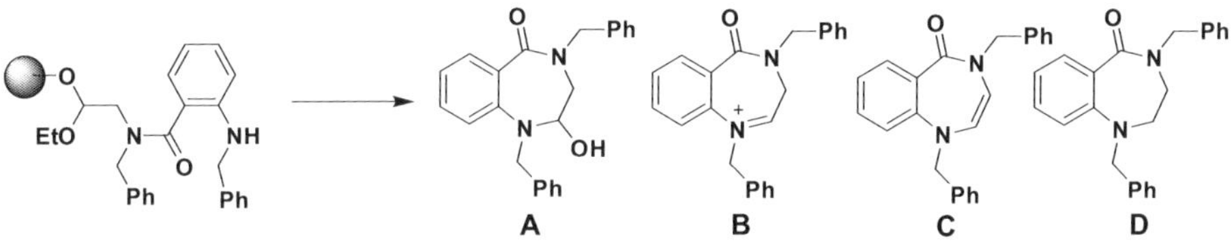

Reagents and condition	Results
30% aq TFA, rt, overnight	Trace A, decomposed
5% TFA in DCE, 60°C, 5 h	B and decomposed side product
Cat. pTSA in DCE, 60°C, 3 h	B and decomposed side product
Neat formic acid, rt, overnight	B
Neat formic acid, 60°C, 3 h	D

Figure 5.9. Reaction condition screening for an optimized procedure.

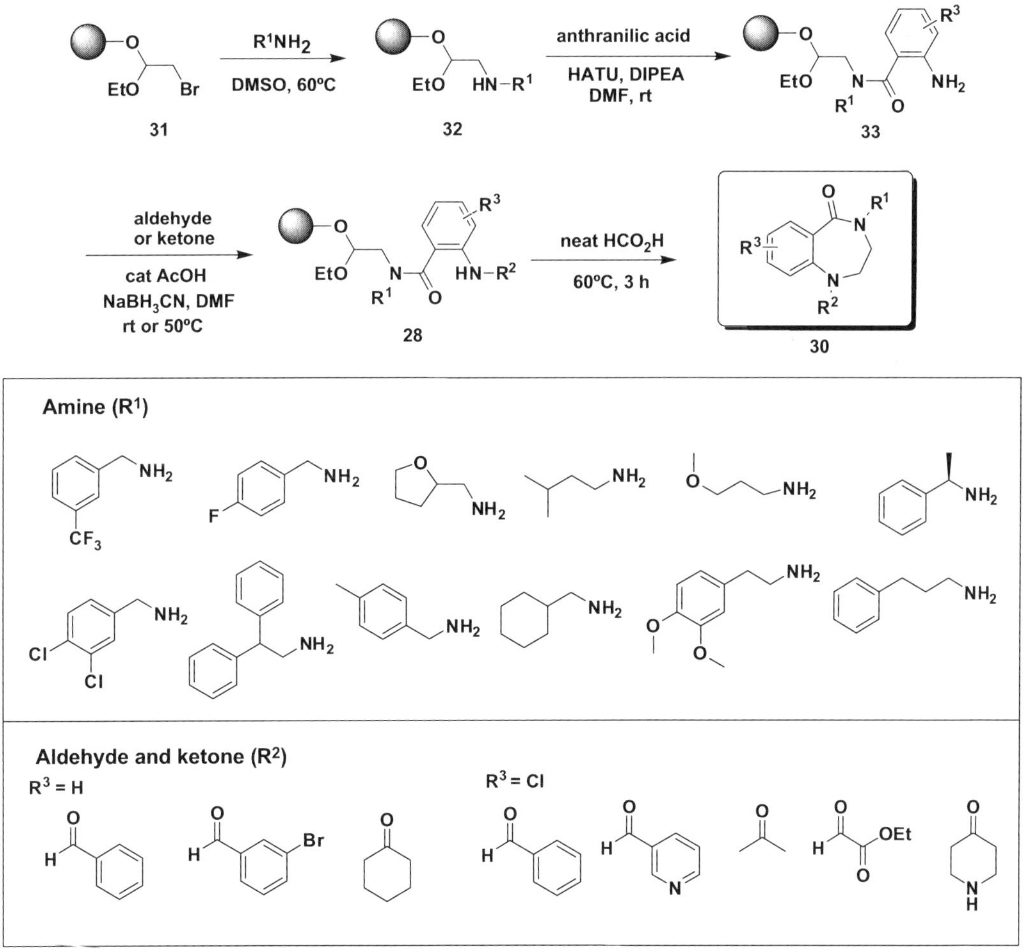

Scheme 5.7. General procedure and set of building blocks for the construction of a 1,4-benzodiazepin-5-one library.

aldehydes and ketones. The LC/MS purities of the crude final products after cleavage from the solid support using neat formic acid were more than 94%.

5.3 CONCLUSIONS

In this chapter, we introduced pDOS strategy as a subclass of DOS strategy to emphasize the importance of privileged structure for the efficient identification of various bioactive small molecules. We also demonstrated the practical application of this approach for the construction of drug-like small molecules containing skeletally diversified heterocyclic core skeletons via solid-phase parallel synthesis. Consequently, we have developed five different types of core skeletons that were derived from bioactive natural products or therapeutic agents, such as diazabicycles (Type I and II), tetrahydro-β-carbolines (Type III), oxopiperazines (Type IV), and benzodiazepines (Type V) along with pDOS strategy. The key transformation in these novel synthetic methods was focused on the *in situ* formation of cyclic iminium as key intermediates via acid-catalyzed cleavage of substrates from solid supports and subsequent intramolecular condensation. The resulting cyclic iminium intermediate can be transformed into five unique core skeletons using Pictet–Spengler-type nucleophilic addition (Types I–III), acid-catalyzed olefin migration (Type IV), and Leuckart–Wallach-type hydride addition (Type V). The robustness and practicality of these methodologies was validated by the successful construction of a pilot library containing five unique core skeletons as a single diastereomer with excellent overall yields and purities. As discussed in this chapter, pDOS approach is proven to be a practical method for the construction of drug-like small molecule library containing diverse core skeletons embedded with privileged substructures and will be utilized for the discovery of novel bioactive small molecule as research tools for chemical biology or as lead compounds for drug discovery.

REFERENCES

1. (a) A. L. Hopkins, C. R. Groom, *Nat. Rev. Drug Discovery* **2002**, *1*, 727–730; (b) R. L. Strausberg, S. L. Schreiber, *Science* **2003**, *300*, 294–295.
2. (a) S. L. Schreiber, *Chem. Eng. News* **2003**, *81*, 51–61. (b) Schreiber, S. L. *Nat. Chem. Biol.* **2005**, *1*, 64–66.
3. X. F. S. Zheng, T.-F. Chan, *Drug Discovery Today* **2002**, *7*, 197–205.
4. (a) C. M. Dobson, *Nature* **2004**, *432*, 824–828; (b) W. Wilk, T. J. Zimmermann, M. Kaiser, H. Waldmann, *Biol. Chem.* **2010**, *391*, 491–497.
5. (a) S. L. Schreiber, *Science* **2000**, *287*, 1964–1969. (b) M. D. Burke, S. L. Schreiber, *Angew. Chem., Int. Ed.* **2004**, *43*, 46–58; (c) T. E. Nielsen, S. L. Schreiber, *Angew. Chem., Int. Ed.* **2008**, *47*, 48–56;
6. (a) D. R. Spring, *Org. Biomol. Chem.* **2003**, *1*, 3867–3870; (b) D. S. Tan, *Nat. Chem. Biol.* **2005**, *1*, 74–84; (c) R. J. Spandl, A. Bender, D. R. Spring, *Org. Biomol. Chem.* **2008**, *6*, 1149–1158; (d) M. Peuchmaur, Y.-S. Wong, *Comb. Chem. High Throughput Screening* **2008**, *11*, 587–601.
7. (a) R. Breinbauer, I. R. Vetter, H. Waldmann, *Angew. Chem., Int. Ed.* **2002**, *41*, 2878–2890; (b) M. A. Koch, L.-O. Wittenberg, S. Basu, D. A. Jeyarai, E. Gourzoulidou, K. Reinecke, A. Odermatt, H. Waldmann, *Proc. Natl. Acad. Sci. USA*, **2004**, 101, 16721–16726; (c) M. Koch, A. Schuffenhauer, M. Scheck, S. Wetzel, M. Casaulta, A. Odermatt, P. Ertl, H. Waldmann, *Proc. Natl. Acad. Sci. USA,* **2005**, 102, 17272–17277; (d) A. Nören-Müller, I. Reis Corrêa, Jr., Rosenbaum, C., Schwalbe, H., Vestweber, D., Prinz, H., Schiewe, H., Waldmann, H. *Proc. Natl. Acad. Sci. USA*, **2006**, 103, 10606–10611.

8. (a) S. K. Ko, H. J. Jang, E. Kim, S. B. Park, *Chem. Commun.* **2006**, *28*, 2962–2964; (b) H. An, S.-J. Eum, M. Koh, S. K. Lee, S. B. Park, *J. Org. Chem.* **2008**, *73*, 1752–1761; (c) Y. Kim, J. Kim, S. B. Park, *Org. Lett.* **2009**, *11*, 17–20; (d) S. O. Park, J. Kim, M. Koh, S. B. Park, *J. Comb. Chem.* **2009**, *11*, 315–326; (e) S. Lee, S. B. Park, *Org. Lett.* **2009**, *11*, 5214–5217; (f) J. Kim, H. Song, S. B. Park, *Eur. J. Org. Chem.* **2010**, *20*, 3815–3822; (g) S. Oh, H. J. Jang, S. K. Ko, Y. Ko, S. B. Park, *J. Comb. Chem.* **2010**, *12*, 548–558.
9. S. Lee, S. B. Park, *J. Comb. Chem.* **2006**, *8*, 50–57.
10. S. Lee, S. Y. Choi, Y. K. Chung, S. B. Park, *Tetrahedron Lett.* **2006**, *47*, 6843–6847.
11. S. Lee, S. B. Park, *J. Comb. Chem.* **2007**, *9*, 828–835.
12. S. Lee, S. B. Park, *Chem. Commun.* **2007**, 3714–3716.
13. (a) T. E. Nielsen, F. Diness, M. Meldal, *Curr. Opin. Drug Discovery Dev.* **2003**, *6*, 801–804; (b) D. Orain, R. Canova, M. Dattilo, R. D. Klöppner, D. Koch, R. Giger, *Synlett* **2002**, *9*, 1443–1446; (c) T. E. Nielsen, M. Meldal, *J. Comb. Chem.* **2005**, *7*, 599–610; (d) B. Danieli, P. Giovanelli, G. Lesma, D. Passarella, A. Sacchetti, A. Silvani, *J. Comb. Chem.* **2005**, *7*, 458–462; (e) J. J. N. Veerman, R. S. Bon, B. T. B. Hue, D. Girones, F. P. J. T. Rutjes, J. H. Maarseveen, H. Hiemstra *J. Org. Chem.* **2003**, *68*, 4486–4494.
14. D. Farzin, N. Mansouri *Eur. Neuropsychopharmacol.* **2006**, *16*, 324–328.
15. R. S. Kusurkar, S. K. Goswami, *Tetrahedron* **2004**, *60*, 5315–5318.
16. X. Li, L. Zhang, W. Zhang, S. E. Hall, J. P. Tam, *Org. Lett.* **2000**, *2*, 3075–3078.
17. (a) P. D. Bailey, I. D. Collier, S. P. Hollinshead, *J. Chem. Soc., Perkin Trans.* **1997**, *1*, 8, 1209–1214; (b) J. Li, J. M. Cook, *J. Org. Chem.* **1998**, *63*, 4166–4167; (c) W. Dai, H. Zhu, X. Hao, *Tetrahedron Lett.* **1996**, *37*, 5971–5974; (d) E. D. Cox, L. K. Hamaker, J. Li, P. Yu, K. M. Czerwinski, L. Deng, D. W. Bennett, J. M. Cook, *J. Org. Chem.* **1997**, *62*, 44–61.
18. J. Ezquerra, C. Lamas, A. Pastor, P. Alvarez, J. J. Vaquero, W. G. Prowse *Tetrahedron Lett.* **1996**, *37*, 5813–5816.
19. K. Rossen, J. Sager, L. M. DiMichele, *Tetrahedron Lett.* **1997**, *38*, 3183–3186.
20. C. Guo, S. Reich, R. Showalter, E. Villafranca, L. Dong, *Tetrahedron Lett.* **2000**, *41*, 5307–5311.
21. K. Nishimura, M. Ban, K. Fujimura, N. Kobayashi, M., Hori, T. Honda, WO0107419 **2001**.
22. Y. Tong, Y. M. Fobian, M. Wu, N. D. Boyd, K. D. Moeller, *J. Org. Chem.* **2000**, *65*, 2484–2493.
23. B. C., Askew, Jr., T. Aya, K. Biswas, G. Cai, J. J. Chen, N. Han, Q. Liu, T. Nguyen, N. Nishimura, R. Nomak, T. Peterkin, W. Qian, K. Yang, C. C. Yuan, J. Zhu, D. C. D'amico, T. Nguyen, W. Qian, WO2006019975 **2006**.
24. H. Peng, D. Carrico, V. Thai, M. Blaskovich, C. Bucher, E. E. Pusateri, S. M. Sebti, A. D. Hamilton, *Org. Biomol. Chem.* **2006**, *4*, 1768–1784.
25. (a) H. Sugihara, H. Fukushi, T. Miyawaki, Y. Imai, Z. Terashita, M. Kawamura, Y. Fujisawa, S. Kita, *J. Med. Chem.* **1998**, *41*, 489–502. (b) D. C. Horwell, R. A. Lewthwaite, M. C. Pritchard, G. S. Ratcliffe, J. R. Rubin *Tetrahedron* **1998**, *54*, 4591–4606. (c) S. Kitamura, H. Fukushi, T. Miyawaki, M. Kawamura, N. Konishi, Z. Terashita, T. Naka *J. Med. Chem.* **2001**, *44*, 2438–2450.
26. E. A. Clark, T. R. Golub, E. C. Lander, R. O. Hynes *Nature* **2000**, *406*, 532–535.
27. J. F. Cheng, M. Chen, T. Arrhenius, A. Nadzan, *Tetrahedron Lett.* **2002**, *43*, 6293–6295.
28. (a) J. J. N. Veerman, R. S. Bon, B. T. B. Hue, D. Girones, F. P. J. T. Rutjes, J. H. Maarseveen, H. Hiemstra, *J. Org. Chem.* **2003**, *68*, 4486–4494; (b) M. H. Todd, C. Ndubaku, P. A. Bartlett, *J. Org. Chem.* **2002**, *67*, 3985–3988; (c) B. E. Maryanoff, H. C. Zhang, H. H. Cohen, I. J. Turchi, M. A. Cynthia, *Chem. Rev.* **2004**, *104*, 1431–1628.
29. L. H. Sternbach, *J. Med. Chem.* **1979**, *22*, 1–7.
30. D. E. Thurston, D. S. Bose, *Chem. Rev.* **1994**, *94*, 433–465.
31. W. Eberlein, G. Schmidt, A. Reuter, E. Kutter, *Arzneim.-Forsch.* **1977**, *27*, 356–359.

32. M. J. Kukla, H. J. Breslin, C. J. Diamond, P. P. Grous, C. Y. Ho, M. Miranda, J. D. Rodgers, R. G. Sherrill, E. De Clercq, R. Pauwels, K. Andries, L. J. Moens, M. A. C. Janssen, P. A. J. Janssen, *J. Med. Chem.* **1991**, *34*, 3187–3197.

33. G. L. James, J. L. Goldstein, M. S. Brown, T. E. Rawson, T. C. Somers, R. S. Mcdowell, C. W. Crowley, B. K. Lucas, A. D. Levinson, J. C. Marsters, *Science* **1993**, *260*, 1937–1942.

34. (a) G. L. Grunewald, V. H. Dahanukar, P. Ching, K. R. Criscione, *J. Med. Chem.* **1996**, *39*, 3539–3946; (b) R. A. Tapia, C. Cesar, *Synth. Commun.* **2004**, *34*, 2757–2765; (c) D. J. Skalitzky, J. T. Marakovits, K. A. Maegley, A. E. Ekker, X. -H. Yu, Z. Hostomsky, S. E. Webber, B. W. Eastman, R. Almassay, J. Li, N. J. Curtin, D. R. Newell, A. H. Calvert, R. J. Griffin, B. T. Golding, *J. Med. Chem.* **2003**, *46*, 210–213; (d) C. Corral, R. Madronero, S. Vega *J. Heterocycl. Chem.* **1977**, *14*, 99–102; (e) C. Bagolini, P. De Witt, L. Pacifici, M. T. Ramacci, *J. Med. Chem.* **1978**, *21*, 476–480; (f) D. Ferraris, R. P. Ficco, D. Dain, M. Ginski, S. Lautar, K. Lee-Wisdom, S. Liang, Q. Lin, M. X.-C. Lu, L. Morgan, B. Thomas *Bioorg. Med. Chem.* **2003**, *11*, 3695–3707; (g) A. N. Osman, A. A. El-Gendy, R. H. Omar, L. Wagdy, A. H. Omar, *Indian J. Chem., Sect. B* **2002**, *41*, 871–874; (h) A. Bauer, K. H. Weber, DE 2165310, **1971**; (i) A. Santili, T. S. Osedene, US Patent 3,457, 258, **1969**; (j) P. Tempest, V. Ma, M. G. Kelly, W. Jones, C. Hulme, *Tetrahedron Lett.* **2001**, *42*, 4963–2968; (k) H. Sashida, A. Fujii, T. Tsuchiya, *Chem. Pharm. Bull.* **1987**, *35*, 3182–3189.

35. T Wang, I. S. Claudsdale, *Synthesis* **2000**, *2*, 265–268.

36. (a) T. Su, H. Yang, D. Volkots, J. Woolfrey, S. Dam, P. Wong, U. Sinha, R. M. Scarborough, B. Y. Zhu *Bioorg. Med. Chem. Lett.* **2003**, *13*, 729–732; (b) Y. Okada, H. Taguchi, T. Yokoi, *Tetrahedron Lett.* **1996**, *37*, 2249–2252. (c) H. Sugihara, Z. Terashita, H. Fukushi, EP 0 643 072 A1 **1995**; (d) J. F. Miller, A. Spaltenstein, *Tetrahedron Lett.* **1996**, *37*, 2521–2523; (e) U. Bhatt, N. Mohamed, G. Just, *Tetrahedron Lett.* **1997**, *38*, 3679–3682; (f) B. E. Maryanoff, H.-C. Zhang, J. H. Cohen, I. J. Turchi, C. A. Maryanoff, *Chem. Rev.* **2004**, *104*, 1431–1628.

6

CHEMO- AND REGIOSELECTIVITY ENHANCEMENT IN SOLID-SUPPORTED REACTIONS

Douglas D. Young and Alexander Deiters

6.1 INTRODUCTION

Solid-supported reactions have been extensively used in organic synthesis, as they provide access to the rapid and efficient preparation of arrays of compounds in high purity.[1] Among the benefits afforded by a solid support are facile workup procedures, the ability to conduct reactions under pseudo-high dilution conditions, and the use of excesses of reagents to drive reactions to full conversion. These benefits have been well documented, and solid-supported chemistry has found numerous applications in drug discovery, as well as peptide and oligonucleotide synthesis.[2] A benefit of the solid support, which is perhaps less recognized, is its potential to confer an enhanced degree of chemo- and regioselectivity in various reactions. Owing to the immobilization of a substrate, undesirable side reactions that occur in solution can readily be washed away during purification providing chemoselectivity. Although controversially discussed,[3] "pseudo-high dilution conditions" caused by spatial separation of reactive sites through the solid support can also prevent undesired side reactions, provided that the resin loading is not too high.[4] In addition, the bulk of the polymeric resin can direct the approach of reactants eliciting a degree of regioselectivity. For example, through imposing selective reactivity in a monofunctional fashion within difunctional molecules (e.g., diamines, dialdehydes, diacids). This chapter will summarize selected examples of these effects, but is not meant to be a comprehensive survey of the prevalence of chemo- and regioselectivity in the solid-phase literature. Consequentially, while the solid support is often employed to elicit a degree of selectivity in the preparation of biopolymers (e.g., peptides,[5] oligonucleotides,[6] and carbohydrates),[7] these reactions will

Solid-Phase Organic Synthesis: Concepts, Strategies, and Applications, First Edition.
Edited by Patrick H. Toy and Yulin Lam.

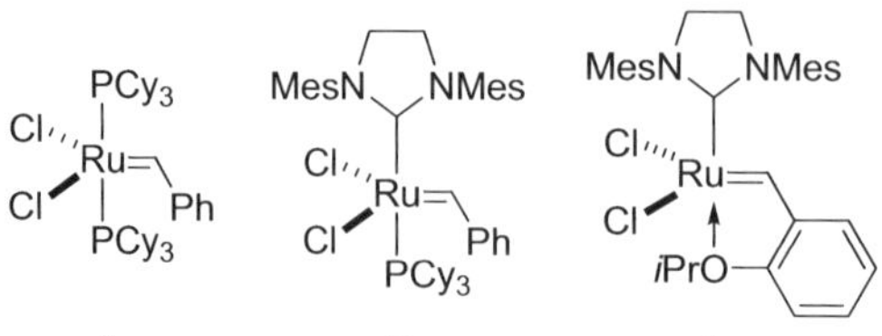

A B C

Figure 6.1. Common olefin metathesis catalysts (Cy = cyclohexyl; Mes = mesitylene).

not be discussed in this chapter due to a focus on small molecules synthesis. Moreover, in many cases the solid support inherently affords a degree of chemo- and regioselectivity as it functions as a protecting group to the site of immobilization. These examples are not discussed either, since these results could be replicated with traditional protecting group under solution-phase conditions.[8]

6.2 TRANSITION METAL-MEDIATED SOLID-SUPPORTED REACTIONS

6.2.1 Olefin Metathesis Reactions

The olefin metathesis reaction has been proven to be an extremely useful tool for carbon–carbon bond formation in organic synthesis.[9,10] Owing to the prevalence of this reaction in the synthetic organic literature, it is not surprising that it has been translated to the solid phase. Several metathesis catalysts, for example, A–C, have been developed and employed on the solid support for various applications (Figure 6.1).[11] Moreover, in several examples, the solid support affords a mechanism for the differentiation of the olefins, allowing the rapid removal of undesirable side products.

6.2.1.1 Olefin Cross Metathesis. Olefin cross metathesis (CM) represents a valuable alternative to traditional carbon–carbon double bond forming reactions; however, its practical utilization is sometimes limited by the generation of multiple homodimerization products (**4** and **5**) when reacting **1** and **2** to obtain the desired heterodimerization product **3** in solution phase (Scheme 6.1a).[12] A solid support can induce chemoselectivity in CM reactions, as the immobilization of one olefin as **6** can effectively prevent its homodimerization (Scheme 6.1b). Moreover, the soluble olefin **2** can be added in excess to drive the reaction to completion. While the soluble olefin is capable of

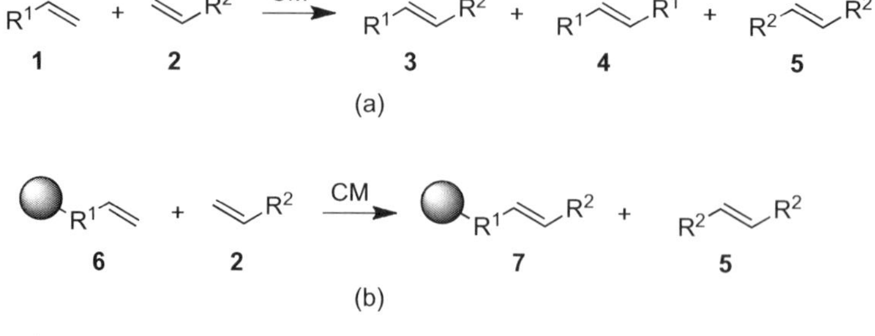

Scheme 6.1. (a) Solution-phase CM reaction lacking chemoselectivity. (b) Chemoselective solid-phase CM reaction.

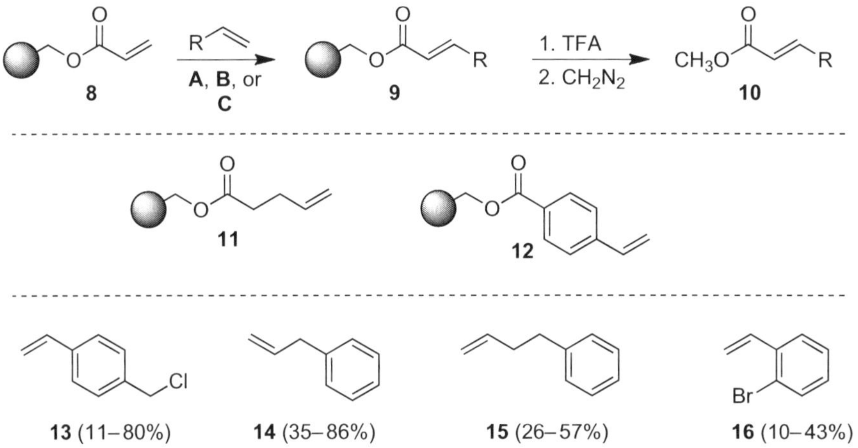

Scheme 6.2. Solid-supported cross-metathesis reaction of immobilized and soluble alkenes.

homodimerization to **5**, the generated side product is easily removed via filtration. However, it should be noted that highly selective CM reactions can be achieved in solution as well.[12]

The chemoselectivity provided by the solid support in the cross-metathesis reaction is evident in ene-ene metathesis reactions, as both olefins are capable of a homodimerization reaction. The Mata laboratory has devoted significant effort toward the understanding and development of cross-metathesis reactions on a solid support. To probe the limits of the reaction, aliphatic, aryl, and acryloyl alkenes **8**, **11**, and **12** were immobilized and reacted with a variety of soluble olefins **13–16** from different classifications based on homodimerization potential (Scheme 6.2).[13] For example, the immobilized acrylate **8** was reacted with several soluble olefins to form the immobilized coupling product **9**. Subsequent cleavage with TFA and esterification with diazomethane delivered the esters **10**. Other immobilized alkenes explored in the same process were **11** and **12**. Among other olefins, they were reacted with the soluble alkenes **13–16**. Modest to good yields (shown in parentheses) were obtained, but, most importantly, no dimerization of the immobilized alkenes **8**, **11**, and **12** was observed, and undesired homocoupling products of the alkenes **13–16** in solution were easily washed away in the workup step.

After investigating the fundamentals of solid-supported CM reactions, Mata and coworkers have extended their solid-supported CM approach toward the preparation of 4-aryl-3-alkenyl-β-lactams (**19**), which have previously been shown to be efficient antibiotics and β-lactamase inhibitors.[14] The β-lactam precursor **17** was prepared on a solid support via the generation of an imine, followed by a Staudinger reaction with Mukaiyama's reagent and crotonic acid to introduce the olefin functionality.[15] A chemoselective cross-metathesis reaction exclusively afforded the product **18** (Scheme 6.3). Gratifyingly, no "intramolecular" metathesis of two immobilized alkenes **17** was observed, as the excess of soluble alkene and low resin loading conditions prevented its formation. In addition, dimerized soluble olefin was removed during the filtration and washing of the resin.

Jeong and coworkers applied a solid-supported CM reaction in the two-step synthesis of (*E*)-stilbenoids **22**. Cross metathesis of substituted styrenes in solution resulted in three products (two homodimers and one heterodimer **22**, see also Scheme 6.1) in an approximate 1:1:1 ratio.[16] However, via immobilization of **20**, the stilbenoids **21** were prepared and cleaved from the solid support in high yields (54–81%) as exclusive heterodimers **22** (Scheme 6.4), and with excellent *E* double bond selectivity (>99:1). Several solid supports

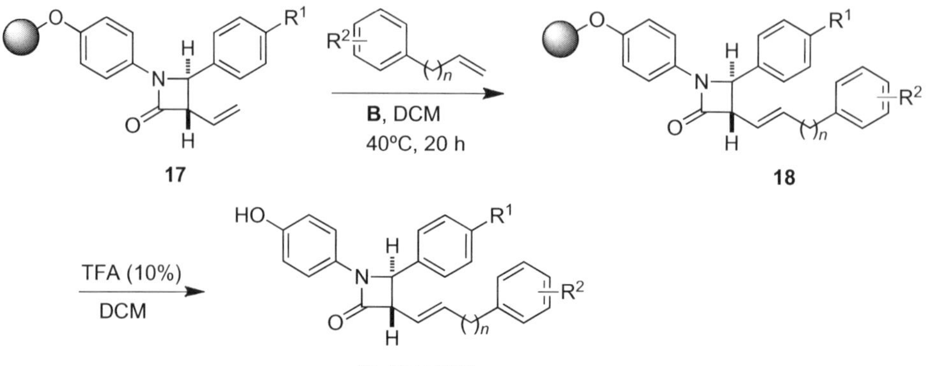

Scheme 6.3. Solid-supported synthesis of 4-aryl-3-alkenyl-β-lactams. The product **19**, was prepared in 35–78%. $n = 0, 1$; $R^1 = $ MeO, HBr; $R^2 = $ 4-OMe, 4-Me, 2-Br, H.

were examined, and Merrifield resin was shown to be the most suitable. This chemoselective approach was applied to the synthesis of resveratrol (**23**), a natural product with anti-inflammatory, HCV inhibitory,[17] and anticarcinogenic activities,[18] in 61% overall yield.

More recently, an alternative approach toward β-lactams was also developed, but with a different application of the cross-metathesis reaction.[19] The synthesis commenced with a chemoselective CM reaction to immobilize the precursor **24**, followed by a similar elaboration to the β-lactam **25** as previously described. However, at this point a second cross-metathesis reaction was conducted to not only afford a new functionality, but also to selectively cleave the molecule from the solid support (**24** → **25**, Scheme 6.5).

Others have also employed this "metathetical cleavage" approach toward the chemoselective cleavage of olefinic substrates. For example, a similar approach was taken using an octenediol-based linker system **26** toward the cleavage of saccharides **27** (Scheme 6.6).[20]

Another unique approach to solid-supported metathesis was examined by Schreiber and coworkers toward the production of homodimeric compounds via an "intra-site" cross metathesis.[21] This strategy used of a highly loaded (1–2 mmol/g), lightly cross-linked (1% DVB) resin **28** to minimize the pseudo-high dilution conditions on the resin surface and facilitate metathesis reactions between immobilized molecules yielding **29** (Scheme 6.7). Again, the *E*/*Z* ratios of the homodimeric products **30** were identical to those prepared in solution, as the solid support had no influence on the ratio of stereoisomers. This approach

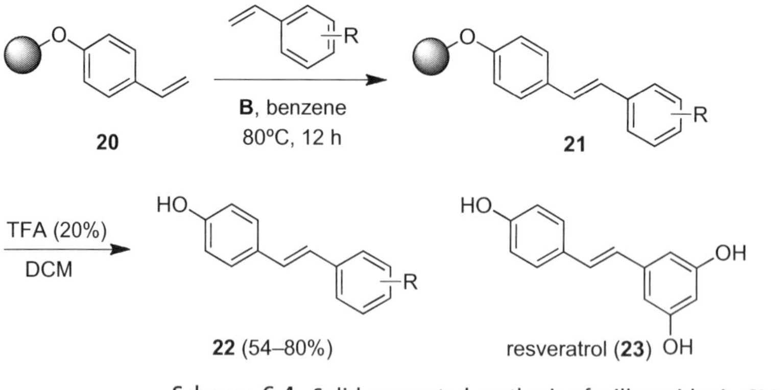

Scheme 6.4. Solid-supported synthesis of stilbenoids via CM.

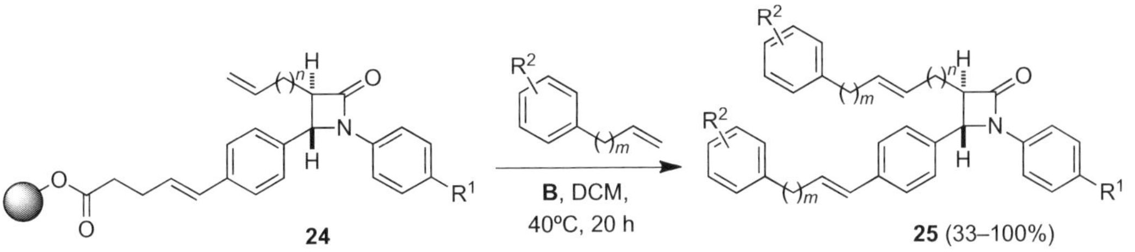

Scheme 6.5. Synthesis of β-lactams followed by resin cleavage via a metathesis reaction. $n = 0, 1$; $m = 0, 1$; $R^1 = F$, OMe; $R^2 = H$, OMe, OAc.

afforded several homodimers that have the potential to be employed as new scaffolds in drug discovery. Liao et al. also adopted this approach toward the preparation of benzo[*b*]furan dimers using high-capacity silica supports. Employing this methodology, they were able to prepare libraries with 80–95% conversion in the metathesis step.[22]

Other examples of CM have been realized by Blechert and Schurer employing ene-yne cross-metathesis reactions to generate 1,3-dienes that can be further elaborated via Diels–Alder reactions or Pd-catalyzed allylic substitutions.[23] Initial studies investigated the immobilization of the olefin as an allylsilane **31**, followed by reaction with a variety of functionalized alkynes in solution.[23] Owing to the nature of the metathesis catalyst, the soluble alkyne (in contrast to a soluble alkene, see Schemes 6.2–6.5) is not capable of a homodimerization reaction and is not required in excess. Moreover, via immobilization of the olefin moiety as **31**, its homodimerization is efficiently restricted, thus affording a chemoselective ene-yne reaction toward **32**. Mild acidic cleavage via a conjugate elimination mechanism yielded the diyne **33** in moderate yields (Scheme 6.8). It should be noted that *E*:*Z* ratios of **33** ranged from 1:1 to 8:1 and appear to be independent of the solid support.

These developments were extended to multistep reactions on the solid support, for example, via immobilization of the alkene or the alkyne component as an ester.[24] The immobilized alkyne **34** created the possibility for a cross metathesis with a soluble olefin; however, as previously noted, undesired CM products generated from soluble alkenes can easily be removed by filtration and washing. After ene-yne cross metathesis to **35**, the resin

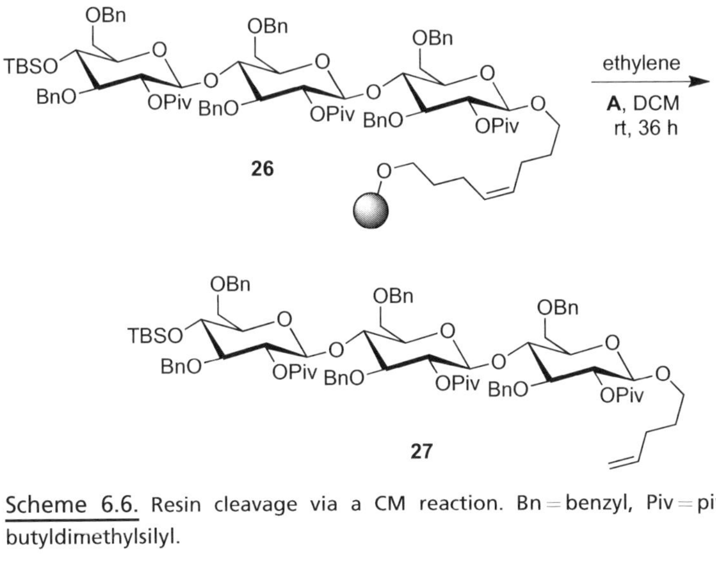

Scheme 6.6. Resin cleavage via a CM reaction. Bn = benzyl, Piv = pivaloyl, and TBS = *tert*-butyldimethylsilyl.

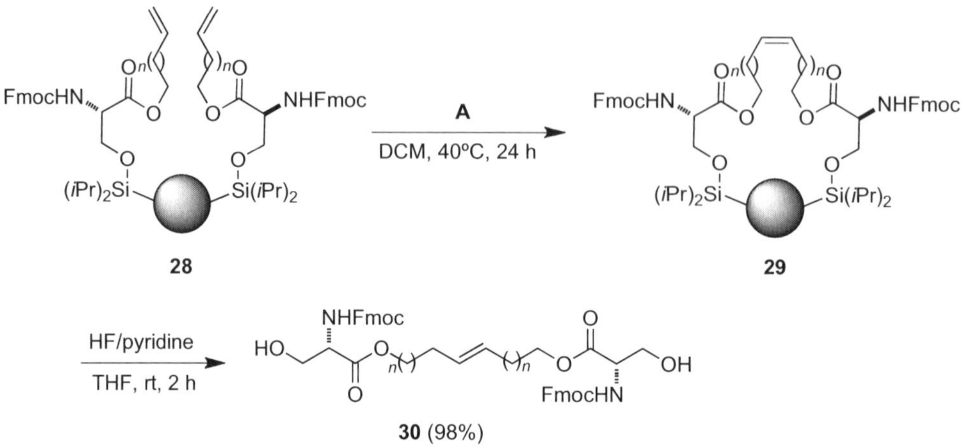

Scheme 6.7. Homodimerization of alkenyl esters of serine immobilized on macrobeads of 500–600 μm consisting of 1% cross-linked polystyrene; $n = 1$–3, 7.

was subjected to a $MeAlCl_2$-catalyzed Diels–Alder reaction. Interestingly, higher yields were obtained with longer spacer units between the diene and the resin. A subsequent reductive amination and lactam formation to **37** leads to a selective release (see Section 6.4.1 for a detailed discussion of cyclizative cleavage approaches) of 1,2,7-trisubstituted octahydrobenzo[*c*]azepin-3-ones (**38**) from the resin (Scheme 6.9). These compounds serve as interesting scaffolds for drug discovery, as octahydrobenzoazepinones have proven to be effective targets of $GABA_A$ receptors in the brain.[25]

6.2.1.2 Ring-Closing Metathesis. Of all olefin metathesis reactions, the ring-closing metathesis (RCM) has found the most widespread application in organic chemistry. Several RCM reactions have been performed on solid supports, occasionally with enhanced chemoselectivity due to immobilization of the RCM precursor, suppressing dimerization/oligomerization side reactions. This strategy has been employed in the preparation of drug-like small molecules, natural products, and macrocyclic peptides.[1a,26] Despite the generation of a carbene on the solid support **40**, these reactions are typically performed with 5–20% catalyst. While standard ring-closing metathesis reactions have been performed on a solid support, the intramolecular nature makes them less interesting from a chemoselectivity perspective.[27] Perhaps more intriguing are approaches that

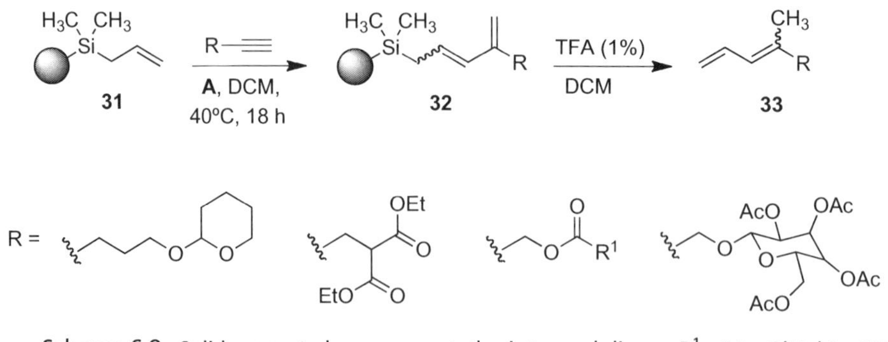

Scheme 6.8. Solid-supported ene-yne metathesis toward dienes; $R^1 = Me$, $C(Me)C=CH_2$.

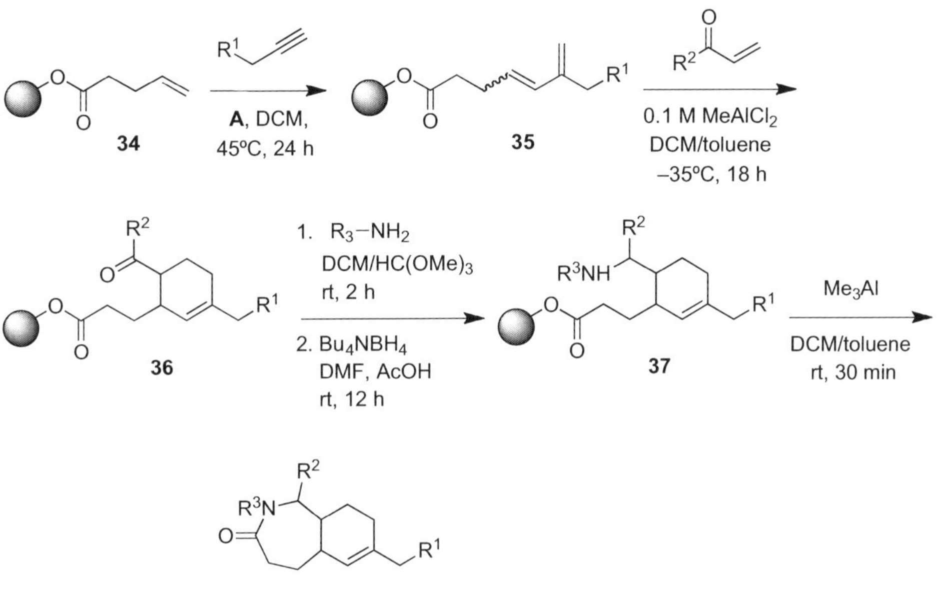

38 (14–28%)

Scheme 6.9. Solid-supported sequence of ene-yne CM and Diels–Alder reactions for the synthesis of octahydropenzoazepinones. $R^1 = CH_2OH$, $(CH_2)_4Me$, OCH_2Ph; $R^2 = Me$, H; $R^3 = CH_2Ph$, $(CH_2)_2Ph$, Me, $(CH_2)_3Me$.

employ immobilized olefins **39**, which are cyclizatively cleaved via an RCM strategy to release the product **41** from the resin (Scheme 6.10). See Section 6.4.1 for a more detailed discussion of cyclizative cleavage approaches and additional examples.

RCM was employed toward the synthesis of a seven-membered lactam **44**.[28] The most pronounced advantage of the solid support is the elimination of undesired metathesis reactions, eliminating the requirement for additional purification. These undesired reactions derive from the residual olefin **45** and its homodimer **46** (Scheme 6.11). However, the reaction progressed more slowly on the solid support requiring 100% catalyst loading and a 54% yield of **44**, compared to a 10% catalyst loading and 97% yield in solution.

Waldmann and coworkers demonstrated this approach in the release of solid-supported products from the resin while preventing undesired side reactions.[29] In the total synthesis of dysidiolide (**48**), a protein phosphatase inhibitor,[30] several model studies were conducted to optimize the metathesis cleavage conditions. The unique linker strategy in **47** enabled an RCM reaction to form an immobilized cyclopentenyl ring **49**, while releasing the product **48** as a terminal olefin (Scheme 6.12). Overall, the metathesis yield was determined to be 82%, and after cleaving the residual compounds from the resin, no cross-linked product was found. However, several reaction intermediates were isolated, demonstrating the benefit of the metathetical cleavage.

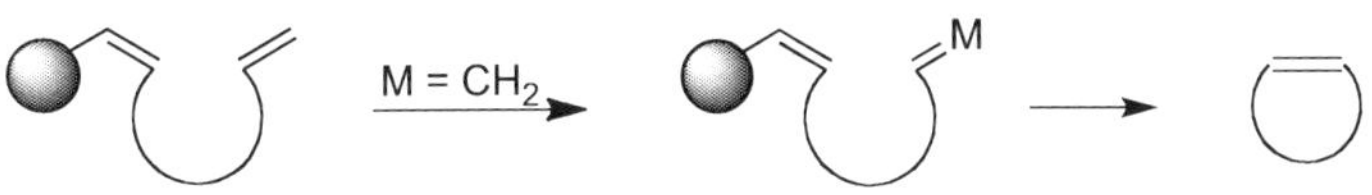

39 40 41

Scheme 6.10. Cyclizative cleavage of a cyclic olefin via an RCM of an immobilized diene.

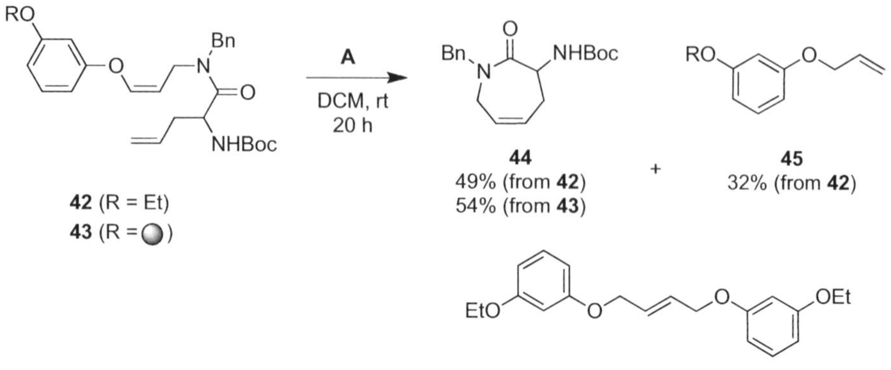

42 (R = Et)
43 (R = ●)

44
49% (from **42**)
54% (from **43**)

\+

45
32% (from **42**)

46
19% (from **42**)

Scheme 6.11. Cyclizative cleavage of the desired product via solid-supported ring-closing metathesis. The undesired byproduct **46** is formed only when the reaction is conducted in solution.

Additional studies have employed this methodology toward the solid-supported preparation of dihydropyrans and pipecolinates, demonstrating the unique ability to afford cleavage from the resin via a carbon–carbon bond formation rather than the traditional carbon–heteroatom cleavage.[26b]

An elegant example of a similar approach was employed toward the total synthesis of epothilone A (**52**) by Nicolaou et al., utilizing the solid support for the synthesis and elaboration of the natural product core scaffold **50** and cleaving it from the resin via RCM to generate the macrolactone **51** (Scheme 6.13).[31] In case of similar solution-phase reactions (without the solid support), high-dilution conditions were necessary to prevent side reactions and obtain the RCM product in high yield.

6.2.1.3 Ring-Opening Metathesis. The ring-opening metathesis (ROM) reaction has also been employed in solid-supported chemistry, but to a lesser extent. However, a solid-supported ROM reaction has been used in the selective formation of bicyclic lactams.[32] Since the bicyclic alkene **53** is immobilized on the resin, any potential side products formed from the solution-phase metathesis homodimerization of the styrene **54** are simply removed in the filtration and washing of the resin (Scheme 6.14). In addition,

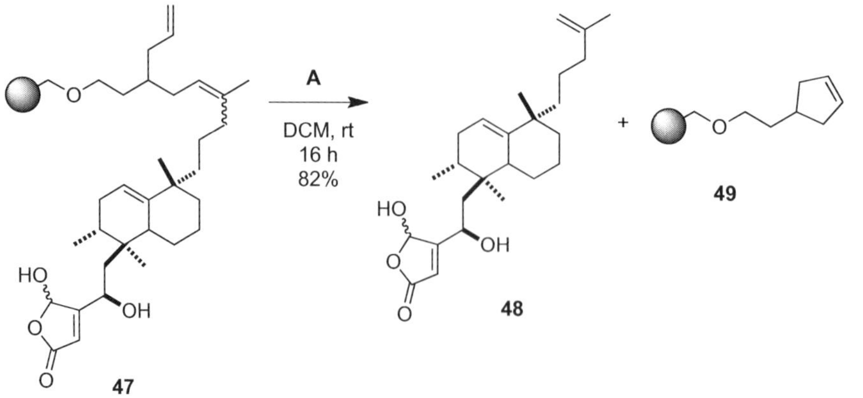

47

48

49

Scheme 6.12. Ring-closing metathesis cleavage toward the synthesis of **48**.

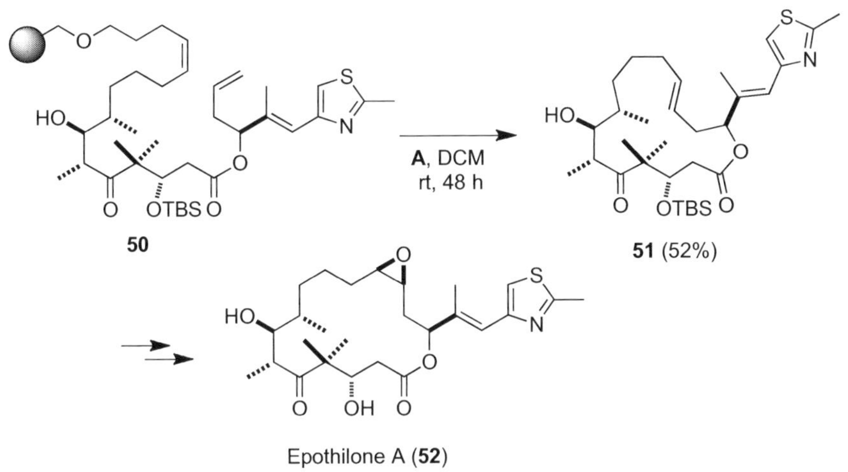

Scheme 6.13. Solid-supported macrolactonization and simultaneous release from the resin via RCM reaction. Product **51** was further elaborated to **52**.

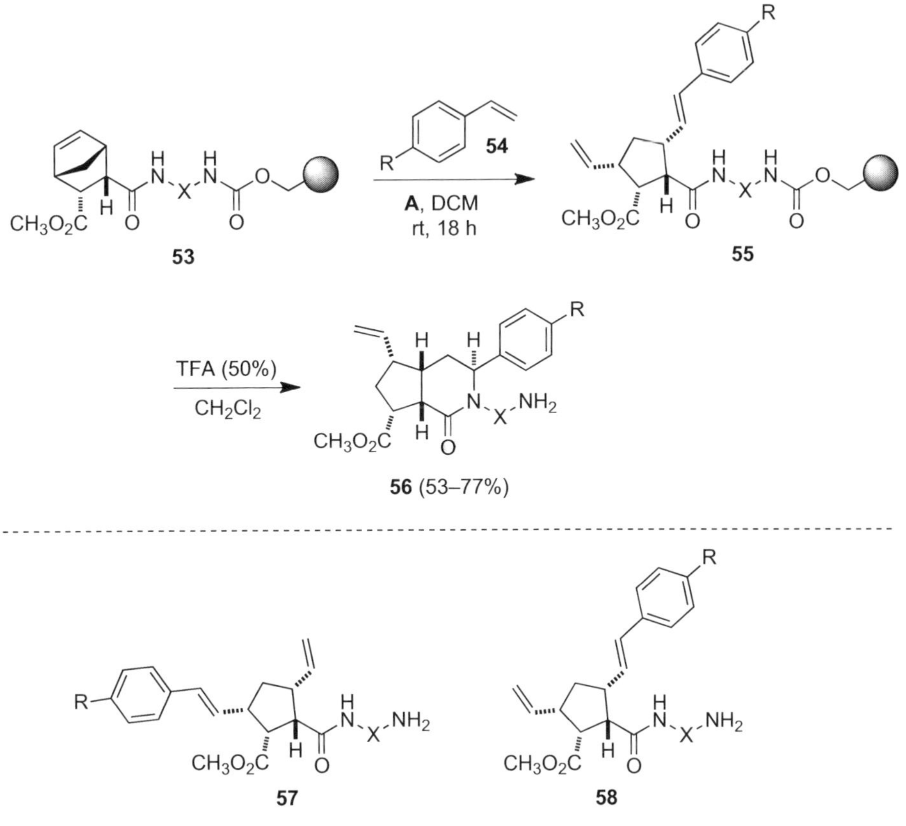

Scheme 6.14. Regioselective solid-supported ring-opening metathesis. X = $(CH_2)_3$, $(CH_2)_2O(CH_2)_2O(CH_2)_2$, *m*-$CH_2C_6H_5CH_2$, piperazine; R = OMe, OPh.

potential dimerization or polymerization of **53** in the ROM step is prevented by the pseudo-high dilution conditions on the solid phase. Perhaps most interesting was the regioselectivity induced by the solid support: based on the orientation of the olefin in the metathesis step, two possible regioisomers, **57** and **58**, are obtained in a ratio of 1.6:1 to 1.3:1 in solution-phase reactions. On the basis of linker choice (linear primary diamines versus cyclic secondary diamines) and resin proximity (Wang resin versus TentaGel resin containing a PEG spacer),[33] exclusive regioselectivity could be obtained in the product **55**. Cleavage of **55** from the resin was performed using 10% TFA to afford the substituted cylopentanes that directly underwent a cyclization reaction to yield the fused bicyclic lactams **56**. Owing to the alterations in regioselectivity with respect to the resin (especially as the TentaGel resin afforded a 3:1 regioisomeric mixture, while the Wang resin afforded exclusively one regioisomer), it was found that the solid support can play a determining factor in the regioselectivity control of a reaction.

6.2.2 The [2 + 2 + 2] Cyclotrimerization Reactions

Due to the chemo- and regioselectivity issues surrounding [2 + 2 + 2] cyclotrimerization reaction, immobilization on a solid support can address chemoselectivity (similar to the case of the metathesis reaction) as the immobilized component can no longer react with itself, and products resulting from side reactions of the soluble components are easily removed in the resin filtration and wash steps. For example, in the case of a crossed [2 + 2 + 2] cyclotrimerization reaction between two different alkynes and a nitrile to yield a pyridine, eight different pyridines can be formed from the incorporation of the different alkynes in different orientations. Of these eight possible pyridines, at least four are typically observed in solution-phase reactions.[34] To address these significant issues, propargyl alcohol was immobilized on a trityl resin (as **59**).[35] This immediately prevents homo-cyclotrimerization reactions due to the pseudo-high dilution conditions of the solid support.[4] [2 + 2 + 2] Cyclotrimerization reactions with different nitriles **61** and different alkynes **60** delivered the pyridines **63** (after cleavage from the solid support **62**) as exclusive chemoisomers, but as mixtures of regioisomers (Scheme 6.15).

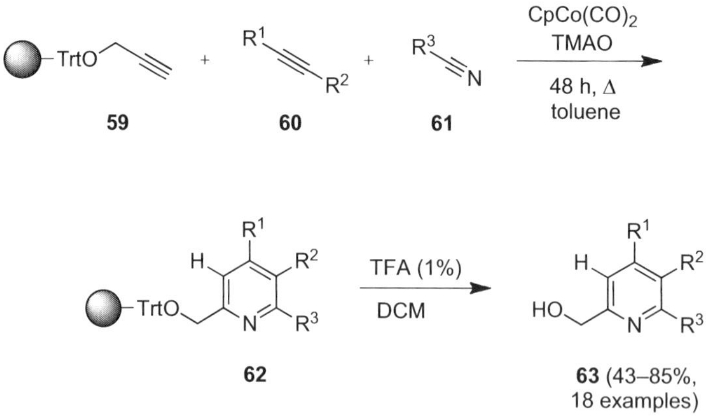

Scheme 6.15. Solid-supported crossed [2 + 2 + 2] cyclotrimerization reaction toward pyridines. R^1 = Bu, Ph, $HOCH_2$, $HOC(CH_3)_2$, $BocNHCH_2$, $MeOCH_2$; R^2 = H, $MeOCH_2$, R^3 = Me, Et, Ph; TMAO = trimethylamine *N*-oxide.

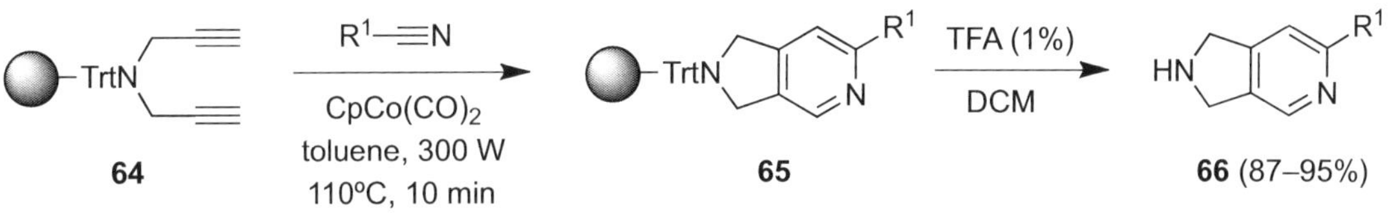

64 **65** **66** (87–95%)

67 (87–96%) **68** (91–93%)

Scheme 6.16. Solid-supported cyclotrimerization reaction to bicyclic pyridines. $R^1 = $ Me, Ph, $CH_2{=}CH$, CH_2 piperazine; $R^2 = $ Ph, Bu, Cy; $R^3 = i$Pr, Cy.

This approach was then extended to the immobilization of diynes, for example, **64**, and their reaction with a soluble nitrile.[36] This immobilization strategy effectively suppressed undesired reactions, including the formation of benzenes via trimerization of the alkynes, and does not require the typically high-dilution and syringe pump conditions required in solution phase.[34b] The reaction of **64** with a variety of nitriles afforded fused pyridines **65** in excellent yields after cleavage from the resin to **66** (Scheme 6.16). In contrast, a solution-phase reaction of trityl-protected dipropargylamine under otherwise identical conditions resulted in only a 46% yield due to both pyridine formation between two diynes and the nitrile, and benzene formation from the reaction of two or three diynes. Similar results were observed in the formation of pyridones **67** (via cyclotrimerization with isocyanates, R^2NCO) and iminopyridines **68** (via cyclotrimerization with carbodiimides, R^3NCNR^3).

The same immobilized diyne **64** was also employed in [2 + 2 + 2] cyclotrimerization reactions with soluble alkynes leading to immobilized benzenes **69**.[37] After cleavage from the resin, the isoindolines **70** were obtained in good to excellent yield and with high purity (Scheme 6.17). A similar chemoselective approach was also accomplished by Martinez and coworkers utilizing microwave irradiation and an iridium catalyst.[38] The selectivity induced by the pseudo-high dilution conditions on the solid support is especially apparent when less reactive internal alkynes are used, for example, 3-hexyne ($R^1 = R^2 = $ Et). In this case, substantial dimerization and trimerization of the diyne starting material, leading to the undesired side products **75** and **76**, was observed.[37,39] Similar approaches have been used for the selective synthesis of indans **71**, phthalans **72**, and indanones **73**, as well as tetrahydroisoquinolines **74** (Scheme 6.17).[40]

Alkynylnitriles are notoriously difficult to react in [2 + 2 + 2] cyclotrimerization reactions, since in the catalytic cycle a reaction between the triple bond of the alkynylnitrile and the external alkyne occurs first, followed by the intramolecular insertion of the tethered cyano group.[34a,41] Thus, these substrates are prone to undesired side reactions with two external alkynes and with another alkynylnitrile. Immobilization on a solid support can alleviate these problems through spatial separation. The immobilized substrate **77** reacted smoothly to the fused pyridines **78** that are subsequently cleaved from the polymeric support to **79** (Scheme 6.18). Control reactions of alkynylnitriles in solution phase under otherwise identical conditions delivered the corresponding fused pyridines in only 8–23% yield.

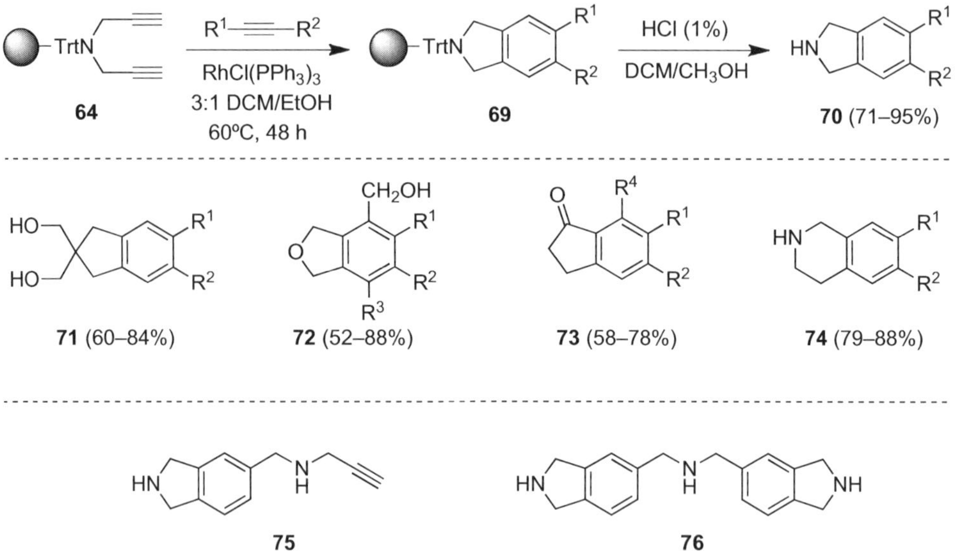

Scheme 6.17. Synthesis of isoindolines and related structures via solid-supported [2 + 2 + 2] cyclotrimerization reactions. R^1 = H, Bu, Ph, CH_2OH, CH_2OBn, CH_2NHBoc, $(CH_2)_3CN$, $SiMe_3$, $(CH_2)_4Cl$, CO_2Me, Et, CH_2OMe; R^2 = H, Et, CH_2OMe; R^3 = H, CH_2OH; R^4 = H, Me, TMS; Trt = trityl linker.

6.2.3 Pauson–Khand Reactions

The Pauson–Khand reaction is a cobalt-mediated cycloaddition of an alkene, an alkyne, and carbon monoxide to yield cyclopentenones.[42] When employing reactive precursors (acetylene or terminal akynes), the reaction proceeds in excellent yield; however, chemoselectivity issues can contribute to decreased reaction yields and unwanted side products in cases where the alkyne component competitively undergoes cyclotrimerization reactions.[43]

Thus, as in case of the [2 + 2 + 2] cyclotrimerization reaction described previously, chemoselectivity could potentially be induced by immobilizing the alkyne reaction partner on a solid support and this prevents its self-trimerization. Schore and coworkers undertook an extensive study to adapt this reaction to the solid phase, optimizing reaction conditions and probing reaction chemoselectivity.[44] The alkyne component was immobilized (**80**) and linker length, resin polymerization (1% versus 2%), alkene substrate, and reaction conditions were altered to identify the optimal parameters (Scheme 6.19). Ultimately, immobilization on a 2% resin led to optimized formation of **81** when using an excess of a soluble

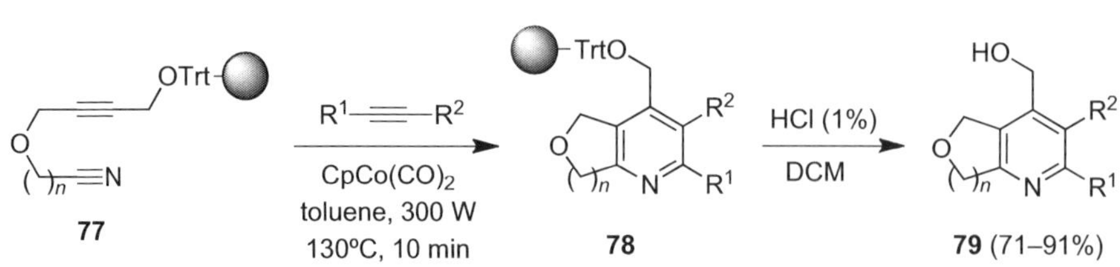

Scheme 6.18. Synthesis of fused pyridines by a chemo- and regioselective solid-supported cyclotrimerization reaction of immobilized alkynylnitriles. n = 1, 2; R^1 = n-Bu, *tert*-Bu, Ph; R^2 = H, Ph; Trt = trityl linker.

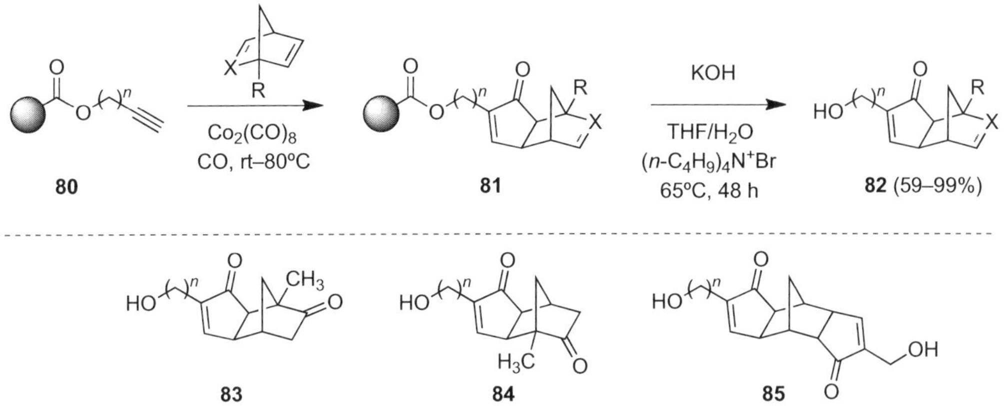

80 **81** **82** (59–99%) **83** **84** **85**

Scheme 6.19. Chemo- and regioselective, solid-supported Pauson–Khand reactions. $n = 1$–3; X = CH, C=O; R = H, Me.

alkene. The Pauson–Khand product **82** was formed in moderate to excellent yields (59–99%) that were dramatically higher than analogous reactions in solution (8–29%). For example, the solution-phase Pauson–Khand reaction between 1-pentynol and norborneone resulted in only an 8% yield with a 3:2 mixture of enones **83** and **84**. The enhanced yield in the case of **80** was attributed to the influence of the solid support, as undesirable side reactions were suppressed via the immobilization of the alkyne. Interestingly, when 1% cross-linked resin was employed, or longer linkers, and stoichiometric quantities of the alkene, the double Pauson–Khand product **85** was observed in 20% yield, as the loss of pseudo-high dilution conditions enabled the reaction of two immobilized alkynes with both double bonds of norbornadiene.

The developed solid-supported, chemoselective Pauson–Khand reaction has been subsequently applied toward the preparation of complex tricyclic scaffolds for diversity-oriented synthesis.[45] The triple bond was introduced into the substrate via a stereoselective Ferrier reaction generating the ene-yne **86**, followed by immobilization on the solid support. The subsequent Pauson–Khand reaction yielded the desired tricyclic product **87** in an efficient fashion (however, no yields were provided), which was further elaborated toward structurally complex small molecule libraries (Scheme 6.20).

6.2.4 Miscellaneous Transition Metal-Mediated Reactions

6.2.4.1 Dötz Benzannulation Reactions. The Dötz benzannulation involves the reaction of a vinylic or aryl chromium Fischer carbene with an alkyne to afford substituted

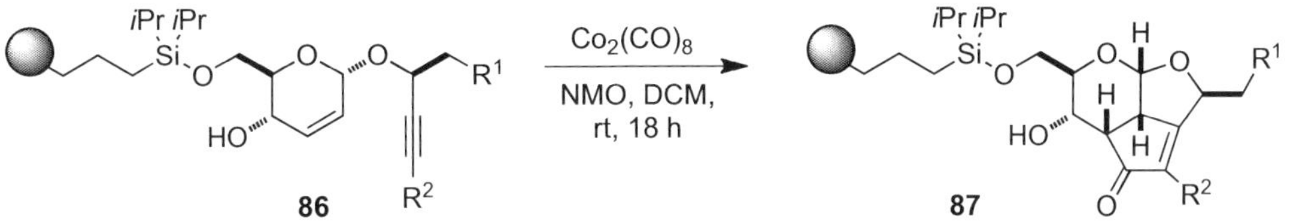

86 **87**

Scheme 6.20. Solid-supported, intramolecular Pauson–Khand reactions in the synthesis of tricyclic compounds **87**. R^1 = Ph, CH_2OH, CH_2NHCH_2Ph; R^2 = H, Ph, piperazine.

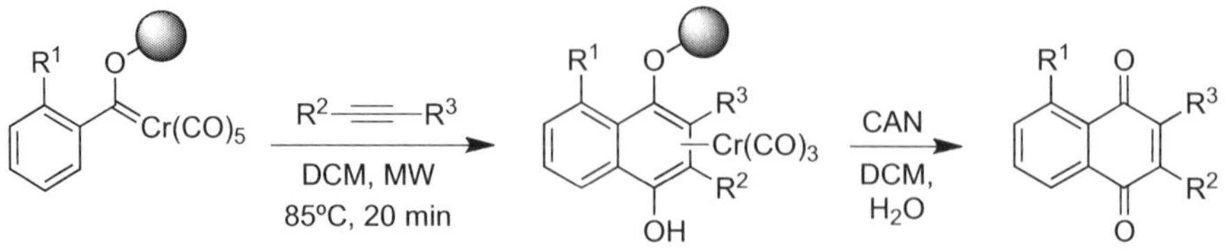

88 **89** **90** (70–99%)

Scheme 6.21. Solid-supported Dötz reaction leading to the chemoselective synthesis of 1,4-naphthoquinones. R^1 = H, OMe; R^2 = H, *n*-Pr, Et, Me, Ph; R^3 = Ph, C_5H_{11}, C_6H_{13}, C_3H_6OH, CO_2Et.

phenols.[46] Based on the mechanism of the reaction, different products and side products, including naphthols, indenes, furans, and cyclobutanones, can be formed. Martinez and coworkers reported the first example of a solid-supported Dötz annulation, by immobilizing the Fisher carbene **88** on a Wang resin.[47] The chromium species was reacted with various alkynes under microwave irradiation to furnish 1,4-naphthoquinones **89** in moderate to good yields, which were successively cleaved from the resin to **90** (Scheme 6.21). While the observed regioselectivity was identical to the corresponding solution-based reactions, exclusive chemoselectivity was observed on the solid support, as no indene, indenone, or cyclobutenone products were observed. This is in contrast to identical reactions in solution phase that led to the formation of all four possible products.[47]

6.2.4.2 Cadiot–Chodkiewicz Coupling Reactions. Conjugated diynes are not only structural motifs found in natural products, but also have become increasingly prevalent in materials chemistry.[48] Owing to the nature of the coupling reaction, chemoselectivity issues can result from a competition between cross-coupling and homocoupling of alkyne starting materials. The Cadiot–Chodkiewicz reaction represents the coupling of a terminal alkyne **92** with a 1-haloalkyne **91** in the presence of a Cu(I) catalyst.[49] While preparation of symmetrical products is relatively trivial, the generation of unsymmetrical diynes **93** is complicated by potential homocouplings of 2 equiv of **91** to **94**. Schore and coworkers have exploited the advantages of the spatial separation of alkynes on a solid support to eliminate these issues and exclusively provide the cross-coupling product **93**.[50] Immobilization of the 1-iodo- or 1-bromoalkyne substrate **95** and reaction with a soluble terminal alkyne yielded exclusively the cross-coupled product **96** after cleavage from the resin (Scheme 6.22). This can most likely be attributed to the spatial separation of the immobilized haloalkanes. Identical reaction conditions in solution phase delivered the undesired homocoupling product **94** in up to 34% yield.

6.2.4.3 Cyclopropanation Reactions. Davies and Nagashima reported the first example of a catalytic asymmetric cyclopropanation of alkenes on a solid support.[51] Carbene dimerization represents a limitation in solution phase, lowering yields and necessitating additional purification steps.[52] Immobilization of the olefin **97** on a polystyrene diethylsilyl resin followed by reaction with various diazoacetates in the presence of a rhodium catalyst generated the cyclopropanes **98** and **99** in high yield and enabled the removal of dimerization products **102** through a simple wash step (Scheme 6.23). The products **100** and **101** were cleaved as a mixture of diastereomers from the resin under mild conditions. The stereoselectivity of the reaction was not influenced by the solid support, but rather by the catalyst selection; most important, >90% ee was observed under these conditions.

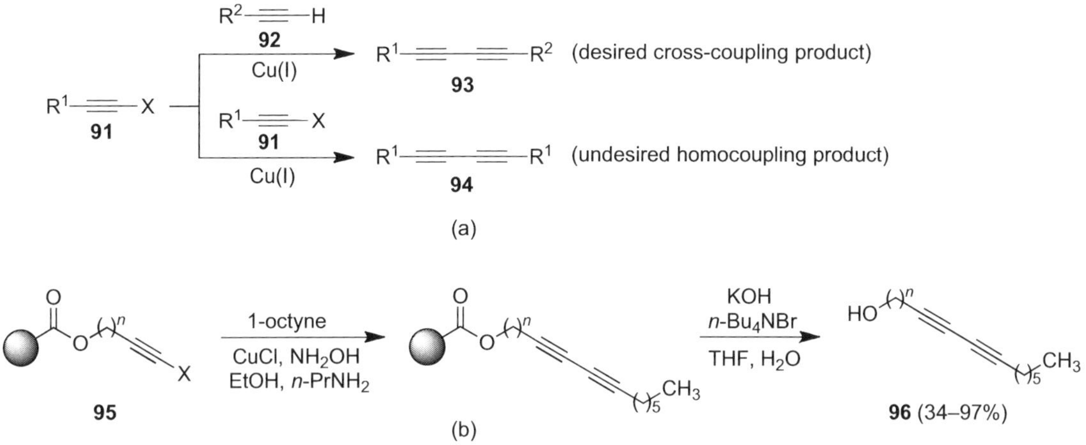

Scheme 6.22. (a) Solution-phase Cadiot–Chodkiewicz reactions deliver undesired homocoupling products. (b) The same reaction can be selectively performed on a solid support. $n = 1–4$; X = Br, I.

6.2.4.4 Heck Reactions. Owing to the nature of transition metal-catalyzed cross-coupling reactions, and extensive development in organic chemistry, chemo- and regioselectivity issues are rarely a problem.[53] However, the solid support has been proven to be advantageous for several palladium cross-coupling reactions, increasing yield and decreasing reaction times.[54] Bolton and Hodges have described the solid-supported preparation of substituted benzazepines.[55] Solution-phase intramolecular Heck cyclization to **105** was achieved only at yields below 40% due to reductive deiodination (toward **106**) or polymerization; however, via immobilization of the precursor **104** as **103**, a

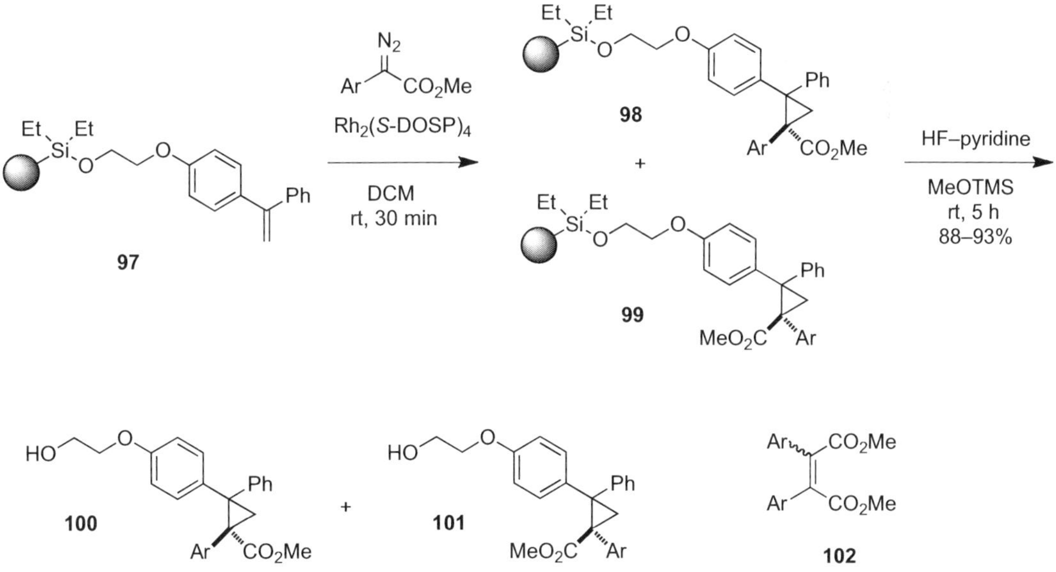

Scheme 6.23. Solid-supported cyclopropanation reaction. The immobilization of the olefin **97** prevented the isolation of the undesired side product **102**. $Rh_2(S\text{-DOSP})_4$ = tetrakis[*N*-[(4-dodecylphenyl)sulfonyl]-(*S*)-prolinato]dirhodium; Ar = Ph, 2-naphthyl, 4-CH_3-C_6H_4, 4-CH_3O-, C_6H_4, 4-CF_3-C_6H_4, 4-Br-C_6H_4, 4-Cl-C_6H_4.

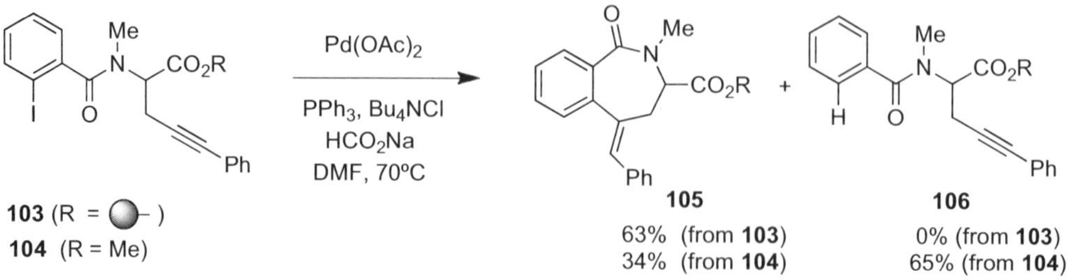

Scheme 6.24. Solid-supported Heck cyclization toward the preparation of benzazepines **105**.

63% overall yield was achieved with no deiodinated material (Scheme 6.24). The basis of this selectivity was not fully elucidated, but Bolton and Hodges hypothesized that either the effective concentrations were approximately five-fold less on the solid support or the inorganic materials from previous solid-supported reactions employed in the synthesis of **103** were trapped within the polymeric support and enhanced the Heck cyclization.

Akaji and Kiso demonstrated another interesting example of a Heck cyclization toward the macrocyclization of peptides.[56] The cyclization reaction was facilitated by the pseudo-high dilution conditions afforded by the solid support to reduce dimer or oligomer formation. An iodobenzylamine was coupled to Fmoc-Asp(O*t*Bu)-OH, which was then deprotected and immobilized on a solid support. Standard Fmoc-based peptide synthesis was then performed and the peptide was capped with an acrylic acid monomer. The Heck macrocyclization was then performed on both the solid support and in solution phase. Interestingly, based on the HPLC analysis of the reaction mixture, the cyclization on the solid support was generally completed within 2 h (affording a 30% overall yield), while the corresponding solution-phase reaction required approximately 8 h (the yield of the solution-based reaction was not reported).

6.3 NON-TRANSITION METAL-MEDIATED SOLID-SUPPORTED REACTIONS

6.3.1 Cycloaddition Reactions

The 1,3-dipolar cycloaddition reaction has been investigated on a solid support primarily toward the construction of diverse small molecule libraries, as the reaction is capable of preparing a variety of privileged five-membered heterocycles.[57] Typically, chemoselectivity is high, and the solid support provides only a mechanism for the rapid preparation of pure compounds in a parallel fashion; however, in select cases the solid support has been demonstrated to enhance both the chemo- and the regioselectivity of the reaction.[58]

The 1,3-dipolar cycloaddition has been employed in the synthesis of tetrahydrofurans via the reaction of a 1,5-diene **107** with a nitrile oxide to generate an isoxazoline intermediate **108**.[58a] In solution phase, the diene **107** can react twice with the nitrile oxide, generating the bis-isooxazoline by-product **109** (Scheme 6.25a). This side reaction could be suppressed by generating the nitrile oxide *in situ* from the solid-supported precursor **110** (Scheme 6.25b). Consequentially, after the first 1,3-cycloaddition reaction, the product **111** was immobilized, preventing the undesired bis-cycloaddition product **109** through the pseudo-high dilution conditions found on the solid support. Release from the resin and simultaneous formation of the tetrahydrofuran **112** was achieved by treatment with ICl. A markedly higher yield of 40% was observed for the solid-supported reaction compared to an 18% yield in solution phase.

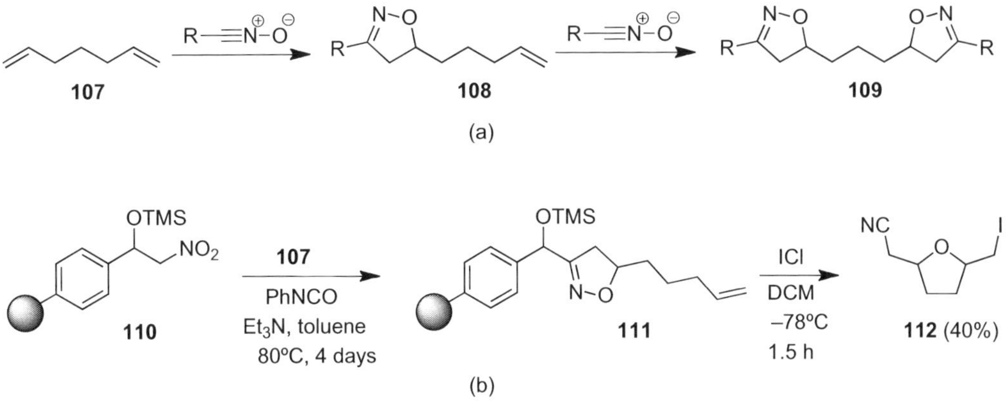

Scheme 6.25. (a) Solution-phase [3 + 2] cycloaddition toward isooxazolines and bis-isooxazolines. (b) Performing the reaction on a solid support enables selective isooxazoline formation followed by cyclization to tetrahydrofurans under simultaneous cleavage from the resin.

Interesting regioselectivity effects in 1,3-dipolar cycloaddition reactions have been observed by several investigators,[58b,58c] as specific substrates appear to be susceptible to the influence of the solid support. An early example was reported by Leznoff and Yedidia in the reaction of benzonitrile oxide with a soluble or an immobilized propiolate ester **113** toward the *meta*- and *ortho*-isoxazoles **114** and **115**, respectively (Scheme 6.26).[58b] In case of the soluble methyl ester ($R = CH_3$), almost exclusive formation of the *ortho*-isomer **115** was observed (Scheme 6.26). In contrast, when the same reaction was conducted on a Wang resin, the regioselectivity was completely reversed, delivering exclusively the *meta*-isomer **114**. A benzyl ester ($R = CH_2Ph$) led to the formation of approximately a 1:1 ratio of **114** and **115** in solution phase. In addition, yields of solid-supported reactions were substantially higher (80–81%), compared to their solution-phase counterparts (30–66%).

Similarly, Back and coworkers reported an increased regioselectivity in the reaction of ethyl diazoacetate with acetylenic sulfones **116** to yield the pyrazoles **117** and **118** (Scheme 6.27).[58c] In solution, the same acetylenic sulfone generated regioisomers in a 4:1 mixture, while a single regiosiomer was afforded on the solid support. Based on these results, it appears that the solid support is capable of directing the approach of the diazoacetate, perhaps due to steric effects.

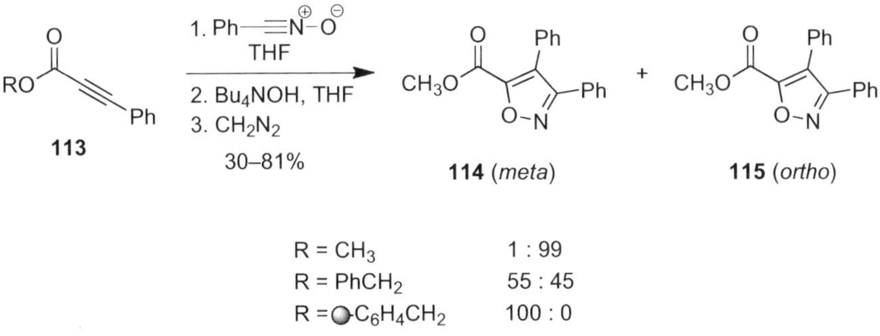

Scheme 6.26. Regioselectivity in isoxazole formation on a solid support compared to solution phase.

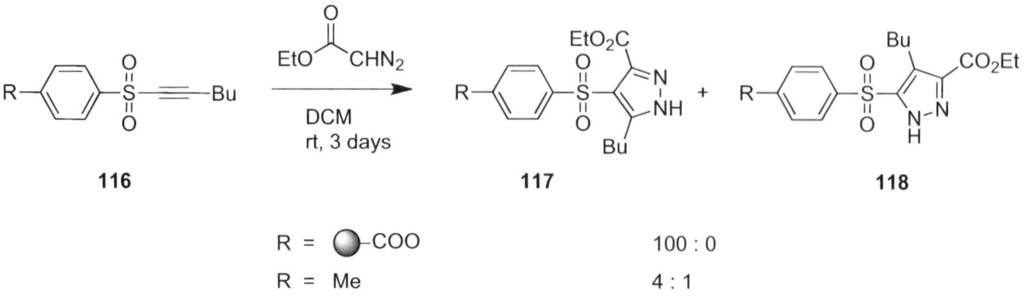

Scheme 6.27. Dipolar cycloadditions of acetylenic sulfones toward the synthesis of pyrazoles.

Enhanced chemoselectivity in solid-supported cycloaddition reactions was also demonstrated by Bilodeau and Cunningham in their preparation of imidazole libraries.[59] Conducting a [3 + 2] cycloaddition on a solid support suppressed self-condensation of the intermediate **120** and significantly increased reaction yields and purities of the obtained imidazoles. Immobilized **119** was reacted with EDC to form the intermediate **120**, which then reacted with a tosylimine to afford the immobilized imidazole **121** (Scheme 6.28). A unique intermediate cleavage step using 90% TFA at room temperature was then performed to remove unreacted materials and undesirable by-products that do not possess an imidizole functionality, rendering them less stable to acidic cleavage. Subsequent resin cleavage in glacial acetic acid at 100°C provided a diverse imidazole library **122** in high yield (63–99%) and with high purity (94–98%).

Diels–Alder cycloadditons have been explored on solid supports, and Winkler and Kwak have demonstrated a chemoselective variant that affords a single reaction of a dienophile containing two reactive double bonds with a bis-diene.[60] Utilizing bis-reactive Diels–Alder reagents in the absence of a solid support is problematic as competing oligomerization can occur. Thus, a dienophile was immobilized as **8** and reacted with the

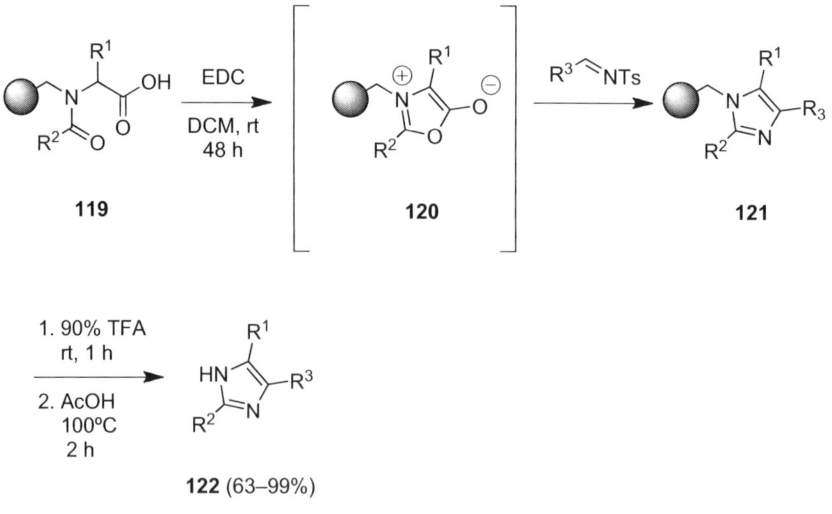

Scheme 6.28. Solid-supported imidazole synthesis via [3 + 2] cycloaddition. $R^1 = Ph$, 4-F-C_6H_4; $R^2 = Ph$, 4-F-C_6H_4, 4-MeO-C_6H_4; $R^3 =$ 3-pyridyl, 4-pyridyl.

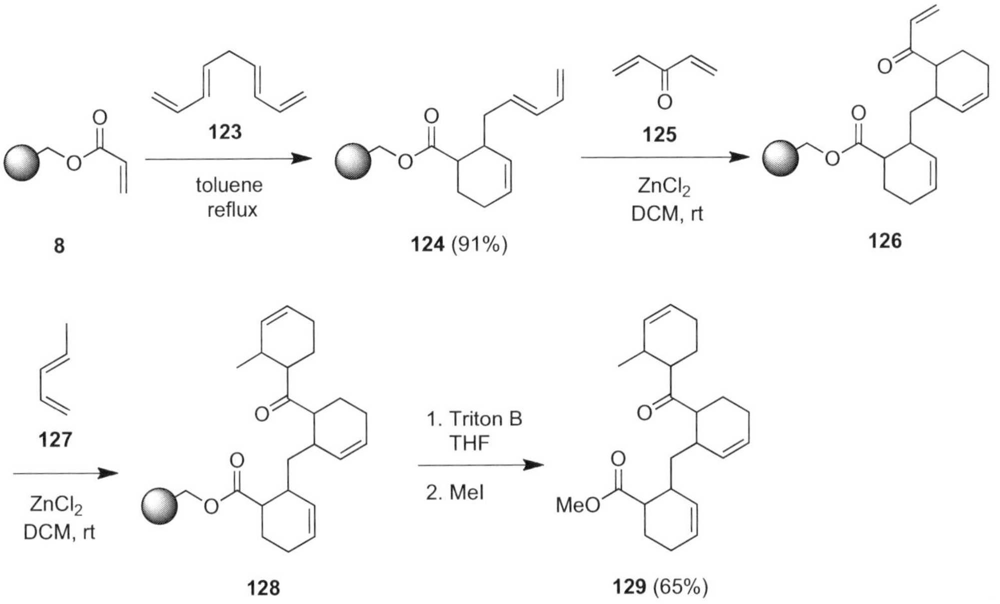

Scheme 6.29. Sequential, controlled Diels–Alder reactions on a solid support enable the selective formation of polycyclic structures.

bis-diene **123**, affording only the single cycloaddition product **124** in 91% yield with no oligo- or polymerization by-product (Scheme 6.29). The comparable solution-phase reaction led to only a 51% yield as a result of undesirable by-products. The product **124** was further elaborated via a second [4 + 2] cycloaddition reaction with the divinyl ketone **125** to generate the double Diels–Alder product **126**, which itself contains a new immobilized dienophile for an additional [4 + 2] cycloaddition reaction. The subsequent reaction with **127** led to the formation of the tricyclic molecule **128**. This iterative process provided a highly controlled oligomerization, to **129** in 65% overall yield after cleavage from the resin, eliminating undesired side reactions. Comparative solution-phase reactions afforded only an overall 24% yield due to polymerization by-products, thus illustrating the enhanced chemoselectivity of solid-supported reactions.

6.3.2 Hydroxylation Reactions

An application of the solid support was investigated toward the stereoselective hydroxylation of vitamin D_3 (Scheme 6.30).[61] The allylic oxidation of **130** with SeO_2 can result in the 1α- and the 1β-product, **134** and **135**, respectively; however, the 1α-product **134** is generally favored. This can be explained by analyzing the bicyclic transition state of the reaction, as positioning the 3-oxy substituent in the axial position results in an unfavorable 1,3-diaxial interaction in **133** compared to **132**. Li and coworkers demonstrated that in the case of an immobilized substrate **131**, a higher degree of stereoselectivity is obtained than in the case of a solution-phase reaction using a simple protected 3-$OSiPh_3$ substrate **130**.[61] In addition, the stereocontrol appears to be influenced by the flexibility of the polymeric support, as increased cross-linking and polar solvents (resulting in nonswelled resin) decreased the resin flexibility and increased the stereoselectivity.

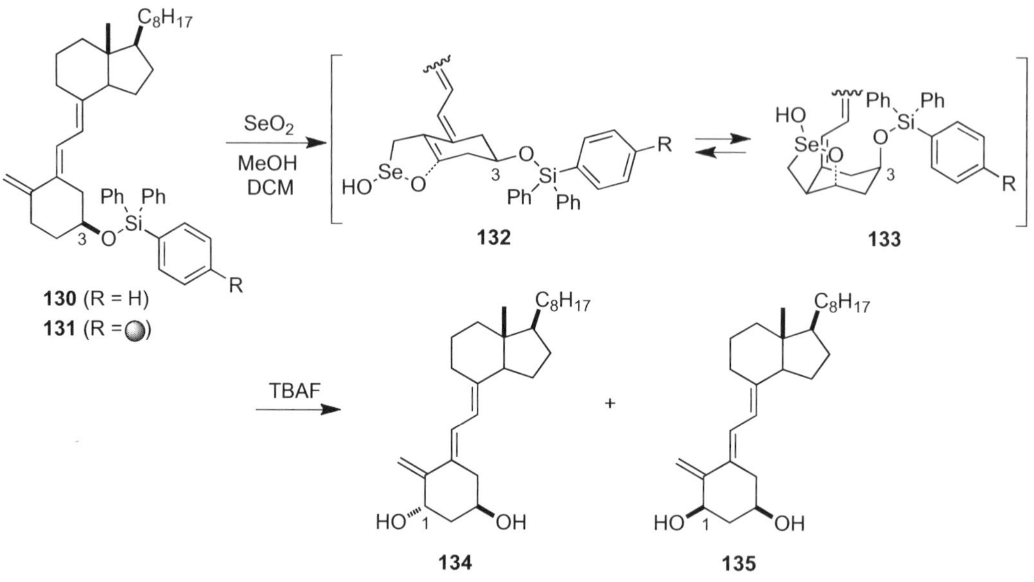

Substrate	MeOH/DCM	134/135
130	1:3	5.3:1
130	3:1	5.2:1
131 (2%)	1:3	6.0:1
131 (2%)	3:1	6.3:1
131 (8%)	1:3	6.4:1
131 (8%)	3:1	6.7:1

Scheme 6.30. Enhanced stereoselectivity in the allylic hydroxylation of immobilized vitamin D_3. The resin cross-linking is shown in parentheses.

6.3.3 Aldol Condensation Reactions

The aldol condensation reaction has been extensively employed in organic chemistry toward the formation of carbon–carbon bonds. Early work by Leznoff and Wong demonstrated the ability to chemoselectively differentiate the carbonyl groups in dialdehydes through selective protection via immobilization on a solid support.[62] In addition, given the presence of an immobilized aldehyde **136**, it was feasible to conduct crossed aldol reactions toward **137** and **138** in high yields (96–100%) without the presence of undesired side products (Scheme 6.31).

6.3.4 Radical Reactions

The Naito laboratory has investigated the applicability of radical reactions in solid-supported chemistry.[63] A pronounced effect of the solid support on the stereochemical outcome of a radical reaction was reported for the addition of an alkyl radical onto the chiral oxime **139**, delivering α-amino acid derivatives **140** in good yield and with high stereoselectivity (Scheme 6.32). Higher diastereoselectivities were observed with the substrate **139** immobilized on the solid support (>90% de), compared to comparable solution-phase reactions of **139** (72–85% de).[63b] This effect was rationalized based on the lower reactivity and reduced reaction rate of the immobilized oxime, inducing an enhanced selectivity.[64]

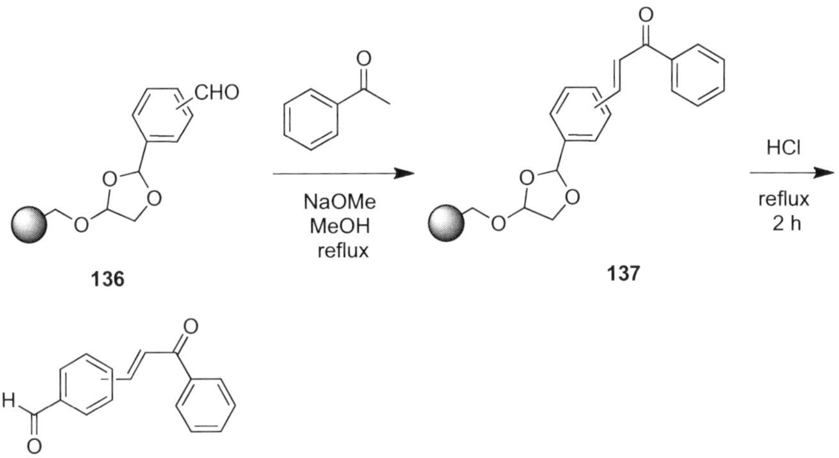

138 (96–99%)

Scheme 6.31. Crossed aldol reaction of an immobilized dialdehyde **136**.

6.3.5 Oxidative Coupling Reactions

The chemoselectivity afforded by the solid support has been exploited toward the oxidative heterocoupling of carpanone-like molecules by the Shair laboratory.[65] The seminal work of Chapman et al. originally synthesized carpanone by a diasteroselcctive oxidative homocoupling of an *ortho*-hydroxystyrene followed by a Diels–Alder cycloaddition reaction.[66] While synthetically elegant, this domino process limits the ability to selectively achieve heterocoupling of different phenols in order to assemble arrays of carpanone analogues. However, this was accomplished via immobilization of the electron-rich phenol **141** onto a solid support. Utilizing a silyl-based linker strategy, heterocoupled/homocoupled product **142** was obtained at a 5.3:1 ratio, achieving a selectivity that could not be achieved in solution. The final product was then rapidly cleaved with HF–pyridine to afford **143** in high purity. To assess the generality of the heterocoupling reaction, a small compound collection

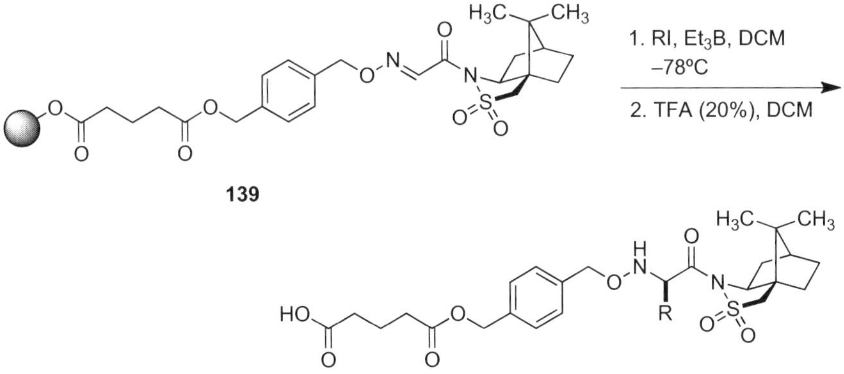

140 (41–74%, >90% de)

Scheme 6.32. Enhanced diastereoselectivity in a solid-supported radical addition onto the oxime **139**. R = Et, *i*-Pr, Cy.

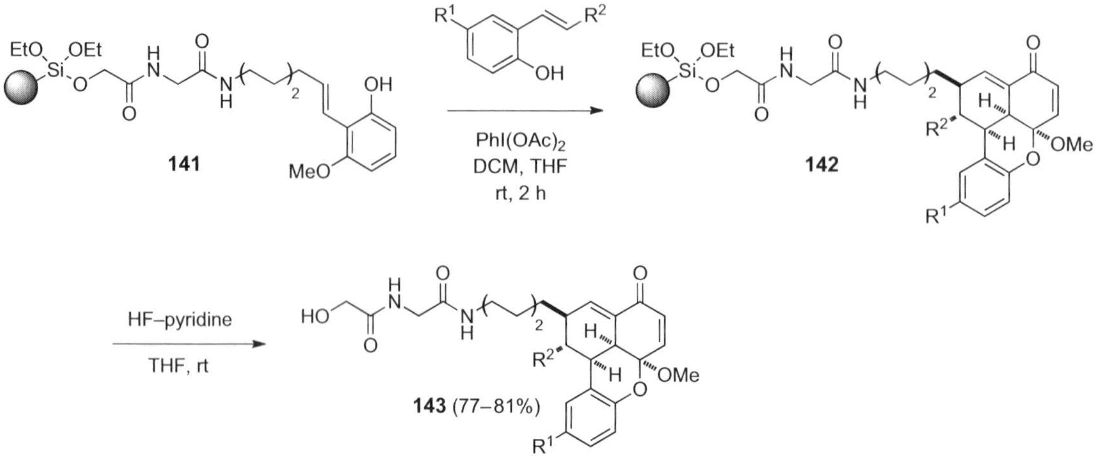

Scheme 6.33. An oxidative heterocoupling on the solid support enables the synthesis of carpanone analogues **143**. $R^1 = CONHCH_2C_6H_4Br$, CO_2Me, CO_2t-Bu; $R^2 = Me$, Et, *i*-Pr, $CH_2CH_2OCH_2Ph$.

(five examples) was prepared affording the heterocoupled products in 77–81% yield (Scheme 6.33).

6.4 TRACELESS CLEAVAGE

6.4.1 Cyclizative Cleavage

It is possible to choose an immobilization strategy that exploits a functional group introduced during the solid-phase chemistry to eventually cleave the synthesized molecule from the resin in a traceless fashion.[67] This is outlined in Scheme 6.34 for a traceless cyclizative cleavage, where the immobilized molecule **144** undergoes a functional group transformation in step a (R $\rightarrow$ Y) to **145**. Even if this transformation is incomplete, only the modified molecule **145** can undergo cyclizative cleavage to **146** in step b, forming a covalent bond between the functional groups X and Y. A simple filtration step can then be used to remove any unreacted starting material **144**, thus greatly facilitating purification of the product **146**. The same approach can be used to remove not only starting materials, but also undesired stereochemical isomers that cannot undergo a cyclization reaction. This approach

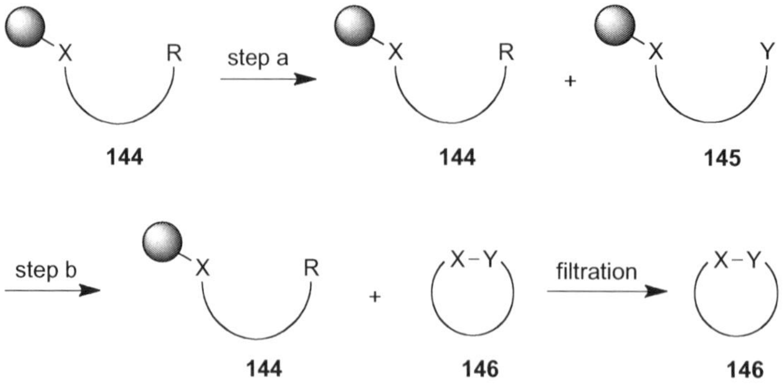

Scheme 6.34. General schematic of a cyclizative cleavage approach.

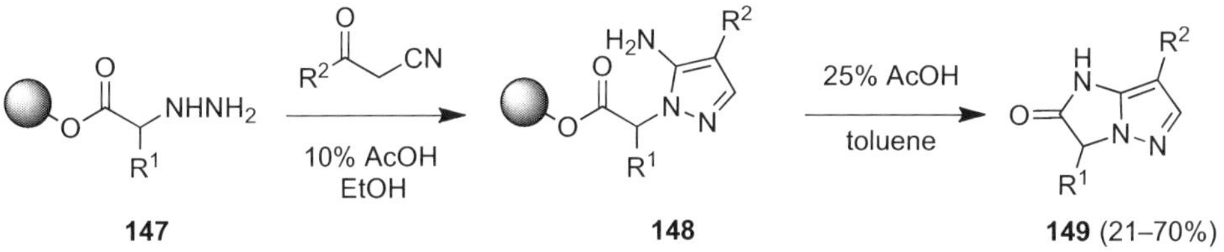

Scheme 6.35. A cyclizative cleavage approach toward the synthesis of imidazo[1,2-*b*]pyrazol-2-ones. R^1 = Bn, $(CH_2)_2Ph$, $CH_2C_6H_4$-4-OCF_3, $CH_2CH(CH_3)_2$; R^2 = *t*-Bu, Ph, 4-MeO-C_6H_4, 4-Cl-C_6H_4.

was previously shown (Section 6.2.1.2 and Schemes 6.11–6.13), as only products that underwent the desired reaction were liberated from the resin, while side-metathesis reactions were unproductive and generated by-products remained immobilized.

For example, this cyclizative cleavage approach was employed toward the solid-supported synthesis of functionalized imidazo[1,2-*b*]pyrazol-2-ones, which have been shown to have antitumor and antiviral activities.[68] Immobilization of racemic amino acid derived hydrazines **147** followed by treatment with malononitriles afforded amino pyrazoles **148** that possessed a nucleophilic nitrogen in the proper orientation to undergo cyclizative cleavage under lactam formation to **149** (Scheme 6.35).[69]

A nucleophilic epoxide ring opening has been employed in the generation of functional groups capable of undergoing a cyclizative cleavage. This approach generates a reactive nucleophile that can undergo an intramolecular nucleophilic attack onto a carbonyl-based linker to liberate the product from the solid support. Owing to the chemoselective cleavage, undesired side products remain immobilized, affording high yielding reactions with high compound purity. This was employed toward the preparation of γ- and δ-lactones, present in pheromones and antifungal substances.[70] Immobilization of an alkenoic acid on a polystyrene support **150** followed by epoxidation yielded the precursor **151**. Nucleophilic ring opening of the epoxide generated **152**, which subsequently underwent lactonization to **153**, releasing the product from the resin in good to high purity and as a 1:1 mixture of diastereoisomers (Scheme 6.36).

In a similar fashion, oxazolidiones were prepared on a solid support via a cyclizative cleavage.[71] The oxazolidinone scaffold can be found in antibacterial agents and MAO inhibitors.[72] As in the previous example, epoxide ring opening was used to generate a

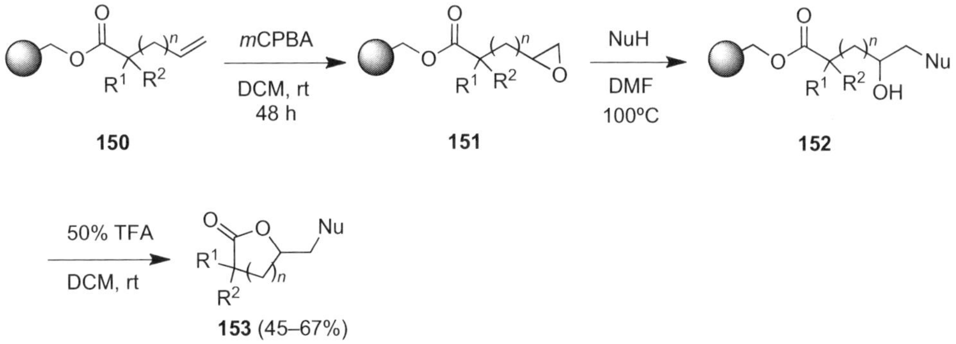

Scheme 6.36. Preparation of lactones on a solid support via cyclizative cleavage. R^1 = H, Me; R^2 = H, Me; *n* = 1, 2; Nu = N_3, PhS, 4-MePhS.

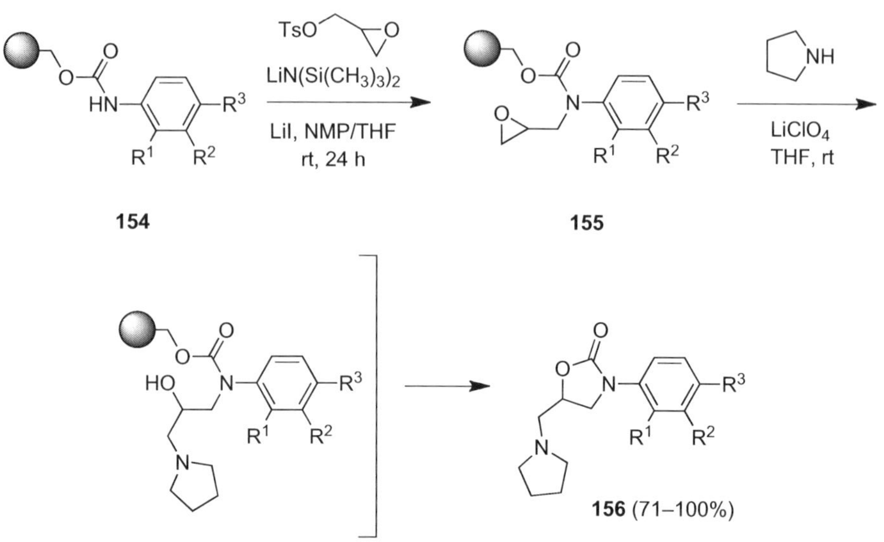

Scheme 6.37. Oxazolidinones synthesis on a solid support via cyclizative cleavage. R^1 = H, OMe, NO_2, Cl; R^2 = H, CN; R^3 = H, CN, OMe, Br, CO_2Et, NO_2; Ts = 4-toluenesulfonyl.

nucleophilic hydroxy group capable of inducing a cyclizative cleavage. Initially, a carbamate was immobilized on the Wang resin **154**, which was alkylated to yield the epoxide **155**. The epoxide was opened with pyrrolidine, leading to spontaneous cyclization and cleavage of the final product **156** from the solid support (Scheme 6.37). This cyclizative cleavage affords the oxazolidinones **156** in high purity, leaving by-products resulting from incomplete reactions immobilized on the solid support.

A cyclizative cleavage approach has been used to efficiently and selectively release final products from solid-supported multistep reactions. This approach eliminates side products produced by incomplete reactions as they do not possess the chemical functionalities in a proper arrangement to undergo the cyclizative cleavage reaction. This has been demonstrated in the synthesis of small molecule libraries (Scheme 6.38). A collection of β-carbolines were prepared on the solid support via immobilization of tryptophan **157** on a

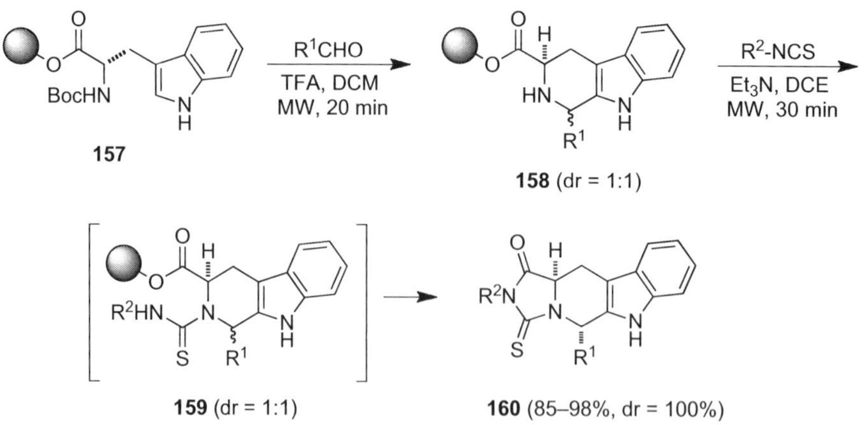

Scheme 6.38. A cyclizative cleavage approach toward the synthesis of carbolines. R^1 = *n*-Bu, *n*-pentyl, *n*-hexyl, piperonyl, $(CH_2)_3Ph$; R^2 = *n*-Bu, Bn, allyl, cyclopentyl, Ph.

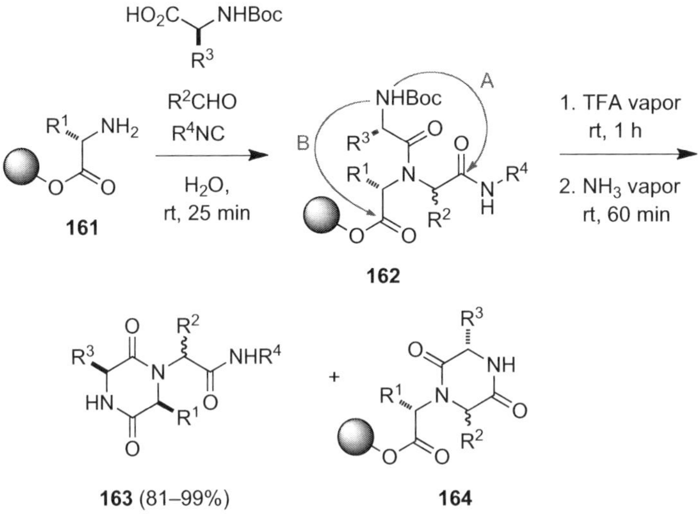

Scheme 6.39. Facile synthesis of two different diketopiperazines by a cyclizative cleavage approach. $R^1 = H$, CH_3, *i*-Bu, Bn, $CH_2C_6H_4$-4-Cl; $R^2 = Cy$, $(CH_2)_5CH_3$, $(CH_2)_2Ph$, CH_2OBn; $R^3 = H$, Me, *i*-Bu, Bn, $CH_2C_6H_4$-4-Cl; $R^4 = t$-Bu, cyclohexyl, pentyl, Bn.

polyethylene glycol resin, followed by a Pictet–Spengler reaction to **158**.[73] Addition of various thiocyanates afforded **159** that directly underwent an intramolecular cyclizative cleavage from the resin, generating the β-carboline scaffold **160** and leaving any unreacted material **157** or **158** still immobilized.[73] The unprecedented 100% *trans*-selective cyclization is a consequence of the orientation of the nitrogen lone pair to reduce electronic repulsions during the thiohydantoin ring formation.

Blackwell and Campbell reported an elegant example of a chemoselective cleavage in the synthesis of diketopiperazine arrays.[74] Based on the site of piperazine formation, two potential products **163** and **164** can be formed; however, the presence of the solid support facilitates their selective isolation via cyclizative cleavage. The cyclization precursor **162** is generated by an Ugi multicomponent reaction of the immobilized amino acid **161**. Diketopiperazine **163** formed by a nucleophilic attack following B is released from the solid support in high purity (81–99%), while the diketopiperazine **164** derived from route A remains immobilized (Scheme 6.39). To obtain **163** as the cyclizative cleavage product, TFA is first utilized to deprotect the nucleophilic amine, followed by base-catalyzed cyclization using ammonia.

Another example of a selectivity inducing cyclizative cleavage reaction has been demonstrated in the synthesis of tricyclic quinoxalizones (Scheme 6.40),[75] small molecules that have been investigated as antibacterials.[76] Here, the precursor **165** was assembled in three steps using a polyethylene glycol (PEG) as a polymeric support and taking advantage of the rate-enhancing effects of microwave irradiation. A nucleophilic aryl substitution under microwave heating was performed toward **166**. Both nitro groups were then reduced to the corresponding amino functionalities; however, only the *ortho*-amino group of **167** was capable of undergoing an intramolecular attack onto the ester linkage between the small molecule and the PEG support, leading to the selective formation of **168** via a cyclizative cleavage reaction and concurrent oxidation. The corresponding benzimidazole in **169** was formed by a condensation with the aldehyde R^3CHO. Surprisingly, no addition of oxidizing agents was required.

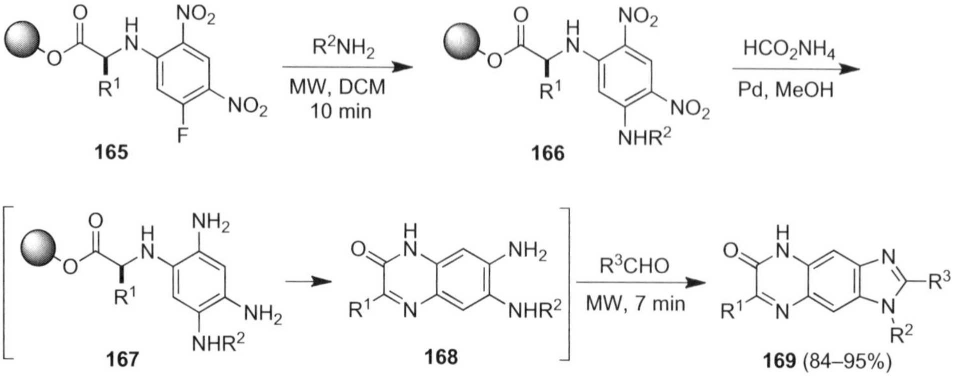

Scheme 6.40. Traceless synthesis of tricyclic quinoxalinones by solid-supported multistep microwave chemistry and a subsequent cyclizative cleavage. R^1 = H, *i*-Pr, *i*-Bu, Bn; R^2 = *i*-Pr, cyc-Pr, *n*-Bu, *i*-Bu, cyc-pentyl, Cy; R^3 = *n*-Bu, Cy, Ph, CH_2CH_2Ph, naphthyl.

Garibay et al. developed a solid-phase methodology for directed *ortho*-lithiations toward the preparation of phthalide libraries.[77] This method uses a linker strategy to direct the regiochemistry and, at the same time, to provide a mechanism for traceless cleavage from the resin. Carboxylic acids were immobilized on an aminomethylated polystyrene resin to generate the secondary amide **170**. This amide is important not only as a linker but also as a directing group for the lithiation. The lithiated compound **171** was trapped with benzaldehydes, benzophenones, and acetophenones to generate **172** that underwent cyclizative cleavage with the amide linker to afford the phthalide compounds **173** (Scheme 6.41). An additional benefit is the chemoselectivity afforded by the cyclizative cleavage, as molecules that were not lithiated remained immobilized on the resin, yielding pure product cleaved from the solid support. Begtrup and coworkers have employed a similar strategy for solid-supported *ortho*-lithiation toward the preparation of substituted hydroxyimidazoles.[78]

As observed in the previous examples, many cyclizative cleavage reactions rely upon a nucleophilic attack onto a carbonyl group; however, several alternative methods exist to accomplish a cyclizative cleavage. Many of these have been previously discussed in

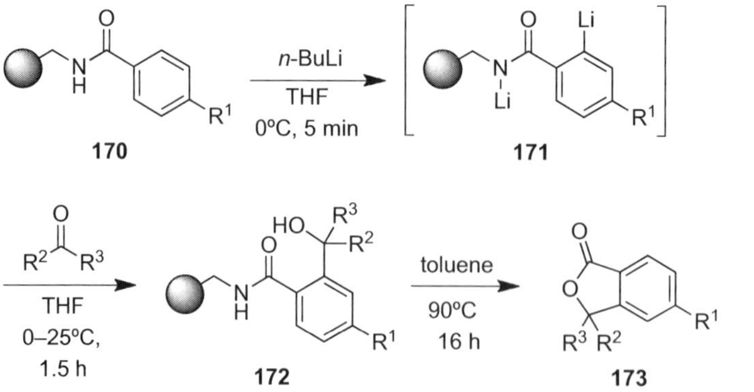

Scheme 6.41. Solid-supported directed *ortho*-lithiation followed by a cyclizative cleavage toward the synthesis of phthalides. R^1 = H, MeO, *t*-Bu, *i*-PrO, CF_3, Ph, Et_2N; R^2 and R^3 = 4-Cl-C_6H_4, Ph, H, 4-OMe-C_6H_4, 4-NO_2-C_6H_4N, 4- $N(Et)_2$-C_6H_4, 4-Ph-C_6H_4,4-OH-C_6H_4.

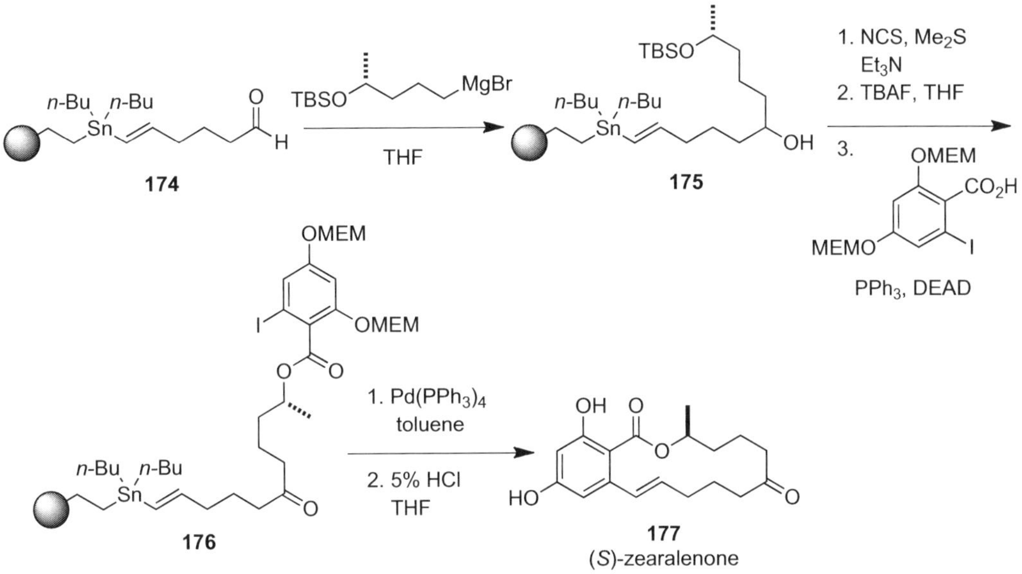

Scheme 6.42. Synthesis of (*S*)-zearalenone (**177**) via a solid-supported Stille cyclizative cleavage. NCS = *N*-chlorosuccinimide; MEM = methoxyethoxymethyl ether; DEAD = diethyl azodicarboxylate.

conjunction with the olefin metathesis reaction (Section 6.2.1.2); however, Nicolaou et al. have developed additional approaches toward cyclizative cleavage. Notably in the synthesis of (*S*)-zearalenone (**177**), a Stille coupling was employed in the release from the solid support (Scheme 6.42).[79] An immobilized aldehyde **174** containing a dibutyltin linker was prepared and reacted with a Grignard reagent to yield the secondary alcohol **175**. This was further elaborated via oxidation, deprotection and coupling to **176**, an immediate precursor to the Stille cyclorelease reaction. Treatment with [$Pd(PPh_3)_4$] produced (*S*)-zearalenone (**177**) as an exclusive *E*-isomer in a 54% yield.

Nicolaou et al. have also utilized a Horner–Wadsworth–Emmons reaction as a cyclizative release step. This approach was employed toward the total synthesis of muscone (**181**), a natural product obtained from the glandular secretion of musk deer (Scheme 6.43).[80] Following the conversion of a standard Merrifield resin to the phosphonate resin **178** (via reaction with $CH_3P(O)(OCH_3)Cl$), the aldehyde **179** was synthesized in three steps on the solid support. A subsequent Horner–Wadsworth–Emmons reaction generated the muscone scaffold **180** in 35–65% yields via simultaneous cleavage from the resin. More importantly, any by-products from incomplete reactions in the solid-phase assembly of **179** remain immobilized on the resin in the cyclizative cleavage step, obviating laborious purification of **180**. Further cuprate addition and reduction afforded muscone (**181**) and structural analogues (12 examples). Nicolaou et al. note that the ease of product isolation, high purity of final products, high-dilution conditions, and absence of dimer formation demonstrate the advantages of the solid support in the assembly of complex natural product-like libraries.[80] This is illustrated when comparing the higher yields and purities from the solid-supported synthesis with similar solution-phase reactions, as 15–20% yield of dimers was obtained even under high dilution conditions.

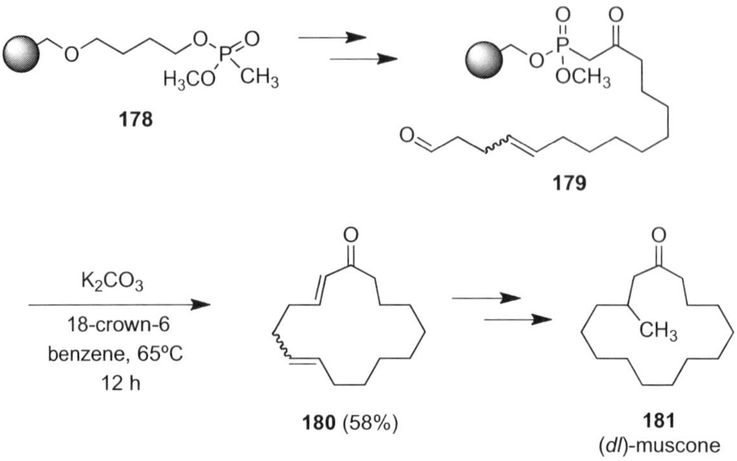

Scheme 6.43. Preparation of muscone (**181**) and analogues (not shown) via Horner–Wadsworth–Emmons cyclizative cleavage.

6.4.2 Cyclizative Immobilization

A final example of a traceless cleavage involves the converse of the cyclizative cleavage, utilizing a unique selenium-based resin that undergoes a "cyclizative loading" reaction with *ortho*-olefinic phenols and anilines (Scheme 6.44).[81] In this case, the resin itself is involved in the chemoselective addition and generation of a new heterocyclic ring. This strategy has

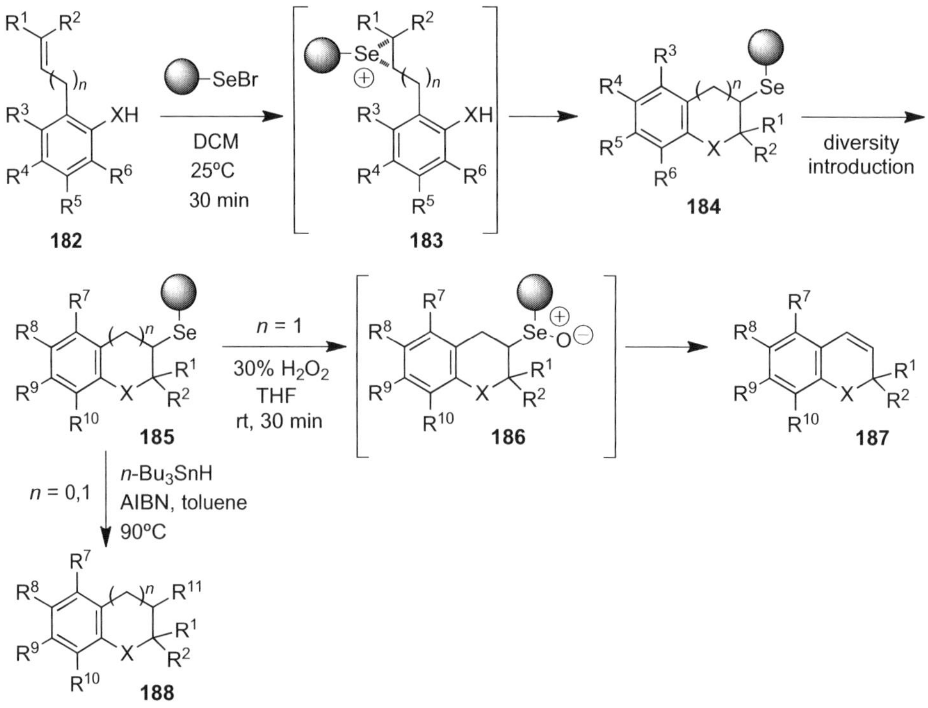

Scheme 6.44. General strategy for selenium-based cycloloading and elaboration to diverse libraries of small molecules. X = O, NH; AIBN = azobisisobutyronitrile.

been employed toward benzopyran and indoline scaffolds, which have been further elaborated into large libraries (<10,000 compounds).[82] Toward a benzopyran scaffold, an *ortho*-prenylated phenol **182** (X = O, $n = 1$) was reacted with a selenium bromide resin (Scheme 6.44). This resulted in selenium electrophilic cyclization to yield an immobilized benzopyran scaffold **184**. This was further elaborated to **185**, or simply cleaved from the solid support via oxidation to the selenoxide **186** followed by *syn*-elimination to **187** in excellent purities and moderate yields.[83] Toward indolines **188** (X = NH_2; $n = 0$), an allylic aniline **182** was immobilized in an analogous fashion, elaborated to introduce novel diversity, and reductively cleaved from the resin to afford additional functionality.[84]

6.4.3 Chemoselective Cleavage

In a similar approach as the cyclizative cleavage, it is feasible to chemoselectively release final products from the resin by using chemical functionalities generated on the solid support without actually forming cyclic compounds. Austin and Savinov demonstrated a chemoselective aminolysis of a compound generated from a 1,3-dipolar cycloaddition reaction of immobilized **189** to intermediate **190**.[85] The recognition element for the cleavage is not generated until the 1,3-cycloaddtion has occurred, and thus only the product **191** is cleaved and not any by-products or starting material (Scheme 6.45). This nucleophilic cleavage is the result of the product properly orienting the primary amine with the ester group using two hydrogen bonding interactions (oriented by the ether and the carbonyl oxygen in closest proximity to the solid support), dramatically increasing the rate of cleavage.

Seminal work by Morphy and coworkers illustrated the selective cleavage methodology, as well as the chemoselective alkylation of tertiary amines.[86] Secondary, primary, and resin-bound "ammonia" were immobilized on an REM (REgenerated after Michael reaction) resin via the Michael addition of **192** to an immobilized acrylate **8** (Scheme 6.46). The resulting solid-supported amine **193** was then alkylated to afford the quaternary amine **194**, which can then undergo a Hofmann elimination to be cleaved from the resin as a tertiary amine **195** and regenerate the resin **8** (Scheme 6.46a). This approach provides a chemoselective alkylation of amines, but it can also provide chemoselectivity in the alkylation step by differentiating between multiple nitrogen centers in the immobilized amine. An example of this was reported in the immobilization of piperazine onto a REM resin to yield the

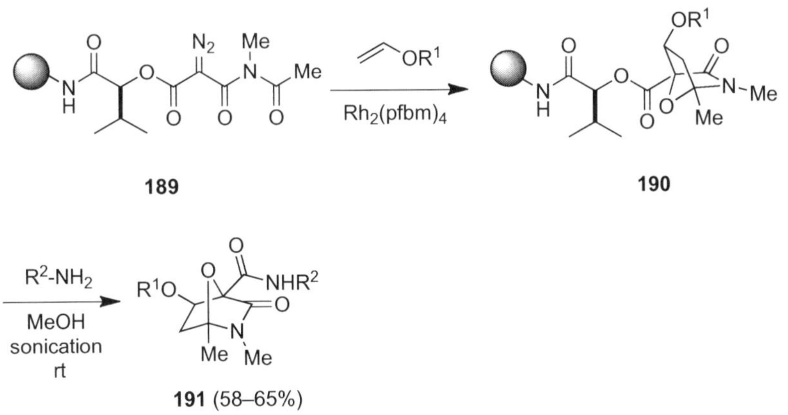

Scheme 6.45. Cycloaddition reaction followed by a chemoselective resin cleavage. R^1 = Et, Cy, *t*-Bu, Bn. R^2 = H, allyl, *i*-Pr, CH_2Cy, OH, $(CH_2)_3OH$, Bn.

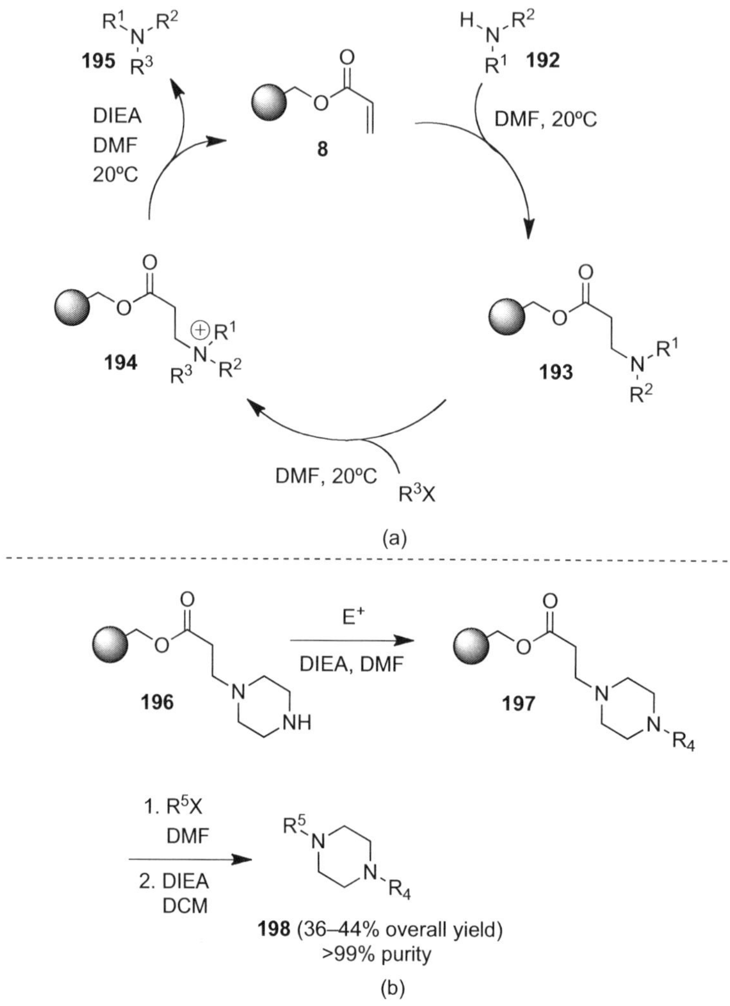

Scheme 6.46. Chemoselective preparation of tertiary amines. (a) Utilization of a REM resin to synthesize multiply substituted tertiary amines. R^1 and R^2 = Me, $(CH_2)_2Ph$, tetrahydroisoquinoline, ethyl isonipecotate, ethyl nipecotate; R^3 = allyl, 4-NO_2-C_6H_4. (b) Selective quaternization of diamines to yield tertiary amines. R^4 = $CHPh_2$, CONHPh; R^5 = Me, allyl.

monoalkylated derivative **196**.[86] The amine **196** was subsequently alkylated at the remaining NH group to **197** in a chemoselective fashion. Moreover, quaternization also occurred with complete chemoselectivity (presumably due to steric hindrance imposed by the $CHPh_2$ group) and elimination afforded **198** in 44% yield with >99% purity. In addition, if quaternization of the undesired nitrogen occurs (which may occur), it is not capable of eliminating and remains immobilized, and only the desired product is cleaved from the resin.

6.5 CONCLUSIONS

It can be concluded that solid supports have been successfully employed in order to enhance the selectivity or organic transformations and to facilitate the isolation of desired products from multistep reaction sequences. Several different approaches to take advantage of these opportunities have been discussed in this chapter, without comprehensively reviewing every reaction that displays an enhanced selectivity when conducted on a solid support. The presented examples should allow researchers to gauge if solid-supported chemistry has the potential to solve selectivity issues encountered in their research.

REFERENCES

1. (a) S. E. Hall, *Mol. Diversity* **1998**, *4*, 131; (b) J. S. Fruchtel, G. Jung, *Angew. Chem., Int. Ed.* **1996**, *35*, 17.
2. (a) S. L. Schreiber, *Science* **2000**, *287*, 1964; (b) H. M. Geysen, F. Schoenen, D. Wagner, R. Wagner, *Nat. Rev. Drug Discovery* **2003**, *2*, 222; (c) M. Verlander, *Int. J. Pept. Res. Ther.* **2007**, *13*, 75.
3. (a) L. T. Scott, J. Rebek, L. Ovsyanko, C. L. Sims, *J. Am. Chem. Soc.* **1977**, *99*, 625; (b) J. Rebek, J. E. Trend, *J. Am. Chem. Soc.* **1979**, *101*, 737; (c) J. I. Crowley, H. Rapoport, *Acc. Chem. Res.* **1976**, *9*, 135; (d) J. I. Crowley, T. B. Harvey, H. Rapoport, *J. Macromol. Sci. Chem.* **1973**, *A7* 1117.
4. (a) B. Yan, Q. Sun, *J. Org. Chem.* **1998**, *63*, 55; (b) M. A. Kraus, A. Patchorn, *Isr. J. Chem.* **1971**, *9*, 269; (c) P. Jayalekshmy, S. Mazur, *J. Am. Chem. Soc.* **1976**, *98*, 6710; (d) W. T. Ford, *ACS Symp. Ser.* **1986**, *308*, 247; (e)S. Mazur, P. Jayalekshmy, *J. Am. Chem. Soc.* **1979** *101*, 677.
5. (a) I. Coin, M. Beyermann, M. Bienert, *Nat. Protoc.* **2007**, *2*, 3247; (b) A. R. Mitchell, *Biopolymers* **2008** *90*, 175.
6. (a) K. Itakura, J. J. Rossi, R. B. Wallace, *Annu. Rev. Biochem.* **1984**, *53*, 323; (b) R. Eritja, *Int. J. Pept. Res. Ther.* **2007** *13*, 53.
7. (a) P. H. Seeberger, *Chem. Soc. Rev.* **2008**, *37*, 19; (b) P. H. Seeberger, S. J. Danishefsky, *Acc. Chem. Res.* **1998** *31*, 685.
8. (a) M. Schelhaas, H. Waldmann, *Angew. Chem., Int. Ed.* **1996**, *35*, 2056; (b) D. Orain, J. Ellard, M. Bradley, *J. Comb. Chem.* **2002**, *4*, 1.
9. (a) S. J. Connon, S. Blechert, *Angew. Chem., Int. Ed.* **2003**, *42*, 1900; (b) T. M. Trnka, R. H. Grubbs, *Acc. Chem. Res.* **2001**, *34*, 18; (c) A. Deiters, S. F. Martin, *Chem. Rev.* **2004**, *104*, 2199; (d) Y. Schrodi, R. L. Pederson, *Aldrichim. Acta* **2007**, *40*, 45.
10. (a) G. C. Vougioukalakis, R. H. Grubbs, *Chem. Rev.* **2010**, *110*, **1746**; (b) R. R. Schrock, A. H. Hoveyda, *Angew. Chem., Int. Ed.* **2003**, *42*, 4592;
11. (a) J. S. Kingsbury, J. P. A. Harrity, P. J. Bonitatebus, A. H. Hoveyda, *J. Am. Chem. Soc.* **1999**, *121*, 791; (b) E. L. Dias, S. T. Nguyen, R. H. Grubbs, *J. Am. Chem. Soc.* **1997** *119*, 3887.
12. A. K. Chatterjee, T. L. Choi, D. P. Sanders, R. H. Grubbs, *J. Am. Chem. Soc.* **2003**, *125*, 11360.
13. A. A. Poeylaut-Palena, S. A. Testero, E. G. Mata, *J. Org. Chem.* **2008**, *73*, 2024.
14. (a) M. Kidwai, P. Sapra, K. R. Bhushan, *Curr. Med. Chem.* **1999**, *6*, 195; (b) S. N. Maiti, O. A. Phillips, R. G. Micetich, D. M. Livermore, *Curr. Med. Chem.* **1998** *5*, 441.
15. S. A. Testero, E. G. Mata, *Org. Lett.* **2006**, *8*, 4783.
16. S. Chang, Y. Na, H. J. Shin, E. Choi, L. S. Jeong, *Tetrahedron Lett.* **2002**, *43*, 7445.
17. J. B. Park, *J. Nat. Prod.* **2001**, *64*, 381.
18. M. S. Jang, E. N. Cai, G. O. Udeani, K. V. Slowing, C. F. Thomas, C. W. W. Beecher, H. H. S. Fong, N. R. Farnsworth, A. D. Kinghorn, R. G. Mehta, R. C. Moon, J. M. Pezzuto, *Science* **1997**, *275*, 218.
19. A. A. Poeylaut-Palena, E. G. Mata, *J. Comb. Chem.* **2009**, *11*, 791.

20. (a) L. G. Melean, W. C. Haase, P. H. Seeberger, *Tetrahedron Lett.* **2000**, *41*, 4329; (b) T. Kanemitsu, P. H. Seeberger, *Org. Lett.* **2003** *5*, 4541.
21. H. E. Blackwell, P. A. Clemons, S. L. Schreiber, *Org. Lett.* **2001**, *3*, 1185.
22. Y. Liao, R. Fathi, Z. Yang, *J. Comb. Chem.* **2003**, *5*, 79.
23. S. C. Schurer, S. Blechert, *Synlett* **1998**, 166.
24. S. C. Schurer, S. Blechert, *Synlett* **1999** 1879.
25. S. Khom, I. Baburin, E. N. Timin, A. Hohaus, W. Sieghart, S. Hering, *Mol. Pharmacol.* **2006**, *69*, 640.
26. (a) K. Koide, J. M. Finkelstein, Z. Ball, G. L. Verdine, *J. Am. Chem. Soc.* **2001**, *123*, 398; (b) A. D. Piscopio, J. F. Miller, K. Koch, *Tetrahedron Lett.* **1997**, *38*, 7143; (c) S. M. Miles, R. J. Leatherbarrow, S. P. Marsden, W. J. Coates, *Org. Biomol. Chem.* **2004**, *2*, 281; (d) S. J. Miller, H. E. Blackwell, R. H. Grubbs, *J. Am. Chem. Soc.* **1996** *118*, 9606.
27. (a) N. Schmiedeberg, H. Kessler, *Org. Lett.* **2002**, *4*, 59; (b) S. Varray, C. Gauzy, F. Lamaty, R. Lazaro, J. Martinez, *J. Org. Chem.* **2000** *65*, 6787.
28. J. H. vanMaarseveen, J. A. J. den Hartog, V. Engelen, E. Finner, G. Visser, C. G. Kruse, *Tetrahedron Lett.* **1996**, *37*, 8249.
29. D. Brohm, N. Philippe, S. Metzger, A. Bhargava, O. Muller, F. Lieb, H. Waldmann, *J. Am. Chem. Soc.* **2002**, *124*, 13171.
30. G. Draetta, J. Eckstein, *Biochim. Biophys. Acta.* **1997**, *1332*, M53.
31. K. C. Nicolaou, N. Winssinger, J. Pastor, S. Ninkovic, F. Sarabia, Y. He, D. Vourloumis, Z. Yang, T. Li, P. Giannakakou, E. Hamel, *Nature* **1997**, *387*, 268.
32. G. D. Cuny, J. R. Cao, J. R. Hauske, *Tetrahedron Lett.* **1997**, *38*, 5237.
33. (a) E. Bayer, M. Dengler, B. Hemmasi, *Int. J. Pept. Protein Res.* **1985**, *25*, 178; (b) M. Delgado, K. D. Janda, *Curr. Org. Chem.* **2002**, *6*, 1031.
34. (a) J. A. Varela, C. Saa, *Chem. Rev.* **2003**, *103*, 3787; (b) A.Naiman, K. P. C. Vollhardt, *Angew. Chem. Int. Ed.* **1977**, 16, 708.
35. R. S. Senaiar, D. D. Young, A. Deiters, *Chem. Commun.* **2006**, 1313.
36. D. D. Young, A. Deiters, *Angew. Chem., Int. Ed.* **2007**, *46*, 5187.
37. D. D. Young, R. S. Senaiar, A. Deiters, *Chem. Eur. J.* **2006**, *12*, 5563.
38. M. Shanmugasundaram, A. L. Aguirre, M. Leyva, B. Quan, L. E. Martinez, *Tetrahedron Lett.* **2007**, *48*, 7698.
39. (a) Y. Yamamoto, T. Arakawa, R. Ogawa, K. Itoh, *J. Am. Chem. Soc.* **2003**, *125*, 12143; (b) R. Grigg, R. Scott, P. Stevenson, *J. Chem. Soc. Perkin Trans. 1 1988*, P 1357.
40. (a) D. D. Young, L. Sripada, A. Deiters, *J. Comb. Chem.* **2007**, *9*, 735; (b) R. S. Senaiar, J. A. Teske, D. D. Young, A. Deiters, *J. Org. Chem.* **2007** *72*, 7801.
41. (a) A. F. Moretto, H. C. Zhang, B. E. Maryanoff, *J. Am. Chem. Soc.* **2001**, *123*, 3157; (b) J. A. Varela, L. Castedo, C. Saa, *J. Org. Chem.* **1997** *62*, 4189.
42. J. Blanco-Urgoiti, L. Anorbe, L. Perez-Serrano, G. Dominguez, J. Perez-Castells, *Chem. Soc. Rev.* **2004**, *33*, 32.
43. N. E. Schore, B. E. Labelle, M. J. Knudsen, H. Hope, X. J. Xu, *J. Organomet. Chem.* **1984**, *272*, 435.
44. J. L. Spitzer, M. J. Kurth, N. E. Schore, *Tetrahedron* **1997**, *53*, 6791.
45. H. Kubota, J. Lim, K. M. Depew, S. L. Schreiber, *Chem. Biol.* **2002**, *9*, 265.
46. K. H. Dotz, P. Tomuschat, *Chem. Soc. Rev.* **1999**, *28*, 187.
47. M. Shanmugasundaram, I. Garcia-Martinez, Q. Y. Li, A. Estrada, N. E. Martinez, L. E. Martinez, *Tetrahedron Lett.* **2005**, *46*, 7545.
48. P. Siemsen, R. C. Livingston, F. Diederich, *Angew. Chem., Int. Ed.* **2000**, *39*, 2633.
49. N. V. Kolotilo, N. I. Tkachenko, T. V. Kulik, A. Y. Ilchenko, *Zh. Org. Khim.* **1989**, *25*, 728.

50. J. M. Montierth, D. R. DeMario, M. J. Kurth, N. E. Schore, *Tetrahedron* **1998**, *54*, 11741.
51. T. Nagashima, H. M. L. Davies, *J. Am. Chem. Soc.* **2001**, *123*, 2695.
52. M. R. Gowravaram, M. A. Gallop, *Tetrahedron Lett.* **1997**, *38*, 6973.
53. (a) I. P. Beletskaya, A. V. Cheprakov, *Chem. Rev.* **2000**, *100*, 3009; (b) G. T. Crisp, *Chem. Soc. Rev.* **1998**, *27*, 427.
54. (a) R. Franzen, *Can. J. Chem.* **2000**, *78, 957;* (b) P. H. H. Hermkens, H. C. J. Ottenheijm, D. Rees, *Tetrahedron* **1996**, *52*, 4527.
55. G. L. Bolton, J. C. Hodges, *J. Comb. Chem.* **1999**, *1*, 130.
56. K. Akaji, Y. Kiso, *Tetrahedron Lett.* **1997**, *38*, 5185.
57. E. J. Kantorowski, M. J. Kurth, *Mol. Diversity* **1997**, *2*, 207.
58. (a) X. Beebe, N. E. Schore, M. J. Kurth, *J. Am. Chem. Soc.* **1992**, *114*, 10061; (b) V. Yedidia, C. C. Leznoff, *Can. J. Chem.* **1980**, *58*, 1144; (c) D. T. Gao, H. M. Zhai, M. Parvez, T. G. Back, *J. Org. Chem.* **2008**, *73*, 8057.
59. M. T. Bilodeau, A. M. Cunningham, *J. Org. Chem.* **1998**, *63*, 2800.
60. J. D. Winkler, Y. S. Kwak, *J. Org. Chem.* **1998**, *63*, 8634.
61. Y. B. Han, J. P. Chen, B. N. Liu, G. Q. Yang, Y. Li, *Chin. Chem. Lett.* **2006**, *17*, 159.
62. C. C. Leznoff, J. Y. Wong, *Can. J. Chem.* **1973**, *51*, 3756.
63. (a) H. Miyabe, K. Fujii, H. Tanaka, T. Naito, *Chem. Commun.* **2001**, 831;(b) H. Miyabe, C. Konishi, T. Naito, *Org. Lett.* **2000** 2, 1443.
64. (a) H. Miyabe, C. Ushiro, T. Naito, *Chem. Commun.* **1997**, P 1789;(b) H. Miyabe, C. Ushiro, M. Ueda, K. Yamakawa, T. Naito, *J. Org. Chem.* **2000**, *65*, 176.
65. C. W. Lindsley, L. K. Chan, B. C. Goess, R. Joseph, M. D. Shair, *J. Am. Chem. Soc.* **2000**, *122*, 422.
66. O. L. Chapman, M. R. Engel, J. P. Springer J. C. Clardy, *J. Am. Chem. Soc.* **1971**, *93*, 6696.
67. P. Blaney, R. Grigg, V. Sridharan, *Chem. Rev.* **2002**, *102*, 2607.
68. (a) H. L. Ennis, L. Moller, J. J. Wang, O. S. Selawry, *Biochem. Pharm.* **1971**, *20*, 2639; (b) J. C. Pelling, C. Shipman, *Biochem. Pharm.* **1976** *25*, 2377.
69. (a) W. J. Chang, M. V. Kulkarni, C. M. Sun, *J. Comb. Chem.* **2006**, *8*, P 141; (b) B. E. Blass, A. Srivastava, K. R. Coburn, A. L. Faulkner, J. J. Janusz, J. M. Ridgeway, W. L. Seibel, *Tetrahedron Lett.* **2004** *45*, 1275.
70. (a) K. Mori, *Tetrahedron* **1989** *45*, 3233; (b) P. Dubs, R. Stussi, *Helv. Chim. Acta* **1978**, *61*, 990; (c) C. LeHetet, M. David, F. Carreaux, B. Carboni, A. Sauleau, *Tetrahedron Lett.* **1997**, *38*, 5153.
71. H. P. Buchstaller, *Tetrahedron* **1998**, *54*, 3465.
72. (a) D. M. Gleave, S. J. Brickner, *J. Org. Chem.* **1996**, *61*, 6470; (b) P. L. Dostert, M. Strolin Benedetti, K. F. Tipton, *Med. Res. Rev.* **1989**, *9*, 45.
73. (a) W. J. Chang, M. V. Kulkarni, C. M. Sun, *J. Comb. Chem.* **2006**, *8*, 141; (b) H. Wang, A. Gansean, *Org. Lett.* **1999**, *10*, 1647.
74. J. Campbell, H. E. Blackwell, *J. Comb. Chem.* **2009**, *11*, 1094.
75. M. J. Lin, C. M. Sun, *J. Comb. Chem.* **2006**, *8*, 455.
76. (a) K. Makino, H. S. Kim, Y. Kurasawa, *J. Heterocycl. Chem.* **1998**, *35*, 489; (b) A. Monge, J. A. Palop, A. O. Deretana, I. Urbasos, E. F. Alvarez, *An. Quim.* **1988**, *84*, 364.
77. P. Garibay, P. Vedso, M. Begtrup, T. Hoeg-Jensen, *J. Comb. Chem.* **2001**, *3*, 332.
78. S. Havez, M. Begtrup, P. Vedso, K. Andersen, T. Ruhland, *J. Org. Chem.* **1998**, *63*, 7418.
79. K. C. Nicolaou, N. Winssinger, J. Pastor, F. Murphy, *Angew. Chem., Int. Ed.* **1998**, *37*, 2534.
80. P. Dowd, S. C. Choi, *Tetrahedron* **1992**, *48*, 4773.
81. (a) K. C. Nicolaou, J. A. Pfefferkorn, G. Q. Cao, *Angew. Chem., Int. Ed.* **2000**, *39*, 734; (b) K. C. Nicolaou, G. Q. Cao, J. A. Pfefferkorn, *Angew. Chem., Int. Ed.* **2000**, *39*, 739; (c) K. C. Nicolaou, A. J. Roecker, J. A. Pfefferkorn, G. Q. Cao, *J. Am. Chem. Soc.* **2000**, *122*, 2966.

82. K. C. Nicolaou, J. A. Pfefferkorn, H. J. Mitchell, A. J. Roecker, S. Barluenga, G. Q. Cao, R. L. Affleck, J. E. Lillig, *J. Am. Chem. Soc.* **2000**, *122*, 9954.
83. K. C. Nicolaou, J. A. Pfefferkorn, A. J. Roecker, G. Q. Cao, S. Barluenga, H. J. Mitchell, *J. Am. Chem. Soc.* **2000**, *122*, 9939.
84. K. C. Nicolaou, J. Pastor, S. Barluenga, N. Winssinger, *Chem. Commun.* **1998**, 1947.
85. S. N. Savinov, D. J. Austin, *Org. Lett.* **2002**, *4*, 1419.
86. A. R. Brown, D. C. Rees, Z. Rankovic, J. R. Morphy, *J. Am. Chem. Soc.* **1997**, *119*, 3288.

Part II

APPLICATIONS

7

ASYMMETRIC SYNTHESIS ON SOLID SUPPORT

Baburaj Baskar and Kamal Kumar

7.1 INTRODUCTION

Solid-phase synthesis refers to the molecular transformation of a substrate that is covalently bound to polymeric and insoluble support and is one of the powerful techniques for generating organic molecules for small or large compound collections.[1] Applications of solid-phase synthesis in generating small- to medium-sized organic molecules take roots from the oligopeptide and oligonucleotide synthesis on polymeric supports.[2] The basic advantage in this strategy is avoiding the cumbersome and tedious workup procedures at each reaction step. However, overall yield and purity of the product are often the decisive criteria for a successful synthesis strategy on a polymeric support. The initial developments in the solid-phase synthesis of compound collections witnessed racemic synthesis of molecules and rarely asymmetric chemical transformations were attempted on solid phase.[3] When required, the chiral building blocks that were synthesized separately in solution-phase synthesis were coupled to solid-phase-bound substrates and the synthesis was further followed to the desired target structures.

Emergence of combinatorial chemistry and parallel synthesis in early nineties and synthesis of small molecules inspired by the structural architectures of natural products in the beginning of this century called for further developments in the solid-phase synthesis protocols. Synthesis of compound libraries possessing enantiopure molecules embodying a complex framework and decorated with more than one stereocenters in a combinatorial and parallel fashion calls for highly feasible solid-phase synthesis methods.[4] In particular, these efforts must address the stereocontrol of the reaction course in order to minimize isomer

Solid-Phase Organic Synthesis: Concepts, Strategies, and Applications, First Edition.
Edited by Patrick H. Toy and Yulin Lam.

formation. These asymmetric transformations should provide a combinatorial platform for the synthesis of target structures and their further diversification that may include both functional group and skeletal modifications.[5]

Although an enantioselective reaction carried out on solid phase, in principle, also includes the reactions wherein a chiral catalyst[6] or a chiral auxiliary[7] is polymer bound, in this chapter we have mainly focused on reactions wherein a substrate is bound to a solid support and the chemical transformation is carried out by chiral reagents. However, we have included few examples of employing solid-supported catalyst and auxiliaries to give readers an idea of these alternative synthesis strategies. Readers are, however, advised to refer to Refs 6 and 7 for a review on the applications of immobilized chiral catalysts and solid-phase-bound chiral auxiliaries, respectively, and Refs 1, 5, and 8 for diverse solid-phase synthesis approaches directed toward small molecules and natural products.[8]

7.2 ASYMMETRIC CHEMICAL TRANSFORMATIONS OF SOLID-SUPPORTED SUBSTRATES

The majority of the reported solid-phase syntheses employ substrates bound to a polymeric support and use different reagents to cause chemical transformations. Excess of reagents are then washed off providing a pure polymer-bound product, assuming the reactions are optimized for negligible side product formation including the unreacted substrates. The pure product is finally released by cleaving the solid support off the product. In principle, most of the reactions that work in solution-phase synthesis could be performed on solid phase too; however, so far the experience has been that not all reactions work well. Among the reactions that mostly work nicely and for which reliable protocols have been established are condensation reactions (Knoevenagel condensation, heterocycle condensations), nucleophile addition to activated carbonyls (amide bond formation, esterification/saponification, reactions with isocyanates/thioisocyanates), nucleotide oligomerization, reductive amination, ozonolysis, and some Pd-catalyzed cross-coupling. Among the lesser explored chemistry on solid phase include cycloaddition, aldol and allylation chemistry, organometallic chemistry, and photochemical transformations.

Asymmetric and enantioselective syntheses on solid support gained attention in the late 1990s with the first solid-phase allylation chemistry reported by Panek and Zhu[9] but remained slow to grow further. Eventually in the beginning of this century, different research groups made successful efforts to transfer the solution-phase asymmetric chemical transformations onto solid supports and their applications in the library synthesis were realized. In the following section, we overview some of these enantioselective reactions developed on solid phase and their role in the synthesis of interesting small molecules.

7.2.1 Asymmetric Aldol Reactions

The aldol reaction is a key carbon–carbon bond forming reaction and is one of the most widely used transformations in organic synthesis that provides the β-hydroxy carbonyl structural unit prevalent in many natural products. Although many elegant protocols for enantioselective aldol reactions have been developed, not all perform equally well on a solid support. Among these protocols, aldol reactions employing boron enolates bearing chiral substituents at the boron atom[10] were successfully used for solid-phase asymmetric synthesis of natural product analogues almost simultaneously by two groups.

Waldmann and coworkers[11] employed the aldol reactions of chiral boron enolates as the key stereoselective transformations toward the solid-phase synthesis of 6,6-spiroketals, a scaffold often found in biologically interesting complex natural products. The polymer-bound aldehyde **1** with a loading of 0.75 mmol/g was treated at −78°C with the preformed (*Z*)-diisopinocampheyl borinate **2** in dichloromethane for 1.5 h (Scheme 7.1). After storing the reaction mixture at −27°C for 16 h, the resin was filtered and the whole process was repeated once. An oxidative workup followed by TBS protection of the secondary alcohol yielded immobilized *syn*-aldol product **3**.

In the crucial second aldol reaction toward the synthesis of advanced precursor for spiroacetal scaffold, the chiral boron enolate (**4**) was generated on the solid support by treating the resin-bound adduct **3** with dicyclohexylboron chloride and triethylamine in diethylether at 0°C. After 6 h the resin was washed and the procedure was repeated once again to yield (*E*)-dicyclohexylboron enolate **4** on the solid support (Scheme 7.1). Boron enolate resin **4** was then treated at −78°C with aldehyde **5** and after oxidative workup to cleave B−O bond, the secondary alcohol (monitored by IR that displayed strong absorption at 3504 cm^{-1} and 1714 cm^{-1} in the spectrum) of the *anti*-aldol adduct was protected with a TBS group to provide aldol adduct **6**. Treating **6** with DDQ in a CH_2Cl_2/buffer (pH 7) mixture resulted in simultaneous cleavage of the PMB ether, release from the Wang resin,

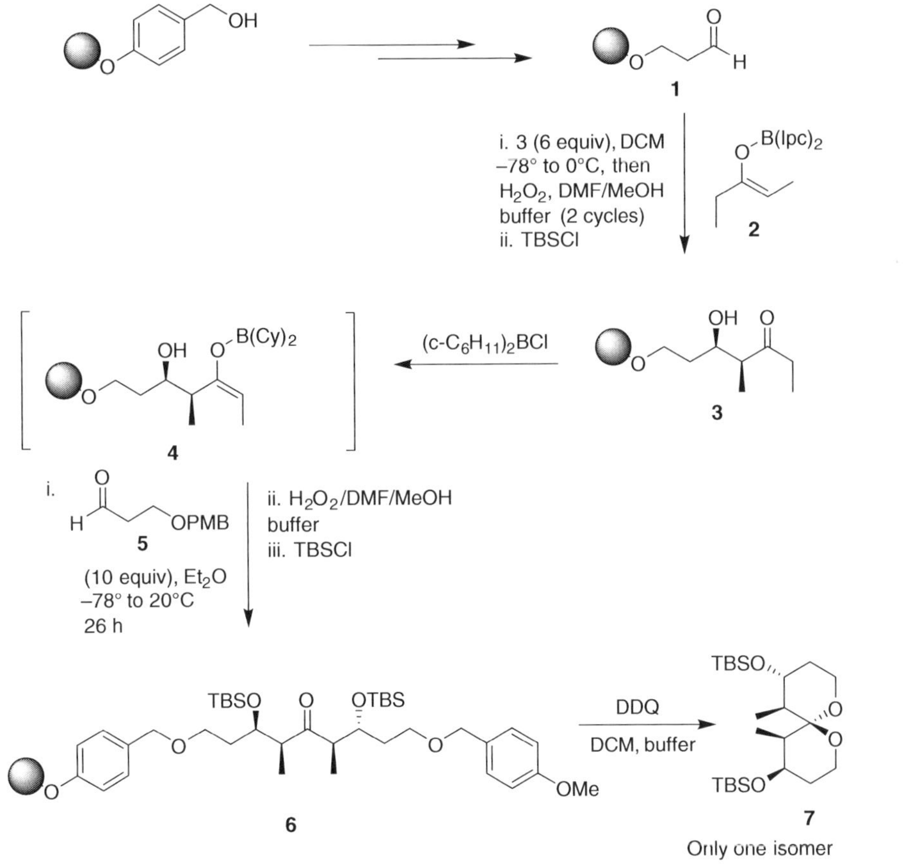

Scheme 7.1. Asymmetric aldol reactions on solid phase for synthesis of spiroketal library.

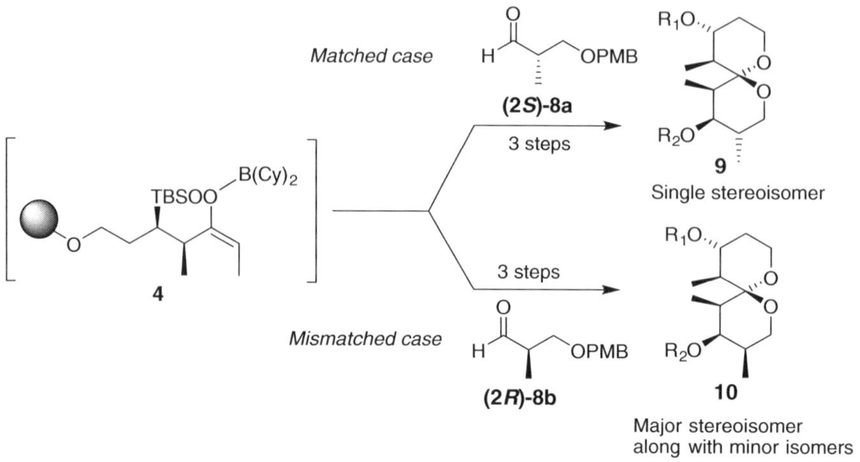

Scheme 7.2. Double diastereodifferentiation in solid-phase aldol reactions toward spiroketal synthesis.

and spiroketalization to give **7**. The compound was found to be identical to the one synthesized in solution phase, which conforms the right stereoselectivities obtained in the solid-phase asymmetric aldol reactions.

Control of stereoselectivity and double diastereodifferentiation was further investigated using resin-bound β-enolate **4** and both enantiomers of chiral aldehyde **8**. After a sequence of aldol reaction as described previously, alcohol protection, and DDQ-mediated spiroketalization, reactions with aldehyde **8a(2*S*)** provided pure single diastereoisomer of the spiroketal **9**. However, the reaction of **4** with aldehyde **8b(2*R*)** yielded **10** as the major diastereomer along with minor inseparable isomers (Scheme 7.2). Thus, although in both aldol reactions of the chiral enolate **4** with the enantiomeric aldehydes **8a** and **8b** the *anti*-aldol adduct is formed as the major product, the combination of **4** and **8a** represents the *matched case* and the combination of **4** and **8b** the *mismatched case*.

Later, Paterson et al.[12] (Scheme 7.3) also reported a similar solid phase synthesis of a spiroketal that is a fragment of the natural product spongistatin. Chiral boron enolate was treated with resin-bound chiral aldehyde **11** to yield the aldol product **12** with 20:1

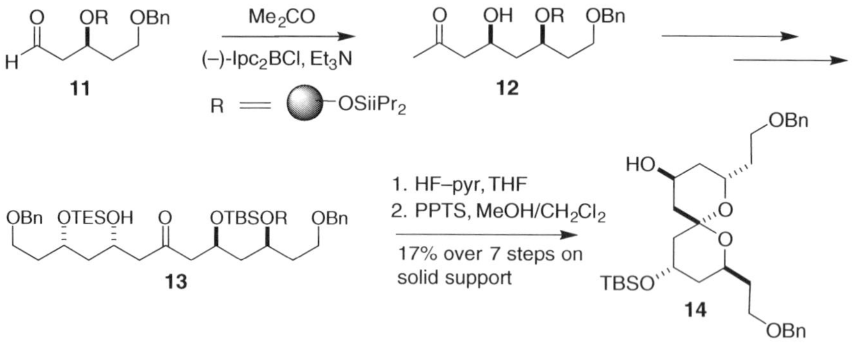

Scheme 7.3. Synthesis of natural product fragment spiroacetal using asymmetric solid-phase aldol strategy.

diastereoselectivity. A second aldol addition was performed with preformed boron enolate of resin-bound methyl ketone to the aldehyde. The final spiroketal **14** formation was realized by cleavage of solid support and silyl group through one-pot reaction. Although solution-phase reaction was comparable to the solid-phase synthesis in terms of stereoselectivity, the solid-supported transformations required 2 equiv of chiral reagents with two cycles to get the complete conversion.

7.2.2 Asymmetric Allylation Reactions

Allylation of carbonyl compounds is another very useful carbon–carbon bond forming asymmetric transformation in organic synthesis. This transformation yields the homoallylic alcohols that have proven to be valuable reagents and intermediates that have found numerous applications in natural product total synthesis. In particular, asymmetric allylboration of aldehydes employing tartrate- and pinane-derived reagents has been widely exploited.[13] Although the asymmetric allylation reaction is well documented and widely used in solution phase, the asymmetric variant of the allylation of carbonyl compounds on the solid support has remained largely unexplored.[9]

In 1997, Panek et al. first reported the asymmetric crotyl transfer reaction to aldehydes on solid support using a chiral (*E*)-crotylsilane reagent in the presence of trimethylsilyl triflate (TMSOTf).[9,14] As observed in solution-phase crotylation reported by the same group before,[15] the solid-supported crotylation of achiral/chiral aldehydes and acetals exhibited *syn*-selectivity in homoallylic ether generation (Scheme 7.4). The solid-supported silane reagent **(*R*)-15** when combined with excess of acetal in the presence of TMSOTf at low temperature followed by basic hydrolysis to cleave the polymer linkage afforded homoallylic ether **16** in up to 74% yield and favored *syn*-adducts in the ratio of 7:1 (Scheme 7.4a). The crotylation reactions of aldehydes *via* reaction with *in situ* generated oxocarbenium ions were also successfully performed with the immobilized silane. In this three-component reaction, immobilized silane **(*R*)-15** with excess of aldehyde and methoxytrimethylsilane (TMSOMe) in CH_2Cl_2 was treated with TMSOTf under similar conditions as those used for the acetal reactions to provide functionalized homoallylic ether **16** with yields up to 92% and *syn*-selectivity up to 30:1 (Scheme 7.4a). The methodology was further extended to the synthesis of polypropionate-like subunits through an iterative crotylation sequence. To this end, after first solid-phase crotylation of the resin-bound aldehyde **17** with silane **(*R*)-18**, the polymer-supported homoallylic ether **19** was subjected to ozonolysis to generate the chiral aldehyde that was further transformed to acetal **21** before the next crotylation was performed with silane reagents in the presence of $BF_3{\cdot}OEt_2$ to yield **22** after cleavage from the polymer support (Scheme 7.4).

Interestingly, it took almost 8 years when the next solid-phase allylation reaction was reported again using a chiral silane. Tan and coworkers[16] reported the asymmetric allylation of an aliphatic polymer-supported aldehyde using a strained allylsilacycle **24** developed by Leighton and coworkers[17] (Scheme 7.4b). The homoallylic alcohol **25** was obtained in good enantiopurity and yields.

Waldmann and coworkers later developed and explored the solid-supported carbonyl allylation reactions[18] for the stereoselective solid-phase synthesis of a collection of NP-inspired δ-lactones.[19] To identify reaction conditions that would give rise to the allylation products with high enantioselectivity and in high yield, immobilized aldehyde **26** was synthesized as model compound and was then subjected to allylation with different chiral allylboron reagents. After some experimentation, it was found that treatment of the resin-bound aldehyde with 4 equiv of D-Ipc_2BAll **28** at −78°C in THF/ether 5:1 (v/v) followed by

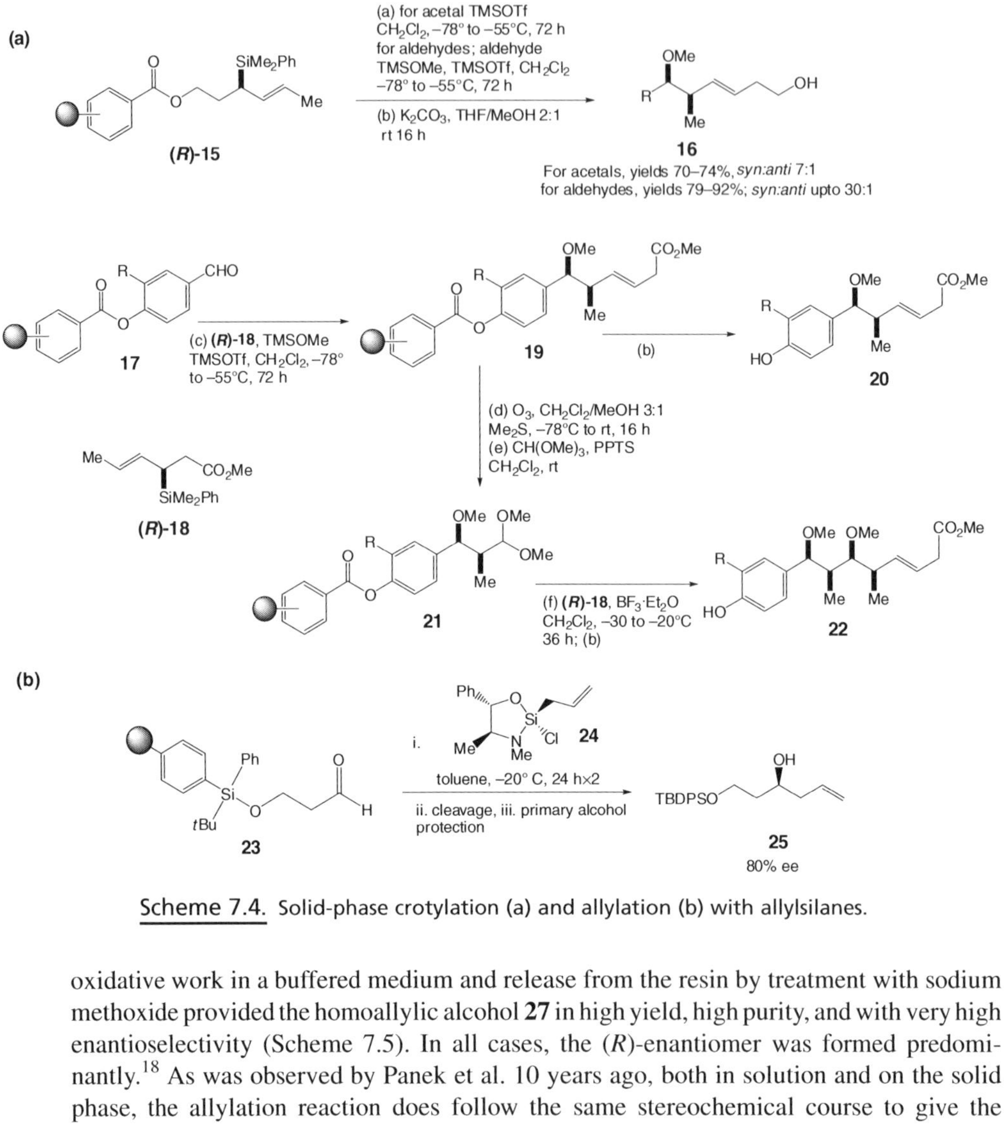

Scheme 7.4. Solid-phase crotylation (a) and allylation (b) with allylsilanes.

oxidative work in a buffered medium and release from the resin by treatment with sodium methoxide provided the homoallylic alcohol **27** in high yield, high purity, and with very high enantioselectivity (Scheme 7.5). In all cases, the (*R*)-enantiomer was formed predominantly.[18] As was observed by Panek et al. 10 years ago, both in solution and on the solid phase, the allylation reaction does follow the same stereochemical course to give the products with very comparable stereoselectivities.

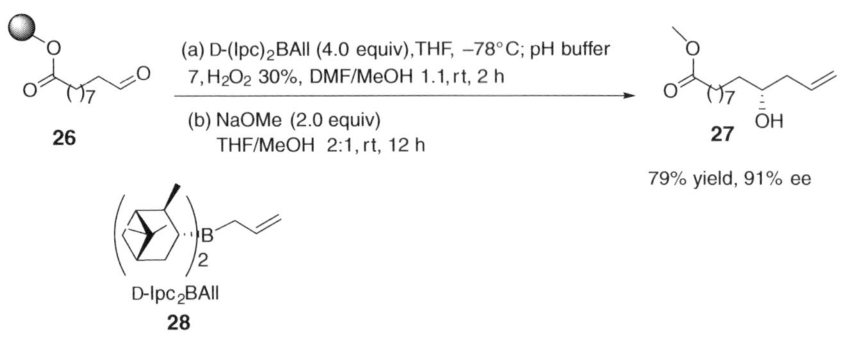

Scheme 7.5. Solid-phase enantioselective allylation of aldehydes.

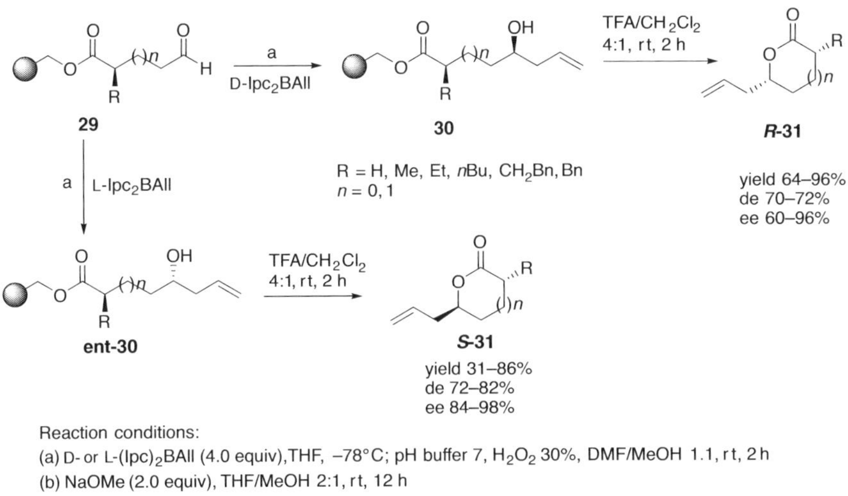

Scheme 7.6. Synthesis of δ-lactones using enantioselective solid-phase allylation reaction.

The usefulness of this methodology was proven for the synthesis of six-membered ring lactones starting from the supported aldehyde **29**, which was derived from the commercially available 5-hexenoic acid. Stereoselective allylation with D-Ipc_2BAll (which occurred with an enantiomeric ratio of 80:20 for *R*:*S*) followed by release from the resin under acidic conditions yielded the lactone ***R*-31** exclusively. Employing L-Ipc_2BAll instead provided the lactone ***S*-31** in good yields and enantioselectivities (Scheme 7.6).

High yields and enantiomeric excess of the solid-phase allylation of aldehydes encouraged exploring the sequence for synthesis of natural product cryptocarya diacetate, an α,β-unsaturated-δ-lactone isolated from *Cryptocarya latifolia* that is representative of a large group of biologically active secondary metabolites and its analogues. The synthesis design included multiple stereocomplementary allylation reactions on the polymeric carrier followed by a ring-closing metathesis to access the NP analogues (Scheme 7.7).[19] Thus, a careful ozonolysis of the resin-bound homoallylic alcohols **32** and **33** obtained after the stereoselective allylation of resin-bound aldehydes by L- and D-Ipc_2BAll reagents, respectively, provided the aldehydes **34** and **35**. The later were subjected to a second allylation with L- and D-Ipc_2BAll and the formed secondary alcohols were converted to acrylic acid esters **36–39**. Ring-closing metathesis employing the Grubbs II catalyst to induce the lactone formation, release from the solid support, with consecutive cleavage of the silyl group by treatment with trifluoroacetic acid and subsequent acetylation yielded natural product (all-*syn* isomer of **40**) and its analogues **40–43** (Scheme 7.7).

To further explore the generality of the method for asymmetric synthesis of homoallylic alcohols on the solid support, the enantio- and diastereoselective crotylation of polymer-bound aldehydes **44** and **45** with (*E*)- and (*Z*)-*B*-crotyldiisopinocampheylborane was investigated (Scheme 7.8). The reactions were carried out under the conditions described above for the allylation with the analogous crotylboron reagents. Diastereomeric homoallylic alcohols **46–49** were obtained in high yields and with high diastereomer ratios. The major diastereomers were formed with high enantiomeric ratios. The direction of the diastereoselection parallels the observations made for analogous crotylation reactions in solution, clearly advocating the solid-phase synthesis planning for complex molecules.

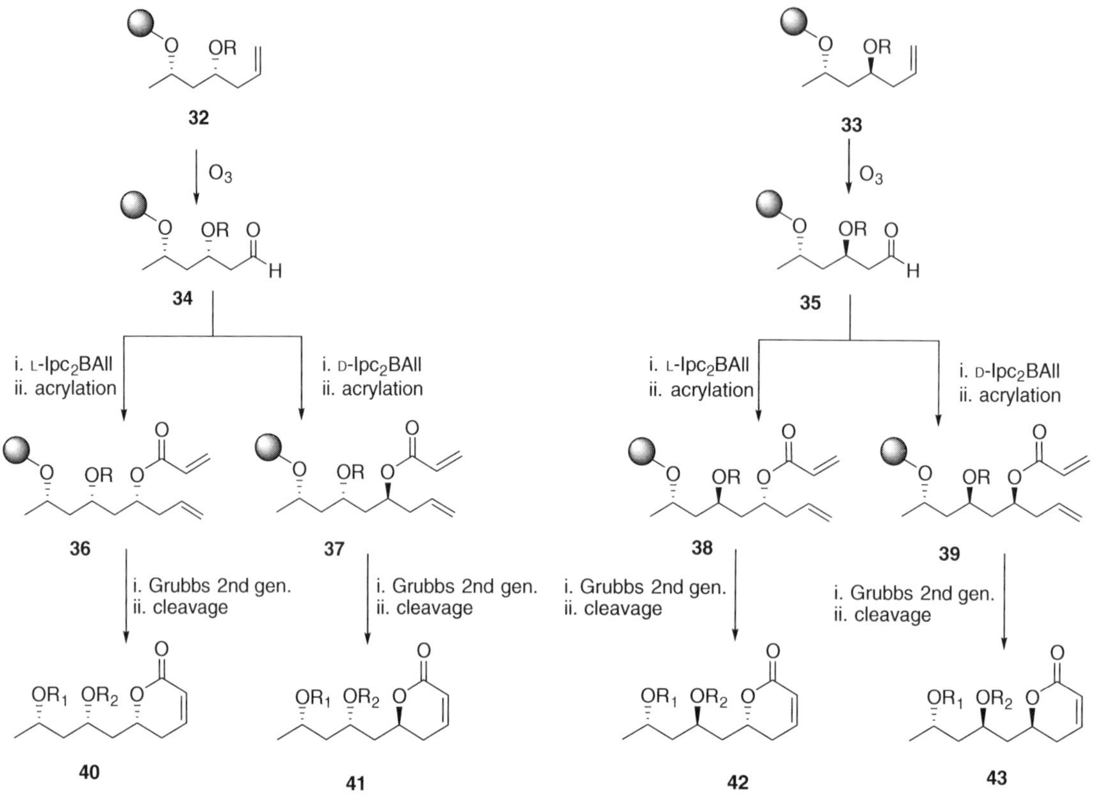

Scheme 7.7. Iterative solid-phase allylation for the synthesis of natural product analogues.

7.2.3 Enantioselective Cycloaddition Reactions

One of the most popular strategies to access different ring systems is the cycloaddition reactions, and a lot of chemistry has been developed for stereoselective cycloadditions, annulations, and cyclization reactions providing diverse carbo- and heterocycles. However, only few reports have emerged where solid-phase synthesis employed the enantioselective cycloaddition or annulation chemistry for a desired target structure. Jiang and Zou[20] reported the enantioselective solid-phase 1,3-dipolar cycloaddition of nitrile oxides to allyl alcohol for the synthesis of isooxazoline **52** (Scheme 7.9). Resin-bound chloroxime (**50**) was treated with (−)-diisopropyl tartrate ((−)-DIPT) as a chiral auxiliary and Grignard reagent as base to give the resin-bound isooxazoline **51**. Resulting product was cleaved from the resin in the presence of acid catalyst to give the compound **52** in 50–70% yield and up to 95% ee. Upon treating the diethyl zinc reagent in solid phase gave the racemic products against optically pure product obtained in the solution phase.[21]

Schreiber and coworkers[22] (Scheme 7.10) described the [3 + 2] cycloaddition of an immobilized aromatic azomethine ylide with *tert*-butyl acrylate in the presence of silver acetate/(*S*)-QUINAP. Reaction gave rise to the *endo* isomer (**53**) with 79% overall yield and 90% ee. However, to achieve the results as good as in solution-phase chemistry, a three times more catalyst loading was required.

Kurosu et al.[23] (Scheme 7.11) employed the polymer-supported vinyl ethers for hetero-Diels–Alder reactions with structurally diverse heterodienes using 1 equiv of

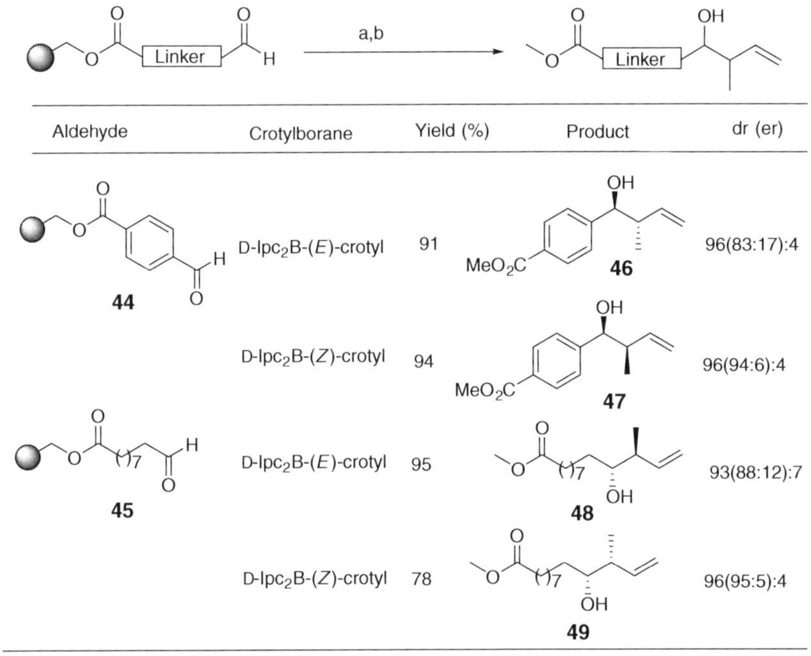

Reaction conditions: (a) D-$(Ipc)_2BAll$ (4.0 equiv), THF, −78°C; pH buffer 7, H_2O_2 30%, DMF/MeOH 1.1, rt, 2 h
(b) NaOMe (2.0 equiv), THF/MeOH 2:1, rt, 12 h

Scheme 7.8. Solid-phase crotylation of aldehydes.

[(1*R*,2*S*)-Inda-Box–$Cu(OTf)_2$] (**56**). The corresponding cycloaddition products, allyl dihydropyrancarboxylates (**57**), were obtained with good diastereoselectivities (12:1) and 86–97% ee.

Waldmann and coworkers[24] (Scheme 7.12) described the enantioselective synthesis of a library of 2,4,6-trisubstituted tetrahydropyrans by an oxa-Diels–Alder reaction. The corresponding pyrane ring is prevalent in a number of natural products. To access it stereoselectively with a solid-phase synthesis strategy, cycloaddition reaction of Danishefsky's diene **59** with resin-bound aldehydes **58** was carried out in the presence of 5 mol% of the chromium catalyst **63**. After the release of resin, the product (**61**) was

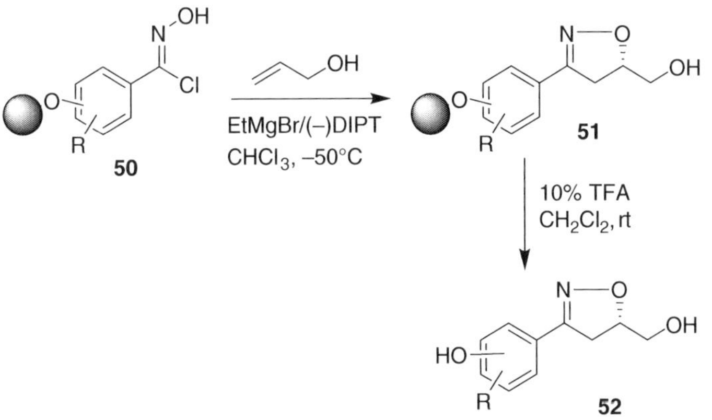

Scheme 7.9. Enantioselective [3 + 2] cycloaddition reaction on solid support.

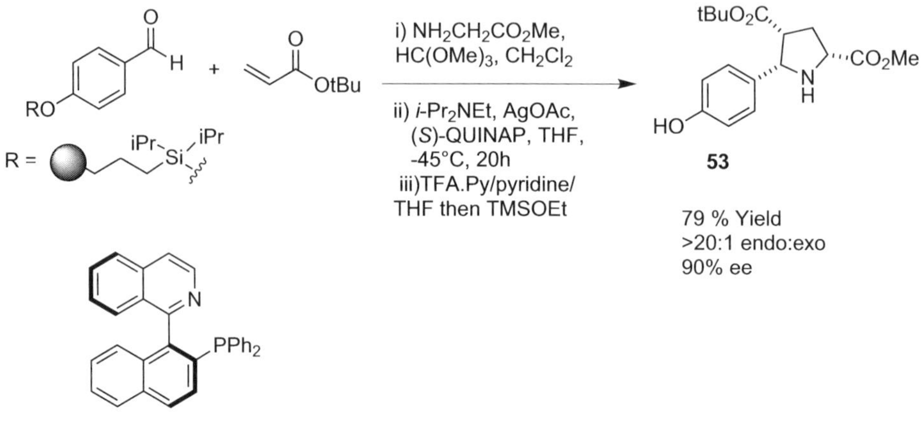

(*S*) - Quinap

Scheme 7.10. Enantioselective [3 + 2] cycloaddition of an immobilized aromatic azomethine ylide.

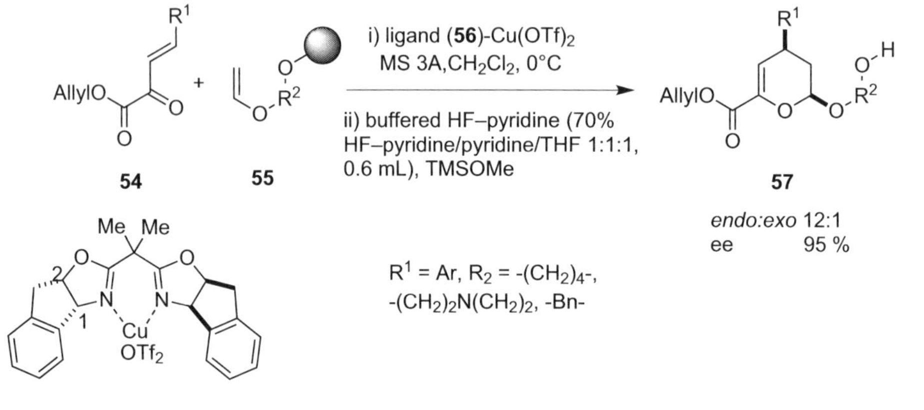

(*1R 2S*) -Inda-Box (**56**)

Scheme 7.11. Enantioselective hetero-Diels–Alder reaction on solid phase.

obtained up to >98% ee. A further transformation of the polymer-bound dihydropyrones **60** was performed by subsequent conjugate cuprate addition and reduction of the ketone to provide highly enantioenriched pyrans **62**.

7.2.4 Stereoselective Epoxide Ring-Opening Reactions

Epoxides present numerous opportunities for functionalization via catalytic asymmetric opening of this ring system with the help of nucleophiles. A solid-phase approach for asymmetric epoxide ring opening was undertaken by Jacobsen and coworkers.[25] They studied the asymmetric ring opening of polymer-bound *meso*-epoxides **64** with trimethylsilyl azide catalyzed by (salen)CrN$_3$ complex ***(R,R)*-(65)**. Resin was removed by the transesterification of resin-bound alcohol **66** and then the resulting soluble product **67** was obtained in 95% yield and 92% ee (Scheme 7.13). Furthermore, products obtained were converted into series of cyclic peptides that contained the Arg-Gly-Asp (RGD) sequence characteristic of integrin ligands.

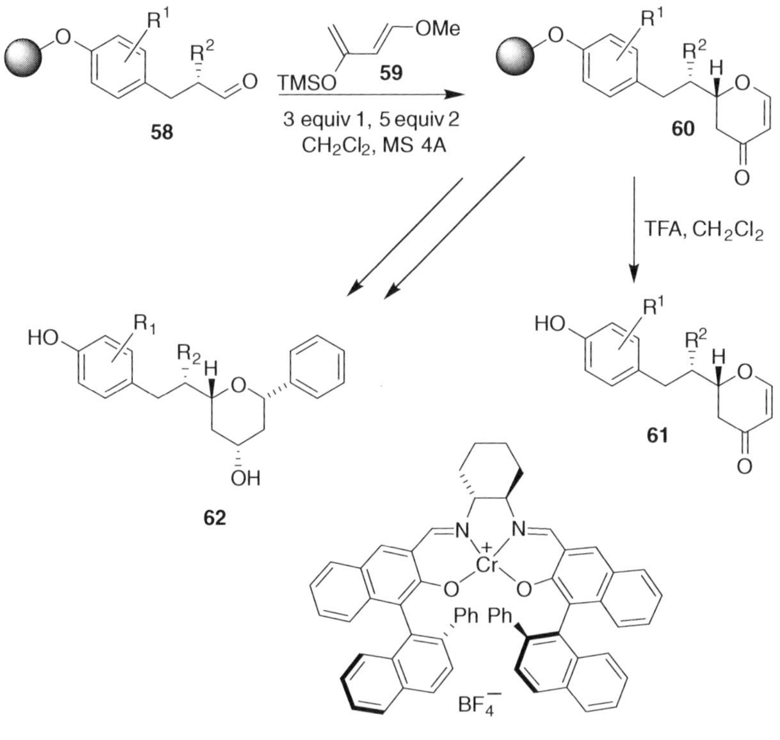

Scheme 7.12. Enantioselective oxa-Diels–Alder reactions on solid phase.

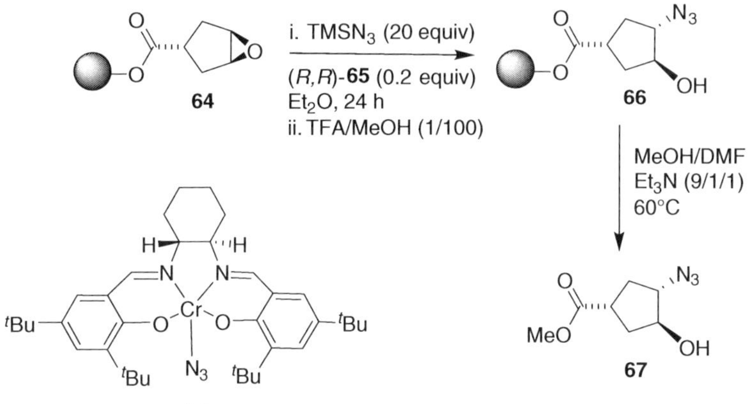

Scheme 7.13. Stereoselective ring opening of immobilized epoxide.

7.2.5 Asymmetric Alkene Cyclopropanation Reactions

Catalytic asymmetric cyclopropanations of alkenes on solid support was studied by Davies and Nagashima (Scheme 7.14).[26] Reaction of an immobilized 1,1-diarylethylene **68** with a set of seven different aryldiazoacetates **69** with 1 mol% of the chiral catalyst

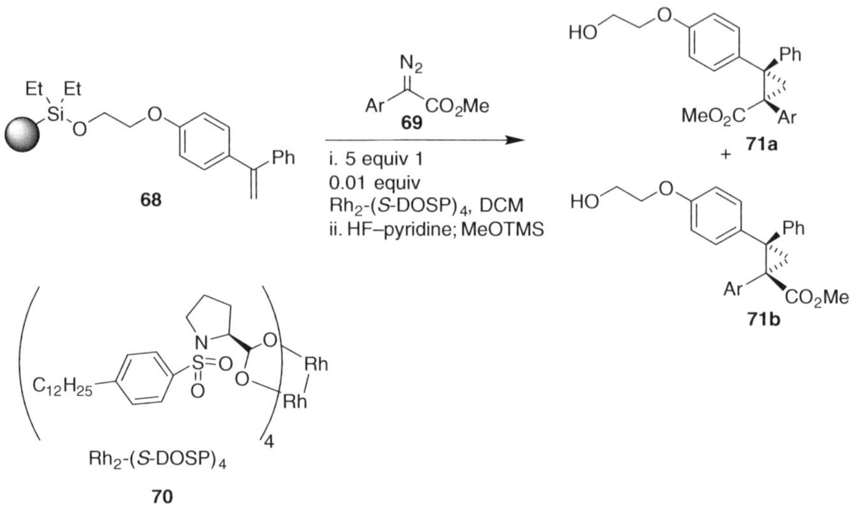

Scheme 7.14. Enantioselective cyclopropanation on solid phase.

Rh_2(*S*-DOSP)$_4$ (**70**) provided the corresponding cyclopropanated product **71** in >80% yield and with high diastereoselectivity (***E*-71a**):(***Z*-71b**) > 3:1 and enantioselectivity (up to 93%).

7.2.6 Enantioselective Alkylation Reactions

The point of attachment of the solid support could play a significant role in steering the stereoselection of the reaction course. In 1989, O'Donnell et al.[27] observed that a tertiary butyl group was required for achieving high enenatioselectivity in the phase transfer alkylation of diphenylmethyleneglycinimine ester. As a consequence, Park et al.[28] (Scheme 7.15) linked the solid support to the substrate (**72**) onto the imine moiety to

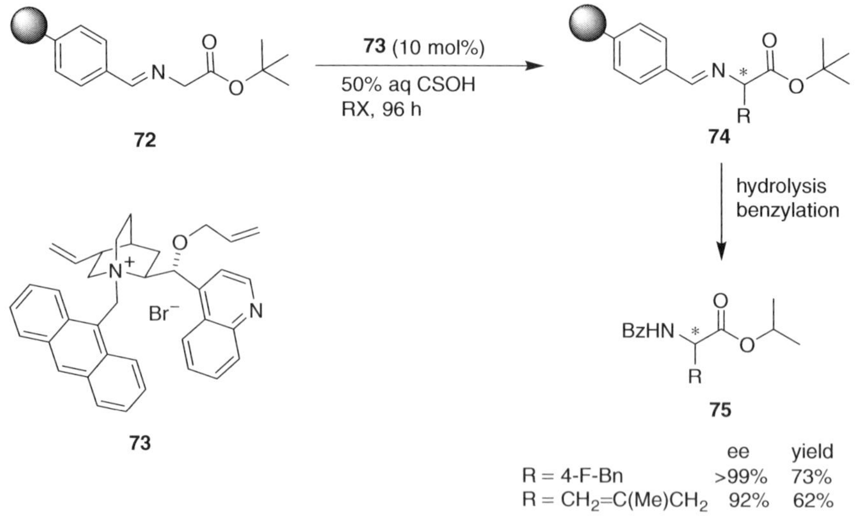

Scheme 7.15. Stereoselective alkylation of immobilized amino ester.

retain the *tert*-butyl esters for desired asymmetric alkylation on solid phase. Employing 10 mol% of benzyl bromide catalyst in the alkylation of **72** provided the alkylated compounds (**75**) with more than 90% ee.

The easy preparation of the solid-supported substrate **72**, the high enantioselectivity, and the very mild reaction conditions make this method very practical for the synthesis of the chiral non-natural *R*-amino acid library *via* combinatorial synthesis or parallel synthesis.

7.3 ASYMMETRIC TRANSFORMATIONS USING RESIN-BOUND CHIRAL CATALYSTS AND AUXILIARIES

Over the years, pharmaceutical industry and academic laboratories involved in the interdisciplinary research of small molecule interactions with biological systems have realized the significance of having enantiopure molecules whenever possible.[29] This definitely has called for further developments in asymmetric catalytic syntheses. Asymmetric catalysis strongly relies on employing catalysts/catalysts complexes based in general on enantiomerically pure and thus highly valuable ligands and transition metals. Therefore, unless a very low loading of the catalyst works efficiently, the cost and manpower factors remain bigger obstacles in applications of these catalytic methods. In such scenario, the immobilization of homogeneous catalysts could be of immense significance and interest.[6] Once optimized for a high efficiency, the resin-bound catalysts could provide very high turnover frequencies and in principle can be recycled for further reactions.

In another approach, a chiral auxiliary can be fixed to solid-supported substrate and help direct the stereoselectivity of the asymmetric reactions to provide enantiopure molecules.[7] In this chapter, we present some representative examples of resin-bound catalysts and chiral auxiliaries in asymmetric synthesis. For detailed reviews about the polymer-bound chiral ligands used in conjunction with metals and metalloids in asymmetric catalysis, chiral organic catalysts attached to polymer supports, and chiral auxiliaries on solid support, readers are referred to Refs 6 and 7.

7.3.1 Catalytic Asymmetric Synthesis with Resin-Bound Chiral Catalysts

Organocatalysts have emerged as an interesting and efficient class of catalysts for a plethora of asymmetric transformations. Particularly, L-proline has been used to carry out a number of reactions providing high yields and high enantioselectivities—for instance, aldol reactions, Michael additions, Robinson annulation, and a variety of α-substitutions of aldehydes and ketones.[30] Contrary to its extensive solution-phase exploration and exploitation, only a limited number of heterogeneous proline-based catalysts have thus far been reported in asymmetric solid-phase synthesis.[31] Proline directly tethered to a support through the carboxylate was initially used to catalyze the aldol reaction of aliphatic aldehydes with acetone that resulted in only moderate enantioselectivity. To overcome the above disadvantages, Portnoy and Kehat[32] reported the preparation of nonpeptidic hydroxyproline-based catalysts on dendronized supports and their exploration in an asymmetric aldol reaction (Scheme 7.16).

The asymmetric transformation for aldol reaction was carried out for the reactions of benzaldehyde and 4-nitrobenzaldehyde with acetone in the presence of dendritic and

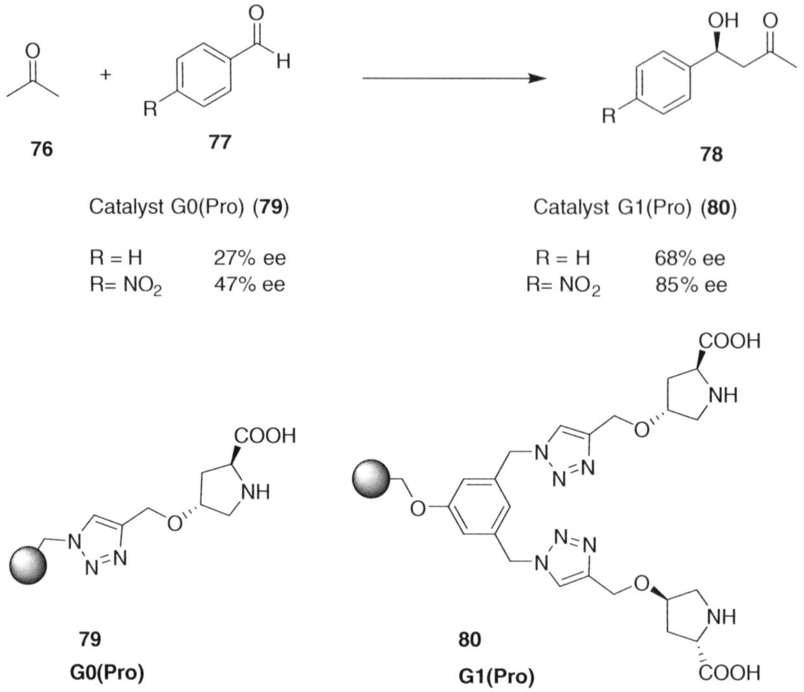

Scheme 7.16. Supported proline catalysts in asymmetric synthesis.

nondendritic catalysts. While the nondendritic G0(Pro) (**79**) could induce only 27% of enantiomeric excess in the aldol products of the reaction between acetone and benzaldehyde, the ee for the same reaction went up to 68% using dendritic catalyst G1(Pro) (**80**). For 4-nitrobenzaldehyde, ee of the reaction with dendritic catalyst (**80**) was increased greatly over the nondendritic one (85% versus 47%). Interestingly, the enantioselectivity observed using G1(Pro) (**80**) is higher than that achieved in solution with L-proline as catalyst.

Itsuno and coworkers[33] reported the asymmetric transfer hydrogenation of aromatic ketones by using the amphiphilic polymer-supported chiral catalyst in water (Scheme 7.17). Among the various polymeric catalysts tested, the sulfonated polymeric catalyst (**83**) prepared from the (*R*,*R*)-diphenylethandiamine was found to possess good catalytic activity for the reduction of acetophenone (**81**) to the corresponding alcohol **82** with high conversion (100%) and high enantioselectivity (***R*-82** with up to 99%). The reaction was performed under mild reaction conditions using 2-propanol or formic acid derivatives as a hydrogen source. Interestingly, in all cases, the enantioselectivities obtained with the polymeric catalysts were superior to those in the model reaction using optically active *N*-toluenesulfonyl-1,2-diphenylethylenediamine (TsDPEN). Moreover, the catalyst was recycled several times without loss of the catalytic activity.

Chromanes[34] are biologically relevant and interesting heterocycles for their important medicinal properties and this scaffold being part of many natural products such as sappone B and robustadial, in addition to being a bioisostere for the hydantoin moiety, presents synthetic challenges for easy accessibility. Literature reports reveal that asymmetric synthesis of this scaffold class[6] using both hetero- and homogeneous catalysis and supported or nonsupported catalysts need high catalyst loading and long reaction times,[31c,35] while delivering varied enantioselectivities with no consistent yields. Kurth

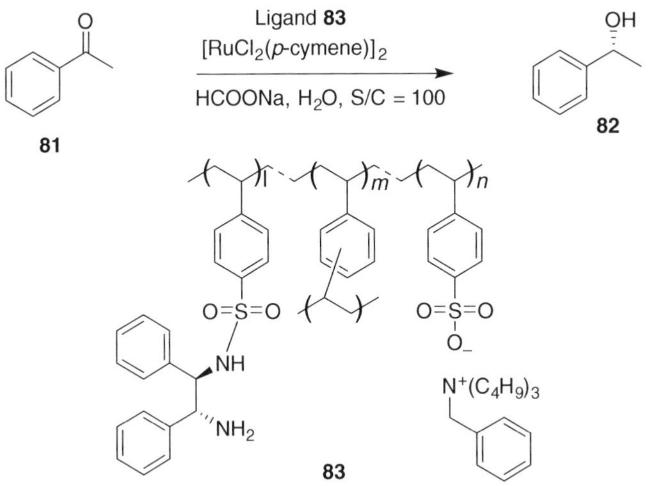

Scheme 7.17. Supported catalysts for hydrogen transfer reaction.

and coworkers[36] reported the synthesis of resin-bound asymmetric pyrrolidine catalysts **84a** and **84b** (Scheme 7.18), which were employed for the preparation of optically active chromanones and other enamine-derived compounds. Reaction of substituted hydroxyacetophenone (**85**) with cycloalkanones (**86**) was carried out using catalysts **84a** or **84b** in MeOH at room temperature and then under microwave irradiation at 110°C for 11 min to yield the spirochromanes **87** in 78–89% yields and enantiomeric excess in the range of

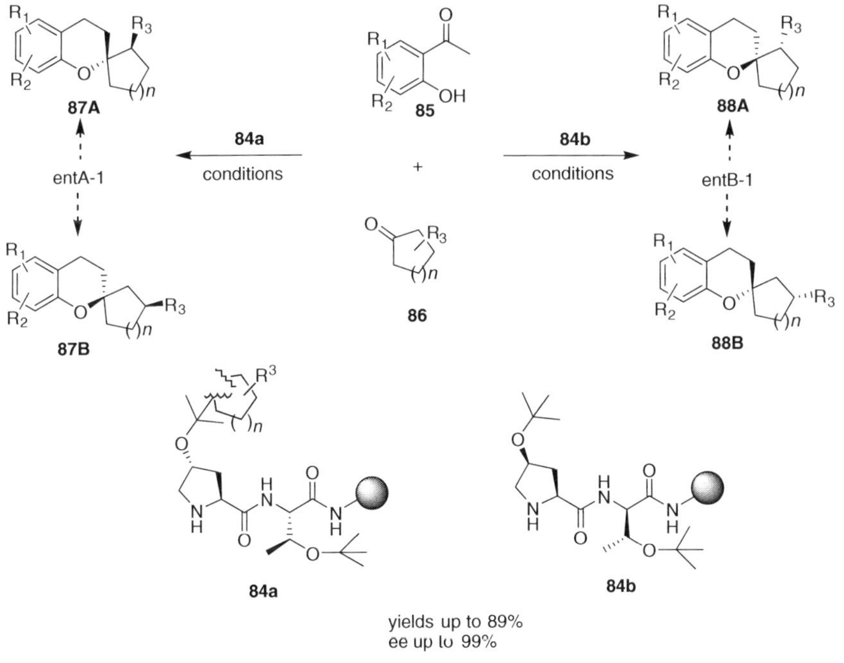

Scheme 7.18. Supported pyrrolidine catalysts in asymmetric synthesis of chromanes.

93–99%. Again, in this case too, the catalysts employed could be recycled several times without any loss of activity.

Synthesis of the chiral catalysts to introduce enantioselectivity in carbene transfer reactions is a subject of great interest.[37] Often copper and rhodium chiral catalysts are of choice for the carbene transfer reactions. In some reports, immobilized chiral dirhodium (II) catalyst were employed successfully in asymmetric cyclopropanation reactions. Ubeda and coworkers[38] reported the immobilization of chiral $Rh_2(O_2CR)_2(PC)_2$ (PC = *ortho*-metalated phosphine) compounds on cross-linked polystyrene (PS) resin by an exchange reaction of carboxylate ligands[39] between metal complexes and carboxyethyl-polystyrene ($PS\text{-}C_6H_4(CH_2)_2CO_2H$) (Scheme 7.19a). The immobilized catalyst **92** (1 mol%) was employed for the asymmetric cyclopropanation of styrene with ethyl diazoacetate in refluxing *n*-pentane (Scheme 7.19b). Although the yields of the cyclopropanation reactions (51–79%) were higher than some corresponding solution-phase reactions, their diastereo- (58 : 42 and 69 : 31) and enantioselectivities (53–57% for **90** and 9–12% for **91**) were lower.

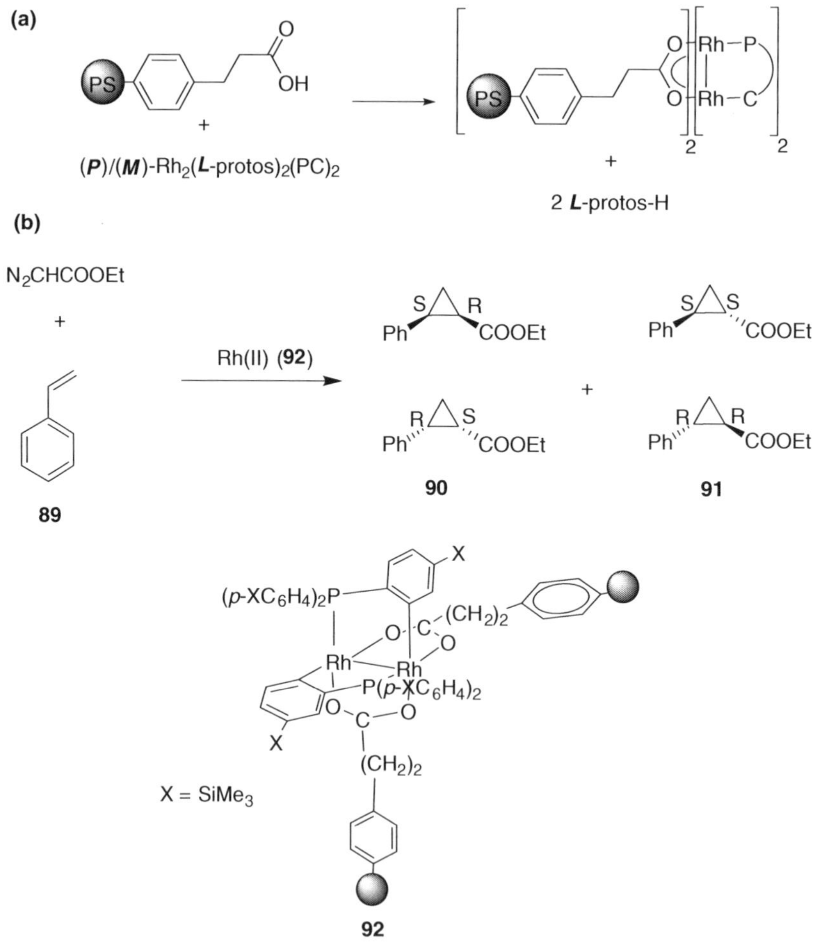

Scheme 7.19. Immobilized rhodium catalyst for asymmetric cyclopropanation.

7.3.2 Asymmetric Synthesis Using Resin-Bound Chiral Auxiliaries

"Glycomics" is a research field that seeks to identify and understand the processes involved in the formation of cell type and oligosaccharide patterns at different developmental stages.[40] To this end, collections of structurally and stereochemically well-defined oligosaccharides are required to develop algorithms for the assignment of oligosaccharide mass spectrometry spectra, for fabricating microarrays, for elucidating the biosynthetic pathways of glycoconjugate assembly, and also as immunogens to produce monoclonal antibodies for glycoprotein visualization and isolation by immunoprecipitation. Solid-phase oligosaccharide synthesis provides a powerful technology for increasing the speed of oligosaccharide assembly, primarily by eliminating intermediate purification steps and by automation.[41] However, the stereoselective installation of 1,2-*cis*-glycosides by solid-phase oligosaccharide synthesis is a major drawback and often a mixture of compounds would be produced if many units need to be attached. Boons and coworkers[42] have successfully demonstrated the solid-supported synthesis of complex, branched oligosaccharides (Scheme 7.20) using the glycosyl donors having a (*S*)-(phenylthiomethyl)benzyl chiral auxiliary (**A** or **B**) at C2 position. Neighboring group participation assists the anomeric control by an (*S*)-(phenylthiomethyl)benzyl chiral auxiliary at C2 of the glucosyl donors.[43] Addition of these glycosides was carried out by replacing the 9-fluorenylmethyloxycarbonyl (Fmoc) and

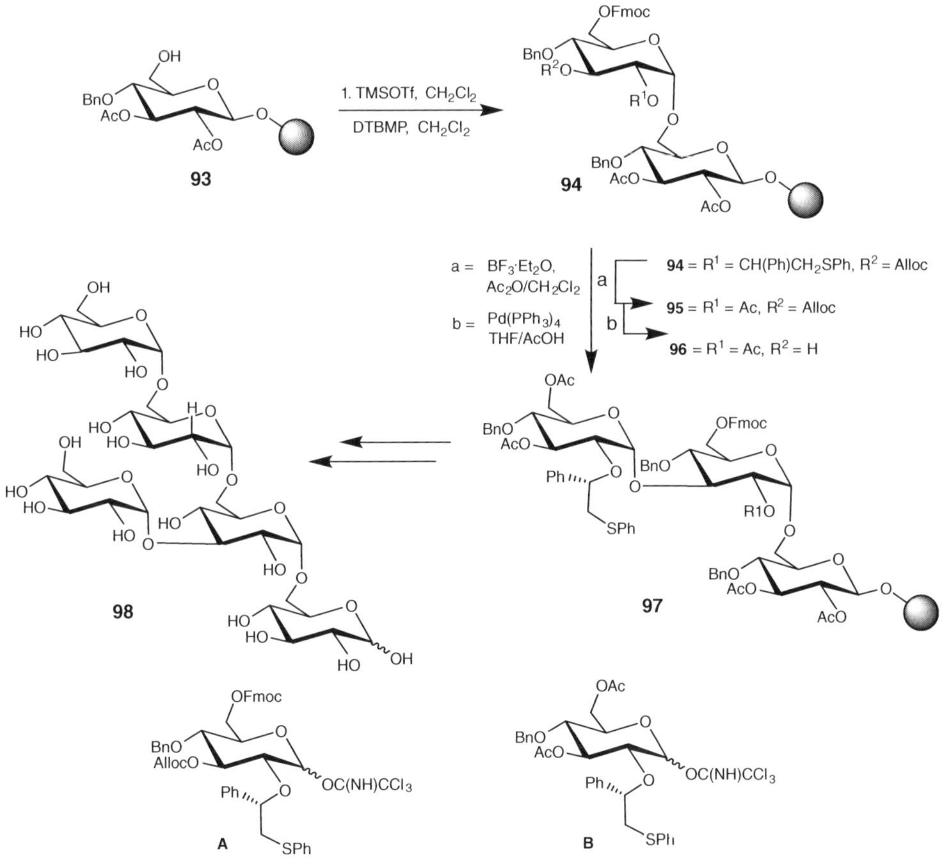

Scheme 7.20. Solid-phase stereoselective synthesis of oligosaccharides using chiral auxiliary.

allyloxycarbonyl (Alloc) protecting groups. The resin-bound acceptor (**93**) was coupled with preactivated auxiliary-containing glucosyl donor **A** using stoichiometric amount of TMSOTf and 2,6-di-*tert*-butyl-4-methylpyridine (DTBMP) in CH_2Cl_2 at −40°C to yield **94**. The Alloc function of **95** was easily removed by treatment with $Pd(PPh_3)_4$ in a mixture of THF and AcOH to give disaccharide acceptor (**96**).

However, solution-phase model studies indicated that C3′ hydroxyl of **96** was sterically shielded owing to the neighboring (*S*)-(phenylthiomethyl)benzyl ether. Therefore, the auxiliary of **94** was converted into acetyl ester (**95**) by treatment with acetic anhydride in the presence of $BF_3{\cdot}OEt_2$. Removal of Alloc in **95** under standard conditions provided glycosyl acceptor **96**, which was coupled with preactivated **B** to form resin-bound trisaccharide **97**. Finally, the required target **98** was achieved after performing few more stereoselective additional reactions.

In another application of sugar-based auxiliary, Kunz and Zech[44] used an immobilized galactose auxiliary for the stereoselective synthesis of chiral piperidones. In this strategy, an amine **99** was treated with an aldehyde in the presence of acetic acid to obtain the polymer-bound imine **100**. The reaction of electron-rich Danishefsky's diene with **100** in the presence of zinc chloride afforded the corresponding resin-bound didehydropiperidinones **102** *via* a domino Mannich–Michael condensation reaction. Finally, the cleavage of the linker with TBAF yielded the 2-substituted *N*-galactosyl-5,6-dehydropiperidin-4-ones (**103**) in 40–81% with diastereomeric ratios ranging from 80:20 to 100:0. It is important to mention here that unlike solution-phase reactions no anomerization was observed during condensation reaction on solid phase. Moreover, the polymer-bound enaminones (**102**) were explored for the conjugate addition reactions by treating it with both methylaluminum bis(2,6-di-*tert*-butyl-4-methyl-phenoxide) (MAD) and cyano-modified Gilman reagents in the presence of Lewis acids (Scheme 7.21). Finally, cleavage of the product obtained from the resin provided 2,6-disubstituted piperidinones (**104**) in 49–78% yield and with *cis*/*trans* ratios ranging from 93:7 to 98:2.

Procter and coworkers[45] have described a Sm(II)-mediated (**106**), asymmetric capture and release approach (Scheme 7.22) to γ-butyrolactones (**107**) that involves intermolecular radical additions to α,β-unsaturated esters (**105**) attached to resin through an ephedrine chiral linker (**108**). Resin capture–release is a hybrid technique that combines elements of traditional solid-phase synthesis and the use of supported reagents. Fukuzawa's Sm(II)-mediated, asymmetric method to γ-butyrolactones was chosen to demonstrate the feasibility of such a process. γ-Butyrolactones (**107**) were obtained by capture of a reactive intermediate from solution through an asymmetric transformation starting from α,β-unsaturated esters (**105**) immobilized on an ephedrine chiral resin. Lactone products were obtained in moderate yields with selectivities up to 96% ee. Nevertheless, the ephedrine resin can be efficiently reused for many cycles although in some cases lower yields were obtained on reuse of the chiral resin.

Ephedrine and pseudoephedrine auxiliaries are cheap commercially available compounds and readily accessible without any further modification. Procter and coworkers developed a polymer-supported pseudoephedrine auxiliary for asymmetric alkylations on solid phase.[46] The resin-bound amide **109** is deprotonated and alkylated with benzyl bromide to provide enantiopure **110**. Treating later with different reagents resulted in the enantioselective formation of various alcohols, acids, and amide in 31–55% yield (overall yields), with ee ranging from 78% to 92% (Scheme 7.23).

Chiral tertiary α-hydroxy carboxylic acids are important building blocks for the syntheses of many biologically active and relevant molecules.[47] These tertiary alcohols

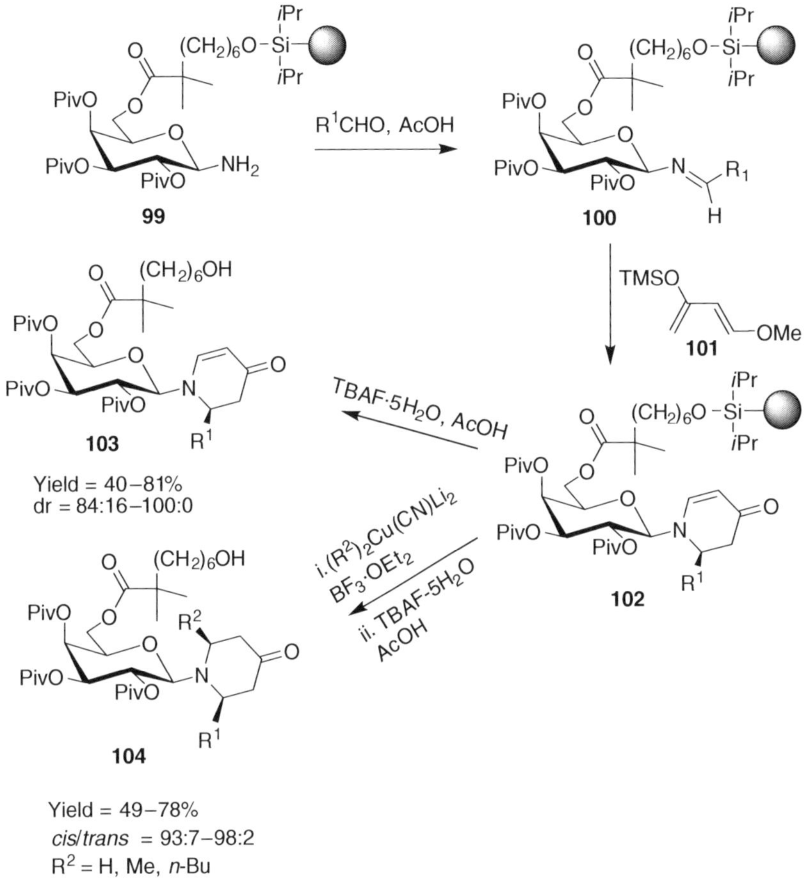

Scheme 7.21. Synthesis of chiral piperidones using immobilized galactose auxiliary.

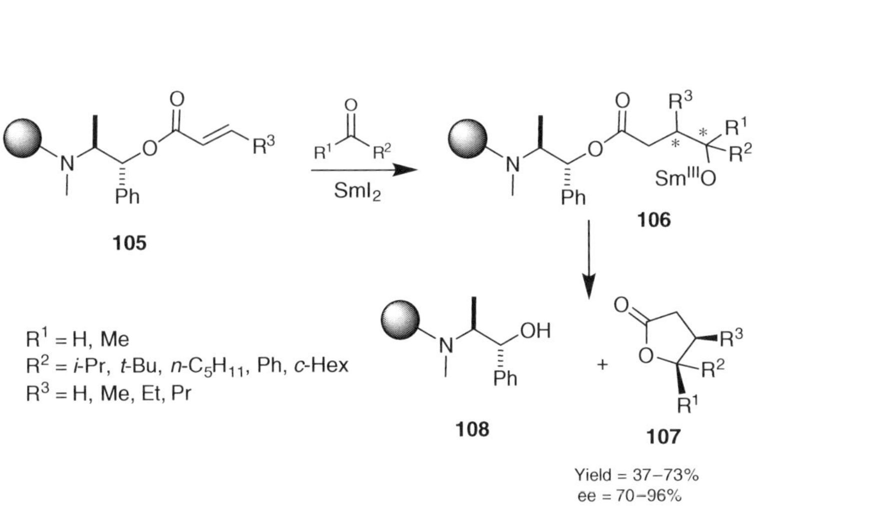

Scheme 7.22. Asymmetric synthesis of butyrolactones using supported ephedrine linker.

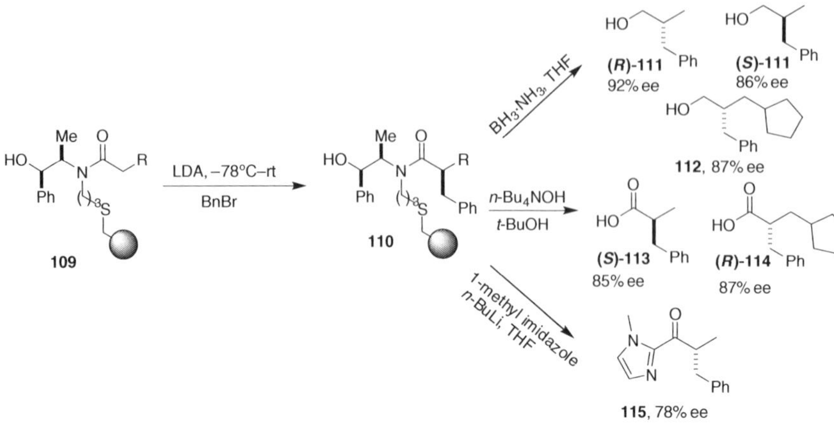

Scheme 7.23. Polymer-supported pseudoephedrine auxiliary for asymmetric alkylation.

can be prepared in principle in high enantiomeric excess by addition of organometallic reagents to a chiral auxiliary coupled α-keto carboxylic acids or its derivatives. Gaertner and coworkers used *m*-hydrobenzoin-derived chiral auxiliaries (**116**) for the L-Selectride-mediated stereoselective reduction of phenylglyoxylates.[48] Noe's anhydrolactols[49] were used for the desymmetrization of *m*-hydrobenzoin and the subsequent diastereoselective addition of organozinc reagents (Scheme 7.24). Nucleophilic addition of several organozinc reagents to resin-bound α-keto esters (**118**)[48] provided the chiral tertiary alcohols **121** with

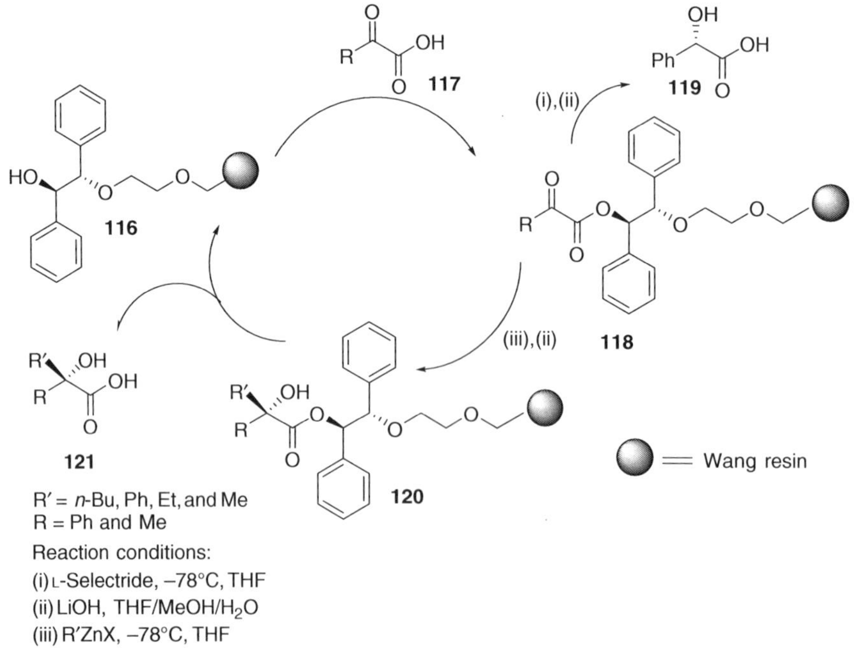

Scheme 7.24. Resin-bound *m*-hydrobenzoin-derived chiral auxiliary for stereoselective reduction of phenylglyoxylates.

good (65:35 and 75:25) to excellent diastereoselectivities (95:5 and 93:5) for the smaller alkyl groups (Me- and Et-) and larger alkyl groups (*n*-Bu- and *c*-hex), respectively, added to phenylglyoxylate.[50]

7.4 CONCLUSIONS

Over the years, it has been experienced that a constant endeavors are required for any discipline to develop and grow. Although organic chemistry can be proud of its 180-year history, often synthesis of a desired small molecule with desired functional and stereochemical decoration proves challenging. In this era of combinatorial chemistry and high-throughput screenings, organic synthesis has different roles to play in providing collections of diverse molecules as pure diastereomers and enantiomers. Solid-phase synthesis and, in particular, solid-phase asymmetric synthesis can be of great help to chemist in such endeavors. The examples summarized in this chapter convincingly demonstrate that asymmetric synthesis on the solid support is possible and comparable to results obtained through solution-phase synthesis efforts. Especially the synthesis of natural products derived and inspired complex molecules can make use of solid-phase protocols for asymmetric synthesis. It should, however, be remembered that in order to avoid the formation of immobilized minor products during the course of solid-phase asymmetric synthesis, only such protocols for asymmetric transformations can be applied to solid-phase organic synthesis that provide high levels of asymmetric induction. Results from solution-phase chemistry give good indications as to which type of reaction can be taken into consideration.

Apparently, there is still a lot to be done to improve and develop the feasible protocols for many chemical transformations that remain challenging and often do not work well on solid phase, for instance, catalytic hydrogenation; addition of organometallic reagents such as Grignard, organolithiums, and so on; reactions involving carbocations (skeletal re-arrangements, etc.) and reactions involving radicals, radical cations, radical anions, and electroorganic synthesis, and so on. This area of research is still very young and thus calls for further investment from all sources for gaining a level of maturity so that it can give back further useful applications to the chemical and biological sciences.

REFERENCES

1. F. Zaragoza-Dörwald, *Organic Synthesis on Solid-phase*, 2nd ed., Wiley, Weinheim, **2002**.
2. B. Merrifield, in Malström, B. G. (Ed.), *Nobel Lectures, Chemistry 1981–1990,* World Scientific Publishing Co., Singapore, **1992**, pp. 149–175.
3. M. Feher, J. M. Schmidt, *J. Chem. Inf. Comput. Sci.* **2003**, *43*, 218–227.
4. (a) K. Kumar, H. Waldmann, *Angew. Chem., Int. Ed.* **2009**, *48*, 3224–3242. (b) J. P. Nandy, M. Prakesch, S. Khadem, P. T. Reddy, U. Sharma, P. Arya, *Chem. Rev.* **2009**, *109*, 1999–2060. (c) W. Wilk, T. J. Zimmermann, M. Kaiser, H. Waldmann, *Biol. Chem.* **2009**, *391*, 491–497.
5. T. Lessmann, H. Waldmann, *Chem. Commun.* **2006**, 3380–3389.
6. (a) E. Framery, B. Andrioletti, M. Lemaire, *Tetrahedron: Asymmetry* **2010**, *21*, 1110–1124. (b) S. Brase, F. Lauterwasser, R. E. Ziegert, *Adv. Synth. Catal.* **2003**, *345*, 869–929.
7. C. W. Y. Chung, P. H. Toy, *Tetrahedron: Asymmetry* **2004**, *15*, 387–399.
8. M. Mentel, R. Breinbauer, *Top. Curr. Chem.* **2007**, *278*, 209–241.

9. J. S. Panek, B. Zhu, *J. Am. Chem. Soc.* **1997**, *119*, 12022–12023.
10. C. J. Cowden, I. Paterson, *Org. React.* **1997**, *51*, 1–200.
11. (a) O. Barun, S. Sommer, H. Waldmann, *Angew. Chem., Int. Ed.* **2004**, *43*, 3195–3199. (b) O. Barun, K. Kumar, S. Sommer, A. Langerak, T. U. Mayer, O. Muller, H. Waldmann, *Eur. J. Org. Chem. 2005*, 4773–4788.
12. I. Paterson, D. Gottschling, D. Menche, *Chem. Commun.* **2005**, 3568–3570.
13. (a) Y. Yamamoto, N. Asao, *Chem. Rev.* **1993**, *93*, 2207–2293. (b) P. V. Ramachandran, *Aldrichim. Acta* **2002** *35*, 23–35.
14. J. S. Panek, M. Yang, F. Xu, *J. Org. Chem.* **1992**, *57*, 5790–5792.
15. (a) J. S. Panek, M. Yang, *J. Org. Chem.* **1991**, *56*, 5755–5758. (b) J. S. Panek, M. Yang, *J. Am. Chem. Soc.* **1991**, *113*, 6594–6600. (c) C. E. Masse, J. S. Panek, *Chem. Rev.* **1995**, *95*, 1293–1316.
16. C. M. DiBlasi, D. E. Macks, D. S. Tan, *Org. Lett.* **2005**, *7*, 1777–1780.
17. J. W. A. Kinnaird, P. Y. Ng, K. Kubota, X. L. Wang, J. L. Leighton, *J. Am. Chem. Soc.* **2002**, *124*, 7920–7921.
18. V. Mamane, A. B. Garcia, J. D. Umarye, T. Lessmann, S. Sommer, H. Waldmann, *Tetrahedron* **2007**, *63*, 5754–5767.
19. (a) A. B. Garcia, T. Lessmann, J. D. Umarye, V. Mamane, S. Sommer, H. Waldmann, *Chem. Commun.* **2006**, 3868–3870. (b) J. D. Umarye, T. Lessmann, A. B. Garcia, V. Mamane, S. Sommer, H. Waldmann, *Chem. Eur. J.* **2007**, *13*, 3305–3319.
20. N. Zou, B. Jiang, *J. Comb. Chem.* **2000**, *2*, 6–7.
21. Y. Ukaji, K. Sada, K. Inomata, *Chem. Lett.* **1993**, 1847–1850.
22. C. Chen, X. D. Li, S. L. Schreiber, *J. Am. Chem. Soc.* **2003**, *125*, 10174–10175.
23. M. Kurosu, J. R. Porter, M. A. Foley, *Tetrahedron Lett.* **2004**, *45*, 145–148.
24. M. A. Sanz, T. Voigt, H. Waldmann, *Adv. Synth. Catal.* **2006**, *348*, 1511–1515.
25. D. A. Annis, O. Helluin, E. N. Jacobsen, *Angew. Chem., Int. Ed.* **1998**, *37*, 1907–1909.
26. T. Nagashima, H. M. L. Davies, *J. Am. Chem. Soc.* **2001**, *123*, 2695–2696.
27. M. J. O'Donnell, W. D. Bennett, S. D. Wu, *J. Am. Chem. Soc.* **1989**, *111*, 2353–2355.
28. H. G. Park, M. J. Kim, M. K. Park, H. J. Jung, J. Lee, S. H. Choi, Y. J. Lee, B. S. Jeong, J. H. Lee, M. S. Yoo, J. M. Ku, S. S. Jew, *J. Org. Chem.* **2005**, *70*, 1904–1906.
29. S. C. Stinson, *Chem. Eng. News* **2001**, *79*, 45–46.
30. S. Mukherjee, J. W. Yang, S. Hoffmann, B. List, *Chem. Rev.* **2007**, *107* 5471–5569.
31. (a) J. D. Revell, D. Gantenbein, P. Krattiger, H. Wennemers, *Biopolymers* **2006**, *84*, 105–113. (b) M. R. M. Andreae, A. P. Davis, *Tetrahedron: Asymmetry* **2005**, *16*, 2487–2492. (c) K. Akagawa, S. Sakamoto, K. Kudo, *Tetrahedron Lett.* **2005**, *46*, 8185–8187. (d) F. Calderon, R. Fernandez, F. Sanchez, A. Fernandez-Mayoralas, *Adv. Synth. Catal.* **2005**, *347*, 1395–1403. (e) G. Szollosi, G. London, L. Balaspiri, C. Somlai, M. Bartok, *Chirality* **2003**. *15*, S90–S96.
32. T. Kehat, M. Portnoy, *Chem. Commun.* **2007**, 2823–2825.
33. Y. Arakawa, A. Chiba, N. Haraguchi, S. Itsuno, *Adv. Synth. Catal.* **2008**, *350*, 2295–2304.
34. H. J. Kabbe, A. Widdig, *Angew. Chem., Int. Ed.* **1982**, *21*, 247–256.
35. M. Benaglia, M. Cinquini, F. Cozzi, A. Puglisi, G. Celentano, *Adv. Synth. Catal.* **2002**, *344*, 533–542.
36. R. D. Carpenter, J. C. Fettinger, K. S. Lam, M. J. Kurth, *Angew. Chem., Int. Ed.* **2008**, *47*, 6407–6410.
37. (a) H. M. L. Davies, R. E. J. Beckwith, *Chem. Rev.* **2003**, *103*, 2861–2903. (b) H. Lebel, J. F. Marcoux, C. Molinaro, A. B. Charette, *Chem. Rev.* **2003**, *103*, 977–1050.
38. J. Lloret, F. Estevan, K. Bieger, C. Villanueva, M. A. Ubeda, *Organometallics* **2007**, *26*, 4145–4151.

39. (a) M. P. Doyle, D. J. Timmons, J. S. Tumonis, H. M. Gau, E. C. Blossey, *Organometallics* **2002**, *21*, 1747–1749. (b) M. P. Doyle, M. Yan, H. M. Gau, E. C. Blossey, *Org. Lett.* **2003**, *5*, 561–563.
40. (a) R. Raman, S. Raguram, G. Venkataraman, J. C. Paulson, R. Sasisekharan, *Nat. Methods* **2005**, *2*, 817–824. (b) K. T. Pilobello, L. K. Mahal, *Curr. Opin. Chem. Biol.* **2007**, *11*, 300–305. (c) M. S. M. Timmer, B. L. Stocker, P. H. Seeberger, *Curr. Opin. Chem. Biol.* **2007**, *11*, 59–65. (d) N. Laurent, J. Voglmeir, S. L. Flitsch, *Chem. Commun.* **2008**, 4400–4412.
41. P. H. Seeberger, *Chem. Soc. Rev.* **2008**, *37*, 19–28.
42. T. J. Boltje, J. H. Kim, J. Park, G. J. Boons, *Nat. Chem.* **2010**, *2*, 552–557.
43. (a) J. H. Kim, H. Yang, G. J. Boons, *Angew. Chem., Int. Ed.* **2005**, *44*, 947–949. (b) J. H. Kim, H. Yang, J. Park, G. J. Boons, *J. Am. Chem. Soc.* **2005**, *127*, 12090–12097.
44. G. Zech, H. Kunz, *Angew. Chem., Int. Ed.* **2003**, *42*, 787–790.
45. N. J. Kerrigan, P. C. Hutchison, T. D. Heightman, D. J. Procter, *Org. Biomol. Chem.* **2004**, *2*, 2476–2482.
46. A. M. McGhee, J. C. Kizirian, D. J. Procter, *Org. Biomol. Chem.* **2007**, *5*, 1021–1024.
47. (a) Y. Nagao, T. Inoue, E. Fujita, S. Terada, M. Shiro, *J. Org. Chem.* **1983**, *48*, 132–133. (b) S. Superchi, M. Contursi, C. Rosini, *Tetrahedron* **1998**, *54*, 11247–11254. (c) P. Scafato, L. Leo, S. Superchi, C. Rosini, *Tetrahedron* **2002**, *58*, 153–159.
48. C. Schuster, M. Knollmueller, P. Gaertner, *Tetrahedron: Asymmetry* **2005**, *16*, 3211–3223.
49. (a) C. R. Noe, *Chem. Ber.-Recl.* **1982**, *115*, 1576–1590. (b) C. R. Noe, M. Knollmuller, G. Steinbauer, E. Jangg, H. Vollenkle, *Chem. Ber.-Recl.* **1988**, *121*, 1231–1239.
50. C. Schuster, M. Knollmueller, P. Gaertner, *Tetrahedron: Asymmetry* **2006**, *17*, 2430–2441.

8

RECENT ADVANCES IN MICROWAVE-ASSISTED SOLID-PHASE SYNTHESIS OF HETEROCYCLES

Prasad Appukkuttan, Vaibhav, P. Mehta, and Erik Van der Eycken

8.1 INTRODUCTION

The ingenious work of Bruce Merrifield in the early 1960s[1] regarding the synthesis of peptides on resin beads composed of a copolymer of styrene and divinylbenzene has paved the way for the synthesis of a plethora of organic molecules on solid support.[2] Solid-supported reactions quickly became very popular due to various advantages, such as the ease of separation of the target molecules from the reactants by simple filtration, the possibility of the use of high concentrations of reagents, the viability of the protocol toward automation, and hence the possibility for the generation of large libraries.[3] Until the resurgence boosted by the invention of combinatorial chemistry techniques in the early 1990s, solid-phase organic synthesis was primarily used for the synthesis of peptides, nucleic acids, and carbohydrates, while the synthesis of small molecules and heterocycles was relatively underexplored.[4] However, the field has witnessed significant advances in the development of new solid supports and interesting linkers to encompass diverse chemistry and allow the attachment of various functionalities to the solid support.[5] Furthermore, advantageous concepts such as cyclative cleavage, traceless linking, and safety-catch linking have reduced the complexity of the solid-phase protocols.[6] As a result, the last decade has seen an exponential increase in the solid-phase synthesis of heterocycles and small molecules with a plethora of interesting biological activities.[7] A major drawback of solid-supported reactions is the requirement of relatively long reaction times in comparison to their solution-phase counterparts. However, the development of microwave-assisted chemistry in the last two decades also had a beneficial effect on solid-phase chemistry: reactions could be sped up

Solid-Phase Organic Synthesis: Concepts, Strategies, and Applications, First Edition.
Edited by Patrick H. Toy and Yulin Lam.

tremendously under microwave irradiation providing cleaner products and sometimes protocols could be performed that were unattainable via the classical conditions. In this chapter, we will provide the reader a selection of examples of recent advances in microwave-assisted solid-phase synthesis of heterocyclic moieties. The literature covered is a selective representation of the vast collection of available literature ranging from 2005 to the early 2010. We deliberately decided not to include fields such as solid-phase synthesis of peptides and carbohydrates as these deserve a chapter on their own.

8.2 FUSED 1,3-OXAZIN-6-ONES

Fused 1,3-oxazin-6-ones feature a multitude of biological activities, such as anticoagulant, antiviral, and herbicidal activities. They have also been described as potential inhibitors of HSV-1 and C1r serine protease.[8] Furthermore, oxazinones fused with other heterocyclic systems such as thieno[2,3-*d*]-1,3-oxazinones, pyrrolo[2,3-*d*]-1,3-oxazin-4-ones, pyrazole [3,4-*d*][1,3]oxazin-4-ones, though scarce in nature, feature potent biological activities and therefore have invited high synthetic interest.[9] Lam and coworkers have recently developed a convenient microwave-assisted traceless solid-phase approach to bi- and tricyclic heteroannulated 1,3-oxazin-6-ones,[10] which involves the synthesis of a five-membered heterocyclic acyl amino ester on solid support and subsequent cyclization to the fused heterocyclic oxazinone derivative (Scheme 8.1). The authors initially established the proof of concept in solution phase, where they used a benzyloxy group as the solid-phase mimic in their six-step protocol. It is noteworthy that the dibromotetrachloroethane and triphenylphosphine-mediated ring closure to a fused heterocyclic oxazinone derivative was carried out under microwave irradiation, even in solution phase. The solid-supported synthesis was then explored using Wang resin, which was first converted to the resin-bound cyanoacetate moiety (Scheme 8.1).

This loaded to the resin and converted to the corresponding resin-bound pyrazole by initially reacting with triethyl orthoformate and acetic anhydride in DMF for 30 min at 140°C, followed by treatment with a suitable hydrazine in ethanol at 120°C for 30 min (Scheme 8.1). Both reactions run exceedingly fast under microwave irradiation, while a prolonged 36 and 12 h, respectively, were required when carried out under classical heating conditions. The authors converted the thus formed resin-bound pyrazole into the corresponding amide by a microwave-assisted diamidation of the amine in DMF at 120°C for

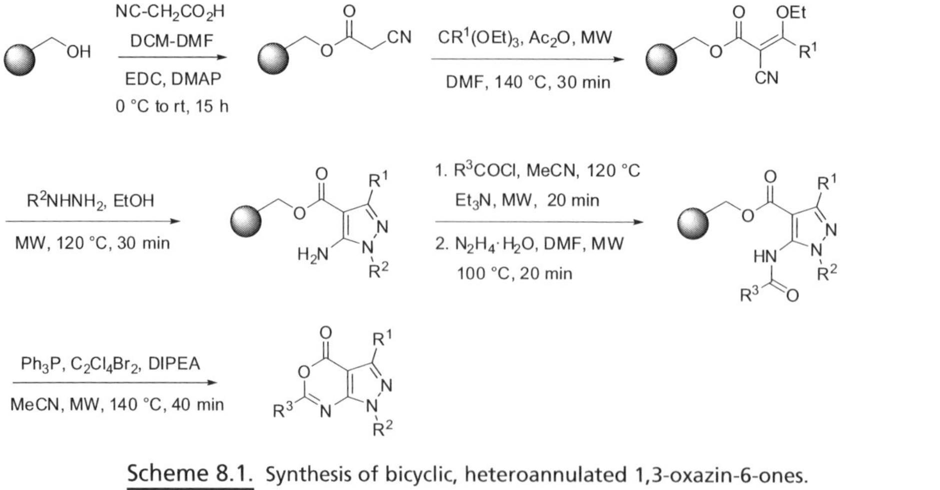

Scheme 8.1. Synthesis of bicyclic, heteroannulated 1,3-oxazin-6-ones.

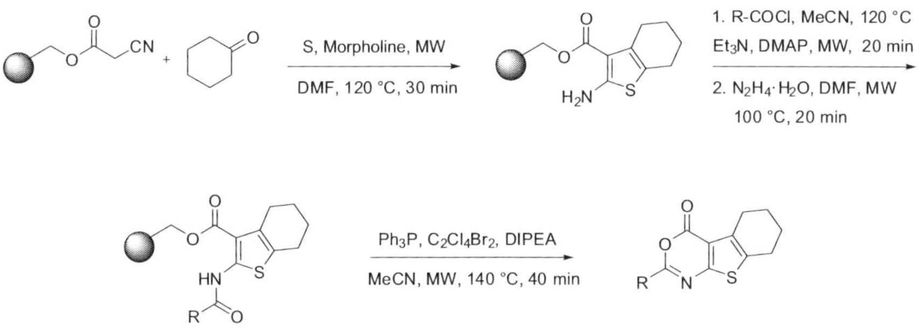

Scheme 8.2. Synthesis of tetrahydrobenzo[*b*]thiophene-annulated 1,3-oxazin-6-ones.

20 min, followed by the cleavage of the diamide to the corresponding monoamide using hydrazine hydrate in DMF at 100°C for 20 min, once again under microwave irradiation. The formation of the target oxazinone and the subsequent cleavage from the resin were accomplished by reacting the resin-bound pyrazole amide with a triphenylphosphine–carbon tetrabromide mixture in MeCN at 140°C for 40 min under microwave irradiation, using DIPEA as the base of choice.

The authors also presented a well-rounded study on heteroannulated 1,3-oxazin-6-ones by proposing three more solid-supported, microwave-assisted protocols, including a solid-supported Gewald thiophene synthesis.[11] A representative library of 20 bi- and tricyclic heteroannulated 1,3-oxazin-6-ones was generated (Scheme 8.2). Thus, the resin-bound α-cyanoketone was reacted with cyclohexanone and sulfur in the presence of morpholine in DMF at 120°C for 30 min under microwave irradiation, yielding the corresponding resin-bound 2-amino-4,5,6,7-tetrahydro-benzo[*b*]thiophene-3-carboxylic acid (Scheme 8.2).

After converting the amine to the diamide, the subsequent cleavage to the monoamide was effected once again using microwave irradiation (Scheme 8.2). The formation of the target oxazinone and the subsequent cleavage of the product was then accomplished by a DIPEA-mediated reaction with Ph_3P and CBr_4 (Scheme 8.2). It is also interesting to note that the reaction time for the three-component Gewald condensation is shortened from 18 h under conventional heating to 30 min under microwave irradiation.

8.3 THIAZOLO[4,5-*d*]PYRIMIDINE-5,7-DIONES

Thiazolo[4,5-*d*]pyrimidine-5,7-dione derivatives exhibit a plethora of interesting biological activities. They have been described as TNF-α inhibitors for rheumatoid arthritis, multiple sclerosis, and asthma, as antidepressant agents acting on the central nervous system, as hepatitis C virus (HCV) polymerase inhibitors, and as antihuman cytomegalovirus (HCMV) agents.[12] On the basis of their experience with traceless linker-based solid-phase strategies for the preparation of thiazoles and fused thiazole scaffolds,[13] Gong and coworkers have recently reported the first solid-phase synthesis of 2,4,6-trisubstituted thiazolo[4,5-*d*]pyrimidine-5,7-diones employing an ingenious six-step strategy using Merrifield resin as the solid support of choice (Scheme 8.3).[14]

The authors initially investigated the synthesis of resin-bound 4-aminothiazole-5-esters, which were generated in two convenient steps starting from Merrifield resin (Scheme 8.3) by the reaction with dipotassium cyanodithioimidocarbonate followed by a condensation reaction with ethyl bromoacetate in DMF at 80°C using TEA as the base. The urea formation with a large number of isocyanates was then explored under microwave irradiation using $i\text{Pr}_2\text{EtN}$ in DMSO at 150°C. After the NaH-mediated cyclization and

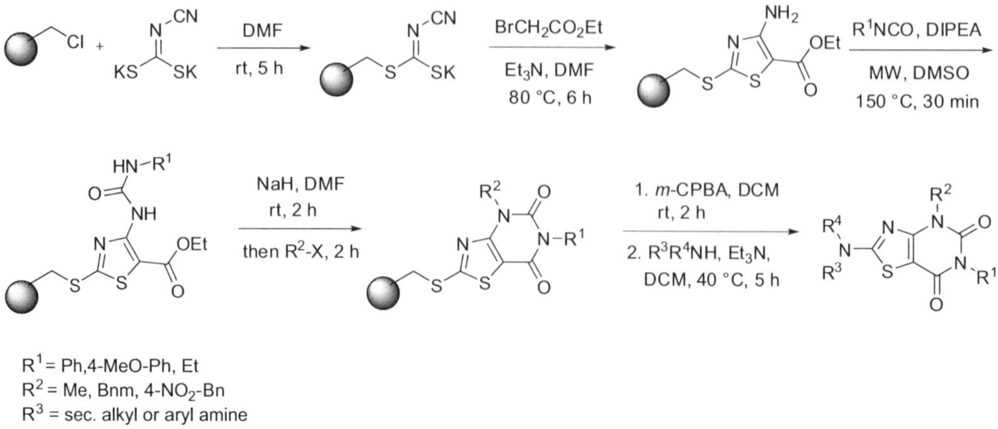

Scheme 8.3. Synthesis of 2,4,6-trisubstituted thiazolo[4,5-*d*]pyrimidine-5,7-diones.

N-alkylation, the resin-bound heterocycle was converted into the corresponding sulfone by treatment with *m*-CPBA. The traceless cleavage of the resin was finally accomplished using a variety of amines (Scheme 8.3), thus generating a small library of 48 compounds with good overall yields of 12–34% over six steps.

8.4 PYRAZOLES

Yli-Kauhaluoma and coworkers have recently elaborated an interesting synthesis of amino acid-derived *N*-unsubstituted pyrazoles using solid-supported, stable, mesoionic sydnones, which react as azomethine imine-type dipoles.[15] This investigation is particularly valuable owing to the abundance of pyrazole-containing compounds such as Celecoxib, Rimonabant, Fomepizole, and Sildenafil that show interesting biological and pharmacological activities.[16] Though sydnones have previously been studied in 1,3-dipolar cycloaddition reactions furnishing pyrazoles, they normally generate *N*-substituted pyrazoles as the *N*-nitrosation reaction proceeds only with an *N*-alkylated amino acid. The authors attached a number of amino acids to the formyl-functionalized Ameba resin[17] using a $NaBH(OAc)_3$-mediated reductive amination protocol, followed by hydrolysis of the ester to generate the free acid (Scheme 8.4). The corresponding nitroso derivatives were then generated by the use of nitrous acid at rt for 20 h.

The cycloaddition reaction was performed in toluene under microwave irradiation at 150°C for 30 min using a water removing agent such as acetic anhydride. The generated resin-linked pyrazole (Scheme 8.4) was easily cleaved from the resin using DCM/TFA (7:3) at rt. It is interesting to note that the yields of the cycloaddition reaction were found to drop drastically when the temperature was lowered. Furthermore, in case of $R_2 = H$, the authors observed that the 3-substituted pyrazole was always the major product.

8.5 HSP70 MODULATORS

Heat shock proteins (HSPs) are members of a family of molecular chaperones playing critical roles in protein folding, intracellular trafficking of proteins, and coping with proteins

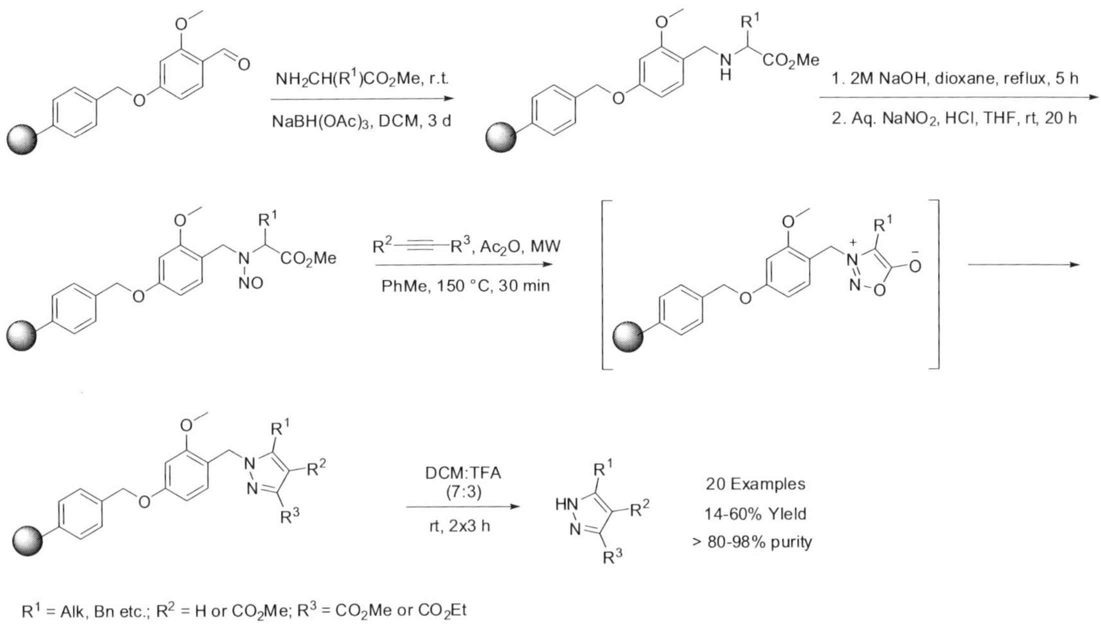

Scheme 8.4. Synthesis of pyrazoles using sydnones.

denatured by heat and other stress factors.[18] Hsp70 is easily triggered by external stress-contributing factors such as infection, inflammation, and exposure to toxins. Small molecule inhibitors of HSPs show promise as anticancer agents,[19] such as the potent Hsp90 inhibitor 17-*N*-allylamino-17-demethoxygeldanamycin (17-AAG), which is currently in clinical trials for treatment of several types of cancer.[20] Most recently, the quest on finding efficient chemical modulators of HSPs has shifted to more "drug-like" compounds that are assembled on a suitable heterocyclic core like dihydropyrimidine.[21] These studies identified SW-02, MAL3-90, and MAL3-101 (Figure 8.1), as potent anticancer agents acting through various biological pathways.

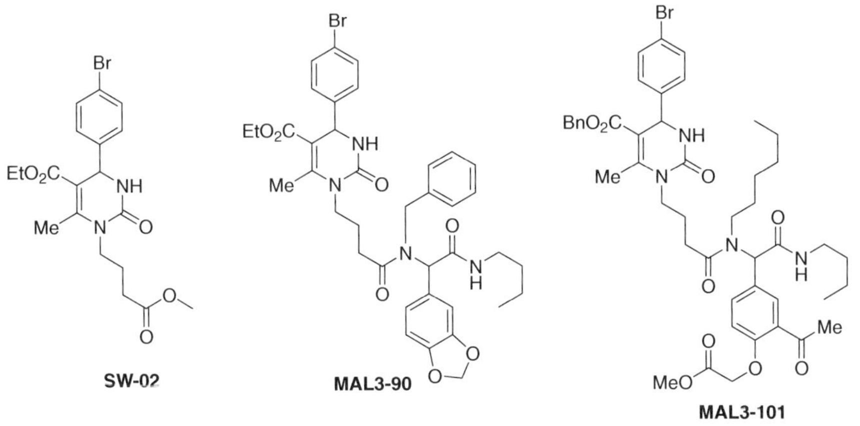

Figure 8.1. Some potent HSP modulators nearing the dihydropyrimidine core.

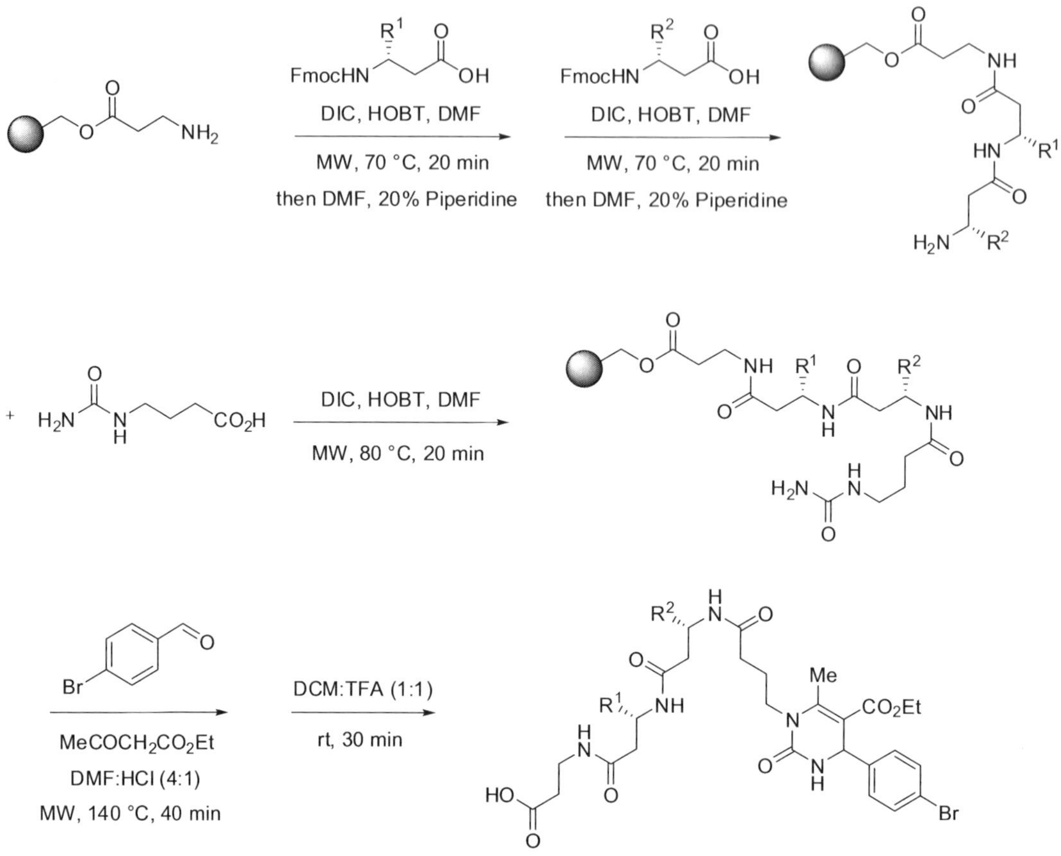

Scheme 8.5. Biginelli reaction in the synthesis of Hsp70 modulators.

In view of generating a focused library of dihydropyrimidines as additional leads, Gestwicki and coworkers investigated a microwave-assisted, solid-supported pathway toward a diverse array of dihydropyrimidine analogues,[22] considering exploring the SAR of the nonheterocyclic part. The route (Scheme 8.5) involves the attachment of suitable amino acids to Wang resin-bound β-alanine using a microwave-assisted, DIC-mediated coupling protocol and subsequent Fmoc deprotection.

This was followed by the attachment of a suitable urea-bearing carboxylic acid to generate the handle for the synthesis of the heterocyclic moiety (Scheme 8.5). The dihydropyrimidine core was then assembled using a standard microwave-assisted Biginelli reaction[23] at 140°C for 40 min using DMF/HCl (4:1), followed by a TFA-mediated cleavage of the target molecule from the resin.

8.6 BENZIMIDAZO[2,1-*b*]QUINAZOLIN-12(5*H*)-ONES

Benzimidazo[2,1-*b*]quinazolin-12(5*H*)-ones are currently receiving increased attention, as they are potent immune suppressors[24] as well as prospective antitumor compounds, owing to the strong DNA intercalating nature of the benzimidazole and quinazolinone moieties. They have been shown to effectively truncate the propagation of human tumor cell lines.[25]

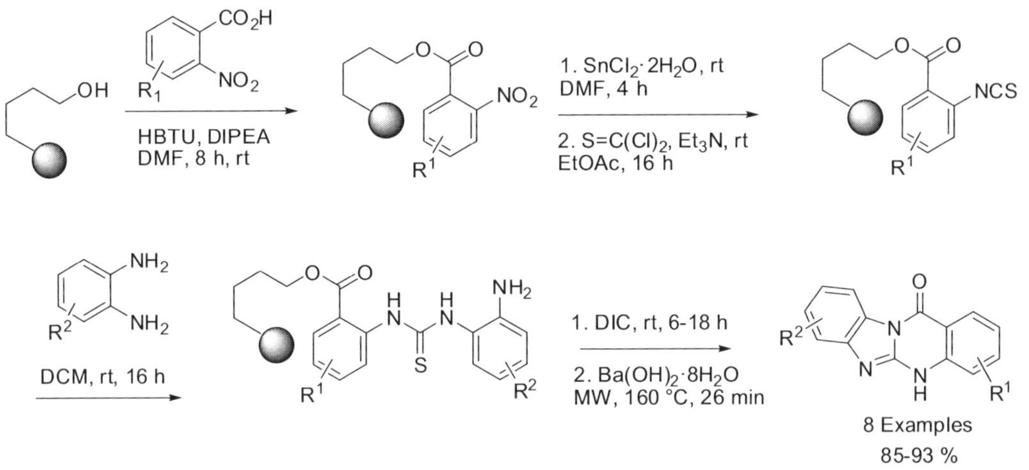

Scheme 8.6. Synthesis of benzimidazo[2,1-*b*]quinazolin-12(5*H*)-ones.

Although there is a large amount of available literature about the synthesis of these interesting heterocycles based on the cyclization of suitably decorated 2-(1*H*-benzoimidazol-2-ylamino) benzoates, these reactions generally suffer from low yields. In view of developing an efficient and high-yielding protocol for the synthesis of benzimidazo[2,1-*b*]quinazolin-12 (5*H*)-ones, Kurth and coworkers have recently developed a microwave-assisted heterocyclization protocol,[26] exploring the chemistry in solution phase as well as on solid support (Scheme 8.6).

The authors used (4-hydroxybutyl)polystyrene resin as the solid support, which was converted into the corresponding isothiocyanate resin by employing an HBTU-mediated coupling of a suitable 2-nitrobenzoic acid, followed by a $SnCl_2$ reduction and a reaction with thiophosgene (Scheme 8.6). After the formation of the corresponding thiourea via reaction of a suitable 1,2-diaminoaryl compound, a DIC-mediated cyclization generated the benzimidazole intermediate. The final heteroannulation with concomitant cleavage of the desired product from the resin in a traceless manner was carried out under microwave irradiation at 160°C for 26 min (Scheme 8.6), using barium hydroxide octahydrate as the reagent of choice. It is noteworthy that this microwave-assisted protocol furnishes excellent overall yields of 85–93% starting from the hydroxybutyl resin with phenomenal purities of >95%, while the formation of hydrolyzed by-products such as 2-(1*H*-benzoimidazol-2-ylamino)benzoic acids was practically negligible.

8.7 IMIDAZOLES

The imidazole core is a common moiety in a variety of natural products and pharmacologically active compounds[27] such as the commercially available drugs etomidate, cimetidine, omeprazole, and lansoprazole.[28] In view of developing an easy and clean protocol for the synthesis of small libraries of imidazole derivatives, Yli-Kauhaluoma and coworkers have recently elaborated a microwave-assisted protocol[29] based on the 1,3-dipolar cycloaddition reaction of 1-(isocyanomethylsulfonyl)-4-methylbenzene (TOSMIC) with suitably functionalized imines immobilized on solid support (Scheme 8.7).

The authors used commercially available Ameba resin[30] for their investigations that upon reaction with trimethyl orthoformate (4 equiv) and a suitably decorated benzylamine (4 equiv)

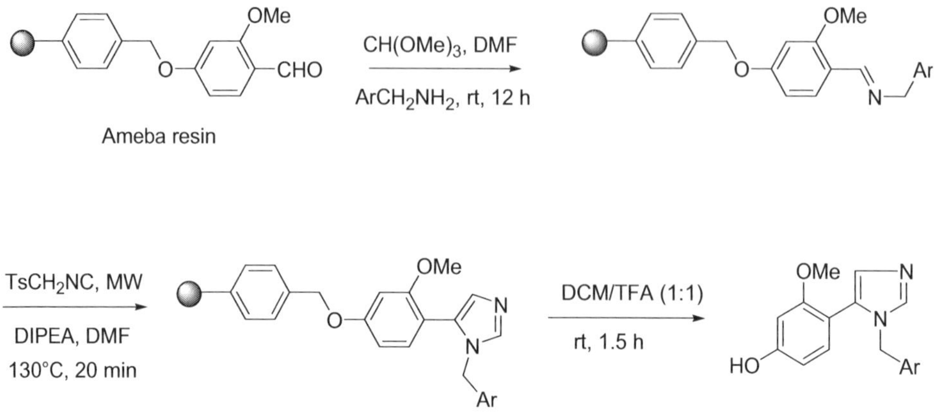

Scheme 8.7. Synthesis of imidazoles.

yielded the immobilized imines for the 1,3-dipolar cycloaddition (Scheme 8.7). This was run in DMF with TOSMIC in the presence of DIPEA at 130°C for 20 min to generate the resin-bound imidazoles, which were finally released into solution by a TFA-mediated resin cleavage (Scheme 8.7). Interestingly, the same reaction conditions applied under conventional heating at 130°C for 1 h completely failed to provide the desired imidazole targets.

8.8 1,4-NAPHTHOQUINONES

Martinez and coworkers[31] investigated the Dötz benzannulation reaction[32] in the synthesis of 2,3-disubstituted-1,4-naphthoquinones. This represents the first focused investigation on the use of this interesting transition metal-mediated carbene-based reaction for the synthesis of a small combinatorial library of these target molecules. The authors initially explored the attachment of the Fischer carbene complex to the resin in a smooth four-step protocol. Reaction of an aryllithium species with $Cr(CO)_6$ generated the metallated intermediate. Subsequent reaction with *N*-tetrabutylammonium bromide (TBAB) furnished the required ionic complex, which was *O*-acylated and attached to the Wang resin (Scheme 8.8).

The authors then investigated the benzannulation reaction of this resin-bound Fischer carbene complex with a number of different alkynes under microwave irradiation. The best conditions were found to be the use of DCM at 85°C for 20 min (Scheme 8.8). Although the same reactions under classical heating conditions furnished compatible yields, the authors noted that the microwave-assisted reactions resulted in a dramatic sixfold shortening of the reaction times. It is particularly noteworthy that the benzannulations proceeded with excellent yields, purities, and regioselectivities in case of an *ortho*-blocked Fischer carbene complex (Scheme 8.8). The target naphthoquinones were finally released into solution by oxidative cleavage of the resin using CAN in a mixture of DCM and water.

8.9 PHTHALOCYANINES

Phthalocyanines are excellent candidates as near-IR fluorizers for tagging biomolecules for a wide variety of bioanalytical applications, such as single gene mutation detection[33] and resonance energy transfer-based assays.[34] They feature a large number of interesting properties such as narrow absorbance and emission spectra and tunable spectral properties

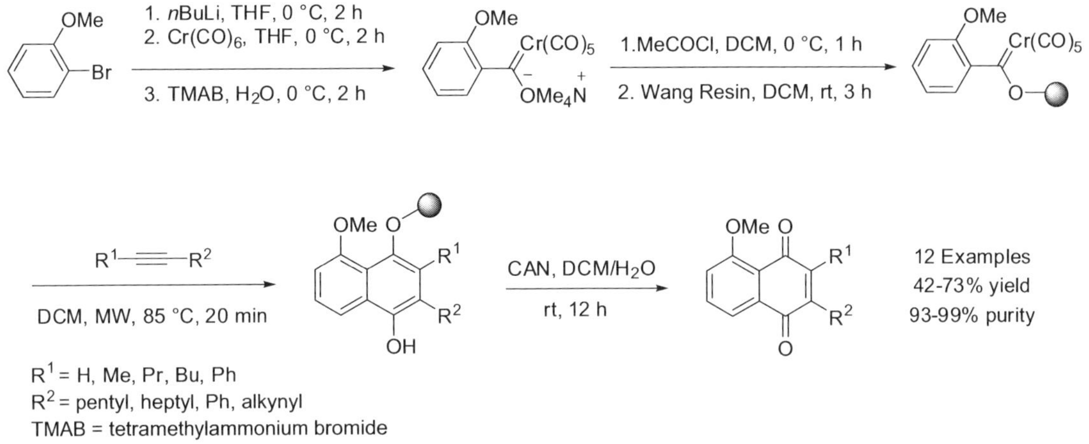

Scheme 8.8. Synthesis of 2,3-disubstituted-1,4-naphthoquinones.

by modifying the substituents on the heterocyclic core as well as by changing the central metal atom.[35] However, the solution-phase chemistry of these molecules is dealing with many problems, hampering their well-deserved applications. On the basis of their previous expertise on the solid-phase synthesis of monohydroxylated analogues,[36] Hammer and coworkers have recently proposed a microwave-assisted, solid-supported synthesis of asymmetrically substituted, monoaminated, AB_3-type phthalocyanines,[37] where they also investigated various protocols for the linkage of the thus generated near-IR fluors with oligonucleotides (Scheme 8.9). The PEG-based Rink amide resin was selected as the solid support. After attaching an α-Fmoc-ε-(Mtt)-Lys-OH to the resin using a standard HBTU- and HOBT-mediated amide coupling reaction at rt, the remaining amine sites of the resin were capped using Ac_2O and DIPEA in DMF as the solvent (Scheme 8.9). After removal of the α-Fmoc group using 4-methyl piperidine in DMF at ambient temperature, the authors attached a suitable carboxyl-bearing phthalonitrile to the solid support using a standard HBTU and HOBT coupling (Scheme 8.9). The phthalocyanine core was then generated via a microwave-assisted cyclotetramerization protocol using a suitably functionalized phthalonitrile in the presence of $Zn(OAc)_2$ with DBU as the base in 1-butanol at 150°C for 40 min (Scheme 8.9).

In this way, the desired metallated phthalocyanines remained on the solid support, while the unwanted by-products such as the symmetrically substituted B4-type moieties were formed in solution and could easily be discarded. After an acid-mediated cleavage from the resin using a TFA/TIPS/DCM mixture at ambient temperature, the ε-amine of the lysine moiety of the metallated phthalocyanine was used for the attachment of suitable oligonucleotides connected with aryl aldehydes or acids (Scheme 8.9). Interestingly, both the reductive amination and the amidation reaction could be dramatically sped up using microwave irradiation, carrying out the reactions at 70–75°C for a mere 30 min. The reductive amination protocol was found to provide a far superior yield, owing to the competitive ester formation and hydrolysis observed in the microwave-assisted amidation protocol.

The authors also demonstrated that the attachment of the oligonucleotide could be performed using a microwave-assisted, Cu(I)-mediated "click-ligation" protocol (Scheme 8.10). The (Mtt) group of the ε-NH of lysine was cleaved under milder conditions using 2% TFA solution so as to leave the zincated phthalocyanine on the resin. The free

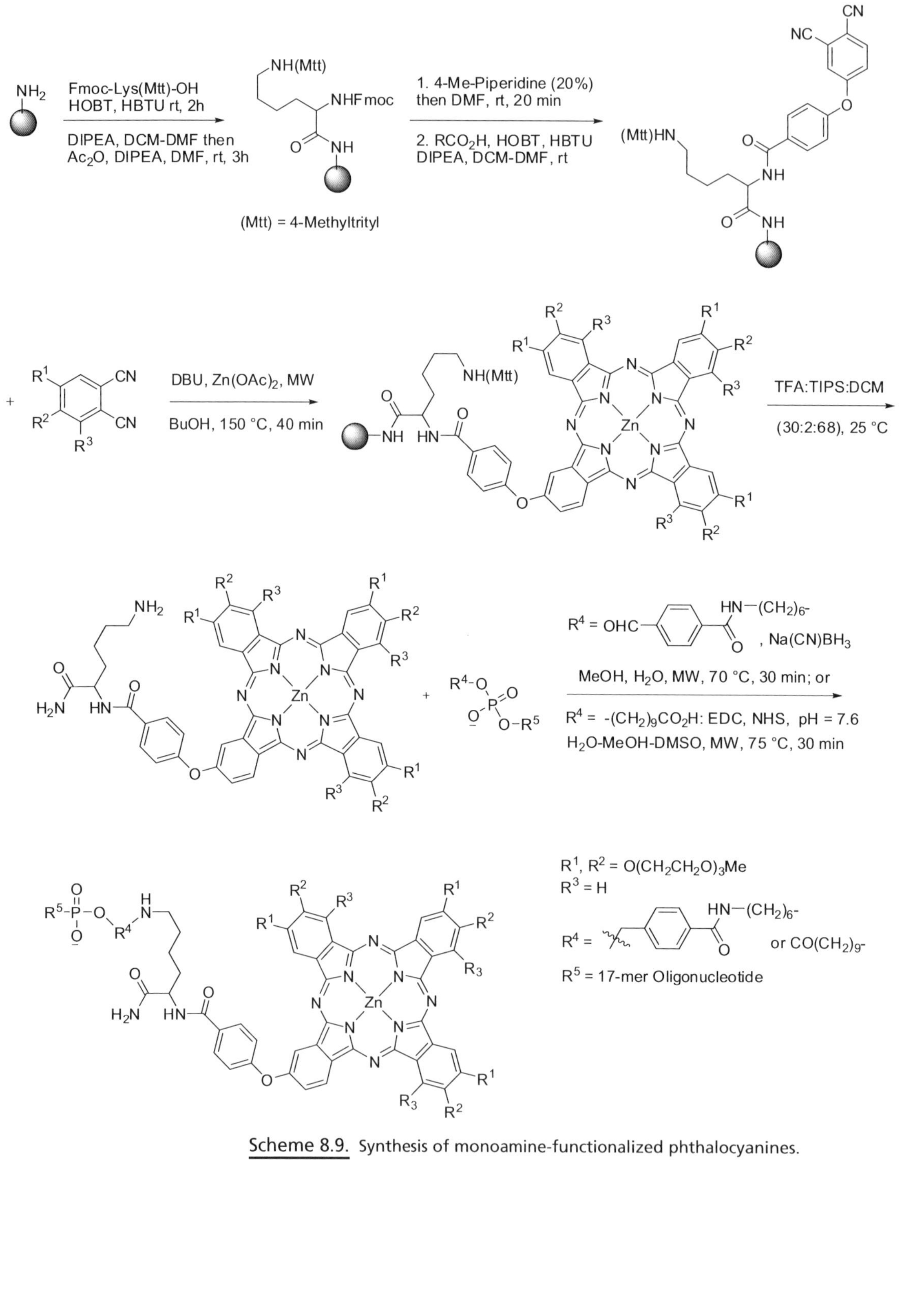

Scheme 8.9. Synthesis of monoamine-functionalized phthalocyanines.

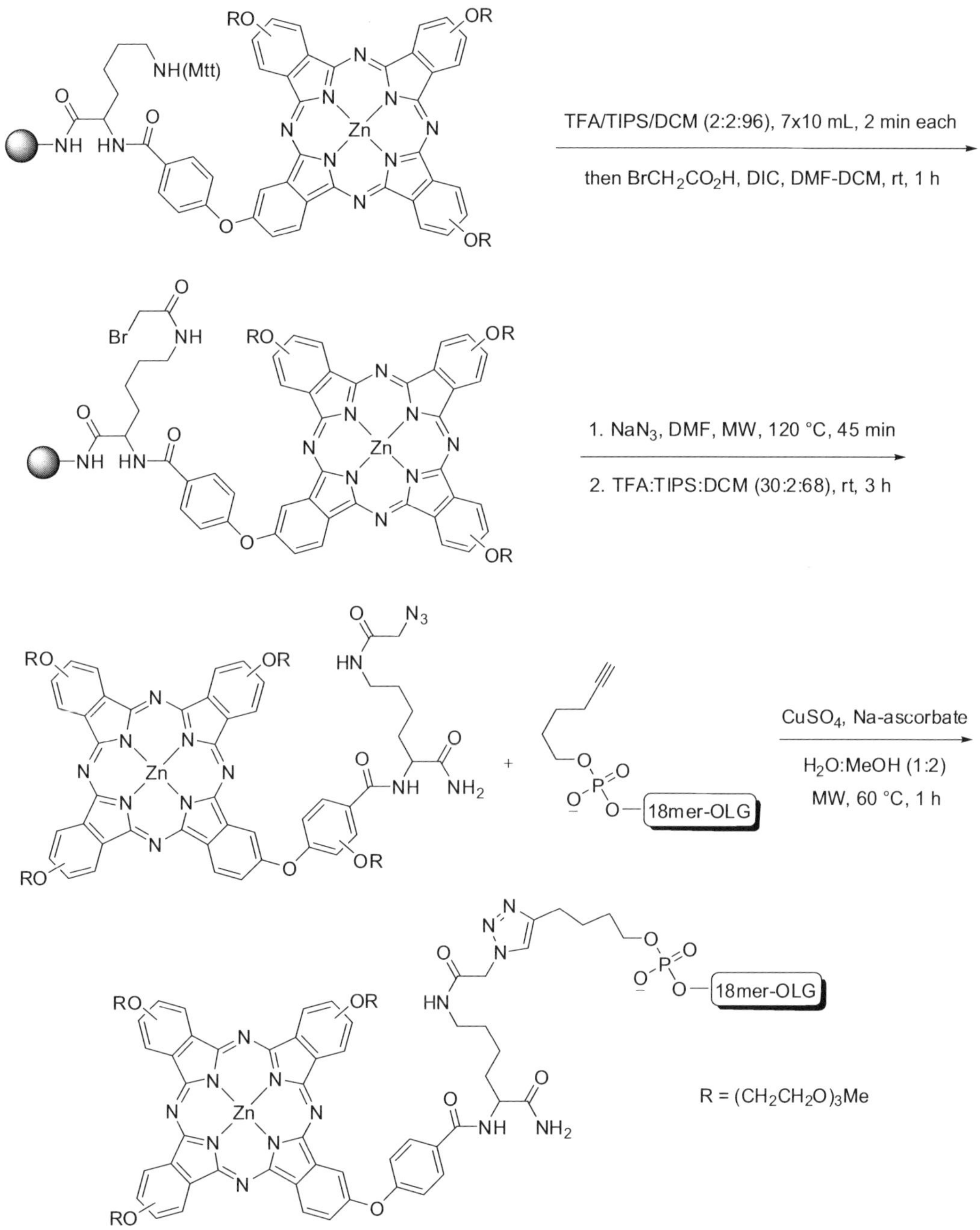

Scheme 8.10. Click ligation of an oligonucleotide with zincated phthalocyanines.

amino group thus generated was then acylated with bromoacetic acid using a standard DIC-mediated coupling (Scheme 8.10), followed by a microwave-assisted azidation reaction using NaN_3 in DMF for 45 min at 120°C and cleavage of the thus generate azide-functionalized phthalocyanine using a TFA/TIPS/DCM mixture at rt (Scheme 8.10).

Finally, the click ligation of the azide-containing phthalocyanine with an oligonucleotide bearing a hexynyl group at the 5′-terminus was performed at 60°C for 1 h under microwave irradiation using a $CuSO_4$/sodium ascorbate system in MeOH/H_2O (Scheme 8.10).

8.10 1,2,3,4-TETRAHYDROQUINOLINES

Fused heterocyclic systems bearing one or more of nitrogen atoms are a very interesting area of research from both synthetic and pharmaceutical points of view as is evident from their pivotal roles in modern drug discovery programs.[38] Silvani and coworkers have recently revealed an interesting solid-phase synthesis of a small library of chiral 1,2,3,4-tetrahydroquinoline derivatives,[39] using a microwave-assisted Pictet–Spengler reaction as the key[40] ring generation step (Scheme 8.11). This investigation is particularly interesting, as the tetrahydroisoquinoline-3-carboxylic acid is the core structural motif in a large number of alkaloids.[41] It is a pharmacophore commonly found in a variety of biologically active molecules such as agonists and antagonists of dopamine receptors, useful therapeutic agents against Parkinson's disease,[42] modified δ-receptor-selective opioid ligands to treat acute and chronic pain or to fight narcotic addition, and so on.[43]

The authors started their synthesis by attaching suitably decorated *N*-Boc-protected amino acids to chloromethyl polystyrene resin under microwave irradiation using Cs_2CO_3 as the base in NMP at 180°C for 15 min. It is noteworthy that the same reaction under classical heating conditions furnished a mere 40–47% loading compared to a superior 95% loading under microwave irradiation (Scheme 8.11).

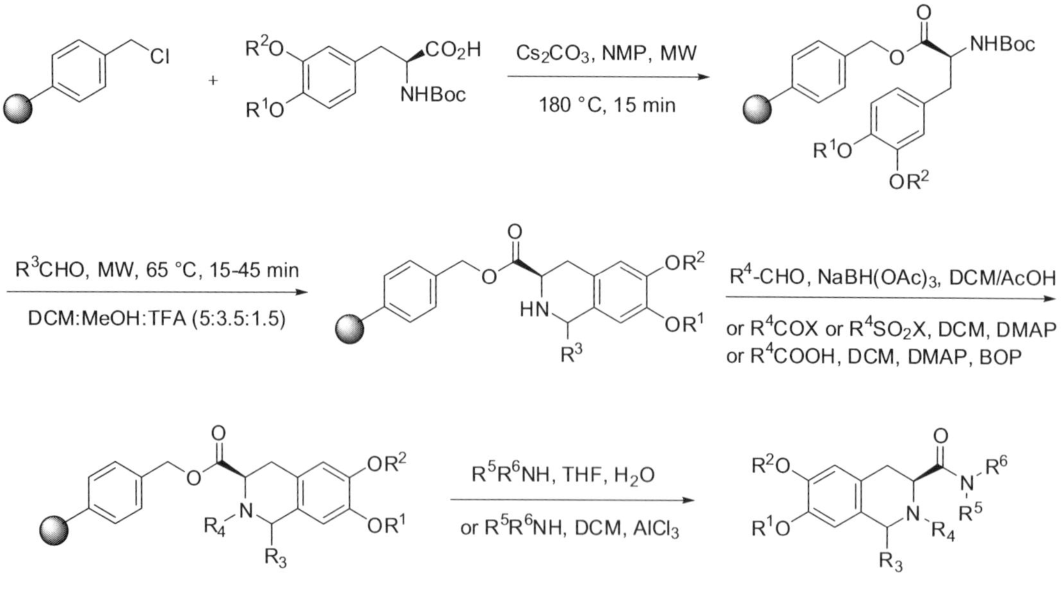

R^1, R^2 = Me or Bn, R^3 = Ar or Alk, R^4 = Boc, COBn, COPh, COAlk, Me, Bn, Ms or SO_2Ar, R^5, R^6 = H, Alk, OMe

Scheme 8.11. Synthesis of 1,2,3,4-tetrahydroquinoline derivatives.

The Silvani group then performed an acid-mediated Boc cleavage and tandem Pictet–Spengler cyclization under microwave irradiation with suitable aldehydes in a 5:3.5:1.5 mixture of DCM/MeOH/TFA at 65°C for 15–45 min (Scheme 8.11). Once again, a marked increase in the yield was observed under microwave irradiation in comparison to classical heating. After variously functionalizing the isoquinoline nitrogen using reductive amination, amidation, or sulfonamidation to impart structural diversity, the desired isoquinolines were cleaved from the resin upon reaction with a suitable amine. The corresponding chiral amides were liberated into solution in excellent overall yields (60–85%) and purities (79–99%) (Scheme 8.11).

8.11 1,2,3-TRIAZOLES

1,2,3-Triazoles have received great synthetic interest in recent years due to their importance as potent pharmacophores.[44] Different triazole-containing heterocycles exhibit submicromolar activity against LPS-induced tumor necrosis factor-α (TNF-α) and some *N*-substituted-phenyl-1,2,3-triazole-4-acylhydrazones possess considerable antiplatelet activity on arachidonic acid- and collagen-induced platelet aggregation.[45] Furthermore, 1,2,3-triazoles have found extensive industrial use as corrosion inhibitors, dyes, photostabilizers, photographic materials, and agrochemicals.[46] Traditionally, 1,2,3-triazoles are generated via 1,3-dipolar cycloaddition of organic azides with alkynes.[47] However, owing to the heavy dependence of the reaction on the steric and electronic nature of the alkynes, this highly useful reaction has found limited use in the solid-phase synthesis of triazole libraries. In particular, the [3 + 2] cycloaddition of azides with electron-poor alkenes has received very little attention due to the demand of harsher reaction conditions in promoting the otherwise sluggish transformations. In view of circumventing these issues, Lam and Gao have recently elaborated a microwave-assisted, fully regioselective and traceless synthesis of di- and trisubstituted 1,2,3-triazoles,[48] which involve resin-linked vinyl sulfones as dipolarophiles (Scheme 8.12).

The authors chose polystyrene/divinylbenzene sodium sulfinate resin as the solid support, as it easily grants access to the vinyl sulfone needed for the cycloaddition reaction. The sulfone was generated via a microwave-assisted reaction with a suitable benzyl halide bearing an electron-withdrawing group (Scheme 8.12).

The resin-bound vinyl sulfone was then generated by a microwave-assisted, base-mediated Knoevenagel condensation of a suitable aldehyde in THF or DMF at 82–111°C for

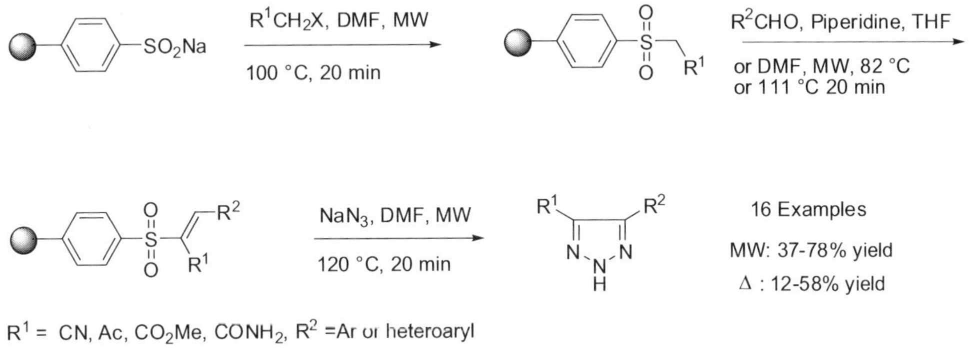

Scheme 8.12. Synthesis of 1,2,3-triazoles on solid support.

20 min (Scheme 8.12). The 1,3-dipolar cycloaddition and subsequent cleavage of the target 1,2,3-triazole from the solid support was effected by reaction of NaN_3 in DMF at 120°C for 20 min under microwave irradiation. Interestingly, in the majority of the cases, the authors observed higher conversions and purities for the microwave-assisted reactions in comparison to their classically heated counterparts.

8.12 2,8-DIAMINOPURINES

The purine nucleus is one of the most well-studied and sought-forward heterocyclic cores, owing to the multitude of physiological and biochemical functions controlled by purine-containing compounds such as nucleic acids.[49] Therefore, novel purine-based compounds are of extreme interest in both synthetic and pharmaceutical chemistry as they offer enormous therapeutic potential, such as for the development of purine-based kinase inhibitors.[50] The importance of purine derivatives and purine mimics in pharmaceutical research necessitates the development of efficient and versatile syntheses of such molecules.[51] In view of developing a robust diversity-oriented protocol for the synthesis of a small combinatorial library of *N*-9-substituted 2,8-diaminopurines where a threefold structural decoration at the N9, C2, and C8 positions is possible, Cole et al. developed a novel microwave-assisted approach (Schemes 8.13 and 8.14).[52] Aminomethyl-Agrogel® resin was chosen as the solid support and a tailor-made linker was developed by acylating the resin with *N*-α-*N*-ε-bis-Fmoc-Lys-OH using standard DIC and HOBT-mediated amidation of the resin in DMF at ambient temperature (Scheme 8.13). After deprotection of the amine ends using Fmoc cleavage mediated by 20% piperidine in DMF at rt, the authors proceeded to attach 4-(4′-formyl-3′-methoxy)phenoxybutyric acid to the free amine terminals to generate the needed linker (Scheme 8.13).

At the beginning of the investigation of their new strategy using the thus generated resin-bound linker, the authors chose to introduce the first point of diversity (N^2-substituent)

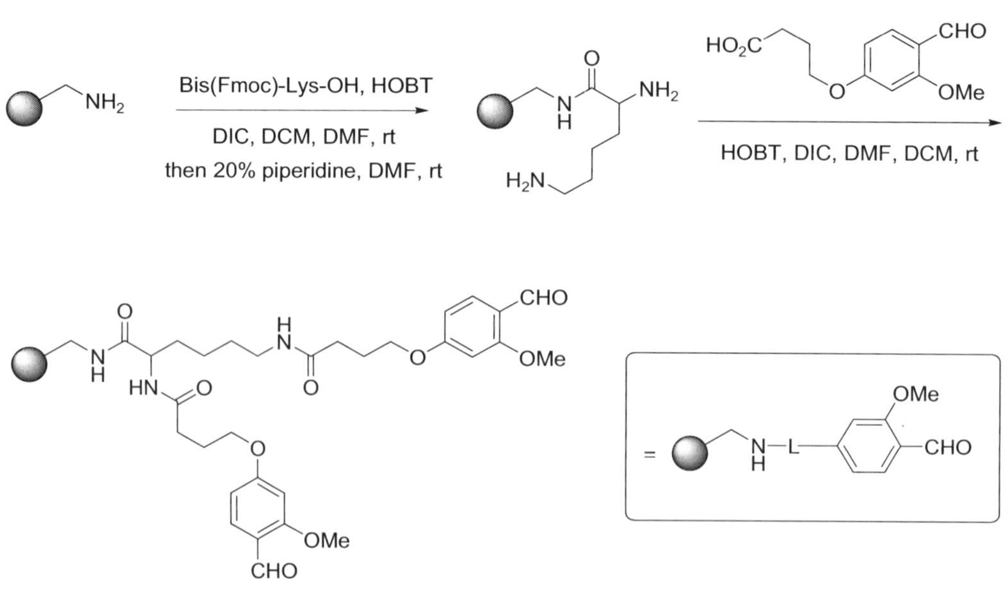

Scheme 8.13. Synthesis of the required linker using aminomethyl-terminated Agrogel.

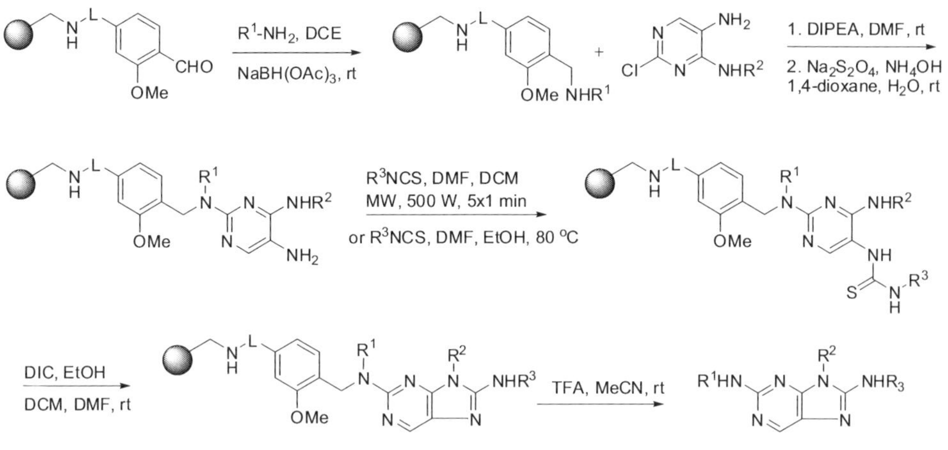

Scheme 8.14. Synthesis of N9-substituted 2,8-diaminopurines.

by performing a reductive amination with a number of amines in 1,2-dichloroethane at rt using $NaBH(OAc)_3$ as the reagent (Scheme 8.14). This monofunctionalized amine was then reacted with various 2-chloro-5-nitropyrimidine-4-amines in the presence of DIPEA in DMF at ambient temperature to incorporate the second point of diversity (substituent at the ninth position) (Scheme 8.14). The nitro group was then converted to the corresponding amino group by a dithionate-mediated reduction protocol using dioxane/water as the solvent system of choice. The third point of diversity (N^8-substituent) was finally introduced by reacting this resin-bound diamine with a variety of aryl isothiocyanates in a mixture of DMF and DCM (Scheme 8.14).

For the generation of the resin-bound thioureas, the transformations were carried out under microwave irradiation at 500 W for 5 × 1 min with a 15 min interval between the different irradiation times (Scheme 8.14). It is particularly noteworthy that the reaction of the resin-bound amine with the less reactive aryl thioureas failed to reach completion under conventional heating conditions, clearly outlining the benefit of microwave irradiation. The cyclization of these intermediates into the corresponding trifunctionalized purine derivatives was carried out using DIC at rt in a mixture of DCM, DMF, and EtOH, followed by a TFA-mediated cleavage of the resin, liberating the target molecules in very good yields (54–89%) and purities (77–89%) (Scheme 8.14).

8.13 IMIDAZOLIDIN-4-ONES

Owing to their promising biological potential, for example, as antibacterials,[53] suitable decorated imidazolidin-4-ones are of high interest for both synthetic and pharmaceutical chemistry.[54] Imidazolidin-4-ones have also been reported to inhibit the binding of VCAM-1 to VLA-4 and are interesting candidates for the treatment of inflammation linked to chronic inflammatory diseases, such as rheumatoid arthritis, multiple sclerosis, asthma, and inflammatory bowel disease.[55] Therefore, the development of robust and versatile syntheses of multifunctionalized imidazolidin-4-ones has attracted considerable interest for drug discovery.[56] In view of developing a general solid-phase protocol for the synthesis of imidazolidin-4-ones that allows the decoration at the N1, C2, and C5 positions, Qin et al.

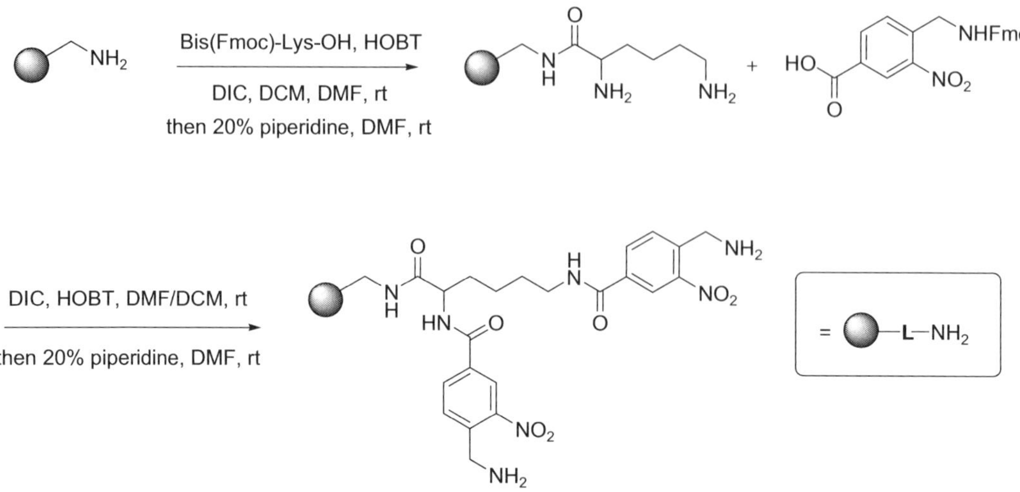

Scheme 8.15. Synthesis of the required amine-terminated linker.

have recently investigated a microwave-assisted protocol.[57] Aminomethyl-Agrogel resin was chosen as the solid support, which was acylated with *N*-α-*N*-ε-bis-Fmoc-Lys-OH using a standard DIC- and HOBT-mediated amidation of the resin in DMF at ambient temperature (Scheme 8.15). Removal of both Fmoc groups upon treatment with 20% piperidine in DMF at rt delivered an amine-terminated resin (Scheme 8.15).

The required bis-aminated linker was then generated from this resin by the DIC- and HOBT-mediated amidation of the free amine-bearing resin with a suitable benzoic acid derivative, followed by removal of both Fmoc groups (Scheme 8.15). Various Fmoc amino acids such as Glu, Ser, and 1,2-diaminopropionic acid were then attached to the free amines (Scheme 8.16). After the piperidine-mediated removal of the Fmoc protection, 2,5-disubstituted imidazolidin-4-ones were generated on the resin by reaction of a suitable aldehyde under microwave irradiation at 180°C in DMF (Scheme 8.16).

The twofold diversity of the desired target bound to the resin was then increased by various reactions such as amidation using BOP-Cl and DIPEA in DCM at rt, or the formation of urea or carbamate derivatives (Scheme 8.16) by the reaction of a suitable isocyanate or chloroformate. The final cleavage of the desired 1,2,5-trisubstituted imidazolidin-4-ones was effected in TFA/water/*i*PrOH (3:20:80) at 50°C (Scheme 8.16).

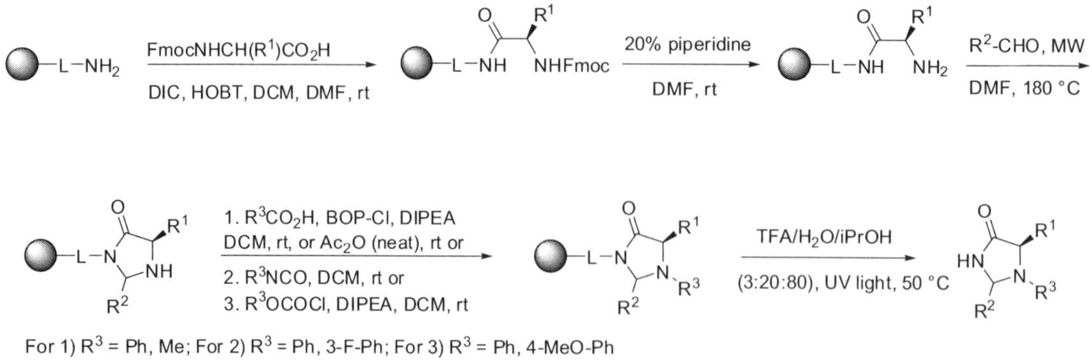

Scheme 8.16. Synthesis of 1,2,5-trisubstituted imidazolidin-4-ones.

8.14 INDOLES

The application of transition metal-mediated reactions on solid support is an area of supreme interest for both peptide chemistry and the synthesis of nonpeptidic "drug-like" molecules. Although many metal-catalyzed reactions have been carried out on resin-bound scaffolds,[58] there are scarce examples of metal-mediated solid-phase reactions promoted by a complexing agent that is covalently bound with the resin in close proximity of the scaffold. Dai and Sun have recently proposed an ingenious route toward the microwave-assisted solid-phase synthesis of a small library of 2-(hetero)aryl indoles.[59] The authors demonstrated the strategic use of a tailor-made linker that features the dual function of a normal linker for scaffold attachment as well as of a promoter for the Cu(II)-mediated heteroannulation of the intermediates (Scheme 8.17).

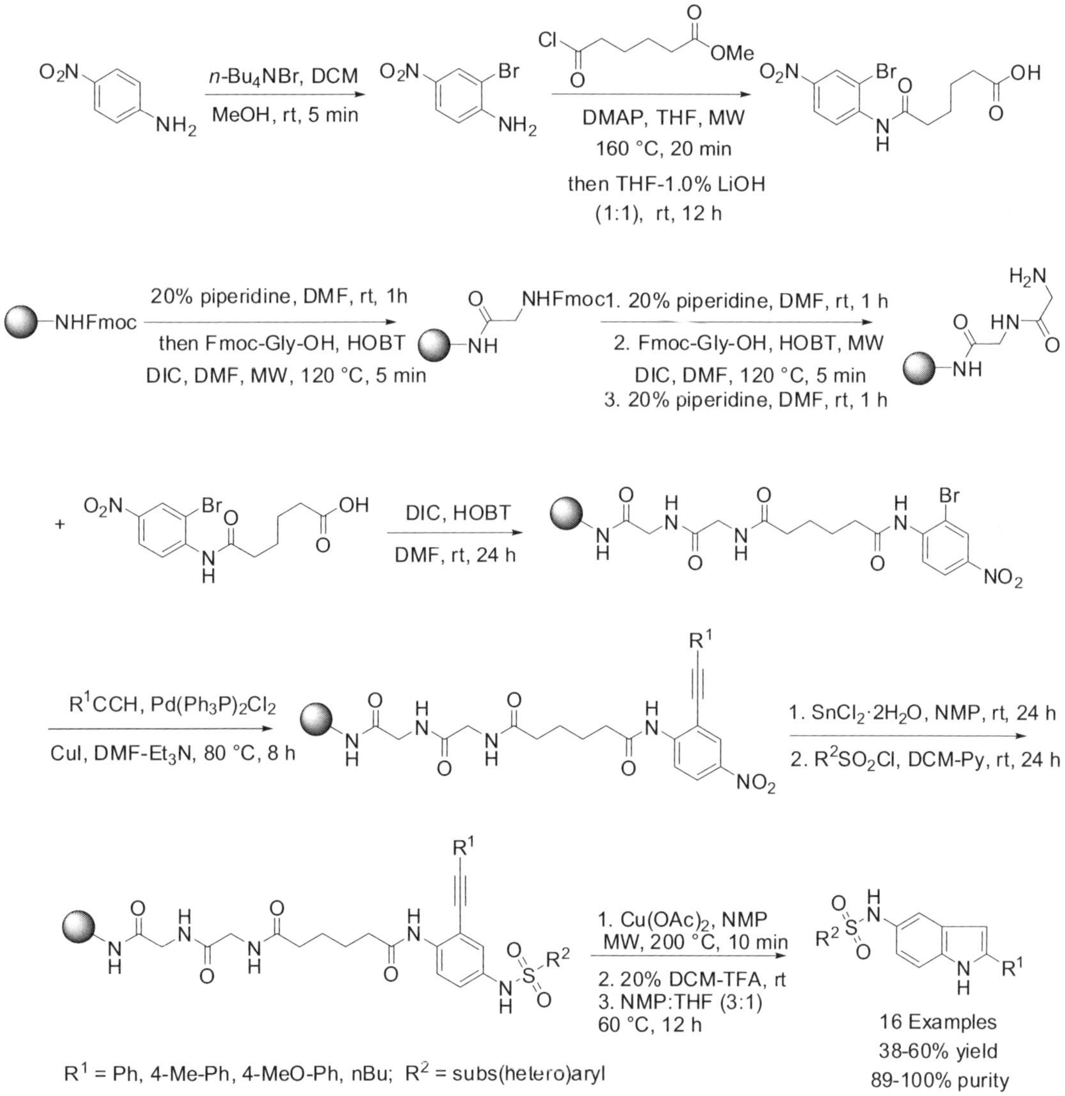

Scheme 8.17. Synthesis of indole derivatives.

Based on their previous experience on the solid-phase synthesis of indole derivatives[60] through metal-catalyzed intramolecular annulation reactions[61] under controlled microwave irradiation, the authors designed a solid-phase synthesis of 2,5-disubstituted indoles by applying Rink amide resin. A tether of two glycine moieties, which was pivotal in promoting the Cu(II)-mediated heteroannulation reaction at the later stages of the synthesis (see below), was attached to the resin using microwave-assisted peptide synthesis (Scheme 8.17). After cleavage of the Fmoc protective group, 2-bromo-4-nitroaniline was attached to the linker using an adipic acid spacer. It is to be noted that the chain length of the diacid was crucial for the synthetic strategy as other chain lengths failed to provide optimum yields or failed altogether to perform the ring closure at the final stage. The precursor for the synthesis of 2,5-disubstituted indoles was then assembled using a three-step sequence involving a Sonogashira coupling with a suitable acetylene, followed by a $SnCl_2$-mediated reduction of the nitro group and subsequent conversion of the thus formed aniline to the corresponding aryl sulfonamide (Scheme 8.17).

The crucial microwave-assisted, Cu(II)-mediated heteroannulation reaction was then investigated in NMP using $Cu(OAc)_2$ at 200°C for 10 min, resulting in the formation of the desired indoles in excellent yields and purities upon cleavage from the resin using 20% TFA in DCM at ambient temperature (Scheme 8.17). The amide carbonyls of the bis-glycine unit were found to play a pivotal role in the success of the reaction, due to complex formation with the Cu(II) catalyst in the near vicinity of the alkyne and thus presenting the active metal catalyst for assisting in the heteroannulation reaction. This was proven by the authors by synthesizing the corresponding mono- and triglycine tethers, where the Cu(II)-mediated cyclization was found to be impossible. The *N*-linked indoles were subjected to a base-mediated *N*-acyl cleavage (Scheme 8.17) to provide the desired indole derivatives as free bases in solution with excellent overall yields (38–60%) and purities (93–100%).

8.15 1,2,3,4-TETRAHYDROQUINOLINES USING A SmI_2-CLEAVABLE LINKER

The development of suitable linkers is a continuous area of interest in solid-supported synthesis. Based on their previously explored traceless linking strategy using HASC (α-heteroatom-substituted carbonyl) linkers[62] employing SmI_2,[63] Procter and coworkers have demonstrated the use of a novel sulfur-based linker for the synthesis of the tetrahydroquinolone framework (Scheme 8.18)[64] that is present in a number of natural and unnatural biologically active compounds.[65] The authors demonstrated that sulfur was an excellent choice not only because of the obvious use in the linker strategy, but also because it is helpful during the C=C bond forming reaction. The authors initially explored the synthesis in solution phase, using an S-benzyl moiety as a mimic to the solid support. The key Heck reaction was run under microwave irradiation at 100°C for 7 h, delivering the addition products in very high yields and purities. They then synthesized the benzyl thiol resin,[66] to which the authors attached the α-bromoamide (Scheme 8.18).

Subsequently, the Heck reaction with *tert*-butyl acrylate was investigated, where the best conditions were found to be the use of $Pd(OAc)_2$ and $(o\text{-Tol})_3P$ as the catalytic system in DMF for an unusually long irradiation time of 20 h at 100°C (Scheme 8.18). The conversion of the sulfur to the corresponding sulfone was achieved upon treatment with *m*-CPBA at rt. This was followed by a K_2CO_3-mediated cyclization of the obtained intermediate to generate the required tetrahydroquinolone still linked to the solid support (Scheme 8.18).

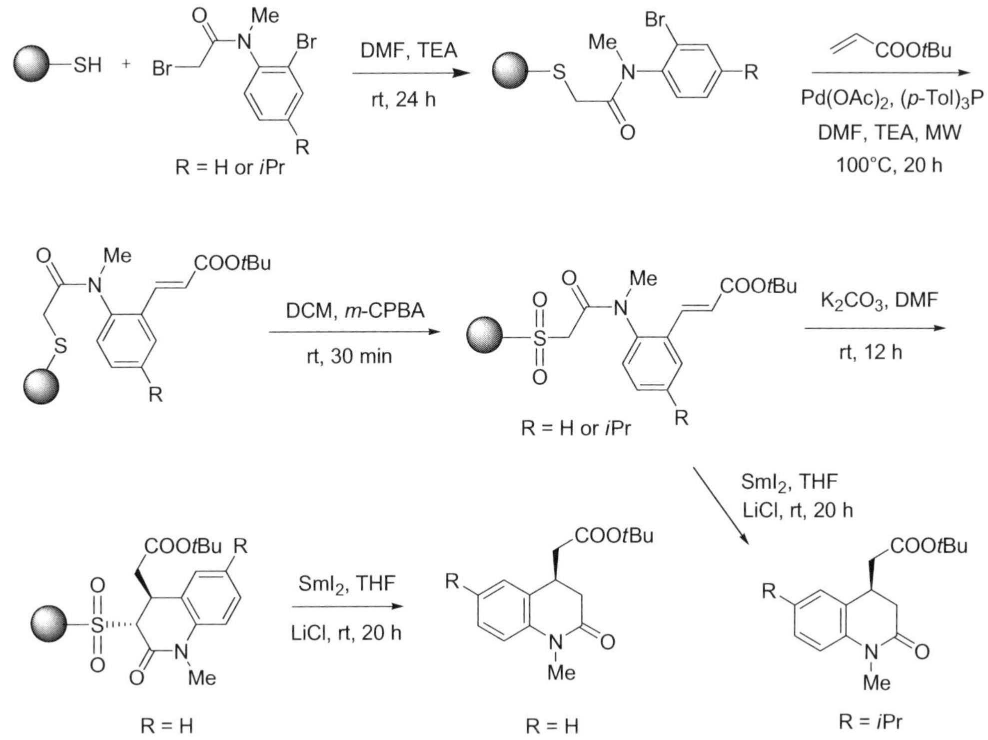

Scheme 8.18. Synthesis of tetrahydroquinolones using a SmI_2-cleavable sulfur-based linker.

Finally, the heterocyclic moiety was cleaved from the resin upon treatment with SmI_2 in THF and LiCl at rt, furnishing the desired product in a respectable 27% overall yield. It is noteworthy that the authors also demonstrated a LiCl/SmI_2-mediated cyclative cleavage delivering the target tetrahydroquinolone in an overall yield of 25% over four steps (Scheme 8.18).

8.16 HYDANTOINS

The hydantoin nucleus could be found in a broad range of biological active compounds displaying versatile activities such as neuroprotective, antihypertensive, anti-inflammatory, analgesic, antidiabetic, antibacterial, antiviral, and diuretic activities.[67] An additional lure of hydantoins to a synthetic chemist is the fact that they are key intermediates in the synthesis of optically pure natural and unnatural amino acids.[68] In view of quickly generating large libraries of hydantoins, Parrot and coworkers have investigated the first microwave-assisted, solid-supported synthesis of hydantoin derivatives (Scheme 8.19).[69] They explored their synthetic methodology on a large number of solid supports including Merrifield, Wang, PAM, and MBHA resins, employing different combinations of linkers and spacers together with various amino acids. The authors chose to start the synthesis using a variety of commercially available resins that were preloaded with suitable amino acids with or without the presence of suitable linkers (Scheme 8.19).

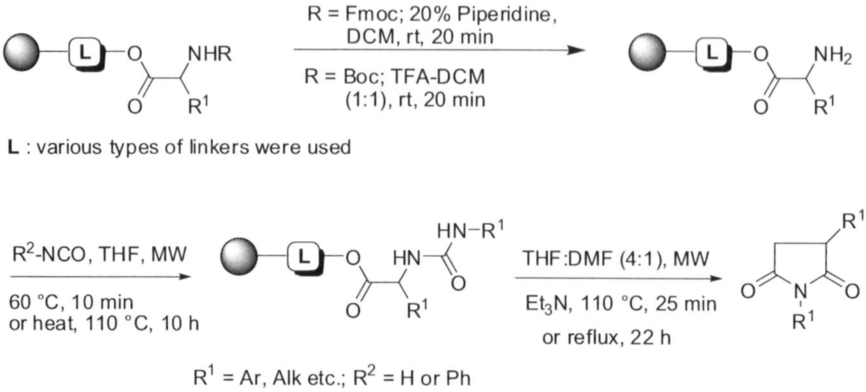

Scheme 8.19. Synthesis of hydantoin derivatives.

The protecting group was then removed by acid-mediated cleavage in the case of Boc protection and subsequent conversion of the amine salt into the corresponding free base, or by base-mediated cleavage in the case of Fmoc protection. The thus generated free amine was then reacted with a suitable isocyanate in THF under microwave irradiation at 60°C for 10 min, resulting in the formation of the corresponding urea (Scheme 8.19). It is noteworthy that the same reaction took a long 10 h for complete conversion under conventional heating conditions at 110°C. The desired hydantoin was then generated and subsequently cleaved from the resin in a traceless manner using a suitable base under microwave irradiation in 25 min at 110°C in a 4:1 mixture of THF and DMF as the solvent (Scheme 8.19). It is of particular interest that the same reactions, when carried out under conventional heating, required a much longer (22 h) reaction time to furnish similar results for the experiments conducted under microwave irradiation.

8.17 IMATINIB

Imatinib (Scheme 8.20), commercially developed by Novartis with the trade name Gleevec/Glivec, is a blockbuster used for the chemotherapy of chronic myeloid leukemia (CML).[70] It is a potent and selective inhibitor of BCR-ABL and c-kit oncogenic tyrosine kinases. Imatinib has become a prototype for small molecule targeted anticancer treatment, featuring exceptional therapeutical efficacy and a very low toxicity profile.[71] Unfortunately, the downside of this wonderful compound is the development of drug resistance in patients with advanced oncological stages after a very promising initial response.[72] In view of generating small libraries of structurally novel, potent, and selective BCR-ABL inhibitors for the treatment of advanced-stage CML and its hazardous mutants, Carotti and Capaldi have recently investigated the microwave-assisted synthesis of Imatinib on solid support (Scheme 8.20).[73]

The AMEBA linker[74] was installed upon NaH-mediated alkylation of Merrifield resin with 4-hydroxy-2-methoxy-benzaldehyde in DMF at 120°C for 5 min under microwave irradiation (Scheme 8.20). The authors then chose to attach 4-methyl-3-nitroaniline to the linker through a previously established two-step reductive amination protocol,[75] involving the imine formation mediated by $Ti(O^iPr)_4$ followed by reduction of the imine using NaBH$(OAc)_3$ (Scheme 8.20). The free amino group was then acylated with 4-(chloromethyl)benzoyl chloride using DIPEA in DMF to form the corresponding amide.

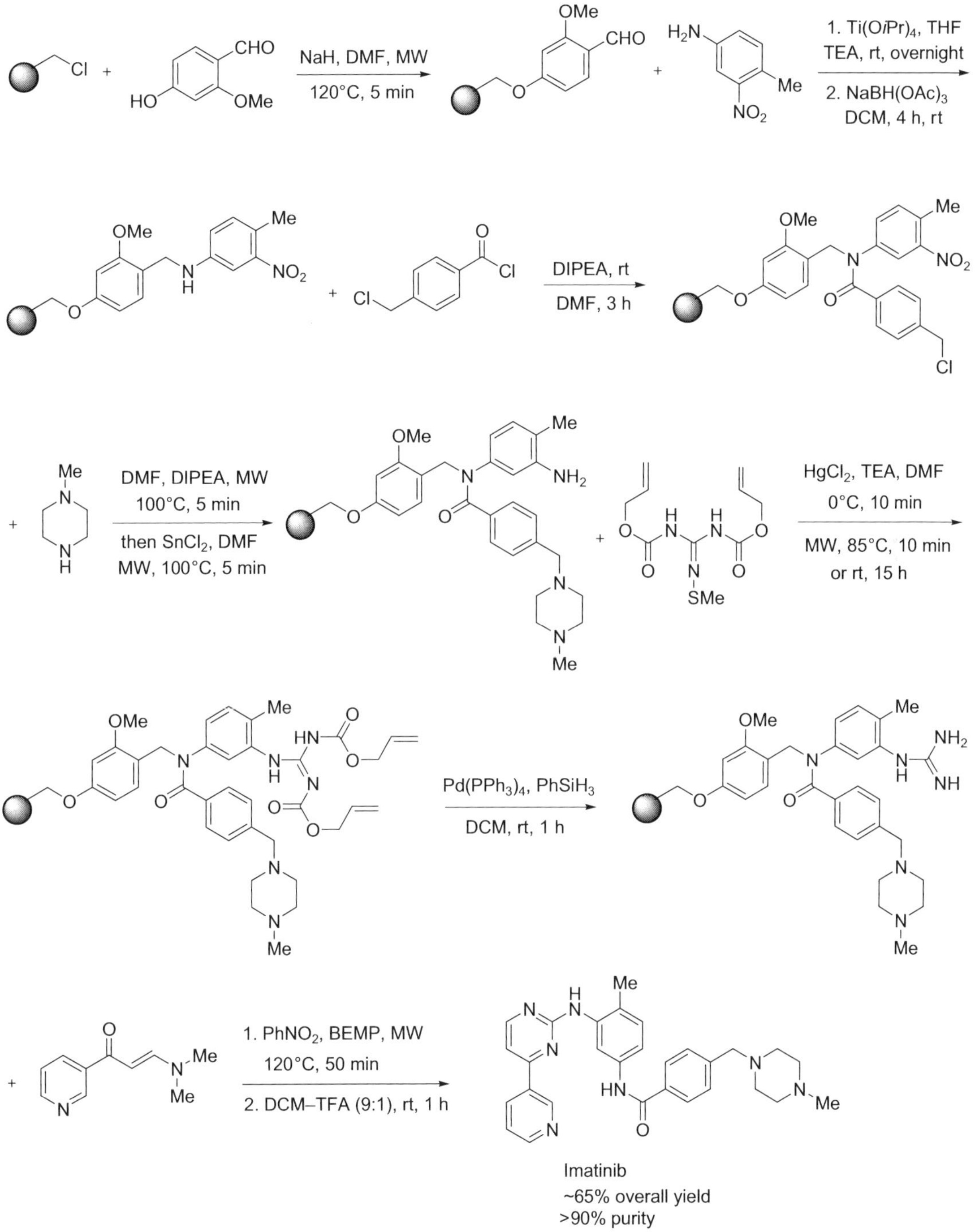

Scheme 8.20. Synthesis of Imatinib.

The DIPEA-mediated nucleophillic displacement of the halogen atom by *N*-methyl piperazine in DMF was effected under microwave irradiation at 100°C for 5 min, followed by a microwave-assisted, $SnCl_2$-mediated reduction of the nitro group in DMF at 100°C to generate the corresponding aniline in 5 min (Scheme 8.20). Introduction of the guanidine group was then carried out with bis-Alloc-protected methylthiopseudourea using $HgCl_2$ and TEA in DMF for 10 min at 85°C under microwave irradiation, followed by removal of the Alloc groups upon treatment with $Pd(PPh_3)_4$ and $PhSiH_3$ in DCM for 1 h at rt. It is noteworthy that the same guanylation reaction when performed at rt required a prolonged 15 h reaction time for obtaining the complete conversion. Finally, the cyclization with 3-dimethylamino-1-pyridin-3-yl-propenone to generate the corresponding 5-substituted-2-arylaminopyrimidine core of Imatinib was carried out using 2-*tert*-butylimino-2-diethylamino-1,3-dimethylperhydro-1,3,2-diazaphosphorine (BEMP)[76] mediated cyclization in nitrobenzene under microwave irradiation at 120°C, followed by a TFA-mediated cleavage of Imatinib from the resin. It should be noted that the cyclization time was dramatically shortened under microwave irradiation from 20 h to a mere 50 min, providing the target molecule in a phenomenal 98% yield (Scheme 8.20).

8.18 ISOINDOLINES

Although the isoindoline ring system is observed in many naturally occurring moieties,[77] it is relatively underexplored in the broad area of heterocycle-based, "drug-like" small molecules. Compounds bearing an isoindoline core features numerous interesting biological properties such as platelet aggregation inhibitory and antitumor activities.[78] On the basis of their previous expertise on microwave-assisted carbocyclizations on solid support,[79] Martinez and coworkers recently explored an interesting solid-phase combinatorial synthesis of isoindolines[80] through the first microwave-assisted, Ir-catalyzed [2 + 2 + 2] cycloaddition reaction (Scheme 8.21). The authors found that the common side products of the cycloaddition reaction, such as trimerization and oligomerization of the diynes, as well as the cyclotrimerization of the alkynes involved in the process, could be suppressed by immobilization of the diynes on solid support.

The resin-bound diyne was then generated by the DIPEA-mediated alkylation of the commercially available tritylchloride resin with dipropargylamine in DMF at ambient temperature. The microwave-assisted cycloaddition reaction was performed using a catalytic system comprising $[Ir(COD)Cl]_2$ and 1,2-diphenylphosphinoethane (dppe) in

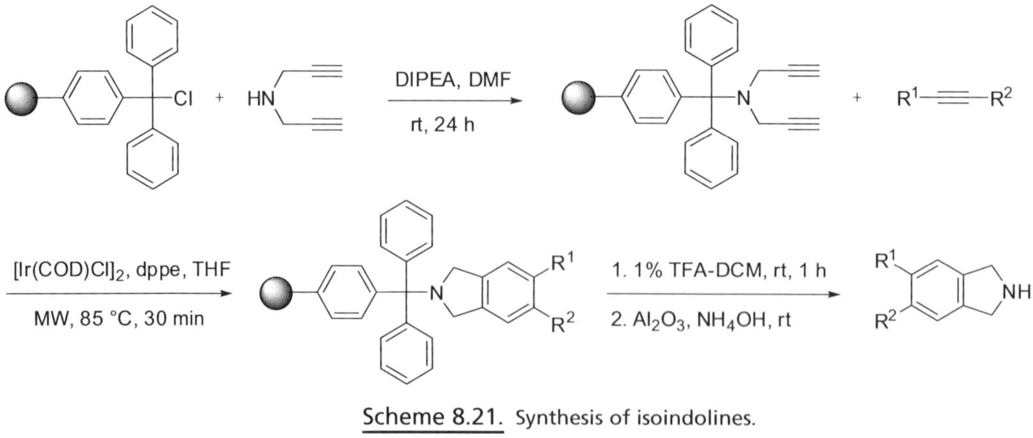

Scheme 8.21. Synthesis of isoindolines.

THF at 80°C for 30 min, followed by the cleavage of the isoindoline from the resin under mild acidic conditions by applying 1% TFA in DCM at rt for 1 h (Scheme 8.21). The authors noted the formation of single isoindoline analogues in very good yields (50–70%; over the two last steps) and purities (97–99%), while no products corresponding to the self-dimerization of the diynes were observed. Interestingly, microwave irradiation dramatically shortened the reaction time to 30 min of the cyclization, in comparison to the prolonged reaction times of 48 h needed when these reactions were run under classical heating conditions (Scheme 8.21).[81] It is of particular interest that the authors performed a catch–release purification strategy to get the salt-free isoindoline using inexpensive alumina and NH_4OH, completely avoiding the use of any expensive polymer-supported reagents or column chromatography.

8.19 2-(BENZYLTHIO)IMIDAZO[1,2*a*]-PYRIMIDIN-5-ONES

Imidazo[1,2*a*]pyrimidin-5-ones have received increased synthetic interest in present-day chemistry owing to their diverse biological potential.[82] This interesting heterocyclic core resembles both purine and benzimidazole ring systems and is found in a variety of pharmaceutically interesting targets such as phosphodiesterase inhibitors,[83] benzodiazepine receptor ligands,[84] gonadotropin-releasing hormone antogonists,[85] and so on.

It is evident from the available literature that both 2-benzylthio-benzimidazoles and 8-benzylthiopurines exhibit stronger biological potential than their parent analogues,[86] although the available literature on the synthesis of the corresponding 2-benzylthio derivatives of imidazo[1,2*a*]-pyrimidin-5-ones is relatively rare. Lam and Soh have recently explored the microwave-assisted, solid-supported synthesis of substituted 2-(benzylthio) imidazo[1,2*a*]pyrimidin-5-ones (Scheme 8.22).[87] They elaborated their synthesis on bromomethyl polystyrene resin, generating the desired target molecules through a traceless cyclative cleavage strategy.

A suitable β-ketoester was first converted into the corresponding 2-mercaptopyrimidine by employing a microwave-assisted, EtONa-mediated condensation with thiourea in EtOH/DMF (10:1) at 130°C for 30 min. The obtained compound was then attached to bromomethyl polystyrene resin at 100°C, once again under microwave irradiation for 10 min (Scheme 8.22). The amide group was then subjected to a TEA-mediated alkylation using 2-bromoacetonitrile in a 10:1 mixture of DMF and EtOH at 120°C for 20 min under

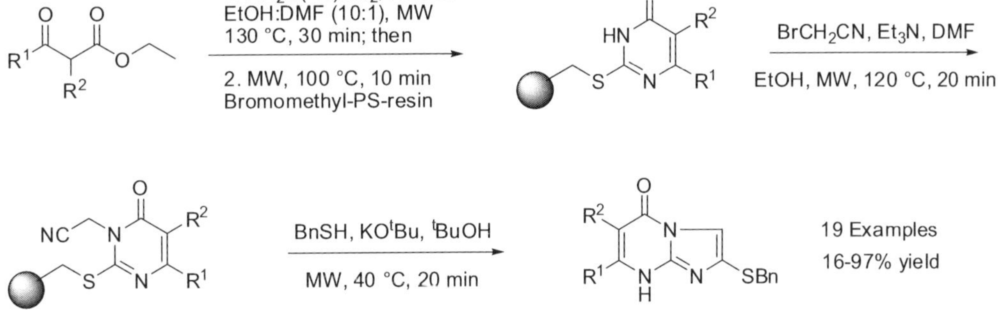

Scheme 8.22. Synthesis of substituted 2-(benzylthio)imidazo[1,2*a*]pyrimidin-5-ones.

microwave irradiation to introduce the desired "handle" for the heterocyclization reaction (Scheme 8.22). The formation of the fused ring system and the subsequent cleavage from the resin was then effected by a microwave-assisted, KO^{t}Bu-mediated reaction with BnSH using ^{t}BuOH as the solvent for 20 min at 40°C (Scheme 8.22). A library of 19 examples was generated with yields ranging 16–97%.

8.20 2-AMINOBENZOTHIAZOLES

The generation of small to medium-sized libraries comprising fused heterocyclic systems still receives great interest from the synthetic community, owing to the high demand for these compounds as versatile scaffolds in drug discovery. The 2-aminobenzothiazole is a valuable addition to the broad family of indole-like heterocycles and can be regarded as an interesting "privileged" structure for medicinal chemistry.[88,89] Despite the existing literature on the solid-supported synthesis of benzothiazoles,[90] no available procedures could be found about the synthesis of 2-aminobenzothiazoles on solid support until the very recent exploration of Ballatore and coworkers.[91] The authors investigated the synthesis of 2-aminobenzothiazoles employing carboxypolystyrene resin, which was converted into the corresponding isothiocyanate according to a two-step literature procedure (Scheme 8.23).[92] Thus, conversion of the resin into the corresponding acid chloride was effected in 1,2-DCE at rt using oxalylchloride. This was followed by the reaction of tetrabutylammonium isothiocyanate in a mixture of THF and 1,2-DCE at rt for 16 h (Scheme 8.23), yielding the resin-bound isothiocyanate. This was reacted with a variety of mono- and polyhalogenated anilines in DMF at rt for 16 h to generate the corresponding acylated thiourea analogues that were then subjected to cyclization under three different conditions, depending on the R_2 substituent in the *ortho*-position of the aniline (Scheme 8.23). The final compounds could easily be cleaved from the resin upon microwave irradiation using $N_2H_4 \cdot H_2O$ in EtOH at 150°C for 30 min.

The authors further explored the chemistry on solid support by decorating the obtained resin-bound 2-aminobenzothiazoles using different palladium-catalyzed cross-coupling reactions (Scheme 8.24). Thus, the reaction product of 3-bromoaniline was subjected to bromine-mediated cyclization in AcOH at rt for 16 h, furnishing a 1:1 mixture of 5- and

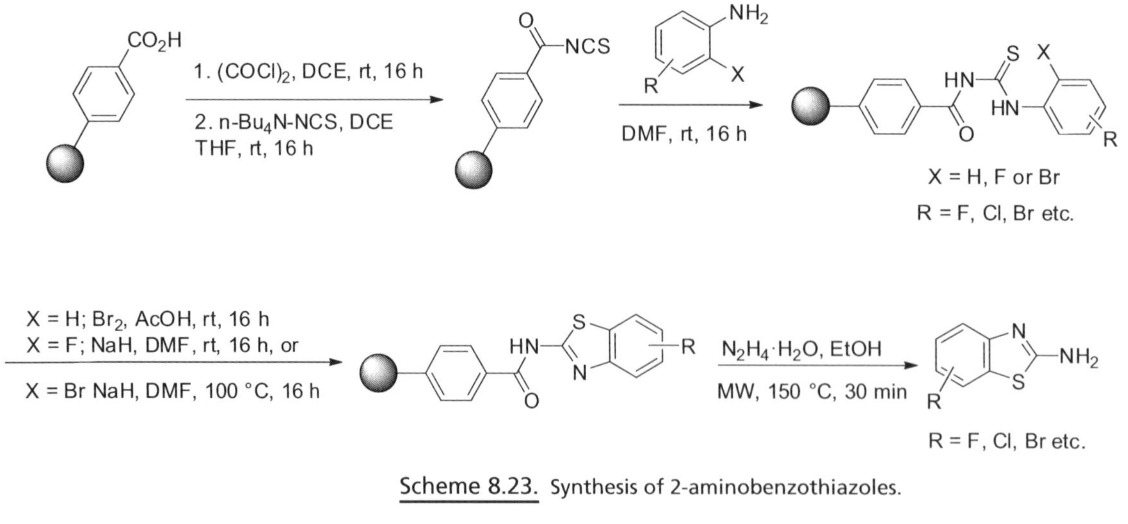

Scheme 8.23. Synthesis of 2-aminobenzothiazoles.

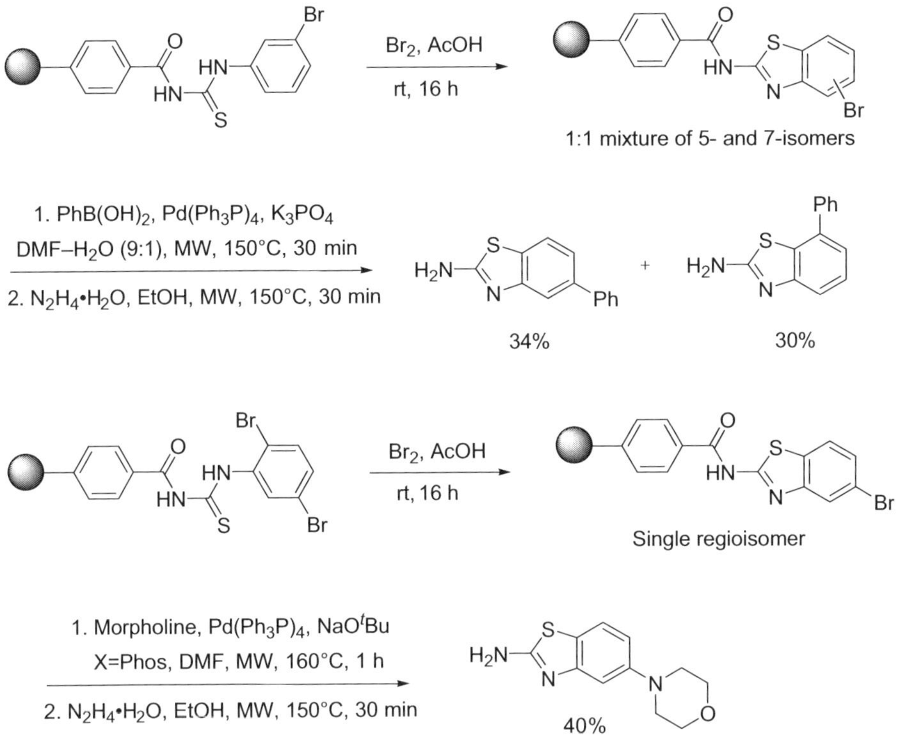

Scheme 8.24. Cross-coupling reactions of 2-aminobenzothiazoles.

7-bromo-2-aminobenzothiazoles (Scheme 8.24). This was subjected to a microwave-assisted Suzuki–Miyaura cross-coupling reaction[93] with phenylboronic acid at 150°C for 30 min, using K_3PO_4 as the base and $Pd(PPh_3)_4$ as the catalyst. The obtained compounds were finally cleaved from the resin upon treatment with $N_2H_4 \cdot H_2O$ under microwave irradiation at 150°C (Scheme 8.24).

In a similar fashion, the resin-bound 5-bromo-2-aminobenzothiazole was subjected to a microwave-assisted Buchwald–Hartwig amination[94] with morpholine using ^{t}BuONa as the base and a catalytic system comprising $Pd(PPh_3)_4$ and X-Phos in DMF at 160°C for 1 h. The product was cleaved from the resin using the previous conditions (Scheme 8.24).

8.21 PYRIMIDINES, PYRAZOLES, AND ISOXAZOLES

In view of synthesizing novel ligands with promising inhibitory potency toward tyrosine kinases[95] (PTKs) or other key target enzymes such as COXs and HIV reverse transcriptase,[96] Leonetti and coworkers have recently proposed a microwave-assisted synthesis of a 40-mer library of pyrimidines, pyrazoles, and isoxazoles on solid support.[97] The authors started their investigation by choosing Rink amide resin as the solid support of choice, which was amidated with 3-hydroxybenzoic acid using EDC in DMF at rt. The free OH-terminus was then alkylated with a number of bromomethyl ketones to impart the first element of structural diversity to the synthesis. The reactions were carried out under microwave irradiation using DBU as the base in a 10% solution of HMPA in DMF at 140°C for 30 min (Scheme 8.25).

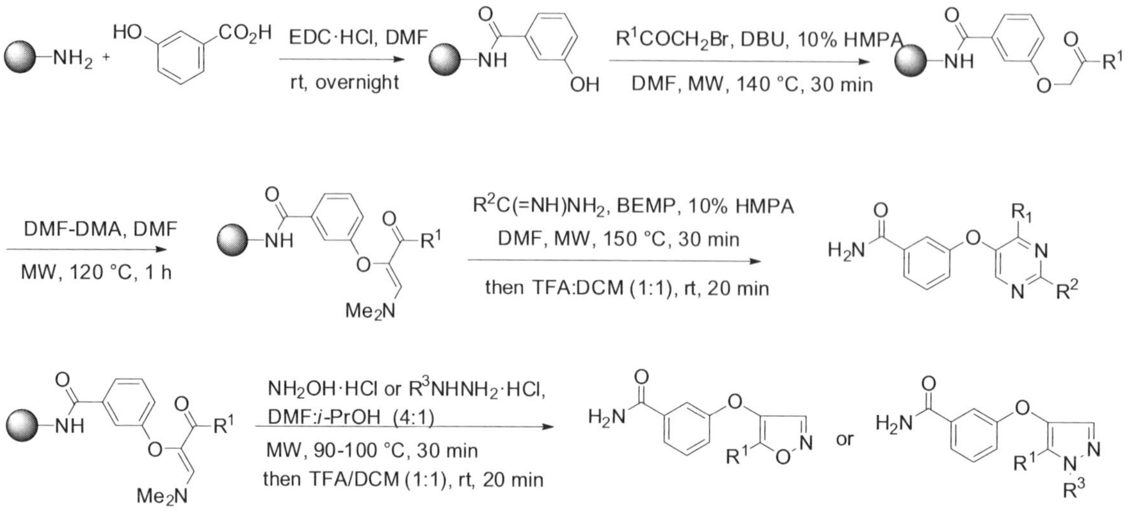

Scheme 8.25. Synthesis of biologically potent pyrimidines, pyrazoles, and isoxazoles.

The required enamino ketone was then generated using a microwave-assisted reaction of the resin-bound ketone with *N*,*N*-dimethylformamide dimethylacetal (DMF-DMA) in DMF at 120°C for 1 h (Scheme 8.25). Generation of the structurally diverse 2,4-disubstituted-5-phenoxy-pyrimidines was achieved by a microwave-assisted, BEMP-mediated cyclization reaction with suitable guanidines at 150°C for 30 min, using a 10% solution of HMPA in DMF as the solvent of choice. The pyrimidine derivatives were then cleaved from the resin upon treatment with TFA/DCM (1:1) at rt and the target molecules were isolated in good yields of 69–73% (Scheme 8.25). Alternatively, the isoxazoles were generated in 50–70% yields by the microwave-assisted condensation of hydroxylamine hydrochloride at 90°C for 30 min, whereas the corresponding pyrazole analogues were synthesized via a microwave-assisted condensation of suitable hydrazines at 100°C for 30 min (Scheme 8.25). Both reactions were carried out in DMF/*i*-PrOH (4:1).

8.22 QUINOLIN-2(1*H*)-ONES AND COUMARINS

Although triphenylphosphine-based building blocks, especially ylides,[98] are very valuable tools in the synthesis of various heterocyclic systems,[99] the difficulty in separating triphenylphosphine oxide from the desired target is often a herculean task for the synthetic chemist. In view of circumventing the problem, Henkel has described an ingenious protocol (Scheme 8.26) that uses polymer-bound triphenylphosphine.[100] The author coupled bromoacetic acid *N*-hydroxysuccinimidyl ester (NHS ester) with the polymer-bound triphenylphosphine,[101] which was converted to the corresponding ylide by the reaction of DCM/TEA (1:1) at rt for 16 h (Scheme 8.26). The thus generated resin-bound ylide was demonstrated to be useful in the synthesis of biologically interesting heterocyclic scaffolds such as coumarins[102] and quinolin-2(1*H*)-ones.[103]

Therefore a cascade protocol was elaborated by employing a Wittig reaction followed by intramolecular cyclization. The best conditions were found when a mixture of the compounds in 1,4-dioxane was irradiated at 180°C for 15 min (Scheme 8.26). The authors further demonstrated the applicability of their strategy by generating a small library

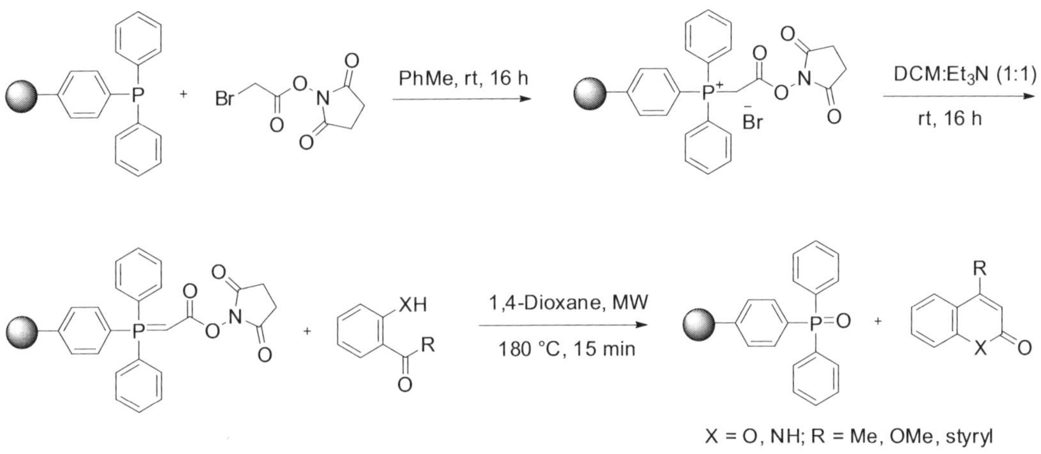

Scheme 8.26. Synthesis of quinolin-2(1*H*)-ones and coumarins.

of various heterocycles with good overall yields of 49–73%, while the triphenylphosphine oxide generated in the reaction was left behind on the resin, which was easily removed from the product by filtration (Scheme 8.26).

8.23 BENZOFURANS

As a result of their broad biological potential, substituted benzofurans are of high interest in present-day organic chemistry.[104] They are of additional interest due to the fact that they are the isoelectronic analogues of indoles, arguably the most well-investigated heterocyclic moiety for generating small molecules with drug-like features. Giacomelli and coworkers have recently proposed an interesting microwave-assisted protocol for the synthesis of diversely functionalized 2-substituted benzofurans (Scheme 8.27).[105] The authors investigated a solid-supported "catch-and-release" strategy to generate the title molecules with great success. An added advantage of the protocol is the easy separation of the formed resin-bound triphenylphosphine oxide from the products. The authors applied triphenylphosphine polystyrene resin as the solid support of choice, which was treated with suitable 2-hydroxybenzyl bromides. The best conditions for this alkylation appeared to be microwave irradiation in DMF at 85°C for two 15 min cycles (Scheme 8.27).

The cyclization to the desired benzofurans of the thus formed phosphonium bromide with a variety of acid chlorides was performed under microwave irradiation at 110°C for two cycles of 30 min in toluene (Scheme 8.27) using triethylamine as the base. The solid-bound triphenylphosphine oxide was easily removed from the product by simple filtration of the resin.

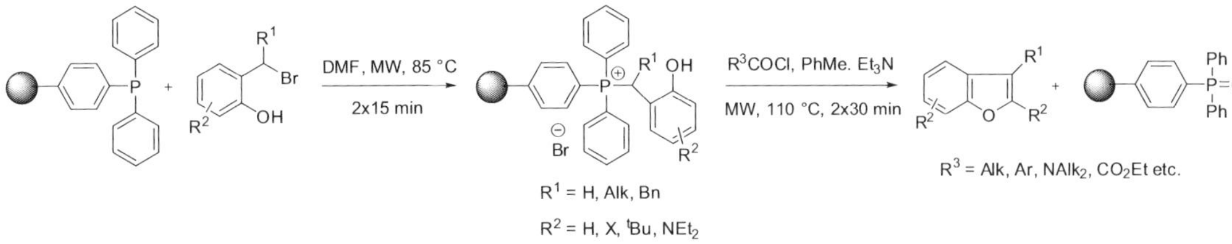

Scheme 8.27. Synthesis of diversely functionalized benzofurans.

8.24 *i*-CONDENSED PURINES

Owing to the presence of a purine core in a plethora of physiologically active molecules, the synthesis and decoration of small molecules containing this unit holds a never-ending interest from the synthetic and pharmaceutical chemistry community. A large number of manuscripts has been published on the synthesis and the decoration of purine-bearing small molecules, for example, purine-based kinase inhibitors.[106] However, most of the available literature deals with the generation of N9 isomers, while the corresponding N7-substituted purines are left relatively untouched,[107] although some of the last class of compounds have been shown to possess interesting antiviral activity[108] or strong antiproliferative activity.[109] Furthermore, N7-substituted guanines are potent telomerase inhibitors and enhance the efficacy of chemotherapeutic agents in the treatment of cancer.[110] On the basis of their previous experience in the regiospecific traceless solid-phase synthesis of N1,N7-disubstituted guanines using 6-chloropurine, Lam and coworkers have recently elaborated an interesting microwave-assisted synthesis of N7-alkylated purines on solid support,[111] as well as of the previously unknown *i*-condensed purines (Scheme 8.28). The authors chose the BOBA (benzyloxybenzylamine) resin as the preferable solid support, which was easily generated from commercially available Wang resin upon Mitsunobu reaction with phthalimide in THF at rt, followed by conversion to the corresponding amine by hydrazinolysis in THF at rt (Scheme 8.28).

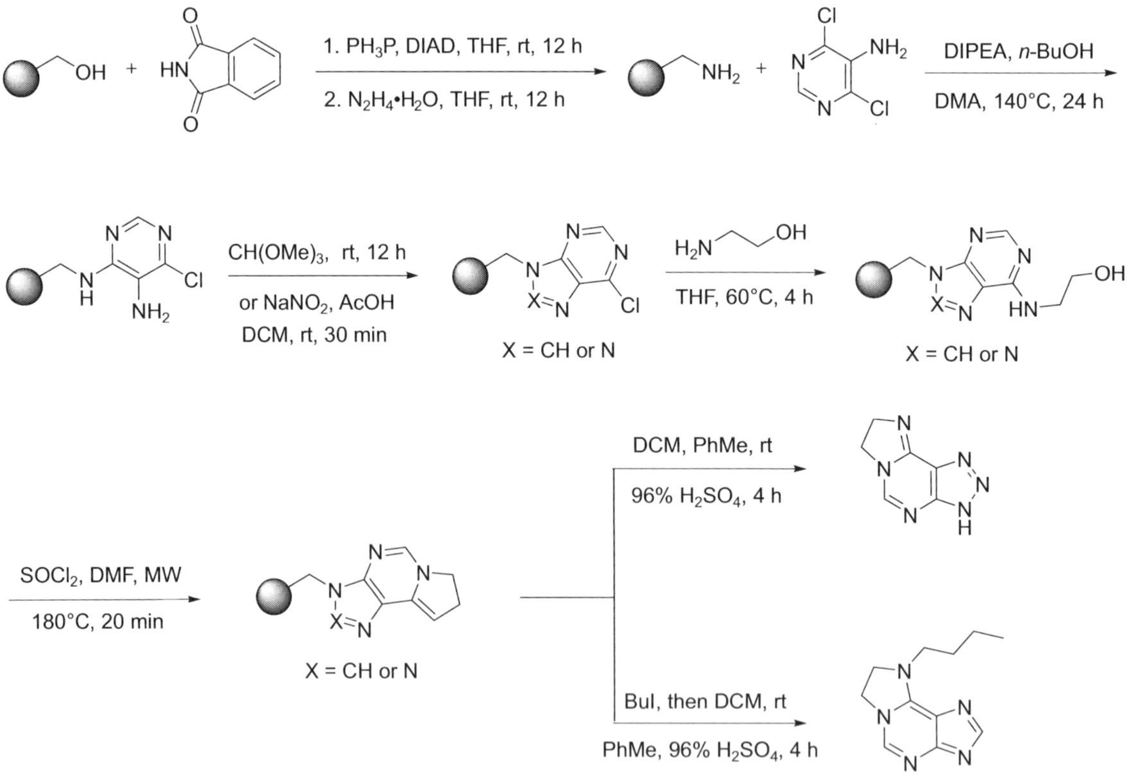

Scheme 8.28. Synthesis of *i*-condensed purines.

They next explored the DIPEA-mediated alkylation of the amino terminus of the resin with 4,6-dichloro-5-aminopyrimidine, which was performed at 140°C for 24 h in a mixture of DMA and *n*-BuOH. Treatment overnight of the thus obtained resin with trimethyl orthoformate in DMF at rt resulted in the formation of the purine imidazole fused ring system, whereas a diazotization reaction using $NaNO_2$ in AcOH and DCM at rt for 30 min provided the corresponding purine–triazole fused ring system (Scheme 8.28). After substitution of the chlorine with ethanolamine in THF at 60°C for 4 h, conversion of the hydroxyl group to the corresponding chlorine and subsequent cyclization to the *i*-condensed purine core was achieved via a microwave-assisted reaction with $SOCl_2$ in DMF at 180°C for 20 min (Scheme 8.28). The cleavage of the product from the resin was finally effected by the treatment of the resin-bound *i*-condensed purine with 96% H_2SO_4 in a mixture of DCM and toluene at rt for 4 h. Alternatively, the resin-bound *i*-condensed purine was first converted into the quaternary salt by reaction with BuI, which was then cleaved from the resin upon treatment with 96% H_2SO_4.

8.25 2(1*H*)-PYRAZINONES

The 2(1*H*)-pyrazinone scaffold allows the incorporation of pharmacologically active groups with the ability to address many biological targets. Pyrazinone derivatives have been shown to feature a diverse array of potent biological and pharmacological activities, such as corticotropin-releasing factor (CRF) receptor antagonism,[112] HIV replication inhibitory acitivity,[113] and so on. They also feature inhibitory action on platelet aggregation, vasodilating activity, and inhibitory action on liperoxide generation.[114] In addition to this, Diels–Alder reactions of 2(1*H*)-pyrazinones afford various bicyclic and heterocyclic products that provide access to different scaffolds of pharmaceutical interest, such as bridged analogues of piperazine drugs[115] or β-turn mimics.[116]

Van der Eycken and coworkers have demonstrated an interesting solid-phase synthesis of 2(1*H*)-pyrazinones[117] based on the Strecker reaction[118] of the resin-bound amine with an appropriate aldehyde and a cyanide, allowing a wide diversity at the C6 position of the pyrazinone ring (Scheme 8.29). The authors investigated a number of interesting decorations on the resin-bound 2(1*H*)-pyrazinones, using a variety of microwave-assisted, transition metal-mediated cross-coupling reactions (Scheme 8.29).

The resin-linked 3,5-dichloropyrazinones[110] were subjected to a number of microwave-assisted, transition metal-mediated reactions for further diversifying the structure (Scheme 8.29). Suzuki–Miyaura cross-coupling reaction with phenylboronic acid was carried out using $Pd(Ph_3P)_4$ and Na_2CO_3 as the catalyst and the base, respectively, at 170°C in toluene for 30 min. The pyrazinone was cleaved from the solid support upon treatment with TFA/DCM (1:2) under microwave irradiation at 120°C for 20 min. Alternatively, Stille cross-coupling with tetraphenyl tin was also investigated by the authors under microwave irradiation at 150°C for 20 min. Further structural diversity was elaborated using a Sonogashira reaction with phenylacetylene, carried out under microwave irradiation at 120°C for 30 min using $Pd(Ph_3P)_2Cl_2$ and CuI (Scheme 8.29). Microwave-assisted Ullmann-type aminations[119] were then explored with 4-anisidine as the coupling partner, using a mixture of Cu(0) and Cu(I) as the catalytic system at 175°C for 30 min, with K_2CO_3 as the base. All products were subsequently subjected to acid-mediated cleavage upon microwave irradiation at 120°C for 20 min (Scheme 8.29).

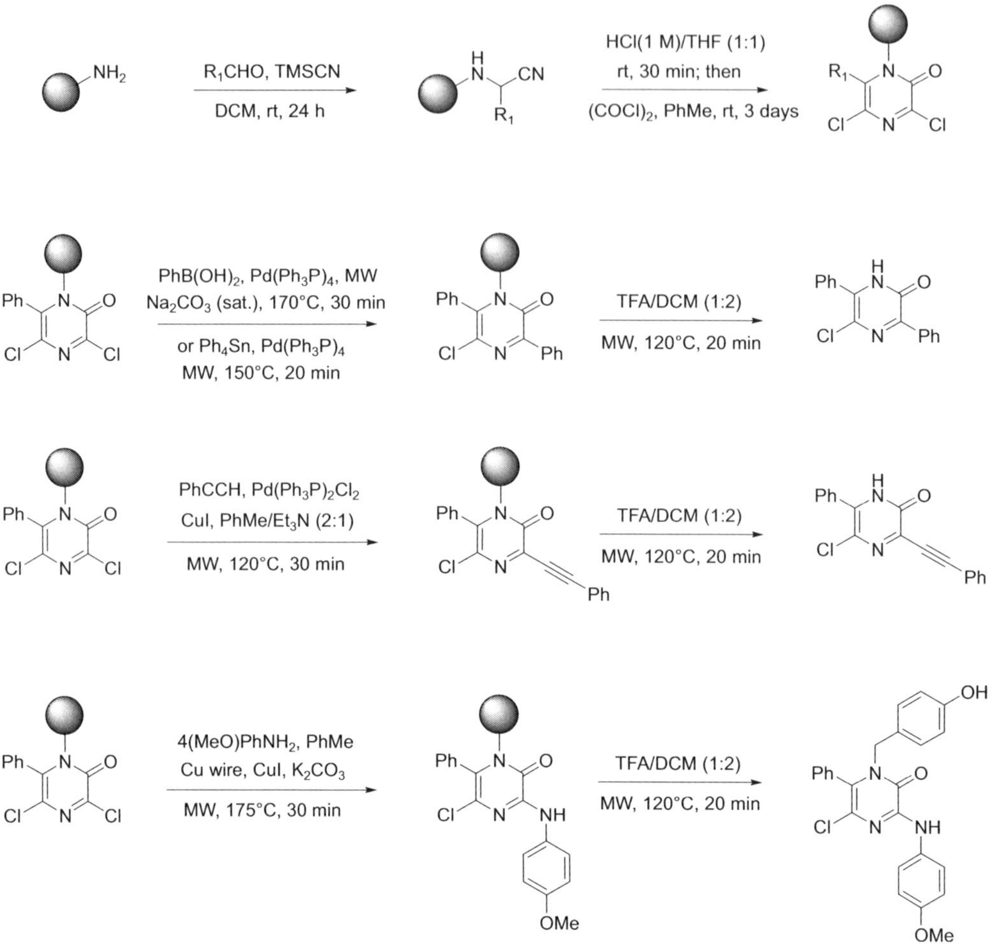

Scheme 8.29. Decoration of 2(1*H*)-pyrazinones.

8.26 CONCLUSIONS

In this chapter, we have attempted to provide a very selective but hopefully interesting set of examples of recent microwave-assisted solid-phase syntheses of heterocycles. It has been clearly demonstrated that the application of microwave irradiation has mostly a beneficial effect on the outcome of the reaction, resulting in dramatically shortened reaction times and higher purities of the target compounds. Moreover, in some cases, protocols could be performed that are nearly impossible under conventional heating conditions. Many different ways of generating small, structurally diverse heterocyclic moieties, where the structural cores were directly generated or manipulated on solid support under microwave irradiation, passed the *revue*, exemplifying the nearly endless creativity in this field. We sincerely hope that this chapter will stimulate the reader to pay a more in-depth focus on the subject, resulting in an increased creative research activity in the field.

REFERENCES

1. R. B. Merrifield, *J. Am. Chem. Soc.* **1963**, *85*, 2149.
2. R. B. Merrifield, *Peptides* **1995**, *16*, 93.
3. For selective earlier literature on solid-phase organic reactions, see (a) C. C. Leznoff, *Chem. Soc. Rev.* **1974**, *3*, 65. (b) C. C. Leznoff, *Acc. Chem. Res.* **1978**, *11*, 327. (c) J. I. Crowley, H. Rapoport, *Acc. Chem. Res.* **1976**, *9*, 135.
4. (a) M. H. Geysen, S. J. Rodda, T. Mason, *J. Mol. Immunol.* **1986**, *23*, 709. (b) K. S. Lam, S. E. Salmon, E. M. Hersh, V. J. Hruby, W. M. Kazmierski, R. J. Knapp, *Nature* **1991**, *354*, 82. (c) R. A. Houghten, C. Pinilla, S. E. Blondelle, J. R. Appel, C. T. Dooley, J. H. Cuervo, *Nature* **1991**, *354*, 84. (d) P. H. Seeberger, W.-C., Haase, *Chem. Rev.* **2000**, *100*, 4349.
5. (a) B. A. Bunin, J. A. Ellman, *J. Am. Chem. Soc.* **1992**, *114*, 10997. (b) S. H. DeWitt, J. S. Kiely, C. J. Stankovic, M. C. Schroeder, D. M. R. Cody, M. R. Pavia, *Proc. Natl. Acad. Sci. USA* **1993**, *90*, 6909.
6. P. Blaney, R. Grigg, V. Sridharan, *Chem. Rev.* **2002**, *102*, 2607.
7. V. Krchňák, M. W. Holladay, *Chem. Rev.* **2002**, *102*, 61.
8. For a selective reading, see (a) P. Jakobsen, P. B. Ritsmar, E. Persson, *Bioorg. Med. Chem.* **2000**, *8*, 2095. (b) E. Persson, P. Jakobsen, H. Worsaae, *PCT Int. Appl.* WO 9948878, **1999**. (c) P. W. Hsieh, F. R. Chang, C. H. Chang, P. W. Cheng, L. C. Chiang, F. L. Zeng, K. H. Lin, Y. C. Wu, *Bioorg. Med. Chem. Lett.* **2004**, *14*, 4751. (d) S. J. Hays, B. W. Caprathe, J. L. Gilmore, N. Amin, M. R. Emmerling, W. Michael, R. Nadimpalli, R. Nath, K. J. Raser, D. Stafford, D. Watson, K. Wang, J. C. Jaen, *J. Med. Chem.* **1998**, *41*, 1060. (e) R. L. Jarvest, M. J. Parratt, C. M. Debouck, J. G. Gorniak, L. J. Jennings, H. T. Serafinowska, J. E. Strickler, *Bioorg. Med. Chem. Lett.* **1996**, *6*, 2463.
9. For a very selective reading on heteroannulated oxazinones, see (a) R. L. Jarvest, I. L. Pinto, S. M. Ashman, C. E. Dabrowski, A. V. Fernandez, L. J. Jennings, P. Lavery, D. G. Tew, *Bioorg. Med. Chem. Lett.* **1999**, *9*, 443. (b) P. Jakobsen, A. M. Hornemann, E. Persson, *PCT Int. Appl.* WO 2000030646, **2000**. (c) P. Jakobsen, A. M. Horneman, E. Persson, *Bioorg. Med. Chem.* **2000**, *8*, 2803. (d) J. Ward, R. Jain, D. James, H. J. Verheij, J. C. C. Schultz, *PCT Int. Appl.* WO 2006044826, **2006**. (e) J. Pato, G. Keri, L. Orfi, F. Waczek, Z. Horvath, P. Banhegyi, G. Z. Szabadkai, A. Choidas, G. Bacher, A. Missio, A. Koul,US Patent 2004171603, **2004** (f) A. Missio, G. Bacher, A. Koul, A. Choidas, *PCT Int. Appl.* WO 2003084947, **2003**. (g) D. Witter, A. L. Castelhano, *PCT Int. Appl.* WO 2003053944, **2003**. M. Linschoten, *PCT Int. Appl.* WO 2004108139, 2004. (h) P. Jakobsen, A. M. Horneman, E. Persson, *Bioorg. Med. Chem.* **2000**, *8*, 2803. (i) P. Jakobsen, A. M. Hornemann, E. Persson, *PCT Int. Appl.* WO 2000030646, **2000**.
10. J. Che, M. S. Raghavendra, Y. Lam, *J. Comb. Chem.* **2009**, *11*, 378.
11. K. Gewald, E. Schinke, H. Bottcher, *Chem. Ber.* **1966**, *99*, 94.
12. For selected literature on biological activities of thiazolo[4,5-d]pyrimidine-5,7-dione derivatives, see (a) D. A. Carson, H. B. Cottam, L. Deng, *PCT Int. Appl.* WO 2000069861, **2000**. (b) A. Berger, E. E. Borgeas, *Ger. Offen.* DE2038922, **1971**. (c) H. Abe, M. Tanaka, K. Sugimoto, A. Suma, M. Yokota, M. Shiozaki, K. Ito, K. Ueyama, D. Motoda, T. Noguchi, T. Adachi, J. Tsuruha, S. Doi, *PCT Int. Appl.* WO 2007119889, **2007**. (d) A. F. Lewis, G. R. Revankar, S. M. Fennewald, M. Susan, J. H. Huffman, R. F. J. Rando, *Heterocycl. Chem.* **1995**, *32*, 547.
13. Lee, J. Y. Lee, H. J. Lee, Y.-D. Gong, *Synlett* **2005**, 2483.
14. T. Lee, J.-H. Park, D.-H. Lee, Y.-D. Gong, *J. Comb. Chem.* **2009**, *11*, 495.
15. K. Harju, J. Vesterinen, J. Yli-Kauhaluoma, *Org. Lett.* **2009**, *11*, 2219.
16. (a) T. D. Penning, J. J. Talley, S. R. Bertenshaw, J. S. Carter, P. W. Collins, S. Docter, M. J. Graneto, L. F. Lee, J. W. Malecha, J. M. Miyashiro, R. S. Rogers, D. J. Rogier, S. S. Yu, G. D. Anderson, E. G. Burton, J. N. Cogburn, S. A. Gregory, C. M. Koboldt, W. E. Perkins, K. Seibert, A. W. Veenhuizen, Y. Y. Zhang, P. C. Isakson, *J. Med. Chem.* **1997**, *40*, 1347. (b) M. Rinaldi-Carmona, F. Barth, M. Héaulme, D. Shire, B. Calandra, C. Congy, S. Martinez, J. Maruani, G.

Néliat, D. Caput, P. Ferrara, P. Soubrié, J. C. Breliére, G. Le Fur, *FEBS Lett.* **1994**, *350*, 240. (c) F. J. Baud, C. Bismuth, R. Garnier, M. Galliot, A. Astier, G. Maistre, M. Soffer, *J Toxicol. Clin. Toxicol.* **1986**, *24*, 463. (d) N. K. Terrett, A. S. Bell, D. Brown, P. Ellis, *Bioorg. Med. Chem. Lett.* **1996**, *6*, 1819.

17. (a) M. T. Bilodeau, A. M. Cunningham, *J. Org. Chem.* **1998**, *63*, 2800. (b) A. F. Abdel-Magid, K. G. Carson, B. D. Harris, C. A. Maryanoff, R. D. Shah, *J. Org. Chem.* **1996**, *61*, 3849.
18. For some selected literature on HSPs, see (a) S. Lindquist, E. A. Craig, *Annu. Rev. Genet.* **1988**, *22*, 631. (b) B. Bukau, J. Weissman, A. Horwich, *Cell* **2006**, *125*, 443. (c) S. W. Fewell, K. J. Travers, J. S. Weissman, J. L. Brodsky, *Annu. Rev. Genet.* **2001**, *35*, 149. (d) R. J. Ellis, *Trends Biochem. Sci.* **2006**, *31*, 395. (e) J. C. Young, J. M. Barral, F. Ulrich Hartl, *Trends Biochem. Sci.* **2003**, *28*, 541. (f) M. J. Schlesinger, *J. Biol. Chem.* **1990**, *265*, 12111. (g) S. Walter, J. Buchner, *Angew. Chem., Int. Ed. Engl.* **2002**, *41*, 1098.
19. (a) R. J. Binder, *Expert Rev. Vaccines* **2008**, *7*, 383. (b) C. Didelot, D. Lanneau, M. Brunet, A.-L. Joly, A. De Thonel, G. Chiosis, C. Garrido, *Curr. Med. Chem.* **2007**, *14*, 2839.
20. (a) D. B. Solit, N. Rosen, *Curr. Top. Med. Chem.* **2006**, *6*, 1205–1214. (b) R. Bagatell, L. Gore, M. J. Egorin, R. Ho, G. Heller, N. Boucher, E. G. Zuhowski, J. A. Whitlock, S. P. Hunger, A. Narendran, H. M. Katzenstein, R. J. Arceci, J. Boklan, C. E. Herzog, L. Whitesell, S. P. Ivy, T. M. Trippett, *Clin. Cancer Res.* **2007**, *13*, 1783.
21. S. W. Fewell, C. M. Smith, M. A. Lyon, T. P. Dumitrescu, P. Wipf, B. W. Day, J. L. Brodsky, *J. Biol. Chem.* **2004**, *279*, 51131.
22. S. Wisén, J. Androsavich, C. G. Evans, L. Chang, J. E. Gestwicki, *Bioorg. Med. Chem. Lett.* **2008**, *18*, 60.
23. (a) D. Dallinger, C. O. Kappe, *Nat. Protoc.*, **2007**, *2*, 1713. (b) C. O. Kappe, A. Stadler, *Methods Enzymol.*, **2003**, *369*, 197.
24. W. H. W. Lunn, R. W. Harper, R. L. Stone, *J. Med. Chem.* **1971**, *14*, 1069.
25. L. Dalla Via, O. Gia, S. Marciani Magno, A. Da Settimo, A. M. Marini, G. Primofiore, F. Da Settimo, S. Salerno, *Il Farmaco* **2001**, *56*, 159.
26. R. D. Carpenter, K. S. Lam, M. J. Kurth, *J. Org. Chem.* **2007**, *72*, 284.
27. For a selective literature, see (a) W. J. Greenlee, P. K. S. Siegl, *Annu. Rep. Med. Chem.* **1992**, *27*, 59. (b) S. C. Shilcrat, M. K. Mokhallalati, J. M. D. Fortunak, L. N. Pridgen, *J. Org. Chem.* **1997**, *62*, 8449. (c) J. P. Rizzi, A. A. Nagel, T. Rosen, S. McLean, T. Seeger, *J. Med. Chem.* **1990**. *33*, 2721. (d) G. Shapiro, B. Gomez-Lor, *J. Org. Chem.* **1994**, *59*, 5524. (e) J. L. Adams, J. C. Boehm, S. Kassis, P. D. Gorycki, E. F. Webb, R. Hall, M. Sorenson, J. C. Lee, A. Ayrton, D. E. Griswold, T. F. Gallagher, *Bioorg. Med. Chem. Lett.* **1998**, *8*, 3111.
28. W. H. Beggs, F. A. Andrews, G. A. Sarosi, *Life Sci.* **1981**, *28*, 111.
29. S. K. Samanta, I. Kylänlahti, J. Yli-Kauhaluoma, *Bioorg. Med. Chem. Lett.* **2005**, *15*, 3717.
30. Ameba resin – PS-MB-CHO, Sigma–Aldrich, catalog no. 51, 644.
31. M. Shanmugasundaram, I. Garcia-Martinez, Q. Li, A. Estrada, N. E. Martinez, L. E. Martinez, *Tetrahedron Lett.* **2005**, *46*, 7545.
32. (a) K. H. Dötz, *Angew. Chem., Int. Ed. Engl.* **1975**, *14*, 644. (b) K. H. Dötz, *Pure Appl. Chem.* **1983**, *55*, 1689. (c) K. H. Dötz, *Angew. Chem., Int. Ed. Engl.* **1984**, *23*, 587. (d) A. de Meijere, H. Schirmer, M. Duetsch, *Angew. Chem., Int. Ed.* **2000**, *39*, 3964. (e) J. W. Herndon, *Tetrahedron* **2000**, *56*, 1257. (f) M. Gómez-Gallego, M. J. Mañcheno, M. A. Sierra, *Acc. Chem. Res.* **2005**, *38*, 44.
33. X. J. Zhao, R. Tapec-Dytioco, K. M. Wang, W. H. Tan, *Anal. Chem.* **2003**, *75*, 3476.
34. I. V. Nesterova, S. S. Erdem, S. Pakhomov, R. P. Hammer, S. A. Soper, *J. Am. Chem. Soc.* **2009**, *131*, 2432.
35. V. T. Verdree, S. Pakhomov, G. Su, M. W. Allen, A. C. Countryman, R. P. Hammer, S. A. Soper, *J. Fluoresc.* **2007**, *17*, 547.
36. S. S. Erdem, I. V. Nesterova, S. A. Soper, R. P. Hammer, *J. Org. Chem.* **2008**, *73*, 5003.

37. S. S. Erdem, I. V. Nesterova, S. A. Soper, and R. P. Hammer, *J. Org. Chem.* **2009**, *74*, 9280.
38. S. Bräse, C. Gil, K. Knepper, *Bioorg. Med. Chem.* **2002**, *10*, 2415.
39. G. Lesma, B. Danieli, F. Lodroni, D. Passarella, A. Sacchetti, A. Silvani, *Comb. Chem. High Throughput Screening* **2006**, *9*, 691.
40. (a) E. D. Cox, J. M. Cook, *Chem. Rev.* **1995**, *95*, 1797.(b) W. M. Whaley, T. R. Govindachari,in Adams, R. (Ed.), *Organic Reactions*, Vol. 6, Wiley, New York, **1951**, p. 151.
41. (a) T. Kametani, The total syntheses of isoquinoline alkaloids, in ApSimon, J. (Ed.), *The Total Synthesis of Natural Products*, Vol. 3, Wiley-Interscience, New York, **1977**, pp. 1–218. (b) M. D. Rozwadowska, *Heterocycles*, **1994**, *39*, 903. (c) T. S. Kaufman, *Tetrahedron: Asymmetry* **2004**, *15*, 1203.
42. (a) Y. Gao, R. Zong, A. Campbell, N. S. Kula, R. J. Baldessarini, L. Neumeyer, *J. Med. Chem.* **1988**, *31*, 1392. (b) J. M. Schaus, R. D. Titus, M. M. Foreman, N. R. Mason, L. L. Truex, *J. Med. Chem.* **1990**, *33*, 600. (c) G. M. Cingolani, A. Di Stefano, B. Mosciatti, F. Napolitani, G. Giorgioni, M. Ricciutelli, F. Claudi, *Bioorg. Med. Chem. Lett.* **2000**, *10*, 1385.
43. (a) P. Virta, J. Rosenberg, T. Karskela, P. Heinonen, H. Lönnberg, *Eur. J. Org. Chem.* **2001**, 3467. (b) E. Mannekens, M. Crisma, S. Van Cauwenberghe, D. Tourwé, *Eur. J. Org. Chem.* **2003**, 3300.
44. (a) Y. Bourne, H. C. Kolb, Z. Radić, K. B. Sharpless, P. Taylor, P. Marchot, *Proc. Natl. Acad. Sci. USA* **2004**, *101*, 1449. (b) W. G. Lewis, L. G. Green, F. Grynszpan, Z. Radić, P. R. Carlier, P. Taylor, M. G. Finn, K. B. Sharpless, *Angew. Chem., Int. Ed.* **2002**, *41*, 1053. (c) N. A. Al-Masoudi, Y. A. Al-Soud, *Tetrahedron Lett.* **2002**, *43*, 4021. (d) K. Dabak, A. Akar, *Heterocyclic Commun.* **2002**, *8*, 385. (e) D. J. Guerin, S. J. Miller, *J. Am. Chem. Soc.* **2002**, *124*, 2134. (f) S. Kamijo, T. Jin, Z. Huo, Y. Yamamoto, *Tetrahedron Lett.* **2002**, *43*, 9707.
45. (a) A. C. Cunha, J. M. Figueiredo, J. L. M. Tributino, A. L. P. Miranda, H. C. Castro, R. B. Zingali, C. A. M. Fraga, M. C. B. V. de Souza, V. F. Ferreira, E. J. Barreiro, *Bioorg. Med. Chem.* **2003**, *11*, 2051. (b) J. S. Tullis, J. C. VanRens, M. G. Natchus, M. P. Clark, B. De, M. J. Janusz, L. C. H. Janusz, *Bioorg. Med. Chem. Lett.* **2003**, *13*, 1665.
46. W.-Q. Fan, A. R. Katritzky,in Katritzky, A. R., Rees, C. W., Scriven, E. F. V. (Eds), *Comprehensive Heterocyclic Chemistry II*, Vol. 4, Elsevier Science, Oxford, **1996**, pp. 1–126.
47. R. Huisgen, *Angew. Chem.* **1963**, *75*, 604.
48. Y. Gao, Y. Lam, *Org. Lett.* **2006**, *8*, 3283.
49. H. Rosemeyer, *Chem. Biodiversity* **2004**, *1*, 361.
50. (a) L. M. Toledo, N. B. Lydon, D. Elbaum, *Curr. Med. Chem.* **1999**, *6*, 775. (b) M. Legraverend, O. Ludwig, E. Bisagni, S. Leclerc, L. Meijer, *Bioorg. Med. Chem. Lett.* **1998**, *8*, 793.
51. N. S. Gray, L. Wodicka, A.-M. Thunnissen, T. C. Norman, S. Kwon, F. H. Espinoza, D. O. Morgan, G. Barnes, S. LeClerc, L. Meijer, S.-H. Kim, D. J. Lockhart, P. G. Schultz, *Science* **1998**, *281*, 533.
52. A. G. Cole, A. Metzger, G. Ahmed, M.-R. Brescia, R. J. Chan, J. Wen, L. O'Brien, L.-Y. Qin, I. Henderson, *Tetrahedron Lett.* **2006**, *47*, 8897.
53. (a) P. Gomes, M. J. Araujo, M. Rodrigues, N. Vale, Z. Azevedo, J. Iley, P. Chambel, J. Morais, R. Moreira, *Tetrahedron* **2004**, *60*, 5551. (b) M. J. Araujo, J. Bom, R. Capela, C. Casimiro, P. Chambel, P. Gomes, J. Iley, F. Lopes, J. Morais, R. Moreira, E. De Oliveira, V. Do Rosario, N. Vale, *J. Med. Chem.* **2005**, *48*, 888.
54. (a) J. P. Lyssikatos, B. V. Yang,US Patent 6,194,438 **2001**. (b) M. P. Gauthier, C. Michaux, S. Rolin, C. Vastersaegher, X. de Leval, F. Julemont, L. Pochet, B. Masereel, *Bioorg. Med. Chem.* **2006**, *14*, 918.
55. K. G. Hull, A. Sidduri, J. Tilley, *PCT Int. Appl.* WO 0048994, **2000**.
56. (a) B. E. Blass, K. Coburn, N. Fairweather, A. Fluxe, S. Hodson, C. Jackson, J. Janusz, W. Lee, J. Ridgeway, R. White, S. Wu, *Tetrahedron Lett.* **2006**, *47*, 7497. (b) R. A. Houghten, J. M. Ostresh, Y. Yu, *Tetrahedron* **2002**, *58*, 3349.

57. L.-Y. Qin, A. G. Cole, A. Metzge, L. O'Brien, X. Sun, J. Wu, Y. Xu, K. Xu, Y. Zhang, I. Henderson, *Tetrahedron Lett.* **2009**, *50*, 419.
58. (a) S. Bräse, J. H. Kirchhoff, J. Klbberling, *Tetrahedron* **2003**, *59*, 885. (b) C. Gil, S. Bräse, *J. Comb. Chem.* **2009**, *11*, 175.
59. L.-P. Sun, W.-M. Dai, *Angew. Chem., Int. Ed.* **2006**, *45*, 7255.
60. W.-M. Dai, D.-S. Guo, L.-P. Sun, X.-H. Huang, *Org. Lett.* **2003**, *5*, 2919.
61. For a selective reading on Pd(II)- or Cu(II)-mediated synthesis of indoles, see (a) B. C. J. van Esseveldt, F. L. van Delft, R. de Gelder, F. P. J. T. Rutjes, *Org. Lett.* **2003**, *5*, 1717 and references therein. (b) M. G. Saulnier, D. B. Frennesson, M. S. Deshpande, D. M. Vyas, *Tetrahedron Lett.* **1995**, *36*, 7841. (c) K. Hiroya, S. Itoh, M. Ozawa, Y. Kanamori, T. Sakamoto, *Tetrahedron Lett.* **2002**, *43*, 1277. (d) K. Hiroya, S. Itoh, T. Sakamoto, *J. Org. Chem.* **2004**, *69*, 1126.
62. (a) F. McKerlie, D. J. Procter, G. Wynne, *Chem. Commun.* **2002**, 584. (b) F. McKerlie, I. M. Rudkin, G. Wynne, D. J. Procter, *Org. Biomol. Chem.* **2005**, *3*, 2805.
63. For a selective literature on the use of SmI_2 in organic synthesis, see (a) J. A. Soderquist, *Aldrichim. Acta* **1991**, *24*, 15. (b) G. A. Molander, *Chem. Rev.* **1992**, *92*, 29. (c) G. A. Molander, *Org. React.* **1994**, *46*, 211. (d) G. A. Molander, C. R. Harris, *Chem. Rev.* **1996**, *96*, 307. (e) A. Krief, A.-M. Laval, *Chem. Rev.* **1999**, *99*, 745. (f) D. J. Edmonds, D. Johnston, D. J. Procter, *Chem. Rev.* **2004**, *104*, 3371.
64. K. L. Turner, T. M. Baker, S. Islam, D. J. Procter, M. Stefaniak, *Org. Lett.* **2006**, *8*, 329.
65. For a selected number of examples on tetrahydroquinolones, see. (a) M. Patel, R. J. McHugh, Jr., B. C. Cordova, R. M. Klabe, L. T. Bacheler, S. Erickson-Viitanen, J. D. Rodgers, *Bioorg. Med. Chem. Lett.* **2001**, *11*, 1943. (b) H. Hayashi, Y. Miwa, I. Miki, S. Ichikawa, N. Yoda, A. Ishii, M. Kono, F. Suzuki, *J. Med. Chem.* **1992**, *35*, 4893. (c) R. W. Carling, P. D. Leeson, K. W. Moore, J. D. Smith, C. R. Moyes, I. M. Mawer, S. Thomas, T. Chan, R. Baker, A. C. Foster, S. Grimwood, J. A. Kemp, G. R. Marshall, M. D. Tricklebank, K. L. Saywell, *J. Med. Chem.* **1993**, *36*, 3397. (d) E. Christopher, E. Bedir, C. Dunbar, I. A. Khan, C. O. Okunji, B. M. Schuster, M. M. Iwu, *Helv. Chim. Acta* **2003**, *86*, **2914**. (e) C. Ito, M. Itoigawa, T. Otsuka, H. Tokuda, H. Nishino, H. Furukawa, *J. Nat. Prod.* **2000**, *63*, 1344.
66. S. Kobayashi, I. Hachiya, S. Suzuki, M. Moriwaki, *Tetrahedron Lett.* **1996**, *37*, 2809.
67. For a very selective literature on the biological activities of hydantoins, see (a) M. Meusel, M. Gutschow, *Org. Prep. Proced. Int.* **2004**, *36*, 391. (b) W. J. Brouillette, V. P. Jestkov, M. L. Brown, M. S. Akhtar, T. M. Delorey, G. B. Brown, *J. Med. Chem.* **1994**, *37*, 3289. (c) R. M. Schelkun, P. W. Yuen, K. Serpa, L. T. Meltzer, L. D. Wise, E. R. Whittemore, R. M. Woodward, *J. Med. Chem.* **2000**, *43*, 1892. (d) H. U. Stilz, W. Guba, B. Jablonka, M. Just, O. Klingler, W. Konig, V. Wehner, G. Zoller, *J. Med. Chem.* **2001**, *44*, 1158. (e) R. Z. Kuang, J. B. Epp, S. M. Ruan, H. Y. Yu, P. Huang, S. He, J. Tu, N. M. Schechter, J. Turbov, C. J. Froelich, W. C. Groutas, *J. Am. Chem. Soc.* **1999**, *121*, 8128. (f) R. Z. Kuang, J. B. Epp, S. Ruan, L. S. Chong, R. Venkataraman, J. Tu, S. He, T. M. Truong, W. C. Groutas, *Bioorg. Med. Chem.* **2000**, *8*, 1005.
68. (a) J. A. Monn, M. J. Valli, S. M. Massey, R. A. Wright, C. R. Salhoff, B. G. Johnson, T. Howe, C. A. Alt, G. A. Rhodes, R. L. Robey, K. R. Griffey, J. P. Tizzano, M. J. Kallman, D. R. Helton, D. D. Schoepp, *J. Med. Chem.* **1997**, *40*, 528. (b) J. A. Monn, M. J. Valli, S. M. Massey, M. M. Hansen, T. J. Kress, J. P. Wepsiec, A. R. Harkness, J. L. Grutsch, R. A. Wright, B. G. Johnson, S. L. Andis, A. Kingston, R. Tomlinson, R. Lewis, K. R. Griffey, J. P. Tizzano, D. D. Schoepp, *J. Med. Chem.* **1999**, *42*, 1027. (c) F. Tellier, F. Acher, I. Brabet, J. P. Pin, R. Azerad, *Bioorg. Med. Chem.* **1998**, *6*, 195.
69. E. Colacino, F. Lamaty, J. Martinez, I. Parrot, *Tetrahedron Lett.* **2007**, *48*, 5317.
70. (a) R. Capdeville, E. Buchdunger, J. Zimmermann, A. Matter, *Nat. Rev. Drug Discovery* **2002**, *1*, 493. (b) A. Arora, E. M. Scholar, *J. Pharmacol. Exp. Ther.* **2005**, *315*, 971.
71. I. Melnikova, J. Golden, *Nat. Rev. Drug Discovery* **2004**, *3*, 993.
72. A. Hochhaus, T. Hughes, *Hematol. Oncol. Clin. North Am.* **2004**, *18*, 641.

73. C. Capaldi, A. Carotti, *Tetrahedron Lett.* **2007**, *48*, 3455.

74. A. M. Fivush, T. M. Wilson, *Tetrahedron Lett.* **1997**, *38*, 7151.

75. F. Leonetti, P. Salvati, C. Caccia, M. Ferappi, A. Carotti,in *Proceedings of the 17th Convegno Nazionale Divisione di Chimica Farmaceutica*, Pisa, Italy, September 6-10 **2004**, p. 97, 163.

76. (a) R. Schwesinger, *Chimia* **1985**, *39*, 269. (b) M. Grøtli, M. Douglas, B. Beijer, R. G. Garsia, R. Eritja, B. Sproat, *J. Chem. Soc., Perkin Trans. 1*, **1997**, 2779.

77. (a) S. Ruchirawatt, P. Sahakitpichan, *Tetrahedron Lett.* **2000**, *41*, 8007. (b) M. Horiuchi, T. Maoka, N. Iwase, K. Ohnishi, *J. Nat. Prod.* **2002**, *65*, 1204. (c) V. Rys, A. Couture, E. Deniau, P. Grandclaudon, *Eur. J. Org. Chem.* **2003**, 231. (d) X. Zhang, W. Ye, S. Zhao, C.-T. Che, *Phytochemistry* **2004**, *65*, 929.

78. (a) T. Berg, S. B. Cohen, J. Deshamals, C. Sonderegger, D. J. Maslyar, J. Goldbers, D. L. Boger, P. K. Vogt, *Proc. Nat. Acad. Sci. USA* **2002**, *99*, 3830. (b) D. L. Boger, J. Desharnais, K. Capps, *Angew. Chem., Int. Ed.* **2003**, *42*, 4138.

79. M. Shanmugasundaram, G. Garcia-Martinez, Q. Li, N. E. Martinez, L. E. Martinez, *Tetrahedron Lett.* **2005**, *45*, 7465.

80. M. Shanmugasundaram, A. L. Aguirre, M. Leyva, B. Quan, L. E. Martinez, *Tetrahedron Lett.* **2007**, *48*, 7698.

81. Q. Sun, X. Zhou, K. Islam, D. J. Kyle, *Tetrahedron Lett.* **2001**, *42*, 6495. (b) D. D. Young, R. S. Senaiar, A. Deiters, *Chem. Eur. J.* **2006**, *12*, 5563.

82. For selective literature on the biologically active imidazo[1,2*a*]pyrimidin-5-ones, see (a) L. Dalla Via, O. Gia, S. M. Magno, A. Da Settimo, A. M. Marini, G. Primofiore, F. Da Settimo, S. I. Salaerno, *IlFarmaco* **2001**, *56*, 159. (b) Y. Rival, G. Grassy, A. Taudon, R. Ecalle, *Eur. J. Med. Chem.* **1991**, *26*, 13. (c) A. Mai, M. Artico, G. Sbardella, S. Quartarone, S. Massa, A. G. Loi, A. D. Montis, F. Scintu, M. Putzolu, P. L. Colla, *J. Med. Chem.* **1997**, *40*, 1447. (d) A. Mai, M. Artico, G. Sbardella, S. Massa, E. Novellino, G. Greco, A. G. Loi, E. Tramontano, M. E. Marongiu, P. L. Colla, *J. Med. Chem.* **1999**, *42*, 619. (h) T. Kappe, *J. Heterocycl. Chem.* **1995**, 1003. (i) H. S. Ahn, A. Bercovic, G. Boykow, A. Bronnenkant, S. Chackalamannil, J. Chow, R. Cleven, J. Cook, M. Czarniecki, C. Domalski, A. Fawzi, M. Green, A. Gundes, G. Ho, M. Laudicina, N. Lindo, K. Ma, M. Manna, B. McKittrick, B. Mirzai, T. Nechuta, B. Neustadt, C. Puchalski, K. Pula, L. Silverman, E. Smith, A. Stamford, R. P. Tedesco, H. Tsai, D. Tulshian, H. Vaccaro, R. W. Watkins, X. Weng, J. T. Witkoski, Y. Xia, H. Zhang, *J. Med. Chem.* **1997**, *40*, 2196.

83. (a) J. Boryski, B. Golankiewicz, E. DeClercq, *J. Med. Chem.* **1988**, *31*, 1351. (b) B. Golankiewicz, T. Ostrowski, P. Januszczyk, J. Zeidler, D. Baranowski, E. De Clercq, *J. Med. Chem.* **2001**, *44*, 4284.

84. G. Trapani, M. Franco, A. Latrofa, G. Genchi, V. Iacobazzi, C. A. Ghiani, E. Maciocco, G. Liso, *Eur. J. Med. Chem.* **1997**, *32*, 83.

85. (a) Y. F. Zhu, A. Q. Guo, T. D. Gross, Y. H. Gao, P. J. Cornnors, Jr., R. S. Struthers, Q. Xie, F. C. Tucci, G. J. Reinhart, D. P. Wu, J. Saunders, C. Chen, *J. Med. Chem.* **2003**, *46*, 1769. (b) Y. F. Zhu, R. S. Struthers, P. J. Cornnors, Jr., Y. H. Gao, T. D. Gross, J. Saunders, K. Wilcoxen, G. J. Reinhart, N. Ling, C. Chen, *Bioorg. Med. Chem. Lett.* **2002**, *12*, 399.

86. (a) R. W. Devivar, E. Kawashima, G. R. Revankar, J. M. Breitenbach, E. D. Kreske, J. C. Drach, L. B. Townsend, *J. Med. Chem.* **1994**, *37*, 2942. (b) J. P. Miller, K. H. Boswell, R. B. Meyer, Jr., L. F. Christensen, R. K. Robins, *J. Med. Chem.* **1980**, *23*, 242. (c) F. Manetti, J. A. Esté, I. C. Codina, M. Armand-Ugón, G. Maga, E. Crespan, R. Cancio, C. Mugnaini, C. Bernardini, A. Togninelli, C. Carmi, M. Alongi, E. Petricci, S. Massa, F. Corelli, M. Botta, *J. Med. Chem.* **2005**, *48*, 8000.

87. C. H. Soh, Y. Lam, *J. Comb. Chem.* **2010**, *12*, 286.

88. B. E. Evans, K. E. Rittle, M. G. Bock, R. M. DiPardo, R. M. Freidinger, W. L. Whitter, G. F. Lundell, D. F. Veber, P. S. Anderson, *J. Med. Chem.* **2002**, *31*, 2235.

89. (a) P. Jimonet, F. Audiau, M. Barreau, J.-C. Blanchard, A. Boireau, Y. Bour, M.-A. Coleno, A. Doble, G. Doerflinger, C. Do Huu, M.-H. Donat, J. M. Duchesne, P. Ganil, C. Guérémy, E. Honoré, B. Just, R. Kerphirique, S. Gontier, P. Hubert, P. M. Laduron, J. Le Blevec, M. Meunier, J.-M. Miquet, C. Nemecek, M. Pasquet, O. Piot, J. Pratt, J. Rataud, M. Reibaud, J.-M. Stutzmann, S. Mignani, *J. Med. Chem.* **1999**, *42*, 2828. (b) R. C. Young, R. C. Mitchell, T. H. Brown, C. R. Ganellin, R. Griffiths, M. Jones, K. K. Rana, D. Saunders, I. R. Smith, N. E. Sore, T. J. Wilks, *J. Med. Chem.* **1988**, *31*, 656.

90. (a) H. J. Lim, D. Myung, I. Y. Lee, M. H. Jung, *J. Comb. Chem.* **2008**, *10*, 501. (b) S. Mourtas, D. Gatos, K. Barlos, *Tetrahedron Lett.* **2001**, *42*, 2201.

91. F. Piscitelli, C. Ballatore, A. B. Smith, III, *Bioorg. Med. Chem. Lett.* **2010**, *20*, 644.

92. L. J. Wilson, *Org. Lett.* **2001**, *3*, 585.

93. N. Miyaura, A. Suzuki, *Chem. Rev.* **2002**, *95*, 2457.

94. A. R. Muci, S. L. Buchwald, *Top. Curr. Chem.* **2002**, *219*, 131.

95. (a) S. Lapenna, A. Giordano, *Nat. Rev. Drug Discovery* **2009**, *8*, 547. (b) J. T. Hartmann, M. Haap, H. G. Kopp, H. P. Lipp, *Curr. Drug Metab.* **2009**, *10*, 470. (c) R. Morphy, *J. Med. Chem.* **2010**, *4*, 1413.

96. (a) M. Anzini, M. Rovini, A. Cappelli, S. Vomero, F. Manetti, M. Botta, L. Sautebin, A. Rossi, C. Pergola, C. Ghelardini, M. Norcini, A. Giordani, F. Makovec, P. Anzellotti, P. Patrignani, M. Biava, *J. Med. Chem.* **2008**, *51*, 4476. (b) G. A. Fitzgerald, C. N. Patrono, *N. Engl. J. Med.* **2001**, *345*, 433. (c) M. Radi, C. Falciani, L. Contemori, E. Petricci, G. Maga, A. Samuele, S. Zanoli, M. Terrazas, M. Castria, A. Togninelli, J. A. Esté, I. Clotet-Codina, M. Armand-Ugón, M. Botta, *Chem. Med. Chem.* **2008**, *3*, 573. (d) C. E. Mowbray, C. Burt, R. Corbau, M. Perros, I. Tran, P. A. Stupple, R. Webster, A. Wood, *Bioorg. Med. Chem. Lett.* **2009**, *19*, 5599.

97. G. Pellegrino, F. Leonetti, A. Carotti, O. Nicolotti, L. Pisani, A. Stefanachi, M. Catto, *Tetrahedron Lett.* **2010**, *51*, 1702.

98. (a) C. N. Matthews, G. H. Birum, *Tetrahedron* **1966** *46*, 5707. (b) H. J. Bestmann, M. Schmidt, R. Schobert, *Synthesis* **1988**, 49. (b) O. Isler, H. Gutmann, M. Montavon, R. Rüegg, G. Ryser, P. Zeller, *Helv. Chim. Acta* **1957**, *40*, 1242. (c) H. J. Bestmann, D. Sandmeier, *Chem. Ber.* **1980**, *113*, 274. (d) H. J. Bestmann, D. Sandmeier, DE 2530329, **1975**; *Chem. Abstr.* **1977**, *87*, 6188. (e) C. N. Matthews, G. H. Birum,US Patent 3,459,804, 1969; Chem. Abstr. **1969**, 71, 815269. (f) C. N. Matthews, G. H. Birum,US Patent 3,655,766, **1972**; *Chem. Abstr.* **1972**, *77*, 19803.

99. (a) R. Schobert, S. Müller, H. J. Bestmann, *Synlett* **1995**, 425. (b) R. Schobert, J. Löffler, *J. Chem. Soc., Perkin Trans. 1* **1996**, 2799. (c) R. Schobert, S. Siegfried, M. Nieuwenhuyzen, W. Millius, F. Hampel, *J. Chem. Soc., Perkin Trans. 1* **2000**, 1723. (d) J. Löffler, R. Schobert, *Synlett* **1997**, 283. (e) R. Schobert, S. Siegfried, G. Gordon, *J. Chem. Soc., Perkin Trans. 1* **2001**, 2393. (f) H. J. Bestmann, *Angew. Chem., Int. Ed. Engl.* **1977**, *16*, 349; *Angew. Chem.* **1977**, *89*, 361. (g) J. Westmann, K. K. Orrling, *Comb. Chem. High Throughput Screening* **2002**, *7*, 571.

100. B. Henkel, *DE 102004014120A1,* 2004; *Chem. Abstr.* **2005**, *143*, 387168.

101. B. Henkel, *Synlett* **2008**, 355.

102. (a) Y. Okada, N. Miyauchi, K. Suzuki, T. Kobayashi, C. Tsutsui, K. Mayuzum, S. Nishibe, T. Okuyama, *Chem. Pharm. Bull.* **1995**, *43*, 1385. (b) R. G. Bell, J. A. Sadowski, J. T. Matschiner, *Biochemistry* **1972** *11*, 1959.

103. (a) J. J. Kulagowski, R. Baker, N. R. Curtis, P. D. Leeson, I. M. Mawer, A. M. Moseley, M. P. Ridgill, M. Rowley, I. Stansfield, A. C. Foster, S. Grimwood, R. G. Hill, J. A. Kemp, G. R. Marshall, K. L. Saywell, M. D. Tricklebank, *J. Med. Chem.* **1994**, *37*, 1402. (b) A. G. Chapmann, N. Dürmüller, B. L. Harrison, B. M. Baron, N. Parvez, B. S. Meldrum, *Eur. J. Pharmacol.* **1995**, *274*, 83. (c) R. W. Carling, P. D. Leeson, K. W. Moore, C. R. Moyes, M. Duncton, M. L. Hudson, R. Baker, A. C. Foster, S. Grimwood, J. A. Kemp, G. R. Marshall, M. D. Tricklebank, K. L. Saywell, *J. Med. Chem.* **1997**, *40*, 754. (d) E. Christopher, E. Bedir, C. Dunbar, I. A. Khan, C. O. Okunji, B. M. Schuster, C. O. Iwu, *Helv. Chim. Acta* **2003**, *86*, 2914. (e) C. Ito, M. Itoigawa, A.

Furukawa, T. Hirano, T. Murata, N. Kaneda, Y. Hisada, K. Okuda, H. Furukawa, *J. Nat. Prod.* **2004**, *67*, 1800.

104. For selected literature on the pharmacological properties of benzofurans, see (a) M. Halabalaki, N. Aligiannis, Z. Papoutsi, S. Mitakou, P. Moutsatsou, C. Sekeris, A.-L. Skaltsounis, *J. Nat. Prod.* **2000**, *63*, 1672. (b) E. Von Angerer, C. Biberger, S. Leitchtl, *Ann. NY Acad. Sci.* **1995**, *761*, 176. (c) C. C. Teo, O. L. Kon, K. Y. Sim, S. C. Ng, *J. Med. Chem.* **1992**, *35*, 1330. (d) R. R. Crenshaw, A. T. Jeffries, G. M. Luke, L. C. Cheney, G. Bialy, *J. Med. Chem.* **1971**, *14*, 1185. (e) G. A. Gfesser, R. Faghih, Y. L. Bennani, M. P. Curtis, T. A. Esbenshade, A. A. Hancock, M. D. Cowart, *Bioorg. Med. Chem. Lett.* **2005**, *15*, 2559. (f) M. Cowart, J. K. Pratt, A. O. Stewart, Y. L. Bennani, T. A. Esbenshade, A. A. Hancock, *Bioorg. Med. Chem. Lett.* **2004**, *14*, 689. (g) C. Hocke, O. Prante, S. Lober, H. Hubener, P. Gmeiner, T. Kuwert, *Bioorg. Med. Chem. Lett.* **2004**, *14*, 3963. (h) Y. Hu, J. S. Xiang, M. J. Di Grandi, X. Du, M. Ipek, L. M. Laakso, J. Li, W. Li, T. S. Rush, J. Schmid, J. S. Skotnicki, S. Tam, J. R. Thomason, Q. Wang, J. I. Levin, *Bioorg. Med. Chem.* **2005**, *13*, 6629. (i) M. W. Khan, M. J. Alam, M. A. Rashid, R. Chowdhury, *Bioorg. Med. Chem.* **2005**, *13*, 4796. (j) K. Kawasaki, M. Masubuchi, K. Morikami, S. Sogabe, T. Aoyama, H. Ebiike, S. Niizuma, M. Hayase, T. Fujii, K. Sakata, H. Shidoh, Y. Stiratori, Y. Aoki, T. Ohtsuka, N. Stimma, *Bioorg. Med. Chem. Lett.* **2003**, *13*, 87.

105. L. De Luca, G. Giacomelli, G. Nieddu, *J. Comb. Chem.* **2008**, *10*, 517.

106. (a) L. M. Toledo, N. B. Lydon, D. Elbaum, *Curr. Med. Chem.* **1999**, *6*, 775. (b) M. Legraverend, O. Ludwig, E. Bisagni, S. Leclerc, L. Meijer, *Bioorg. Med. Chem. Lett.* **1998**, *8*, 793.

107. (a) J. Liu, Q. Dang. Z. Wei, F. Shi, X. Bai, *J. Comb. Chem.* **2006**, *8*, 410. (b) C. Gaulon, H. Dijkstra, C. Springer, *Synthesis* **2005**, *13*, 2227. (c) H. Fu, Y. Lam, *J. Comb. Chem.* **2006**, *7*, 734.

108. (a) L. Naesens, L. Lenaerts, G. Andrei, R. Snoeck, D. Van Beers, A. Holy, J. Balzarini, E. De Clercq, *Antimicrob. Agents Chemother.* **2005**, *49*, 1010. (b) J. Neyts, J. Balzarini, G. Andrei, C. Zhu, R. Snoeck, A. Zimmermann, T. Mertens, A. Karlssen, E. De Clercq, *Mol. Pharmacol.* **1998**, *53*, 157. (c) G. H. Hakimelahi, T. W. Ly, A. A. Moosavi-Movahedi, M. L. Jain, M. Zakerinia, H. Davari, H.-C. Mei, T. Sambaiah, A. A. Moshfegh, S. Hakimelahi, *J. Med. Chem.* **2001**, *44*, 3710. (d) G. Jähne, H. Kroha, A. Müller, M. Helsberg, I. Winkler, G. Gross, T. Scholl, *Angew. Chem., Int. Ed. Engl.* **1994**, *33*, 562.

109. M. C. Núñez, M. G. Pavani, M. Díaz-Gavilán, F. Rodríguez-Serrano, J. A. Gómez-Vidal, J. A. Marchal, A. Aránega, M. A. Gallo, A. Espinosa, J. M. Campos, *Tetrahedron* **2006**, *62*, 11724.

110. A. Bargiotti, A. Ermoli, M. Menichincheri, E. Vanotti, L. Bonomini, A. Fretta,US Patent 138212, **2004**.

111. H. Fu, M. K. C. Tan, Y. Lam, *J. Comb. Chem.* **2007**, *9*, 804.

112. (a) T. M. Sielecki-Dzurdz, A. G. Arvanitis, C. D. Dzierba,PCT-US 0330570; WO 04031189, **2002**; *Chem. Abstr.* **2004**, *40*, 321385. (b) A. Arvanitis, P. Gilligan, R. Hartz,PCT-US 02: 15493; WO 02:092090, **2002**; *Chem. Abstr.* **2002**, *137*, 370110.

113. P. A. J. Janssen, K. J. A. Van Aken, P. J. Lewi, L. M. H. Koymans, M. R. De Jonge, J. Heeres, F. F. D. Daeyaert, G. J. C. Hoornaert, F. J. C. Compernolle, A. Kilonda, *PCT Int. Appl.* WO 02078708, **2002**; *Chem. Abstr.* **2002**, *137*, 294978.

114. M. Yaso, Y. Suzuki, K. Shibata, E. Hayashi,Jpn. Kokai Tokkyo Koho JP 62198671, **1987**;*Chem. Abstr.* **1988**, *108*, 167500.

115. P. K. Loosen, M. F. Khorasani, S. M. Toppet, G. J. Hoornaert, *Tetrahedron* **1991**, *47*, 9269.

116. W. M. De Borggraeve, F. J. R. Rombouts, E. Van der Eycken, S. M. Toppet, G. J. Hoornaert, *Tetrahedron Lett.* **2001**, *42*, 4693.

117. N. Kaval, W. Dehaen, E. Van der Eycken, *J. Comb. Chem.* **2005**, *7*, 90.

118. (a) H. M. Taylor, C. R. Hauser, *Org. Synth., Coll.* **1973**, *5*, 437. (b) B. B. Corson, R. A. Dodge, S. A. Harris, J. S. Yeaw, *Org. Synth., Coll.* **1932**, *1*, 336.

119. P. Moser, A. Sallman, I. Wiesenberg, *J. Med. Chem.* **1990**, *33*, 2358.

9

SOLID-PHASE SYNTHESIS OF HETEROCYCLES FROM PEPTIDES AND AMINO ACIDS

Zhi Li, Marc Giulianotti, Wenteng Chen, Richard A. Houghten, and Yongping Yu

9.1 INTRODUCTION

The solid-phase synthesis of peptides was first described by Merrifield in 1963.[1] Since that time, many advances have been made utilizing Merrifield's core technology. For example, peptidomimetics consisting of unnatural amino acid or other dipolar chemicals have been synthesized on the solid support. These peptides and peptidomimetics have further been modified through amide bond alkylation or reduction generating peralkylated peptidomimetics or polyamines, respectively.[2] In addition, resin-bound peptides and peptidomimetics have been designed in a manner that conveniently affords small molecule or heterocycle libraries with huge molecular diversity.[3] In this manner, these various classes of compounds are all derived from previously established ones. This concept of deriving new compounds from previously established ones enables researchers to expand the synthetic diversity of existing libraries by employing different synthetic methodologies to existing compounds. In this chapter, we focus our attention on heterocyclic libraries derived from solid-phase peptides and peptidomimetics.

9.2 SYNTHESIS OF VARIOUS HETEROCYCLES

9.2.1 Three-Membered Ring Heterocycles

The principles of the Gabriel–Cromwell reaction were applied to a solid-phase synthesis by Filigheddu's group (**Scheme 9.1**).[4] Compound **1** was prepared from Wang resin and

Solid-Phase Organic Synthesis: Concepts, Strategies, and Applications, First Edition.
Edited by Patrick H. Toy and Yulin Lam.

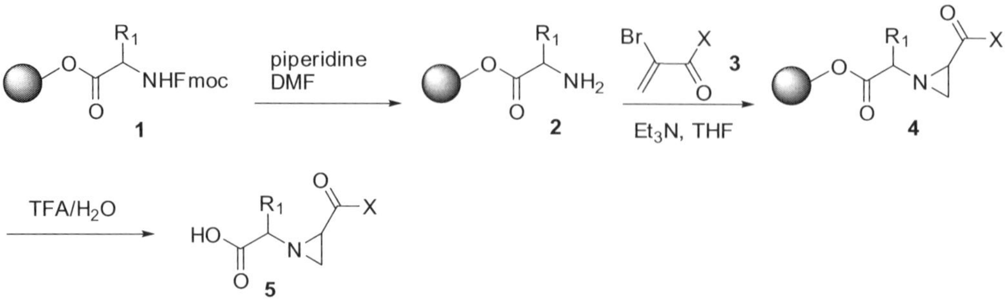

X = O-*t*-Bu, OH, NH-Leu-OMe, NH-Val-Gly-OMe

Scheme 9.1. Synthesis of an aziridine core.

N-Fmoc-amino acids. After removal of the Fmoc group with piperidine in DMF (*N,N*-dimethylformamide), the resin was washed several times with DMF to afford the resin-bound amino acid **2**. The amine group of **2** underwent cyclization with synthon **3** in the presence of triethylamine in THF overnight at room temperature. The resin-bound aziridine (**4**) was treated with TFA:H_2O (95:5) to give the final product **5**.

9.2.2 Four-Membered Ring Heterocycles

In the attempt to obtain β-lactams through solid-phase approaches, there are two main reported strategies. One utilizes an imine group as the key component. In this approach, the imine is acylated to generate a zwitterionic intermediate before the final cyclization step. The other strategy uses 2-chloroacetyl chloride or 2-chloroacetic acid to form an amide group before the ring closure. Notably, thionation of the acetyl group could be achieved by the use of Lawesson's reagent.

Preloaded Fmoc-glycine Wang resin **6** was treated with 30% piperidine in DMF and then coupled with an R_1-substituted aldehyde (**7**) in 1% acetic acid in DMF to afford an imine group **8**. With the presence of triethylamine, the R_2-substituted acetyl chloride **9** formed a ketene structure that underwent a nucleophilic attack by imine **8**, giving a zwitterionic intermediate **10** before the formation of β-lactam ring **11**. Cleavage from the resin was carried out with 10% trifluoroacetic acid in dichloromethane (DCM), followed by esterification with diazomethane, affording the β-lactam product **12** (**Scheme 9.2**).[5] Similar approaches to obtain β-lactams starting from resin-bound amino acids were reported by Delpiccolo and coworkers.[6,7]

β-Thiolactams can be obtained by employing Lawesson's reagent in the solid-phase synthesis process.[8] Starting from resin-bound Fmoc-amino acids **13** and using the synthetic

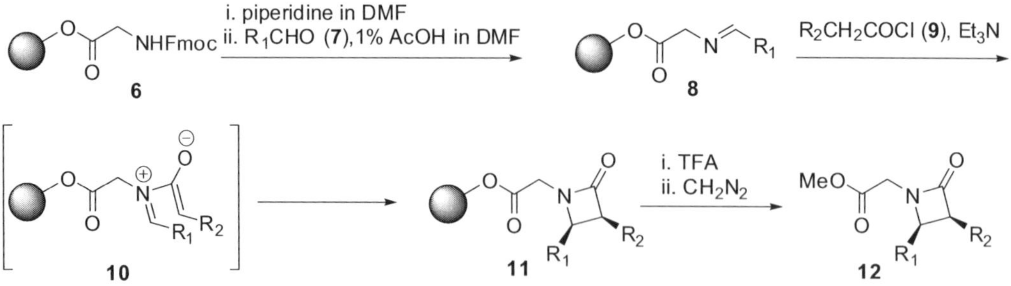

Scheme 9.2. Synthesis of beta-lactams.

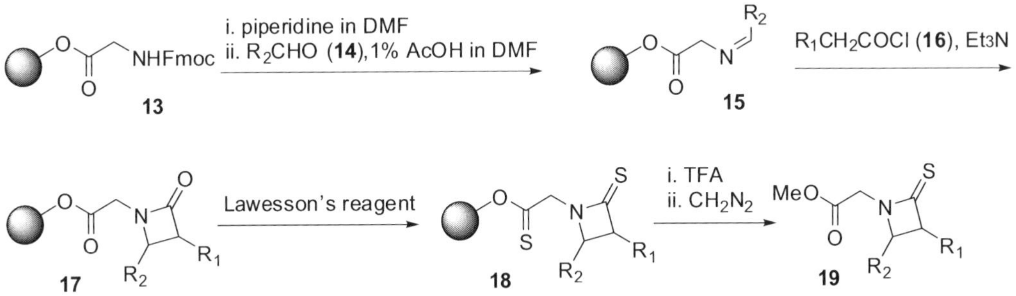

Scheme 9.3. Thionation by Lawesson's reagent.

route described above, polymer-supported β-lactam **17** was stirred at 95°C with 1.5 equiv Lawesson's reagent for 2 h, which afforded the β-thiolactam product **18** (**Scheme 9.3**).

The second common approach for the solid-phase synthesis of β-lactams employs 2-chloroacetyl chloride or 2-chloroacetic acid as the key synthons (Scheme 9.4).[9,10] Starting from Fmoc-Phe loaded Wang resin **22**, and after deprotection of the amine group with piperidine, treatment of 6 equiv *p*-methoxybenzaldehyde in $(MeO)_3CH$ afforded the corresponding imine group that was reduced by $NaBH_3CN$ to give the resin-bound amine product **31**. Then 2-chloroacetyl chloride **28** was coupled to **31** in the presence of propylene oxide in DMF, producing the key intermediate **32**. Cyclization was achieved with BTPP (*tert*-butylimino-tri(pyrrolidino)phosphorane) or BEMP (2-*tert*-butylimino-2-diethylamino-1,3-dimethylperhydro-1,3,2-diazaphosphorine) serving as the base in NMP (*N*-methylpyrrolidone) to give the resin-bound β-lactam **33**. Cleavage from the resin was carried out by exposure to TFA:H_2O (19:1) (**34**).

An alternative method to form an amine bond at the N-terminal of resin-bound Phe was reported by González-Muñiz and coworkers (**Scheme 9.4**).[9] Starting with Fmoc-Phe-HMPB-Leu-MBHA-PS resin, the Fmoc group is removed followed by protection of a nosyl group to afford **25**. The product then underwent alkylation with benzyl bromide to give **26**. After removal of the nosyl group (**27**) and condensation with 2-chloroacetyl chloride **28**, the resulting compound **29** underwent cyclization, producing the desired resin-bound β-lactam **30**.

9.2.3 Synthesis of Five-Membered Ring Heterocycles

9.2.3.1 Five-Membered Ring Heterocycles Containing One Nitrogen Atom. To synthesize pyrrolidine-2-ones, amine **36** was attached to FDMP (2-(3,5-dimethoxy-4-formylphenoxy)ethoxymethyl polystyrene) resin **35** by employing sodium triacetoxyborohydride in the presence of acetic acid in DCM at room temperature overnight (**Scheme 9.5**).[11] The resulting resin-bound amine **37** was shaken at room temperature overnight in a solution containing Fmoc-amino acid **38**, HOBt (*N*-hydroxybenzotriazole), and DIC (1,3-diisopropylcarbodiimide) to afford the corresponding compound **39**. After removal of Fmoc group with 20% piperidine in DMF, the resin-bound amino acid **40** was treated with 16 equiv of monomethyl itaconate **41** in MeOH:toluene (4:1) at 60°C for 3 days to give the pyrrolidinone methyl ester **42**. After removal of methyl group with KOH in dioxane:H_2O (3:1), the amine **44** was added to a mixture of resin-bound pyrrolidinone **43**, HOBt, and DIC in DMF and shaken at room temperature overnight, resulting in compound **45**. Subsequent cleavage from the resin was carried out in 10% TFA in DCM to give the final product **46**.[11]

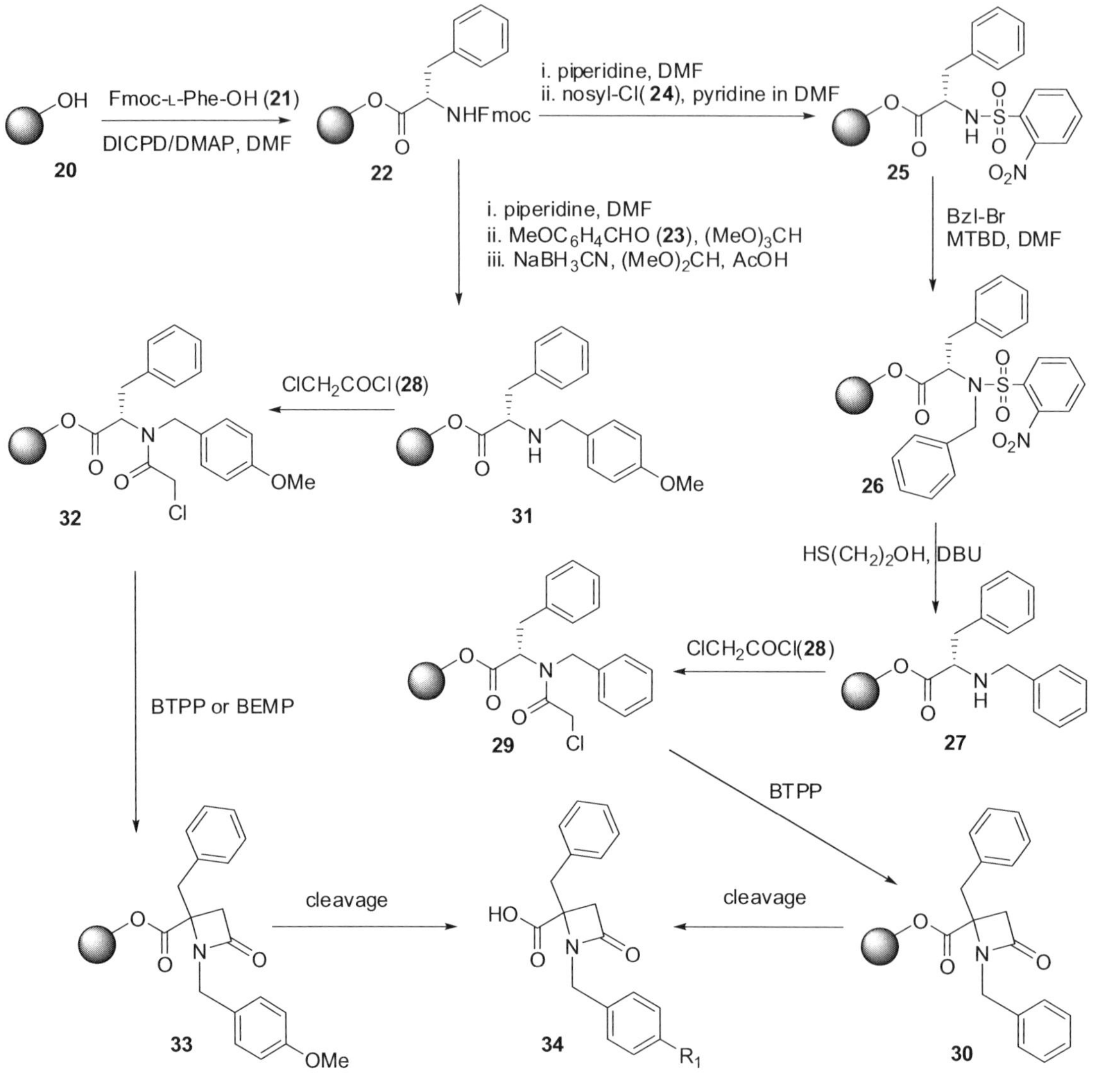

Scheme 9.4. A second method to obtain β-lactams.

An alternative method to produce β-lactam involves the use of a Dieckmann-like reaction. This solid-phase synthesis was developed by the Ganesan and coworkers (**Scheme 9.6**).[12] Amino acid loaded Wang resin was treated with an aldehyde and sodium triacetoxyborohydride in DCM to generate the *N*-alkylated resin-bound amino acid **47**. A mixture of compound **47**, HOBt, R_3-substituted acetic acid **48** in DCM was prepared and cooled to 0°C, followed by the slow addition of diisopropylcarbodiimide. The reaction was then allowed to continue for 18 h after warming to room temperature, affording the key intermediate **49**. Cyclization and cleavage from the resin was carried out in a single step by mixing THF, **49** and tetrabutylammonium hydroxide or lithium bis(trimethylsilyl)amide in

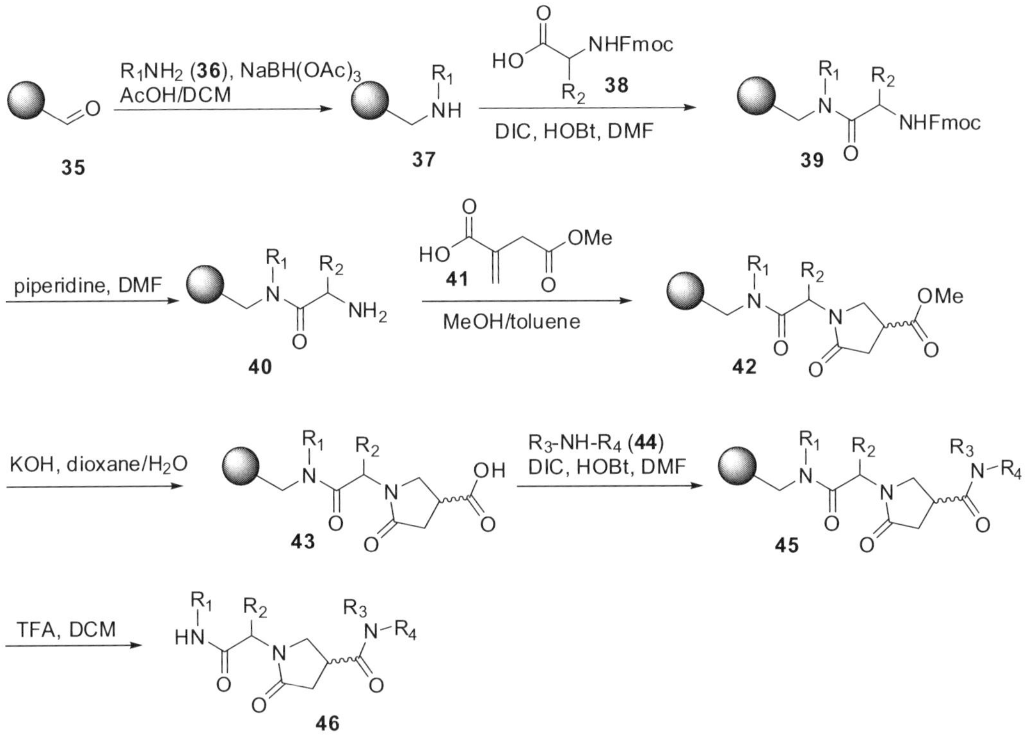

Scheme 9.5. Synthesis of pyrrolidine-2-ones.

methanol, resulting in the tetramic acid product **50**. Similar work employing the Dieckmann condensation was also reported by Duffy and coworkers.[13]

Oxazoles are prepared from tryptophan loaded Wang resin **53** (**Scheme 9.7**).[14] The deprotected N-terminal was condensed with carboxylic acids **54** or carboxylic acid anhydrides **55** to give *N*-acetyltryptophans **56**. The key step involved oxidation of **56** with 2 equiv of DDQ (dichlorodicyanoquinone) in THF:H_2O (9:1) at room temperature for 15 min, producing the key intermediate **57**. Compound **57** underwent cyclocondensation in the presence of triethylamine, CCl_4, and triphenylphosphine in acetonitrile at room temperature for 2 h to afford the oxazole **58**. After cleavage from the resin with 20% TFA in DCM, esterification was carried out using TMS diazomethane to give the final product **59**.

Using a similar strategy, oxazoles and thiazoles were prepared starting from dipeptide-bound Wang resin **60** (**Scheme 9.8**).[15] The hydroxyl group on the threonine residue was oxidized with 3 equiv of Dess–Martin periodinane to give the corresponding β-ketoamide intermediate **61**. Treatment of **61** with triphenylphosphine in the presence of iodine and diisopropylethylamine (DIEA) in DCM for 12 h afforded the resin-bound oxazoles **62**,

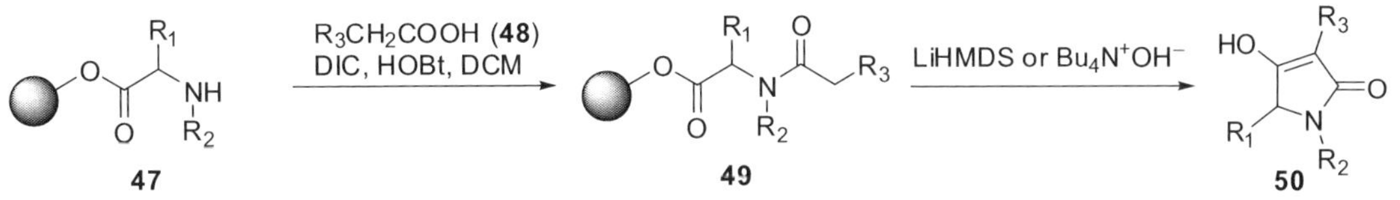

Scheme 9.6. Dieckmann-like condensation on solid support.

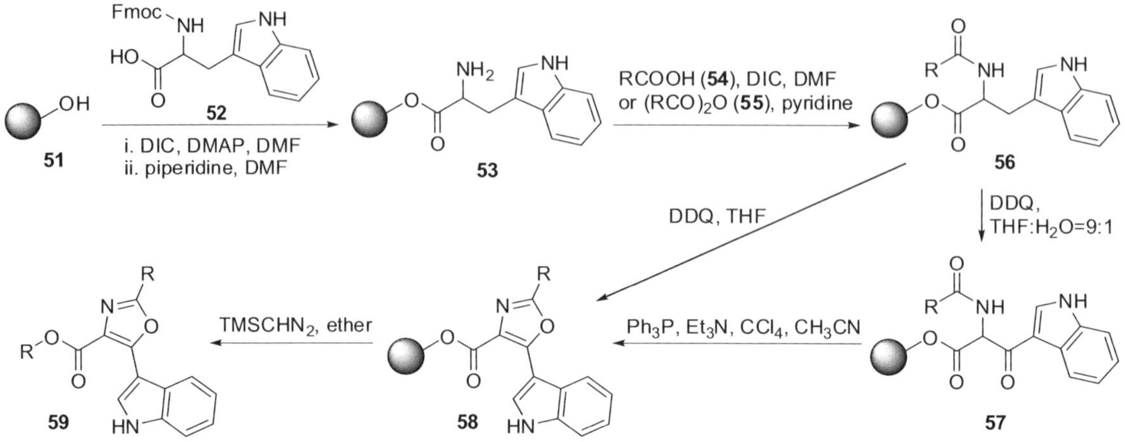

Scheme 9.7. Synthesis of oxazoles.

whereas excess of Lawesson's reagent reacted with **61** at 55°C for 5 h resulting in a thiazole core **63**. Cleavage from the resin in both cases was achieved by using trifluoroacetic acid.

The synthesis of an isoxazole requires the use of a special Boc-amino acid **67** prepared by Kurth and Park (**Scheme 9.9**).[16] Starting from Merrifield resin **65**, 1,3-dihydroxypropane **64** was attached as a linker. The resin-bound hydroxyl compound **66** was coupled with **67** to form the key intermediate **68**. The propyne group of **68** is prone to nucleophilic reactions. Subsequent cycloaddition with the nitryl compound **69** afforded the isoxazole core **70**. Further treatment with TFA in DCM removed the Boc group, providing the free amine of **71** that was then treated with aldehyde **72** in the presence of trimethyl orthoformate (TMOF) in THF and reduced with $NaBH_3CN$ in THF to complete the alkylation (**73**). This reductive alkylation of an amine group is very useful and will be covered many times in the following sections. Compound **73** was treated with isocyanate **74** in THF affording the urea ester intermediate **75**, which under heating at 60°C resulted in the final product **76**, containing an isoxazole and a hydantoin ring.

9.2.3.2 Five-Membered Ring Heterocycles Containing Two Nitrogen Atoms

Synthesis of Hydantoin Cores from Urea/Thiourea-Like Intermediates. There are many reports on the solid-phase synthesis of five-atom heterocycles having two nitrogen

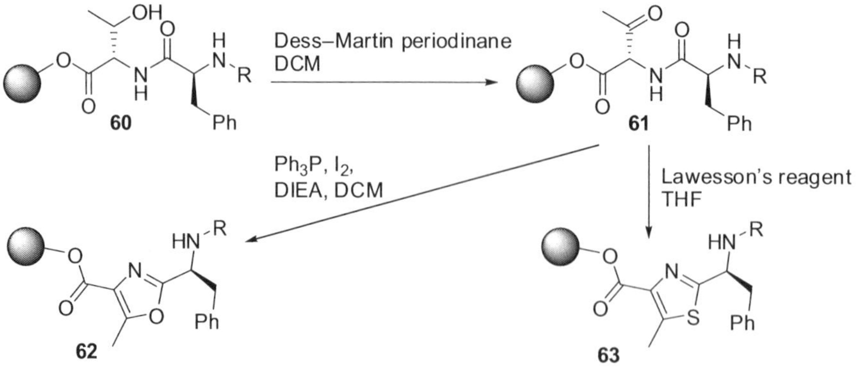

Scheme 9.8. Synthesis of oxazoles and thiazoles.

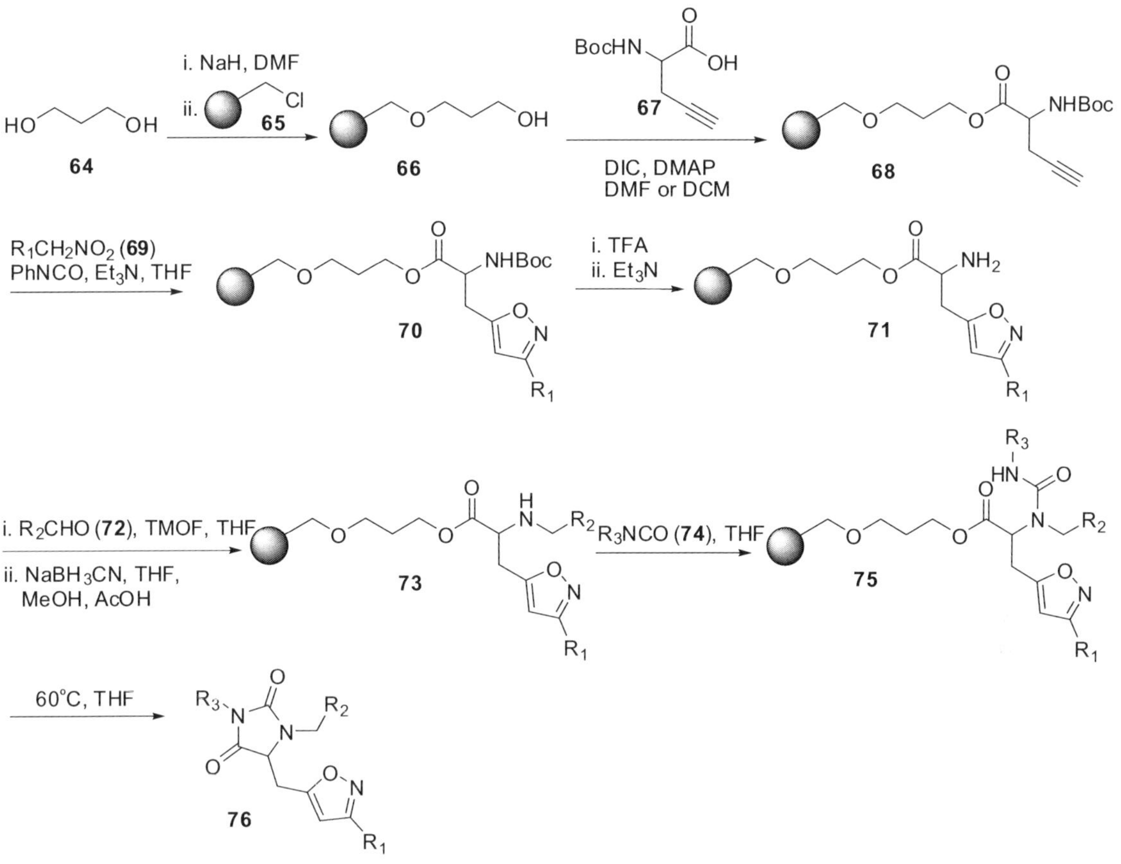

Scheme 9.9. Synthesis of isoxazoles.

atoms in the core. The most commonly reported are hydantoins. Generally, there are two main approaches to obtain a hydantoin core starting from resin-bound peptides. One method is to treat the N-terminal with isocyanates (R-NCO) or isothiocyanates (R-NCS) to form the corresponding ureas or thioureas, respectively. Cyclization can be achieved usually under basic conditions. The other approach to synthesize hydantoins is to treat resin-bound dipeptides directly with electrophilic reagents to generate isocyanate-like intermediates, which ultimately undergo cyclocondensation.

A typical way to synthesize a hydantoin core starting from a resin-bound amino acid is shown in **Scheme 9.10**. Fmoc-amino acid loaded polystyrene resin **77** was deprotected in 20% piperidine in NMP, which then underwent a reductive alkylation process in an NMP solution of 6 equiv of isocyanate (**78**) and 1 equiv of DIEA for 3 h, affording the urea derivatives **79**. Cyclocleavage of **79** was carried out in 10% triethylamine in methanol, shaking for 3 h, to give the desired hydantoin ring **80**.[17]

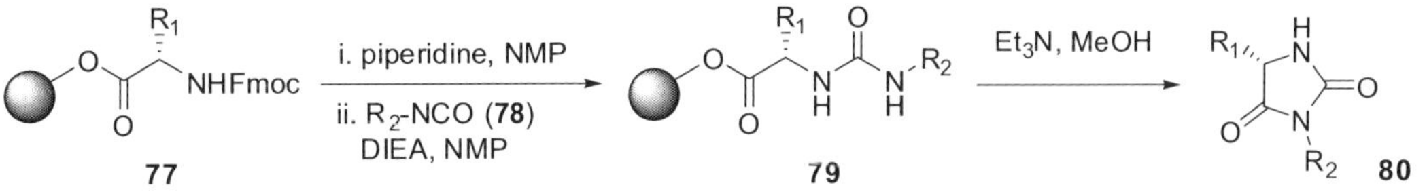

Scheme 9.10. Synthesis of hydantoins.

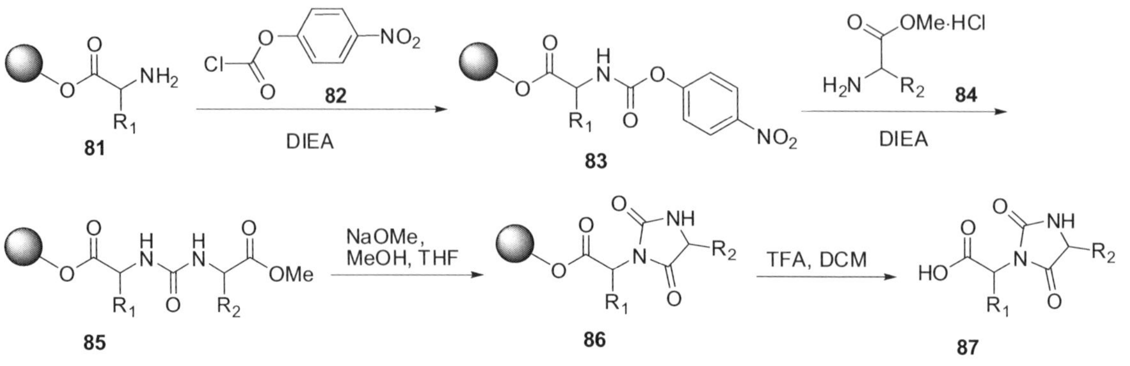

Scheme 9.11. Jung's synthesis of hydantoins.

Using the same strategy, but different reagents and cleavage position (**Scheme 9.11**), Jung and coworkers employed trityl-linked polystyrene–1% divinylbenzene resin to tether an amino acid (**81**).[18] Acylation of the N-terminal of **81** was achieved by using 10 equiv of *p*-nitrophenyl chloroformate **82** and 10 equiv of DIEA in DMF, affording intermediate **83**. Compound **83** reacted with 10 equiv of amino acid ester **84** and 10 equiv DIEA in DMF for 16 h to form the urea intermediate **85**. Cyclization was facilitated by using 10 equiv of sodium methoxide in THF:methanol (1:1) for 3 h (**86**). For cleavage from the resin, 10% TFA in DCM was used, affording the hydantoin peptide derivative **87**.

Alternatively, Boc-amino acid **89** was attached to PS-PO resin **88** under typical coupling conditions to form compound **90** (**Scheme 9.12**).[19] After deprotection of the amine group using TFA in DCM, the resin-bound amino acid **91** was condensed with 6 equiv ketimine **92** in the presence of acetic anhydride and *N*-ethyl-*N,N*-diisopropylamine in DCM to give the imine intermediate **93**, which underwent reduction by $NaBH_3CN$ and acetic acid in DMA (*N,N*-dimethylacetamide) resulting in **94**. Afterward, 10 equiv of isocyanate **95** was used to form the urea ester intermediate **96**. Cyclocleavage was achieved by using a basic suspension of isopropylamine, giving the desired hydantoin core **97**.

Cellulose membranes can also be used to synthesize hydantoins (**Scheme 9.13**). Wenschuh and coworkers used amino-derivatized cellulose membranes as solid support and Rink linker or photolinker to bind subsequent synthons.[20] Support **98** was initially treated with $BrCH_2COODnp$ (2,4-dinitrophenyl) **99** and then with 5 equiv of *n*-butylamine **100** to form the peptide-like compound **101**. This was again treated with **99** and then amino acid **102**, giving a dipeptide intermediate **103**. Three equivalents of *N*-benzylimidazole was used as a catalyst and added to the NMP suspension of **103** 15 min before the addition of 1 equiv of isocyanate **104**, giving the key intermediate **105**. Cyclization and cleavage were carried out in a single step producing **106**. For Rink linker, 95% TFA in H_2O at 60°C was used. For photolinker, cyclization was achieved with 95% TFA in H_2O at 60°C, followed by photocleavage with 365 nm UV light.

Microwave reactions can also be used in the formation of hydantoin scaffolds (**Scheme 9.14**).[21] After deprotection of the N-terminal of the resin-bound amino acid **107** with piperidine in DCM, 4 equiv of isocyanate **109** was reacted with **108** in THF in a microwave reactor at 60°C for 10 min, generating the urea ester intermediate **110**. Cyclocleavage was achieved by using triethylamine in THF:DMF (4:1) at 110°C with the microwave for 15 min, affording the hydantoin core **111**.

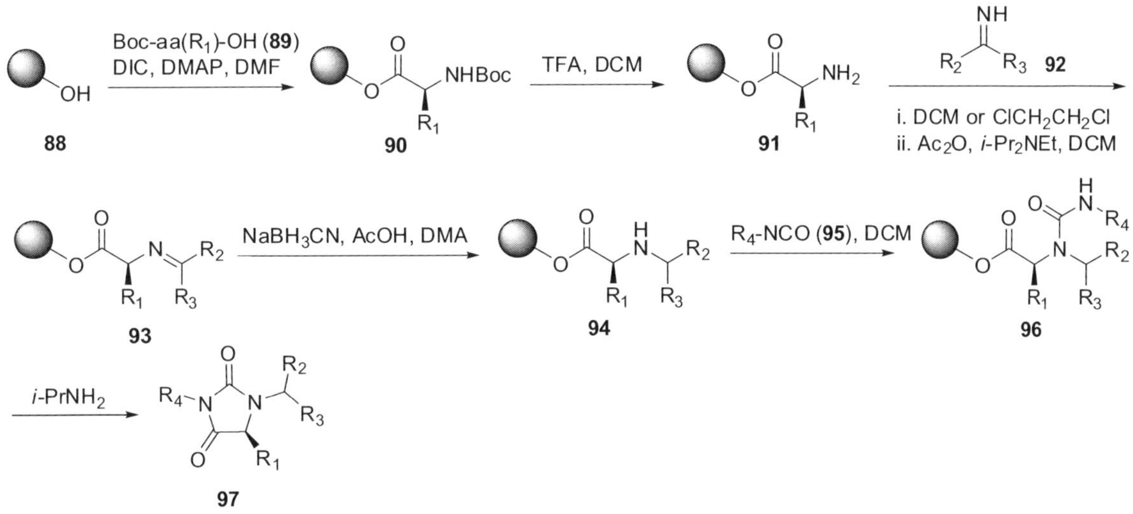

Scheme 9.12. Another route to hydantoins.

Another example of using the microwave in hydantoin scaffold synthesis was reported by Sun and Lin.[22] Polymer HO-PEG-OH **112** was employed as the solid support (**Scheme 9.15**). Fmoc-amino acid **113** loading was carried out in a DCM suspension in the presence of DCC (*N*,*N*-dicyclohexylcarbodiimide) and DMAP (4-(*N*,*N*-dimethylamino) pyridine), and using 150 W microwave for 14 min, affording compound **114** that underwent deprotection in 10% piperidine in DCM to give the polymer-bound diamine intermediate **115**. Treatment of isothiocyanate **116** in DCM with 150 W microwave heating for 7 min afforded the thiourea ester intermediate **117**. Cyclization and cleavage from the polymer was achieved with K_2CO_3 in DCM in a 150 W microwave heating for 7 min generating **118**. Notably, this method does not have a conformational impact on the chiral characteristics of the original amino acids.

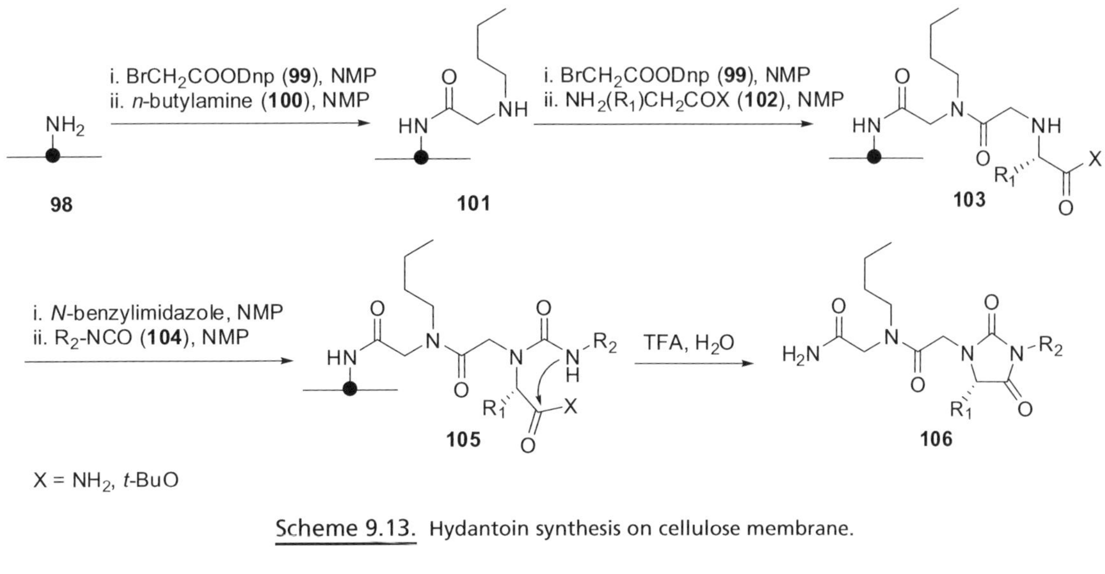

Scheme 9.13. Hydantoin synthesis on cellulose membrane.

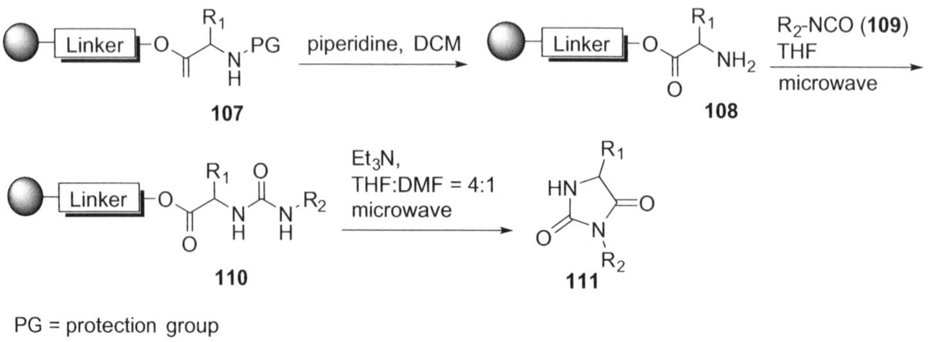

Scheme 9.14. Microwave heating in hydantoin synthesis.

Synthesis of Hydantoin Cores Using Electrophilic Reagents. Starting from Fmoc-amino acid loaded Wang resin **119**, after a deprotection treatment of 20% piperidine in DMF, Fmoc-amino acid **120** was coupled to the free amine to form dipeptide **121** (**Scheme 9.16**).[23] After another deprotection procedure, diphosgene (20% in toluene) and DIEA were used to generate the key isocyanate intermediate **122**. Conversion of **122** into the hydantoin core **123** was achieved by heating the anhydrous toluene suspension to 70°C for 6 h. Cleavage from resin was facilitated by stirring in TFA:DCM:H_2O (90:5:5) for 1 h, affording the desired hydantoin **124**. Thiohydantoins can be obtained by using thiophosgene in the step of disphosgene. A similar synthetic strategy was reported by Houghten and coworkers, with some alterations in resin and reagents.[2]

Oxime resin **125** in DMF was treated with 5 equiv of Boc-amino acid **126** and 5 equiv of DCC in DCM solution prepared at 0°C (**Scheme 9.17**)[24] and then shaking the mixture at room temperature for 24 h afforded resin-bound Boc-amino acid **127**. After removal of the Boc group with 25% TFA in DCM for 30 min, the deprotected amino acid was swollen in anhydrous DCM and treated with 4 equiv of DIEA and *t*-butyl chlorosulfonylcarbamate that was prepared by adding *t*-butyl alcohol in DCM to chlorosulfonyl isocyanate (CSI) **128** in DCM at 0°C. Treatment of **129** with 25% TFA in DCM for deprotection followed by

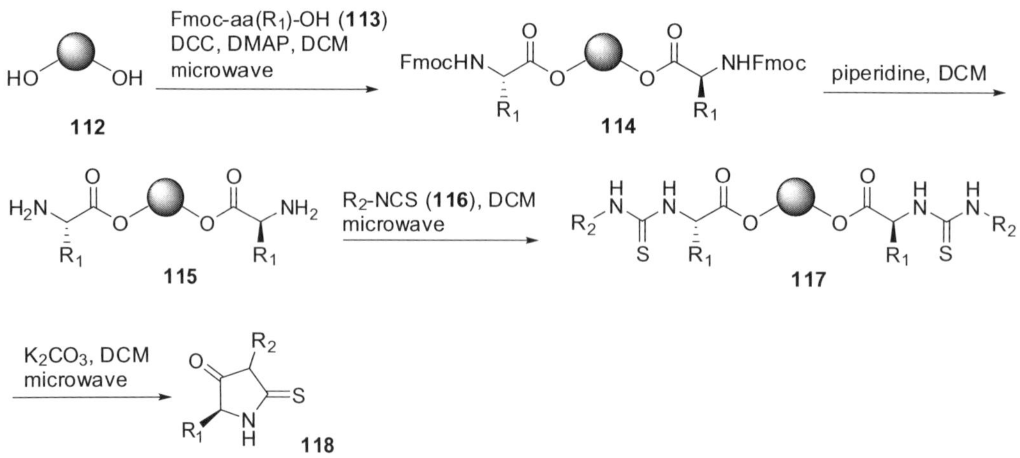

Scheme 9.15. Synthesis of thiohydantoins.

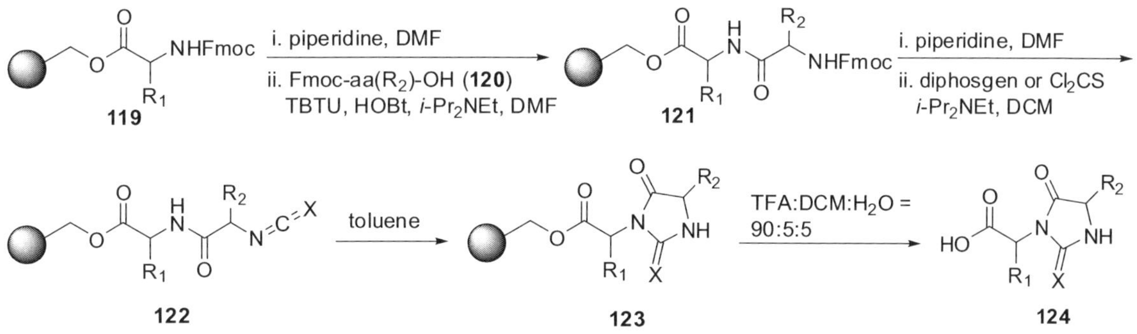

P = Fmoc, H
X = O, S

Scheme 9.16. Synthesis of hydantoins using electrophilic reagents.

treatment with 20 equiv triethylamine in anhydrous DCM under N_2 for 3 h facilitated the sulfahydantoin **132** ring closure. Treatment of **128** with Mitsunobu reaction reagents afforded an alkylated *N*-Boc sulfamide **130** that after deprotection and cyclization provided the substituted sulfahydantoin **131**. Similar work was reported by Tice and coworkers.[25,26]

SYNTHESIS OF IMIDAZOLONES. In general, the synthesis of imadazolones is achieved by the treatment of resin-bound amino acids with isocyanates or isothiocyanates to form the corresponding ureas and thioureas that then undergo an oxidation reaction resulting in a carbodiimide intermediate. Mukaiyama's reagent or other oxidative reagents are employed. The carbodiimide intermediate is treated with primary or secondary amines to form the imidazolone scaffolds.

To a DMF suspension of aminomethyl resin **133**, 3 equiv of Boc-amino acid **134** was added together with 3 equiv of DIC, 3 equiv of HOBt, and 3 equiv of DMAP in DMF (**Scheme 9.18**).[27] The mixture was then shaken at room temperature for 15 h to afford the resin-bound compound **135**. After removal of the Boc group with 30% TFA in DCM, the resin was neutralized with 10% triethylamine in DCM. Then addition of 5 equiv of

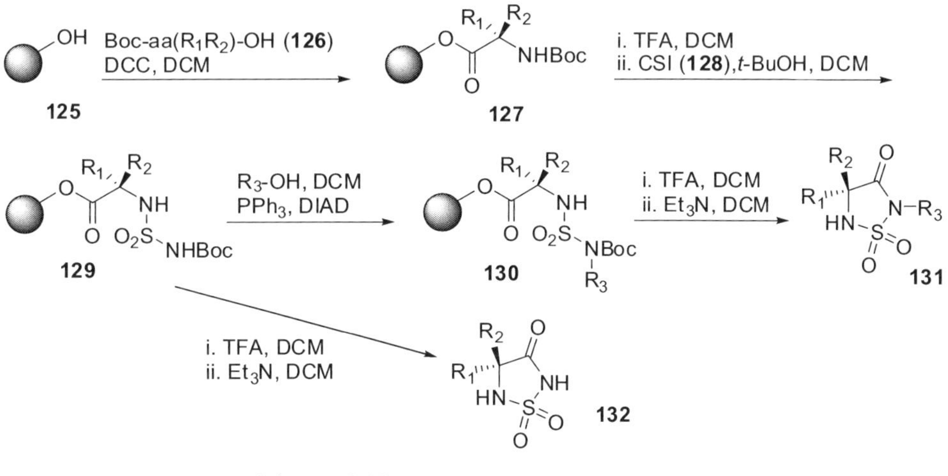

Scheme 9.17. Synthesis of sulfahydantoins.

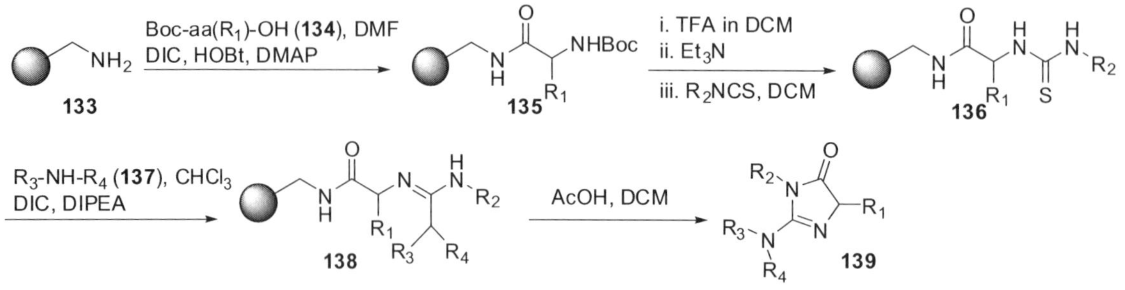

Scheme 9.18. Synthesis of imidazolones.

isothiocyanate in DCM and shaking for 15 h gave the thiourea ester intermediate **136**. Compound **136** was then treated with $CHCl_3$, 5 equiv of secondary amine **137**, 5 equiv of DIC, and 5 equiv of DIPEA and shaken at 50°C for 2 days to obtain the guanidine derivative **138**. Cyclocleavage was achieved by using 10% acetic acid in DCM at room temperature overnight, giving the imidazolone core **139**.

MBHA (4-methylbenzhydrylamine) resin **140** was treated with 6 equiv of Boc-amino acid **141** in the presence of 6 equiv of both DIC and HOBt in anhydrous DMF for 2 h, followed by treatment with 55% TFA in DCM for 30 min, to give resin-bound amino acid **142** (**Scheme 9.19**).[28] Treatment of **142** with 6 equiv of isothiocyanate **143** afforded thiourea **144**, which was reacted with 6 equiv of Mukaiyama's reagent and 6 equiv of triethylamine in DCM to form the carbodiimide intermediate **145**. After an overnight reaction, compound **146** was obtained, which underwent cleavage from the resin with anhydrous HF in the presence of anisole at 0°C for 1.5 h, affording the imidazolone product **149**. Resin-bound **146** can be further alkylated by adding 1 M lithium *t*-butoxide (10 equiv) in THF and shaking for 30 min. After removal of excessive base by decantation, 5 equiv R_3X was added and shaken for 4 h to complete the alkylation **147**. Alkyl-substituted product **148** was obtained using the same HF cleavage method. Lange and Udo also reported a similar

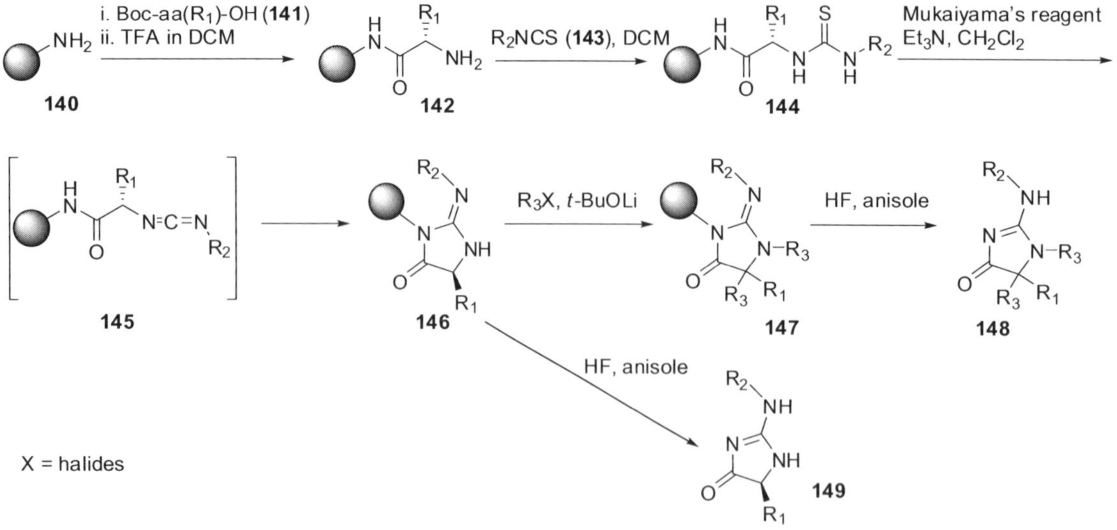

Scheme 9.19. Method 2 for synthesis of imidazolones.

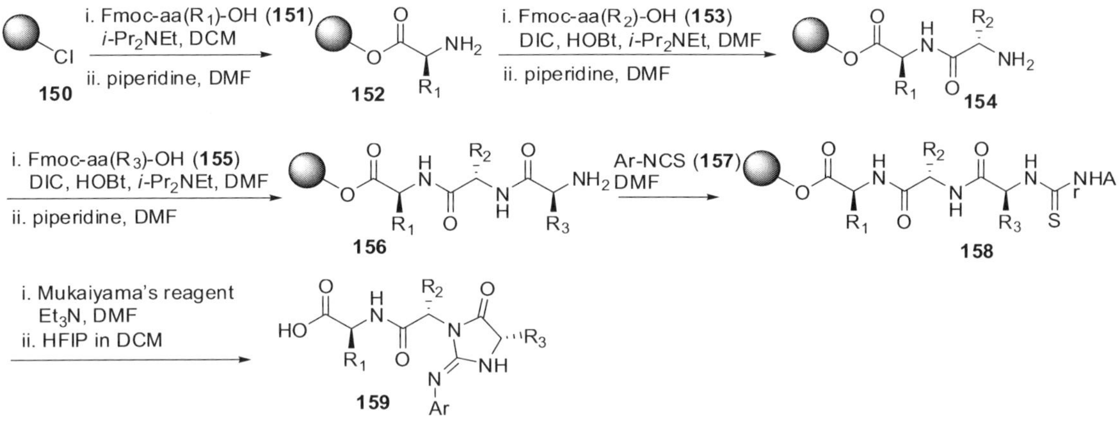

Scheme 9.20. Method 3 for synthesis of imidazolones.

strategy to obtain imidazolones, using isocyanates as the urea generating reagent and Burgess reagent to form the carbodiimide intermediate.[29]

Starting from polystyrene-supported 2-chlorotrityl chloride resin **150**, a solution of 3 equiv of Fmoc-amino acid **151** and 5 equiv of Hünig's base in anhydrous DCM was added (**Scheme 9.20**).[30] The solution was shaken at room temperature for 3 h to produce the resin-bound Fmoc amino acid, which was deprotected with 20% piperidine in DMF for 30 min to give amino acid **152**. Coupling of amino acids **153** and **155** was achieved using the same method: 8 equiv of Hünig's base, 3 equiv of amino acid (**153** or **155**), 3 equiv of HOBt, and 3 equiv of DIC were added into the DMF suspension of **152** or **154** to obtain the tripeptide intermediate **156**. After removal of the Fmoc group, 5 equiv of isothiocyanate **157** were added to the resin suspension of **156** in DMF and shaken at room temperature for 20 min, resulting in the thiourea intermediate **158**, which was then treated with 10 equiv of Mukaiyama's reagent and 10 equiv of triethylamine in DMF to form the imidazolone (or iminohydantoin) core. After cleavage from the resin using HFIP (hexafluoroisopropanol): DCM (1:4), the final product **159** was obtained.

Starting from MBHA resin **160**, coupling with Boc-amino acid **161** and deprotection with 55% TFA in DCM afforded the resin-bound amino acid **162** (**Scheme 9.21**).[31] The thiourea derivative **164** was synthesized using thioisocyanate **163**. Yu et al. describe a method to obtain guanidine **166** from resin-bound thiourea **164** using $HgCl_2$ and amine **165** and shaking at room temperature overnight. Cyclocleavage from the resin was achieved using HF in the presence of anisole at 0°C. Notably, guanidines formed with a secondary amine resulted in product **167** while those formed with a primary amine generated product **168**.

An example of the stereo effect of guanidines in the cyclocleavage step was reported by Ghiron and Drewry (**Scheme 9.22**).[32] Starting with Fmoc-Phe-Wang resin **169**, deprotection of the amine group was carried out with 30% piperidine in DMF. Treatment with 10 equiv of phenylisothiocyanate **170** in THF gave the thiourea intermediate **171**, which was treated with 4.8 equiv of Mukaiyama's reagent and 9 equiv of triethylamine in DCM and heated at 45°C for 3.5 h to give the carbodiimide intermediate **172**. Then 1.1 equiv of primary amine **173** was reacted with **172** in THF for 24 h. Afterward, treatment of the guanidine intermediate **175** with isocyanate scavenger resin for 6 h generated product **177**. When the secondary amine **174** was mixed with **172**, product **178** was correspondingly generated with **176** as the guanidine intermediate.

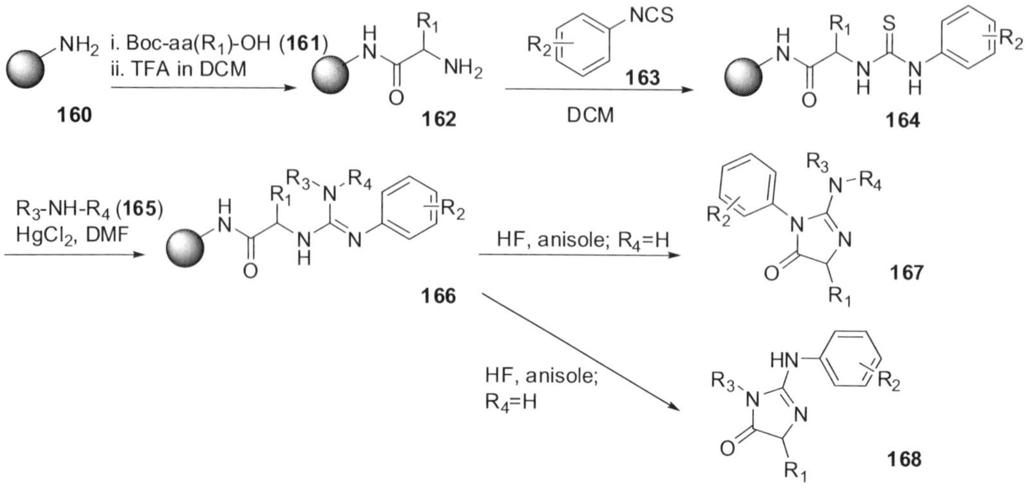

Scheme 9.21. Method 4 for the synthesis of imidazolones.

Fan's group reported another method to obtain the guanidine intermediate for the synthesis of imidazolones (**Scheme 9.23**).[33] Starting from Fmoc-MBHA resin **179**, removal of the Fmoc group was carried out with 20% piperidine in DMA. The free resin-bound amine was added to a solution of 10 equiv of DIEA, 5 equiv of Fmoc-amino acid **180**, and 5 equiv of PyBop (benzotriazo-1-oxytripyrrolidinophosphonium hexafluorophosphate) and shaken overnight. Following another deprotection step, the resin-bound amino acid **181** was obtained. Then **181** underwent a reductive alkylation reaction by shaking overnight with 50 equiv of aldehyde **182**, 50 equiv of $NaBH_3CN$, and 1% acetic acid in trimethyl orthoformate, affording the secondary amine intermediate **183**. The key step involved the treatment of **183** with a solution of 10 equiv of DIPEA in DMF, followed by another solution of 3 equiv of Mukaiyama's reagent and 9 equiv of the thiourea derivative **184** in DMF, shaken at room temperature for 48 h, giving the guanidine intermediate **185**. Cyclocleavage was achieved using TFA:H_2O:triisopropylsilane (TIPS) (94:3:3) for 1 h, affording the imidazolone **186**.

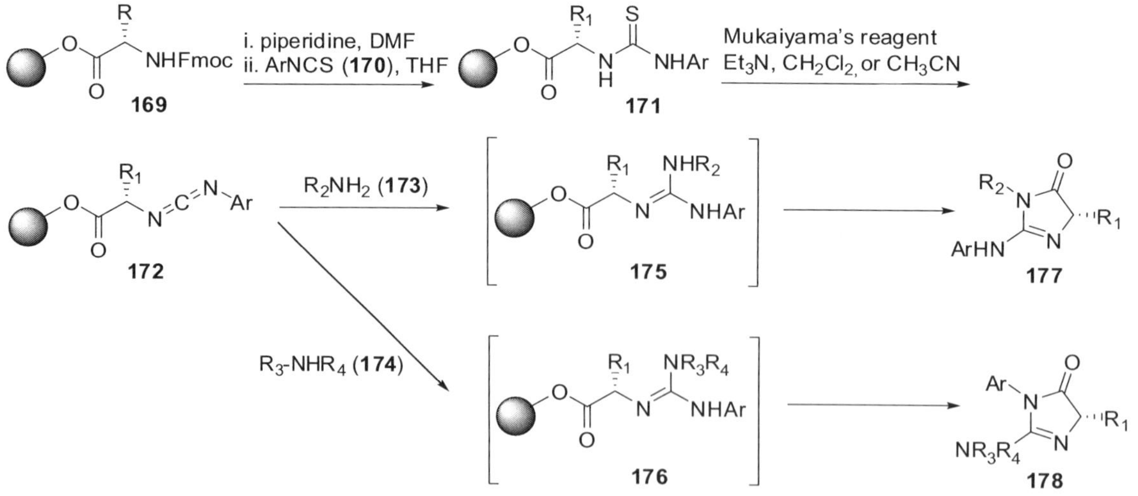

Scheme 9.22. Method 5 for the synthesis of imidazolones.

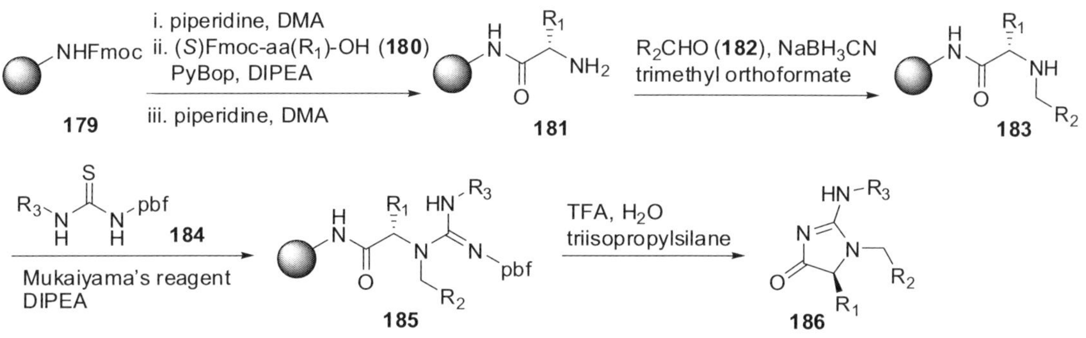

Scheme 9.23. Method 6 for the synthesis of imidazolones.

SYNTHESIS OF IMIDAZOLIDINONE DERIVATIVES. Generally, the key step to obtain imidazolidinones involves using an electrophilic reagent to connect two adjacent amine groups on the resin-bound peptide derivatives. Starting with amino acid-bound MBHA resin **187** and treating it with 10 equiv of aldehyde **188**, 5 equiv of trimethyl orthoformate, 10 equiv of acetic acid, and 10 equiv of $NaBH_3CN$ and shaking in anhydrous DMF:DCM: MeOH (1:8:2) at room temperature for 1 h provided the resin-bound secondary amine **189** (**Scheme 9.24**).[34] An anhydrous benzene suspension of resin **189** was added to 10 equiv of aldehyde **190** and benzotriazole (Bt) and refluxed for 16 h (4 Å sieves were employed to ensure the water free condition) to obtain the intermediate **191**. Then **191** underwent cyclization to form the imidazolidinone core **192**. Cleavage from resin was achieved by using HF at −5°C for 1.5 h, resulting in the imidazolidinone **193**.

To complete the synthesis of imidazolidinimines, five possible schemes are shown. In the first (**Scheme 9.25**),[35] 6 equiv of Boc amino acid **195** was coupled to MBHA resin **194** in the presence of 6 equiv of both DIC and HOBt in DMF at room temperature for 2 h. Then removal of the Boc group was facilitated by 55% TFA in DCM followed by neutralization with 5% DIEA in DCM, giving the resin-bound amino acid **196**. Next, **196** was treated with 10 equiv of carboxylic acid **197** and 10 equiv of both DIC and HOBt in DMF overnight to give the amide intermediate **198**. Exhaustive reduction of resin **198** was carried out with 12 equiv boric acid and 12 equiv trimethyl borate, and then a slow addition of 1 M borane–THF solution (40 equiv), followed by heating at 65°C for 72 h. After quenching with MeOH, the resin was treated with 65°C piperidine for 20 h, giving the triamine intermediate **199**, which

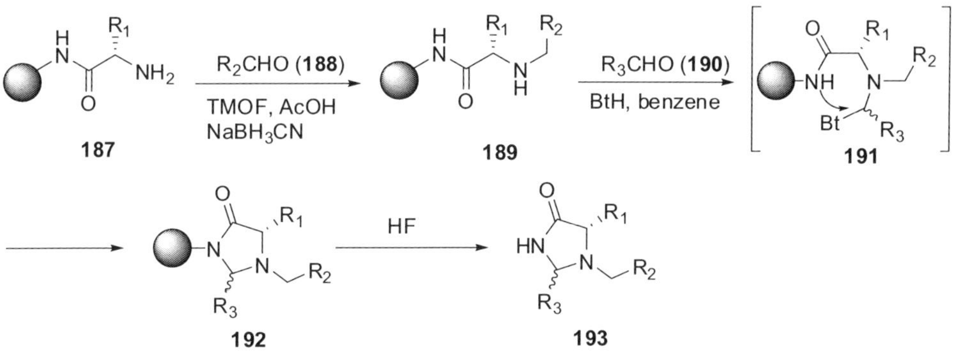

Scheme 9.24. Synthesis of imidazolidinones.

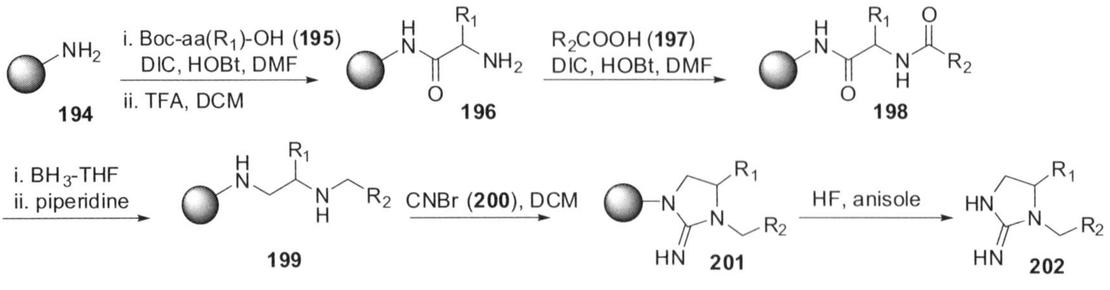

Scheme 9.25. Method 1 for the synthesis of imidazolidinimines.

underwent cyclization with 10 equiv cyanogen bromide **200** in DCM overnight. The imidazolidinone loaded resin **201** was treated with HF and anisole at 0°C for 1.5 h to afford cleavage from the resin, yielding the imidazolidimine product **202**.

An alternative method for the formation of imidazolidinimines is outlined in **Scheme 9.26.**[36] To MBHA resin **203** was attached Boc-amino acid **204** under proper coupling and deprotection conditions to give **205**, which underwent two coupling/deprotection reactions to form the resin-bound tripeptide **208**. Through exhaustive reduction by borane–THF, the tetraamine intermediate **209** was afforded. After treatment with 2.2 equiv of cyanogen bromide **210** in *m*-xylene suspension under N_2 overnight, the double ring imidazolidinimine **211** was obtained. Then **221** underwent cleavage from the resin with HF in the presence of anisole at 0°C, giving the final product **212**.

Fmoc-Sieber amide resin **213** was employed in the third method (**Scheme 9.27**).[37] After removal of the Fmoc group with 20% piperidine in DMF, the resin was treated with 3 equiv of Fmoc-amino acid **214** and 3 equiv of both DIC and DIEA in DMF to form resin **215**. After another deprotection procedure, 10 equiv of *ortho*-fluoro-nitroarene **216** and 10 equiv of DIEA in DMSO were added and the solution was left at room temperature overnight, facilitating the formation of **217**. Next, **217** was treated with 1 M $SnCl_2 \cdot 2H_2O$ in DMF to reduce the nitro group producing aniline **218**. Cyclization was carried out using 10 equiv of

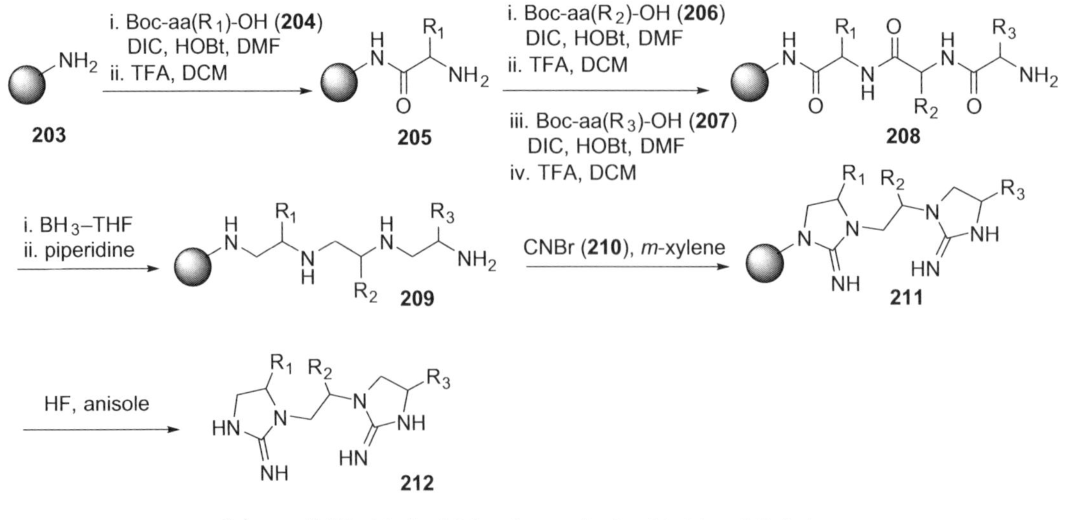

Scheme 9.26. Method 2 for the synthesis of imidazolidinimines.

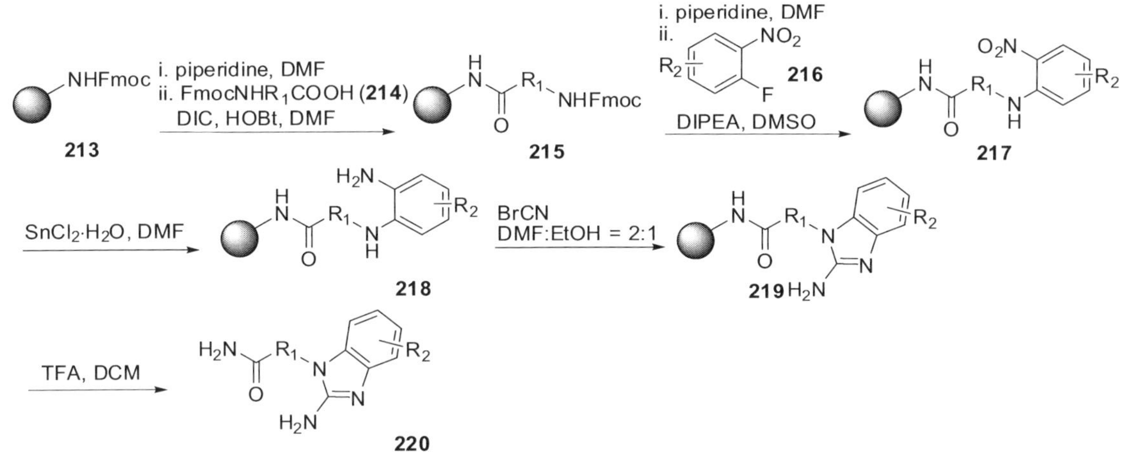

Scheme 9.27. Method 3 for the synthesis of imidazolidinimines.

cyanogen bromide in ethanol:DMF (1:2) at rt overnight, providing the resin-bound benzoimidazolamine core **219**. Cleavage from the resin was achieved with 5% TFA in DCM, producing **220**.

The fourth method is outlined in **Scheme 9.28**.[38] MBHA resin **221**, 6 equiv of Fmoc-amino acid **222**, and 6 equiv of both DIC and HOBt were employed for the first coupling reaction. After removal of the Fmoc group, the resin was treated with 10 equiv of both trityl chloride and DIC in DCM:DMF (9:1), generating the trityl-protected amino acid, which was treated with 1 M lithium *tert*-butoxide (20 equiv) and then 20 equiv of alkylating reagent **223** in anhydrous DMSO to give the alkyl-substituted intermediate **224**. Removal of the trityl group was carried out with 2% TFA in DCM, followed by neutralization with 5% DIEA in DCM. Addition of a second amino acid **225** under proper coupling and deprotection conditions gave the dipeptide **226**, which was acylated with 10 equiv of

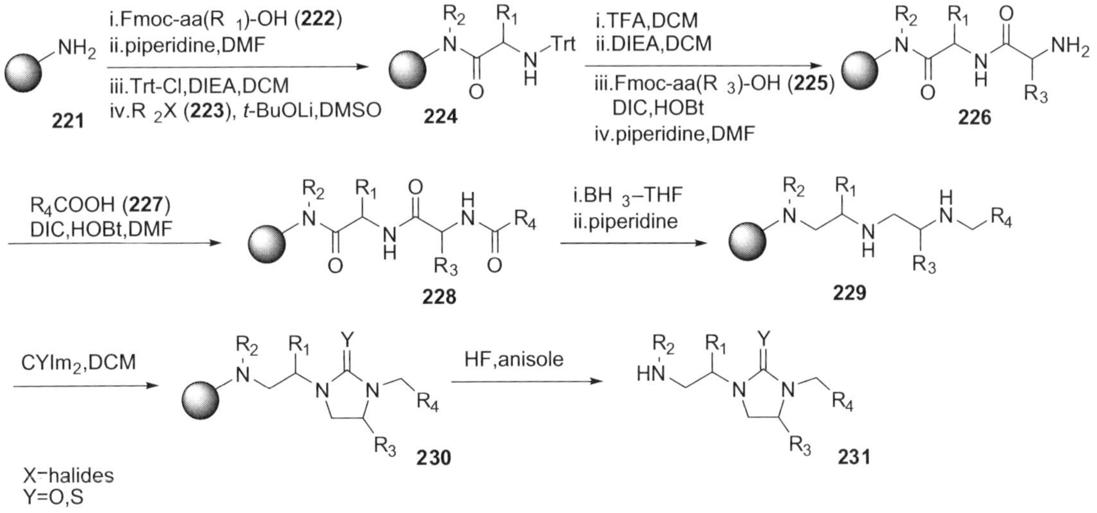

Scheme 9.28. Synthesis of imidazolidinones and imidazolidinthiones.

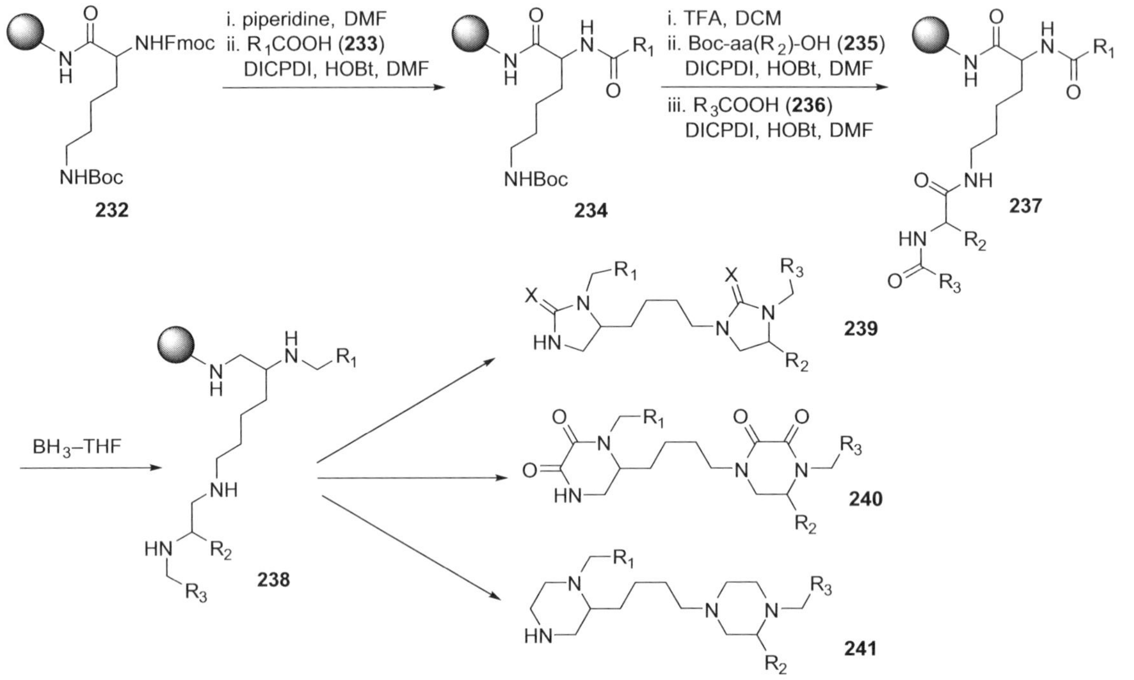

Scheme 9.29. Another method for the synthesis of imidazolidinones and derivatives X = O, S.

carboxylic acid **227** in the presence of 10 equiv of DIC and HOBt in anhydrous DMF to afford **228**. After exhaustive reduction with borane–THF complex, the triamine resin **229** was treated with 20 equiv of carbonyldiimidazole (or thiocarbonyldiimidazole) to form the urea-like compounds **230**. Cleavage from resin in HF and anisole at 0°C afforded the imidazolidinone (or imidazolidinthione) derivatives **231**.

The fifth method is outlined in **Scheme 9.29**.[39] MBHA resin loaded with Fmoc-lysine (ξ-Boc) **232** and after removal of the Fmoc group with 20% piperidine in DMF, the resin-bound amino acid was treated with 10 equiv of carboxylic acid **233** and 10 equiv of both DIPCDI and HOBt in anhydrous DMF overnight to afford the acylated compound **234**. After treatment with 50% TFA in DCM to remove the ξ position Boc group, a Boc-amino acid **235** and carboxylic acid **236** were coupled to the resin to form intermediate **237**, which went through exhaustive reduction with borane–THF complex, giving the tetraamine intermediate **238**. Next, **238** was reacted with an electrophilic reagent to afford the ring closure. Different reagents were used for this ring closure: 5 equiv of carbonyldiimidazole or thiocarbonyldiimidazole in anhydrous DCM, or oxalyldiimidazole in anhydrous DMF, was shaken with resin **238** overnight, followed by treatment with HF at 0°C, affording products **239** (X = O or S) and **240**, respectively. Notably, after resin **238** reacted with oxalyldiimidazole, the double diketopiperazine product can be further treated with borane–THF complex, generating the double piperazine derivative **241**.

Another example of using carbonyldiimidazole derivatives for cyclization was reported by Houghten and coworkers (**Scheme 9.30**).[40] Starting with Boc-amino acid loaded MBHA resin **242** and after deprotection with TFA in DCM and neutralization with DIEA, 4-fluoro-3-nitrobenzoic acid **243** was attached to the resin-bound amino acid, giving resin **244**. The fluoro group was substituted by different symmetrical diamines in DMF overnight to afford corresponding resin **245**, the free amine of which reacted with Boc-amino acid **246** under

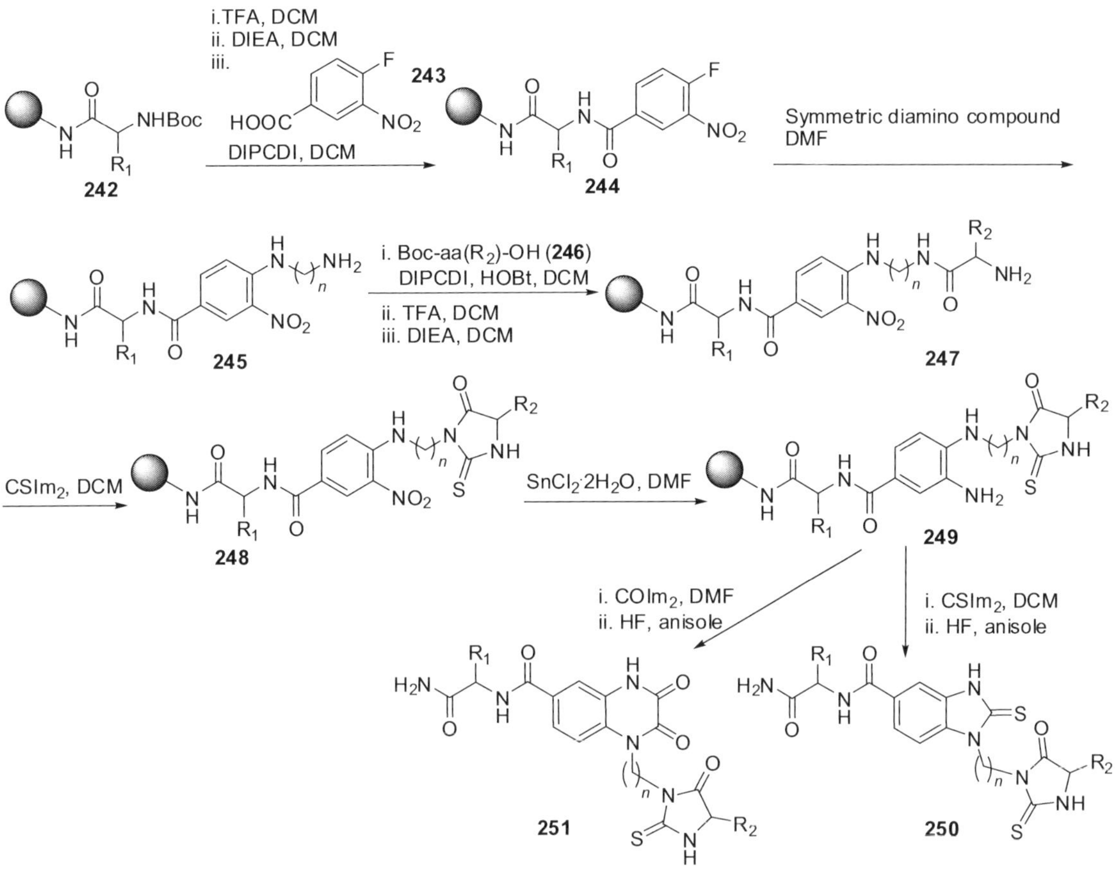

Scheme 9.30. Synthesis using $CSIm_2$ and $COIm_2$.

proper coupling and deprotection conditions to give intermediate **247**. After treatment with 5 equiv of thiocarbonyldiimidazole in DCM overnight to produce the thiohydantoin core **248**, the nitrogen group was reduced with 2 M $SnCl_2{\cdot}2H_2O$ in DMF overnight, giving resin **249**. This was treated with thiocarbonyldiimidazole or oxalydiimidazole, followed by HF/anisole cleavage from the resin, generating double heterocycle products **250** and **251**, respectively.

Synthesis of Imidazoles and Imidazole-Like Cores. The synthesis of imidazoles starting from resin-bound amino acids usually involves the linkage of adjacent amine groups of a peptide bond through the treatment of phosphoryl chloride. The phosphoryl chloride plays the role of the dehydration reagent in the conversion of an *in situ* formed imidoyl chloride intermediate to the desired imidazole scaffold.

Starting with MBHA resin **252**, 6 equiv of Boc-amino acid **253** was used as well as 6 equiv of both DIC and HOBt in DMF at room temperature for 2 h for the first coupling procedure (**Scheme 9.31**).[41] After deprotection of the amine group with 55% TFA in DCM for 30 min, followed by neutralization with 5% DIEA in DCM, resin **254** was acylated with 10 equiv of carboxylic acid **255** in the presence of 10 equiv of both DIC and HOBt in DMF to give the resin-bound amide **256**. Then **256** underwent exhaustive reduction upon treatment

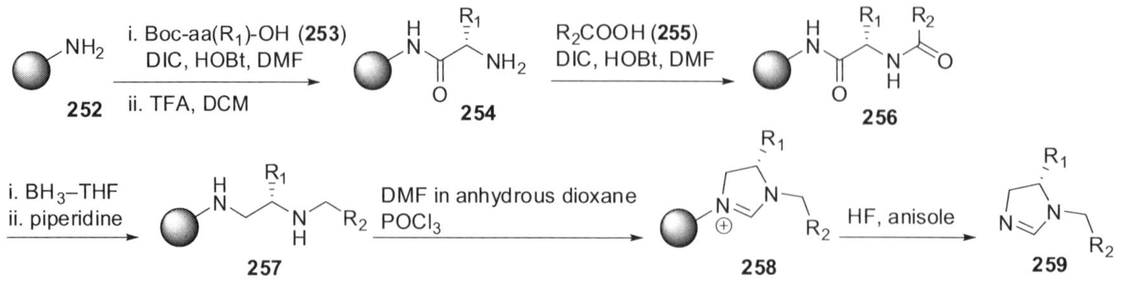

Scheme 9.31. Synthesis of imidazoles using Vilsmeier's reagent.

with 12 equiv of boric acid and 12 equiv of trimethyl borate, followed by the slow addition of 40 equiv borane–THF solution and heating at 65°C for 72 h. After decantation of the reaction solution and quenching with MeOH, treatment of piperidine at 65°C for 20 h afforded the diamine intermediate **257**. An *in situ* formed Vilsmeier reagent of DMF and phosphoryl chloride in anhydrous dioxane facilitated the sequential nucleophilic attacks on the secondary amines of resin **257**, generating product **258**. Treatment of **258** with anhydrous HF in the presence of anisole at 0°C for 7 h afforded the imidazole product **259**.

Another method of synthesizing imidazoles involves the coupling of 2.5 equiv of diamino acid **261** to MBHA resin **260** in the presence of 2.5 equiv of both DIC and HOBt (**Scheme 9.32**).[42] After removal of the Boc group with 50% TFA in DCM for 30 min, resin **262** was condensed with amino acid **263** and then carboxylic acid **264** under proper coupling conditions. Further treatment with 20% piperidine in DMF at room temperature for 30 min facilitated the removal of the Boc group of the original resin-bound diamine, giving intermediate **265**, which underwent exhaustive reduction with borane–THF to afford **266**. Protection of the free primary amine was carried out with 1.5 equiv of Dde-OH (2-acetyldimedone) at room temperature for 3 h, followed by treatment with 7 equiv of oxalyldiimidazole in DMF or 10 equiv of carbonyldiimidazole or thiocarbonyldiimidazole in DCM, forming the corresponding diketopiperazine or imidazolidinones **267**. Removal of the Dde group was achieved by treating with 2% hydrazine in DMF at room temperature for 1 h, resulting in **268**. Treatment of **268** with 3 equiv of carboxylic acid **269**, 3 equiv of HTBU, and 6 equiv of DIEA in DMF for 3.5 h facilitated the subsequent acylation to afford **270**. Cyclodehydration was carried out with 10 equiv of phosphoryl chloride in anhydrous dioxane and heated at 110°C for 2.5 h to give the imidazole core **271**. After cleavage from the resin with HF/anisole at 0°C, products **272–274** were obtained.

Another example of synthesizing multiple heterocycle cores was reported by Houghten and coworkers (**Scheme 9.33**).[43] Starting again with MBHA resin **275**, Boc-amino acid **276** was coupled and then deprotected under proper conditions, giving resin-bound amino acid **277**, which underwent exhaustive reduction with borane–THF complex to form the diamine intermediate **278**. Next, 3 equiv of 4-fluoro-3-nitrobenzoic acid **279** was coupled to resin **278** in the presence of 3 equiv of HBTU (2-(1*H*-benzotriazole-1-yl)-1,1,3,3-tetramethyluronium hexafluorophosphate) and 6 equiv of DIEA at room temperature for 3 h, giving the *N*-acylated intermediate **280**. Next, **280** was treated with 10 equiv of $POCl_3$ in anhydrous dioxane at 110°C for 2.5 h to form the imidazole derivative **281**. The fluoro group was then replaced by amine **282** in the presence of 20 equiv of DIEA in DMF at rt for 18 h yielding **283**. Treatment with 20 equiv of $SnCl_2{\cdot}2H_2O$ in DMF at room temperature for 14 h afforded the benzene diamine intermediate **284**, which was then treated with thiocarbonyldiimidazole to form the imidazolidinthione core **285**. Further treatment with 20 equiv of alkyl halide

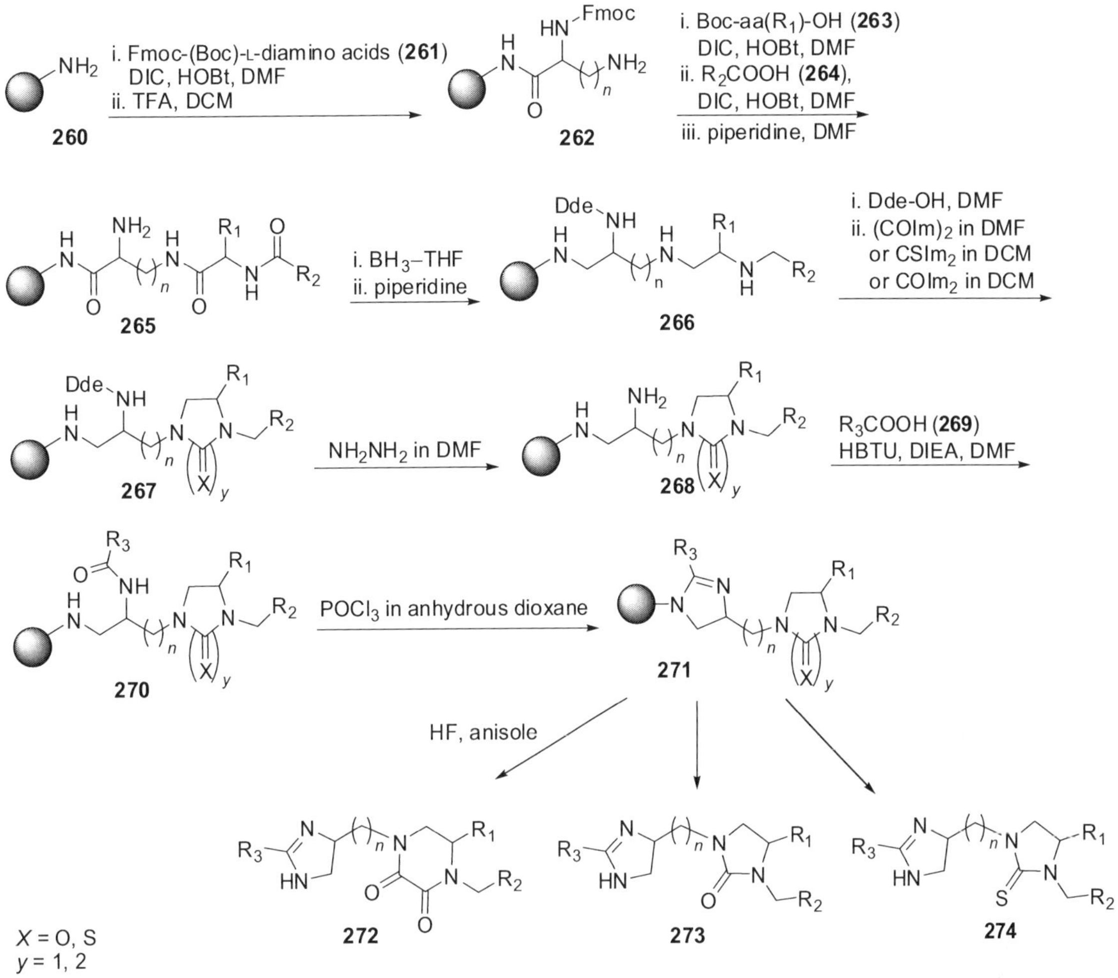

Scheme 9.32. Synthesis of imidazoles along with other heterocycles.

286 and 10 equiv of 1-methylimidazole in DCM at rt for 20 h facilitated alkylation, producing the thiourea **287**. Cleavage from resin was achieved with anhydrous HF in the presence of anisole at 0°C for 7 h producing imidazole **288**.

To begin the synthesis, Fmoc-lysine **290** was coupled to the MBHA resin **289** (**Scheme 9.34**).[44] After removal of the Boc group and neutralization, the Boc-amino acid **292** was coupled to the ξ-amine of lysine, producing compound **293**. After removal of the Boc group and exhaustive reduction with borane–THF, resin **294** was treated with excessive carboxylic acid **295** to acylate both primary amines and afford **296**. Two nonadjacent imidazole rings in **297** were obtained by using $POCl_3$ in anhydrous dioxane. Cleavage from resin was achieved by the HF/anisole method, giving the final product **298**. Similar synthetic strategies to obtain a dual imidazole core compound have been reported by Houghten and coworkers.[45]

Tosyl-protected dipeptide loaded Wang resin **299** was prepared and treated with 3 equiv of Ph_3PO and 1.5 equiv of trifluoromethanesulfonic anhydride in DCM for 4 h and then oxidized with 2.6 equiv of DBU (1,8-diazabicyclo[5.4.0]undec-7-ene) and 2.2 equiv of

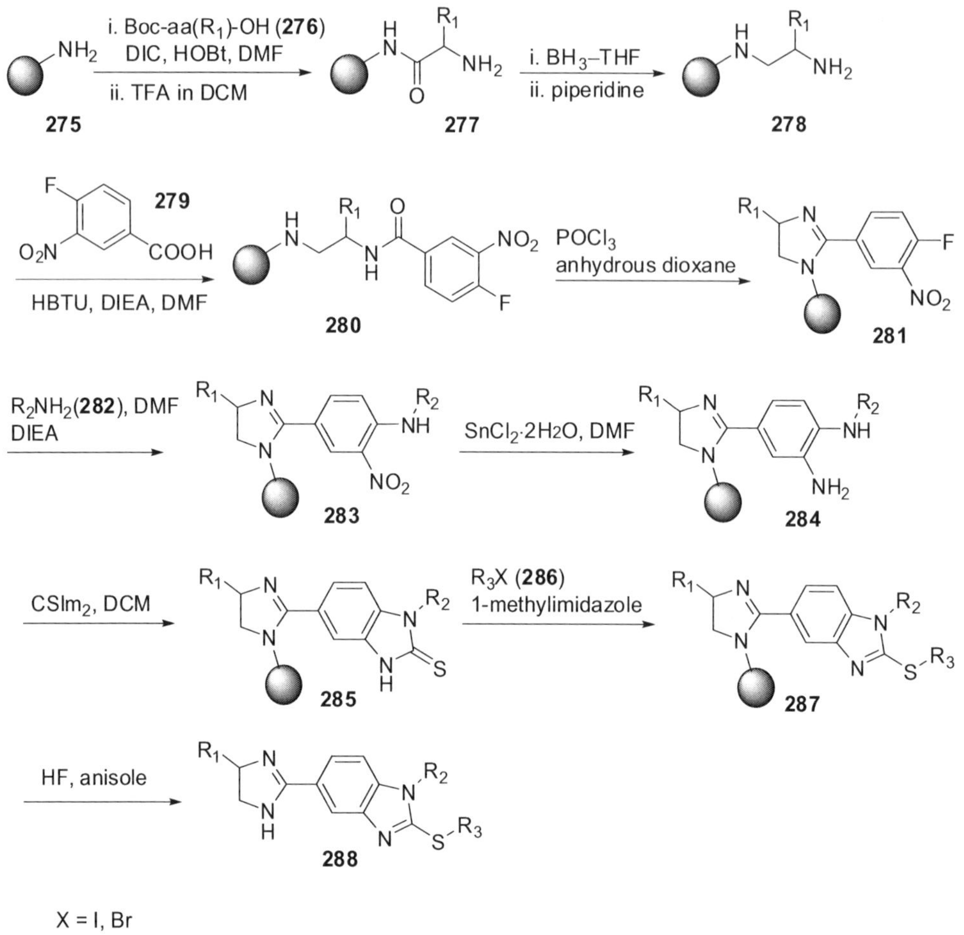

Scheme 9.33. Synthesis of imidazole cores.

bromotrichloromethane for 12 h, giving the imidazole core **300** (**Scheme 9.35**).[15] Cleavage from the resin was carried out using a mixture of TFA:H_2O:TIS (95:2.5:2.5), resulting in **301**.

Cunningham and Bilodeau reported a method to synthesis imidazoles through a münchnone intermediate (**Scheme 9.36**).[46] Starting from polystyrene-poly(ethylene glycol) graft copolymer resin **302**, amino acid ester **303** was attached using proper reductive alkylation conditions. The corresponding resin **304** was treated with 10 equiv of carboxylic acid chloride **305** in the presence of Hünig's base in DCM in order to acylate the resin. Treatment with KOH in H_2O:dioxane (1:3) facilitated the formation of the carboxylic acid intermediate **306**, which was subjected to 10 equiv of tosylimine **307** and 10 equiv of EDC (1-ethyl-3-(3-dimethylaminopropyl)carbodiimide hydrochloride) in DCM at room temperature for 24–48 h to obtain intermediate **308**, which was converted to resin-bound imidazole **309**. Cleavage from the resin was achieved by using 90% TFA in H_2O, resulting in the final product **310**.

A library of imidazo[1,2-α]pyridines was prepared by Shankaraiah and coworkers (**Scheme 9.37**).[47] Starting from 4-hydroxyphenylsulfide resin **311**, 3 equiv of Boc-amino nicotinic acid **312** was added along with 3 equiv of EDC in DCM:DMF(1:1), followed by

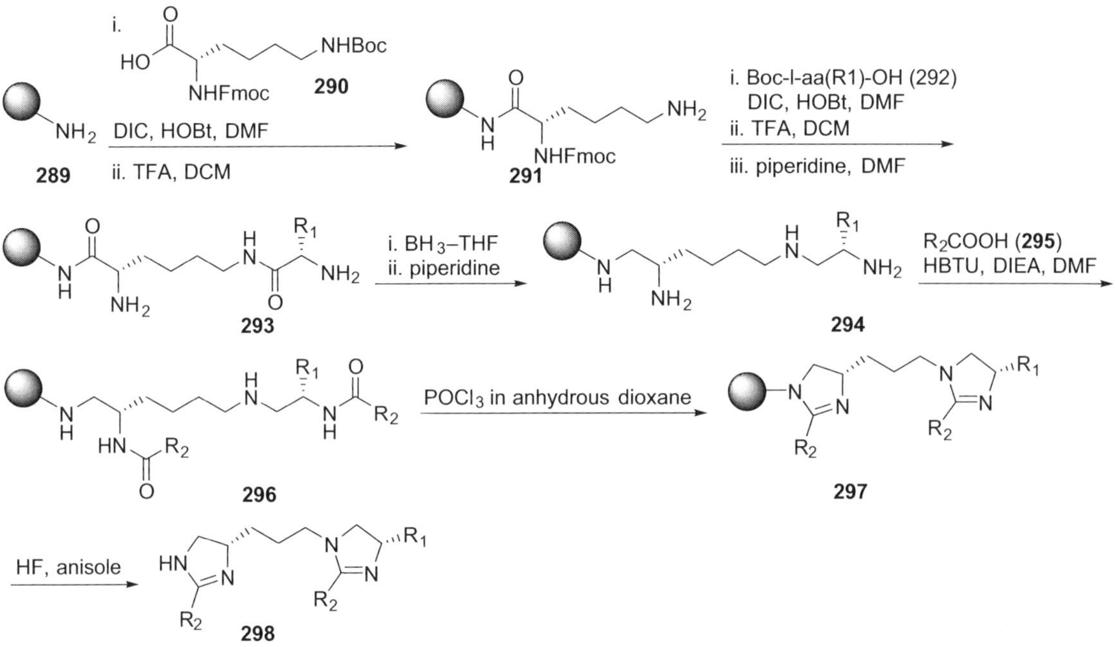

Scheme 9.34. Synthesis of dual imidazole core compounds.

addition of 10% DMAP in DCM, forming resin **313**. Removal of the Boc group was achieved with a solution of 4 N HCl in dioxane to afford **314**. Resin **314** was treated with 3 equiv of α-haloketone **315** in ethanol and refluxed for 24 h to give resin-bound imidazo[1,2-α]pyridine **316**. This was treated with 5 equiv of amine **317** in pyridine, shaken for 18–24 h, and treated with a solid-supported liquid–liquid extraction (SLE) procedure to obtain the final product **318**.

Synthesis of Multiple Rings. Starting with MBHA resin **319**, dipeptide **322** was obtained through the coupling of Boc-amino acids **320** and **321** under proper coupling and deprotection conditions (**Scheme 9.38**).[48] Then 10 equiv of carboxylic acid **323** along with DICI and HOBt in DMF suspension was used for the acylation of the free amine of resin **322**. Intermediate **324** was treated with 15 equiv of phosphorus oxylchloride in dioxane, protected by N_2, and heated at 100°C for 18 h to give the resin-bound bicyclic product **325**, which underwent cleavage from the resin with HF/anisole at 0°C, affording **326**.

MBHA resin-bound dipeptide **330** was synthesized through two successive coupling/deprotection reactions starting from resin **327** (**Scheme 9.39**).[49] A selective acylation was

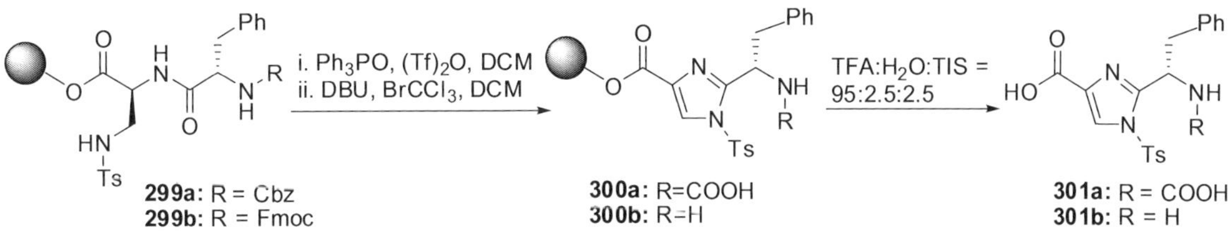

Scheme 9.35. Synthesis of tosyl-protected imidazoles.

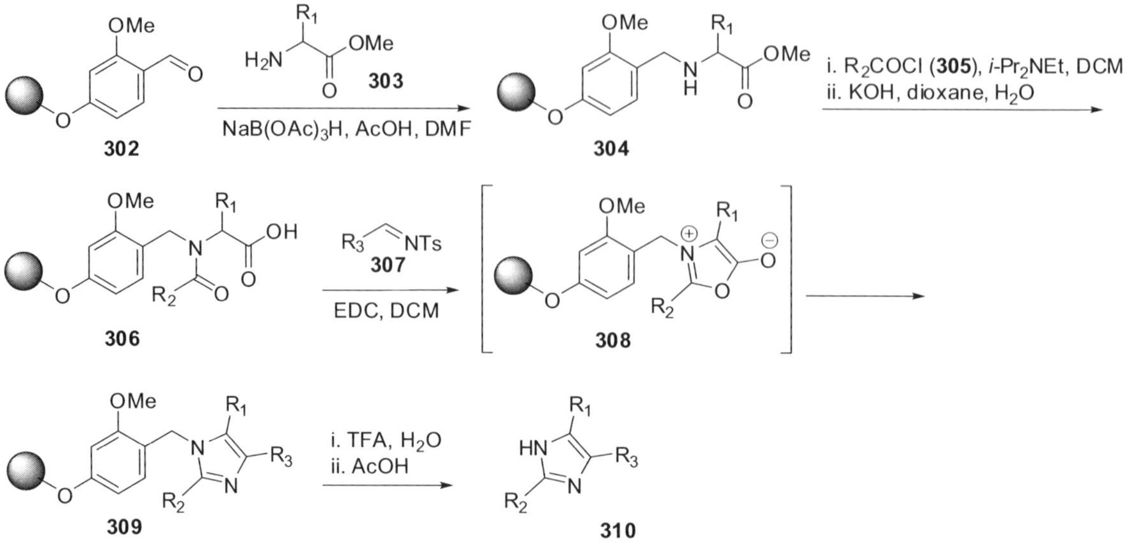

Scheme 9.36. Münchnone reactions.

carried out using 10 equiv of carboxylic acid **331** and 10 equiv of both DIC and HOBt in DMF overnight to give resin **332**. Resin **332** then underwent exhaustive reduction with borane–THF complex, affording the tetraamine intermediate **333**. The free primary amine was treated with 10 equiv of trityl chloride in the presence of 25 equiv of DIEA in DCM for 2 h, generating **334**. The Trt-protected intermediate **334** was treated with 24 equiv of thiocarbonyldiimidazole in anhydrous DCM under N_2 overnight to form a thioimidazolidinone intermediate. The intermediate was then treated with 10 equiv of mercuric acetate in anhydrous DMF overnight, forming the bicyclic product. Treatment with 5% TFA in DCM removed the trityl group, giving resin **335**, which was coupled with another amino acid **336**, resulting in **337** that was then reacted with an isocyanate **338** to afford the urea ester product **339**. Final product **340** was obtained by using anhydrous HF in the presence of anisole at 0°C for 7 h. A similar strategy employing thiocarbodyldiimidazole for a bicyclic synthesis was reported by Houghten and coworkers.[50]

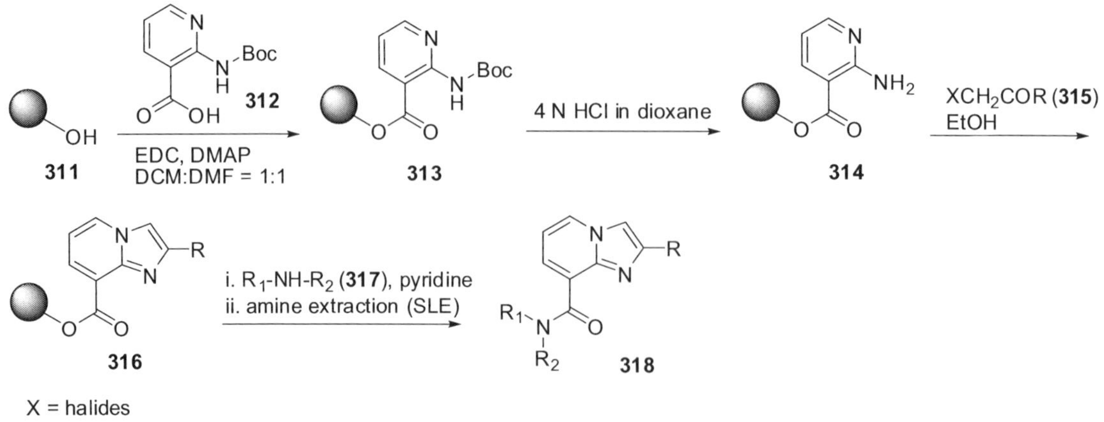

Scheme 9.37. Synthesis of imidazo[1,2-α]pyridines.

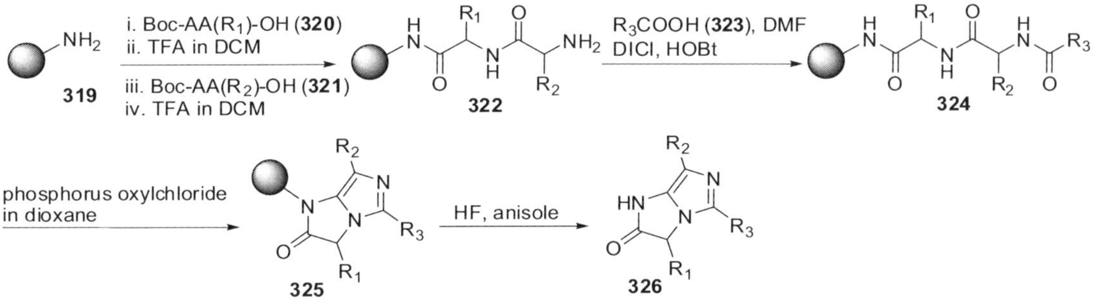

319 322 324 325 326

Scheme 9.38. Synthesis of bicyclic compounds.

9.2.3.3 *Five-Membered Ring Heterocycles Containing Three Nitrogen Atoms.* Starting with Rink amide resin **341**, 4 equiv of Fmoc-amino acid **342** was used in the presence of 4 equiv of EDAC (1-ethyl-3-(3-dimethylaminopropyl)carbodiimide) and HOBt in DMF overnight (**Scheme 9.40**).[51] After removal of the Fmoc group with 20%

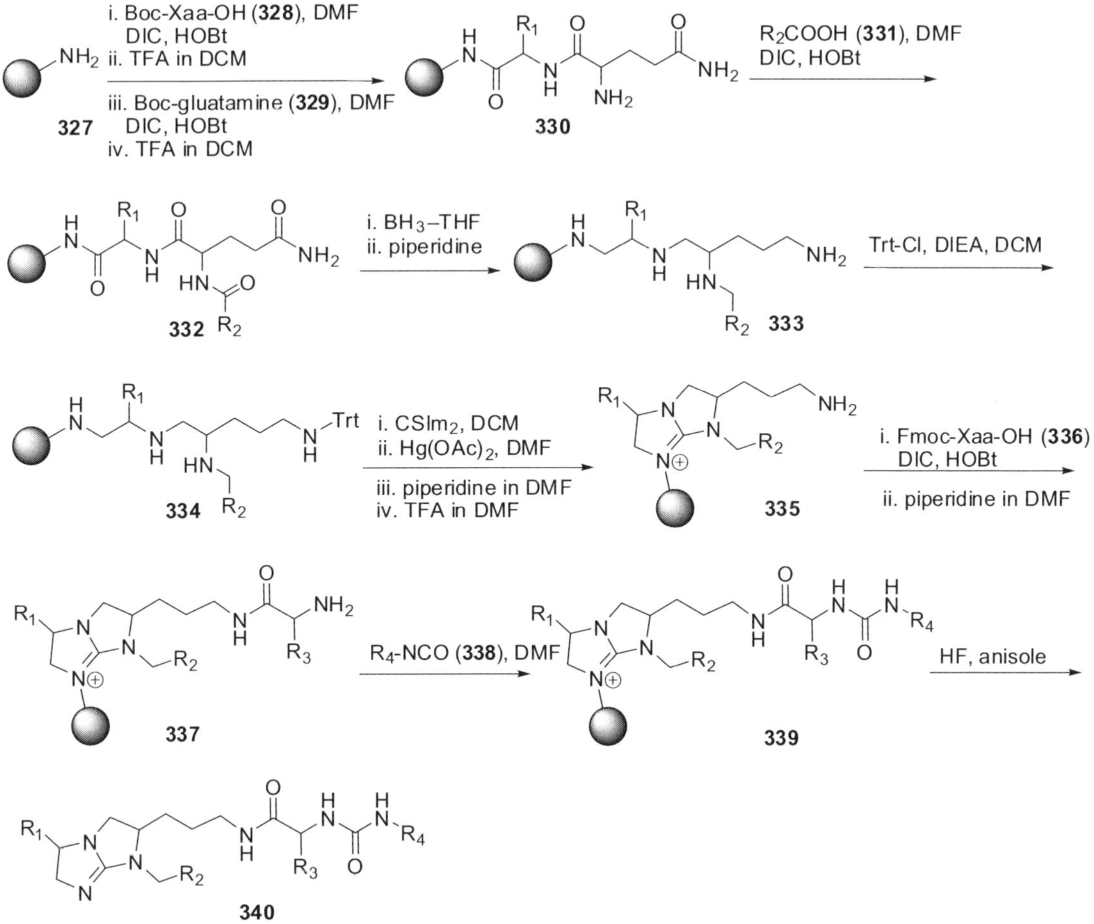

327 330 332 333 334 335 337 339 340

Scheme 9.39. Synthesis of more bicyclic compounds.

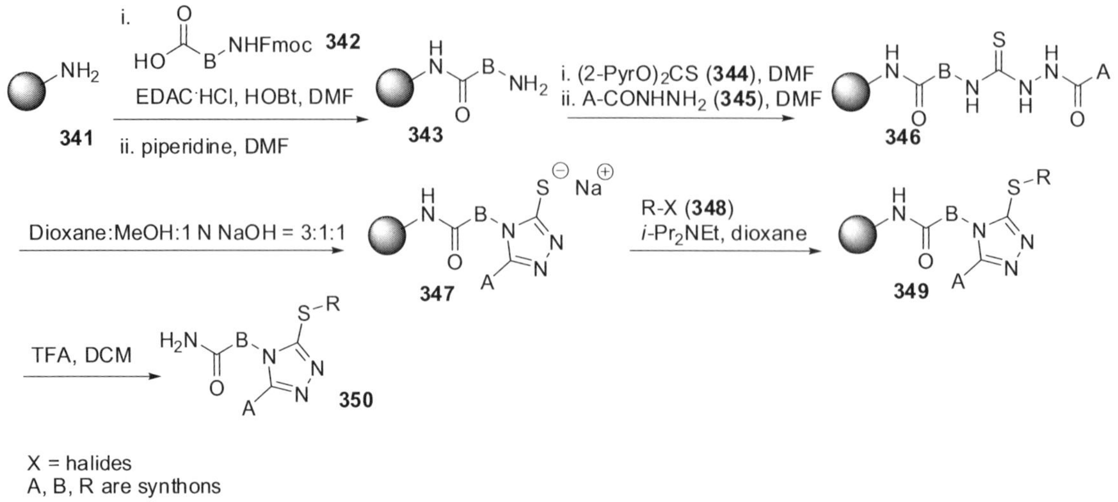

Scheme 9.40. Method 1 for the synthesis of triazoles.

piperidine in DMF, resin **343** was treated with 10 equiv of di-(2-pyridyl)thionocarbonate **344** in DMF at room temperature for 1 h and then treated with 20 equiv of hydrazine **345** in DMF overnight, giving the hydrazinecarbothioamide **346**. Then **346** was suspended in a solution of dioxane:methanol:1 N NaOH and heated at 85°C for 4 h to afford the sodium salt **347**. After resin **347** was placed in a 0.2 M alkyl halide **348** dioxane solution, 2 drops of *N*,*N*-diisopropylethylamine were added. The combined solution was shaken for 1 h, affording product **349**, which was then cleaved from the resin with 2% TFA in DCM to give the triazole derivative **350**.

Kauhaluoma and Samanta used münchnone as the key intermediate to obtain a triazole core (**Scheme 9.41**).[52] Starting with Ameba resin **351**, 3.84 equiv of amino acid ester **352** and 25 equiv of $NaBH_3CN$ were employed for the reductive alkylation in 1% acetic acid in DMF at room temperature for 12 h. Resin **353** was treated with 10 equiv of carboxylic acid chloride **354** in the presence of 12 equiv of *N*,*N*-diisopropylethylamine in DCM at room temperature for 12 h to form amide **355**, which was treated with 5% KOH in dioxane:H_2O (3:1) for 4 h to give the corresponding carboxylic acid. Resin **356** was treated with 23.5 equiv of acetic acid anhydride, 3 equiv of DEAD (diethyl azodicarboxylate) **357**, and 3 equiv of 4-phenyl-4*H*-1,2,4-triazoline-3,5-dione **358** in DCM at room temperature for 5–10 h, giving the münchnone intermediate **359**, which reacted with compound **357** or **358** to form resin **360**. Treatment of this with 30% TFA in DCM at room temperature for 1.5 h resulted in cleavage from the resin, producing **361**.

Meldal and coworkers employed azides and copper(I) catalyst in the formation of a 1,2,3-triazole (**Scheme 9.42**).[53] $PEGA_{800}$ resin was treated with 3 equiv of HMBA (4-(hydroxymethyl)benzoic acid), 2.9 equiv of TBTU (*N*-[(1*H*-benzotriazol-1-yl)(dimethylamino)methylene]-*N*-methylmethanaminium tetrafluoroborate *N*-oxide), and 4 equiv of NEM (*N*-ethylmorpholine) in DCM, and then with 3 equiv of Fmoc-Gly-OH, 3 equiv of MSNT (mesitylenesulfon-1-yl-3-nitro-1,2,4-triazole), and 6 equiv of methylimidazole to form the first resin-bound amino acid. By alternating the use of 3 equiv of Fmoc-Phe-OPfp and 3 equiv of Fmoc-Gly-OPfp with 1 equiv of Dhbt-OH (3-hydroxy-3,4-dihydro-4-oxo-1,2,3-benzotriazine) as the catalyst in DMF, followed by Fmoc removal, the tetrapeptide was formed. Acylation with 3 equiv of propargylic acid in the presence of 3.1 equiv of EEDQ (2-ethoxy-1-methoxycarbonyl-1,2-dihydroquinoline) in DCM for 16 h generated

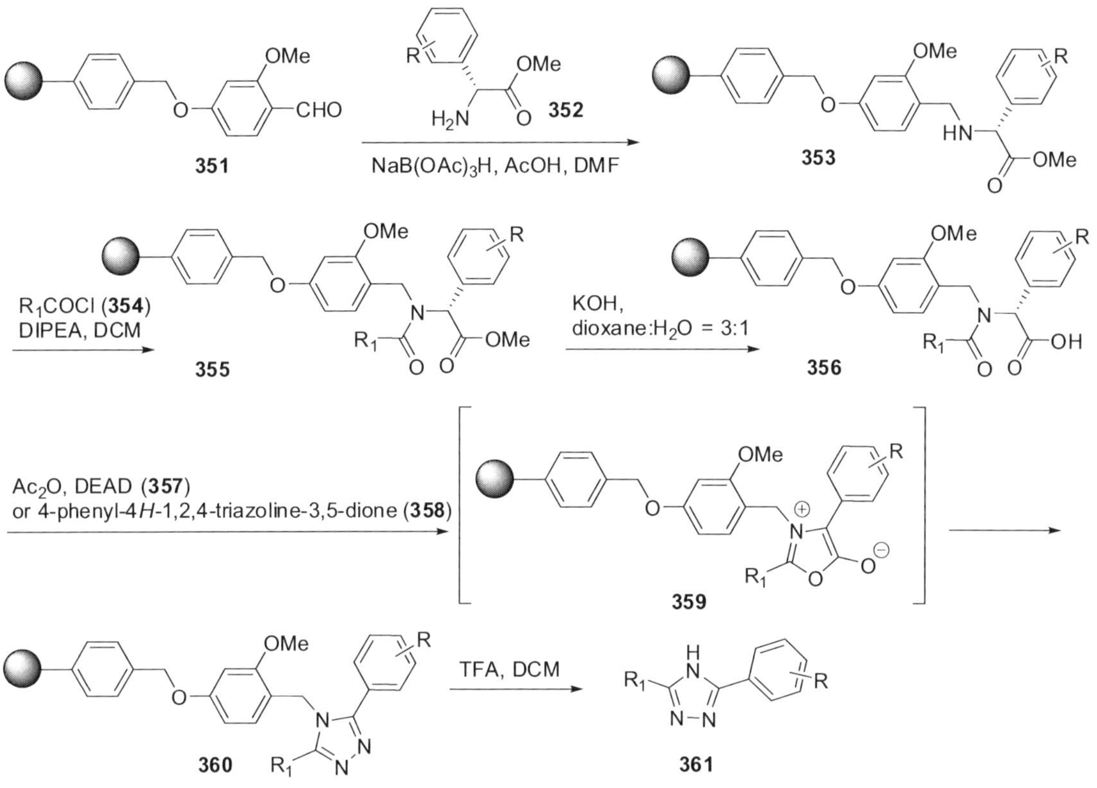

Scheme 9.41. Method 2 for the synthesis of triazoles.

compound **362**. Cyclization was achieved by using 50 equiv of DIEA, 2 equiv of CuI, and 2 equiv of azide **363** in a THF suspension of **362** at 25°C for 16 h. Cleavage from the resin was carried out using 1 N NaOH, resulting in the 1,2,3-triazole **364**.

9.2.3.4 Five-Membered Ring Heterocycles Containing Four Nitrogen Atoms. A suspension of 1.5 equiv of Fmoc-proline **366** and 5 equiv of Hünig's base in anhydrous DCM was prepared before addition of resin **365** (**Scheme 9.43**).[54] The mixture was stirred at room temperature for 3 h. Then methanol was added and stirred for 15 min to cap the remaining 2-chlorotrityl groups on resin **365**, resulting in the resin-bound Fmoc-amino acid **367**. After removal of the Fmoc groups with 20% piperidine in DMF, 3 equiv of Fmoc-glycine **368** and 8 equiv of Hünig's base in DMF were used for the formation of dipeptide **369**. After deprotection using 20% piperidine in DMF, resin **369** was treated with 5 equiv of isothiocyanate **370** in DMF to give the thiourea intermediate **371**, which was swollen in a DMF solution containing 10 equiv of triethylamine and 5 equiv of

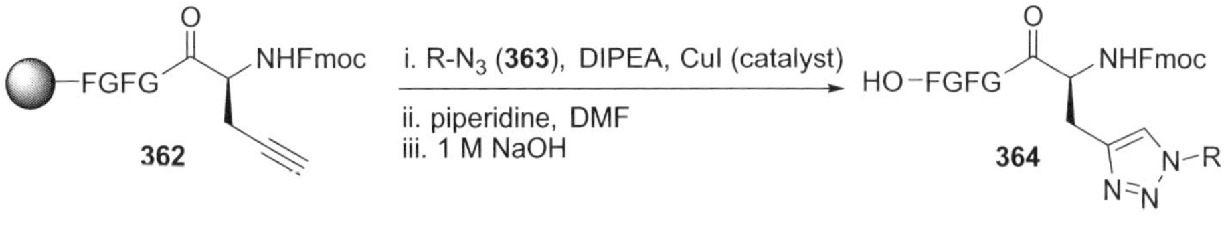

Scheme 9.42. Solid-phase click reactions.

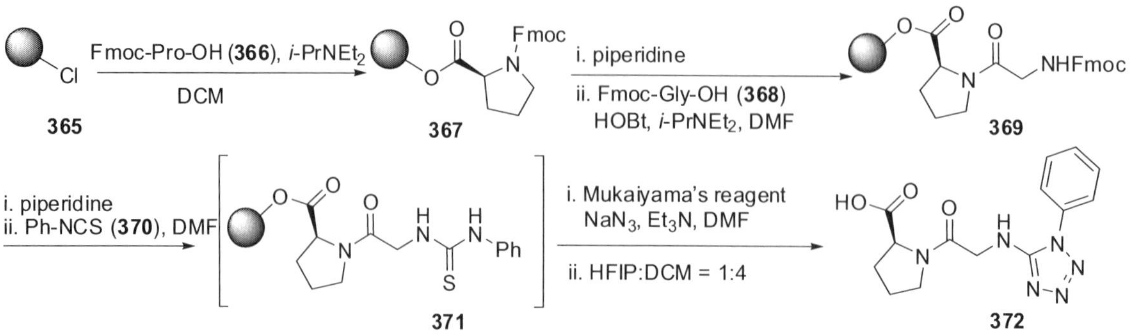

Scheme 9.43. Synthesis of tetrazoles.

Mukaiyama's reagent. The solution was shaken for 3 h, then 10 equiv of sodium azide was added, and the solution was again shaken overnight to get the resin-bound tetrazole. Cleavage from the resin was achieved by using HFIP:DCM (1:4) for 30 min, generating **372**.

9.2.4 Six-Membered Ring Heterocycles

9.2.4.1 Six-Membered Ring Heterocycles Containing One Nitrogen Atom. Fmoc-Cys(Trt)-OH **374** was coupled to MBHA resin **373** to form **375** (**Scheme 9.44**).[55] This was treated with TFA:(i-Bu)$_3$SiH:DCM (5:5:90) in order to

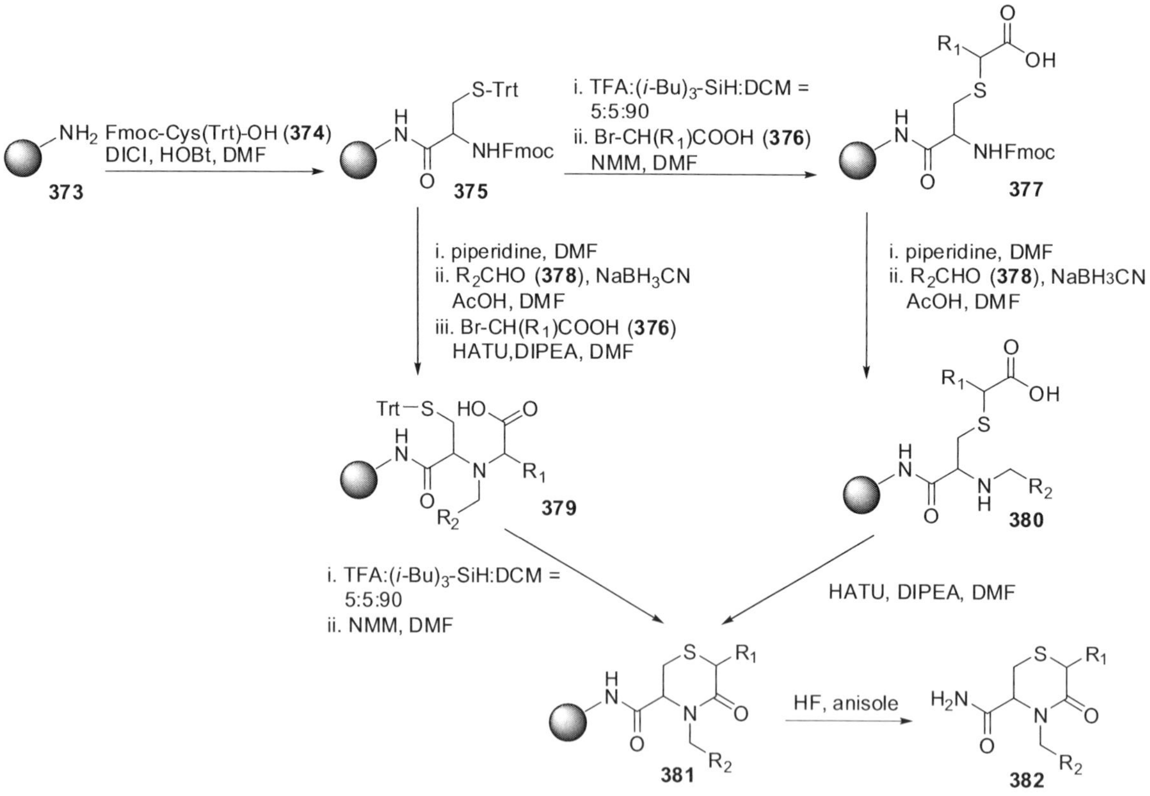

Scheme 9.44. Synthesis of thiomorpholines.

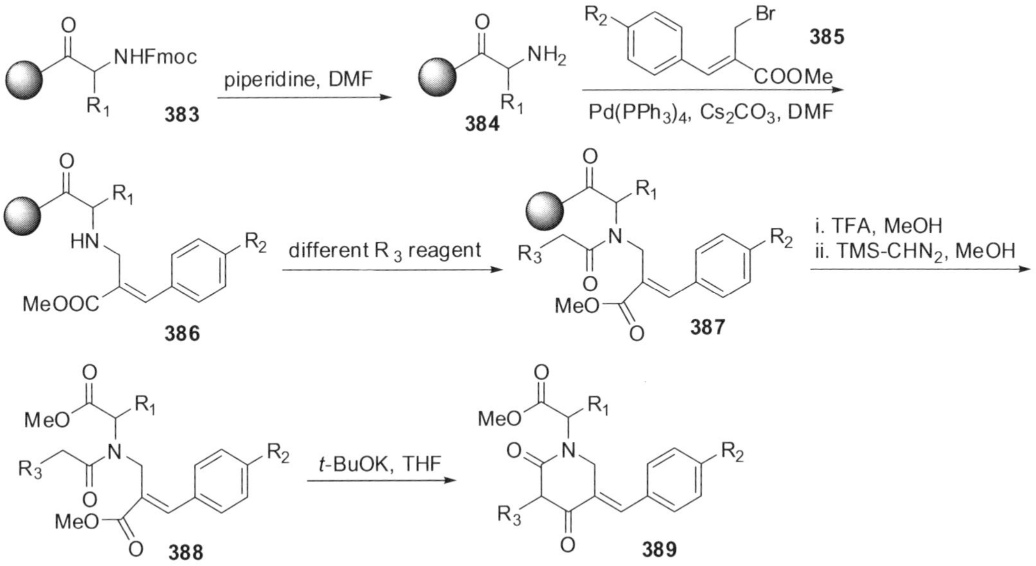

Scheme 9.45. Dieckmann condensation reactions.

remove the trityl groups, and then it was treated with 15 equiv of bromoacetic acid **376** in the presence of NMM (*N*-methylmorpholine) in DMF to give resin **377**. After removal of the Fmoc group with 20% piperidine in DMF, a reductive alkylation was carried out using 5 equiv of aldehyde **378** in 1% acetic acid in DMF followed by the addition of 5 equiv of $NaBH_3CN$, affording **380**. Cyclization to afford **381** was achieved by treatment of **380** with 5 equiv of HATU (*N*-[(dimethylamino)-1*H*-1,2,3-triazolo[4,5-*b*]pyridin-1-ylmethylene-*N*-methylmethanaminium hexafluorophosphate *N*-oxide) and 5 equiv of DIEA in DMF and shaking overnight. The thiomorpholine derivative **382** was obtained through cleavage with anhydrous HF and anisole at 0°C. Another synthetic path involves altering the protection group removal order of resin **375**. Removal of the Fmoc group with piperidine/DMF and subsequent reductive alkylation of the primary amine and then nucleophilic alkylation of the corresponding secondary amine afforded intermediate **379**. After removal of the trityl group from thiol with TFA:$(i\text{-Bu})_3SiH$:DCM (5:5:90), cyclization was achieved using the NMM/DMF method, giving the resin-bound thiomorpholine **381**. A similar strategy as described above was reported by Campbell's group.

Starting from Fmoc-amino acid loaded Wang resin **383**, treatment of piperidine:DMF (1:1) facilitated removal of the Fmoc group (**Scheme 9.45**).[57] Subsequent alkylation of the free amine **384** carried out with 1 equiv of bromo-synthon **385** in the presence of 0.01 equiv of $Pd(PPh_3)_4$ and 1 equiv of Cs_2CO_3 in DMF at room temperature for 16 h afforded **386**. Resin **386** was treated with different acylation reagents (ethyl malonyl chloride, cyanoacetic acid, or diketene) under various conditions to form the different amides **387**. Cleavage from resin was achieved using TFA at 25°C for 2 h. Methylation of the carboxylic acid was carried out by using diazomethyltrimethylsilane in methanol, giving product **388**, which underwent a Dieckmann condensation with 2.7 equiv of *t*-BuOK in THF and stirring at −78°C for 3 h to afford the piperidinone derivative **389**.

Starting with Fmoc-4-nitro-phenylalanine loaded Wang resin **390** and after removal of the Fmoc group with piperidine:DMF (4:1), resin **391** was acylated with carboxyl acid **393** or carboxyl acid chloride **392** to give resin **394** (**Scheme 9.46**).[58] Reduction of the nitro

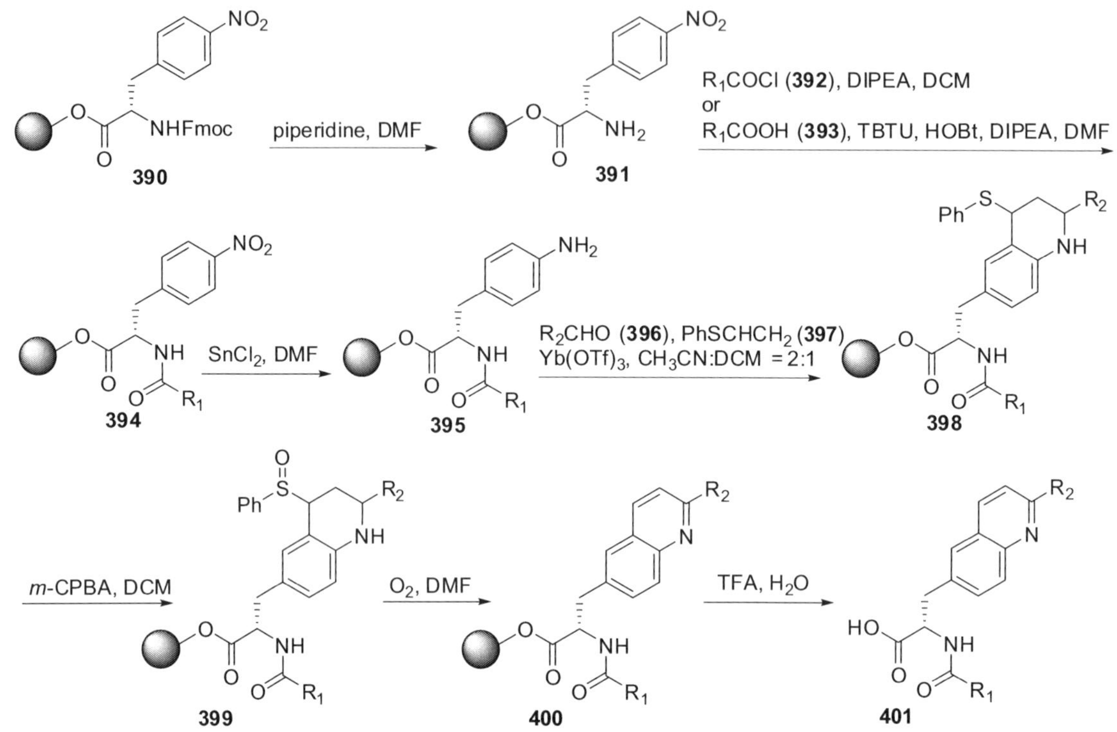

Scheme 9.46. Yb-catalyzed cyclization reactions.

group was achieved using a solution of 2 M $SnCl_2$ in DMF at room temperature for 16 h, affording **395**. The key step involved treating resin **395** with 10 equiv of aldehyde **396** in DCM:acetonitrile (1:2) and a catalytic amount of $Yb(OTf)_3$ in DCM:acetonitrile (1:2) with 10 equiv of phenyl vinyl sulfide **397** and shaking the solution at room temperature for 18 h to generate the tetrahydroquinoline derivative **398**. Oxidation of the benzenethiol group was carried out with 1.3 equiv of *m*-CPBA (*meta*-chloroperoxybenzoic acid) in DCM shaken at room temperature for 4 h to yield **399**. Resin **399** was swollen in DMF and shaken at 80°C for 16 h to give the resin-bound quinoline **400**. Treatment of this with 95% TFA in H_2O facilitated cleavage from the resin, yielding the six-membered ring heterocycle **401**.

Starting with amino acid-bound SASRIN resin **402**, treatment with either phosgene in the presence of 2,6-lutidine in a toluene–DCM solution with progressive warming from 0°C to room temperature, or phosgene (or triphosgene) and 2,6-di-*t*-butylpyridine in toluene–DCM at rt for 1–10 h, produced isocyanate **403** (Scheme 9.47).[59] Resin **403** was treated with anthranilic acid **404** in 10% pyridine in DMF at rt to give the urea intermediate **405**. Cyclization employing DIC, acetic anhydride in DMF, or tosylchloride in pyridine, both at room temperature overnight, gave 1*H*-benzo[*d*][1,3]oxazin-4(2*H*)-one derivative **406**. Cleavage from the resin was achieved with 1% TFA in DCM, forming **407**.

9.2.4.2 Six-Membered Ring Heterocycles Containing Two or More Nitrogen Atoms

Synthesis of 1,4-Diketopiperazine Derivatives. The standard approach to obtain 1,4-diketopiperazines (DKPs) is through the intramolecular aminolysis reaction between the N-terminal and the C-terminal of a resin-bound dipeptide. However, there are other

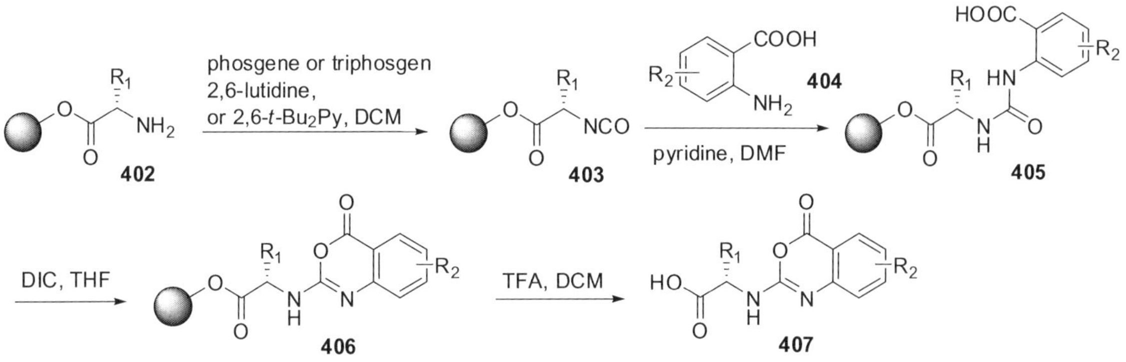

Scheme 9.47. Synthesis of 1*H*-benzo[*d*][1,3]oxazin-4(2*H*)-ones.

approaches, such as using an electrophilic reagent with a resin-bound diamine for cyclization.

Starting from Kaiser oxime resin **408**, 5 equiv of Boc-amino acid **409** and 5 equiv of DIC in DCM or DMF for 16 h completed the first coupling (**Scheme 9.48**).[60] Resin **410** was treated with 25% TFA in DCM for 30 min, then 5 equiv of Boc-amino acid **411** and 5 equiv of both DIC and DIEA in DCM (or DMF) for 4 h generated dipeptide **412**. After another deprotection using 25% TFA in DCM, the subsequent resin-bound dipeptide was treated with 2.2 equiv of DIEA and 5 equiv of acetic acid in DCM for 16 h to induce the intramolecular coupling, resulting in the cyclocleavage product **413**.

Starting from bromoacetyl resin **414**, which was prepared from the reaction of polystyrenc and bromoacetic bromide with a Lewis base in nitrobenzene:DCM (1:1) (**Scheme 9.49**),[61] coupling of 2 equiv of Boc-amino acid **415** to resin **414** was carried out with 2.2 equiv of ethylamine in DMF at room temperature for 24 h to afford **416**. After removal of the Boc group using 3.5 HCl and acetic acid for 30 min, a second Boc-amino acid **417** (3 equiv) was coupled to resin **416** using 3 equiv of HOBt, 3.2 equiv of DIC, and 1.2 equiv of *N*-methylmorpholine in DMF for 4–6 h to give intermediate **418**. Treatment with HCl/AcOH for Boc removal and then 5% triethylamine in THF:H_2O (8:1) afforded diketopiperazine **419** along with a HCl·Et_3N salt that was difficult to remove for purification. A better method was to remove the Boc group of resin **418**, and then neutralize with 10% DIEA in EtOAc before cyclocleavage to give the final product **419** with better purity.

In another synthetic pathway, Fmoc-L-Hyp-OMe **422** was coupled to DHP-HM (3,4-dihydro-2*H*-pyran-2-ylmethoxymethyl polystyrene) resin **421** in the presence of PPTS

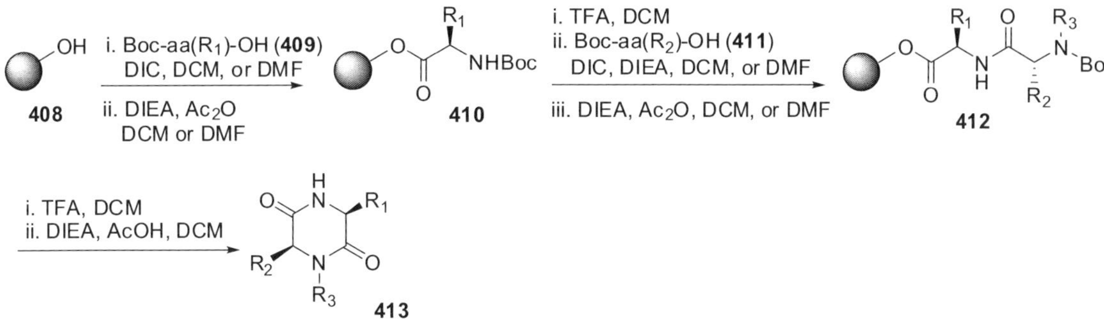

Scheme 9.48. Method 1 for the synthesis of 1,4-diketopiperazine derivatives.

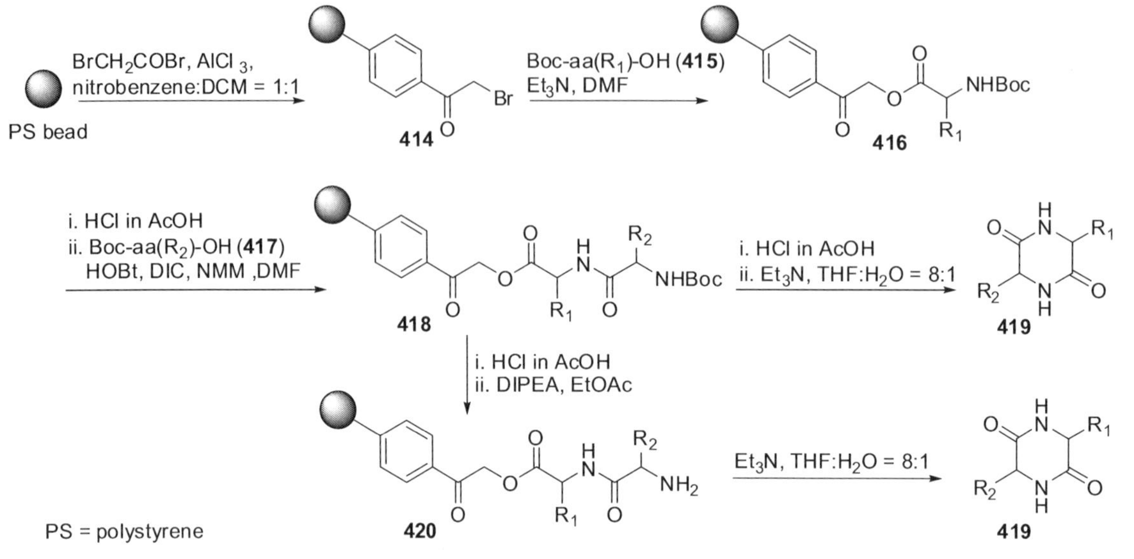

Scheme 9.49. Method 2 for the synthesis of 1,4-diketopiperazine derivatives.

(pyridinium *p*-toluenesulfonate) in 1,2-dichloroethane (1,2-DCE) (**Scheme 9.50**).[62] Resin **423** was treated with 25% piperidine in DMF for removal of the Fmoc group. Then a solution of 3 equiv of Teoc (trimethylsilylethoxycarbonyl)-OBt in DMF at room temperature for 24 h was used to protect the secondary amine against LHMDS (lithium bis(trimethylsilyl)amide) and afford **424**. To a suspension of resin **424** in anhydrous THF under N_2 was added 12 equiv

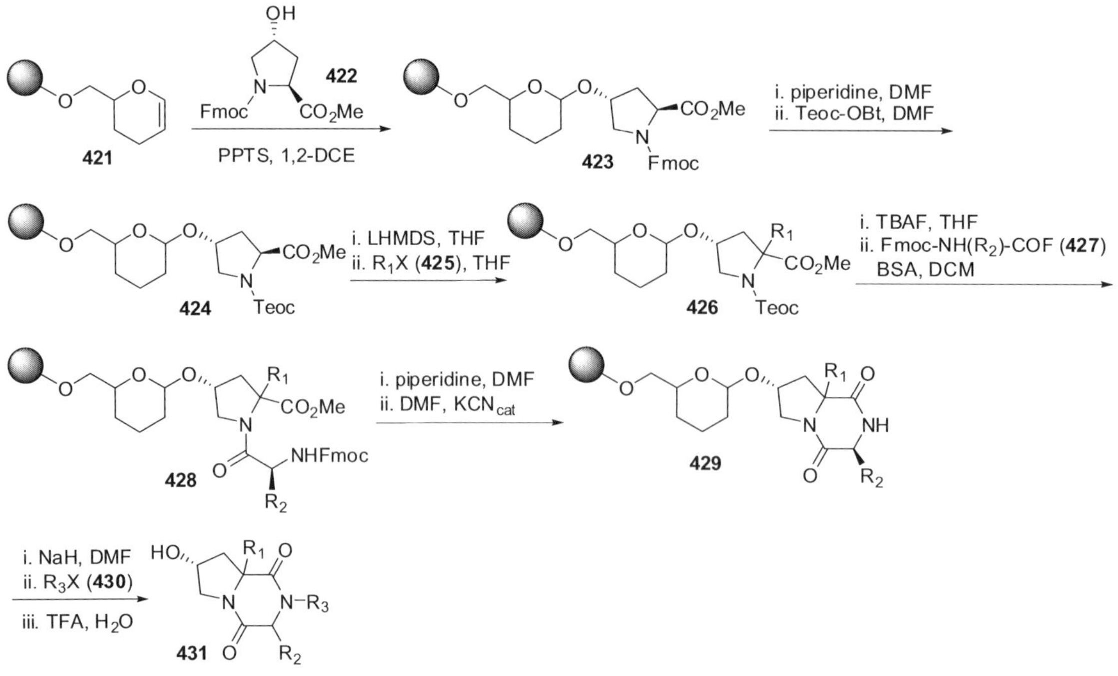

Scheme 9.50. Method 3 for the synthesis of 1,4-diketopiperazine derivatives.

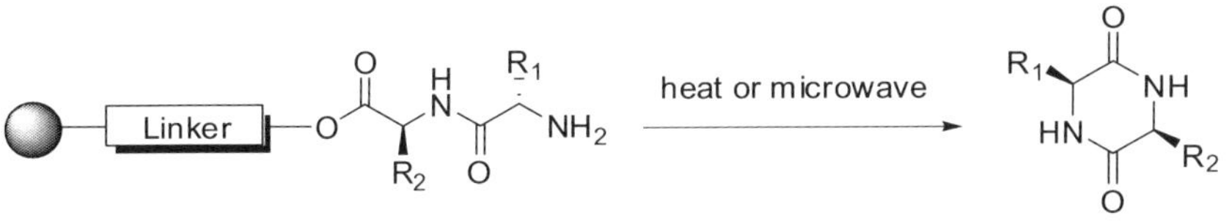

Scheme 9.51. Method 4 for the synthesis of 1,4-diketopiperazine derivatives.

of alkylation reagent **425** in THF and stirred for 5 min. Then 12 equiv of 1 M LHMDS THF solution was added and stirred at room temperature for 6 h to give resin **426**. Cleavage of the Toec group was achieved using 15 equiv of 1 M TBAF (tetrabutylammonium fluoride) in THF for 1 h, followed by treatment first with 5 equiv of BSA (*N,O*-bis(trimethylsilyl) acetamide) in anhydrous DCM for 5 h and then with 5 equiv of Fmoc-amino acid **427** in DCM shaken at rt for 24 h. This gave the key intermediate **428**. After removal of the Fmoc group with 25% piperidine in DMF, resin **428** was heated at 50–55°C with a catalytic amount of KCN for 18 h, affording product **429**. Alkylation of the secondary amine of the resin-bound diketopiperazine was achieved by swelling resin **429** in anhydrous DMF with 20 equiv of NaH and stirring for 7 h, followed by the addition of 20 equiv of alkylation reagent **430** at rt for 16 h. Cleavage from the resin in TFA:H_2O (4:1) gave the final product **431**.

Grøtli and coworkers used various resin supports and linkers and employed thermal heating or microwaves to turn dipeptides into DKPs (**Scheme 9.51**).[63] They found that PEGA-Ser resin is a good choice with water as the solvent under thermal heating for the formation of DKPs, while for microwave reactions, other solid supports in toluene:*t*-butanol (1:4) or *t*-BuOH, or toluene at 120°C, also performed very well.

Aldehyde **433** was mixed with a 2 M Fmoc-amino acid **435** DMF solution and the amino acid loaded cellulose support **432** was treated for 10 min three times with this mixture in H_2O at 25°C for maximum loading (**Scheme 9.52**).[64] Then isocyanide **434** was added to complete the Ugi-like reaction, generating dipeptide **436**. A solution of 10% acetyl chloride in methanol was used at 80°C with microwave irradiation for 30 min to facilitate the formation of the carboxylic methyl ester **437**. Resin **437** then underwent cyclization after the removal of the Fmoc group with 20% piperidine in DMF at room temperature for 1 h to give the DKP loaded resin **438**. After capping the secondary amine of DKP **438** with acetic anhydride in the presence of DIEA in DMF at room temperature for 20 min, photocleavage from the resin was carried out with 366 nm UV light in methanol at room temperature for 16 h, generating the final product **439**.

Starting from $PEGA_{800}$ support with a HMBA linker, Fmoc-Gly-OH was coupled to the linker in the presence of *N*-methylimidazole and MSNT in DCM (**Scheme 9.53**).[65] After removal of the Fmoc group, Fmoc-Thr-OH was coupled with NEM and TBTU in DMF to generate the resin-bound dipeptide **440**. Oxidation of the hydroxyl group of the threonine was achieved through the use of 7.5 equiv of Dess–Martin periodinane in anhydrous DCM for 16 h, producing **441**. After removal of the Fmoc group of **441** with piperidine/DMF, treatment with 95%TFA:TIPS (40:1) facilitated the formation of intermediate **442**. After treatment with 0.1 M NaOH for cleavage, oxidation occurred immediately to form the pyrazinone core. Neutralization with 0.1 M HCl gave the final product **443**.

Starting from MBHA resin **444**, coupling of Boc-amino acid **445** in the presence of DIPCDI and HOBt in DMF, followed by removal of the Boc group in 55% TFA in DCM, afforded the resin-bound amino acid **446** (Scheme 9.54).[66] Acylation with carboxylic acid **447** was carried out under the same coupling condition to give amide **448**, which underwent exhaustive reduction with borane–THF complex to give diamine **449**. Treatment with 10

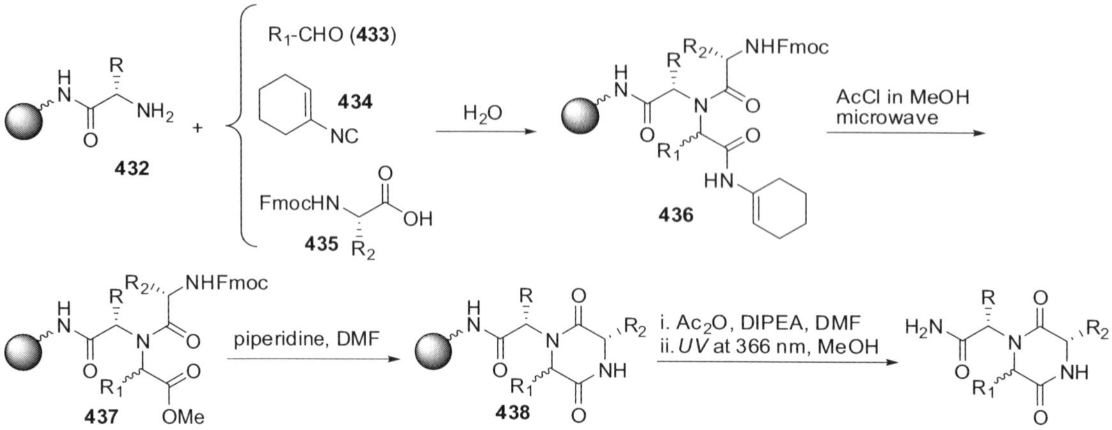

Scheme 9.52. Ugi-like reactions.

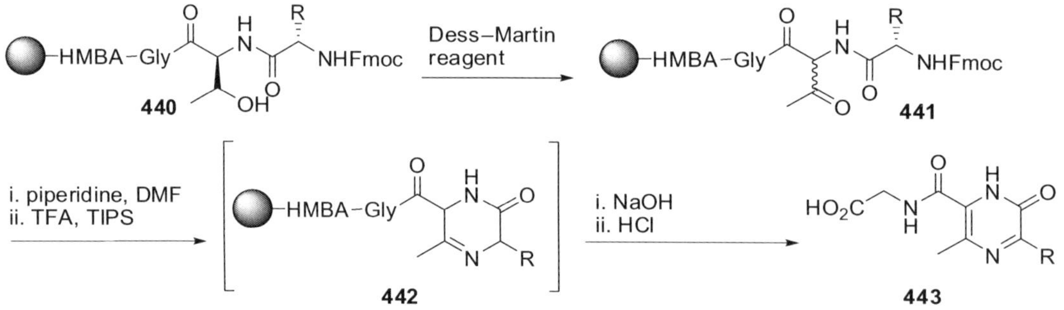

Scheme 9.53. Synthesis of pyrazinones.

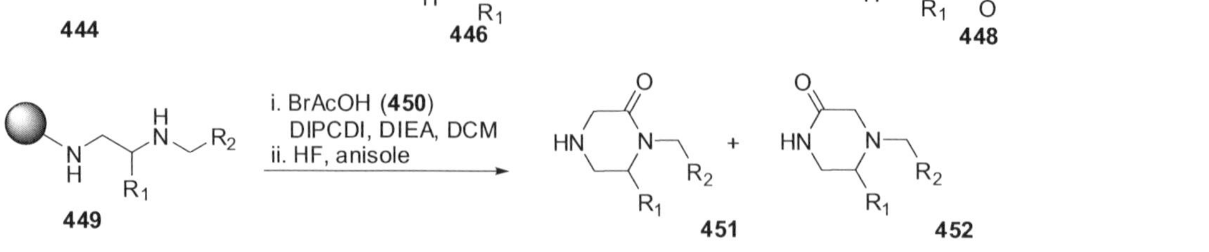

Scheme 9.54. Synthesis of piperazinones.

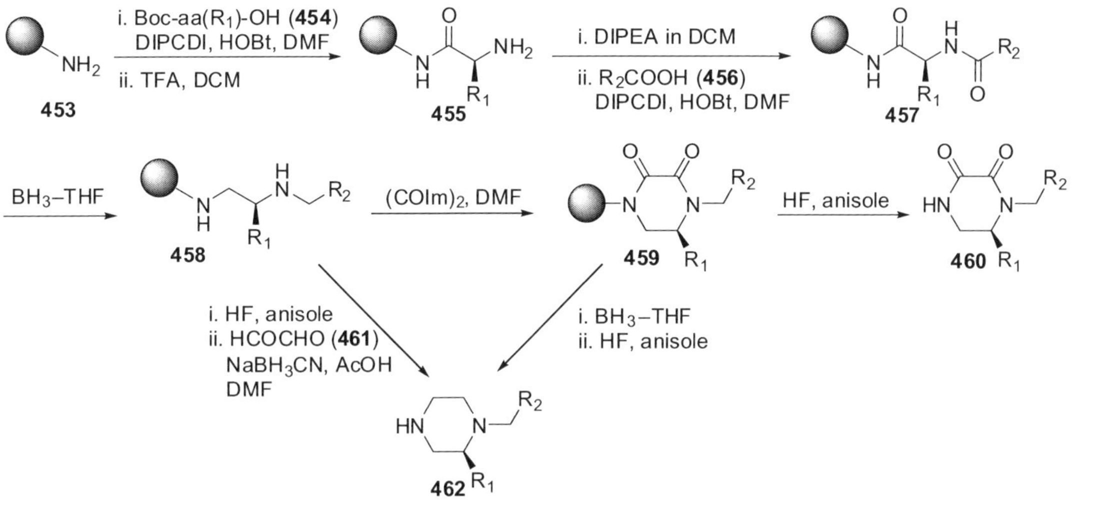

Scheme 9.55. Method 1 for the synthesis of piperidines.

equiv of bromoacetic acid **450**, 10 equiv of DIPCDI, and 5 equiv of DIEA in DCM facilitated the formation of the piperazinone core. Cleavage from resin was carried out by anhydrous HF with anisole at 0°C for 7 h, giving the piperazinone derivatives **451** and **452**, with **451** as the major isomer.

Starting with resin-bound diamine **458**, which was prepared by the same method as in Scheme 9.44, treatment with 0.1 M oxalylcarbodiimidazole in anhydrous DMF afforded the diketopiperazine **459** (**Scheme 9.55**).[67] Cleavage from the resin was achieved using HF/anisole method, forming **460**. Treatment of resin **459** with exhaustive reduction followed by HF/anisole cleavage gave the piperazine derivative **462**. Alternatively, treatment of resin **458** with HF/anisole, generating the free diamine, followed by treatment with oxalaldehyde **461** in the presence of $NaBH_3CN$ in 1% AcOH in DMF also produced the product **462**.

Starting from dipeptide loaded MBHA resin **463**, the free amine was protected by a trityl group using Trt-Cl in DCM:DMF (9:1) (**Scheme 9.56**).[68] Methylation of the remaining

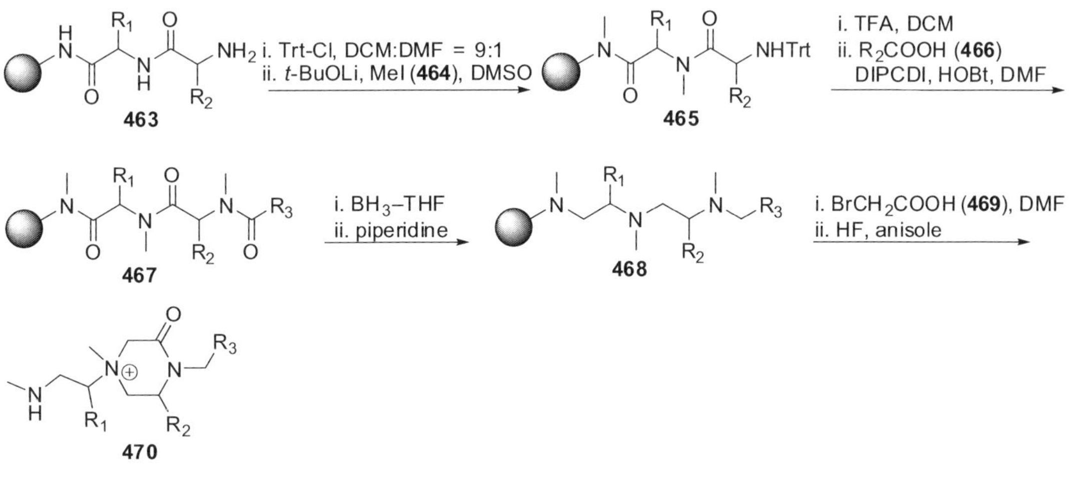

Scheme 9.56. Another method for the synthesis of piprazinones.

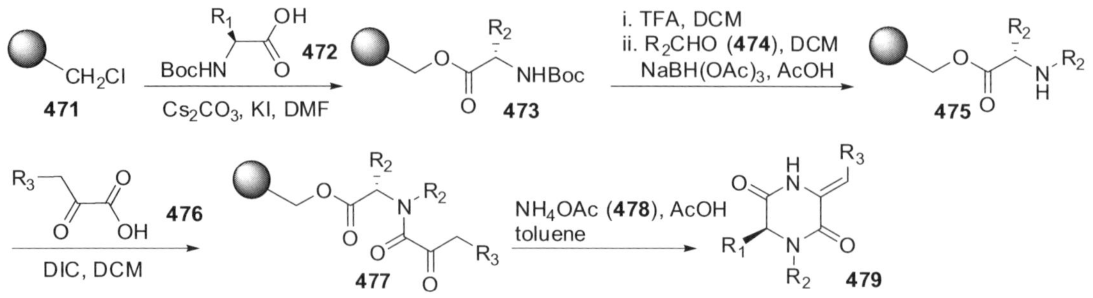

Scheme 9.57. Another method for the synthesis of 1,4-diketopiperazines.

secondary amines was carried out through treatment of the Trt-protected resin with 20 equiv of 0.5 M lithium *t*-butoxide in THF, and after decantation of excessive base, addition of 20 equiv of methyl iodide **464** in anhydrous DMSO afforded **465**. Removal of the trityl group was achieved using 2% TFA in DCM. This was followed by acylation with 10 equiv of carboxylic acid **466** in the presence of 10 equiv of both DIPCDI and HOBt in anhydrous DMF to give the acylated dipeptide **467**. After exhaustive reduction by the borane–THF method, the corresponding triamine **468** was treated with 10 equiv of bromoacetic acid **469** and 10 equiv of DIPCDI in anhydrous DMF, inducing the ring closure. Cleavage from the resin using the HF/anisole method afforded the piprazinone derivative **470**.

Boc-amino acid **472** was attached to **471** using Cs_2CO_3 and KI in DMF to afford **473** (**Scheme 9.57**).[69] After removal of the Boc group using the TFA/DCM method, treatment with aldehyde **474** and $NaBH(OAc)_3$ in the presence of acetic acid in DCM facilitated the reductive alkylation of the free amine. Resin **475** was then treated with oxoacetic acid derivative **476** and DIC in DCM, forming the key intermediate **477**, which was then treated with ammonium acetate **478** and acetic acid in toluene for cyclocleavage to generate the *E*- and *Z*-isomers of **479**.

Starting from Wang resin **480**, activated with *p*-nitrophenyl chloroformate **481** in the presence of DIEA in DCM, the resulting resin **482** was treated with lithium amino acid salt **483** to form salt **484** (**Scheme 9.58**).[70] Resin **484** then underwent alkylation with an alkyl iodide to give resin **485**. Alternatively, resin **485** could be directly obtained through treatment with a lithium *N*-alkyl amino acid salt **486** in DMF. Afterward, treatment with *N*-hydroxysuccinimide ester, HBTU, and DIEA in DMF, followed by addition of lithium glycinate **487**, generated resin-bound dipeptide **488**. Resin **488** in hot DCM was added to 3 equiv of EDC to form oxazolone derivative **489**, which was treated with aldehyde **490** and triethylamine in toluene at 80°C for 2 h to afford the *Z*-azalactone **491**. After cleavage from resin using TFA/DCM, treatment with triethylamine in toluene and heating at 80°C gave the final product **492**.

Synthesis of Benzopiperazinone Derivatives. Employing *ortho*-fluoro nitrobenzene derivatives as the main scaffold, the fluoro group is substituted by amino acids or amines. Subsequent reduction of the nitro group on the benzene ring triggers the cyclization. The key step in obtaining benzopiperazinones is the intramolecular formation of an amide bond.

MBHA resin **493** was treated with 6 equiv of Boc amino acid **494** and 6 equiv of both DIC and HOBt in DMF at room temperature for 2 h, followed by 55% TFA in DCM for 30 min, to give resin **495**, which underwent exhaustive reduction with borane–THF complex, affording the diamine **496** (**Scheme 9.59**).[71] Three equivalents of 4-fluoro-3-nitrobenzoic acid **497** and 3 equiv of HBTU as well as 6 equiv of DIEA in DMF at room

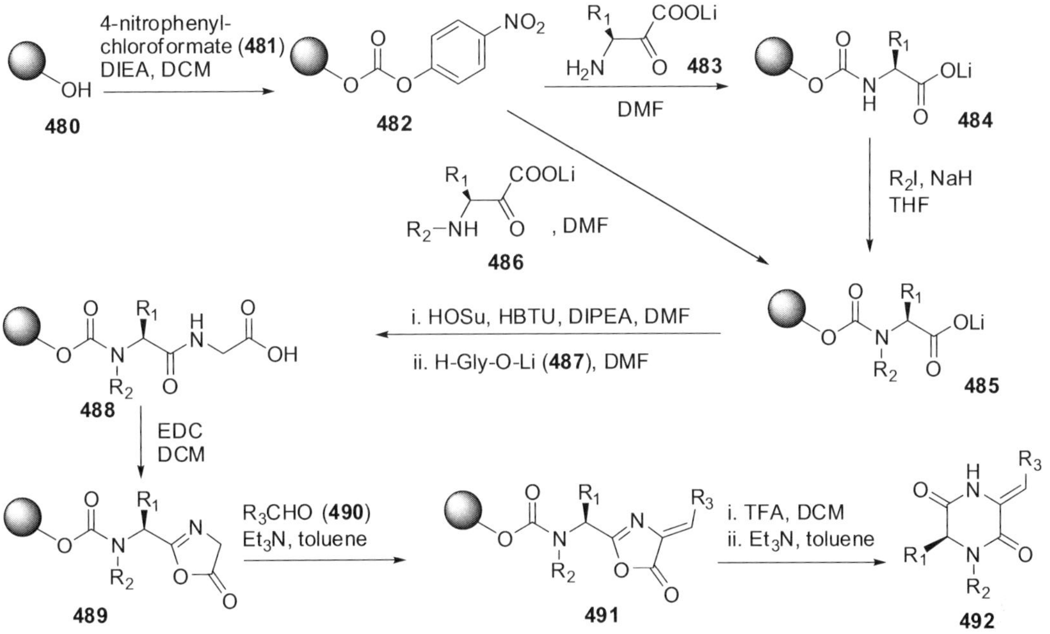

Scheme 9.58. A final method for the synthesis of 1,4-diketopiperazines.

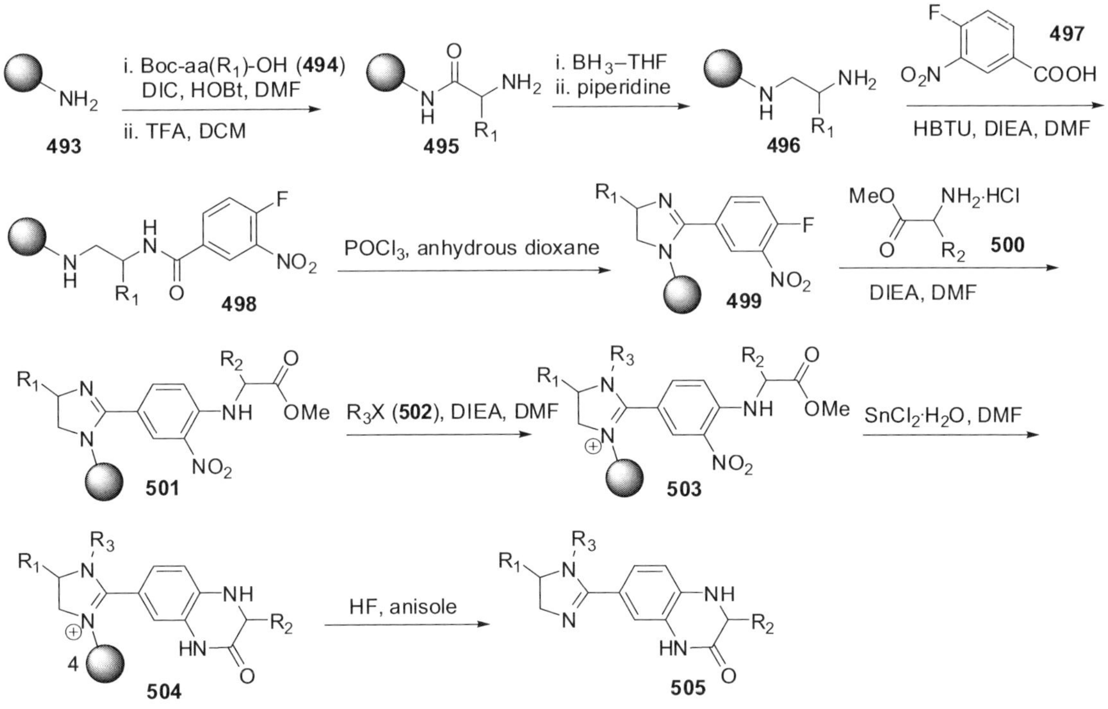

Scheme 9.59. Method 1 for the synthesis of benzopiperazinone derivatives.

temperature for 3 h were used to form the amide **498**. Formation of an imidazole ring was achieved by treating resin **498** with 10 equiv of phosphoryl chloride in anhydrous dioxane under N_2 and heating at 110°C for 2.5 h. Resin **499** was treated with 20 equiv of amino acid ester **500** and 25 equiv of DIEA for 20 h to give resin **501**. Adding 20 equiv of alkyl halide **502** to 10 equiv of DIEA in DMF at rt for 20 h facilitated alkylation of the tertiary amine on the imidazole ring in **503**. Reduction of the nitro group was carried out with 20 equiv of $SnCl_2{\cdot}2H_2O$ in DMF for 15 h, forming the benzopiperazinone product **504**. Cleavage from resin was achieved in anhydrous HF in the presence of anisole at 0°C for 7 h to afford **505**.

In the work of Jarman and coworkers, a more reliable reductive cyclization method was studied (**Scheme 9.60**).[72] Starting from resin-bound *ortho*-amino acid nitro benzene derivative **506** and subsequent cleavage from the resin using TFA/DCM method provides intermediate **507**. Repeated treatment with zinc powder in acetic acid was used for reduction to form **508**.

Starting from 4-(4-formyl-3-methoxyphenoxy)butyryl AM resin **509**, the reductive alkylation method was applied using amino acid ester **510** (**Scheme 9.61**).[73] Resin **511** was treated with *ortho*-fluoro nitrobenzene **512** in DMSO at 75°C for 1–3 days to give intermediate **513**. Reductive cyclization was carried out in a 2 M $SnCl_2{\cdot}2H_2O$ NMP solution under argon at room temperature for 2 h to obtain the resin-bound benzopiperazinone **514**, which was akylated by alkyl halide **515** in the presence of BEMP in DMF at rt for 2 h to yield **516**. Cleavage from the resin was achieved using TFA, gaseous HCl, or gaseous HF at rt for 2 h. Air oxidation of product **517** was carried out in methanol at rt overnight, generating the quinoxalinone derivative **518**.

Starting from Fmoc-amino acid loaded Wang resin **519** and after deprotection of the Fmoc group with 20% piperidine in DMF, 10 equiv of 4-fluoro-3-nitrobenzoic acid **520** was used along with 10 equiv of DIEA and catalytic amount of DMAP at room temperature for 24 h to afford the resin-bound benzoic acid **521** (note the different reaction position from that in Scheme 9.59) (Scheme 9.62).[74] Amidation of resin **521** was achieved by swelling the resin in DMF with 4 equiv of HBTU and DIEA and a catalytic amount of DMAP (the mixture was stirred for 15 min before addition of amine **522**), followed by stirring at rt for 4 h to generate resin **523**. Reductive cyclocleavage with 10 equiv of $SnCl_22H_2O$ in DMF at rt for 24 h gave benzopiperidinone **524**, which was either alkylated with 5 equiv of alkyl halide **525** in the presence of 3 equiv of Cs_2CO_3 in acetone at rt for 8 h to obtain **526**, or oxidized with 1.25 equiv of tetrachloro-1,4-benzoquinone in DMF at rt for 3 h to obtain quinoxalinone **527**.

Starting from Fmoc amide MBHA resin **528** and after removal of Fmoc group with 25% piperidine in DMF, treatment with 3 equiv of Alloc-β-amino acid **529** and 3.5 equiv of both DIC and HOBt in DMF, followed by shaking for 2 h, formed resin **530** (**Scheme 9.63**).[75] Resin **530** was treated with 1 equiv of ethylenediamine **531** and 1.1 equiv of DIEA in DMF for 36 h to afford resin **532**. Treatment of this with 2 equiv of 1 M lithium *tert*-butoxide in

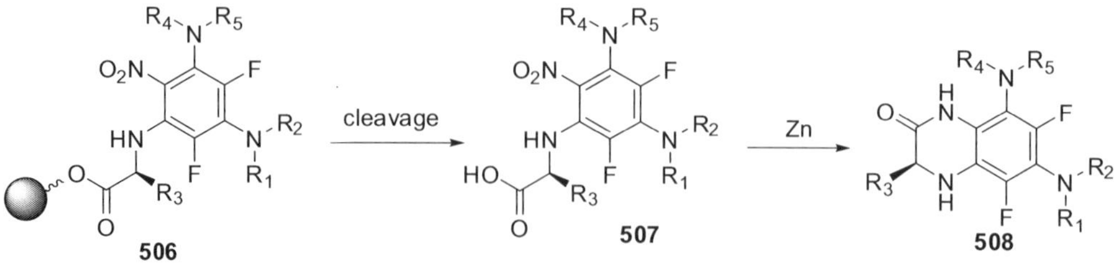

Scheme 9.60. Method 2 for the synthesis of benzopiperazinone derivatives.

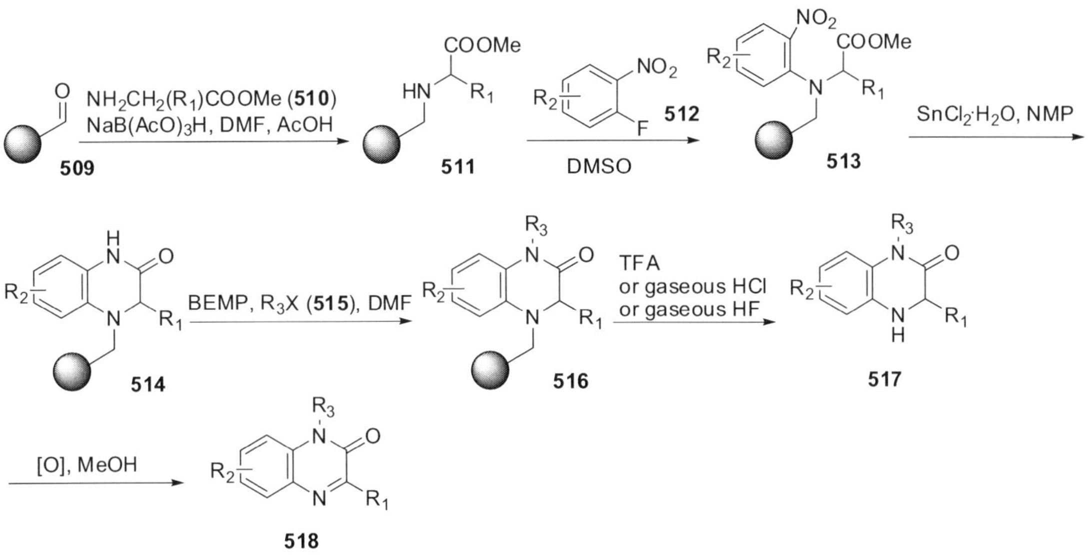

Scheme 9.61. Method 3 for the synthesis of benzopiperazinone derivatives.

THF and a small amount of DMF resulted in the key benzyne intermediate **533**, which was prone to nucleophilic attack on the primary amine. After formation of the tetrahydroquinoxaline ring, cleavage from resin using 95% TFA in H_2O for 2 h gave the final product **534**.

Synthesis of Pyrimidine Derivatives. Starting from Wang resin **535**, treatment with 12.5 equiv of Fmoc-amino acid **536**, 14 equiv of DIC, and 0.5 equiv of DMAP in DMF:DCM (9:1) at 40°C overnight gave resin **537** (**Scheme 9.64**).[76] After deprotection of the amine group with 20% piperidine in DMF to afford **538**, reductive alkylation of the free amine was achieved with 12.5 equiv of 2-nitrobenzaldehyde **539** with 1% AcOH in DMF:MeOH(4:1) stirred at room temperature for 1 h, followed by the addition of 12.5 equiv of $NaBH_3CN$

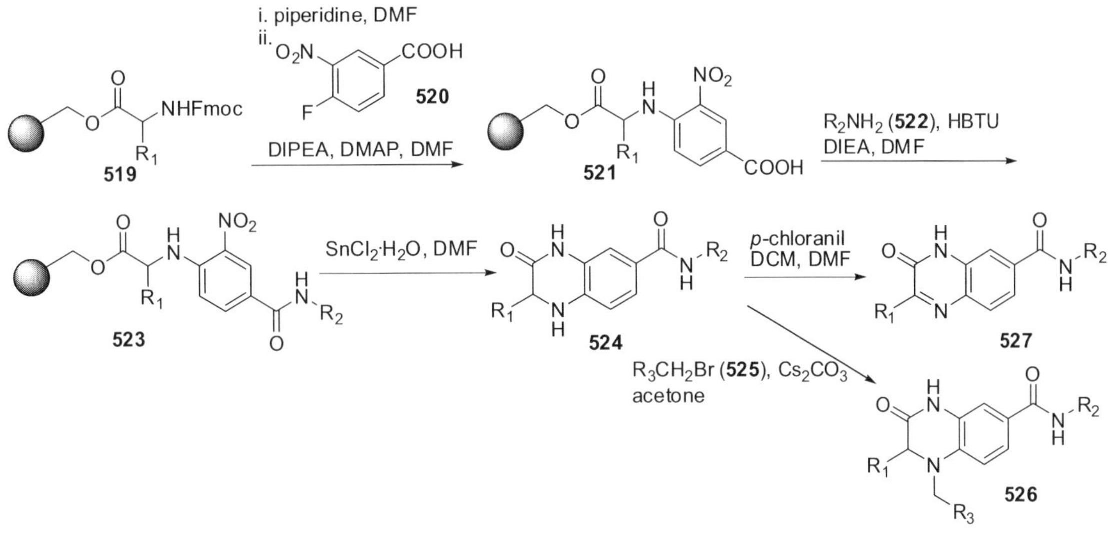

Scheme 9.62. Method 4 for the synthesis of benzopiperazinone derivatives.

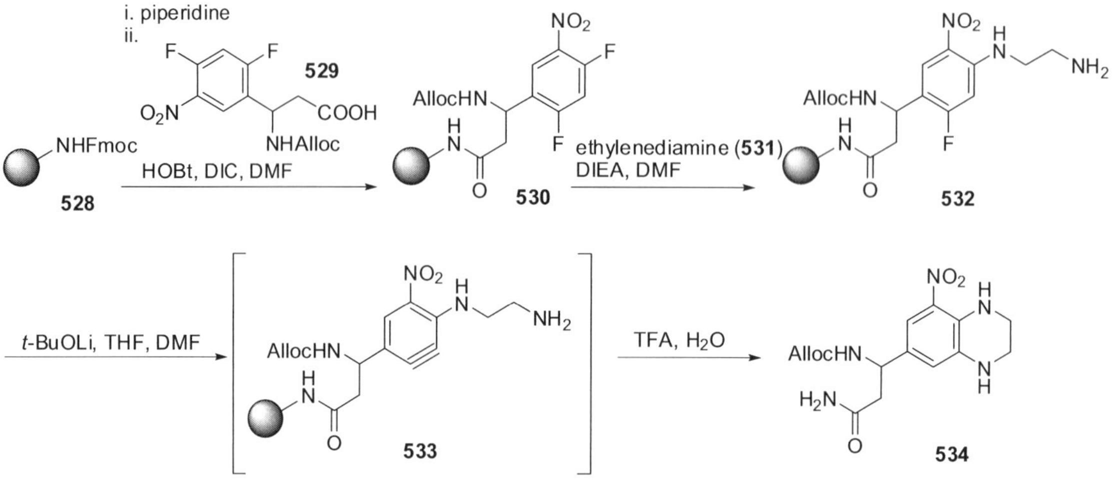

Scheme 9.63. Method 5 for the synthesis of benzopiperazinone derivatives.

stirred at 65°C overnight to yield **540**. Treatment of this with 3 M of $SnCl_2H_2O$ in NMP at 60°C overnight gave the resin-bound diamine **541**. Intermediate **541** was treated with 12.5 equiv of *N,N′*-disuccinimidyl carbonate (DSC) **542** in NMP at room temperature overnight, affording the quinazolinone derivative **545**. Intermediate **541** was also treated with TMOF (**543**):AcOH:NMP (32:1:8) at 65°C for 20 h, providing the dihydroquinazoline derivative **546**. Treatment of **541** with 4.5 equiv of aldehyde **544** in the presence of 2.25 equiv of DDQ in NMP at room temperature overnight gave the 2-R_3-dihydroquinazolinone derivative **547**.

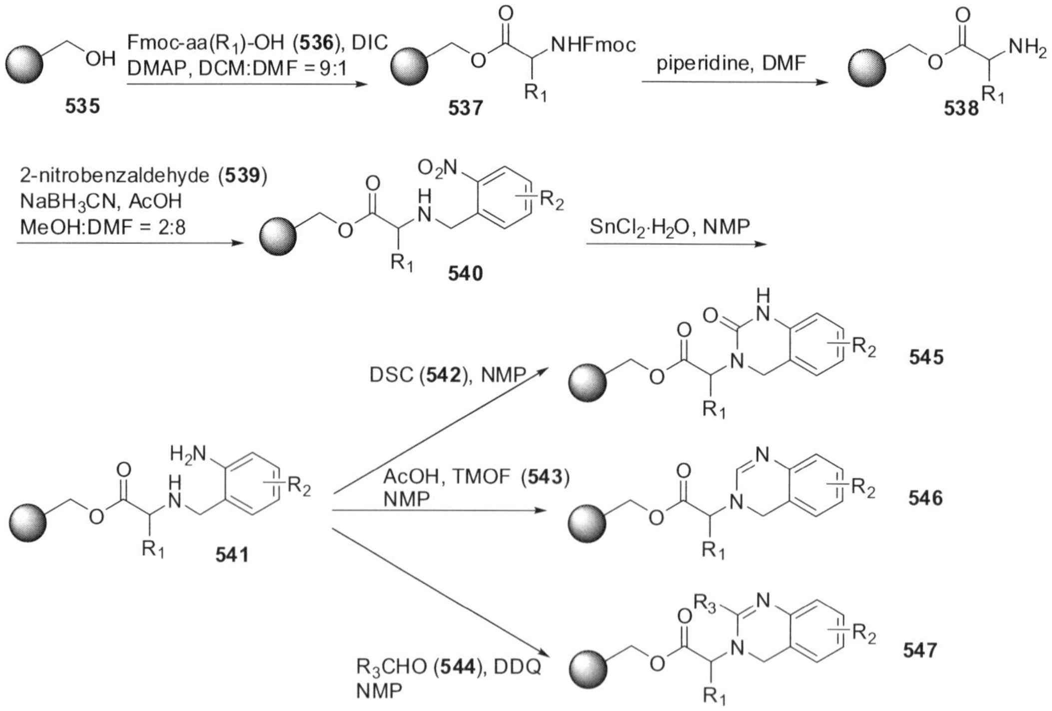

Scheme 9.64. Method 1 for the synthesis of pyrimidine derivatives.

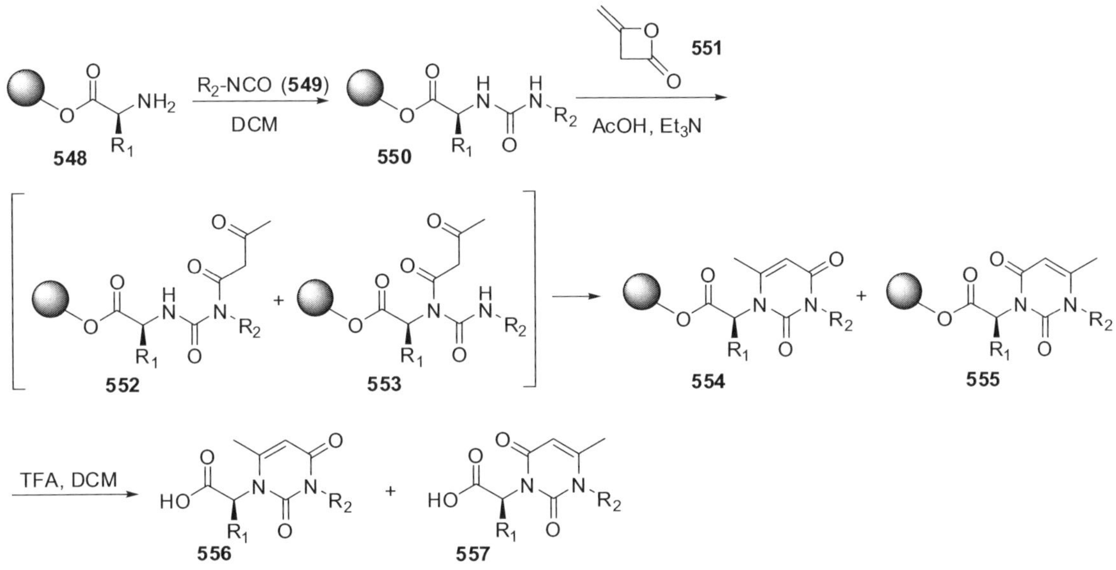

Scheme 9.65. Method 2 for the synthesis of pyrimidine derivatives.

Starting from amino acid loaded Wang resin **548**, treatment with 10 equiv of isocyanate **549** in DCM at room temperature for 16 h afforded the urea ester derivative **550** (**Scheme 9.65**).[77] Resin **550** was added to a mixture of AcOH:Et_3N:diketene (**551**) (200:5:7) and heated at 99°C for 3 h to obtain *in situ* intermediates **552** and **553**, which underwent cyclocondensation to give products **554** and **555**, respectively. Cleavage from resin was carried out with 20% TFA in DCM to form pyrimidinediones **556** and **557**.

Starting from Wang resin **558**, 5 equiv of Fmoc-amino acid **559**, 5 equiv of DIC, and 0.1 equiv of DMAP in DMF shaken at room temperature for 12 h facilitated the initial loading (**Scheme 9.66**).[78] After deprotection with 20% piperidine in DMF, resin **560** was treated with 10 equiv of diketene **561** and 0.1 equiv of DMAP in DCM at 0°C for 4 h to afford the amide resin **562**. This was then exposed to 10 equiv of 4-methyl benzaldehyde **563** in the presence of 0.5 equiv of piperidine and acetic acid (both 1 M in toluene) and heated at 80°C for 4 h to afford intermediate **564**. Cyclocondensation was achieved with 5 equiv of isothiourea hydrochloride **565** and 6 equiv of Cs_2CO_3 in NMP heated at 80°C for 24 h to produce the pyrazine core **566**. Cleavage from the resin was carried out in TFA:H_2O:TIPS (95:2.5:2.5) to produce **567**.

TentaGel-grafted polystyrene resin **568** was treated with 2,6-dimethoxy-4-hydroxybenzaldehyde **569** in the presence of Cs_2CO_3 in NMP to form linker **570** (**Scheme 9.67**).[79] A conventional reductive alkylation procedure was applied to resin **570** using benzylamine **571** to give resin **572**, which was coupled to Fmoc-amino acid **573**, with DIC, HOBt, and triethylamine serving as the activation reagents. Resin-bound amino acid **574** was first deprotected with 20% piperidine in DMF and then treated with 5 equiv of guanidine **575** in the presence of triethylamine for 22 h, providing resin **576**. Removal of the Alloc group with 20 equiv of phenylsilane and a catalytic amount of Pd$(PPh_3)_4$ in DCM for 3 h afforded intermediate **577**, which then underwent cyclocondensation with 10 equiv of acetoacetate **578** and 10 equiv of 6 M NaOMe in methanol for 22 h to afford the pyrazinone ring. Cleavage from the resin in neat TFA for 2 h provided the desired product **579**.

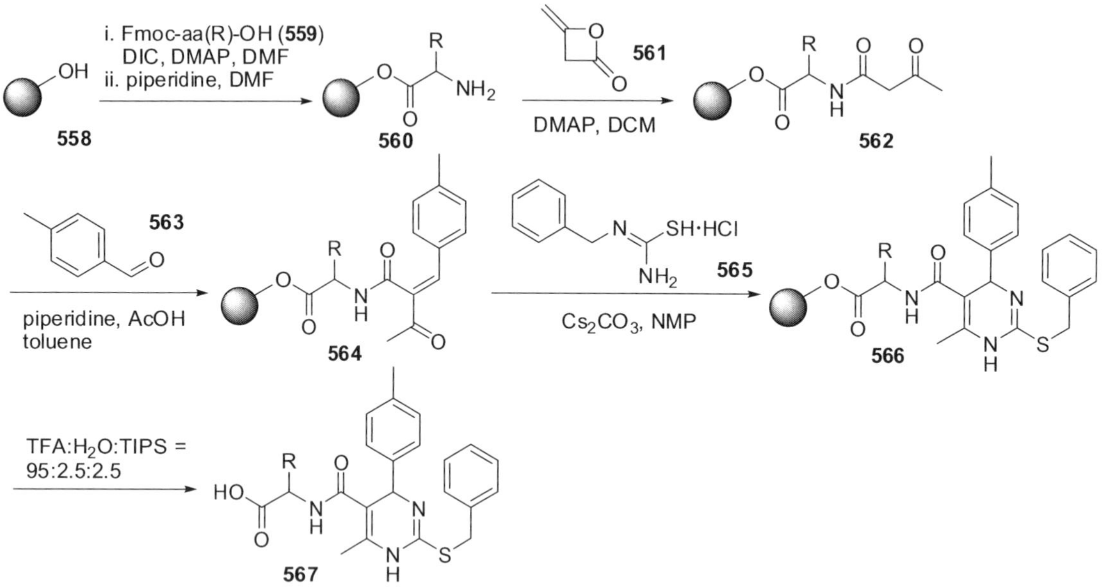

Scheme 9.66. Method 3 for the synthesis of pyrimidine derivatives.

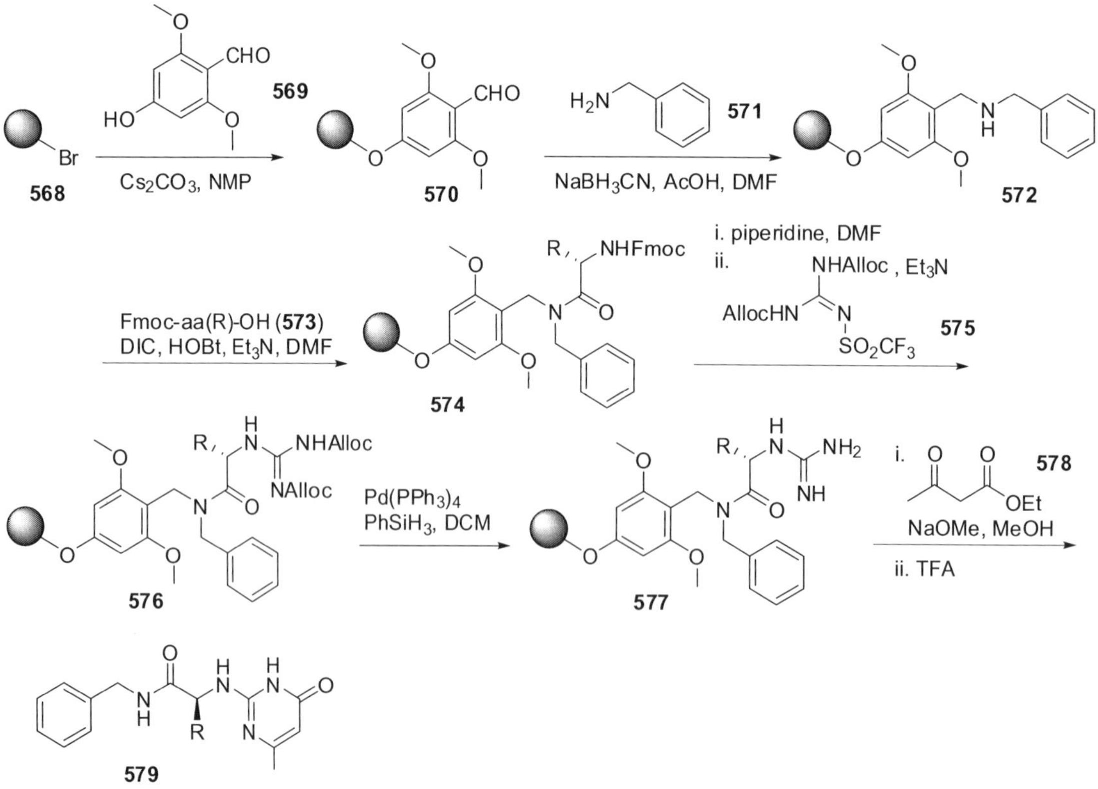

Scheme 9.67. Synthesis of pyrazinone derivatives.

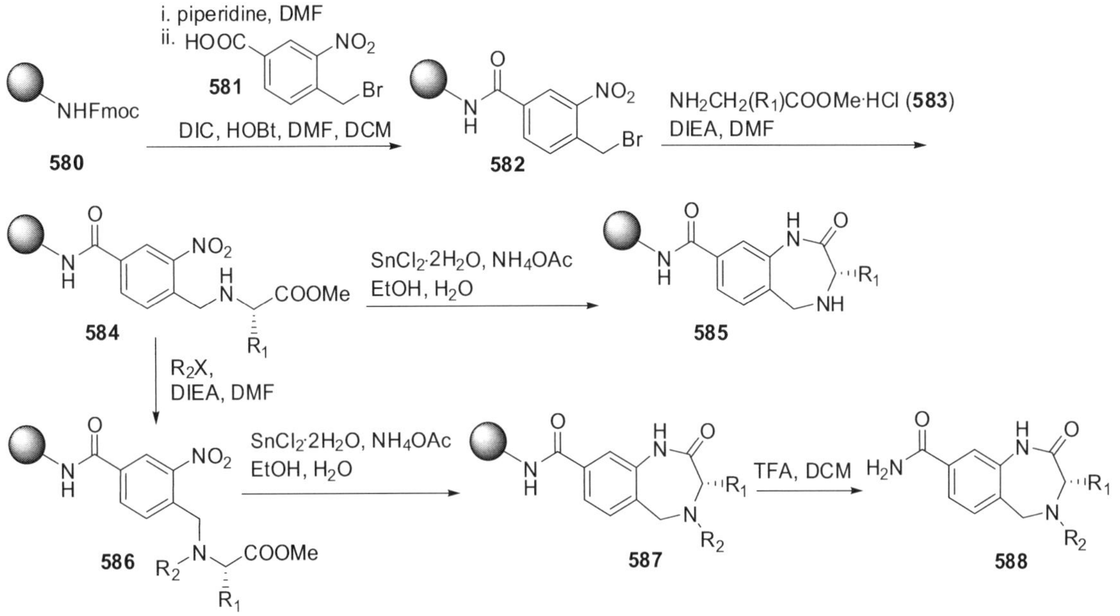

Scheme 9.68. Method 1 for the synthesis of benzodiazepines.

9.2.5 Seven-Membered Ring Heterocycles

Nitrobenzoic acid and its derivatives are commonly used in the synthesis of benzodiazepines and benzothiazepines. Formation of the desired seven-membered ring is usually achieved through an intramolecular amidation.

Fmoc-protected Rink amide polyamide D-series lantern **580** was treated with 20% piperidine in DMF for 40 min to remove the Fmoc group (**Scheme 9.68**).[80] Then treatment with a DMF:DCM (1:4) solution of 0.5 M 4-bromomethyl-3-nitrobenzoic acid **581**, 0.5 M DIC, and 0.005 M HOBt at room temperature for 16 h gave the benzoic amide **582**. Compound **582** was exposed to a DMF solution of 0.5 M amino acid ester **583** and 1 M DIEA in DMF at room temperature for 48 h to afford compound **584**. Reductive cyclization was achieved in a H_2O:EtOH (1:1) solution of 2 M $SnCl_2 \cdot 2H_2O/NH_4OAc$, heated at 90°C for 24 h, to give the benzodiazepine derivative **585**. Further treatment of compound **584** with 1 M alkyl halide and 1 M DIEA in DMF at 100°C for 24 h afforded the tertiary amine **586**, which underwent the same reductive cyclization step to give **587**. Cleavage from the resin was carried out with 20% TFA in DCM for 1 h to give **588**.

Fmoc-amino acid loaded Wang resin **591** was prepared through a conventional coupling method from resin **589** (Scheme 9.69).[81] After removal of the Fmoc group with 20% piperidine in DMF, 2-fluoro-5-nitrobenzoic acid **592** was coupled to the free amine with water-soluble carbodiimide (WSC), HOBt, and DMAP at room temperature for 16 h to give resin **593**. Replacement of the aryl-bound fluoro group was achieved by treatment with a secondary amine **594** in DMSO at room temperature, producing the intermediate **595**. Reduction of the nitro group was carried out with $SnCl_2 2H_2O$ in DMF for 16 h, giving compound **596**. Cyclocleavage was activated by using *t*-BuOLi or *t*-BuONa in THF at 50°C to generate the desired benzodiazepine **597**.

After removal of the Fmoc group from resin **598** with 20% piperidine in DMF for 30 min, the amino acid loaded Wang resin **599** was treated with 2-aminobenzoimine **600** in

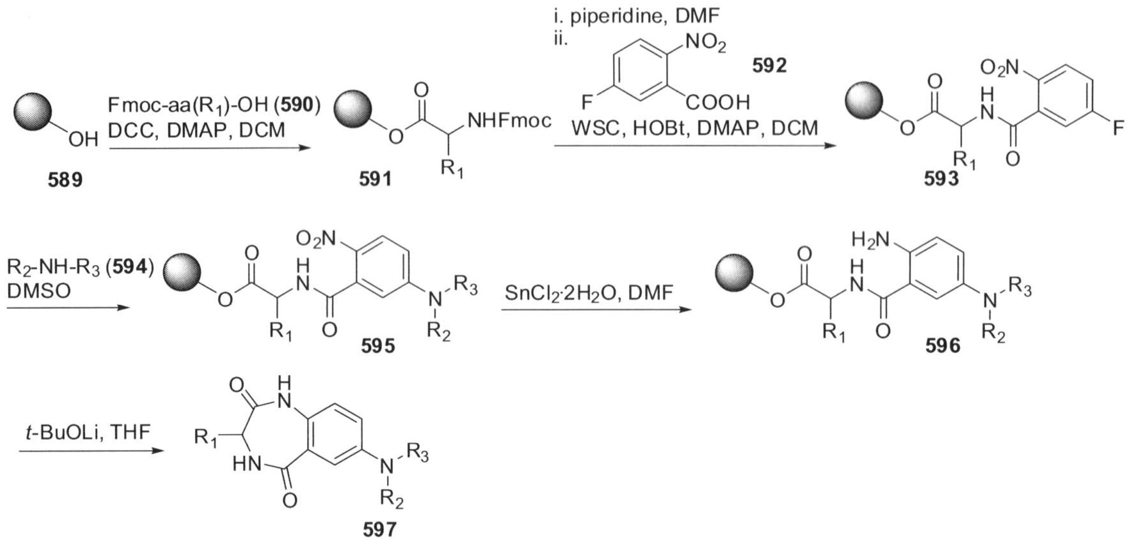

Scheme 9.69. Method 2 for the synthesis of benzodiazepines.

NMP and shaken at 140°C overnight to produce resin **601 (Scheme 9.70)**.[82] Cyclocleavage was facilitated by the use of TFA:1,2-DCE (1:9) for 2 h, resulting in the final product **602**.

Starting from polystyrene-supported Merrifield resin or TentaGel S Br resin **603**, 2,6-dimethoxy-4-hydroxybenzaldehyde **604** in DMF was mixed with 0.97 equiv of NaH then added to **603** and reacted at 50°C for 44–69 h to form **605 (Scheme 9.71)**.[83] Resin **605** was first treated with 2.1 equiv $NaBH(OAc)_3$ in 1% acetic acid/DMF solution for 30 min, then added to 2.1 equiv cyclohexylalanine ester **606**, and finally stirred at room temperature for 2 h to produce resin **607**. Formation of the key intermediate **609** was achieved by treating **607** in a NMP suspension with 5 equiv of EDC·HCl in NMP, stirring for 30 min, and then slowly adding 4 equiv of 5-nitroanthranilic acid **608** and stirring at rt for 20 h. Two different strategies were applied to resin **609**. Treatment with excess (20–30 equiv) lithium acetanilide (prepared by mixing acetanilide in THF and *n*-BuLi in dioxane under −78°C, followed by stirring for 30 min, and finally adding DMF to dissolve precipitate while warming to rt), stirring at rt for 25–67 h, then adding methyl iodide and further stirring for 22 h (for polystyrene support, formation of R group), or adding acetic acid and further stirring for 1 h (for TentaGel) generated the resin-bound benzodiazepine **610**. Resin **610** was treated with 20–30 equiv of $SnCl_2$ in DMF in order to reduce the aryl-bound nitro group and

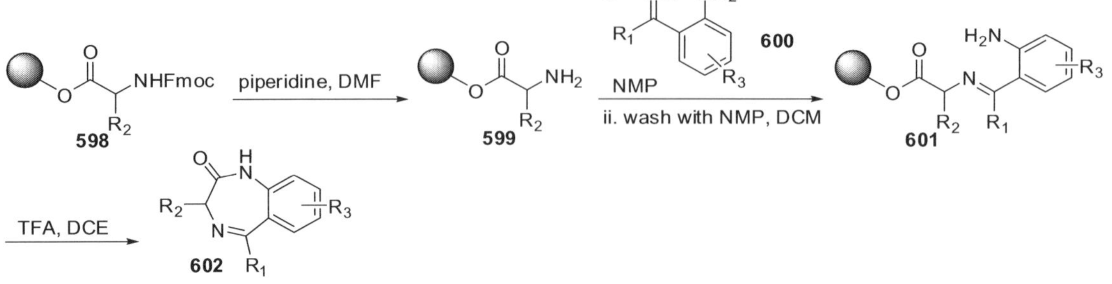

Scheme 9.70. Method 3 for the synthesis of benzodiazepines.

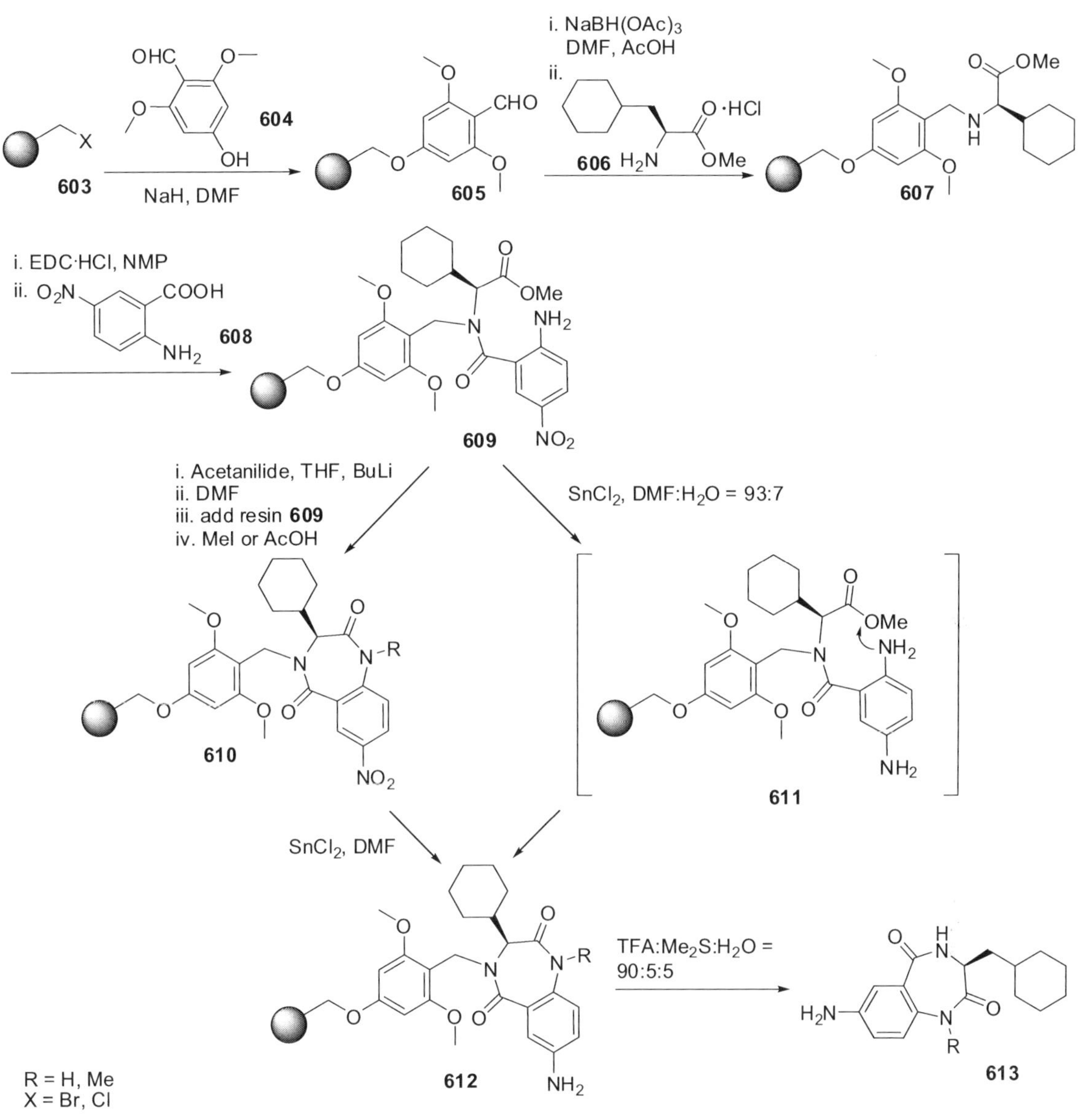

Scheme 9.71. Method 4 for the synthesis of benzodiazepines.

afford **612**. Another strategy to produce **612** from **609** involved treating **609** with $SnCl_2$ in DMF:H_2O (93:7) at 80°C for 70 h. Cleavage from the resin was carried out in TFA:Me_2S:MeOH (90:5:5) at rt for 16 h to afford **613**.

Starting with Fmoc-Cys(Trt)-OH loaded MBHA resin **614**, removal of trityl group was performed in TFA:*t*-BuSiH_3:DCM (10:5:85) for 30 min (**Scheme 9.72**).[84] The resulting product was then treated with 10 equiv of 2-fluoro-5-nitrobenzoic acid **615** overnight, generating resin **616**. After removal of the Fmoc group with 25% piperidine in DMF, a reductive alkylation on the free amine was carried out with 5 equiv of aldehyde **617** and 5 equiv of $NaBH_3CN$ in 1% acetic acid in anhydrous DMF solution for 1 h. Treatment of resin **618** under a coupling condition with 3 equiv of both HBTU and DIEA in DMF overnight

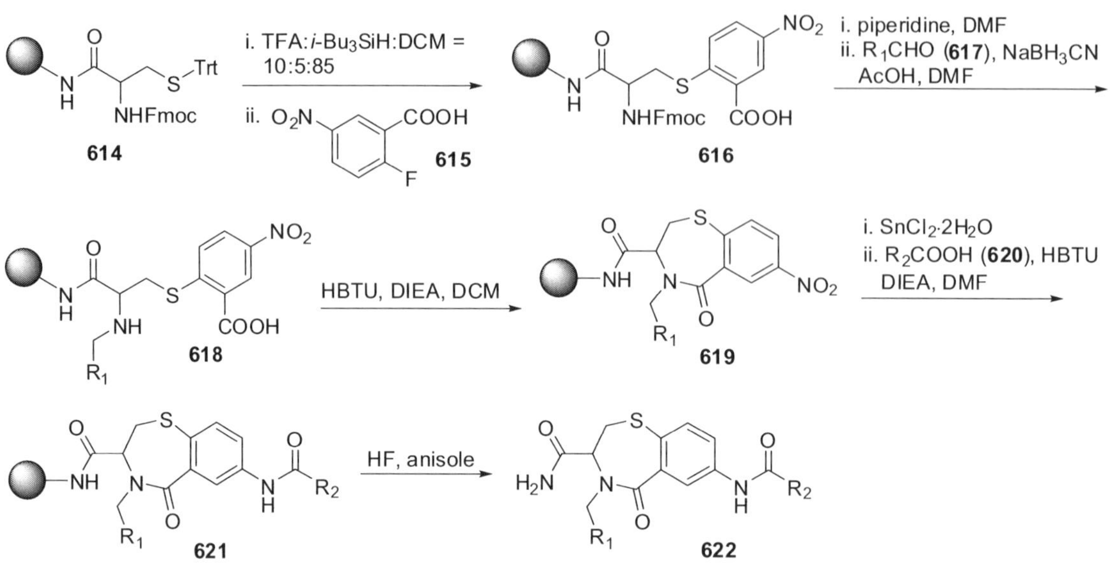

Scheme 9.72. Method 1 for the synthesis of benzothiazepine derivatives.

formed the benzothiazepine core **619**. Further treatment with a 2 M anhydrous DMF solution of $SnCl_2 2H_2O$ overnight generated the aryl-bound primary amine, which was coupled to the carboxylic acid **620** (15 equiv) in the presence of 15 equiv HBTU and 3 equiv DIEA to give product **621**. Exposure to anhydrous HF with anisole at 0°C delivered the desired benzothiazepine product **622**.

A similar strategy to obtain benzothiazepine derivatives was reported by Herpin and coworkers (**Scheme 9.73**).[85] To nitrophenyl carbonate-modified Wang resin **623** in DMF, 5 equiv of Cys-OH **624** and 15 equiv of BSA solution were added under N_2 at room temperature and left overnight, generating the resin **625**. Treatment with 10 equiv of *ortho*-nitrophenyl halides **626** and 10 equiv of DBU in DMF under N_2 at room temperature overnight gave the key intermediate **627**. After reduction of the nitro group with 10 equiv of $SnCl_2 2H_2O$ in DMF at 50°C overnight, resin **628** was treated with 5 equiv of EDC as the coupling reagent in NMP to give the benzothiazepine ring **629**. Further treatment with 12 equiv of *m*-CPBA at room temperature for 6 h in DCM afforded the oxidized sulfone derivative **630**, which underwent alkylation with 10 equiv of alkyl halide **631** in the presence of 10 equiv DBU in DMF at room temperature overnight to afford **632**. Cleavage from the resin with 50% TFA in DCM gave the desired product **633**.

Starting from MBHA resin **634**, resin **636** was obtained under conventional coupling and deprotection procedures used in other synthesis schemes by the Houghten and coworkers (**Scheme 9.74**).[86] After acylation of the primary amine using carboxylic acid **637** along with DIC and HOBt in DMF to afford **638**, exhaustive reduction with borane–THF complex was performed to give the diamine **639**. Treatment of this with 5 equiv of phenyl isocyanatoformate **640** in DMF under N_2 and heating at 60°C for 24 h accomplished the cyclocondensation to give the triazepine derivative **641**. Cleavage from the resin was achieved with the HF/anisole method to afford product **642**.

Starting from dipeptide loaded MBHA resin **643** and after exhaustive reduction with the borane–THF method, triamine **644** was treated with Dde-OH in DMF or Trt-Cl along with DIEA in DCM:DMF (9:1) to protect the primary amine in **645** (**Scheme 9.75**).[87] Treatment of this with 10 equiv of BrCN under N_2 in DCM gave the guanidine intermediate **646**, which

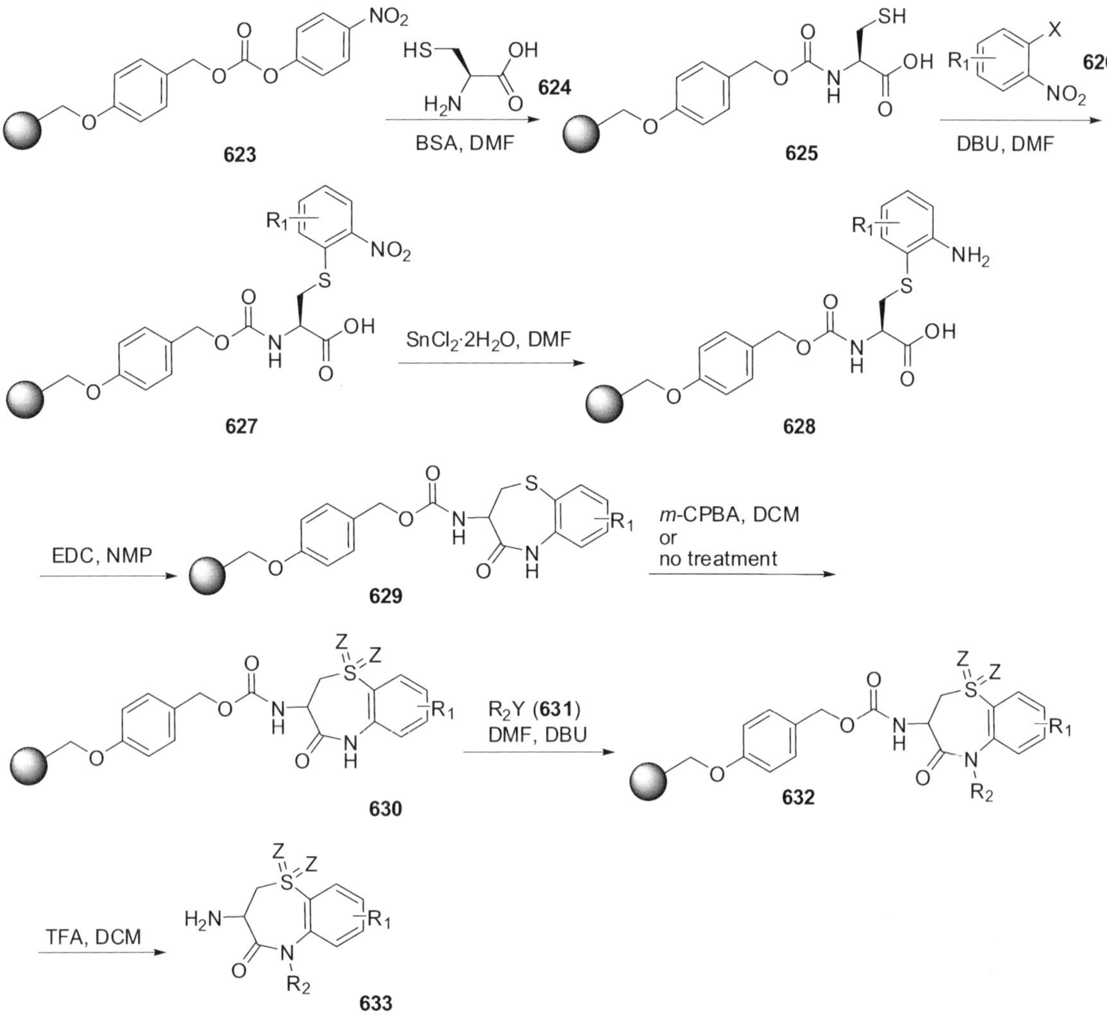

X = Cl, Br, F
Y = Br, Cl
Z = O, electron pairs

Scheme 9.73. Method 2 for the synthesis of benzothiazepine derivatives.

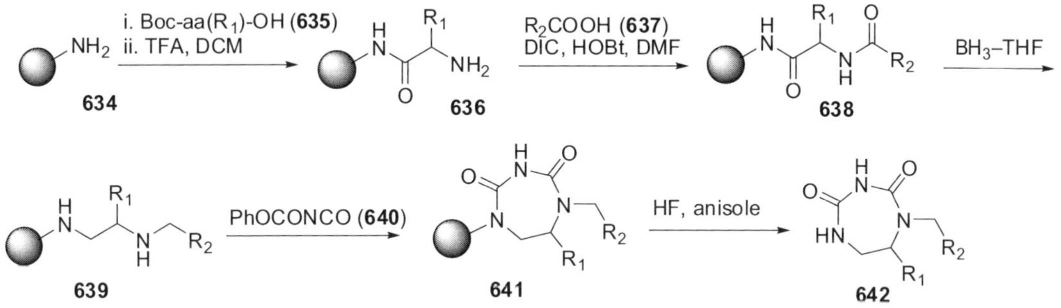

Scheme 9.74. Synthesis of seven-membered heterocycles with three nitrogen atoms.

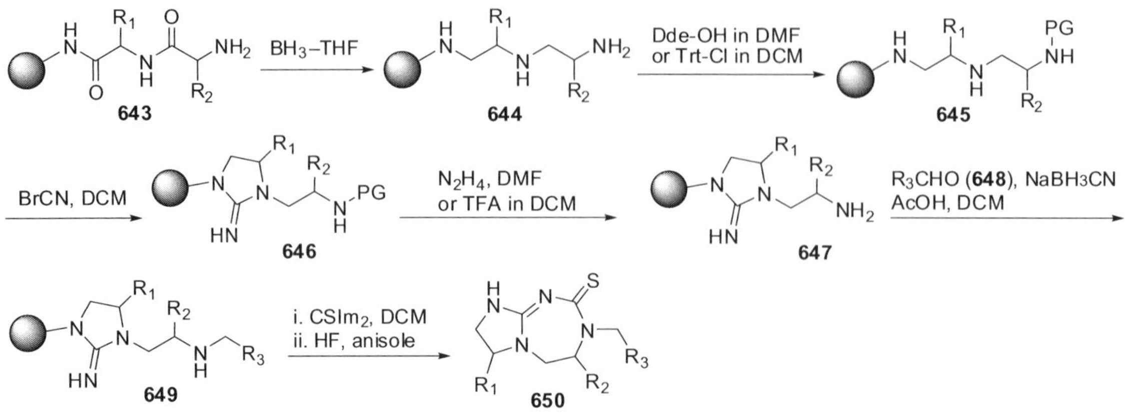

Scheme 9.75. Another synthesis of seven-membered heterocycles with three nitrogen atoms.

was treated with 1% TFA in DCM repeatedly for trityl removal or with 2% hydrazine in DMF for 1 h for removal of the Dde group, generating resin **647**. Reductive alkylation of the free amine was carried out using aldehyde **648** in the presence of $NaBH_3CN$ in 1% AcOH in DMF to afford **649**, which was exposed to 20 equiv of thiocarbonyldiimidazole in DCM under N_2 at room temperature for 15 h for ring closure. Treatment with anhydrous HF and anisole at 0°C gave the final product **650**.

REFERENCES

1. R. B. Merrifield, *J. Am. Chem. Soc.* **1963**, *85*, 2149–2154.
2. A. Nefzi, C. Dooley, J. M. Ostresh, R. A. Houghten, *Bioorg. Med. Chem. Lett.* **1998**, *8*, 2273–2278.
3. A. Nefzi, J. M. Ostresh, Y. Yu, R. A. Houghten, *J. Org. Chem.* **2004**, *69*, 3603–3609.
4. S. N. Filigheddu, S. Masala, M. Taddei, *Tetrahedron Lett.* **1999**, *40*, 6503–6506.
5. C. M. L. Delpiccolo, E. G. Mata, *Tetrahedron: Asymmetry* **2002**, *13*, 905–910.
6. C. M. L. Delpiccolo, E. G. Mata, *Tetrahedron Lett.* **2004**, *45*, 4085–4088.
7. C. M. L. Delpiccolo, L. Méndez, M. A. Fraga, E. G. Mata, *J. Comb. Chem.* **2005**, *7*, 331–334.
8. L. Méndez, C. M. L. Delpiccolo, E. G. Mata, *Synlett* **2005**, *10*, 1563–1566.
9. G. Gerona-Navarro, M. Royo, M. T. García-López, F. Albericio, R. González-Muñiz, *Mol. Diversity* **2003**, *6*, 75–84.
10. P. Pérez-Faginas, M. T. Aranda, L. Coady, M. T. García-López, R. González-Muñiz, *Adv. Synth. Catal.* **2008**, *350*, 2279–2285.
11. A. L. Vergnon, R. S. Pottorf, M. P. Winters, M. R. Player, *J. Comb. Chem.* **2004**, *6*, 903–910.
12. B. A. Kulkarni, A. Ganesan, *Tetrahedron Lett.* **1998**, *39*, 4369–4372.
13. D. M. Fitch, K. A. Evans, D. Chai, K. J. Duffy, *Org. Lett.* **2005**, *7*, 5521–5524.
14. A. Nishida, M. Fuwa, S. Naruto, Y. Sugano, H. Saito, M. Nakagawa, *Tetrahedron Lett.* **2000**, *41*, 4791–4694.
15. E. Biron, J. Chatterjee, H. Kessler, *Org. Lett.* **2006**, *8*, 2417–2420.
16. K.-H. Park, M. J. Kurth, *Tetrahedron Lett.* **1999**, *40*, 5841–5844.
17. A. Boeijen, J. A. W. Kruijtzer, R. M. J. Liskamp, *Bioorg. Med. Chem. Lett.* **1998**, *8*, 2375–2380.

18. M. Bauser, M. Winter, C. A. Valenti, K.-H. Wiesmüller, G Jung, *Mol. Diversity* **1998**, *3*, 257–260.
19. S.-H. Lee, S.-H. Chung, Y.-S. Lee, *Tetrahedron Lett.* **1998**, *39*, 9469–9472.
20. N. Heine, L. Germeroth, J. Schnider-Mergener, H. Wenschuh, *Tetrahedron Lett.* **2001**, *42*, 227–230.
21. E. Colacinl, F. Larnaty, J. Martinez, I. Parrot, *Tetrahedron Lett.* **2007**, *48*, 5317–5320.
22. M.-J. Lin, C.-M. Sun, *Tetrahedron Lett.* **2003**, *44*, 8739–8742.
23. G. Bhalay, D. Cowell, N. D. Hone, M. Scobie, A. D. Baxter, *Mol. Diversity* **1998**, *3*, 195–198.
24. M. Tremblay, N. Voyer, S. Boujabi, G. F. Dewynter, *J. Comb. Chem.* **2002**, *4*, 429–435.
25. F. Albericio, J. Garcia, E. L. Michelotti, E. Nicolás, C. M. Tice, *Tetrahedron Lett.* **2000**, *41*, 3161–3163.
26. F. Albericio, L. M. Bryman, J. Garcia, E. L. Michelotti, E. Nicolás, C. M. Tice, *J. Comb. Chem.* **2001**, *3*, 290–300.
27. M. Li, L. J. Wilson, *Tetrahedron Lett.* **2001**, *42*, 1455–1458.
28. Y. Yu, J. M. Ostresh, R. A. Houghten, *Tetrahedron* **2002**, *58*, 3349–3353.
29. U. E. W. Lange, *Tetrahedron Lett.* **2002**, *43*, 6857–6860.
30. J. I. Gavrilyuk, G. Evindar, R. A. Batey, *J. Comb. Chem.* **2006**, *8*, 237–246.
31. Y. Yu, J. M. Ostresh, R. A. Houghten, *J. Comb. Chem.* **2001**, *3*, 521–523.
32. D. H. Drewry, C. Ghiron, *Tetrahedron Lett.* **2000**, *41*, 6989–6992.
33. J. Li, Z. Zhang, E. Fan, *Tetrahedron Lett.* **2004**, *45*, 1267–1269.
34. M. Rinnová, A. Vidal, A. Nefzi, R. A. Houghten, *J. Comb. Chem.* **2002**, *4*, 209–213.
35. A. N. Acharya, J. M. Ostresh, R. A. Houghten, *J. Comb. Chem.* **2001**, *3*, 578–589.
36. A. N. Acharya, J. M. Ostresh, R. A. Houghten, *Tetrahedron* **2001**, *57*, 9911–9914.
37. J. Lee, A. Doucette, N. S. Wilon, J. Lord, *Tetrahedron Lett.* **2001**, *42*, 2635–2638.
38. A. Nefzi, J. M. Ostresh, M. Giulianotti, R. A. Houghten, *J. Comb. Chem.* **1999**, *1*, 195–198.
39. A. Nefzi, M. A. Giulianotti, R. A. Houghten, *J. Comb. Chem.* **2001**, *3*, 68–70.
40. A. Nefzi, M. A. Giulianotti, R. A. Houghten, *Tetrahedron Lett.* **2000**, *41*, 2283–2287.
41. A. N. Achyara, C. Thai, J. M. Ostresh, R. A. Houghten, *J. Comb. Chem.* **2002**, *4*, 496–500.
42. A. N. Acharya, J. M. Ostresh, R. A. Houghten, *J. Comb. Chem.* **2001**, *3*, 612–623.
43. A. N. Acharya, J. M. Ostresh, R. A. Houghten, *J. Comb. Chem.* **2002**, *4*, 214–222.
44. A. N. Acharya, J. M. Ostresh, R. A. Houghten, *J. Org. Chem.* **2001**, *66*, 8673–8676.
45. A. N. Acharya, J. M. Ostresh, R. A. Houghten, *Tetrahedron Lett.* **2002**, *43*, 1157–1160.
46. M. T. Bilodeau, A. M. Cunningham, *J. Org. Chem.* **1998**, *63*, 2800–2801.
47. A. Kamal, V. Devaiah, K. L. Reddy, Rajenda, R. V. C. R. N. C. Shetti, N. Shankaraiah, *J. Comb. Chem.* **2007**, *9*, 267–274.
48. Y. Yu, H. M. El Abdellaoui, J. M. Ostresh, R. A. Houghten, *Tetrahedron Lett.* **2001**, *42*, 623–625.
49. A. N. Acharya, A. Nefzi, J. M. Ostresh, R. A. Houghten, *J. Comb. Chem.* **2001**, *3*, 189–195.
50. J. M. Ostresh, C. C. Schoner, V. T. Hamashin, A. Nefzi, J.-P. Meyer, R. A. Houghten, *J. Org. Chem.* **1998**, *63*, 8622–8623.
51. M. W. Wilson, A. S. Hernádez, A. P. Calvet, J. C. Hodges, *Mol. Diversity* **1998**, *3*, 95–112.
52. S. K. Samanta, J. Yli-Kauhaluoma, *J. Comb. Chem.* **2005**, *7*, 142–146.
53. C. W. Tornøe, C. Christensen, M. Meldal, *J. Org. Chem.* **2002**, *67*, 3057–3064.
54. J. I. Gavrilyuk, G. Evindar, J. Y. Chen, R. A. Batey, *J. Comb. Chem.* **2007**, *9*, 644–651.
55. A. Nefzi, M. Giulianotti, R. Houghten, *Tetrahedron Lett.* **1998**, *39*, 3671–3674.
56. R. Mortezaei, S. Ida, D. A. Campbell, *Mol. Diversity* **1999**, *4*, 143–148.
57. W. Chai, W. V. Murray, *Tetrahedron Lett.* **1999**, *40*, 7185–7188.
58. T. Demaude, L. Knerr, P. Pasau *J. Comb. Chem.* **2004**, *6*, 768–775.

59. Mikhail F. Gordeev, *Biothechnol. Bioeng. (Comb. Chem.)* **1998**, *61*, 13–16.
60. R. A. Smith, M. A. Bobko, W. Lee, *Bioorg. Med. Chem. Lett.* **1998**, *8*, 2369–2374.
61. D.-X. Wang, M.-T. Liang, G.-J. Tian,H. L., H.-Q. Liu, *Tetrahedron Lett.* **2002**, *43*, 865–867.
62. A. Bianco, C. P. Snoksen, P. Roepstorff, J.-P. Briand, *J. Org. Chem.* **2000**, *65*, 2179–2187.
63. M. Tullberg, K. Luthman, M. Grøtli, *J. Comb. Chem.* **2006**, *8*, 915–922.
64. Q. Lin, H. E. Blackwell, *Chem. Commun.* **2006**, 2884–2886.
65. C. Christensen, C. W. Tornøe, M. Medal, *QSAR Comb. Sci.* **2004**, *23*, 109–116.
66. A. Nefzi, R. A. Mimna, R. A. Houghten, *J. Comb. Chem.* **2002**, *4*, 542–545.
67. A. Nefzi, M. A. Giulianotti, R. A. Houghten, *Tetrahedron Lett.* **1999**, *40*, 8539–8542.
68. A. Nefzi, R. T. Santos, *Tetrahedron Lett.* **2006**, *47*, 3819–3822.
69. W.-R. Li, S.-Z. Peng, *Tetrahedron Lett.* **1998**, *39*, 7373–7376.
70. W.-R. Li, J. H. Yang, *J. Comb. Chem.* **2002**, *4*, 106–108.
71. A. N. Acharya, J. M. Ostresh, R. A. Houghten, *Tetrahedron* **2002**, *58*, 221–225.
72. R. J. Holland, I. R. Hardcastle, M. Jarman, *Tetrahedron Lett.* **2002**, *43*, 6435–6437.
73. V. Krchňák, L. Szabo, J. Vágner, *Tetrahedron Lett.* **2000**, *41*, 2835–2838.
74. E. Laborde, B. T. Peterson, L. Robinson, *J. Comb. Chem.* **2001**, *3*, 572–577.
75. S. Dixon, X. Wang, K. S. Lam, M. J. Kurth, *Tetrahedron Lett.* **2005**, *46*, 7443–7446.
76. A. Ortega, M. Erra, E. Navarro, R. S. Roberts, D. Fernández-Forner, *QSAR Comb. Sci.* **2006**, *25*, 598–604.
77. A. Wahhab, J. Leban, *Tetrahedron Lett.* **2000**, *41*, 1487–1490.
78. L. Zhang, T. M. Rana, *J. Comb. Chem.* **2004**, *6*, 457–459.
79. C. W. Zapf, M. Goodman, *J. Org. Chem.* **2003**, *68*, 10092–10097.
80. Z. Wu, F. Ercole, M. FitzGerald, S. Perera, P. Riley, R. Chempbell, Y. Pham, P. Rea, S. Sandanayake, M. N. Mathieu, A. M. Bray, N. J. Ede, *J. Comb. Chem.* **2003**, *5*, 166–171.
81. C. Migihashi, F. Sato, *J. Heterocyclic Chem.* **2003**, *40*, 143.
82. L. S. Lautsen, C. K. Sams, *J. Comb. Chem.* **2007**, *9*, 1094–1103.
83. P. Ettmayer, S. Chloupek, K. Weigand, *J. Comb. Chem.* **2003**, *5*, 253–259.
84. A. Nafzi, N. A. Ong, M. A. Giulianotti, J. M. Ostresh, R. A. Houghten, *Tetrehedron Lett.* **1999**, *40*, 4939–4942.
85. G. C. Morton, J. M. Salvino, R. F. Labaudinière, T. F. Herpin, *Tetrahedron Lett.* **2000**, *41*, 3029–3033.
86. Y. Yu, J. M. Ostresh, R. Houghten, *Org. Lett.* **2001**, *3*(18), 2797–2799.
87. C. E. Hoesl, J. M. Ostresh, R. A. Houghten, A. Nefzi, *J. Comb. Chem.* **2006**, *8*, 127–131.

10

GENERATION OF DRUG-LIKE FIVE-MEMBERED HETEROCYCLIC LIBRARIES USING CARBON DISULFIDE AND MERRIFIELD RESIN

Young-Dae Gong and Taeho Lee

10.1 INTRODUCTION

Combinatorial chemistry has become an extremely powerful technique for the rapid generation of small drug-like organic molecule libraries for the purpose of medicinal chemistry programs within the pharmaceutical industry.[1–3] Well over half of all therapeutic agents contain heterocyclic skeletons that serve as ideal scaffolds on which pharmacophores can be appended to yield potent and selective drugs.[4,5] In combinatorial synthesis, solid-phase organic synthesis (SPOS) is now routinely used to prepare a large number of small heterocyclic drug-like molecules and is especially useful in creating massive numbers of hit and lead compounds as part of high-throughput screening technologies.[5,6] This is especially true for five-membered heterocyclic compounds that are core components of a large number of substances that possess a wide range of interesting biological activities.[7]

Substituted five-membered ring heterocyclic compounds and their fused counterparts offer a high degree of structural diversity and have proven to be broadly useful as therapeutic agents. In this respect, various approaches for the preparation of these privileged structures with drug-like properties have been developed on solid-phase strategies.[8,9]

The sulfur atom is the fundamental element of various five-membered ring heterocycles,[10] such as thiophene, thiazole, isothiazole, 1,2,3-thiadiazole, 1,2,4-thiadiazole, and 1,3,4-thiadiazole, and is the key element of the reaction intermediate and linkers for solid support.[8,11] Several well-known sulfur source reagents, such as carbon disulfide, thiourea,

Solid-Phase Organic Synthesis: Concepts, Strategies, and Applications, First Edition.
Edited by Patrick H. Toy and Yulin Lam.

thiophosgene, phosphorus sulfide (P_2S_5), Lawesson's reagent, 1,1-thiocarbonyldiimidazole, di-(2-pyridyl)thionocarbonate (DPT), and isothiocyanate, have been used to synthesize various heterocycles.

Thus, many synthetic methods have been investigated for the synthesis of five-membered ring heterocycles and their fused counterparts using sulfur-containing reagents as a sulfur source, linker, substituted site, or intermediate. Among recent examples, many research groups reported the synthesis of 2,3,4,5-tetrasubstituted thiophenes,[12] 1,3,4-thiadiazoles,[13] and 1,3,4-triazolo[3,4-*b*][1,3,4]thiadiazoles and 1,3,4-triazolo[3,4-*b*][1,3,4]thiadiazines[14] using carbon disulfide as a sulfur source and substituted site in solution phase. The carbon disulfide was used as a sulfur source in the solution-phase synthesis of arylthiobenzothiazoles,[15] 4,6-dithiopyrazolo[3,4-*d*]pyrimidines,[16] and 1,2,3,4-tetrahydro-7-thia-6,9,10,11-tetraaza-benzo[*c*]fluorenes.[17] The solution-phase synthesis of 3-alkyldithiocarbonyl-oxazolidines[18] and 5-thia-1,3,8-triaza-cyclopenta[*b*]naphthalen-7-ones[19] was developed by using carbon disulfide as a substituted site. The carbon disulfide was used as a useful reagent in the solution-phase synthesis, such as an intermediate in the synthesis of 2-arylamino-2-imidazolines,[20] as an intermediate in the formation of isothiocyanate for the synthesis of substituted thienopyrimidine-4-ones,[21] and as a ring formation agent in the synthesis of benzimidazoles.[22] In addition, some research groups established the synthesis of 2-amino-5-sulfanylthiazoles using isothiocyanate as a sulfur source,[23] and the synthesis of 1,3,4-thiadiazoles using DPT as a sulfur source,[24] or using isothiocyanate as a sulfur source and diversity element[25] on a solid support. The solid-phase synthesis of 5-arylalkylidene rhodanines using 1,1-thiocarbonyldiimidazole as a sulfur source and linker was reported.[26] The flow chemical synthesis of 4,5-disubstituted thiazoles and imidazoles in combination with immobilized base was developed by using isothiocyanate as a sulfur source and substituted site.[27]

Among these sulfur-containing reagents, *carbon disulfide* is a facile, cheap, and versatile reagent and a starting material for the synthesis of various sulfur and heterocyclic compounds.[28] Recently, solid-phase syntheses of the five-membered ring heterocycles using carbon disulfide and a solid support were developed.[29] For example, thiophenes (from Wang resin),[30] thiazolines (from Merrifield resin),[31] 2-thioxothiazolidin-4-ones (rhodamines) (from Wang resin),[32] 2-thioxoimidazolidin-4-ones (thiohydantoins) (from PEG 3400 (poly(ethylene glycol))),[33] benzimidazoles (from Rink amide resin),[34] imidazoloquinazolines (from AM resin),[35] and 1,3,4-oxadiazoles (from Merrifield resin)[36] were synthesized on solid support with carbon disulfide.

As part of our current interest in SPOS, we examined the efficient and rapid synthesis of nitrogen- or sulfur-containing five-membered ring heterocycles and their fused counterparts by solid-phase strategies using Merrifield resin **1** and carbon disulfide (Figure 10.1).[37–46] In this chapter, we review the synthesis of these five-membered ring heterocycles and their fused counterparts on solid support with carbon disulfide and Merrifield resin **1** as a solid support.

10.2 SOLID-PHASE SYNTHESIS OF RELATED THIAZOLE COMPOUNDS

10.2.1 Solid-Phase Synthesis of 2,4,5-Trisubstituted Thiazoles

Thiazoles are useful heterocycles and building blocks and a prominent structural element of compounds used to treat cancer, bacterial, fungal, and viral infections.[47] Because the thiazole derivatives exhibit a wide range of important biological activities, they serve as

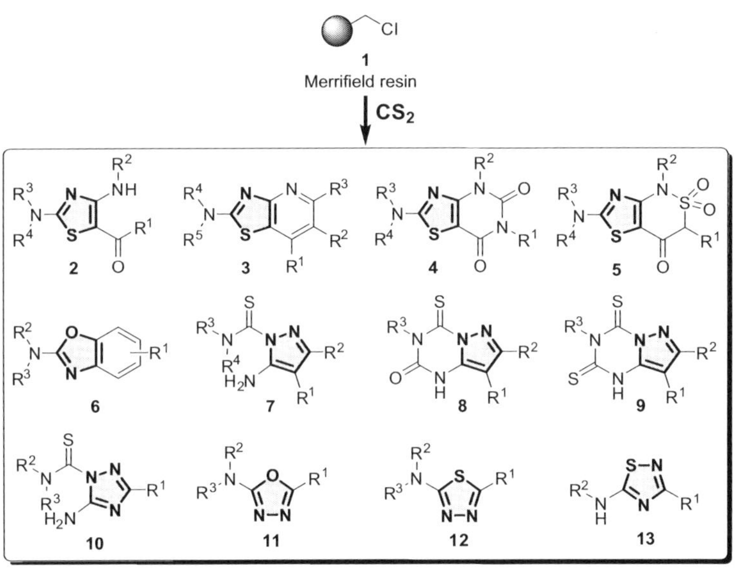

Figure 10.1. Drug-like five-membered heterocycles and their fused heterocycles from Merrifield resin and carbon disulfide.

attractive targets for combinatorial library construction. We have previously described an efficient, facile, and rapid solid-phase synthetic strategy for 2,4,5-trisubstituted thiazoles **2** using a traceless cleavage[48] of the 2-sulfonyl linker of thiazole.[37,49]

The general solid-phase synthetic procedure for thiazole derivatives started with the reaction of commercial Merrifield resin **1** with dipotassium cyanodithioimidocarbonate **14**,[50] which was prepared from carbon disulfide, cyanamide, and KOH in aqueous ethanol. When DMF (*N,N*-dimethylformamide) was used as a solvent, solid-supported cyanocarbonimidodithioate **15** was obtained with a good loading capacity (Scheme 10.1). In addition, the one-pot, three-component reaction of Merrifield resin **1** with carbon disulfide and cyanamide for resin **15** showed a loading capacity reduced by about 40% compared to the stepwise pathway. The resin **15** was treated with 2-bromoacetophenone, monosubstituted 2-bromoacetophene, or ethyl 2-bromoacetate (R^1 diversity element) and triethylamine at 80 °C to afford the corresponding thiazole resin **16** via Thorpe–Ziegler cyclization. After the sulfonyl resin **16** was oxidized to form sulfone resin **17** by treatment with *m*-chloroperoxybenzoic acid (*m*CPBA),[51,52] the desired thiazoles **18** were liberated from resin **17** by nucleophilic addition of various amines (R^3R^4N diversity elements: primary and secondary amines, and anilines). The isolated overall yields for thiazoles **18** (11 examples, >95% purities, Table 10.1) ranged from 25% to 36% from the Merrifield resin **1**.

The acylation with acid chloride and the urea formation with isocyanate of intermediate resin **16** afforded other substituent groups onto 4-aminothiazole. Under microwave (MW) irradiation reaction[53] with isocyanate and acylation reaction with acid chloride, R^2-substituted thiazole resin **19** was obtained. Following conversion of sulfonyl resin **19** to sulfonyl resin **20** (*m*CPBA/CH_2Cl_2), substitution reactions promoted by treatment with appropriate amines (R^3R^4N diversity elements) furnished the 2,4,5-trisubstituted thiazoles **2** (29 examples; 36–25% isolated yields from Merrifield resin **1**, >95% purities, Table 10.1).

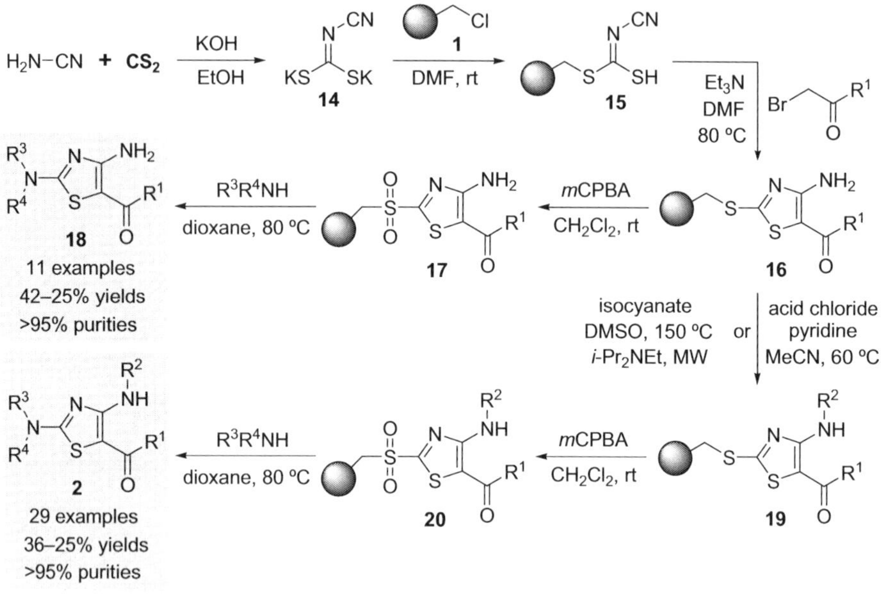

Scheme 10.1. Solid-phase synthesis of 2,4,5-trisubstituted thiazoles.

TABLE 10.1. 2,4,5-Trisubstituted Thiazoles **2** and **18** and Their Yields[a]

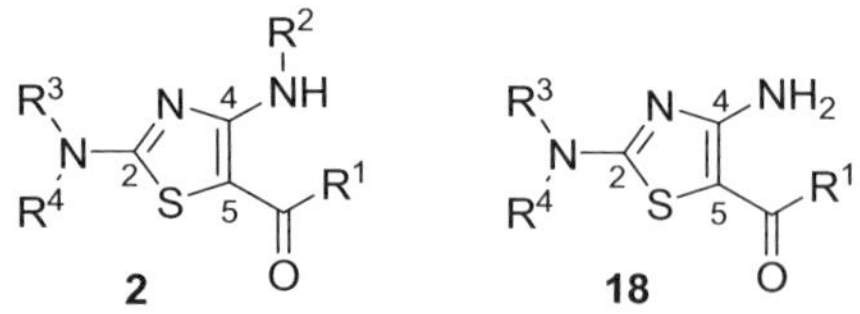

Entry	Products	R^1	R^2	R^3R^4N	Yield (%)[b,c]
1	**18a**	Ph (phenyl)	–	*n*-PrNH	35
2	**18b**	Ph	–	BnNH	28
3	**18c**	Ph	–	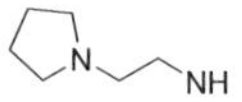	32
4	**18d**	Ph	–	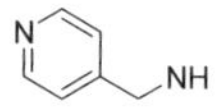	28
5	**18e**	Ph	–	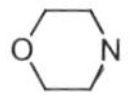	38
6	**18f**	Ph	–	Et_2N	32
7	**18g**	Ph	–	$(i\text{-}Pr)_2N$	42
8	**18h**	Ph	–	PhNH	28
9	**18i**	Ph	–	4-F-PhNH	25
10	**18j**	OEt	–	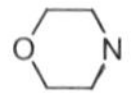	38

TABLE 10.1. (*Continued*)

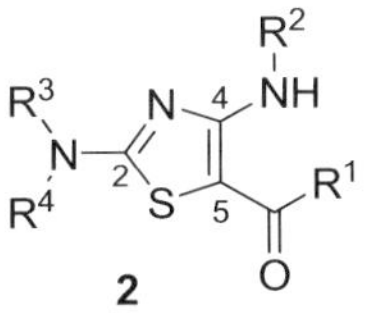

2

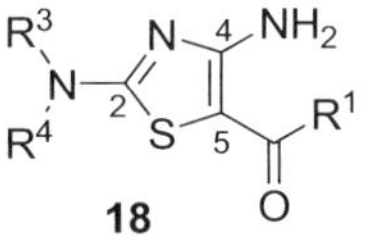

18

Entry	Products	R^1	R^2	R^3R^4N	Yield (%)[b,c]
11	**18k**	OEt	–	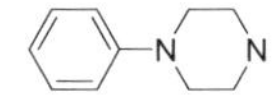	41
12	**2a**	Ph	PhNHCO	*n*-PrNH	25
13	**2b**	Ph	PhNHCO	Et_2N	36
14	**2c**	Ph	PhNHCO	PhNH	28
15	**2d**	Ph	BnNHCO	*n*-PrNH	28
16	**2e**	Ph	BnNHCO	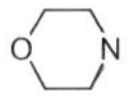	25
17	**2f**	Ph	Ac (acetyl)	*n*-PrNH	30
18	**2g**	Ph	Bz (benzoyl)	*n*-PrNH	29
19	**2h**	Ph	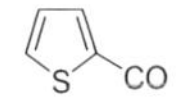	Et_2N	32
20	**2i**	OEt	PhNHCO	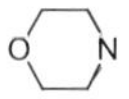	30
21	**2j**	OEt	Bz	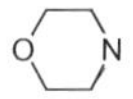	29
22	**2k**	OEt	*t*-BuCO	BnNH	31
23	**2l**	4-F-Ph	*t*-BuCO	BnNH	26
24	**2m**	4-MeO-Ph	Ac	BnNH	34
25	**2n**	4-MeO-Ph	Bz	BnNH	34
26	**2o**	3-MeO-Ph	*n*-BuCO	BnNH	33
27	**2p**	3-MeO-Ph	Bz	BnNH	31
28	**2q**	$4-NO_2-Ph$	$C_6H_{11}CO$	BnNH	28
29	**2r**	$4-NO_2-Ph$	Bz	BnNH	27
30	**2s**	$3-NO_2-Ph$	*t*-BuCO	BnNH	29

[a] All reactions were performed on 200–250 mg scale of resin **17** (for **18**) or **20** (for **2**).
[b] Four-step (for **18**) or five-step (for **2**) overall isolated yield from Merrifield resin **1** (loading capacity = 0.94 mmol/g).
[c] Purities of **2** and **18** were over 95% as judged from LC–MS traces or ^1H NMR spectroscopy.

10.2.2 Solid-Phase Synthesis of 2,5,6,7-Tetrasubstituted Thiazolo[4,5-*b*]pyridines

The fused-thiazole heterocycles are important structural components of bioactive molecules and, as a result, serve as attractive targets for combinatorial library construction

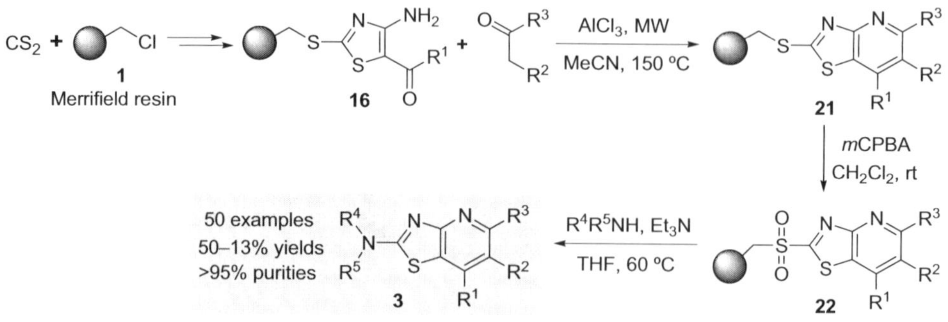

Scheme 10.2. Solid-phase synthesis of 2,5,6,7-tetrasubstituted thiazolo[4,5-*b*]pyridines.

via solution- and solid-phase synthesis.[1,54] Among the fused thiazoles, thiazolo[4,5-*b*] pyridine derivatives **3** exhibit a wide range of biological properties.[55] Thus, many solution-phase synthetic methods have been documented for thiazolo[4,5-*b*]pyridine derivatives using carbon disulfide as a starting material.[51,56,57] Kirsch and coworkers reported a solution-phase synthesis of 7-aminothiazolo[4,5-*b*]pyridines and thiazolo[4,5-*d*][1,2,3] triazines using carbon disulfide as a starting material for 4-amino-1,3-thiazole-5-carbonitrile intermediate.[56] Also, Johnson et al. developed a synthesis of 7-chloro-2-methylsulfanyl-thiazolo[4,5-*b*]pyridine-6-carbonitrile using carbon disulfide as a starting material for 4-aminothiazole.[51] We represented the first solid-phase synthesis of 2,5,6,7-tetrasubstituted thiazolo[4,5-*b*]pyridines **3** using thiazole resin **16** from carbon disulfide and Merrifield resin **1**.[38]

The solid-phase synthetic route for the preparation of thiazolo[4,5-*b*]pyridines uses appropriate 2-bromoacetophenones, ketones, and amines as key building blocks and diversity elements. The sequence began on the basis of successful solution-phase synthetic conditions.[38] The known thiazole resin **16** with R^1 diversity element was reacted under optimized Friedländer reaction conditions ($AlCl_3$ and MW irradiation)[56,58] with ketones ($R^2CH_2COR^3$). This process efficiently produced the thiazolo[4,5-*b*] pyridine resin **21** and introduced the second potential diversity elements R^2 and R^3. Treatment of resin **21** with *m*CPBA in CH_2Cl_2 provided the resin-bound sulfone intermediate **22** (Scheme 10.2).

Finally, the sulfone group on resin **22** was displaced by a desulfonative substitution reaction with the corresponding amines (R^4R^5N diversity elements) in THF (tetrahydrofuran). This process, which is accompanied by concurrent cleavage from the resin, furnished the final thiazolo[4,5-*b*]pyridine derivatives **3** (50 examples; 50–13% five-step overall yields from Merrifield resin **1**, Table 10.2), which were purified by column chromatography (>95% purities as judged from LC–MS traces (integration of diode array 200–400 nm traces)).

10.2.3 Solid-Phase Synthesis of 2,4,6-Trisubstituted Thiazolo[4,5-*d*]pyrimidine-5,7-diones

Thiazolo[4,5-*d*]pyrimidine-5,7-dione derivatives **4**, congeners of xanthine and uracil (pyrimidinedione), exhibit a wide range of important biological properties[59] and, as a result, are attractive targets for combinatorial library construction. In view of these

TABLE 10.2. 2,5,6,7-Tetrasubstituted Thiazolo[4,5-*b*]pyridines **3** and Their Yields[a]

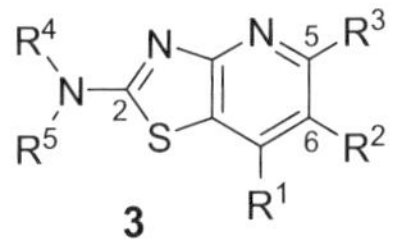

3

Entry	Products	R^1	R^2	R^3	R^4R^5N	Yield (%)[b,c]
1	**3a**	Ph	$-(CH_2)_4-$		BnNH	32
2	**3b**	Ph	$-(CH_2)_4-$		4-MeO-BnNH	29
3	**3c**	Ph	$-(CH_2)_4-$		*n*-PrNH	35
4	**3d**	Ph	$-(CH_2)_4-$		$C_6H_{11}CH_2NH$	27
5	**3e**	Ph	$-(CH_2)_4-$		Et_2N	27
6	**3f**	Ph	$-(CH_2)_4-$			29
7	**3g**	Ph	$-(CH_2)_4-$			33
8	**3h**	Ph	$-(CH_2)_4-$		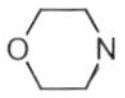	31
9	**3i**	Ph	$-(CH_2)_3-$		BnNH	16
10	**3j**	Ph	$-(CH_2)_3-$		4-MeO-BnNH	21
11	**3k**	Ph	$-(CH_2)_3-$		*n*-PrNH	19
12	**3l**	Ph	$-(CH_2)_3-$			13
13	**3m**	Ph	$-(CH_2)_3-$		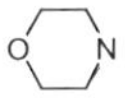	17
14	**3n**	Ph	$-(CH_2)_5-$		BnNH	50
15	**3o**	Ph	$-(CH_2)_5-$		4-MeO-BnNH	43
16	**3p**	Ph	$-(CH_2)_5-$		*n*-PrNH	48
17	**3q**	Ph	$-(CH_2)_5-$			34
18	**3r**	Ph	$-(CH_2)_5-$		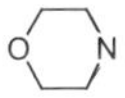	37
19	**3s**	Ph	Me	Et	BnNH	24
20	**3t**	Ph	Me	Et	4-MeO-BnNH	31
21	**3u**	Ph	Me	Et	*n*-PrNH	39
22	**3v**	Ph	Me	Et		33
23	**3w**	Ph	Me	Et	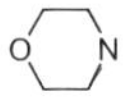	26
24	**3x**	Ph	$-CO(CH_2)_3-$		BnNH	23
25	**3y**	Ph	$-CO(CH_2)_3-$		4-MeO-BnNH	24
26	**3z**	Ph	$-CO(CH_2)_3-$		*n*-PrNH	20
27	**3aa**	Ph	$-CO(CH_2)_3-$			18

(*continued*)

TABLE 10.2. *(Continued)*

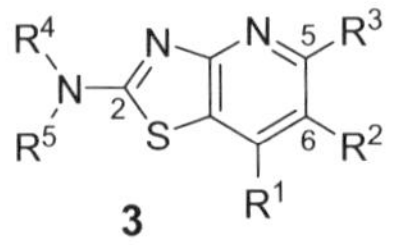

3

Entry	Products	R^1	R^2	R^3	R^4R^5N	Yield (%)[b,c]
28	**3ab**	Ph	$-CO(CH_2)_3-$		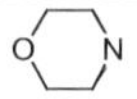	16
29	**3ac**	4-MeO-Ph	$-(CH_2)_4-$		BnNH	37
30	**3ad**	4-MeO-Ph	$-(CH_2)_4-$		*n*-PrNH	48
31	**3ae**	4-MeO-Ph	$-(CH_2)_4-$			32
32	**3af**	4-MeO-Ph	$(CH_2)_4-$		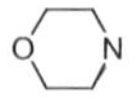	34
33	**3ag**	4-MeO-Ph	$-(CH_2)_5-$		BnNH	40
34	**3ah**	4-MeO-Ph	$-(CH_2)_5-$		*n*-PrNH	49
35	**3ai**	4-MeO-Ph	$-(CH_2)_5-$		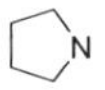	44
36	**3aj**	4-MeO-Ph	$-(CH_2)_5-$		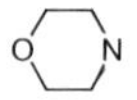	38
37	**3ak**	4-MeO-Ph	Me	Et	BnNH	30
38	**3al**	4-MeO-Ph	Me	Et	*n*-PrNH	38
39	**3am**	4-MeO-Ph	Me	Et		33
40	**3an**	4-MeO-Ph	Me	Et	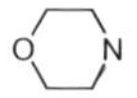	29
41	**3ao**	$4-NO_2-Ph$	$-(CH_2)_4-$		BnNH	39
42	**3ap**	$4-NO_2-Ph$	$-(CH_2)_4-$		4-MeO-BnNH	45
43	**3aq**	$4-NO_2-Ph$	$-(CH_2)_4-$		*n*-PrNH	41
44	**3ar**	$4-NO_2-Ph$	$-(CH_2)_4-$		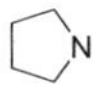	35
45	**3as**	$4-NO_2-Ph$	$-(CH_2)_4-$		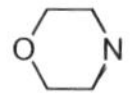	37
46	**3at**	$4-NO_2-Ph$	$-(CH_2)_3-$		BnNH	24
47	**3au**	$4-NO_2-Ph$	$-(CH_2)_3-$		4-MeO-BnNH	27
48	**3av**	$4-NO_2-Ph$	$-(CH_2)_3-$		*n*-PrNH	31
49	**3aw**	$4-NO_2-Ph$	$-(CH_2)_3-$			28
50	**3ax**	$4-NO_2-Ph$	$-(CH_2)_3-$		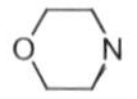	19

[a] All reactions were performed on 150–200 mg scale of resin **22**.
[b] Five-step overall isolated yield from Merrifield resin **1** (loading capacity = 0.94 mmol/g).
[c] Purities of **3** were over 95% as judged from LC–MS traces.

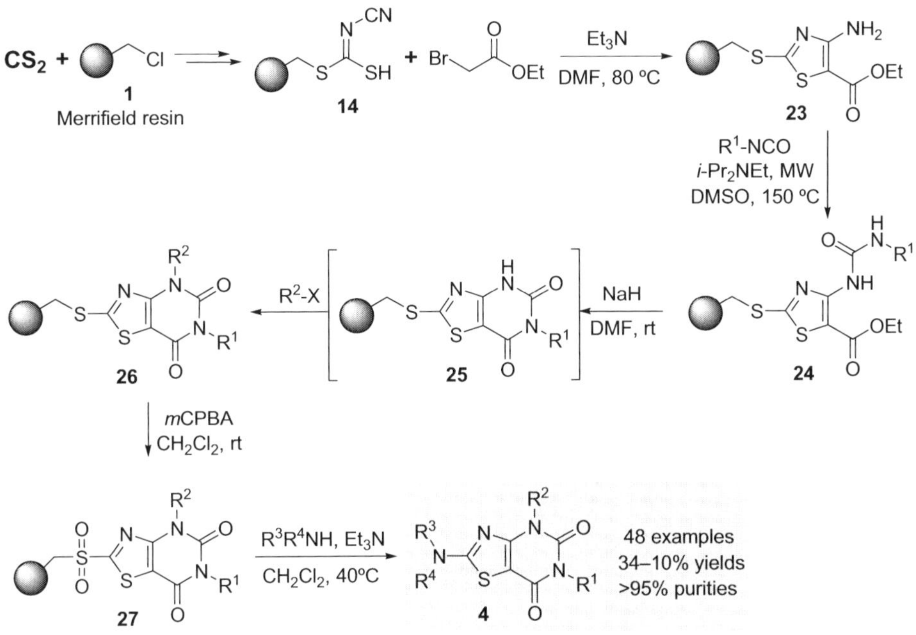

Scheme 10.3. Solid-phase synthesis of 2,4,6-trisubstituted thiazolo[4,5-*d*]pyrimidine-5,7-diones.

diverse properties, an expedient, traceless, solid-phase synthesis of 2,4,6-trisubstituted thiazolo[4,5-*d*]pyrimidine-5,7-dione derivatives was developed.[39] The key step of a solid-phase synthetic route is urea formation by the microwave irradiation[53] promoted reaction of a thiazole amino ester resin **23** with an isocyanate and base-catalyzed cyclization, which were optimized in solution-phase chemistry (Scheme 10.3).

The solid-phase synthesis of thiazolo[4,5-*d*]pyrimidine-5,7-diones **4** utilizes appropriate isocyanates, alkyl halides, and amines as key building blocks and diversity elements. The sequence was started with the formation of thiazole amino ester resin **23** by reacting the solid-supported cyanocarbonimidodithioate **14** with ethyl 2-bromoacetate. The amino ester resin **23** was treated under MW irradiation conditions with isocyanate (R^1 diversity element) to afford the corresponding thiazolourea resin **24**. The one-pot cyclization/*N*-alkylation of thiazolourea resin **24**, using sodium hydride as a base and alkyl halide (R^2 diversity element), was carried out in DMF at room temperature. Accordingly, treatment of resin **24** with NaH in DMF provided the intermediate **25**, which underwent *in situ* *N*-alkylation with alkyl halide to provide the desired thiazolo[4,5-*d*]pyrimidine-5,7-dione resin **26** containing two diversity elements at N4 and N6. After the oxidation of resin **26** to form the sulfone group on resin **27**, nucleophilic C2 substitution with the corresponding amines (R^3R^4N diversity elements) afforded the target 2,4,6-trisubstituted thiazolo[4,5-*d*]pyrimide-5,7-dione derivatives **4**. This process, which was accompanied by concurrent cleavage from the resin, furnished the final thiazolo[4,5-*d*]pyrimidine-5,7-diones **4** (48 examples, Table 10.3). The isolated overall yields for thiazolo[4,5-*d*]pyrimidine-5,7-diones ranged from 10% to 34% for the six-step linear pathway (the average yield for each step was 69% to 84%) from the Merrifield resin **1**.

Lipinski's rule[60] and similar formulations serve as guides for estimating the physicochemical properties of the 2,4,6-trisubstituted thiazolo[4,5-*d*]pyrimidine-5,7-dione

TABLE 10.3. 2,4,6-Trisubstituted Thiazolo[4,5-*d*]pyrimidine-5,7-diones **4** and Their Yields[a]

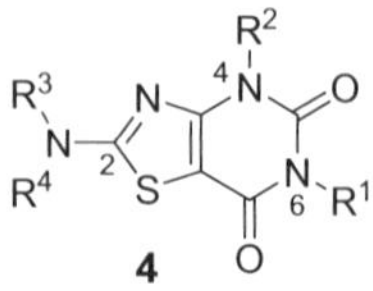

4

Entry	Products	R^1	R^2	R^3R^4N	Yield (%)[b,c]
1	**4a**	Ph	Me	BnNH	34
2	**4b**	Ph	Me	4-MeO-BnNH	31
3	**4c**	Ph	Me	*n*-PrNH	30
4	**4d**	Ph	Me	$C_6H_{11}CH_2NH$	31
5	**4e**	Ph	Me	Et_2N	29
6	**4f**	Ph	Me	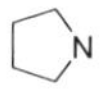	25
7	**4g**	Ph	Me	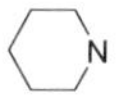	30
8	**4h**	Ph	Me	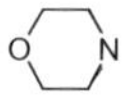	31
9	**4i**	Ph	Bn (benzyl)	BnNH	32
10	**4j**	Ph	Bn	4-MeO-BnNH	32
11	**4k**	Ph	Bn	*n*-PrNH	33
12	**4l**	Ph	Bn	$C_6H_{11}CH_2NH$	29
13	**4m**	Ph	Bn	Et_2N	31
14	**4n**	Ph	Bn	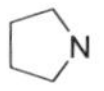	27
15	**4o**	Ph	Bn	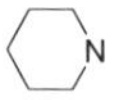	26
16	**4p**	Ph	Bn	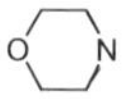	30
17	**4q**	Ph	4-NO_2-Bn	BnNH	27
18	**4r**	Ph	4-NO_2-Bn	4-MeO-BnNH	24
19	**4s**	Ph	4-NO_2-Bn	*n*-PrNH	28
20	**4t**	Ph	4-NO_2-Bn	$C_6H_{11}CH_2NH$	23
21	**4u**	Ph	4-NO_2-Bn	Et_2N	22
22	**4v**	Ph	4-NO_2-Bn	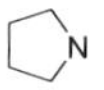	21
23	**4w**	Ph	4-NO_2-Bn		24
24	**4x**	Ph	4-NO_2-Bn	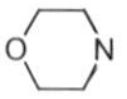	22
25	**4y**	4-MeO-Ph	Me	BnNH	29
26	**4z**	4-MeO-Ph	Me	4-MeO-BnNH	30
27	**4aa**	4-MeO-Ph	Me	*n*-PrNH	29
28	**4ab**	4-MeO-Ph	Me	$C_6H_{11}CH_2NH$	31

TABLE 10.3. (*Continued*)

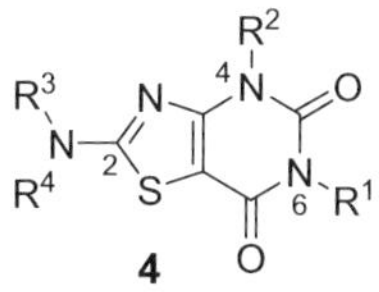

4

Entry	Products	R^1	R^2	R^3R^4N	Yield (%)[b,c]
29	**4ac**	4-MeO-Ph	Me	Et_2N	23
30	**4ad**	4-MeO-Ph	Me		21
31	**4ae**	4-MeO-Ph	Me	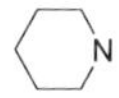	26
32	**4af**	4-MeO-Ph	Me	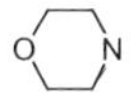	23
33	**4ag**	Et	Me	BnNH	13
34	**4ah**	Et	Me	4-MeO-BnNH	12
35	**4ai**	Et	Me	*n*-PrNH	15
36	**4aj**	Et	Me	$C_6H_{11}CH_2NH$	14
37	**4ak**	Et	Me	Et_2N	12
38	**4al**	Et	Me		15
39	**4am**	Et	Me		15
40	**4an**	Et	Me	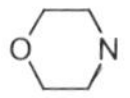	19
41	**4ao**	Et	Bn	BnNH	18
42	**4ap**	Et	Bn	4-MeO-BnNH	17
43	**4aq**	Et	Bn	*n*-PrNH	20
44	**4ar**	Et	Bn	$C_6H_{11}CH_2NH$	16
45	**4as**	Et	Bn	Et_2N	19
46	**4at**	Et	Bn		10
47	**4au**	Et	Bn		21
48	**4av**	Et	Bn	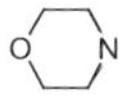	18

[a] All reactions were performed on 150–200 mg scale of resin **13** and R^2X (MeI, BnBr, and 4-NO_2-BnBr) were used.
[b] Six-step overall isolated yield from Merrifield resin **1** (loading capacity = 0.94 mmol/g).
[c] Purities of **4** were over 95% as judged from LC–MS traces.

derivatives **4**. Most of the key parameters (molecular weight, AlogP, number of hydrogen donors and acceptors, polar surface area, and number of rotatable bonds) for members of the library fall within the range of those predicted for reasonable oral bioavailable drugs by using the commonly known guidelines.[39]

10.2.4 Solid-Phase Synthesis of 1,3,6-Trisubstituted 1*H*-Thiazolo[4,5-*c*][1,2]thiazin-4(3*H*)one-2,2-dioxides

We previously described the first solid-phase route for the preparation of 1,3,6-trisubstituted 1*H*-thiazolo[4,5-*c*][1,2]thiazin-4(3*H*)one-2,2-dioxide derivatives.[40] The initial solid-phase synthetic route we developed to prepare substances containing the thiazole scaffold involved the formation of the intermediate thiazole resin **23** (Scheme 10.4).

Resin **23** was first swollen in CH_2Cl_2 and, in a manner that parallels the route employed in the solution-phase synthesis, was reacted with a selected benzylsulfonyl chloride and *t*-BuOLi as a base to afford the corresponding sulfonamide resin **28**, containing the first diversity element R^1. The sulfonamide resin **28** was then reacted under Mitsunobu conditions (PPh_3 (triphenyl phosphine), DIAD (diisopropyl azodicarboxylate), THF, room temperature)[61,62] with the appropriate alcohols. This process efficiently produced resin **29** and introduced the second diversity element R^2. Cyclization reaction of resin **29** was promoted by sodium hydride in DMF and led to the formation of the desired thiazolo[4,5-*c*][1,2]thiazine resin **30**. Treatment of resin **30** with *m*CPBA in CH_2Cl_2 generated the resin-bound cyclic sulfonamide **31**. Finally, the thiazolo[4,5-*c*][1,2]thiazine derivatives **5** were formed and cleaved from the resin (in a traceless manner[48]) by treatment of resin **31** with the corresponding amines (R^3R^4N diversity elements) in respectable yields (34 examples, from 11% to 29% for seven linear steps starting with Merrifield resin **1**, Table 10.4).

When the R^1 on resin **29** was hydrogen or ethyl, the reaction of intermediate resins **29** under the optimal conditions failed to produce precursors of the respective target compounds and only noncyclized products were isolated. This problem is probably due to the lower acidity of the C3 α-protons in resins **29** ($R^1 = H$ or Et). In cases where the R^2 substituent was an electron-withdrawing benzyl (4-NO_2-Bn), the sequence also failed to produce the desired target. It appeared that the acidic proton present in **29** ($R^2 = 4$-NO_2-Bn) renders ineffective the cyclization to produce **30**.[61]

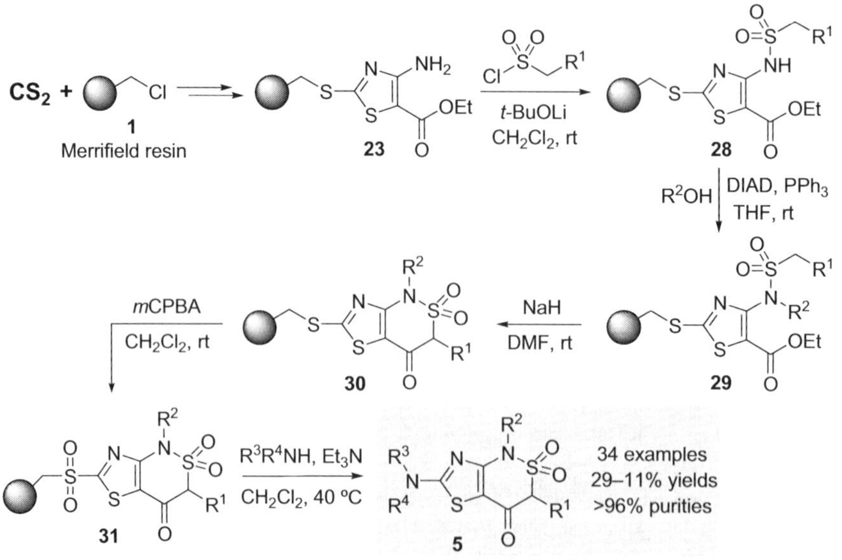

Scheme 10.4. Solid-phase synthesis of 1,3,6-trisubstituted 1*H*-thiazolo[4,5-c][1,2]-4(3*H*)one-2,2-dioxides.

TABLE 10.4. 1,3,6-Trisubstituted 1*H*-thiazolo[4,5-*c*][1,2]-4(3*H*)one-2,2-dioxides **5** and Their Yields[a]

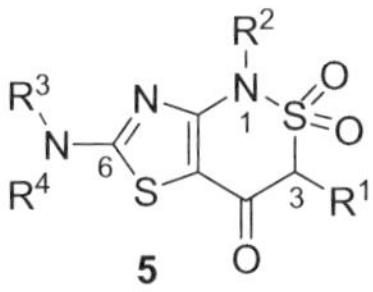

Entry	Products	R^1	R^2	R^3R^4N	Yield (%)[b]	Purity (%)[c]
1	**5a**	Ph	Bn	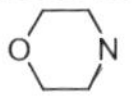	29	99
2	**5b**	Ph	Bn	BnNH	24	99
3	**5c**	Ph	Bn	4-MeO-BnNH	22	99
4	**5d**	Ph	Bn	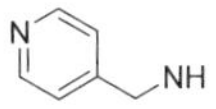	16	99
5	**5e**	Ph	Bn	*n*-PrNH	22	99
6	**5f**	Ph	Bn	$C_6H_{11}CH_2NH$	20	98
7	**5g**	Ph	Bn	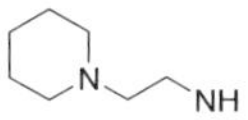	19	96
8	**5h**	Ph	Bn	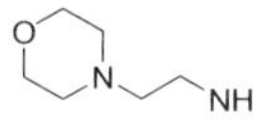	20	98
9	**5i**	Ph	Bn	Et_2N	19	99
10	**5j**	Ph	Bn		18	98
11	**5k**	Ph	Bn		19	97
12	**5l**	H	Bn	BnNH	N.C.[d]	–
13	**5m**	H	Bn	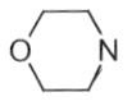	N.C.[d]	–
14	**5n**	Et	Bn	BnNH	N.C.[d]	–
15	**5o**	Et	Bn	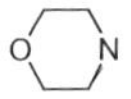	N.C.[d]	–
16	**5p**	Ph	4-MeO-Bn	BnNH	18	99
17	**5q**	Ph	4-MeO-Bn	4-MeO-BnNH	18	99
18	**5r**	Ph	4-MeO-Bn	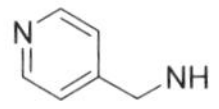	17	98
19	**5s**	Ph	4-MeO-Bn	*n*-PrNH	17	98
20	**5t**	Ph	4-MeO-Bn	$C_6H_{11}CH_2NH$	18	99
21	**5u**	Ph	4-MeO-Bn	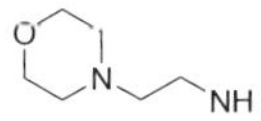	22	99

(continued)

TABLE 10.4. *(Continued)*

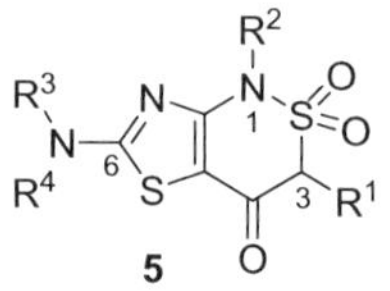

5

Entry	Products	R^1	R^2	R^3R^4N	Yield (%)[b]	Purity (%)[c]
22	**5v**	Ph	4-MeO-Bn	Et_2N	16	97
23	**5w**	Ph	4-MeO-Bn	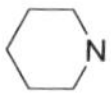	21	99
24	**5x**	Ph	4-MeO-Bn	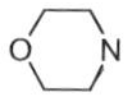	20	99
25	**5y**	Ph	4-Me-Bn	BnNH	28	99
26	**5z**	Ph	4-Me-Bn	4-MeO-BnNH	23	97
27	**5aa**	Ph	4-Me-Bn	*n*-PrNH	23	98
28	**5ab**	Ph	4-Me-Bn	$C_6H_{11}CH_2NH$	22	97
29	**5ac**	Ph	4-Me-Bn		19	98
30	**5ad**	Ph	4-Me-Bn	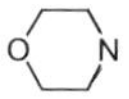	20	97
31	**5ae**	Ph	$4-NO_2-Bn$	BnNH	N.R.[e]	–
32	**5af**	Ph	$4-NO_2-Bn$	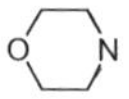	N.R.[e]	–
33	**5ag**	Ph	Et	BnNH	21	99
34	**5ah**	Ph	Et	*n*-PrNH	19	99
35	**5ai**	Ph	Et	$C_6H_{11}CH_2NH$	18	98
36	**5aj**	Ph	Et	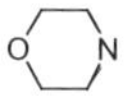	17	99
37	**5ak**	Ph	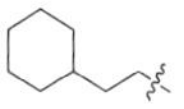	BnNH	11	99
38	**5al**	Ph	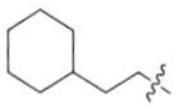	*n*-PrNH	13	99
39	**5am**	Ph	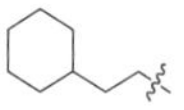	$C_6H_{11}CH_2NH$	12	96
40	**5an**	Ph	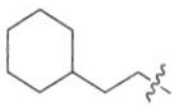	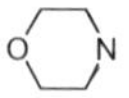	12	98

[a] All reactions were performed on a 150–200 mg scale of resin **31**.
[b] Seven-step overall isolated yield from Merrifield resin **1** (loading capacity = 0.94 mmol/g).
[c] Determined on the basis of LC–MS spectrum (integration of diode array 200–400 nm traces) of isolated product.
[d] Not completed and obtained noncyclized product in about 20% yield.
[e] No reaction and decomposed.

10.3 SOLID-PHASE SYNTHESIS OF BENZOXAZOLES

In SPOS, the choice of the linker that serves to attach the library scaffold to the polymer support is critical. As a result, a variety of elegant linking methods have been developed that enable the introduction of additional diversity into the products during the cleavage reactions. The sulfone linker is an example of a safety-catch linker that can be cleaved from resins by using nucleophilic substitution reactions with amines.

Benzoxazoles are privileged structures of particular interest in medicinal chemistry[63] and, consequently, have been the target of a number of solution- and solid-phase synthetic studies.[64] We have previously employed a carbon disulfide-mediated thioether linker methodology (i.e., safety-catch linker methodology[65]) in a procedure for efficient solid-phase synthesis of 2-aminobenzoxazole derivatives.[41] The sequence used to prepare the target 2-aminobenzoxazole derivatives **6** started with the Merrifield resin **1** as the polymer support. The benzyl chloride groups can be used to produce thioether linkages by reaction with the thiol formed in the cyclization reaction of carbon disulfide and aminophenol (Scheme 10.5).

Specifically, the intermediate benzoxazole resin **32** was generated by treatment of aminophenol **33** with carbon disulfide and Merrifield resin **1** in the presence of triethylamine in acetonitrile, but the resin was obtained in very low yield under this condition. The inefficiency of this process is due to the rapid release of 2-mercaptobenzoxazole **34** from the intermediate **35** during the cyclization reaction (Scheme 10.6, path a). We attempted to circumvent this cleavage process by performing 2-mercaptobenzoxazole **34**; however, reaction of aminophenol with carbon disulfide for 6 h at 80 °C afforded 2-mercaptobenzoxazole **34** in a low yield (Scheme 10.6, path b).

We reasoned that a possible solution to this problem would be to accelerate the cyclization reaction. Therefore, we examined the effects of various additives, including

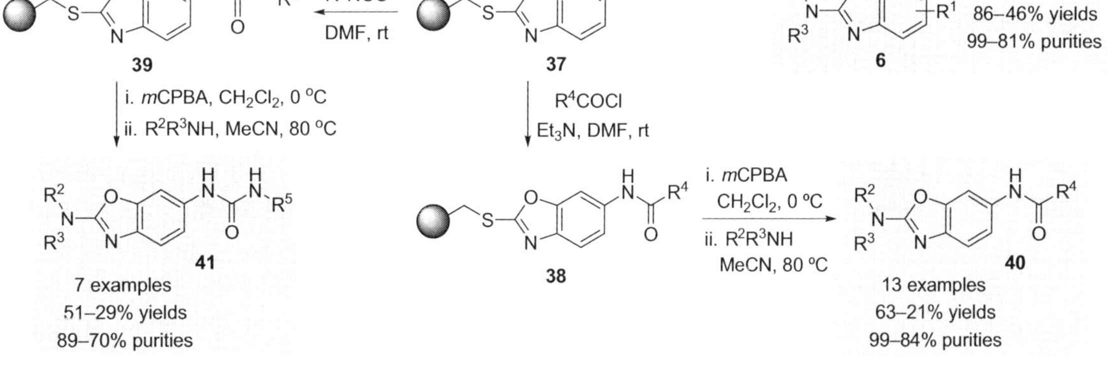

Scheme 10.5. Solid-phase synthesis of 2-aminobenzoxazoles.

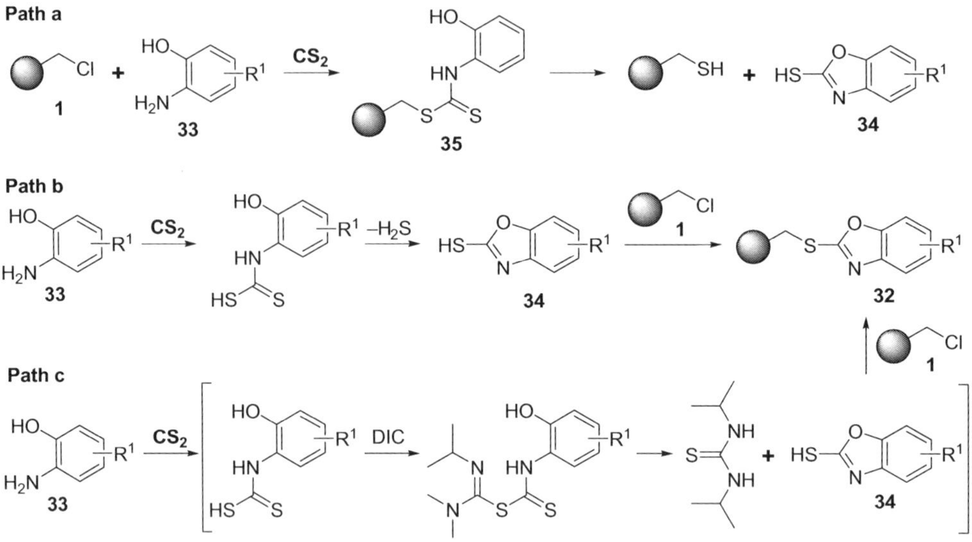

Scheme 10.6. Plausible mechanism for the formation of benzoxazole resin **32**.

N,N′-diisopropylcarbodiimide (DIC), *N,N′*-dicyclohexylcarbodiimide (DCC), *N*-(3-dimethylaminopropyl)-*N′*-ethylcarbodiimide (EDCI), and TsCl (*p*-toluenesulfonyl chloride), and found that the addition of DIC markedly increased the yield of the desired resin **32**. This intermediate is subsequently cyclized to yield 2-mercaptobenzoxazole **34**, which combines with the Merrifield resin **1** to afford the polymer-bound benzoxazole **32** (Scheme 10.6, path c).

The resins **32** were transformed to the corresponding sulfone derivatives **36** by treatment with *m*CPBA in CH_2Cl_2 at 0 °C. To explore the versatility of this methodology, various amines (R^3R^4N diversity elements) were reacted with the sulfone-containing resins to generate the corresponding benzoxazole derivatives **6**. These cleavage reactions generally proceeded with high yields and high purities (20 examples, 86–46% yields, 99–81% purities, Table 10.5).

To introduce additional diversification, the nitro group containing resin **32** (R^1 = 6-NO_2) was reduced to generate 6-aminobenzoxazole resin **37** by treatment with $SnCl_2{\cdot}2H_2O$. Functionalization of the amino group was promoted with acid chlorides (R^4COCl) or isocyanates (R^5NCO) to generate the corresponding amide resins **38** and urea resins **39**. In each case, the desired products **40** and **41** were cleaved by sequential treatment with *m*CPBA and various amines (R^2R^3N diversity elements) (for **40**: 13 examples, 63–21% yields, 99–84% purities; for **41**: 7 examples, 51–29% yields, 89–70% purities, Table 10.5).

10.4 SOLID-PHASE SYNTHESIS OF RELATED PYRAZOLE COMPOUNDS AND 1,3,4-TRIAZOLES VIA A DITHIOCARBAZATE LINKER

10.4.1 Synthesis of a Dithiocarbazate Linker on Solid Support

The heterocyclic-oriented thioureas have scarcely been reported in the research field of drug-like library construction by solid-phase synthesis, compared to their ureas and simple aromatic thiourea analogues. We have previously described the dithiocarbazate linker **42** and its application to drug-like heterocycle formation.[42–44] The key intermediate, the

TABLE 10.5. 2-Aminobenzoxazoles **6**, **40**, and **41** and Their Yields[a]

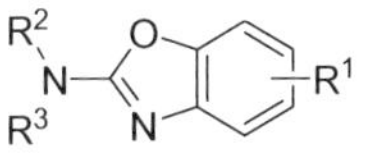
6

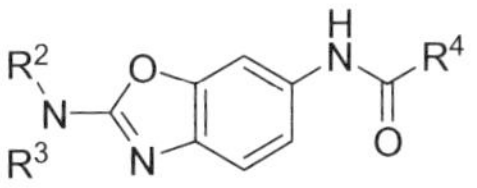
40

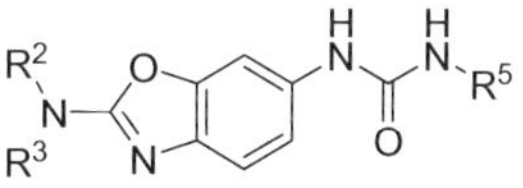
41

Entry	Products	R^1	R^4	R^5	R^2R^3N	Yield (%)[b]	Purity (%)[c]
1	**6a**	H	–	–		70	>99
2	**6b**	H	–	–	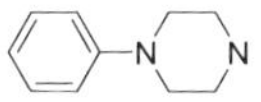	68	>99
3	**6c**	H	–	–	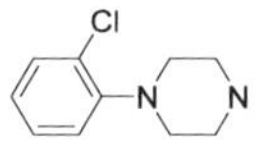	53	>99
4	**6d**	H	–	–	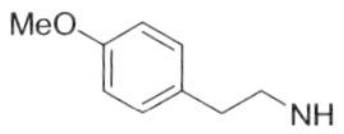	61	89
5	**6e**	4-Cl	–	–	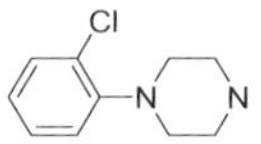	53	>99
6	**6f**	4-Cl	–	–	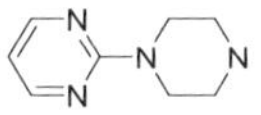	49	95
7	**6g**	4-Cl	–	–	4-Cl-BnNH	47	81
8	**6h**	4-Cl	–	–	PhNH	75	98
9	**6i**	5-NO_2	–	–	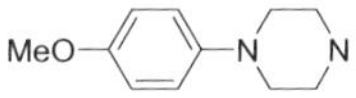	66	74
10	**6j**	5-NO_2	–	–	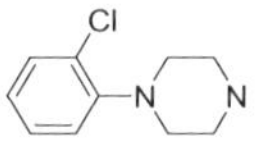	53	95
11	**6k**	5-NO_2	–	–	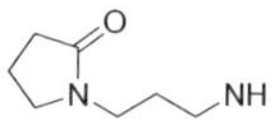	50	89
12	**6l**	4-*t*-Bu	–	–		73	99
13	**6m**	4-*t*-Bu	–	–	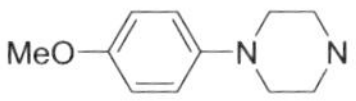	62	93
14	**6n**	4-*t*-Bu	–	–	PhNH	60	>99
15	**6o**	4-*t*-Bu	–	–	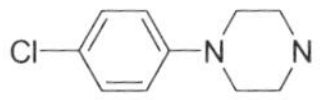	46	99
16	**6p**	5-Me	–	–	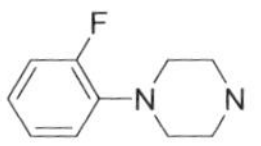	86	96
17	**6q**	5-Me	–	–	4-F-PhNH	56	99
18	**6r**	5-Me	–	–	4-MeO-BnNH	71	97

(continued)

TABLE 10.5. (*Continued*)

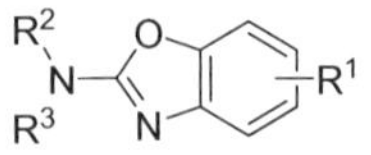
6

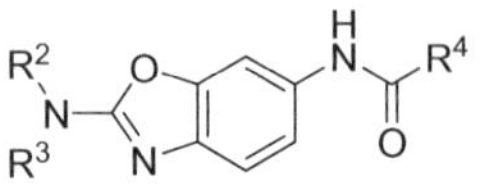
40

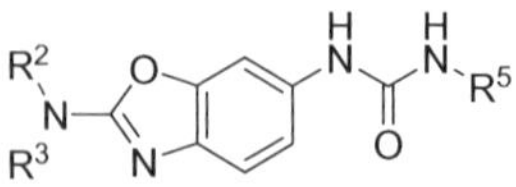
41

Entry	Products	R[1]	R[4]	R[5]	R[2]R[3]N	Yield (%)[b]	Purity (%)[c]
19	**6s**	5-Me	–	–	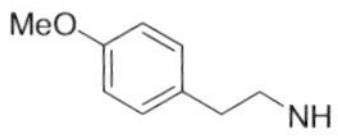	79	99
20	**6t**	5-Me	–	–	4-Me-PhNH	69	>99
21	**40a**	–	Ph	–	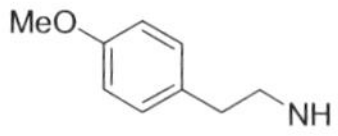	51	95
22	**40b**	–	Ph	–	4-Me-PhNH	48	84
23	**40c**	–	Ph	–	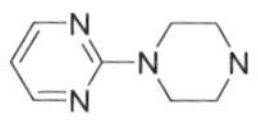	20	87
24	**40d**	–		–	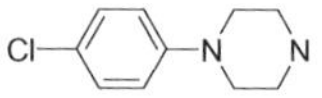	23	>99
25	**40e**	–		–	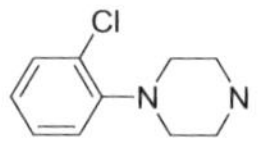	59	99
26	**40f**	–		–	4-Cl-BnNH	32	>99
27	**40g**	–		–	4-F-PhNH	21	96
28	**41a**	–	–	4-MeO-Ph	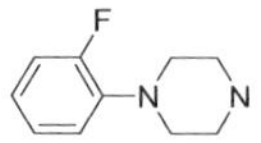	52	95
29	**41b**	–	–	4-MeO-Ph	4-F-BnNH	33	94
30	**41c**	–	–	2-Cl-Ph	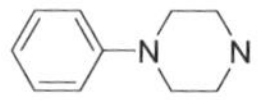	52	95
31	**41d**	–	–	2-Cl-Ph	4-MeO-BnNH	48	93
32	**41e**	–	–	2-Cl-Ph	PhNH	63	97
33	**41f**	–	–	2-Cl-Ph	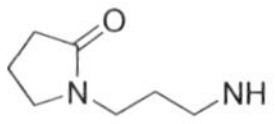	45	95
34	**41g**	–	–	Ph	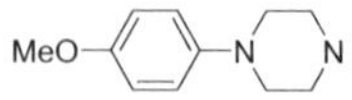	46	83
35	**41h**	–	–	Ph	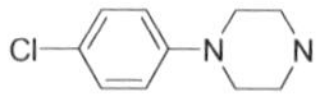	41	82
36	**41i**	–	–	Ph	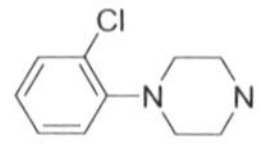	51	87

TABLE 10.5. (*Continued*)

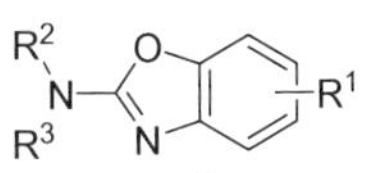
6

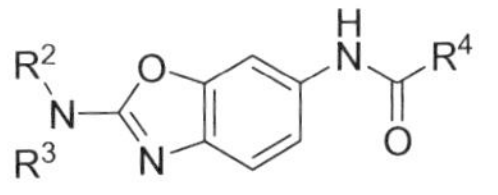
40

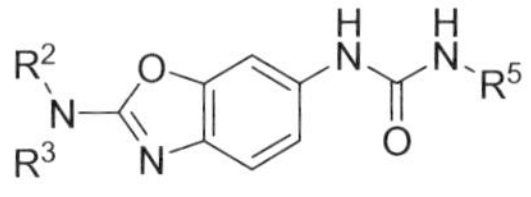
41

Entry	Products	R^1	R^4	R^5	R^2R^3N	Yield (%)[b]	Purity (%)[c]
37	**41j**	–	–	Ph	4-MeO-BnNH	38	87
38	**41k**	–	–	4-MeO-Ph	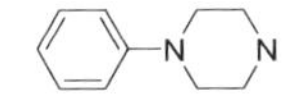	45	89
39	**41l**	–	–	*n*-Bu	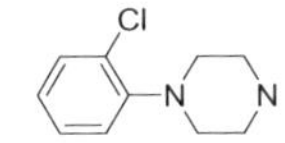	37	82
40	**41m**	–	–	*n*-Bu	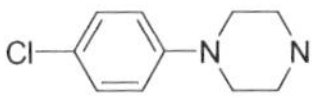	29	70

[a] All reactions were performed on 150–200 mg scale of resins **36** (for **6**), **38** (for **40**), and **39** (for **41**).
[b] Three-step (for **6**) or four-step (for **40** and **41**) overall yield from Merrifield resin **1** (loading capacity = 0.94 mmol/g).
[c] Determined on the basis of LC–MS spectrum (integration of diode array 200–400 nm traces).

polymer-bound dithiocarbazate **42**, was prepared in a one- or two-step procedure starting from the Merrifield resin **1** as shown in Scheme 10.7.

In the first step, Fmoc (9-fluorenylmethoxycarbonyl)-protected dithiocarbazate resin **43** was prepared by a three-component reaction of Merrifield resin **1** and carbon disulfide with Fmoc-protected hydrazine in the presence of sodium hydride in DMF at room temperature. Deprotection of the Fmoc group of resin **43** with 5% piperidine produced the corresponding free dithiocarbazate resin **42**.[42] In this step, the use of 5% piperidine was essential because a higher concentration caused loss of the desired substrate from resin **43**. In addition, we have developed a more convenient synthetic route to the resin **42**, compared to the previous report using hydrazine monohydrate and carbon disulfide with potassium hydroxide in ethanol solvent.[44] Under these reaction conditions, we obtained the polymer-bound dithiocarbazate **42** without the Fmoc protection step of the hydrazine.

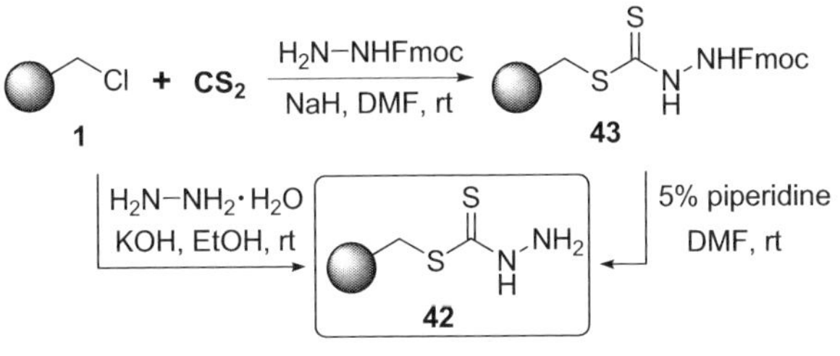

Scheme 10.7. Synthesis of the dithiocarbazate linker.

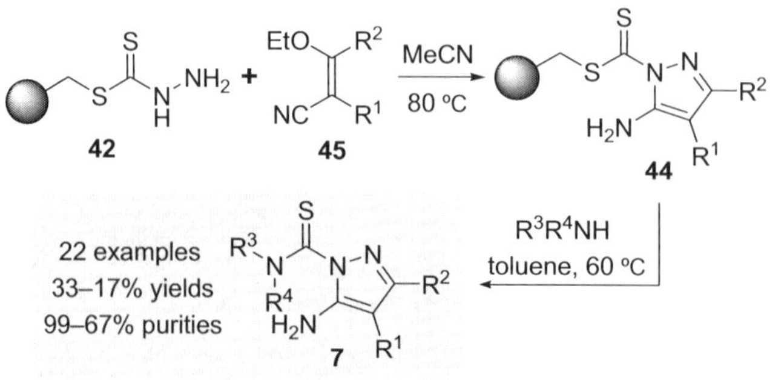

Scheme 10.8. Solid-phase synthesis of pyrazoles via a dithiocarbazate linker.

10.4.2 Solid-Phase Synthesis of Pyrazoles via a Dithiocarbazate Linker

For heterocycle diversification of the hydrazine in the dithiocarbazate system, 5-aminopyrazoles **44** on dithiocarbazate resin **42** (by two-step sequence from Fmoc-hydrazine) were introduced by nucleophilic cyclization reactions with substituted 3-ethoxyacrylonitriles **45** (R^1 and R^2 diversity elements) in acetonitrile (Scheme 10.8).[42]

The desired 5-amino-1-(substituted-thiocarbamoyl)pyrazoles **7** were finally liberated from resin **44** using various amines (see Table 10.6) by thermal cleavage reaction in toluene at 60 °C. Various types of amine (R^3R^4N diversity elements) afforded the 5-amino-1-(substituted-thiocarbamoyl)pyrazole derivatives **7** in good four-step overall yields (22 examples, 33–17% yields, 99–67% purities, Table 10.6) from Merrifield resin **1** with high purities, except for sterically hindered secondary amines, such as diisopropylamine and diisobutylamine.

10.4.3 Solid-Phase Synthesis of Pyrazolo[1,5-*a*][1,3,5]-2-oxo-4-dithioxotriazines

We have described studies that led to the development of an efficient procedure for the synthesis of novel 7,8-functionalized pyrazolo[1,5-*a*][1,3,5]-2-oxo-4-thioxotriazine derivatives **8** (Scheme 10.9) and that involved solid-phase cyclization reactions of resin-bound 3,4-functionalized 5-amino-1-dithiocarboxy-pyrazoles **44** with various isocyanates.[43] These key intermediates then serve as precursors for the target 7,8-functionalized pyrazolo[1,5-*a*][1,3,5]-2-oxo-4-thioxotriazines **8**.

When R^1 was a nitrile group on the 3-ethoxyacrylonitriles **45** ($R^1 = CN$), the pyrazole ring formation of hydrazine dithiocarbazate resin **42** (by a two-step sequence from Fmoc-hydrazine) was accomplished in acetonitrile. In contrast, cyclization reactions of resin **42** with 3-ethoxyacrylonitriles **45** ($R^1 = CO_2Et$) did not run smoothly in acetonitrile. Instead, among various solvents, 1,4-dioxane was used to generalize the process leading to resin **44**. The concurrent cyclization–resin cleavage reactions of the 5-aminopyrazole resins **44** were explored in more detail to determine the optimal conditions. The results showed that reactions of the 4-cyano-5-aminopyrazole resin **44** ($R^1 = CN$) with isocyanates take place in the presence of triethylamine (THF, 40 °C). On the other hand, the reactions of the 4-ethylcarboxy-5-aminopyrazole resin **44** ($R^1 = CO_2Et$) required the strong base sodium hydride (THF, 40 °C). Variously substituted aryl isocyanates (R^3 diversity element) reacted to generate the target 7,8-functionalized pyrazolo[1,5-*a*][1,3,5]-2-oxo-4-thioxotriazine

TABLE 10.6. 5-Amino-1-(substituted-thiocarbamoyl)pyrazole **7** and Their Yields[a]

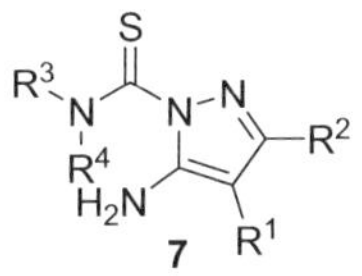

Entry	Products	R^1	R^2	R^3R^4N	Yield (%)[b]	Purity (%)[c]
1	**7a**	H	CO_2Et	*i*-BuNH	33	89
2	**7b**	H	CO_2Et	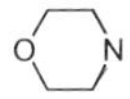	28	>99
3	**7c**	H	CO_2Et	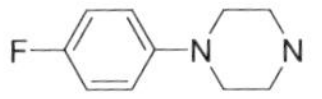	21	91
4	**7d**	H	CO_2Et	4-F-BnNH	26	96
5	**7e**	H	CO_2Et	4-Cl-BnNH	21	67
6	**7f**	H	CO_2Et	4-NO_2-BnNH	27	>99
7	**7g**	H	CO_2Et	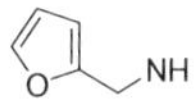	26	93
8	**7h**	H	CO_2Et	$C_3H_7CH_2NH$	22	96
9	**7i**	H	CN	*i*-BuNH	19	89
10	**7j**	H	CN	4-MeO-BnNH	21	97
11	**7k**	H	CN	4-Cl-BnNH	24	90
12	**7l**	Me	CO_2Et	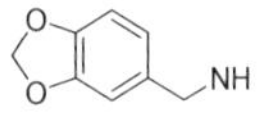	18	91
13	**7m**	Me	CO_2Et	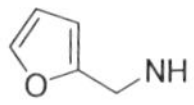	27	92
14	**7n**	Me	CO_2Et	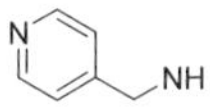	19	96
15	**7o**	Me	CO_2Et	4-MeO-BnNH	22	94
16	**7p**	Me	CO_2Et	2-Cl-BnNH	26	93
17	**7q**	Me	CO_2Et	2-Me-BnNH	20	98
18	**7r**	Me	CN	2-Cl-BnNH	23	79
19	**7s**	Me	CN	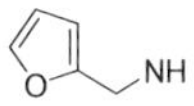	17	>99
20	**7t**	Me	CN	*i*-PrNH	22	96
21	**7u**	Me	CN		24	97
22	**7v**	Me	CN	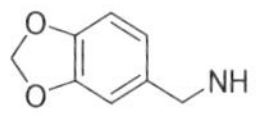	20	81

[a] All reactions were performed on a 150–200 mg scale of resin **44**.

[b] Four-step overall yield from Merrifield resin **1** (loading capacity = 1.60 mmol/g).

[c] Determined on the basis of LC–MS spectrum (integration of diode array 200–400 nm traces) after short-passed silica gel column chromatography.

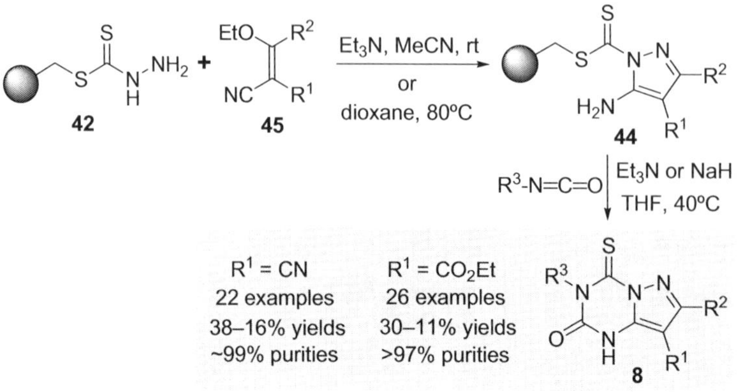

Scheme 10.9. Solid-phase synthesis of pyrazolo[1,5-*a*][1,3,5]-2-oxo-4-thioxotriazines via a dithiocarbazate linker.

derivatives **8** in good five-step overall yields starting from the Merrifield resin **1** and with high purities (R^1 = CN: 22 examples, 38–16% yields; R^1 = CO_2Et: 26 examples, 30–11% yields, Table 10.7).

Having established a flexible method for the solid-phase synthesis of 7,8-functionalized pyrazolo[1,5-*a*][1,3,5]-2-oxo-4-thioxotriazine derivatives **8**, we next evaluated the potential drug properties of members of this family. Most of the key parameters (molecular weight, AlogP, number of hydrogen donors and acceptors, polar surface area, and number of rotatable bonds) for members of the constructed library fall within the range of those predicted for reasonable oral bioavailability of drugs by using the commonly known guidelines.[43,60]

10.4.4 Solid-Phase Synthesis of Pyrazolo[1,5-*a*][1,3,5]-2,4-dithioxotriazines

We have also described the development of an efficient procedure for the synthesis of novel 7,8-functionalized pyrazolo-[1,5-*a*][1,3,5]-2,4-dithioxotriazine derivatives **9** (Scheme 10.10) using a method similar to the synthesis of pyrazolo[1,5-*a*][1,3,5]-2-oxo-4-thioxotriazines **8**, except for the cyclization by various isothiocyanates.[44]

The polymer-bound 5-amino-1-dithiocarboxypyrazole resin **44** was obtained by the reaction of dithiocarbazate resin **42** (by a one-step sequence from hydrazine monohydrate) with 3-ethoxyacrylonitriles **45** (R^1 and R^2 diversity elements) using previously reported methods. Finally, the desired target, 7,8-functionalized pyrazolo[1,5-*a*][1,3,5]-2,4-dithioxotriazine derivatives **9**, was obtained from the respective 5-aminopyrazole resins **44** through a reaction sequence involving the intermediate formation of a solid-supported thiourea with various isothiocyanates (R^3 diversity element) and the intramolecular cyclization reaction of the resulting intermediate.

We could obtain various 7,8-functionalized pyrazolo[1,5-*a*][1,3,5]-2,4-dithioxotriazine derivatives **9** by the concurrent reaction of 5-aminopyrazole resin **44** with aryl isothiocyanates in good three-step overall yields starting from the Merrifield resin **1** with high purities (R^1 = CN: 14 examples, 46–19% yields; R^1 = CO_2Et: 20 examples, 55–10% yields, Table 10.8).

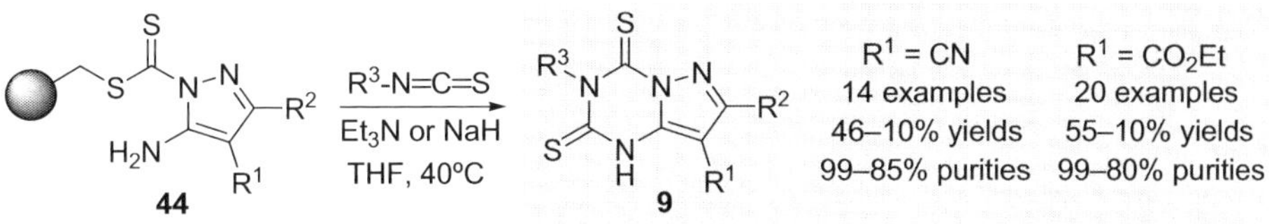

Scheme 10.10. Solid-phase synthesis of pyrazolo[1,5-*a*][1,3,5]-2,4-dithioxotriazines.

TABLE 10.7. 7,8-Functionalized Pyrazolo[1,5-*a*][1,3,5]-2-oxo-4-thioxotriazines **8** and Their Yields[a]

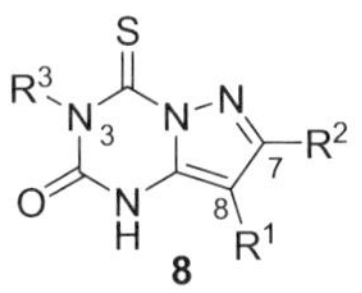

8

Entry	Products	R^1	R^2	R^3	Yield (%)[b]	Purity (%)[c]
1	**8a**	CN	Me	Ph	32	99
2	**8b**	CN	Me	4-Me-Ph	28	99
3	**8c**	CN	Me	4-Et-Ph	29	99
4	**8d**	CN	Me	4-*t*-Bu-Ph	20	99
5	**8e**	CN	Me	3,5-di-Me-Ph	30	99
6	**8f**	CN	Me	4-MeO-Ph	29	99
7	**8g**	CN	Me	4-NO_2-Ph	32	99
8	**8h**	CN	Me	4-F-Ph	34	99
9	**8i**	CN	Me	3-CF_3-Ph	33	99
10	**8j**	CN	Me	2-Cl-Ph	33	99
11	**8k**	CN	Me	Bn	25	99
12	**8l**	CN	Ph	Ph	30	99
13	**8m**	CN	Ph	4-Me-Ph	37	99
14	**8n**	CN	Ph	4-Et-Ph	37	99
15	**8o**	CN	Ph	4-*t*-Bu-Ph	31	99
16	**8p**	CN	Ph	3,5-di-Me-Ph	38	99
17	**8q**	CN	Ph	4-MeO-Ph	38	99
18	**8r**	CN	Ph	4-NO_2-Ph	16	99
19	**8s**	CN	Ph	4-F-Ph	38	99
20	**8t**	CN	Ph	3-CF_3-Ph	31	99
21	**8u**	CN	Ph	2-Cl-Ph	34	99
22	**8v**	CN	Ph	Bn	33	99
23	**8w**	CO_2Et	H	Ph	27	99
24	**8x**	CO_2Et	H	4-Me-Ph	20	99
25	**8y**	CO_2Et	H	4-Et-Ph	23	99
26	**8z**	CO_2Et	H	4-*t*-Bu-Ph	22	99
27	**8aa**	CO_2Et	H	4-MeO-Ph	28	99
28	**8ab**	CO_2Et	H	4-NO_2-Ph	20	99
29	**8ac**	CO_2Et	H	4-F-Ph	11	99
30	**8ad**	CO_2Et	H	3-CF_3-Ph	23	99
31	**8ae**	CO_2Et	Me	Ph	25	99

(continued)

TABLE 10.7. (*Continued*)

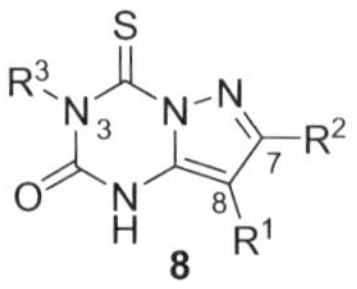

Entry	Products	R^1	R^2	R^3	Yield (%)[b]	Purity (%)[c]
32	**8af**	CO_2Et	Me	4-Me-Ph	23	99
33	**8ag**	CO_2Et	Me	4-Et-Ph	29	97
34	**8ah**	CO_2Et	Me	4-*t*-Bu-Ph	27	98
35	**8ai**	CO_2Et	Me	4-MeO-Ph	26	99
36	**8aj**	CO_2Et	Me	4-NO_2-Ph	27	99
37	**8ak**	CO_2Et	Me	4-F-Ph	22	99
38	**8al**	CO_2Et	Me	3-CF_3-Ph	18	99
39	**8am**	CO_2Et	Me	Bn	25	99
40	**8an**	CO_2Et	Ph	Ph	25	99
41	**8ao**	CO_2Et	Ph	4-Me-Ph	26	99
42	**8ap**	CO_2Et	Ph	4-Et-Ph	30	99
43	**8aq**	CO_2Et	Ph	4-*t*-Bu-Ph	29	97
44	**8ar**	CO_2Et	Ph	4-MeO-Ph	25	99
45	**8as**	CO_2Et	Ph	4-NO_2-Ph	28	99
46	**8at**	CO_2Et	Ph	4-F-Ph	28	99
47	**8au**	CO_2Et	Ph	3-CF_3-Ph	26	99
48	**8av**	CO_2Et	Ph	Bn	24	99

[a] All reactions were performed on a 150–200 mg scale of resin **44**.
[b] Five-step overall isolated yield from Merrifield resin **1** (loading capacity = 2.00 mmol/g).
[c] Determined on the basis of LC–MS spectrum (integration of diode array 200–400 nm traces) of isolated product.

10.4.5 Solid-Phase Synthesis of 1,3,4-Triazoles

A general method has been reported for the parallel solid-phase synthesis of 1,2,4-triazole derivatives based on the cyclization of polymer-bound dithiocarbazate **42** (by a two-step sequence from Fmoc-hydrazine) with electrophiles using a similar route to the formation of pyrazole **7**.[42] The linker **42** served as a nucleophile for the cyclization reactions with electrophiles such as cyanocarboimidates **46** (R^1 diversity elements) (Scheme 10.11).

Further nucleophilic substitution on these polymer-bound 1,2,4-triazoles **47** with various amines (R^2R^3N diversity elements) under thermal cleavage conditions produced the desired 5-amino-1-(substituted thiocarbamoyl)-1,2,4-triazoles **10**. The final compounds **10** were obtained in good four-step overall yields and with high purities upon cleavage from the resins (13 examples, 28–18% yields, 99–81% purities, Table 10.9).

10.5 SOLID-PHASE SYNTHESIS OF 1,3,4-OXADIAZOLES AND 1,3,4-THIADIAZOLES VIA SELECTIVE CYCLIZATION

In the five-membered ring heterocyclic compound family, 1,3,4-oxadiazoles and 1,3,4-thiadiazoles have been used as privileged scaffolds to produce substances of interest in numerous therapeutic areas.[66] In addition, these heterocycles serve as intermediates in

TABLE 10.8. 7,8-Functionalized Pyrazolo[1,5-*a*][1,3,5]-2,4-dithioxotriazines **9** and Their Yields[a]

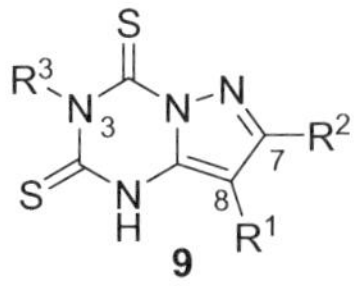

9

Entry	Products	R^1	R^2	R^3	Yield (%)[b]	Purity (%)[c]
1	**9a**	CN	Me	Ph	37	98
2	**9b**	CN	Me	CO_2Et	10	85
3	**9c**	CN	Me	2-MeO-Ph	30	90
4	**9d**	CN	Me	3-MeO-Ph	28	98
5	**9e**	CN	Me	4-MeO-Ph	36	90
6	**9f**	CN	Me	4-F-Ph	35	95
7	**9g**	CN	Me	4-NO_2-Ph	34	99
8	**9h**	CN	Ph	Ph	17	92
9	**9i**	CN	Ph	CO_2Et	40	98
10	**9j**	CN	Ph	2-MeO-Ph	39	92
11	**9k**	CN	Ph	3-MeO-Ph	26	97
12	**9l**	CN	Ph	4-MeO-Ph	42	99
13	**9m**	CN	Ph	4-F-Ph	46	99
14	**9n**	CN	Ph	4-NO_2-Ph	44	92
15	**9o**	CO_2Et	H	Ph	53	98
16	**9p**	CO_2Et	H	2-MeO-Ph	46	98
17	**9q**	CO_2Et	H	3-MeO-Ph	31	99
18	**9r**	CO_2Et	H	4-MeO-Ph	55	99
19	**9s**	CO_2Et	H	4-F-Ph	50	99
20	**9t**	CO_2Et	H	4-NO_2-Ph	49	95
21	**9u**	CO_2Et	Me	Ph	36	90
22	**9v**	CO_2Et	Me	2-MeO-Ph	30	99
23	**9w**	CO_2Et	Me	3-MeO-Ph	29	99
24	**9x**	CO_2Et	Me	4-MeO-Ph	39	95
25	**9y**	CO_2Et	Me	4-F-Ph	40	99
26	**9z**	CO_2Et	Me	4-NO_2-Ph	39	99
27	**9aa**	CO_2Et	Ph	Ph	46	92
28	**9ab**	CO_2Et	Ph	2-MeO-Ph	36	90
29	**9ac**	CO_2Et	Ph	3-MeO-Ph	33	90
30	**9ad**	CO_2Et	Ph	4-MeO-Ph	45	95
31	**9ae**	CO_2Et	Ph	4-F-Ph	49	99
32	**9af**	CO_2Et	Ph	4-NO_2-Ph	47	95
33	**9ag**	CO_2Et	Ph	Ac	10	85
34	**9ah**	CO_2Et	Ph	*n*-Pr	16	80

[a] All reactions were performed on a 150–200 mg scale of resin **44**.

[b] Five-step overall isolated yield from Merrifield resin **1** (loading capacity = 2.00 mmol/g).

[c] Determined on the basis of LC–MS spectrum (integration of diode array 200–400 nm traces) of isolated product.

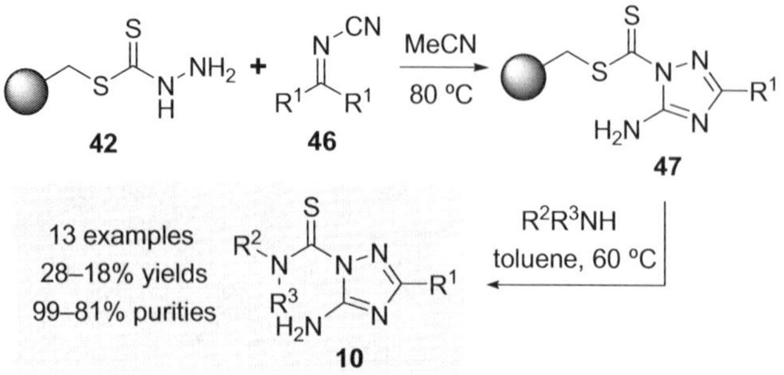

Scheme 10.11. Solid-phase synthesis of 1,3,4-triazoles via a dithiocarbazate linker.

TABLE 10.9. 5-Amino-1-(substituted-thiocarbamoyl)-1,2,4-triazoles **10** and Their Yields[a]

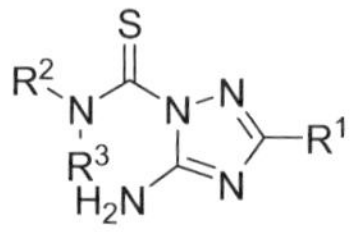

10

Entry	Products	R^1	R^2R^3N	Yield (%)[b]	Purity (%)[c]
1	**10a**	SMe	2-Me-BnNH	26	98
2	**10b**	SMe	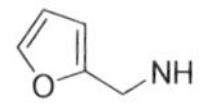	23	>99
3	**10c**	SMe	4-MeO-BnNH	27	97
4	**10d**	SMe	*i*-PrNH	20	98
5	**10e**	SMe	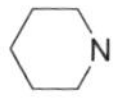	28	96
6	**10f**	SMe	2-Cl-BnNH	22	93
7	**10g**	SMe	Ph_2CHCH_2NH	19	81
8	**10h**	SMe	*i*-BuNH	27	98
9	**10i**	SMe	$C_3H_7CH_2NH$	24	>99
10	**10j**	OPh	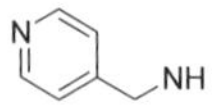	18	94
11	**10k**	OPh	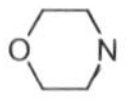	23	92
12	**10l**	OPh	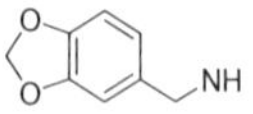	25	85
13	**10m**	OPh	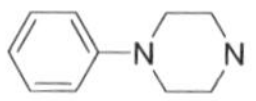	22	87

[a] All reactions were performed on a 150–200 mg scale of resin **47**.
[b] Four-step overall yield from Merrifield resin **1** (loading capacity = 1.60 mmol/g).
[c] Determined on the basis of LC–MS spectrum (integration of diode array 200–400 nm traces) after short-passed silica gel column chromatography.

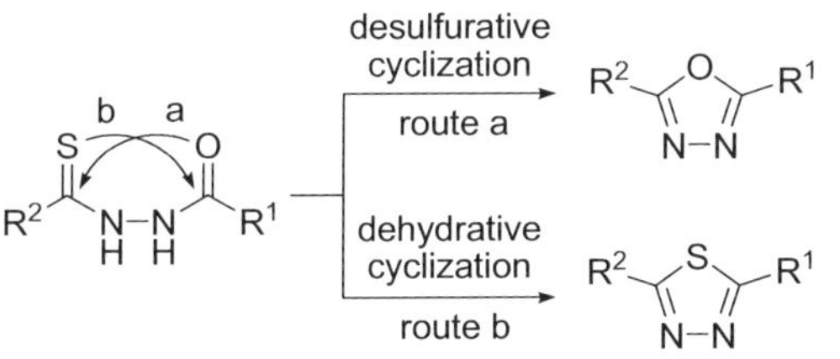

Figure 10.2. Synthesis of 1,3,4-oxadiazoles and 1,3,4-thiadiazoles using an acyldithiocarbazate.

the preparation of various biologically important compounds.[67] As a result of these applications, 1,3,4-oxadiazole and 1,3,4-thiadiazole derivatives have been the targets of a number of solution- and solid-phase synthetic studies.[24a,25]

In a recent investigation (described below), we expanded the utility of the carbazate resin by applying it to the efficient solid-phase syntheses of 1,3,4-oxadiazole derivatives **11** and 1,3,4-thiadiazole derivatives **12** via selective, reagent-based cyclization of an acyldithiocarbazate.[45] The process employs an acyldithiocarbazate resin, from which the respective targets are generated by cyclodesulfurization (Figure 10.2, route a) or cyclodehydration (Figure 10.2, route b).

The sequence used to prepare the acyldithiocarbazate resins **48** (Scheme 10.12) employs the Merrifield resin **1** as a polymer support. Treatment of **1** with carbon disulfide and various hydrazides **49** in the presence of sodium hydride at room temperature led to production of the corresponding acyldithiocarbazate resins **48**.[68]

To investigate suitable methods for the reagent-based, skeletal, diversity-oriented synthesis of 1,3,4-oxadiazoles or 1,3,4-thiadiazoles, cyclization reactions of the acyldithiocarbazate resin **48** were investigated by using various reagents, including EDCI, DCC, TMSCl (trimethylsilyl chloride), TsCl, PPh_3, $SOCl_2$, PCl_5, and diphenyl chlorophosphate. The desired products **11** (1,3,4-oxadiazoles) and **12** (1,3,4-thiadiazoles) were cleaved from the resins **50** and **51**, respectively, by sequential treatment with *m*CPBA and NaOH in aqueous dioxane (producing the sulfones **52** and **53**) and piperidine in 1,4-dioxane at 100 °C.

Reactions of the acyldithiocarbazate resin **48** with both EDCI and DCC afforded 1,3,4-oxadiazole **11a** (R^1 = 4-CF_3-Ph, R^2R^3NH = piperidine) as the major product, but in a low yield, whereas the use of $SOCl_2$ promoted the desulfurative cyclization process that produced a 1,3,4-oxadiazole as the major product with high chemoselectivity (99:1) and in a moderate yield (28%) via route a. The reaction of the resin **48** with TsCl as a reagent for cyclization via route a in the presence of triethylamine generated the 1,3,4-oxadiazole **11a** in a high yield (50%) and high chemoselectivity (98:2). The 1,3,4-thiadiazole **12a** (R^1 = 4-CF_3-Ph, R^2R^3NH = piperidine) was produced in a high yield (53%) and excellent chemoselectivity (99:1) by the dehydrative cyclization upon treatment of the resin **48** with TMSCl as a reagent for cyclization via route b. Similarly, reaction of the resin **48** with diphenyl chlorophosphate also afforded **12a** in a good yield (51%) and high chemoselectivity (98:2) via route b. However, PCl_5 and PPh_3 treatment of **48** produced **12a**, but with low chemoselectivity.

To explore the diversity of this methodology, various amines (R^2R^3N diversity elements) were used to liberate the 1,3,4-oxadiazoles **11** and 1,3,4-thiadiazoles **12** from the functionalized sulfone-containing resin. Cleavage reactions with amines generally afforded the desired products in high yields (22 examples, 72–30% yields, Table 10.10).

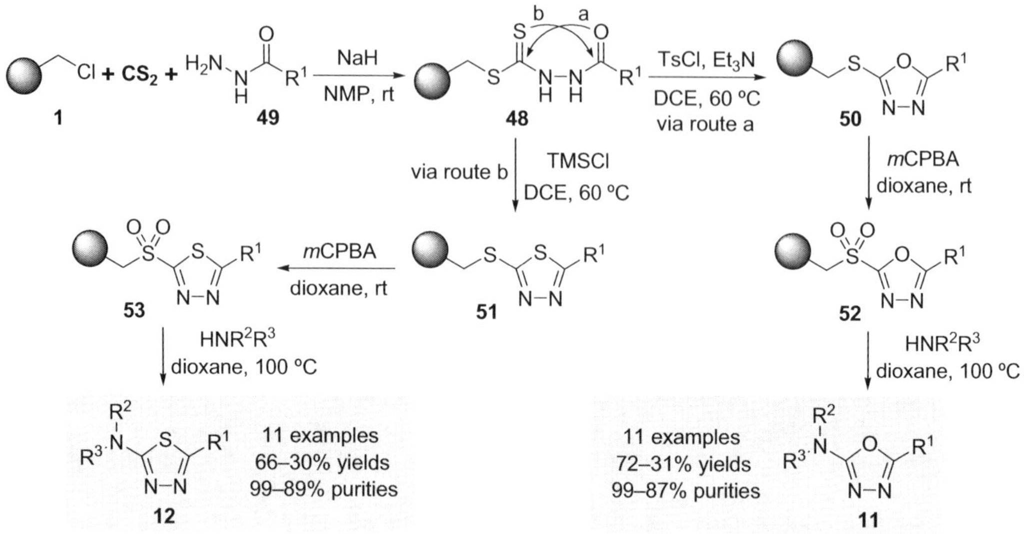

Scheme 10.12. Solid-phase synthesis of 1,3,4-oxadiazoles and 1,3,4-thiadiazoles.

TABLE 10.10. 1,3,4-Oxadiazoles **11** and 1,3,4-Thiadiazoles **12** and Their Yields[a]

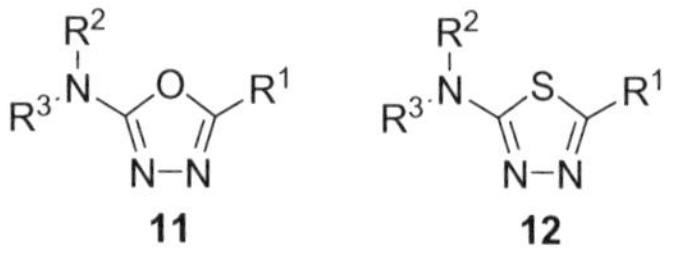

Entry	Products	R^1	R^2R^3N	Yield (%)[b]	Purity (%)[c]
1	**11a**	4-CF_3-Ph		48	>99
2	**11b**	4-CF_3-Ph	*i*-BuNH	34	93
3	**11c**	4-CF_3-Ph	4-MeO-BnNH	37	92
4	**11d**	3-F-Ph	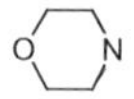	49	>99
5	**11e**	3-F-Ph	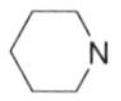	45	92
6	**11f**	Ph	*i*-BuNH	45	>99
7	**11g**	Ph	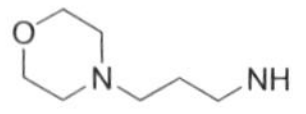	72	>99
8	**11h**	4-Me-Ph	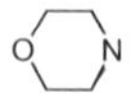	54	>99
9	**11i**	4-*t*-Bu-Ph	4-Cl-BnNH	54	89
10	**11j**	4-*t*-Bu-Ph	3,5-di-MeO-PhNH	31	86
11	**11k**	4-*t*-Bu-Ph	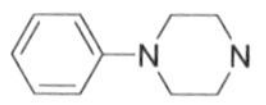	54	99

TABLE 10.10. *(Continued)*

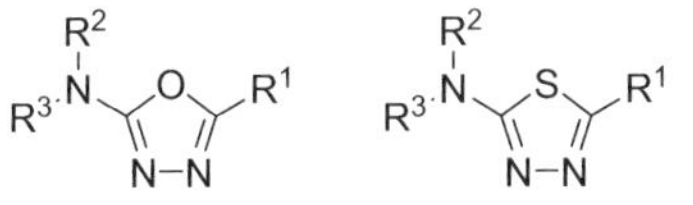

11 **12**

Entry	Products	R^1	R^2R^3N	Yield (%)[b]	Purity (%)[c]
12	**11l**	4-*t*-Bu-Ph	Bn_2NH	trace	–
13	**12a**	4-CF_3-Ph		47	92
14	**12b**	4-CF_3-Ph	*i*-BuNH	37	>99
15	**12c**	4-CF_3-Ph	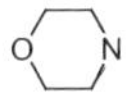	43	99
16	**12d**	4-CF_3-Ph	4-MeO-BnNH	49	97
17	**12e**	3-F-Ph	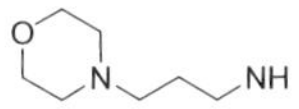	66	99
18	**12f**	3-F-Ph		30	91
19	**12g**	Ph		35	97
20	**12h**	Ph	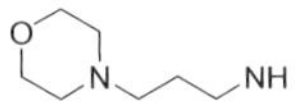	46	99
21	**12i**	4-Me-Ph	*i*-BuNH	32	89
22	**12j**	4-*t*-Bu-Ph	$C_6H_{11}CH_2NH$	34	90
23	**12k**	4-*t*-Bu-Ph		40	94
24	**12l**	4-*t*-Bu-Ph	Bn_2NH	trace	–

[a] All reactions were performed on a 150–200 mg scale of resins **52** (for **11**) and **53** (for **12**).
[b] Four-step overall yield from Merrifield resin **1** (loading capacity = 0.94 mmol/g).
[c] Determined on the basis of LC–MS spectrum (integration of diode array 200–400 nm traces) after short-passed silica gel column chromatography.

10.6 SOLID-PHASE SYNTHESIS OF 1,2,4-THIADIAZOLES

As a member of the thiadiazole family, which is a privileged structure for the generation of drug-like libraries, 1,2,4-thiadiazoles have been used as the basic framework for substances of interest in numerous therapeutic areas.[69] Recently, we reported the results of a study that led to the development of a solution-phase parallel synthesis of various drug-like 5-amino-1,2,4-thiadiazoles via a three-component nucleophilic substitution reaction between carbon disulfide, benzamidine, and benzyl chloride, using a key cyclization reaction of a carboxamidine dithiocarbazate induced by TsCl.[70] We also developed a simple and efficient solid-phase parallel synthetic method capable of facilitating the production of a variety of 5-amino- and 5-amido-1,2,4-thiadiazoles derived from a common intermediate.[46]

The strategy for the efficient solid-phase synthesis of various 5-amino- and 5-amido-functionalized 1,2,4-thiadiazole derivatives is given in the sequence illustrated in Scheme 10.13. Resin-bound carboxamidine thioureas **54** were used as key intermediates and underwent cyclization to produce the 1,2,4-thiadiazole resin **55**. *N*-Alkylation and *N*-acylation reactions of the resin **55** then yielded the respective resins **56** and **57**, which were transformed to the 5-amino- and 5-amido-1,2,4-thiadiazoles **13** and **58**, respectively.

The isothiocyanate-terminated resin **59** was prepared from the amine resin **60** (the BOMBA resin was synthesized by a two-step sequence from Merrifield resin **1**) by the reaction with thiophosgene in the presence of triethylamine or with carbon disulfide and TsCl in the presence of diisopropylethylamine (DIPEA) in CH_2Cl_2. The resin-bound isothiocyanate **59** reacted with carboxamidine **61** (R^1 diversity element) in the presence of DBU (1,8-diazabicyclo[5.4.0]undec-7-ene) in dichloroethane (DCE) at 60 °C to afford the resin-bound carboxamidine thiourea **54**. To develop methods for the solid-phase synthesis of various 5-functionalized 1,2,4-thiadiazoles, cyclization of the carboxamidine thiourea resin **54** was investigated using a number of different activating agents, including EDCI, DCC, TMSCl, TsCl, Ph_3P, $SOCl_2$, PCl_5, and diphenyl chlorophosphate. This investigation demonstrated that the best cyclization condition involved the use of TsCl in the presence of triethylamine in dichloroethane at 60 °C. This process led to the formation of the 3-substituted 5-amino-1,2,4-thiadiazole resin **55**.

Alkylation reactions of 5-amino resin **55** with alkyl halides (R^3 diversity element) provided the desired 3-substituted 5-(*N*-alkylamino)-1,2,4-thiadiazole resin **56**. In a similar manner, 3-substituted 5-(*N*-acylamino)-1,2,4-thiadiazole resin **57** was produced by acylation reactions of resins **55** with acid chlorides (R^3 diversity element). More important, resins **56a** ($R^1 = Ph$, $R^2 = Bn$) and **57a** ($R^1 = R^2 = Ph$) underwent smooth

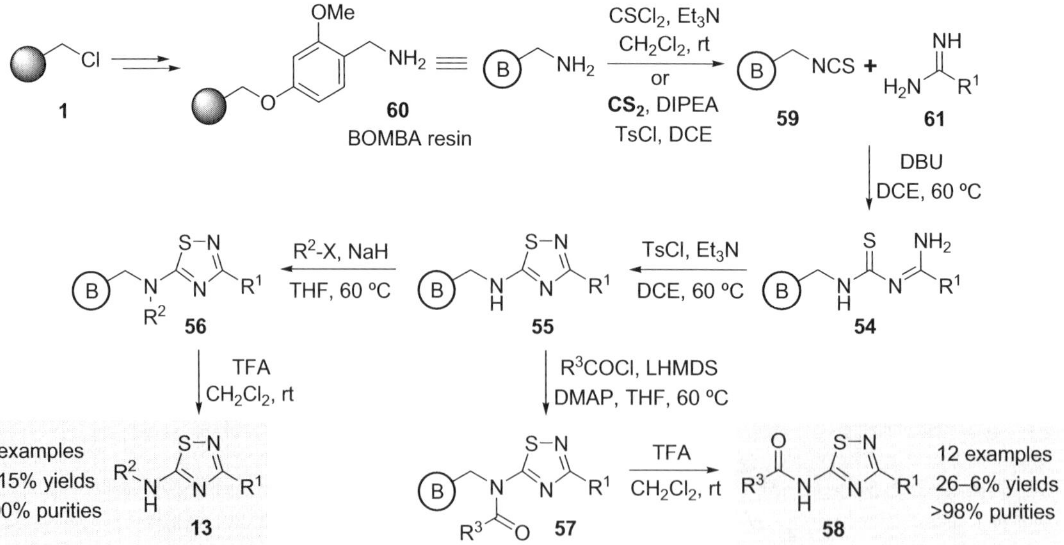

Scheme 10.13. Solid-phase synthesis of 1,2,4-thiadiazoles.

TABLE 10.11. 5-Amino- and 5-Amido-1,2,4-thiadiazoles **13** and **58** and Their Yields[a]

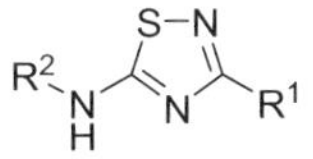

13

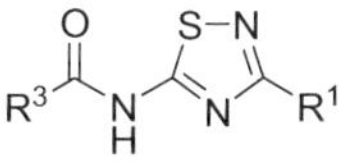

58

Entry	Products	R^1	R^2	R^3	Yield (%)[b]	Purity (%)[c]
1	**13a**	Ph	Bn	–	28	90
2	**13b**	Ph	4-MeO-Bn	–	31	96
3	**13c**	Ph	3,5-di-MeO-Bn	–	19	98
4	**13d**	Ph	4-F-Bn	–	31	>99
5	**13e**	Ph	4-Me-Bn	–	29	98
6	**13f**	Ph	2,5-di-Me-Bn	–	26	93
7	**13g**	Ph	4-CN-Bn	–	27	>99
8	**13h**	Ph	4-NO_2-Bn	–	17	>99
9	**13i**	Ph	4-Cl-Bn	–	30	97
10	**13j**	Ph	2-Cl-Bn	–	23	95
11	**13k**	Ph	3-F-Bn	–	23	>99
12	**13l**	Ph	4-CF_3-Bn	–	15	96
13	**13m**	Ph	4-*t*-Bu-Bn	–	24	93
14	**13n**	Ph	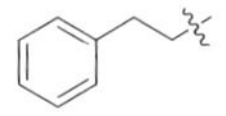	–	15	94
15	**13o**	Ph	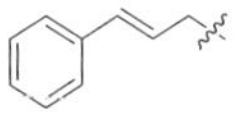	–	19	93
16	**13p**	Ph	$C_6H_{11}CH_2$-	–	34	91
17	**13q**	Ph	Allyl	–	27	>99
18	**13r**	Ph	*i*-Bu	–	33	96
19	**13s**	Ph	Me	–	29	>99
20	**13t**	3-NO_2-Ph	Bn	–	20	95
21	**13u**	3-NO_2-Ph	4-MeO-Bn	–	20	>99
22	**13v**	3-NO_2-Ph	4-CF_3-Bn	–	20	>99
23	**58a**	Ph	–	Ph	26	–[d]
24	**58b**	Ph	–	2-Cl-Ph	7	–[d]
25	**58c**	Ph	–	2-F-Ph	23	–[d]
26	**58d**	Ph	–		20	–[d]
27	**58e**	Ph	–	2,6-di-F-Ph	21	–[d]
28	**58f**	Ph	–	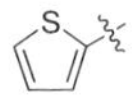	6	–[d]
29	**58g**	Ph	–	4-Cl-Ph	6	–[d]
30	**58h**	Ph	–	4-F-Ph	14	–[d]
31	**58i**	Ph	–	3-CF_3-Ph	6	–[d]
32	**58j**	Ph	–	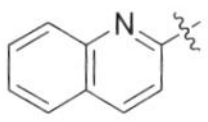	13	–[d]

(continued)

TABLE 10.11. *(Continued)*

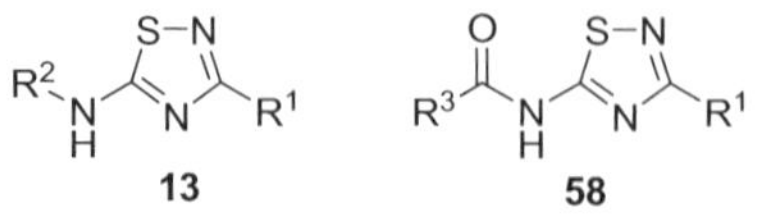

Entry	Products	R^1	R^2	R^3	Yield (%)[b]	Purity (%)[c]
33	**58k**	3-NO_2-Ph	–	Ph	21	–[d]
34	**58l**	3-NO_2-Ph	–	3-CF_3-Ph	6	–[d]

[a] All reactions were performed on a 150–200 mg scale of resins **56** (for **13**) and **57** (for **58**).
[b] Five-step overall isolated yield from BOMBA resin **60** (loading capacity = 1.20 mmol/g).
[c] Determined on the basis of LC–MS spectrum (integration of diode array 200–400 nm traces) of isolated product.
[d] Purities of **58** were over 98% as judged from LC–MS traces.

reactions to yield 5-amino-functionalized 1,2,4-thiadiazoles **13a** ($R^1 = Ph$, $R^2 = Bn$) or 5-amido-functionalized 1,2,4-thiadiazoles **58a** ($R^1 = R^2 = Ph$), respectively, in high yields and purities when treated with TFA (trifluoroacetic acid) in CH_2Cl_2 at room temperature. Various 5-amino-1,2,4-thiadiazoles **13** and 5-amido-1,2,4-thiadiazoles **58** could be produced by this five-step route in high overall yields and purities (for **13**: 22 examples, 34–15% yields, 99–90% purities; for **58**: 12 examples, 26–6% yields, 89–70% purities, Table 10.11).

10.7 SUMMARY

The combinatorial synthesis of small heterocyclic organic molecules plays a significant role in drug discovery. Especially, substituted and fused five-membered ring heterocycles acting as bioactive molecules have proven to be broadly useful as therapeutic agents because of their high degree of structural diversity. In this respect, many synthetic methods for synthesizing these privileged five-membered heterocyclic ring structures with drug-like properties using carbon disulfide have been developed using solid-phase strategies. Carbon disulfide is a facile, cheap, and versatile reagent capable of acting as a sulfur source, traceless linker, substitution site, diversity element, and intermediate in the synthesis of five-membered ring heterocycles and their fused counterparts on a solid support. In this chapter, we have introduced the preparation of diverse and drug-like five-membered ring heterocycles such as thiazoles, benzoxazoles, pyrazoles, triazoles, oxadiazoles, thiadiazoles, and their related compounds using carbon disulfide and Merrifield resin. Following further studies in this field, more experimental conditions with carbon disulfide and Merrifield resin as a solid support will be reported for medicinal chemistry and drug discovery.

REFERENCES

1. (a) R. E. Dolle, B. Le Bourdonnec, A. J. Goodman, G. A. Morales, C. J. Thomas, W. Zhang, *J. Comb. Chem.* **2009**, *11*, 739–790; (b) R. E. Dolle, B. Le Bourdonnec, A. J. Goodman, G. A.

Morales, C. J. Thomas, W. Zhang, *J. Comb. Chem.* **2008**, *10*, 753–800; (c) R. E. Dolle, B. Le Bourdonnec, A. J. Goodman, G. A. Morales, J. M. Salvino, W. Zhang, *J. Comb. Chem.* **2007**, *9*, 855–902.

2. J. P. Kennedy, L. Williams, T. M. Bridges, R. N. Daniels, D. Weaver, C. W. Lindsley, *J. Comb. Chem.* **2008**, *10*, 345–354.
3. K. C. Nicolaou, R. Hanko, W. Hartwig (Eds), *Combinatorial Chemistry: Drugs, Catalysts, Materials*, Wiley-VCH, Weinheim, **2002**.
4. D. Lednicer, *Strategies for Organic Drug Synthesis and Design*, 2nd ed., Wiley, Hoboken, NJ, **2009**, pp. 239–318.
5. V. Krchnňák, M. W. Holladay, *Chem. Rev.* **2002**, *102*, 61–91.
6. D. A. Horton, G. T. Bourne, M. L. Smyth, *Chem. Rev.* **2003**, *103*, 893–930.
7. For a recent review, see A. R. Katritzky, C. A. Ramsden, E. F. V. Scriven, R. J. K. Taylor, (Eds), *Comprehensive Heterocyclic Chemistry III*, Vols 3–6, Elsevier, Oxford, **2008**.
8. W. Bannwarth, B. Hinzen (Eds), *Combinatorial Chemistry: From Theory to Application*, 2nd ed., Wiley-VCH, Weinheim, **2006**.
9. (a) C. Gil, S. Bräse, *J. Comb. Chem.* **2009**, *9*, 175–197; (b) K. Rück-Braun, T. H. E. Freysoldt, F. Wierschem, *Chem. Soc. Rev.* **2005**, *34*, 507–516; (c) R. G. Franzén, *J. Comb. Chem.* **2000**, *2*, 195–214.
10. T. Eicher, S. Hauptmann, A. Speicher, *The Chemistry of Heterocycles*, 2nd ed. , Wiley-VCH, Weinheim, **2003**, pp. 52–221.
11. S. Bräse (Ed.), Combinatorial Chemistry on Solid Supports, Springer, Heidelberg, **2007**.
12. Y. Wang, J. Huang, Y. Chai, Q. Liu, Y. Liang, D. Dong, *J. Comb. Chem.* **2008**, *10*, 511–516.
13. Y.-P. Luo, G.-F. Yang, *Bioorg. Med. Chem.* **2007**, *15*, 1716–1724.
14. E. S. H. El Ashry, A. A. Kassem, H. Abdel-Hamid, F. F. Louis, Sh. A. N. Khattab, M.R. Aouad, *ARKIVOC* **2006**, *xiv*, 119–132.
15. S. Murru, H. Ghosh, S. K. Sahoo, B. K. Patel, *Org. Lett.* **2009**, *11*, 4254–4257.
16. D. A. Ibrahim, A. M. El-Metwally, E. E. Al-Arab, *ARKIVOC* **2009**, *vii*, 12–25.
17. J. C. A. Hunt, E. Briggs, E. D. Clarke, W. G. Whittingham, *Bioorg. Med. Chem. Lett.* **2007**, *17*, 5222–5226.
18. F.-B. Han, Z.-M. Ge, T.-M. Cheng, R.-T. Li, *Synlett* **2009**, 648–650.
19. L. Li, Z. Wang, Y. Chen, Y. Yuan, G. Liu, *J. Comb. Chem.* **2007**, *9*, 959–972.
20. M. Genc, S. Servi, *Heteroatom Chem.* **2005**, *16*, 142–147.
21. A. Ivachtchenko, S. Kovalenko, O. V. Tkachenko, O. Parkhomenko, *J. Comb. Chem.* **2004**, *6*, 573–583.
22. L. Poitout, V. Brault, C. Sackur, S. Bernetière, J. Camara, P. Plas, P. Roubert, *Bioorg. Med. Chem. Lett.* **2007**, *17*, 4464–4470.
23. M. Grimstrup, M. Zaragoza, *Eur. J. Org. Chem.* **2002**, 2953–2960.
24. (a) R. Severinsen, J. P. Kilburn, J. F. Lau, *Tetrahedron* **2005**, *61*, 5565–5575; (b) J. P. Kilburn, J. Laua, R. C. F. Jones, *Tetrahedron Lett.* **2003**, *44*, 7825–7828.
25. I. R. Baxendale, S. V. Ley, M. Martinelli, *Tetrahedron* **2005**, *61*, 5323–5349.
26. C. L. Lee, M. M. Sim, *Tetrahedron Lett.* **2000**, *41*, 5729–5732.
27. I. R. Baxendale, S. V. Ley, C. D. Smith, L. Tamborini, A.-F. Voica, *J. Comb. Chem.* **2008**, *10*, 851–857.
28. (a) Y. Tominaga, *J. Heterocycl. Chem.* **1989**, *26*, 1167–1204; (b) L. A. Paquette (Ed.), *Encyclopedia of Reagents for Organic Synthesis*, Wiley, Chichester, **1995**.
29. Y.-D. Gong, T. Lee, *J. Comb. Chem.* **2010**, *12*, 393–409.
30. (a) H. Stephensen, F. Zaragoza, *J. Org. Chem.* **1997**, *62*, 6096–6097; (b) H. Stephensen, F. Zaragoza, Resin-bound isothiocyanates as intermediates for the solid-phase synthesis of

substituted thiophenes, in Czarnik, A. W. (Ed.), *Solid-Phase Organic Syntheses*, Vol. 1, Wiley, New York, **2001**, pp. 149–157.

31. L. Gomez, F. Gellibert, A. Wagner, C. Mioskowski, *Tetrahedron Lett.* **2008**, *49*, 2726–2729.
32. O. A. Attanasi, L. De Crescentini, G. Favi, P. Filippone, G. Giorgi, F. Mantellini, G. Moscatelli, M. S. Behalo, *Org. Lett.* **2009**, *11*, 2265–2268.
33. F. Xiang, S. Zhang, C. Lu, Z. Chen, G. Yang, *Synth. Commun.* **2008**, *38*, 953–960.
34. A. D. Roy, S. Sharma, R. K. Grover, B. Kundu, R. Roy, *Org. Lett.* **2004**, *6*, 4763–4766.
35. Y. Zhang, Q. Zhou, R. A. Houghten, Y. Yu, *Tetrahedron Lett.* **2007**, *48*, 7042–7045.
36. Z. Liu, J. Zhao, X. Huang, *Bioorg. Med. Chem. Lett.* **2006**, *16*, 1828–1830.
37. I. Y. Lee, J. Y. Lee, H. J. Lee, Y.-D. Gong, *Synlett* **2005**, 2483–2485.
38. T. Lee, D. Lee, I. Y. Lee, Y.-D. Gong, *J. Comb. Chem.* **2010**, *12*, 95–99.
39. T. Lee, J.-H. Park, D.-H. Lee, Y.-D. Gong, *J. Comb. Chem.* **2009**, *11*, 495–499.
40. T. Lee, J.-H. Park, M.-K. Jeon, Y.-D. Gong, *J. Comb. Chem.* **2009**, *11*, 288–293.
41. J. Y. Hwang, Y.-D. Gong, *J. Comb. Chem.* **2006**, *8*, 297–303.
42. J. Y. Hwang,; H.-S. Choi, D.-Y. Lee, S.-E. Yoo, Y.-D. Gong, *J. Comb. Chem.* **2005**, *7*, 136–141.
43. Y.-D. Gong, I. A. Ryu, *J. Comb. Chem.* **2009**, *11*, 626–630.
44. Y.-D. Gong, J. G. Kim, *Bull. Korean Chem. Soc.* **2009**, *30*, 3085–3088.
45. J. Y. Hwang, H.-S. Choi, D.-Y. Lee, Y.-D. Gong, *J. Comb. Chem.* **2005**, *7*, 816–819.
46. I. A. Ryu, J. Y. Park, H. C. Han, Y.-D. Gong, *Synlett* **2009**, 999–1003.
47. (a) Z. Jin, *Nat. Prod. Rep.* **2003**, *20*, 584–605; (b) J. R. Lewis, *Nat. Prod. Rep.* **1999**, *16*, 389–416.
48. P. Blaney, R. Grigg, V. Sridharan, *Chem. Rev.* **2002**, *102*, 2607–2624.
49. Unpublished results of solid-phase synthesis of 2,4,5-trisubstituted thiazoles in our laboratory.
50. (a) W. A. Thaler, J. R. McDivitt, *J. Org. Chem.* **1971**, *36*, 14–18; (b) C. D'Amico, *J. Org. Chem.* **1967**, *32*, 2567–2570; (c) R. J. Timmons, L. S. Wittenbrook, *J. Org. Chem.* **1967**, *32*, 1566–1572.
51. S. G. Johnson, P. J. Connolly, W. V. Murray, *Tetrahedron Lett.* **2006**, *47*, 4853–4856.
52. (a) C. Alvarez-Ibarra, R. Fernández-Granda, M. L. Quiroga, A. Carbonell, F. Cádenas, E. Giralt, *J. Med. Chem.* **1997**, *40*, 668–676; (b) H. Yamanaka, S. Ohba, T. Sakamoto, *Heterocycles* **1990**, *31*, 1115–1127.
53. (a) J. P. Tierney, P. Lidström (Eds), *Microwave Assisted Organic Synthesis*, Blackwell, Oxford, UK, **2005**; (b) V. Molteni, D. A. Ellis, *Curr. Org. Synth.* **2005**, *2*, 333–375; (c) A. Lew, P. O. Krutzik, M. E. Hart, R. Chamberin, *J. Comb. Chem.* **2002**, *4*, 95–105.
54. (a) J. Lee, X. Wu, M. Pasca di Magliano, E. C. Peters, Y. Wang, J. Hong, M. Hebrok, S. Ding, C. Y. Cho, P. G. Schultz, *ChemBioChem* **2007**, *8*, 1916–1919; (b) J. C. Milne, P. D. Lambert, S. Schenk, D. P. Carney, J. J. Smith, D. J. Gange, L. Jin, O. Boss, R. B. Perni, C. B. Vu, J. E. Bemis, R. Xie, J. S. Disch, P. Y. Ng, J. J. Nunes, A. V. Lynch, H. Yang, H. Galonek, K. Israelian, W. Choy, A. Iffland, S. Lavu, O. Medvedik, D. A. Sinclair, J. M. Olefsky, M. R. Jirousek, P. J. Elliott, C. H. Westphal, *Nature* **2007**, *450*, 712–716; (c) J. Das, P. Chen, D. Norris, R. Padmanabha, J. Lin, R. V. Moquin, Z. Shen, L. S. Cook, A. M. Doweyko, S. Pitt, S. Pang, D. R. Shen, Q. Fang, H. F. de Fex, K. W. McIntyre, D. J. Shuster, K. M. Gillooly, K. Behnia, G. L. Schieven, J. Wityak, J. C. Barrish, *J. Med. Chem.* **2006**, *49*, 6819–6832.
55. For recent examples, see; (a) R. Lin, S. G. Johnson, P. J. Connolly, S. K. Wetter, E. Binnun, T. V. Hughes, W. V. Murray, N. B. Pandey, S. J. Moreno-Mazza, M. Adams, A. R. Fuentes-Pesquera, S. A. Middleton, *Bioorg. Med. Chem. Lett.* **2009**, *19*, 2333–2337; (b) S. S. Kulkarni, A. H. Newman, *Bioorg. Med. Chem. Lett.* **2007**, *17*, 2987–2991; (c) S. Komoriya, S. Kobayashi, K. Osanai, T. Yoshino, T. Nagata, N. Haginoya, Y. Nakamoto, A. Mochizuki, T. Nagahara, M. Suzuki, T.

Shimada, K. Watanabe, Y. Isobe, T. Furugoori, *Bioorg. Med. Chem.* **2006**, *14*, 1309–1330; (d) K. Walczyński, O. P. Zuiderveld, H. Timmerman, *Eur. J. Med. Chem.* **2005**, *40*, 15–23.

56. D. Thomae, E. Perspicace, S. Hesse, G. Kirsch, P. Seck, *Tetrahedron* **2008**, *64*, 9309–9314.

57. (a) V. A. Artyomov, V. L. Ivanov, A. M. Shestopalov, V. P. Litvinov, *Tetrahedron* **1997**, *53*, 13351–13360; (b) V. L. Ivanov, V. A. Artemov, A. M. Shestopalov, V. N. Nesterov, Y. T. Struchkov, V. P. Litvinov, *Chem. Heterocycl. Compd. (NY)* **1996**, *32*, 413–419.

58. D. Thomae, G. Kirsch, P. Seck, *Synthesis* **2008**, 1600–1606.

59. (a) D. A. Carson, H. B. Cottam, L. Deng,*PCT Int. Appl.* WO2000069861, **2000**; (b) A. Berger, E. E. Borgeas,*Ger. Offen.* DE2038922, **1971**; (c) H. Abe, M. Tanaka, K. Sugimoto, A. Suma, M. Yokota, M. Shiozaki, K. Ito, K. Ueyama, D. Motoda, T. Noguchi, T. Adachi, J. Tsuruha, S. Doi, *PCT Int. Appl.* WO2007119889, **2007**; (d) A. F. Lewis, G. R. Revankar, S. M. Fennewald, M. Susan, J. H. Huffman, R. F. Rando, *J. Heterocycl. Chem.* **1995**, *32*, 547–556.

60. C. A. Lipinski, F. Lombardo, B. W. Doming, P. J. Feeney, *Adv. Drug Delivery Rev.* **1997**, *23*, 3–25.

61. M.-K. Jeon, M.-S. Kim, J.-J. Kwon, Y.-D. Gong, D.-H. Lee, *Tetrahedron* **2008**, *64*, 9060–9072.

62. (a) L. Banfi, A. Basso, G. Guanti, N. Kiellamd, C. Repetto, R. Riva, *J. Org. Chem.* **2007**, *72*, 2151–2160; (b) T. Ohtani, Y. Kawano, K. Kitano, J. Matsubara, M. Komatsu, M. Uchida, F. Tabusa, Y. Nagao, *Heterocycles* **2005**, *66*, 481–502.

63. For recent examples, see; (a) S. K. Tipparaju, S. Joyasawal, M. Pieroni, M. Kaiser, R. Brun, A. P. Kozikowski, *J. Med. Chem.* **2008**, *51*, 7344–7347; (b) S. Aiello, G. Wells, E. L. Stone, H. Kadri, R. Bazzi, D. R. Bell, M. F. G. Stevens, C. S. Matthews, T. D. Bradshaw, A. D. Westwell, *J. Med. Chem.* **2008**, *51*, 5135–5139; (c) M. A. Siracusa, L. Salerno, M. N. Modica, V. Pittalà, G. Romeo, M. E. Amato, M. Nowak, A. J. Bojarski, I. Mereghetti, A. Cagnotto, T. Mennini, *J. Med. Chem.* **2008**, *51*, 4529–4538; (d) R. E. Martin, L. G. Green, W. Guba, N. Kratochwil, A. Christ, *J. Med. Chem.* **2007**, *50*, 6291–6294; (e) J. Easmon, G. Pürstinger, K.-S. Thies, G. Heinisch, J. Hofmann, *J. Med. Chem.* **2006**, *49*, 6343–6350; (f) S. Yoshida, S. Shiokawa, K.-I. Kawano, T. Ito, H. Murakami, H. Suzuki, Y. Sato, *J. Med. Chem.* **2005**, *48*, 7075–7079.

64. (a) A. Kumar, R. A. Maurya, P. Ahmad, *J. Comb. Chem.* **2009**, *11*, 198–201; (b) H.-J. Lim, D. Myung, I. Y. Choi Lee, M. H. Jung, *J. Comb. Chem.* **2008**, *10*, 501–503; (c) G. Evindar, R. A. Batey, *J. Org. Chem.* **2006**, *71*, 1802–1808; (d) Z. Tian, D. J. Plata, S. J. Wittenberger, A. V. Bhatia, *Tetrahedron Lett.* **2005**, *46*, 8341–3843; (e) C. Chen, Y.-J. Chen, *Tetrahedron Lett.* **2004**, *45*, 113–115.

65. F. Guillier, D. Orain, M. Bradley, *Chem. Rev.* **2000**, *100*, 2091–2157.

66. (a) A. A. El-Emam, O. A. Al-Deeb, M. Al-Omar, J. Lehmann, *Bioorg. Med. Chem.* **2004**, *12*, 5107–5113; (b) T. Akbarzadeh, S. A. Tabatabai, M. J. Khoshnoud, D. Shafaghi, A. Shafiee, *Bioorg. Med. Chem.* **2003**, *11*, 769–773; (c) H. N. Dogan, A. Duran, S. Rollas, G. Sener, M. K. Uysal, D. Gülen, *Bioorg. Med. Chem.* **2002**, *10*, 2893–2898; (d) M. Tyagi, A. Kumar, *Orient. J. Chem.* **2002**, *18*, 125–130; (e) L. Labanauskas, V. Kalcas, E. Udrenaite, P. Gaidelis, A. Brukstus, A. Dauksas, *Pharmazie* **2001**, *56*, 617–619; (f) B. S. Hollar, R. Gonsalves, S. Shenoy, *Eur. J. Med. Chem.* **2000**, *35*, 267–271.

67. (a) J. Y. Chou, S. Y. Lai, S. L. Pan, G. M. Jow, J. W. Cherm, J. H. Guh, *Biochem. Pharmacol.* **2003**, *66*, 115–124; (b) P. Brown, D. J. Best, J. J. P. Broom, R. Casels, P. J. O'Hanlon, T. J. Mitchell, N. F. Osborne, J. M. Wilson, *J. Med. Chem.* **1997**, *40*, 2563–2570.

68. L. Gomez, F. Gellibert, A. Wagner, C. Mioskowski, *J. Comb. Chem.* **2000**, *2*, 75–79.

69. (a) L. Shen, Y. Zhang, A. Wang, E. Sieber-McMaster, X. Chen, P. Pelton, J. Z. Xu, M. Yang, P. Zhu, L. Zhou, M. Reuman, Z. Hu, R. Russell, A. C. Gibbs, H. Ross, K. Demarest, W. V. Murray, G.-H. Kuo, *Bioorg. Med. Chem.* **2008**, *16*, 3321 3341; (b) A. Castro, A. Encinas, C. Gil, S. Bräse, W. Porcal, C. Pérez, F. J. Moreno, A. Martínez, *Bioorg. Med. Chem.* **2008**, *16*, 495–510; (c) R. J.

Watson, D. R. Allen, H. L. Birch, G. A. Chapman, D. R. Hannah, R. L. Knight, J. W. G. Meissner, D. A. Owen, E. J. Thomas, *Bioorg. Med. Chem. Lett.* **2007**, *17*, 3321–3341; (d) L. Shen, Y. Zhang, A. Wang, E. Sieber-McMaster, X. Chen, P. Pelton, J. Z. Xu, M. Yang, P. Zhu, L. Zhou, M. Reuman, Z. Hu, R. Russell, A. C. Gibbs, H. Ross, K. Demarest, W. V. Murray, G.-H. Kuo, *J. Med. Chem.* **2007**, *50*, 3954–3963; (e) A. Castro, T. Castaño, A. Encinas, W. Porcal, C. Gil, *Bioorg. Med. Chem.* **2006**, *14*, 1644–1652.

70. J. Y. Park, I. A. Ryu, J. H. Park, D. C. Ha, Gong, Y. D. *Synthesis* **2009**, 913–920.

11

RECENT ADVANCES IN SOLID-PHASE 1,3-DIPOLAR CYCLOADDITION REACTIONS

Kirsi Harju and Jari Yli-Kauhaluoma

11.1 INTRODUCTION

This chapter discusses the solid-phase synthesis of five-membered nitrogen-containing heterocycles via 1,3-dipolar cycloaddition. 1,3-Dipolar cycloaddition reactions belong to the family of thermally occurring pericyclic reactions, which lead to five-membered heterocyclic rings.[1] Huisgen[2] carried out the major pioneering studies of 1,3-dipolar cycloadditions in the 1960s. He classified and defined cycloadditions in general[3] and also studied the mechanism of cycloadditions.[4]

The chemistry of solid-phase synthesis has been substantially developed since 1963, when Merrifield[5] introduced solid-phase synthesis and reported the synthesis of a tetrapeptide. Attesting to the novelty and significance of the solid-supported method, Merrifield was awarded the Nobel Prize in Chemistry in 1984.[6] Leznoff[7] studied extensively the solid-phase synthesis of small organic compounds in the 1970s and 1980s. Since the 1990s, the growth of solid-phase organic synthesis has been rapid.[8] Removal of peptides from the solid support is straightforward, as the amide bond is easily cleaved. Cleavage of diversely linked small organic compounds from a solid support presents a greater challenge. Originally, the resin was used as a protecting group for a specific functionality; currently, more sophisticated, traceless methods to cleave organic compounds from the resin are available.[9]

Solid-phase 1,3-dipolar cycloadditions have led to various heterocyclic compounds during recent years. Yedidia and Leznoff[10] reported solid-phase synthesis of isoxazoles via 1,3-dipolar cycloaddition in 1980. Alkynes were attached to the resin with an ester linkage

Solid-Phase Organic Synthesis: Concepts, Strategies, and Applications, First Edition.
Edited by Patrick H. Toy and Yulin Lam.

and the cycloaddition with benzonitrile oxide produced isoxazoles. The effect of the resin on the regioselectivity of the cycloaddition was also investigated. Since then, the 1,3-dipolar cycloadditions on solid supports have been studied extensively. The most recent review covers the literature up to December 2003.[11] The review is divided into three parts dealing with cycloadditions to resin-bound dipolarophiles, cycloadditions to resin-bound dipoles, and intramolecular cycloadditions. Reactions of dipolarophiles, such as alkenes, alkynes, and imines, with dipoles, such as azomethine ylides, azomethine imines, nitrile imines, azides, nitrones, and nitrile oxides, are described. In addition, Kantorowski and Kurth[12] reviewed 1,3-dipolar cycloaddition reactions of nitrile oxides and azomethine ylides on solid support. Solid-phase 1,3-dipolar cycloadditions are now discussed on the basis of the compound types prepared, discussing the linkers and cleavage strategies of various five-membered heterocycles, such as pyrroles, pyrazoles, 1,2,3-triazoles, and isoxazoles.

11.2 SOLID-PHASE SYNTHESIS OF PYRROLIDINES, PYRROLINES, AND PYRROLES

Five-membered heterocycles with one nitrogen atom can be prepared from azomethine ylide-type dipoles and alkynes or alkenes. Several solid-supported cycloadditions with maleimide as a dipolarophile have been reported. Trityl resin-bound maleimide captured azomethine ylides that were generated *in situ* from amino acid methyl esters and aldehydes, and substituted resin-bound pyrrolidines were obtained (Scheme 11.1).[13] Traceless cleavage of the C–N bond between acid-sensitive trityl resin and the *N*-unsubstituted cycloadduct was achieved with 50% trifluoroacetic acid.

Resin-bound maleimide has been used as a dipolarophile in the synthesis of *N*-substituted pyrrolidines as well (Scheme 11.2).[14] Cycloaddition with α-imino esters, which were generated from aryl aldehydes and amino acid esters, yielded resin-bound pyrrolidines. The compounds were bound to the acid-cleavable SASRIN resin with an ester linkage and then cleaved from the resin as carboxylic acids with highly diluted trifluoroacetic acid.

Similar maleimide-derived compounds were obtained by Bicknell and Hird.[15] The reaction of a Wang resin-linked amino acid with an aldehyde produced the resin-bound imine that reacted with the maleimides in solution (Scheme 11.3). Ester-linked products

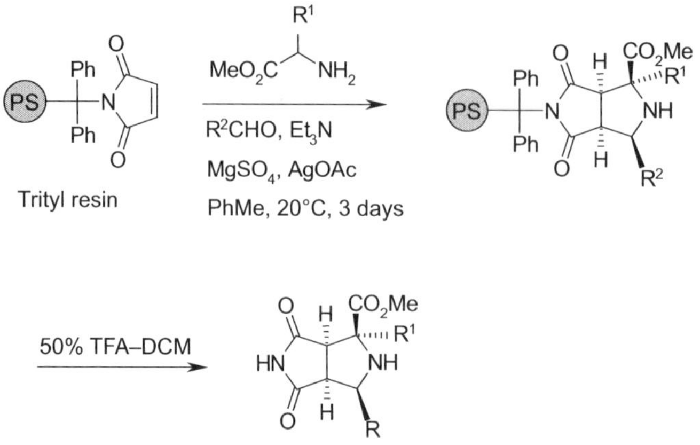

Scheme 11.1. Traceless cleavage of the cycloadducts from trityl resin.

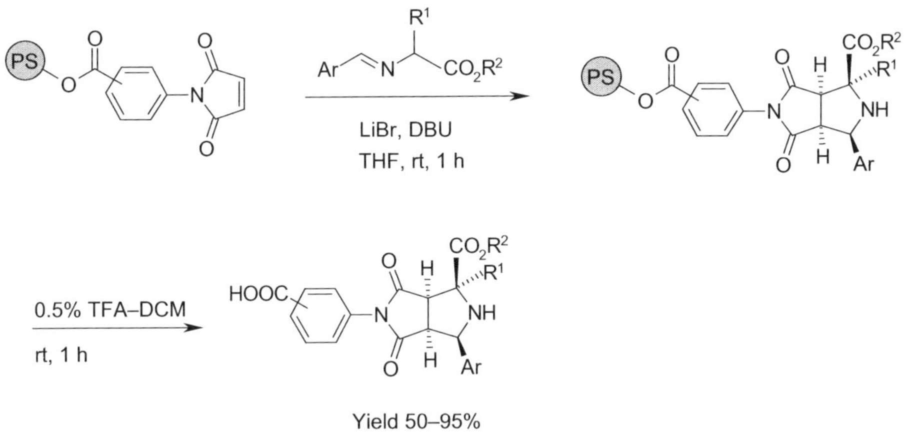

Scheme 11.2. SASRIN resin-bound maleimide with an ester linker.

were cleaved from the resin with 50% trifluoroacetic acid. *N*-Acylation of the intermediates offered further possibilities to diversify the chemical structures of the products obtained from readily available starting materials.

Nine substituted hydroxybenzaldehydes were coupled to a Wang resin with a Mitsunobu coupling (Scheme 11.4).[16] Treatment of the aldehydes with amino acid methyl esters produced resin-bound azomethine ylides that reacted with maleimide in a 1,3-dipolar cycloaddition. Ether-linked cycloadducts were cleaved from the resin as phenols with 50% trifluoroacetic acid. The compounds were retained as a mixture and were not isolated.

Acrylates have also been commonly used as dipolarophiles. Acrylates were attached to the Wang resin with a phenolic linker (Scheme 11.5).[17] Highly substituted pyrrolidines were obtained by a 1,3-dipolar cycloaddition of a resin-bound alkene and azomethine ylide, and the products were cleaved from the resin as phenols with 50% trifluoroacetic acid.

A Wang resin-bound acrylate was attached to the resin with an ester bond (Scheme 11.6).[18] The subsequent reaction of the dipolarophile with imines produced

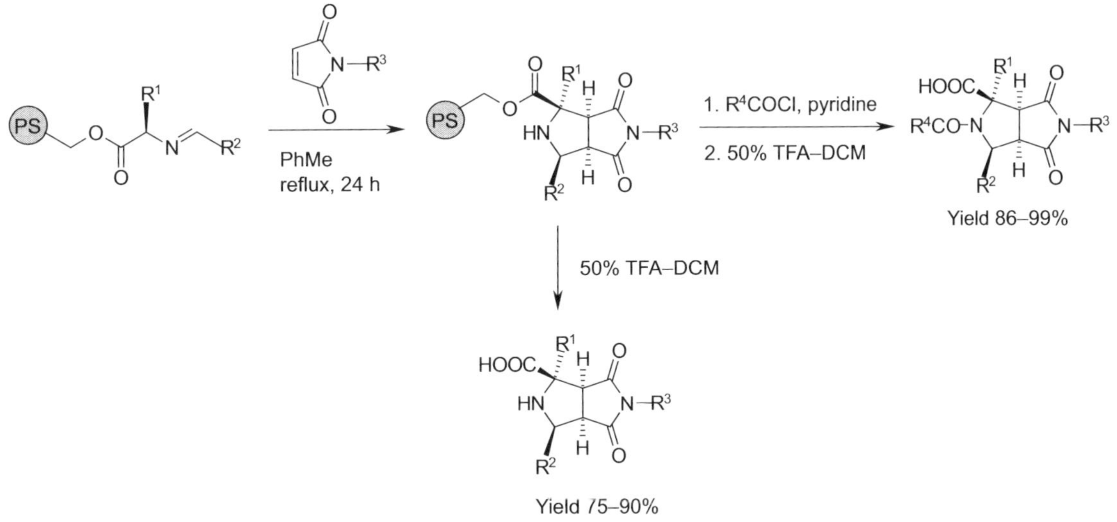

Scheme 11.3. Azomethine ylides attached to the Wang resin with an ester linker.

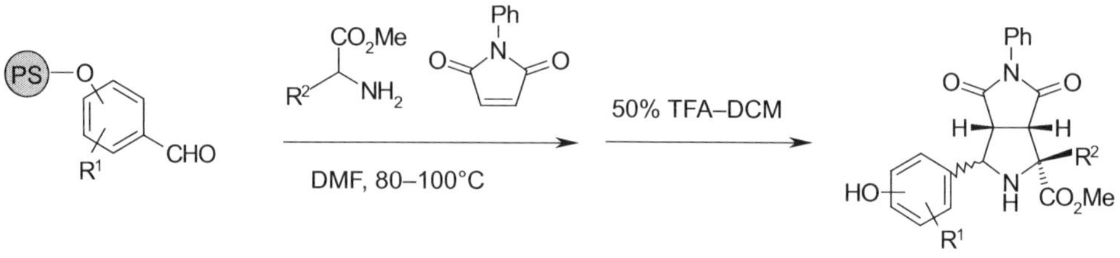

Scheme 11.4. Resin-bound azomethine ylides with a phenolic linker.

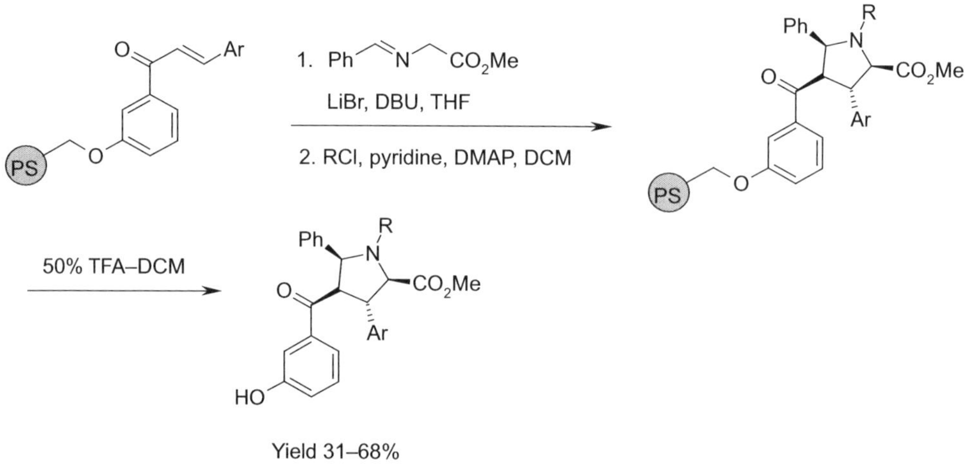

Scheme 11.5. Resin-bound alkene with a phenolic linker.

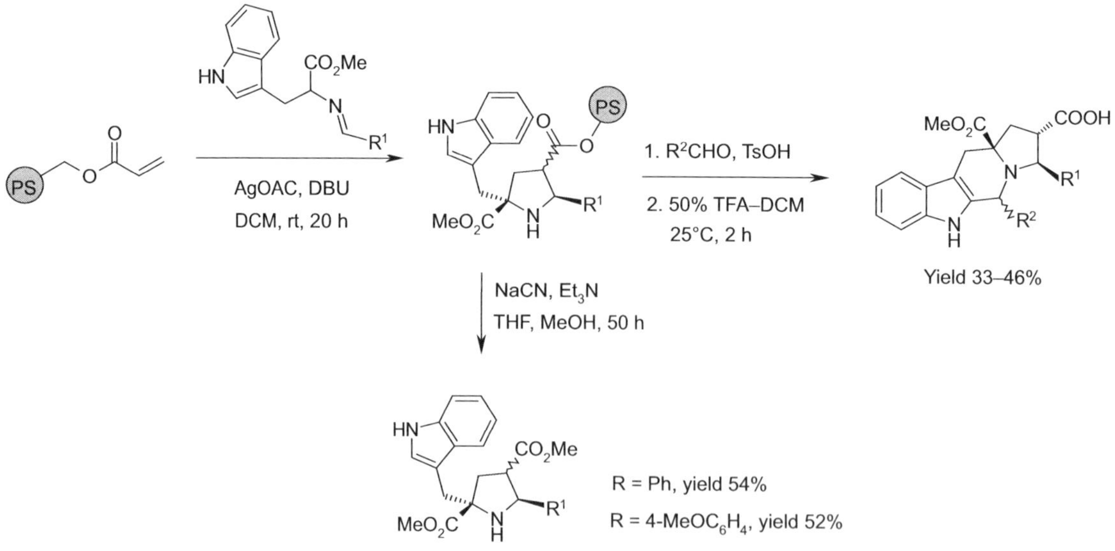

Scheme 11.6. Resin-bound alkene with an ester linker.

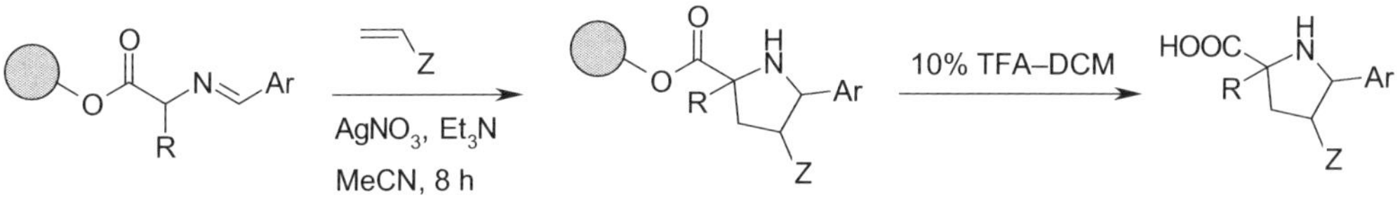

Scheme 11.7. Resin-bound azomethine ylide with an ester linker.

ester-linked cycloadducts that were cleaved from the resin as methyl esters by transesterification. Moreover, the cycloadducts were then used in Pictet–Spengler reactions with various aldehydes, producing fused rings. Treatment of the resin with trifluoroacetic acid released the products as carboxylic acids.

A library of highly functionalized pyrrolidines was prepared from resin-bound azomethine ylides and electron-deficient olefins (Scheme 11.7).[19] Four amino acids, four aldehydes, five olefins, and three acyl chlorides were used as building blocks to produce a library of diverse cycloadducts for various enzymatic assays. Ester-linked products were cleaved from TentaGel or SASRIN resins as carboxylic acids with diluted trifluoroacetic acid. Structural diversity of the proline derivatives was enhanced with *N*-acylation of the pyrrolidines.

A 1,2-silatropic shift was a key step when generating resin-bound azomethine ylides from α-silylimines (Scheme 11.8).[20] A 1,3-dipole was captured with maleimide, and a polymer-bound pyrrolidine was obtained. Acidic or alkylative cleavage of the N–Si bond released *N*-unsubstituted or *N*-substituted pyrrolidines, respectively, from the resin in a traceless manner.

Azomethine ylides were also obtained via a 1,2-silatropic shift of a trimethylsilyl group (Scheme 11.9).[21] Traceless cleavage of the C–Si bond was achieved with tetrabutylammonium fluoride (TBAF), releasing phenyl-substituted pyrrolidine from the resin.

Resin-bound pyridinium betaine acted as an azomethine ylide that reacted with phenyl vinyl sulfone giving various tropane derivatives (Scheme 11.10).[22] Cleavage of benzylic C–N bond was obtained with acyl chlorides in the presence of potassium iodide, and

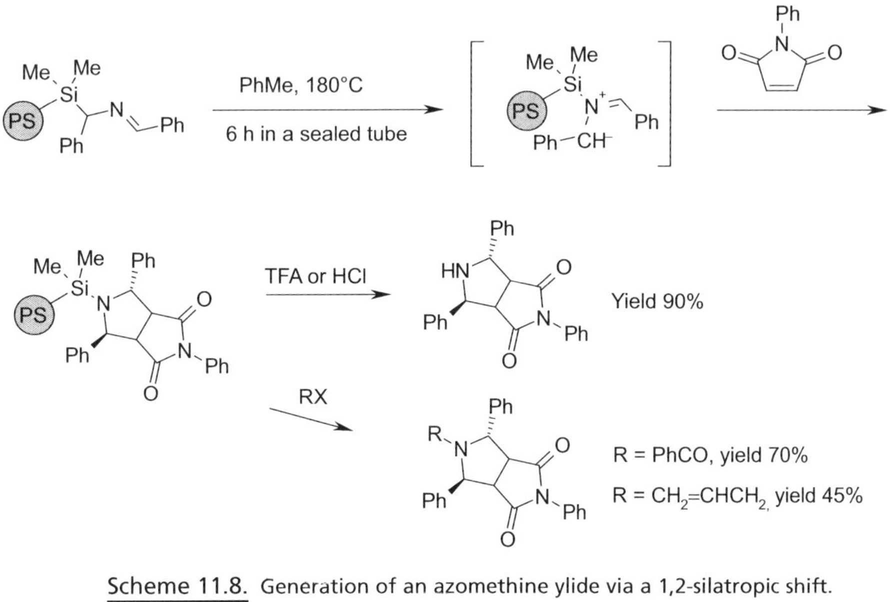

Scheme 11.8. Generation of an azomethine ylide via a 1,2-silatropic shift.

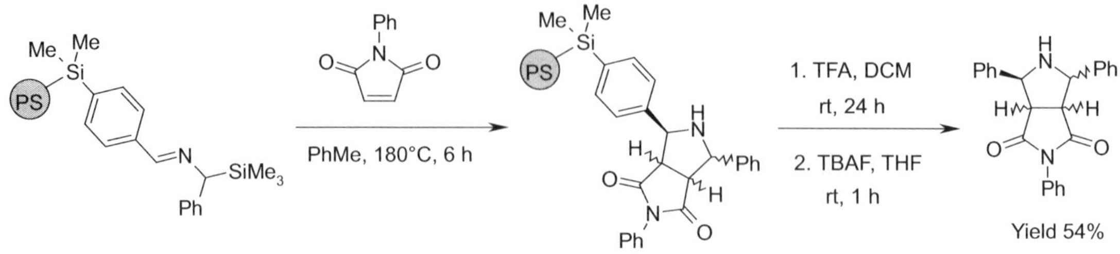

Scheme 11.9. Traceless cleavage of pyrrolidine from the silyl resin.

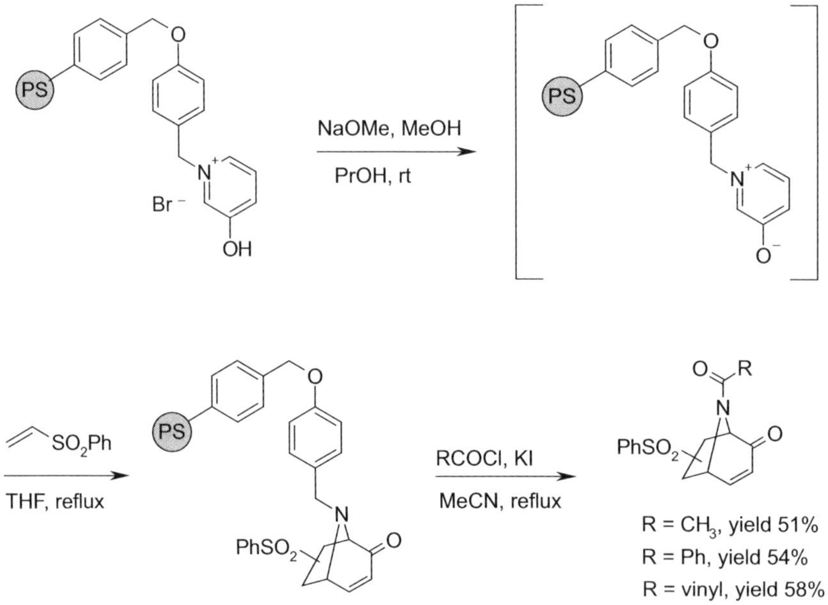

Scheme 11.10. Cleavage of the benzylic C–N bond of tropane derivatives.

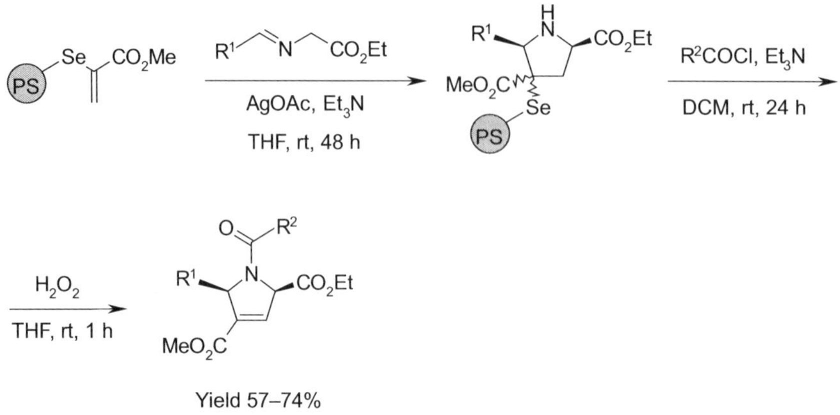

Scheme 11.11. Traceless cleavage from the organoselenium resin.

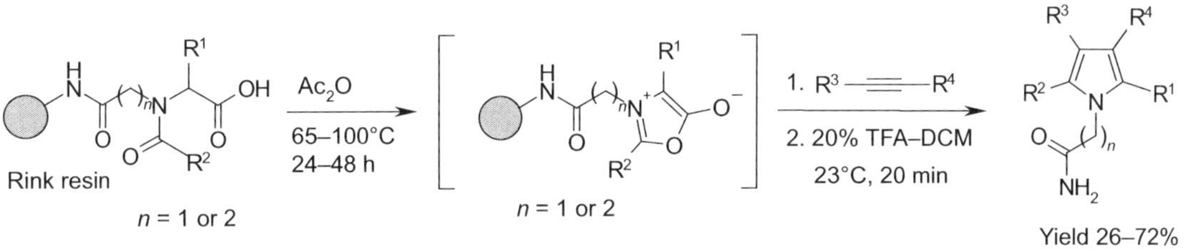

Scheme 11.12. Solid-phase synthesis of pyrroles via münchnones.

pyrrolidines were released from the resin as amides. The use of potassium iodide was essential, otherwise the cleavage was inefficient. Several attempts to cleave the benzylic bond with trifluoroacetic acid, oxidative 2,3-dichloro-5,6-dicyano-1,4-benzoquinone (DDQ), and ceric ammonium nitrate (CAN) were unsuccessful.

Traceless cleavage has also been achieved with an organoselenium resin.[23] Resin-bound methyl 2-seleno acrylate reacted with an excess of azomethine ylide at room temperature (Scheme 11.11). *N*-Acylation of the pyrrolidines gave diverse resin-bound products that were released from the support by oxidative cleavage with hydrogen peroxide. A total of 13 substituted 2,5-dihydro-1*H*-pyrroles were produced with 57–74% yield.

Solid-phase synthesis of pyrroles was reported by Mjalli et al.[24] Pyrroles were prepared via Rink resin-bound mesoionic münchnones (Scheme 11.12). Münchnones behave as azomethine ylides in 1,3-dipolar cycloaddition. Cycloadditions with alkynes give pyrroles after aromatization and spontaneous release of carbon dioxide. A set of compounds was obtained from simple starting materials, aldehyde, amine, carboxylic acid, and isocyanide via the Ugi four-component condensation. Trifluoroacetic acid released highly substituted pyrroles as amides from the Rink resin in high overall yield and purity. A related method via the münchnone pathway was reported by Strocker et al.[25], and two pyrroles were obtained in 4% and 17% yields.

11.3 SYNTHESIS OF PYRAZOLINES AND PYRAZOLES

1,3-Dipolar cycloaddition of azomethine imine or nitrile imine-type dipoles with alkenes or alkynes yields nitrogen heterocycles containing two nitrogen atoms in the ring. MeOPEG resin was applied in the solid-phase synthesis of pyrazolines (Scheme 11.13).[26] Resin-bound

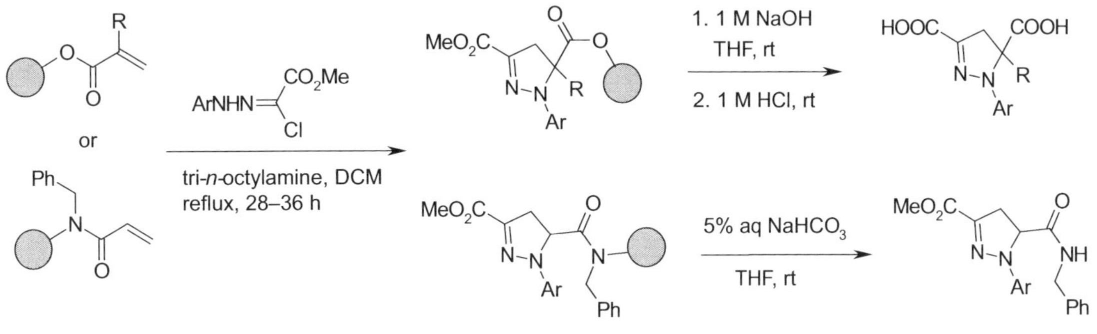

Scheme 11.13. Solid-phase synthesis of pyrazolines from MeOPEG resin-bound dipolarophiles and nitrile imines.

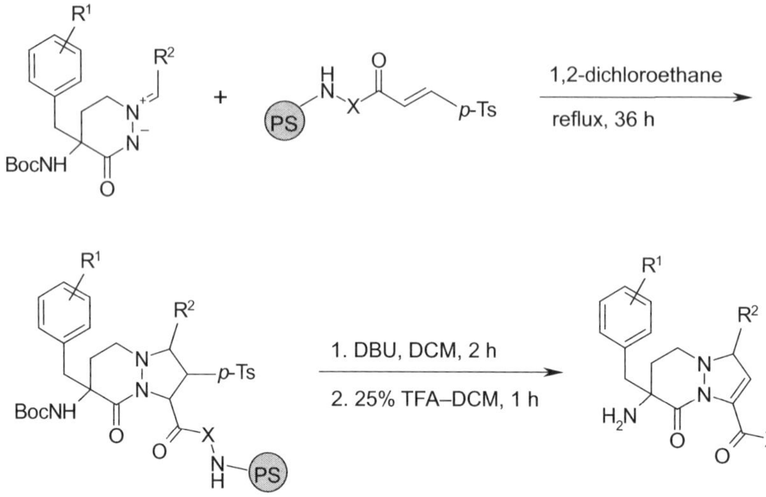

Scheme 11.14. Solid-phase synthesis of pyrazolines from Rink resin-bound vinyl sulfone.

acrylates and acryl amides reacted with nitrile imine generated *in situ* from hydrazonoyl chloride. The products were released from the resin as carboxylic acids or amides.

Rink resin-bound vinyl sulfone has been reported to act as a dipolarophile in the synthesis of pyrazolines from azomethine imines (Scheme 11.14).[27] Elimination of the *p*-toluenesulfonyl group by treatment with 1,8-diazabicyclo[5.4.0]undec-7-ene (DBU) and cleavage from the resin with trifluoroacetic acid gave a library of β-strand mimicking pyrazolines.

A similar method was later applied to the solid-phase synthesis of pyrazoles.[28] A library of 25 compounds was prepared from a Rink resin-bound vinyl sulfone and nitrile imines generated *in situ* from hydrazonoyl chlorides (Scheme 11.15). The pyrazoles were obtained as a mixture of regioisomers.

Traceless cleavage of products from the resin can be achieved through an elimination reaction. Aromatization of the product can be "the driving force" for the elimination. Piperazine resin-bound enamine reacted regioselectively in a 1,3-dipolar cycloaddition with nitrile imines generated *in situ* (Scheme 11.16).[29] Elimination of the piperazine linker with highly diluted trifluoroacetic acid gave diverse 1,4-diarylpyrazoles in a traceless manner.

N-Unsubstituted pyrazoles have been obtained via resin-bound sydnones (Scheme 11.17).[30] Sydnones are mesoionic compounds that react as azomethine imine-type 1,3-dipoles in 1,3-dipolar cycloaddition. The cycloaddition of sydnones with electron-withdrawing alkynes gave pyrazoles. 2-Methoxy-substituted resin acted as a

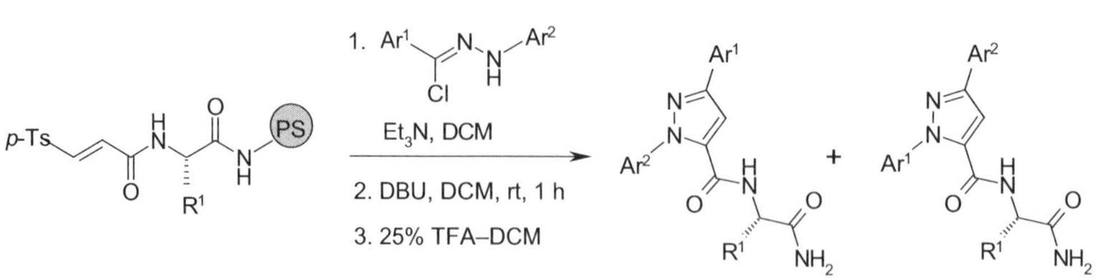

Scheme 11.15. Solid-phase synthesis of pyrazoles from Rink resin-bound vinyl sulfone.

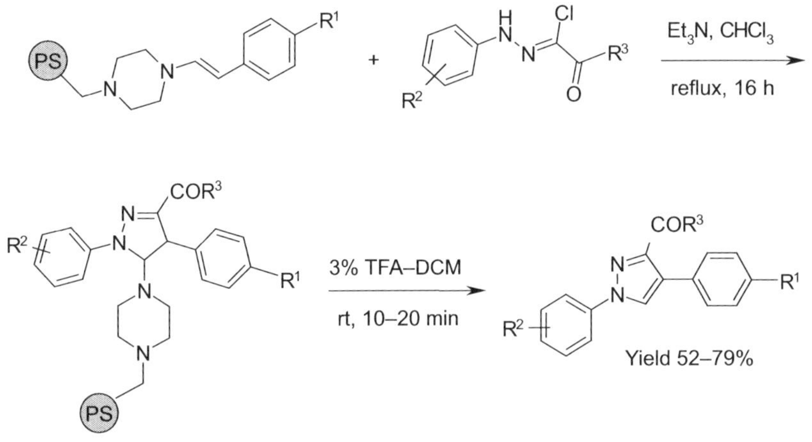

Scheme 11.16. Traceless synthesis of pyrazoles from resin-bound enamines.

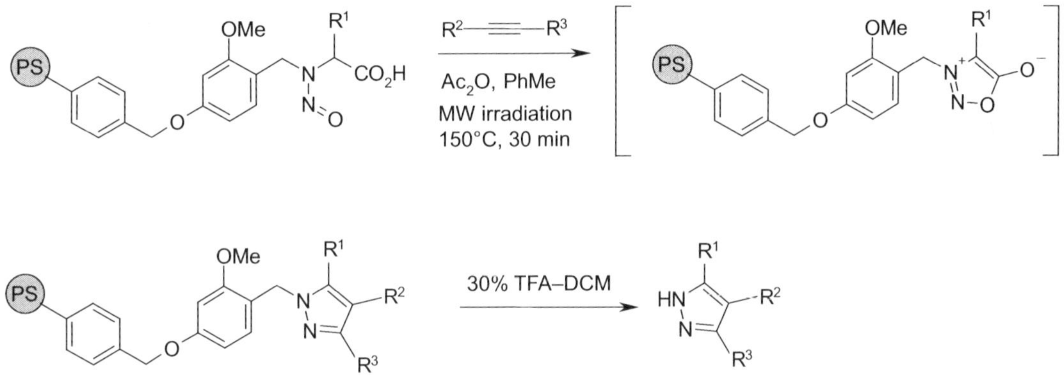

Scheme 11.17. Traceless synthesis of pyrazoles via sydnones.

protecting group for the amino acid and enabled *N*-nitrosation of the secondary amino acid that was bound to the resin. The resin-bound *N*-nitrosated amino acid was dehydrated to form sydnones that reacted *in situ* with alkynes under microwave irradiation. Aromatization of the cycloadduct and release of carbon dioxide led to resin-bound pyrazoles. Diversity of the products was obtained with various amino acids, and traceless cleavage with trifluoroacetic acid gave *N*-unsubstituted pyrazoles.

In addition, pyrazoles have been synthesized in a traceless manner from resin-bound azomethine imines that were generated via a 1,4-silatropic shift (Scheme 11.18).[31]

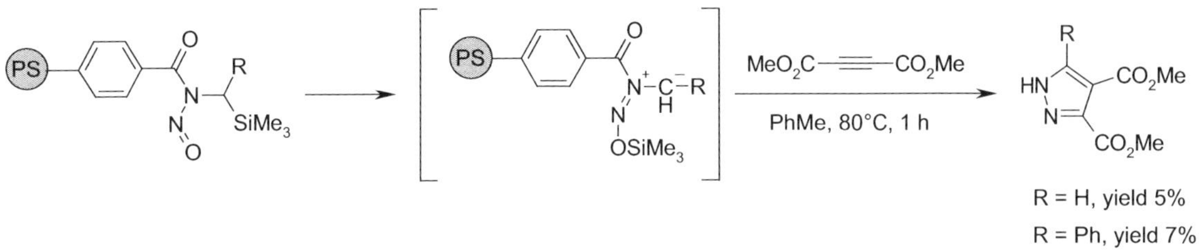

Scheme 11.18. Traceless synthesis of pyrazoles via a 1,4-silatropic shift.

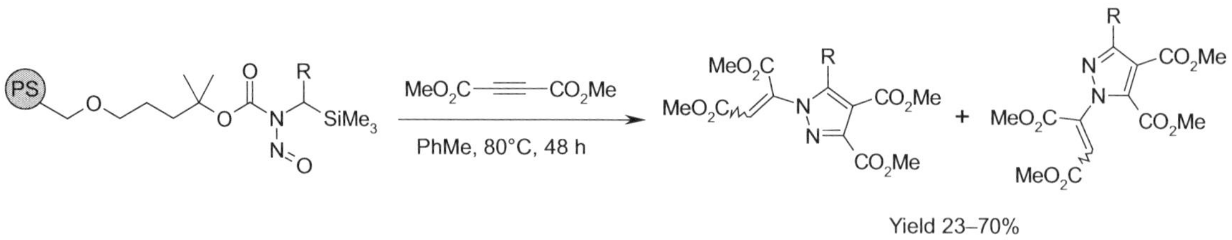

Scheme 11.19. Solid-phase synthesis of pyrazoles with an extended linker.

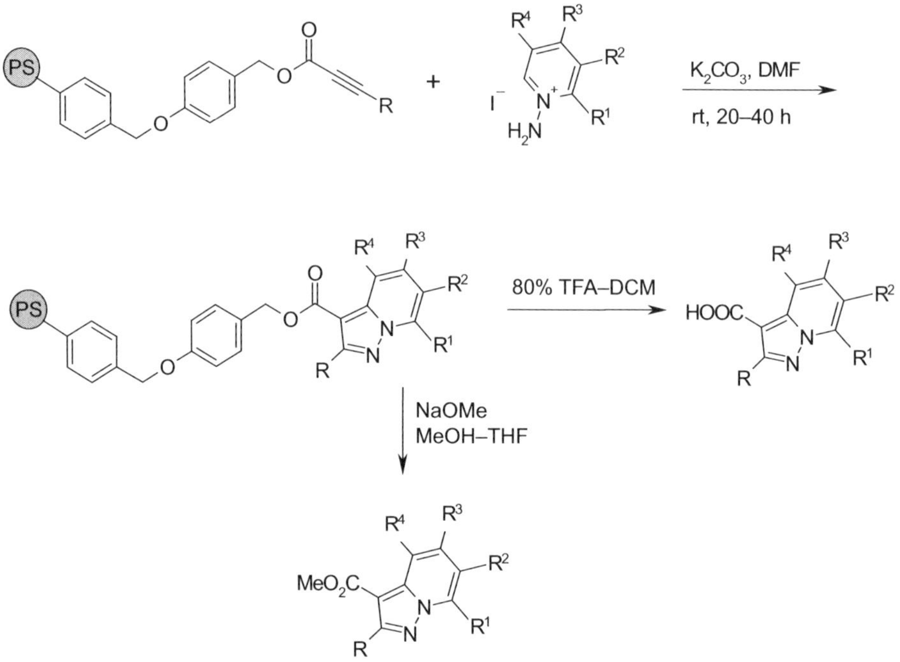

Scheme 11.20. Resin-bound alkynes with an ester linker.

Cycloaddition of the dipoles with dimethyl acetylenedicarboxylate (DMAD) gave *N*-unsubstituted pyrazoles when released spontaneously from the carboxypolystyrene resin. An extended linker improved the yields: however, an excess of DMAD gave Michael adducts and *N*-substituted pyrazoles were obtained instead (Scheme 11.19).

Pyrazolopyridines were synthesized from various resin-bound alkynes and azomethine imines generated *in situ* in solution (Scheme 11.20).[32] Ester-linked products were cleaved with trifluoroacetic acid or sodium methoxide, and the products were obtained as carboxylic acids or methyl esters, respectively.

11.4 SOLID-PHASE SYNTHESIS OF IMIDAZOLES, 1,2,4-TRIAZOLES, AND 1,2,3-TRIAZOLES

Various nitrogen-containing heterocycles have been prepared in a traceless manner using acid-sensitive 2-methoxy-substituted resin. Traceless synthesis of imidazoles was reported

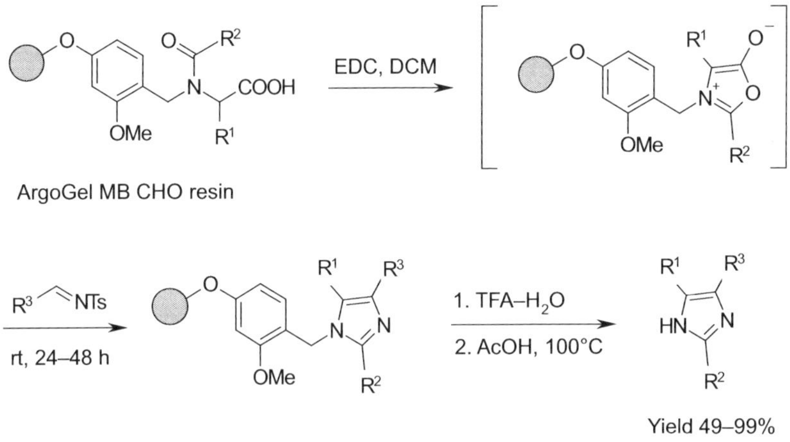

Scheme 11.21. Traceless cleavage of imidazoles.

by Bilodeau and Cunningham.[33] Resin-bound acylated amino acids were dehydrated to mesoionic münchnones that reacted as azomethine ylide-type dipoles (Scheme 11.21). Cycloaddition with aryltosylimines gave resin-bound imidazoles. The electron-donating effect of the 2-methoxy group promoted the traceless cleavage of the benzylic C–N bond, and 12 diverse products were obtained in high yields with higher than 90% purities. Interestingly, treatment with trifluoroacetic acid did not release the imidazoles from the polymer support; instead, it removed the major side products before efficient cleavage with hot acetic acid.

A related method with a resin-bound münchnone intermediate was applied to the synthesis of 1,2,4-triazoles. Cycloaddition of resin-bound münchnones with the N=N double bond of the dipolarophilic diethyl diazocarboxylate (DEAD) or 4-phenyl-4*H*-1,2,4-triazoline-3,5-dione gave resin-bound 1,2,4-triazoles. Traceless cleavage of the products from the 2-methoxy-substituted resin was achieved with 30% trifluoroacetic acid (Scheme 11.22).[34]

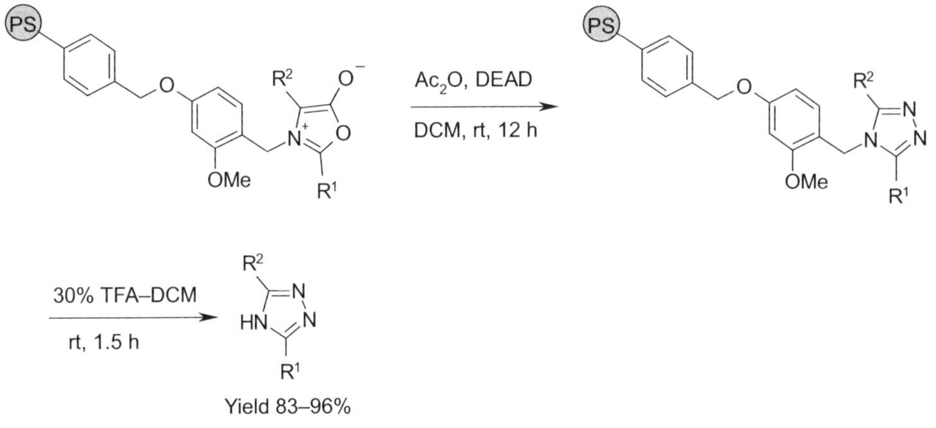

Scheme 11.22. Traceless cleavage of 1,2,4-triazoles.

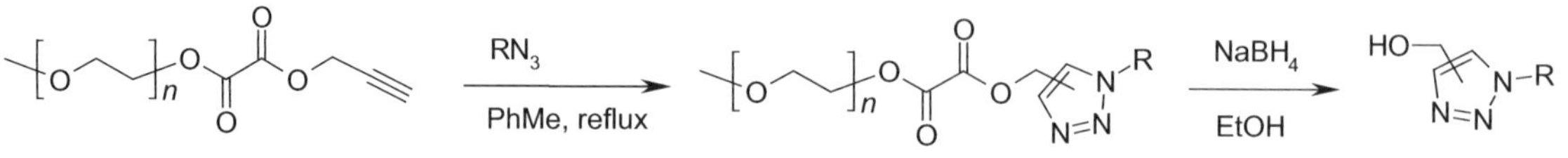

Scheme 11.23. Reductive cleavage of 1,2,3-triazoles.

The reaction of solid-supported imines and *p*-toluenesulfonylmethyl isocyanide (TOSMIC) gave resin-bound imidazoles.[35] In this case, the cleavage of the products occurred with the 2-methoxy-substituted linker because the imidazole core was attached directly to the aryl ring with a carbon–carbon bond.

1,2,3-Triazoles are obtained via 1,3-dipolar cycloadditions between azides and alkynes, and they are probably the most common nitrogen heterocycles prepared on solid supports via 1,3-dipolar cycloaddition. A wide variety of 1,2,3-triazoles have been prepared on solid supports, with an alkyne or azide attached to the resin. In addition, various linkers have been studied. Alkyne-functionalized alcohol was attached to the MeOPEG resin with an oxalyl chloride linkage (Scheme 11.23).[36] Cycloaddition of alkynes with carbohydrate-derived azides gave resin-bound 1,2,3-triazoles, and reductive cleavage with sodium borohydride released the products as primary alcohols.

Resin-bound azide was used in the cycloaddition with various alkynes (Scheme 11.24).[37] *N*-Unsubstituted 1,2,3-triazoles can be cleaved from the 2-methoxy resin in a traceless manner, whereas the Wang resin-linked 1,2,3-triazoles are cleaved from the resin with a linker.

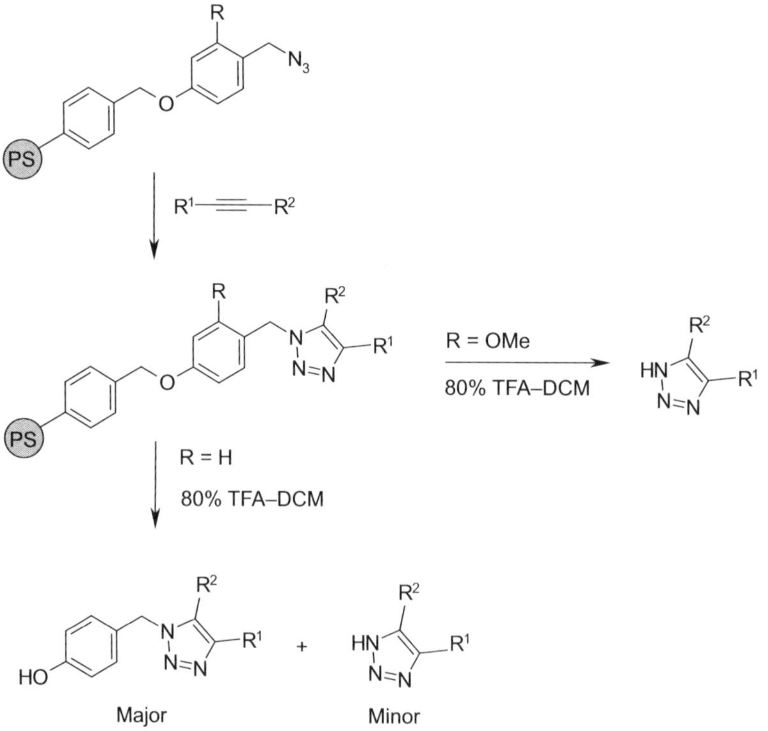

Scheme 11.24. Solid-phase synthesis of *N*-substituted and *N*-unsubstituted 1,2,3-triazoles.

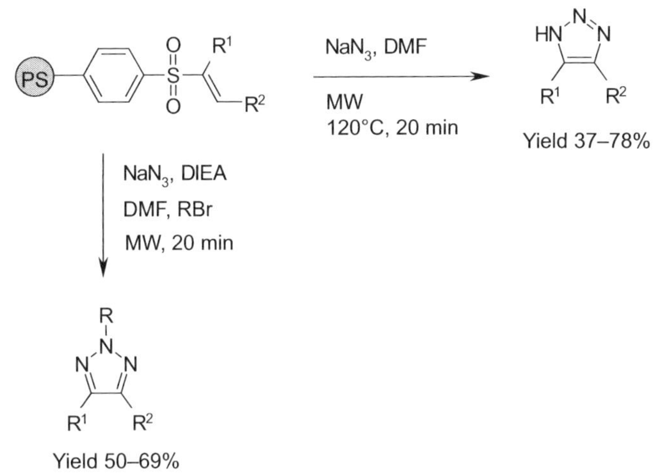

Scheme 11.25. Traceless sulfone linker.

Traceless cleavage of 1,2,3-triazoles has also been accomplished with a sulfone linker. Resin-bound vinyl sulfone was reacted with sodium azide under microwave irradiation, and 1,2,3-triazoles were obtained in a traceless manner (Scheme 11.25).[38] A strongly electron-withdrawing sulfone that is eliminated in the reaction promotes the regioselective cycloaddition in the case of *N*-substituted 1,2,3-triazoles. Microwave irradiation reduced the reaction times and the products were obtained in higher purity than that with conventional heating.

Cyclization is a facile method to cleave compounds from the resin. 1,2,3-Triazoles obtained from a resin-bound azide and methyl propiolate or DMAD were released from the support with spontaneous lactone formation (Scheme 11.26).[39]

Typically, cycloaddition of azides with nonactivated alkynes leads to a regioisomeric mixture of 1,2,3-triazoles. Dramatic improvement in the methods was achieved with a copper catalyst discovered in 2002 by two independent groups.[40,41] Since then, the number of studies concerning 1,3-dipolar cycloadditions of azides and terminal alkynes has increased dramatically. The introduction of the copper catalyst solved many problems, including long reaction times, high reaction temperatures, and poor

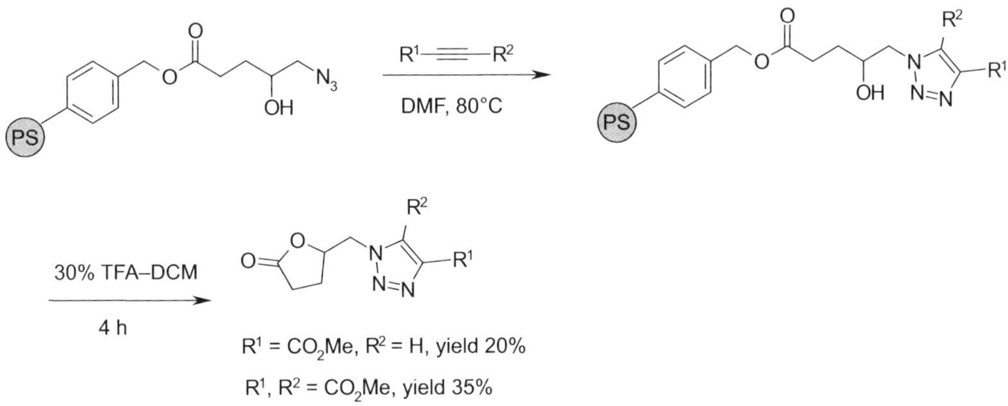

Scheme 11.26. Traceless cyclization cleavage of 1,2,3-triazoles.

Figure 11.1. 1,2,3-Triazole click linker with aldehyde and Michael acceptor functionalities.

regioselectivity, and provided a facile way to produce the heterocyclic ring under mild reaction conditions. Copper(I)-catalyzed synthesis of 1,2,3-triazoles has been used in many applications, and an excellent review covering the field has been written by Meldal and Tornøe.[42]

The first solid-phase copper(I)-catalyzed formation of peptidotriazoles was reported by Tornøe et al.[40] The method was later applied to the synthesis of a library of peptidotriazoles against a recombinant *Leishmania mexicana* cysteine protease.[43] The advantage of the 1,2,3-triazole ring over the peptide bond is the stability of the aromatic ring, which cannot be hydrolyzed by proteases. Recently, the copper-catalyzed method has been exploited in various biomolecular applications. Peptides have been modified with 1,2,3-triazole ligation of azido- and alkyne-functionalized peptide chains.[44] Unexpected peptide cyclodimerization with 1,2,3-triazoles was first reported by Punna et al.[45] Owing to the potential of the cyclodimerization, the reaction was studied further and it was found that the cyclodimerization was favored on solid supports and peptides with at least six residues.[46] Cyclic peptides linked with 1,2,3-triazoles have also been prepared on solid supports with an intramolecular click reaction of azides and alkynes.[47] Moreover, synthesis of peptidomimetics in which amino acid residues are linked with 1,2,3-triazole moieties has been studied.[48] A 1,2,3-triazole linkage has also been used in the solid-phase synthesis of cyclic oligonucleotides[49] and sugar modification of oligonucleotides.[50]

A 1,2,3-triazole-based solid-phase click linker was developed with an aldehyde functionality[51] or a regenerative Michael acceptor (REM) functionality[52] (Figure 11.1). In addition, a chiral pyrrolidine catalyst was grafted to the resin with a 1,2,3-triazole linker, enabling enantioselective Michael addition of ketones to nitroolefins (Figure 11.2).[53]

The copper(I) catalyst promotes the regioselective formation of 1,4-substituted 1,2,3-triazoles. The other regioisomer, 1,5-substituted 1,2,3-triazole, has also been targeted, but without a similar dramatic breakthrough as the copper(I) catalyst in the case of 1,4-substituted 1,2,3-triazoles. A regioselective method, in which Wang resin-bound azide reacts with trimethylsilyl alkyne yielding 1,5-disubstituted 1,2,3-triazole, has been reported (Scheme 11.27).[54] The regioselectivity is controlled by the steric hindrance of the trimethylsilyl (TMS) group and by the ability of silicon to stabilize a partial positive

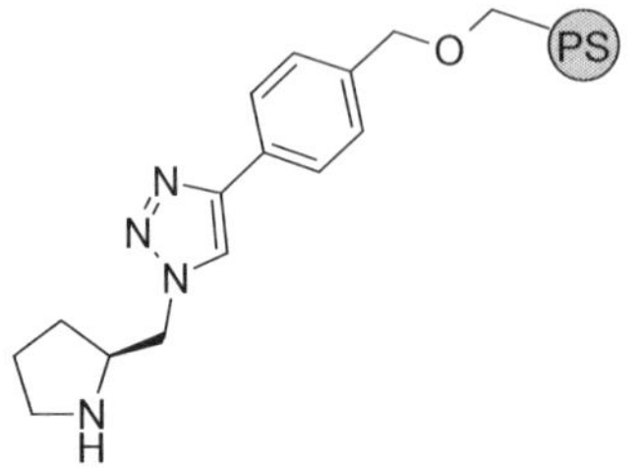

Figure 11.2. Polymer-supported chiral catalyst with a 1,2,3-triazole linker.

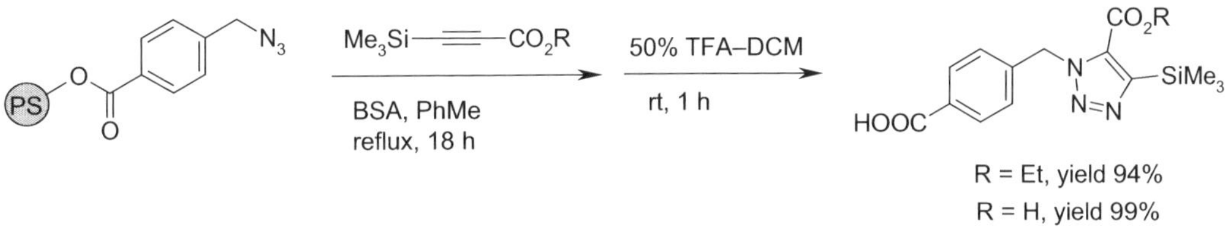

Scheme 11.27. Trimethylsilyl-directed 1,3-dipolar cycloaddition of a resin-bound azide.

charge on the acetylene β-carbon in the transition state. Desilylation and decarboxylation of the products can be prevented with bis(trimethylsilyl)acetamide (BSA). After the cleavage from the resin, the trimethylsilyl group can be removed with 50% aqueous HF in THF.

11.5 SOLID-PHASE SYNTHESIS OF ISOXAZOLIDINES, ISOXAZOLINES, AND ISOXAZOLES

1,3-Dipolar cycloaddition of nitrones and alkenes gives isoxazolidines. Diverse products can be obtained in a one-pot reaction with aldehydes, *N*-substituted hydroxylamines, and alkenes. Haap et al. have studied three different linker strategies for the solid-phase synthesis of isoxazolidines from solid-supported nitrones or alkenes.[55] The reaction of 2-chlorotrityl resin-bound acrylates with nitrone generated *in situ* from 2-bromoaldehyde and *N*-methylhydroxylamine gave resin-bound isoxazolidines (Scheme 11.28). The products were attached to the resin with an ester linkage and cleaved as carboxylic acids under mildly acidic conditions. In the second pathway, the nitrone was bound to the resin with an ether linker, and the reaction with alkenes gave the products with good yields (Scheme 11.29). The third strategy was to enhance the diversity of the products. Rink resin-bound hydroxylamines were condensed with various aldehydes, and the resulting nitrones were trapped with olefins. The isoxazolidines were released from the Rink resin as amides (Scheme 11.30).

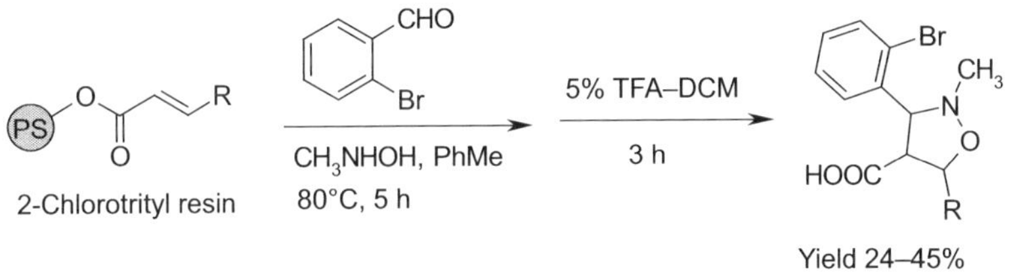

Scheme 11.28. Isoxazolidines from ester-linked alkenes.

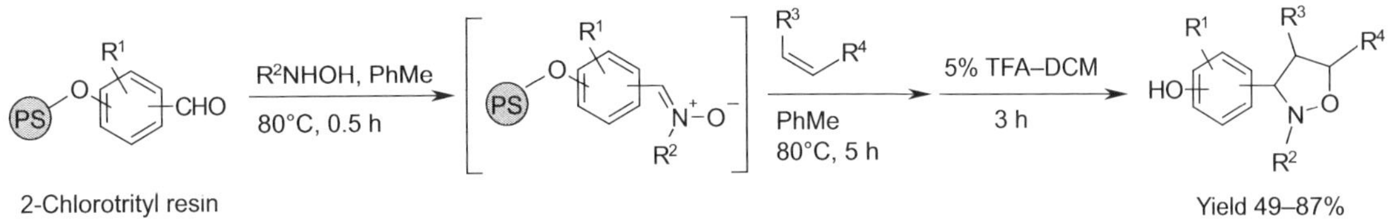

Scheme 11.29. Isoxazolidines from resin-bound aldehydes with a phenolic linkage.

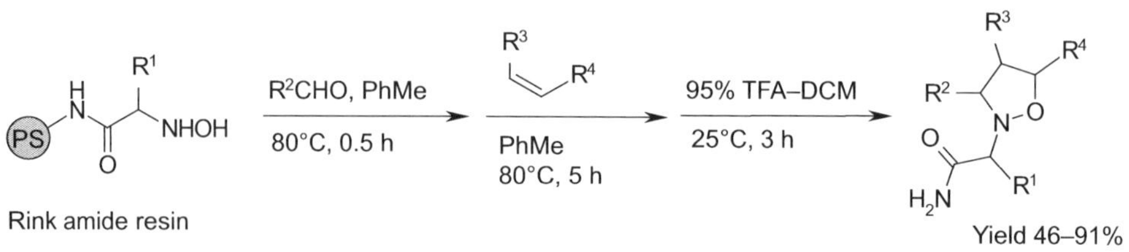

Scheme 11.30. Isoxazolidines from Rink resin-bound hydroxylamines.

Cycloaddition of polymer-supported nitrones with alkenes gave polymer-supported isoxazolidines as intermediates (Scheme 11.31).[56] A catalytic amount of ytterbium(III) triflate activated the alkenes, and the cycloaddition proceeded smoothly at room temperature. Oxidative cleavage of the benzylic C—N bond was achieved with DDQ, releasing 2-isoxazolines in a traceless manner.

1,3-Dipolar cycloadditions of nitrile oxides with alkenes or alkynes give isoxazolines or isoxazoles, respectively. Typically, nitrile oxides are generated *in situ*, and the cycloaddition reaction occurs smoothly at room temperature. A wide variety of nitrile oxide cycloadditions have been studied on solid support since Yedidia and Leznoff published their studies in 1980.[10] Dimerization of the nitrile oxides has limited the utility of nitrile oxide cycloadditions in solution. Solid-supported reactions enabled the formation of resin-bound nitrile oxide *in situ* and trapped the reactive 1,3-dipole with an excess of the dipolarophile in solution (Scheme 11.32).[57] Trityl resin-bound cycloadducts were then cleaved with highly diluted trifluoroacetic acid and the products were obtained as phenols.

An asymmetric solid-phase synthesis of isoxazolines was reported by Zou and Jiang.[58] Cycloaddition of resin-bound nitrile imines with an allyl alcohol in the presence of a chiral diisopropyl tartrate (DIPT) gave isoxazoles with good regio- and enantioselectivity (Scheme 11.33). Ether-linked cycloadducts were released from the Wang resin as phenols with 10% trifluoroacetic acid.

Faita et al. have studied various resin-bound nitrile oxides and their stability on the Wang resin.[59] They found the increased stability of the resin-bound nitrile oxides and storage of the nitrile oxides was possible for at least 1 day in a cool, dry place. The nitrile oxides were generated from Wang resin-linked hydroximoyl chloride derivatives with

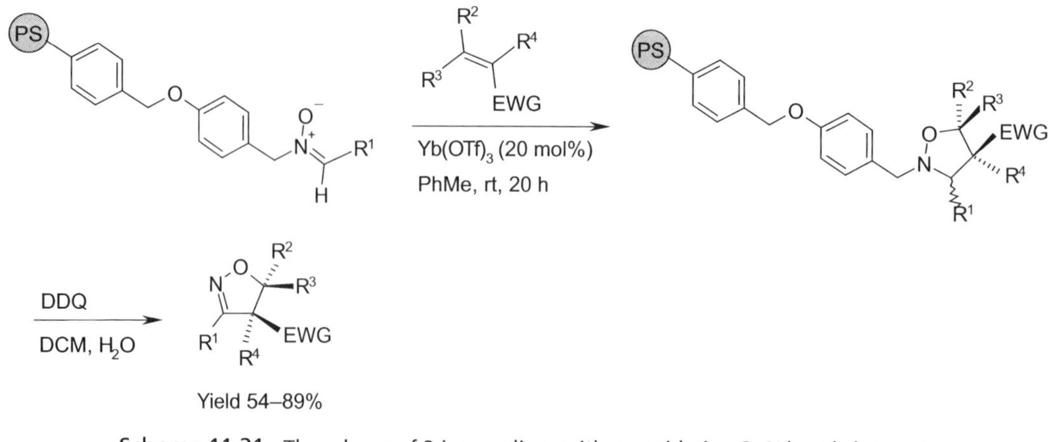

Scheme 11.31. The release of 2-isoxazolines with an oxidative C—N bond cleavage.

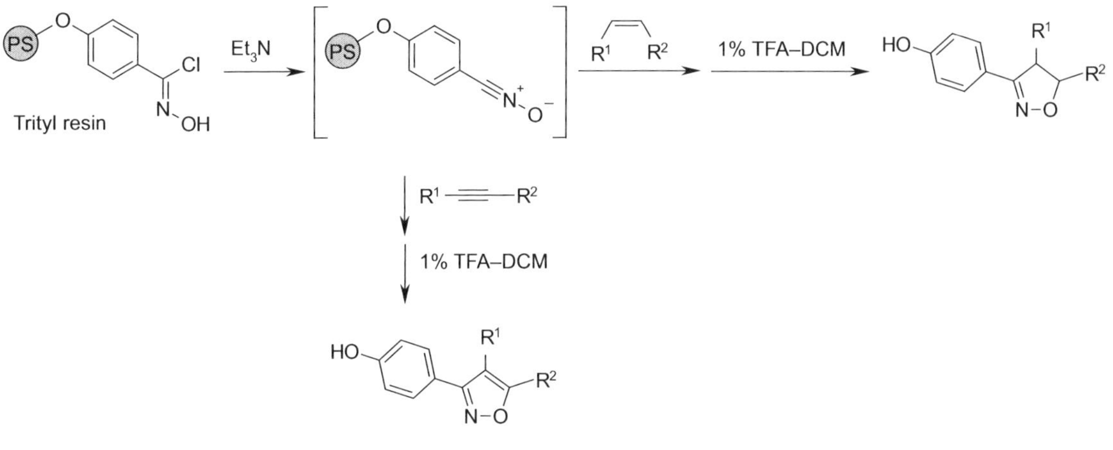

Scheme 11.32. Resin-bound nitrile oxide with a phenolic linker.

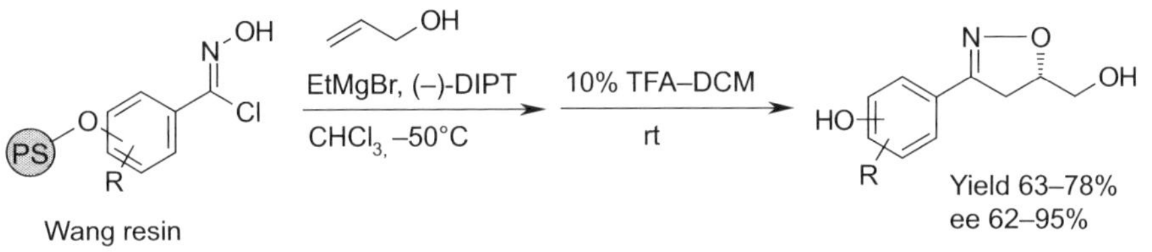

Scheme 11.33. Asymmetric solid-phase synthesis of isoxazolines.

triethylamine (Scheme 11.34). Ester- or phenol-linked nitrile oxides reacted with various dipolarophiles at room temperature. The 2-isoxazolines were cleaved as phenols or carboxylic acids with 20% trifluoroacetic acid. 1,2-Disubstituted electron-poor alkenes gave a mixture of regioisomers.

Various 2-isoxazolines were synthesized from resin-bound nitrile oxide (Scheme 11.35) or alkene (Scheme 11.36).[60] Oxidative treatment of aldoximes with bleach generated nitrile oxides that reacted with various alkenes. The ester-linked products were cleaved from the Wang resin as carboxylic acids using trifluoroacetic acid. Cycloaddition of terminal alkenes

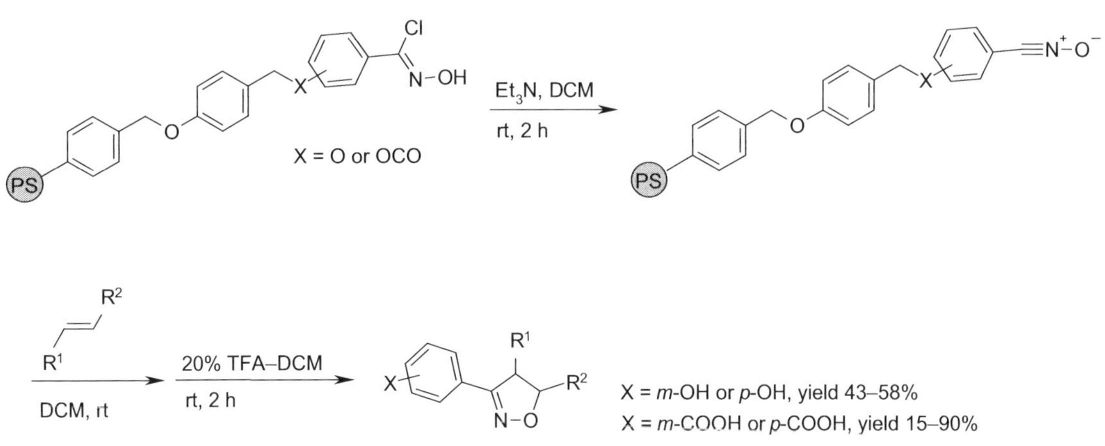

Scheme 11.34. 1,3-Dipolar cycloaddition of resin-bound nitrile oxides.

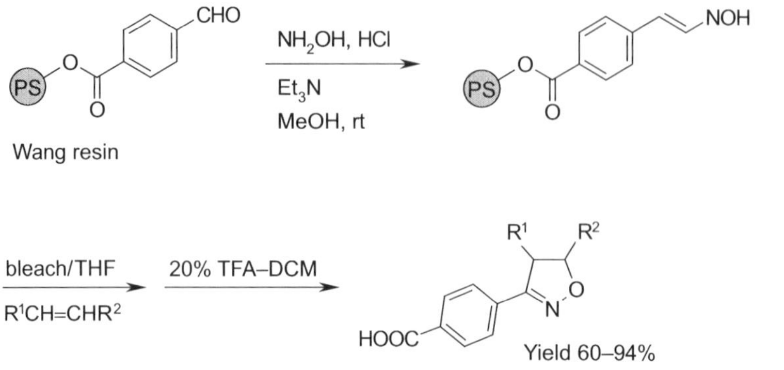

Scheme 11.35. Solid-phase synthesis of 2-isoxazolines from resin-bound nitrile oxide.

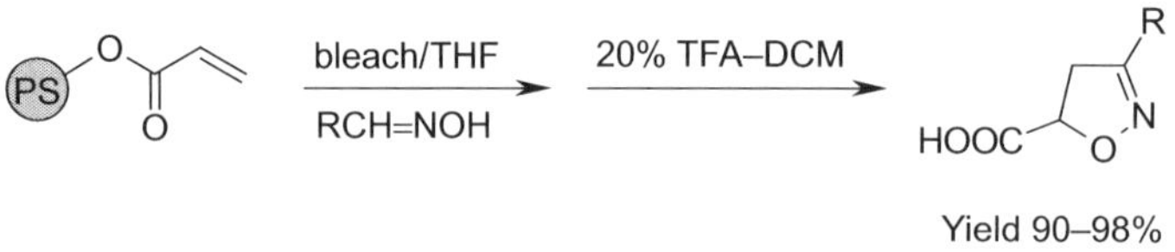

Scheme 11.36. Solid-phase synthesis of 2-isoxazolines from resin-bound acrylate.

produced 5-substituted 2-isoxazolines as single products, whereas 1,2-disubstituted alkene gave a 1:1 mixture of regioisomers.

Lorsbach et al. have studied the synthesis of isoxazoline-functionalized isoquinolines (Scheme 11.37).[61,62] A resin-bound isoxazoline ring was obtained by reacting resin-bound alkenes with nitrile oxides. The traceless cleavage of the heterocycles was then carried out under alkaline conditions. Elimination of the resin released aromatized isoquinoline from the solid support.

Soluble PEG resin has been used in the synthesis of isoxazoles from polymer-bound alkyne (Scheme 11.38).[63] The alkyne was attached to the resin with an ester linkage, and the cycloaddition with nitrile oxides generated *in situ* gave resin-bound isoxazoles. Treatment with sodium hydroxide released the 5-hydroxymethyl-substituted isoxazoles from the resin in high yields and purities.

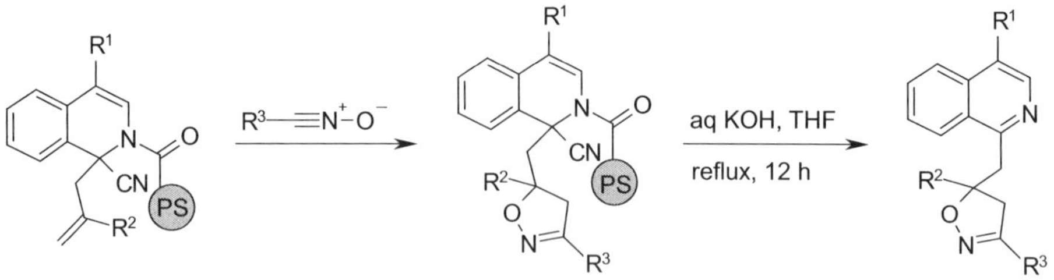

Scheme 11.37. Traceless cleavage of isoquinoline derivatives.

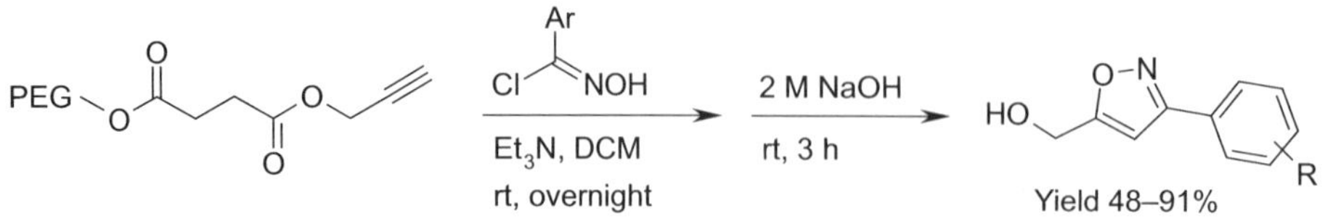

Scheme 11.38. Isoxazoles from resin-bound alkynes.

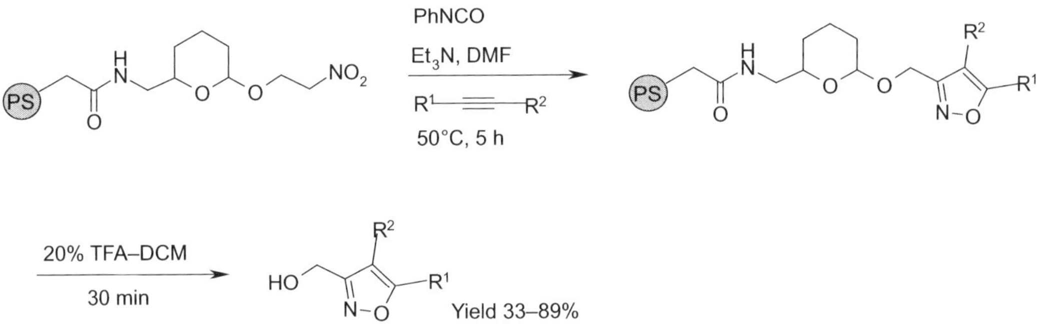

Scheme 11.39. Tetrahydropyranyl-linked isoxazoles.

A tetrahydropyranyl linker is an acid-sensitive linker for alcohols. Nitrile oxides were generated *in situ* from tetrahydropyranyl-linked nitro alkanes and phenyl isocyanate under Mukaiyama conditions, and reactions with various alkynes gave resin-bound isoxazoles (Scheme 11.39).[64] Cleavage with diluted trifluoroacetic acid gave isoxazoles as primary alcohols in a traceless manner. A library of 3-hydroxymethyl-4,5-disubstituted isoxazoles was prepared in a parallel and automated fashion by a 96-well plate synthesizer with an average yield of 60%.

Isoxazoles have also been obtained through an elimination reaction (Scheme 11.40).[65] Cycloaddition of nitrile oxide to resin-bound vinyl ether gave resin-bound 2-isoxazoline, which aromatized to isoxazole after the elimination of the polymer support. A regioselective reaction gave isoxazoles in high purity and with 36–83% yield. Enhanced diversity was obtained with the Suzuki coupling reactions to the R group prior to the cycloadditions.

A selenium linker has been used in the preparation of various nitrogen-containing heterocycles. Cycloaddition of nitrile oxides to resin-bound alkynes gave resin-bound isoxazoles (Scheme 11.41).[66] α-Alkylation of the resin-bound cycloadducts with halides under basic conditions gave additional diversity for the products. Traceless oxidative cleavage of the isoxazoles was achieved via the elimination of the resin with hydrogen peroxide. The diversity of the products was still expanded with a second cascade of 1,3-dipolar cycloadditions to alkenes attached to the resin.

A related method based on an organoselenium resin has been applied to the synthesis of a wide variety of nitrogen-containing heterocycles. Reactions of resin-bound alkenes or alkynes with nitrile oxides and azides gave a variety of isoxazolines, isoxazoles, and 1,2,3-triazoles in high yields and purities (Schemes 11.42 and 11.43).[67,68]

The developed methods were applied to the synthesis of a wide variety of related compounds such as 1,2,4-oxadiazole derivatives (Scheme 11.44)[69] and substituted uracils (Scheme 11.45).[70] The organoselenium resin was also used in the solid-phase synthesis of pyrrolidines from azomethine ylides (Scheme 11.11).[23]

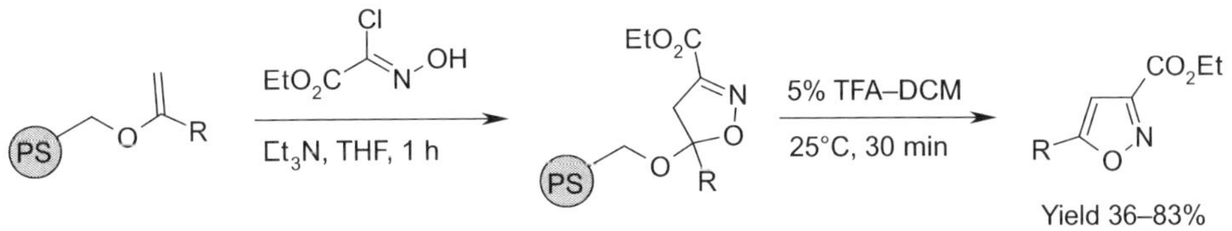

Scheme 11.40. Traceless cleavage of isoxazoles.

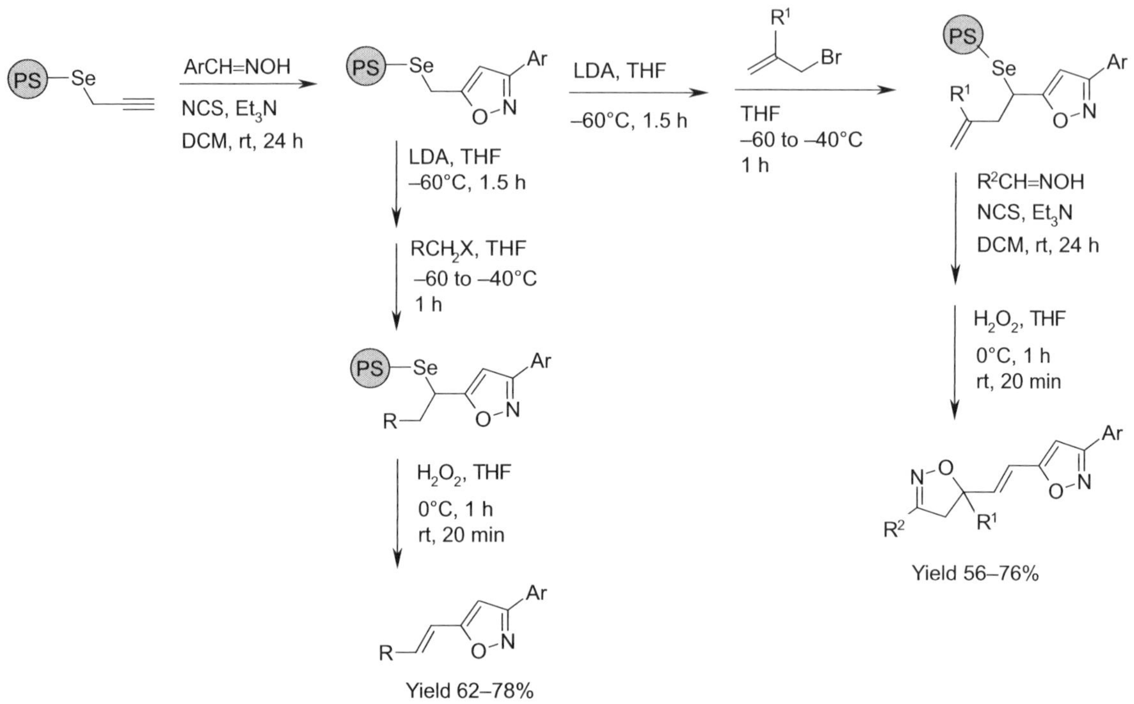

Scheme 11.41. Solid-phase synthesis of isoxazoles with a traceless selenium linker.

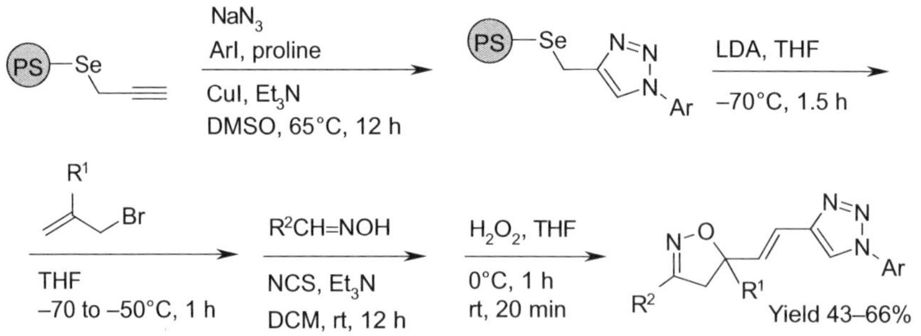

Scheme 11.42. Organoselenium resin in the synthesis of triazolyl- and isoxazolinyl-substituted olefins.

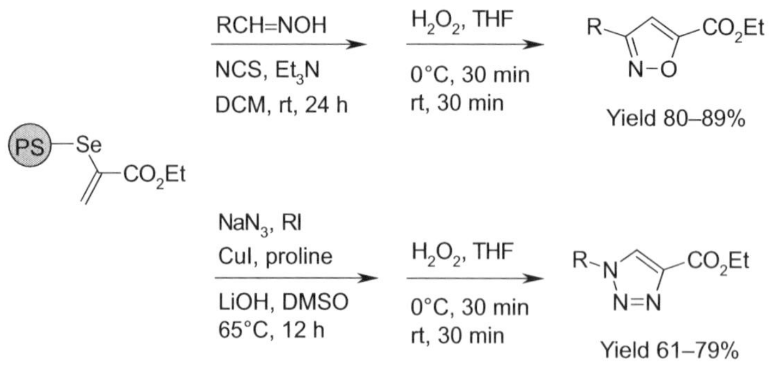

Scheme 11.43. Organoselenium resin in the synthesis of 1,2,3-triazoles and isoxazoles.

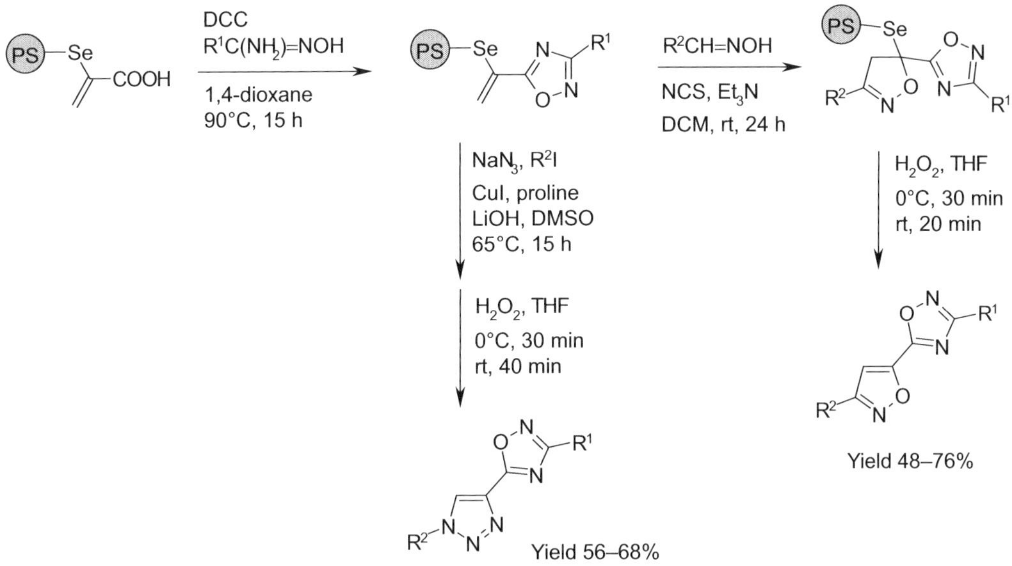

Scheme 11.44. Organoselenium resin in the synthesis of 1,2,4-oxadiazole derivatives.

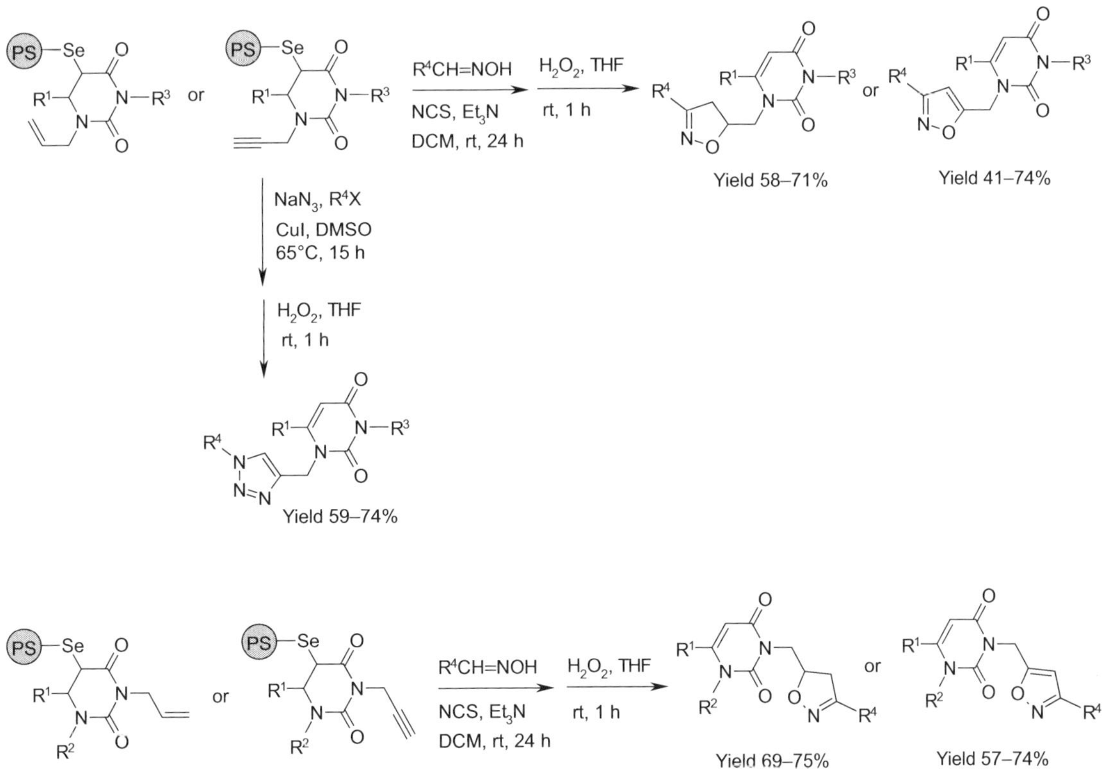

Scheme 11.45. Organoselenium resin in the synthesis of uracil derivatives.

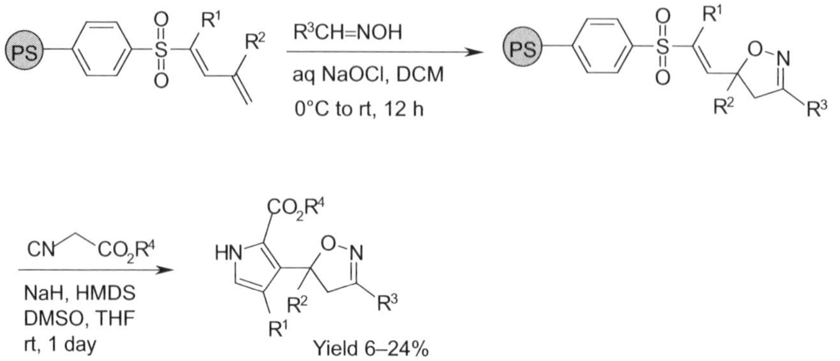

Scheme 11.46. Traceless cleavage of isoxazoline derivatives with a sulfone linker.

Traceless cleavage with a sulfone linker has also been utilized in the solid-phase synthesis of isoxazolines[71] and isoxazoles.[72] Nitrile oxides generated *in situ* reacted regioselectively to the solid-supported diene (Scheme 11.46). Subsequent formation of pyrroles released the cycloadducts from the resin in a traceless manner.

Acetylenic sulfones are reactive dipolarophiles. 1,3-Dipolar cycloaddition with nitrile oxide gave ester-linked isoxazoles (Scheme 11.47).[73] However, the cycloadducts were cleaved from the resin with the sulfone moiety under alkaline hydrolysis. The method was applied later to the 1,3-dipolar cycloaddition reaction to prepare a wide selection of heterocycles, such as pyrroles, pyrazoles, 1,2,3-triazoles, and isoxazoles.[74]

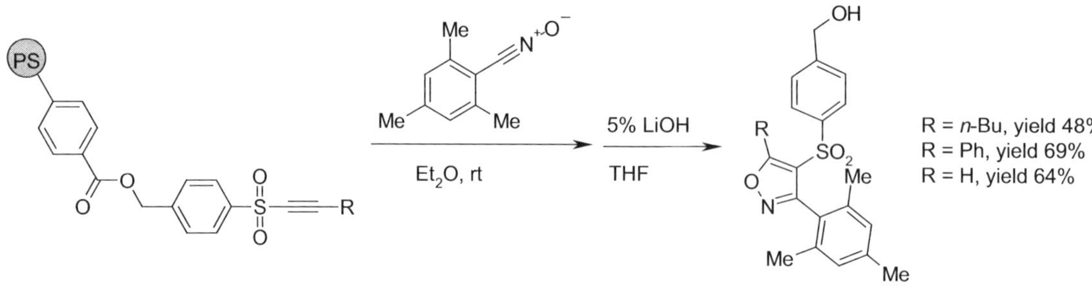

Scheme 11.47. Acetylenic sulfones as dipolarophiles.

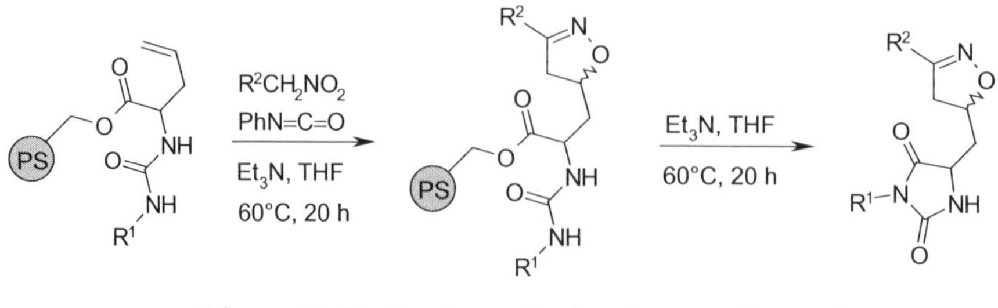

Scheme 11.48. Traceless cyclization cleavage of isoxazolines.

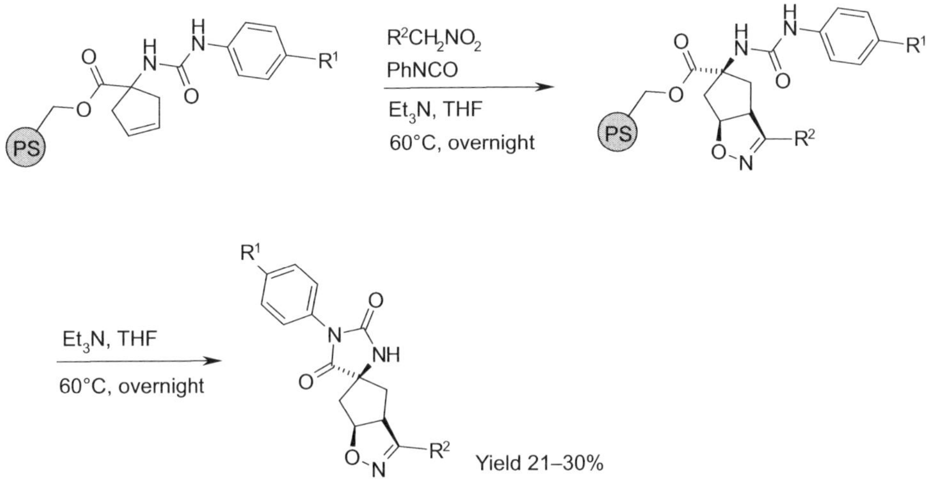

Scheme 11.49. Diastereoselective synthesis of isoxazoline derivatives.

Cyclization is known to release resin-bound compounds without a linker. Hydantoin ring formation releases resin-bound compounds smoothly, producing interesting hydantoin derivatives with potential medicinal or biological activity. Resin-bound alkenes reacted with nitrile oxides that were generated with Mukaiyama's method from nitroalkane (Scheme 11.48).[75] The isoxazolines were cleaved from the resin via a hydantoin formation upon heating under basic conditions.

The method was developed further for the diastereoselective synthesis of isoxazoline derivative (Scheme 11.49).[76] Later, a similar method was applied to the preparation of a diverse 990-member library of hydantoin and isoxazoline heterocycles with multipin technology.[77]

Solid-supported isoxazoles were obtained from resin-bound alkynes and nitrile oxides generated *in situ* from nitroalkanes and isocyanate (Scheme 11.50).[78] The isoxazoles were cleaved from the resin via hydantoin formation upon heating and 18 diverse products were obtained in high purities.

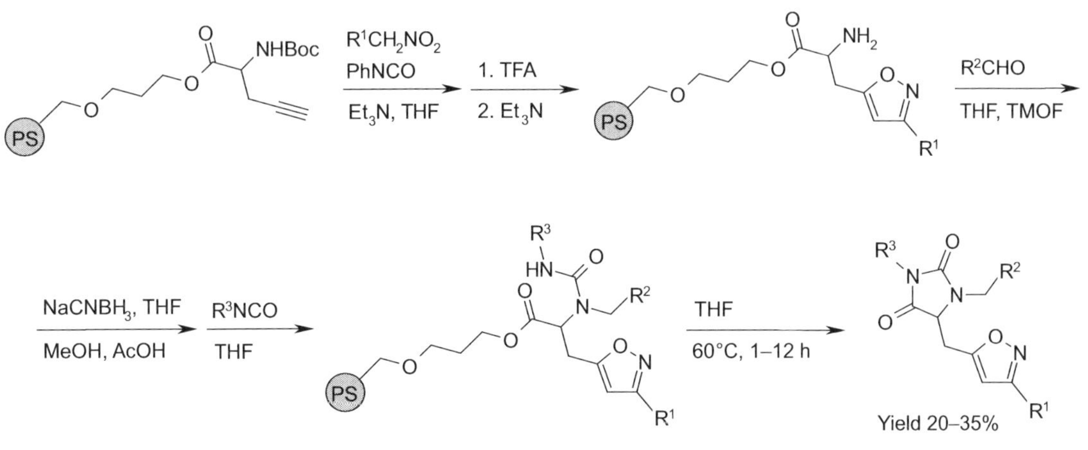

Scheme 11.50. Traceless cyclization cleavage of isoxazolohydantoins.

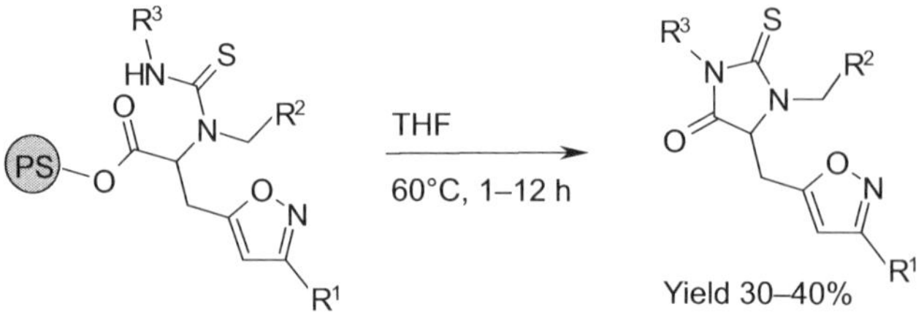

Scheme 11.51. Traceless cyclization cleavage of isoxazolothiohydantoins.

A related method was applied to the synthesis of 18 isoxazolothiohydantoins (Scheme 11.51).[79] Hydantoin formation was also applied to the release of proline derivatives that were obtained from the intramolecular cycloadditions of resin-bound azomethine ylides and alkenes.[80]

11.6 CONCLUSIONS

1,3-Dipolar cycloadditions on solid supports have been widely studied. A diverse amount of nitrogen-containing five-membered heterocycles has been obtained, and various linkers and cleavage strategies have been exploited. Esters and ethers have been the most common linkers. In addition, much effort has been put into the development of traceless methods, and various linker strategies have been studied. To obtain products in high yields and purities from solid supports, efficient synthetic methods are necessary.

Regiochemistry of the 1,3-dipolar cycloadditions is still an interesting research topic. The formation of two regioisomers is not usually the desired result, and it often reduces the yield of the targeted compound. The development of regioselective methods would expand the utility of 1,3-dipolar cycloadditions with unsymmetrical starting materials. The significant improvement in the preparation of 1,2,3-triazoles achieved by a copper(I) catalyst has recently been demonstrated, and the remarkable breakthrough within the field of 1,3-dipolar cycloadditions has been utilized in a variety of applications.

In summary, solid-phase 1,3-dipolar cycloadditions of various dipoles and dipolarophiles provides a great variety of nitrogen-containing heterocycles that can be useful in a wide range of applications.

REFERENCES

1. A. Padwa (Ed.), *1,3-Dipolar Cycloaddition Chemistry*, Vols *1 and 2*, Wiley, New York, **1984**.
2. R. Huisgen, 1,3-Dipolar cycloadditions. Past and future. *Angew. Chem. Int. Ed. Engl.* **1963**, *2*, 565–598.
3. R. Huisgen, Cycloadditions—definition, classification, and characterization. *Angew. Chem. Int. Ed. Engl.* **1968**, *7*, 321–328.
4. R. Huisgen, Kinetics and mechanism of 1,3-dipolar cycloadditions. *Angew. Chem. Int. Ed. Engl.* **1963**, *2*, 633–645.
5. R. B. Merrifield, Solid-phase peptide synthesis. I. The synthesis of a tetrapeptide. *J. Am. Chem. Soc.* **1963**, *85*, 2149–2154.

6. R. B. Merrifield, Solid-phase synthesis (Nobel lecture). *Angew. Chem. Int. Ed. Engl.* **1985**, *24*, 799–810.
7. C. C. Leznoff, 1999 Alfred Bader Award Lecture. From early developments in multi-step organic synthesis on solid phases to multi-nuclear phthalocyanines *Can. J. Chem.* **2000**, *78*, 167–183.
8. (a) Some representative examples of literature reviews concerning the progress of solid-phase organic synthesis: J. S. Früchtel, G. Jung, Organic chemistry on solid supports. *Angew. Chem. Int. Ed. Engl.* **1996**, *35*, 17–42; (b) P. H. H. Hermkens, H. C. J. Ottenheijm, D. Rees, Solid-phase organic reactions: a review of the recent literature. *Tetrahedron* **1996**, *52*, 4527–4554; (c) P. H. H. Hermkens, H. C. J. Ottenheijm, D. C. Rees, Solid-phase organic reactions II: a review of the literature Nov 95–Nov 96. *Tetrahedron* **1997**, *53*, 5643–5678; (d) S. Booth, P. H. H. Hermkens, H. C. J. Ottenheijm, D. C. Rees, Solid-phase organic reactions III: a review of the literature Nov 96–Dec 97. *Tetrahedron* **1998**, *54*, 15385–15443; (e) R. G. Franzen, Recent advances in the preparation of heterocycles on solid support: a review of the literature. *J. Comb. Chem.* **2000**, *2*, 195–214.
9. C. Gil, S. Bräse, Traceless and multifunctional linkers for the generation of small molecules on solid supports. *Curr. Opin. Chem. Biol.* **2004**, *8*, 230–237.
10. V. Yedidia, C. C. Leznoff, Regioselectivity in cycloaddition reactions on solid phases. *Can. J. Chem.* **1980**, *58*, 1144–1150.
11. K. Harju, J. Yli-Kauhaluoma, Recent advances in 1,3-dipolar cycloaddition reactions on solid supports. *Mol. Diversity* **2005**, *9*, 187–207.
12. E. J. Kantorowski, M. J. Kurth, Dipolar cycloadditions in solid-phase organic synthesis (SPOS). *Mol. Diversity*, **1996**, *2*, 207–216.
13. A. G. M. Barrett, R. J. Boffey, M. U. Frederiksen, C. G. Newton, R. S. Roberts, Pyrrolidine synthesis on polystyrene supports: development of a 'one-pot' dipolar cycloaddition strategy. *Tetrahedron Lett.* **2001**, *42*, 5579–5581.
14. H. R. Hoveyda, D. G. Hall, Solid-phase synthesis of cleavable *N*-arylmaleimides: applications in 1,3-dipolar cycloaddition and in thiol scavenging. *Org. Lett.* **2001**, *3*, 3491–3494.
15. A. J. Bicknell, N. W. Hird, Synthesis of a highly functionalized rigid template by solid phase azomethine ylide cycloaddition. *Bioorg. Med. Chem. Lett.* **1996**, *6*, 2441–2444.
16. B. C. Hamper, D. R. Dukesherer, M. S. South, Solid-phase synthesis of proline analogs *via* a three component 1,3-dipolar cycloaddition. *Tetrahedron Lett.* **1996**, *37*, 3671–3674.
17. S. P. Hollinshead, Stereoselective synthesis of highly functionalised pyrrolidines *via* 1,3-dipolar cycloaddition reactions on a solid support. *Tetrahedron Lett.* **1996**, *37*, 9157–9160.
18. H. A. Dondas, R. Grigg, W. S. MacLachlan, D. T. MacPherson, J. Markandu, V. Sridharan, S. Suganthan, Solid phase sequential 1,3-dipolar cycloaddition-Pictet–Spengler reactions. *Tetrahedron Lett.* **2000**, *41*, 967–970.
19. M. M. Murphy, J. R. Schullek, E. M. Gordon, M. A. Gallop, Combinatorial organic synthesis of highly functionalized pyrrolidines: identification of a potent angiotensin converting enzyme inhibitor from a mercaptoacyl proline library. *J. Am. Chem. Soc.* **1995**, *117*, 7029–7030.
20. M. Komatsu, H. Okada, T. Akaki, Y. Oderaotoshi, S. Minakata, Generation and cycloaddition of polymer-supported azomethine ylide via a 1,2-silatropic shift of α-silylimines: traceless synthesis of pyrrolidine derivatives. *Org. Lett.* **2002**, *4*, 3505–3508.
21. H. Okada, T. Akaki, Y. Oderaotoshi, S. Minakata, M. Komatsu, Generation and cycloaddition of polymer-supported azomethine ylide by utilizing the characteristics of silicon: a facile route to pyrrolidines and pyrroles from α-silylimines bound to resin. *Tetrahedron* **2003**, *59*, 197–205.
22. S. Caix-Haumesser, I. Hanna, J.-Y. Lallemand, J.-F. Peyronel, Solid-phase synthesis of functionalized tropane derivatives via 1,3-dipolar cycloaddition. *Tetrahedron Lett.* **2001**, *42*, 3721–3723.

23. X. Huang, J.-F. Xu, Solid-phase synthesis of 2,5-dihydro-1*H*-pyrroles,1,3-dioxo-2,3,5,7*a*-tetrahydro-1*H*-pyrrolo[1,2-*c*]imidazoles and 1,4-dioxo-1,2,3,4,6,8*a*-hexahydropyrrolo[1,2-*a*]pyrazines using a supported selenium resin. *J. Comb. Chem.* **2009**, *11*, 350–354.

24. A. M. M. Mjalli, S. Sarshar, T. J. Baiga, Solid phase synthesis of pyrroles derived from a four component condensation. *Tetrahedron Lett.* **1996**, *37*, 2943–2946.

25. A. M. Strocker, T. A. Keating, P. A. Tempest, R. W. Armstrong, Use of a convertible isocyanide for generation of Ugi reaction derivatives on solid support: synthesis of α-acylaminoesters and pyrroles. *Tetrahedron Lett.* **1996**, *37*, 1149–1152.

26. L. Garanti, G. Molteni, P. Casati, Nitrilimine cycloadditions to MeOPEG-bounded alkenyl dipolarophiles. *J. Chem. Soc. Perkin Trans.* **2002**, *1* 2504–2508.

27. N. Fuchi, T. Doi, B. Cao, M. Kahn, T. Takahashi, The solid-phase parallel synthesis of β-strand mimetic templates via 1,3-dipolar cycloaddition with resin-bound vinylsulfone. *Synlett*, **2002**, 285–289.

28. N. Fuchi, T. Doi, T. Takahashi, A library synthesis of pyrazoles by azomethine imine cycloaddition to the polymer-supported vinylsulfone. *Chem. Lett.* **2005**, *34*, 438–439.

29. A. C. Donohue, S. Pallich, T. D. McCarthy, Cycloaddition of nitrile imines to resin-bound enamines: a solid phase synthesis of 1,4-diarylpyrazoles. *J. Chem. Soc. Perkin Trans.* **2001**, *1*, 2817–2822.

30. K. Harju, J. Vesterinen, J. Yli-Kauhaluoma, Solid-phase synthesis of amino acid derived *N*-unsubstituted pyrazoles via sydnones. *Org. Lett.* **2009**, *11*, 2219–2221.

31. K.-I. Washizuka, K. Nagai, S. Minakata, I. Ryu, M. Komatsu, Generation and cycloaddition of polymer-supported azomethine imines: traceless synthesis of pyrazole derivatives from α-silylnitrosoamide derivatives bound to resin. *Tetrahedron Lett.* **2000**, *41*, 691–695.

32. K. Harju, I. Kylänlahti, T. Paananen, M. Polamo, J. Nielsen, J. Yli-Kauhaluoma, Solid-phase synthesis of pyrazolopyridines from polymer-bound alkyne and azomethine imines. *J. Comb. Chem.* **2006**, *8*, 344–349.

33. M. T. Bilodeau, A. M. Cunningham, Solid-supported synthesis of imidazoles: a strategy for direct resin-attachment to the imidazole core. *J. Org. Chem.* **1998**, *63*, 2800–2801.

34. S. K. Samanta, J. Yli-Kauhaluoma, Polymer-supported 1,3-oxazolium-5-olates: synthesis of 1,2,4-triazoles. *J. Comb. Chem.* **2005**, *7*, 142–146.

35. S. K. Samanta, I. Kylänlahti, J. Yli-Kauhaluoma, Microwave-assisted synthesis of imidazoles: reaction of *p*-toluenesulfonylmethyl isocyanide and polymer-bound imines. *Bioorg. Med. Chem. Lett.* **2005**, *15*, 3717–3719.

36. M. Moore, P. Norris, Dipolar cycloaddition reactions on a soluble polymer-supported dipolarophile: synthesis of sugar-derived triazoles. *Tetrahedron Lett.* **1998**, *39*, 7027–7030.

37. K. Harju, M. Vahermo, I. Mutikainen, J. Yli-Kauhaluoma, Solid-phase synthesis of 1,2,3-triazoles via 1,3-dipolar cycloaddition. *J. Comb. Chem.* **2003**, *5*, 826–833.

38. Y. Gao, Y. Lam, [3 + 2] Cycloaddition reactions in the solid-phase synthesis of 1,2,3-triazoles. *Org. Lett.* **2006**, *8*, 3283–3285.

39. N. Gouault, J. -F. Cupif, A. Sauleau, M. David, γ-Methyl-substituted-γ-butyrolactones: solid-phase synthesis employing a cyclisation–cleavage strategy *Tetrahedron Lett.* **2000**, *41*, 7293–7297.

40. C. W. Tornøe, C. Christensen, M. Meldal, Peptidotriazoles on solid phase: [1,2,3]-triazoles by regiospecific copper(I)-catalyzed 1,3-dipolar cycloadditions of terminal alkynes to azides. *J. Org. Chem.* **2002**, *67*, 3057–3064.

41. V. V. Rostovtsev, L. G. Green, V. V. Fokin, K. B. Sharpless, A stepwise Huisgen cycloaddition process: copper(I)-catalyzed regioselective "ligation" of azides and terminal alkynes. *Angew. Chem. Int. Ed.* **2002**, *41*, 2596–2599.

42. M. Meldal, C. W. Tornøe, Cu-catalyzed azide–alkyne cycloaddition. *Chem. Rev.* **2008**, *108*, 2952–3015.

43. C. W. Tornøe, S. J. Sanderson, J. C. Mottram, G. H. Coombs, M. Meldal, Combinatorial library of peptidotriazoles: identification of [1,2,3]-triazole inhibitors against a recombinant *Leishmania mexicana* cysteine protease. *J. Comb. Chem.* **2004**, *6*, 312–324.

44. R. Franke, C. Doll, J. Eichler, Peptide ligation through click chemistry for the generation of assembled and scaffolded peptides. *Tetrahedron Lett.* **2005**, *46*, 4479–4482.

45. S. Punna, J. Kuzelka, Q. Wang, M. G. Finn, Head-to-tail peptide cyclodimerization by copper-catalyzed azide–alkyne cycloaddition. *Angew. Chem. Int. Ed.* **2005**, *44*, 2215–2220.

46. R. Jagasia, J. M. Holub, M. Bollinger, K. Kirshenbaum, M. G. Finn, Peptide cyclization and cyclodimerization by Cu^{I}-mediated azide–alkyne cycloaddition. *J. Org. Chem.* **2009**, *74*, 2964–2974.

47. V. Goncalves, B. Gautier, A. Regazzetti, P. Coric, S. Bouaziz, C. Garbay, M. Vidal, N. Inguimbert, On-resin cyclization of peptide ligands of the vascular endothelial growth factor receptor 1 by copper(I)-catalyzed 1,3-dipolar azide–alkyne cycloaddition. *Bioorg. Med. Chem. Lett.* **2007**, *17*, 5590–5594.

48. N. G. Angelo, P. S. Arora, Solution- and solid-phase synthesis of triazole oligomers that display protein-like functionality. *J. Org. Chem.* **2007**, *72*, 7963–7967.

49. J. Lietard, A. Meyer, J.-J. Vasseur, F. Morvan, New strategies for cyclization and bicyclization of oligonucleotides by click chemistry assisted by microwaves. *J. Org. Chem.* **2008**, *73*, 191–200.

50. G. Pourceau, A. Meyer, J.-J. Vasseur, F. Morvan, Synthesis of mannose and galactose oligonucleotide conjugates by bi-click chemistry. *J. Org. Chem.* **2009**, *74*, 1218–1222.

51. S. Löber, P. Rodriguez-Loaiza, P. Gmeiner, Click linker: efficient and high-yielding synthesis of a new family of SPOS resins by 1,3-dipolar cycloaddition. *Org. Lett.* **2003**, *5*, 1753–1755.

52. S. Löber, P. Gmeiner, Click chemistry on solid support: synthesis of a new REM resin and application for the preparation of tertiary amines. *Tetrahedron*, **2004**, *60*, 8699–8702.

53. E. Alza, X. C. Cambeiro, C. Jimeno, M. A. Pericàs, Highly enantioselective Michael additions in water catalyzed by a PS-supported pyrrolidine. *Org. Lett.* **2007**, *9*, 3717–3720.

54. S. J. Coats, J. S. Link, D. Gauthier, D. J. Hlasta, Trimethylsilyl-directed 1,3-dipolar cycloaddition reactions in the solid-phase synthesis of 1,2,3-triazoles. *Org. Lett.* **2005**, *7*, 1469–1472.

55. W. J. Haap, D. Kaiser, T. B. Walk, G. Jung, Solid phase synthesis of diverse isoxazolidines *via* 1,3-dipolar cycloaddition. *Tetrahedron*, **1998**, *54*, 3705–3724.

56. S. Kobayashi, R. Akiyama, Lanthanide triflate-catalyzed 1,3-dipolar cycloaddition reactions of polymer-supported nitrones with alkenes for the preparation of diverse 2-isoxazoline derivatives. *Tetrahedron Lett.* **1998**, *39*, 9211–9214.

57. B. B. Shankar, D. Y. Yang, S. Girton, A. K. Ganguly, One pot solid phase synthesis of isoxazolines. *Tetrahedron Lett.* **1998**, *39*, 2447–2448.

58. N. Zou, B. Jiang, Solid phase asymmetric synthesis of isoxazolines. *J. Comb. Chem.* **2000**, *2*, 6–7.

59. G. Faita, M. Mella, A. Mortoni, A. Paio, P. Quadrelli, P. Seneci, Solid-supported nitrile oxides as stable and valuable reactive intermediates. *Eur. J. Org. Chem.* **2002**, 1175–1183.

60. J.-F. Cheng, A. M. M. Mjalli, Solid-phase synthesis of Δ^2-isoxazolines. *Tetrahedron Lett.* **1998**, *39*, 939–942.

61. B. A. Lorsbach, R. B. Miller, M. J. Kurth, Reissert-based "traceless" solid-phase synthesis: isoquinoline, and isoxazoline-containing heterocycles. *J. Org. Chem.* **1996**, *61*, 8716–8717.

62. B. A. Lorsbach, J. T. Bagdanoff, R. B. Miller, M. J. Kurth, Isoxazolinoisoquinoline heterocycles via solid-phase Reissert and Suzuki reactions. *J. Org. Chem.* **1998**, *63*, 2244–2250.

63. Y.-J. Shang, Y.-G. Wang, Soluble polymer-supported synthesis of isoxazoles. *Tetrahedron Lett.* **2002**, *43*, 2247–2249.

64. E. Cereda, A. Ezhaya, M. Quai, W. Barbaglia, Solid-phase synthesis of 3-hydroxymethyl isoxazoles via resin bound nitrile oxides. *Tetrahedron Lett.* **2001**, *42*, 4951–4953.

65. A. G. M. Barrett, P. A. Procopiou, U. Voigtmann, Solid-phase synthesis of isoxazoles using vinyl ethers as chameleon catches. *Org. Lett.* **2001**, *3*, 3165–3168.
66. X. Huang, W.-M. Xu, Use of selenium-bound resin for the solid-phase synthesis of substituted isoxazolyl-substituted (*E*)-olefins. *Org. Lett.* **2003**, *5*, 4649–4652.
67. W.-M. Xu, X. Huang, E. Tang, Solid-phase synthesis of 1,2-diheterocyclic-substituted (*E*)-olefins from a supported selenium resin. *J. Comb. Chem.* **2005**, *7*, 726–733.
68. Y.-G. Wang, W.-M. Xu, X. Huang, An efficient solid-phase synthesis of substituted isoxazole, triazole, and cycloalkadiene derivatives using supported selenium resin. *Synthesis* **2007**, 28–32.
69. X. Huang, Y.-G. Wang, Solid-phase synthesis of linked heterocycles from a selenopolystyrene resin. *J. Comb. Chem.* **2007**, *9*, 121–130.
70. J. Cao, X. Huang, Solid-phase synthesis of heterocyclic nucleoside analogues: substituted uracils tethered to isoxazoles, isoxazolines, and triazoles from a selenopolystyrene resin. *J. Comb. Chem.* **2008**, *10*, 526–533.
71. S. H. Hwang, M. J. Kurth, Versatile "traceless" sulfone linker for SPOS: preparation of isoxazolinopyrrole 2-carboxylates. *J. Org. Chem.* **2002**, *67*, 6564–6567.
72. W.-C. Cheng, M. Wong, M. M. Olmstead, M. J. Kurth, Solid-phase synthesis of novel isoxazolocyclobutanones and isoxazolinocyclobutenones. *Org. Lett.* **2002**, *4*, 741–744.
73. T. G. Back, H. Zhai, Cyclizations and cycloadditions of acetylenic sulfones on solid supports. *Chem. Commun.* **2006**, 326–328.
74. D. Gao, Z. Huimin, M. Parvez, T. G. Back, 1,3-Dipolar cycloadditions of acetylenic sulfones in solution and on solid supports. *J. Org. Chem.* **2008**, *73*, 8057–8068.
75. K.-H. Park, E. Abbate, S. Najdi, M. M. Olmstead, M. J. Kurth, Solution- and solid-phase synthesis of novel hydantoin and isoxazoline-containing heterocycles. *Chem. Commun.* **1998**, 1679–1680.
76. K.-H. Park, M. M. Olmstead, M. J. Kurth, Diastereoselective solid-phase synthesis of novel hydantoin- and isoxazoline-containing heterocycles. *J. Org. Chem.* **1998**, *63*, 6579–6585.
77. K.-H. Park, J. Ehrler, H. Spoerri, M. J. Kurth, Preparation of a 990-member chemical compound library of hydantoin- and isoxazoline-containing heterocycles using multipin technology. *J. Comb. Chem.* **2001**, *3*, 171–176.
78. K.-H. Park, M. J. Kurth, An uncatalyzed cyclo-elimination process for the release of N_3-alkylated hydantoins from solid-phase: synthesis of novel isoxazoloimidazolidinediones. *Tetrahedron Lett.* **1999**, *40*, 5841–5844.
79. K.-H. Park, M. J. Kurth, Solid-phase synthesis of novel heterocycles containing thiohydantoin and isoxazole rings. *J. Org. Chem.* **1999**, *64*, 9297–9300.
80. Y.-D. Gong, S. Nadji, M. M. Olmstead, M. J. Kurth, Solid-phase synthesis: Intramolecular azomethine ylide cycloaddition (→ proline) and carbanilide cyclization (→ hydantoin) reactions. *J. Org. Chem.* **1998**, *63*, 3081–3086.

12

SULFONES IN SOLID-PHASE HETEROCYCLE SYNTHESIS

Chai Hoon Soh and Yulin Lam

12.1 INTRODUCTION

The concept of solid-phase synthesis was first raised by Merrifield in 1963 for efficient peptide synthesis.[1] Today, it has been applied to the synthesis of biooligomers, such as oligosaccharides[2] and oligonucleotides,[3] and a wide spectrum of small organic, organometallic, and inorganic compounds.[4] Solid-phase synthesis is a methodology whereby the reactions take place on the molecule attached to an insoluble material referred to as a solid support. The solid support comprises a polymer bead (generally a cross-linked, insoluble, polymeric material inert to the conditions of synthesis) and a linker (a bifunctional chemical moiety that joins the polymer and the molecule to be synthesized). A building block is first coupled to the solid support via the linker and modification steps are subsequently performed to achieve the solid support bound final molecule that is eventually cleaved from the solid support (Figure 12.1). The main advantage of solid-phase synthesis is the convenience of purification. Since the compounds are bound to the solid support, only simple filtration and washing of the resin to remove the excess reagents are needed. Other benefits include the possibility of using excess reagents to force the chemical reaction to completion and the ease of straightforward automation. However solid-phase synthesis needs large amounts of reagents and solvents, and, depending on the synthetic strategy, extra attachment and cleavage steps are at times also required.

Solid-Phase Organic Synthesis: Concepts, Strategies, and Applications, First Edition.
Edited by Patrick H. Toy and Yulin Lam.

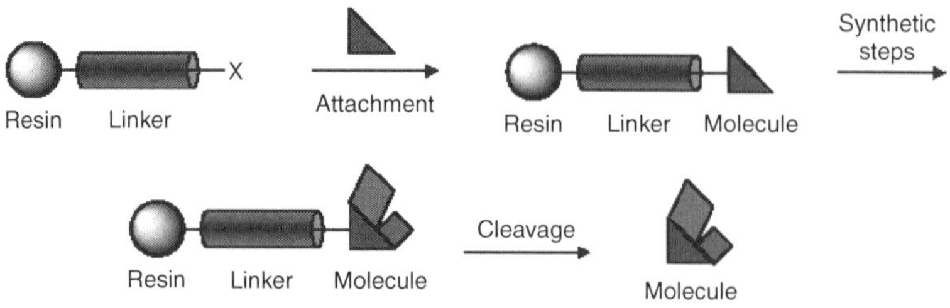

Figure 12.1. Illustration of a solid-phase synthesis.

12.2 LINKERS

The linker is a vital component in solid-phase synthesis. It should be stable enough to tolerate all the reaction conditions and, on the other hand, be sensitive enough to be cleaved after the reaction is completed. Hence, to accommodate the increasing number of reaction conditions that are being adapted for solid-phase synthesis, a large number of sophisticated linker units have been developed and employed in solid-phase synthesis.[5] In this chapter, we will focus exclusively on the sulfone linker unit and its applications in the synthesis of heterocycles.

12.2.1 Sulfone Chemistry

The sulfone functional group **1-1** owes its name to its formal resemblance to the carbonyl group and is best represented as the resonance between two canonical forms **1-2** (Figure 12.2), which demonstrates its ability to function as a proton acceptor; that is, it possesses essentially basic properties. Three important properties of the sulfonyl group mainly determine the overall properties of organic sulfonyl compounds. They are (i) nonenolizability of the sulfonyl group, (ii) its electron attracting (−I) effect, and (iii) the resultant negative charge on the oxygen atoms. The sulfone group renders acidic the hydrogen atoms attached to the α-carbon atoms. Hence, sulfonyl group plays an important role in both organic synthetic and pharmaceutical aspects.[6]

The use of sulfone as an auxiliary group is an important synthetic strategy, especially for carbon–carbon bond formation. The sulfone moiety can modify the polarity of the molecule by acting as a leaving group or as an electron-withdrawing moiety to stabilize carbanions. It can also be exchanged by a hydrogen (reductive desulfonation), an alkyl group (alkylative desulfonation), a carbonyl functionality (oxidative desulfonation), or a nucleophile (nucleophilic displacement), as well as or could be displaced via α,β-elimination or a sulfur dioxide excursion process.[7]

12.2.2 Sulfone Linker Units

12.2.2.1 Preparation of Sulfone Linkers. The common means for preparing sulfone linkers are (i) oxidation of the corresponding thioether linker, (ii) using resin-bound benzenesulfinate linker, and (iii) using resin-bound vinyl sulfone.

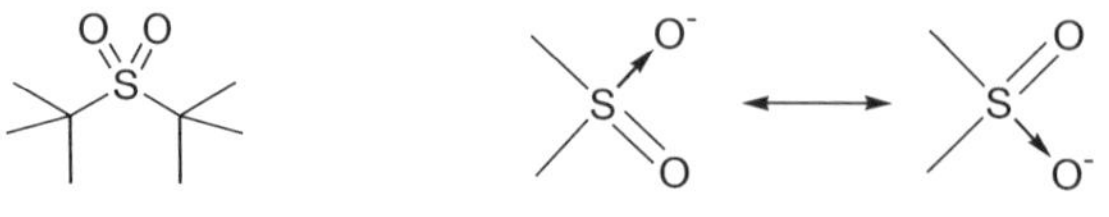

1-1 **1-2**

Figure 12.2. Structure of sulfones.

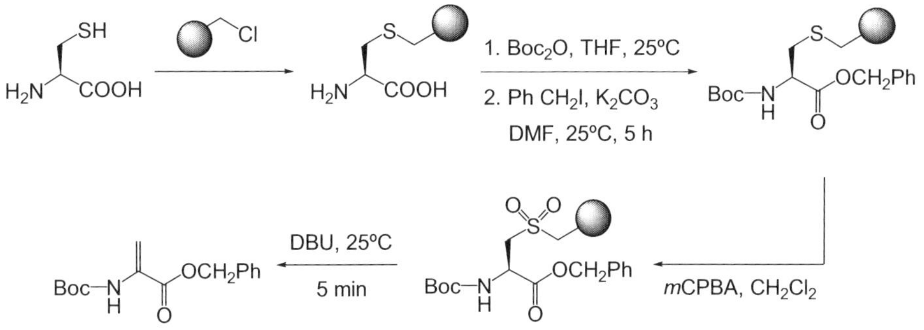

Scheme 12.1. Sulfone displacement via base-promoted β-elimination.

Oxidation of Thioether-Based Linker. The oxidation of thioethers on solid phase is a common reaction that leads to the formation of resin-bound sulfones. Typical oxidizing agents include *meta*-chloroperbenzoic acid (*m*CPBA),[8] H_2O_2,[9] Oxone®,[10] and sodium periodate ($NaIO_4$).[11] The activation of sulfide into sulfone enhances the leaving group ability and allows the sulfone to be displaced via β-elimination (Scheme 12.1) or with a wide variety of nucleophiles. In this context, the sulfide linkage is a typical example of safety-catch linker because the oxidation can be considered as an activation of the linker for the final cleavage. Meanwhile, it is also regarded as a multifunctional linker because the final cleavage introduces various amino groups at the linkage site. This strategy has been widely employed in a diversity approach to synthesize heterocycles and will be highlighted in Sections 12.2.2.2 and 12.2.2.3.

Via Resin-Bound Benzenesulfinate Linker. The use of resin-bound benzenesulfinate in solid-phase synthesis was first reported in 1989 by Huang and coworkers[12] for the synthesis of substituted acrylates (Scheme 12.2). The synthesis was performed using resin-bound sodium benzenesulfinate, which was prepared by bubbling sulfur dioxide into a suspension of cross-linked polystyrene, $AlCl_3$/HCl, and carbon disulfide to obtain the resin-bound benzenesulfinic acid that in turn was treated with sodium hydroxide to yield the resin-bound sodium benzenesulfinate (Scheme 12.2).

The benzenesulfinate resin can also be presented as resin-bound lithium benzenesulfinate. The preparation of such a resin was first described by Hagen et al. where the sulfinate moiety was introduced onto the cross-linked polystyrene via lithiation reaction, followed by

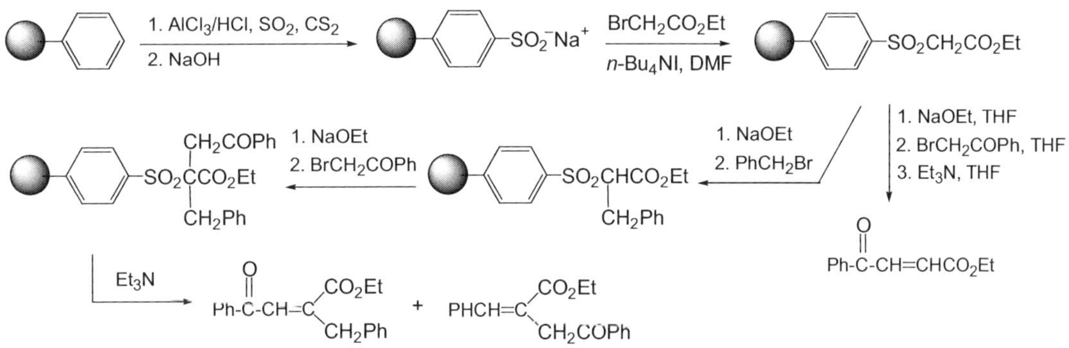

Scheme 12.2. Huang's synthesis and application of resin-bound sodium benzenesulfinate.

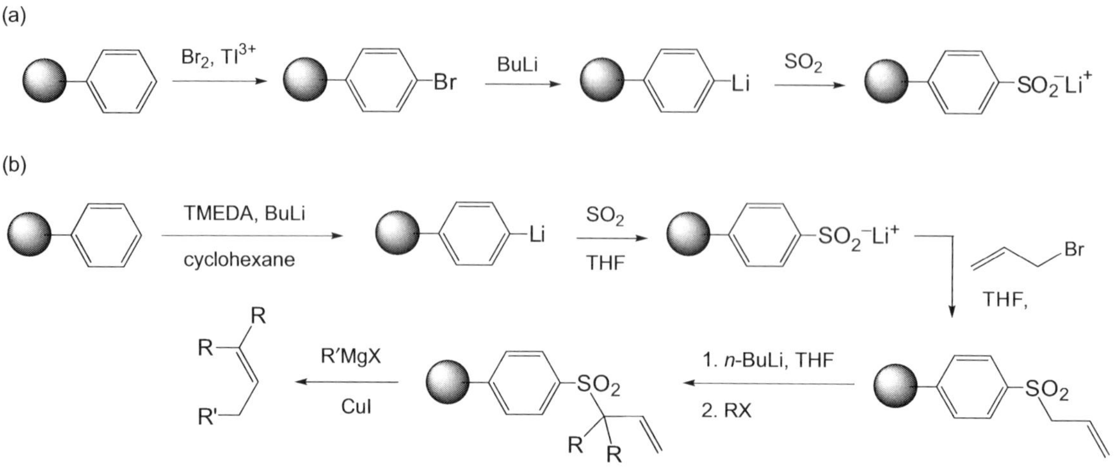

Scheme 12.3. Synthesis and application of resin-bound lithium benzenesulfinate.

quenching with sulfur dioxide (Scheme 12.3a).[13] Using this work and the pioneering work of Leznoff,[14] Kurth and coworkers[15] later reported a slightly modified procedure for the preparation of the resin-bound lithium benzenesulfinate and applied the resin to the synthesis of trisubstituted olefins (Scheme 12.3b).

USING RESIN-BOUND VINYL SULFONE. Resin-bound vinyl sulfones have also been used as an alternative sulfone linker unit since they have been shown to be stable to a wider range of conditions compared to the REM benzyl ester system. The vinyl sulfone group reacts efficiently with nucleophilic substrates, such as secondary amines, via conjugate addition. Quaternization of the resin-bound tertiary amine followed by subsequent deamination provided the respective tertiary amine and the regenerated resin-bound vinyl sulfone (Scheme 12.4).

12.2.2.2 Cleavage of Sulfone Linkers. This section describes the synthesis of *N*-heterocycles using different sulfone linker cleavage techniques.

NUCLEOPHILIC DISPLACEMENT. Sulfones can be displaced by a variety of nucleophiles, the most common of which are the amines. Using this cleavage technique, a range of libraries, including pyrimidines, pyrimido[4,5-*d*]pyrimidines, purines, and triazines, each possessing a diverse variety of amine functionalities, have been prepared.

2-Aminopyrimidine. Early representative examples of a sulfone traceless linker were demonstrated by Villalgordo et al.[17] and Gayo and Suto[8e] in the synthesis of 2-aminopyrimidines. The sulfur linkage was obtained by treating Merrifield resin with thiourea to provide the isothiouronium salts **1** (Scheme 12.5). Condensation of **1** with acetylenic ketones **2** or malononitrile derivatives **3** yielded the pyrimidine skeleton **4**. Oxidation of the alkylthio linkage with *m*CPBA provided the corresponding sulfone derivatives that underwent nucleophilic substitution with various amines to form 2-aminopyrimidines **5**.

Combinatorial synthesis of substituted pyrimidines using the polymer-bound thiouronium salt **1** has also been described by Srivastava et al. (Scheme 12.5c).[18] The reaction

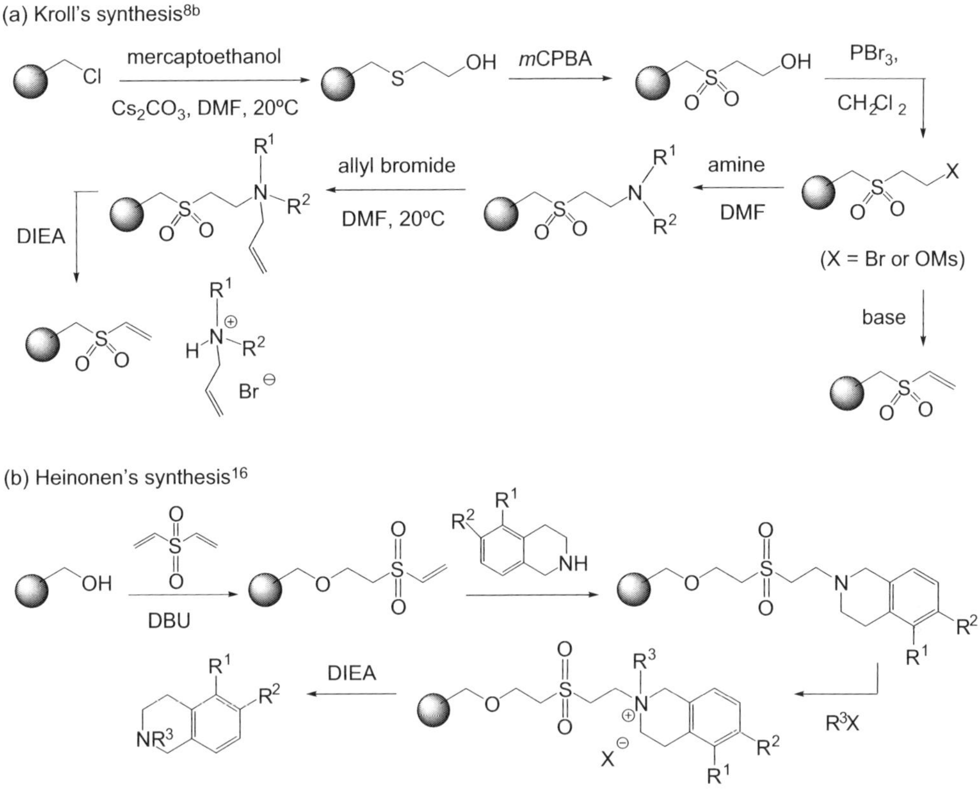

Scheme 12.4. Synthesis and application of resin-bound vinyl sulfone.

between **1** and ethoxymethylene malononitrile (EMMN) yielded the resin-bound pyrimidine **6**. Subsequent oxidation followed by reaction with primary amines yielded trisubstituted pyrimidines **7**.

Pyrimido[4,5-d]pyrimidines. Building on their works on substituted pyrimidines (Scheme 12.5c), Srivatava et al. have reported the use of resin-bound pyrimidine **6** for the synthesis of condensed heterocycles **11** and **14** (Scheme 12.6).[18] Fusion of **6** with urea/thiourea afforded the corresponding pyrimido[4,5-*d*]pyrimidine resin **12**. Oxidation of the resin **12** with Oxone afforded the sulfone-bound derivative **13**, which was then treated with cyclohexylamine to provide the respective pyrimido[4,5-*d*]pyrimidine **14**. Alternatively, resin **6** was also treated with a strong base KOH/EtOH to yield **8**, which then underwent cyclization with phenyl isocynate or phenyl thioisocynate to yield resin **9**. Activation of **9** to the resin-bound sulfone **10** followed by nucleophilic displacement with cyclohexylamine yielded pyrimido[4,5-*d*]pyrimidine derivative **11** as the final product.

The linking of pyrimidines to a resin support has also been applied to the synthesis of pteridine **20** (Scheme 12.7).[19] Oxidative cleavage with dimethyldioxirane followed by nucleophilic substitution with amines, azides, or water provided a traceless synthesis of the compound.

2,6,9-Trisubstituted Purines. Schultz[20a] and Brun[20b] have reported a traceless resin capture and release strategy for the solid-phase synthesis of 2,6,9-trisubstituted purines

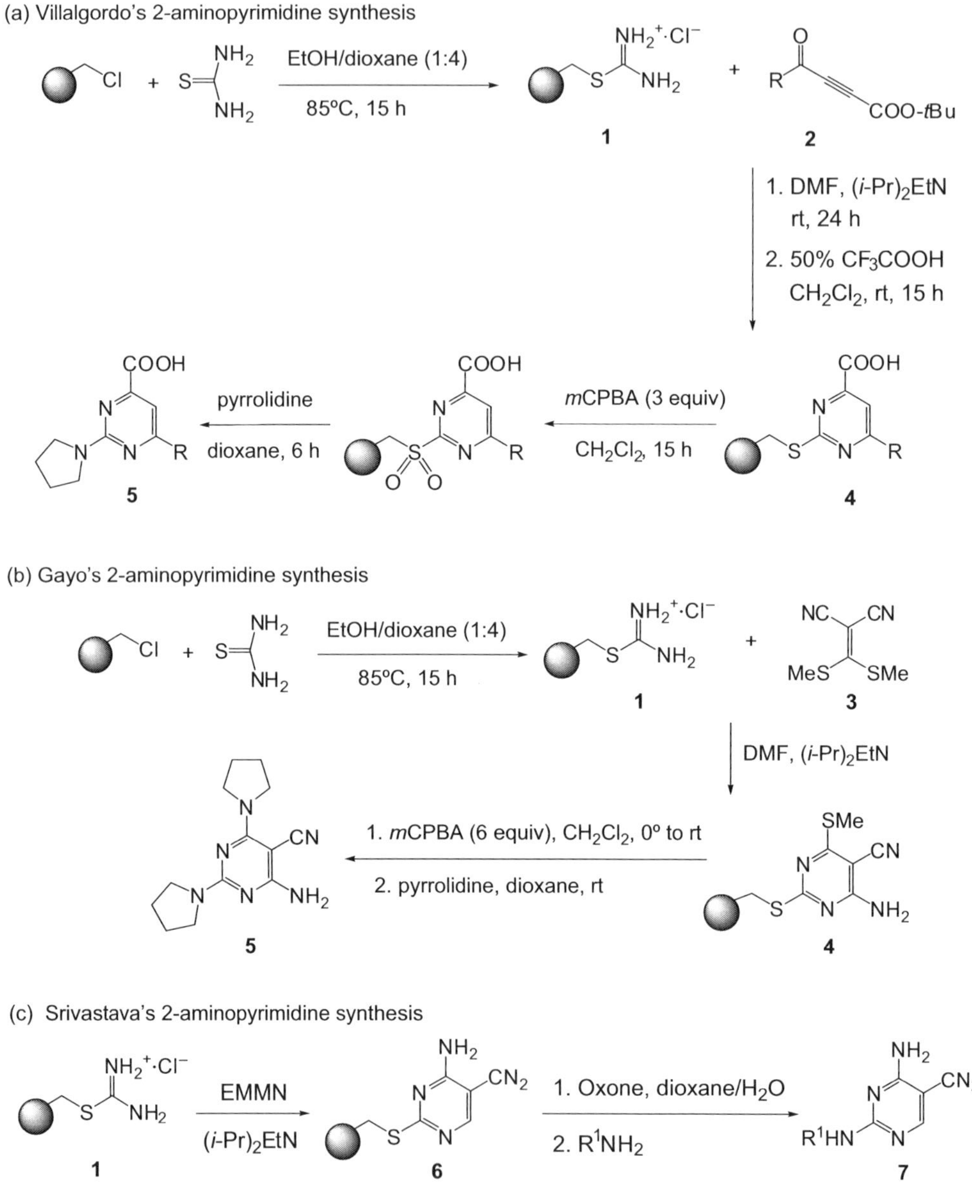

Scheme 12.5. Synthesis of 2-aminopyrimidines.

(Scheme 12.8). In the reaction, N9-substituted 2-halo-6-chloropurine **22** or **28** was captured by a resin at the C6 position. Subsequent substitution at the C2 fluoro/iodo position formed resin **24** or **30**. The thiol group on **24** and **30** was then oxidized to give sulfone **25** and **31**. 2,6,9-Trisubstituted purine derivatives **26** and **32** were then released by C6 substitution with amines and anilines. This approach allows diverse substitution with primary, secondary, and cyclic amines at the C2 position. With the oxidation of the thiol linker just before the cleavage, only the activated polymer-bound purine intermediates **25** and **31** will be released, thus resulting in high purity of the final product.

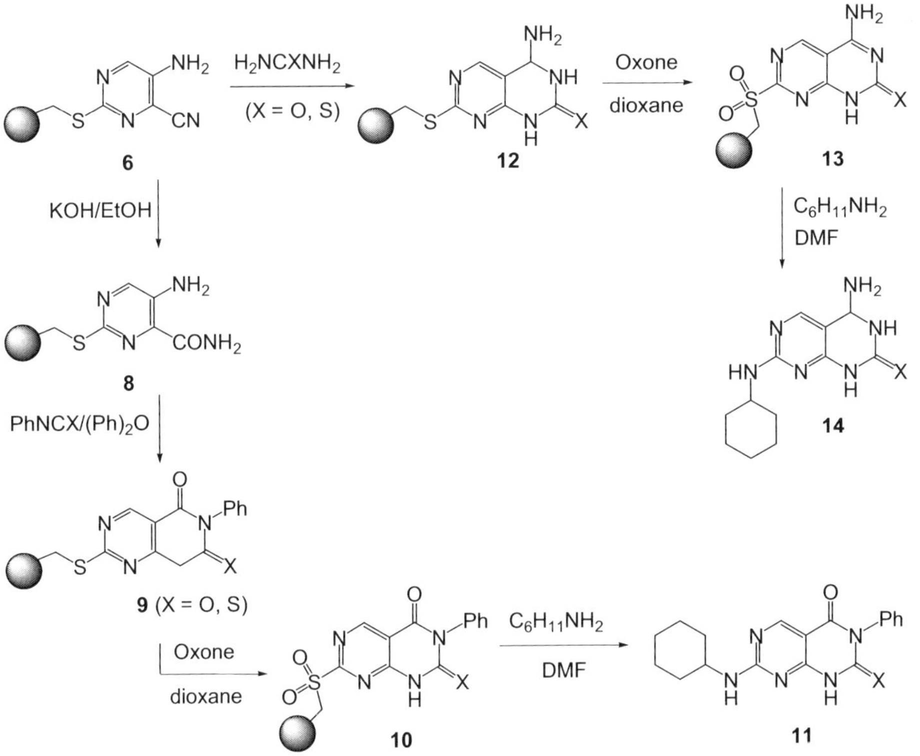

Scheme 12.6. Synthesis of pyrimido[4,5-*d*]pyrimidines.

Triazines. The application of sulfone as a safety-catch linker has also been demonstrated by Chang and coworkers in their orthogonal synthesis of highly pure trisubstituted triazines **37** (Scheme 12.9).[21] Polystyrene thiophenol resin was first coupled with the triazine **33** to yield the polymer-bound resin **34**. The chlorinated site

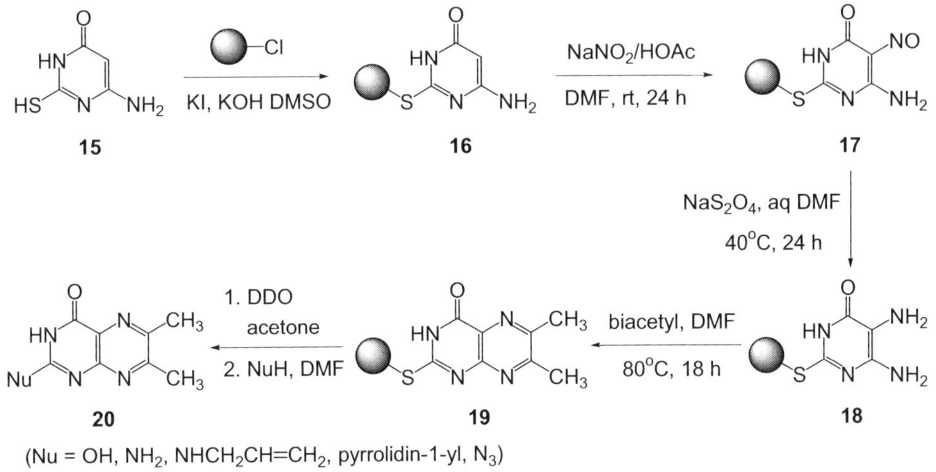

Scheme 12.7. Synthesis of pteridine.

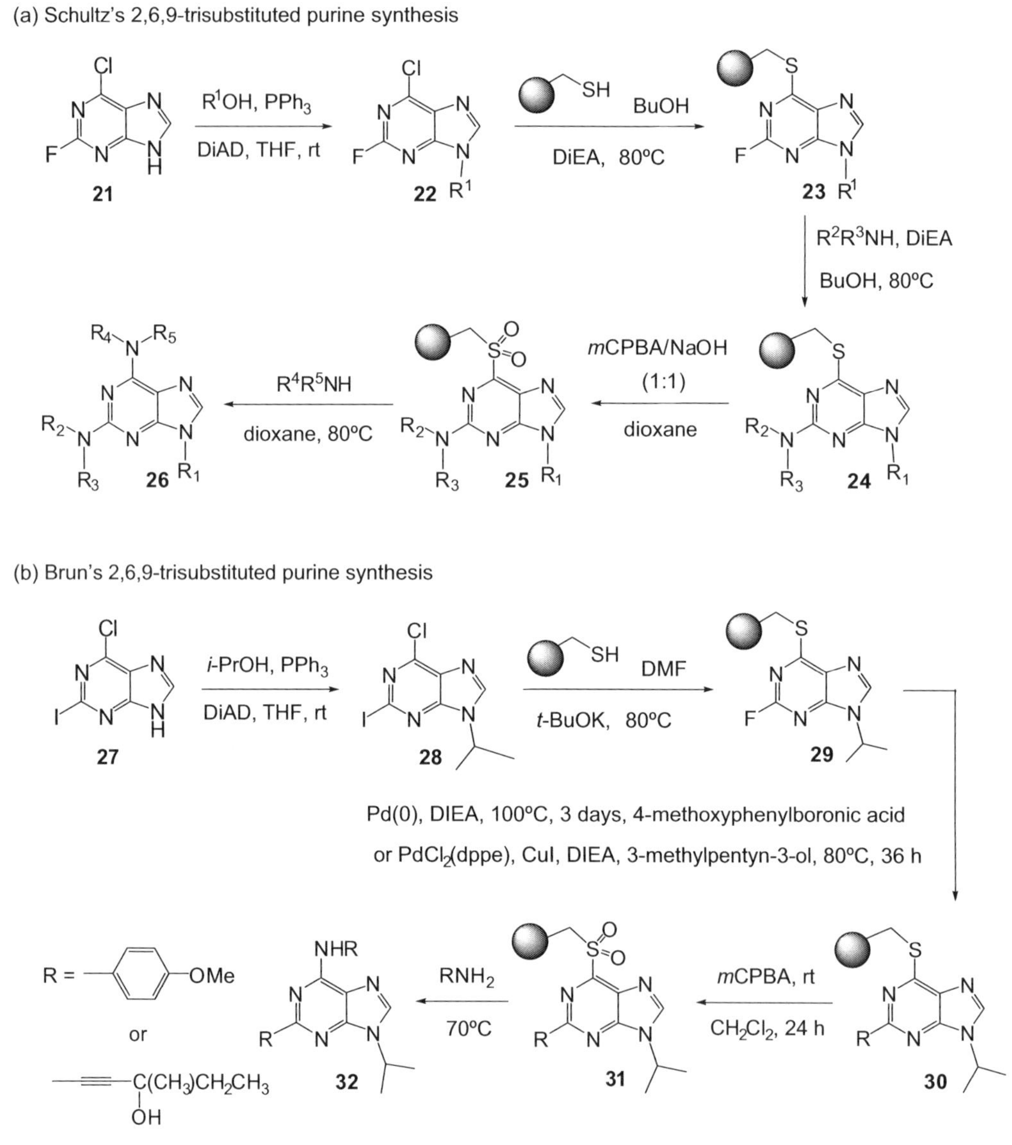

Scheme 12.8. Synthesis of 2,6,9-trisubstituted purines.

of **34** was then aminated using DIEA and a Boc-protected amine to give **35**, which was then subjected to trifluoroacetic acid followed by treatment with a variety of acyl chlorides to remove Boc and yield the polymer intermediate **36**. For the oxidation of **36** with *m*CPBA to form the activated sulfone, the pH of the reaction mixture was kept at around 4 with sodium hydroxide to protect any acid liable groups present on the scaffold. Finally, the activated sulfone linker was displaced by amines in anhydrous acetonitrile to release triazine derivatives **37** in high yields. Chang has reported that with this synthetic strategy, sterically hindered secondary and tertiary amines provided low product purity and phenol did not result in a satisfactory yield of the corresponding product.[21]

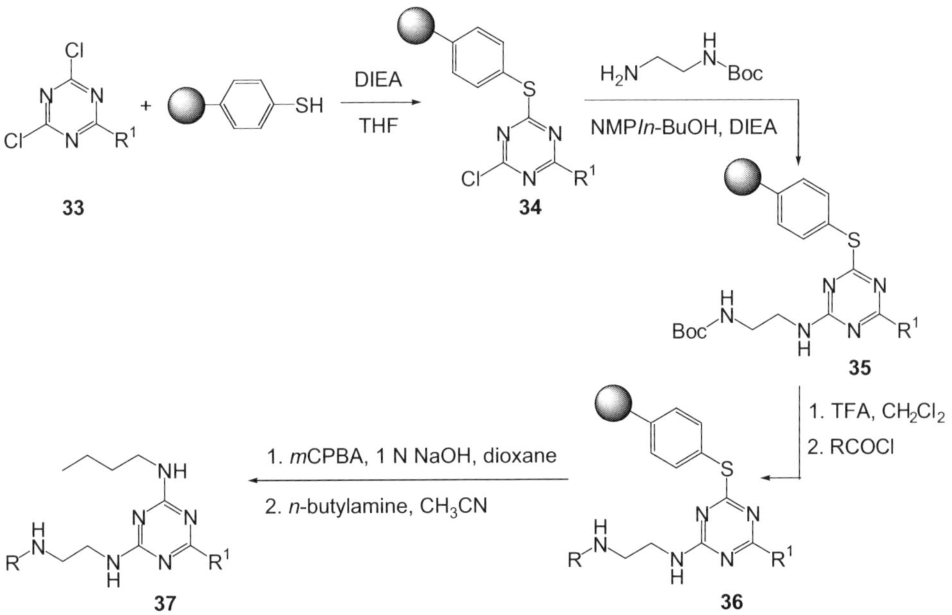

Scheme 12.9. Synthesis of trisubstituted triazines.

Masquelin et al. have also reported the production of pyrazolo[1,5-*a*][1,3,5]triazines and pyrozolo[1,5-*a*][1,3,5]triazin-4-ones from polymer-bound 3*H*-pyrozolo[1,5-*a*][1,3,5] triazin-4-one intermediates **38** (Scheme 12.10).[22] Resin **38** was prepared by condensing 4,5-disubstituted 2*H*-pyrazol-3-ylamine with ethoxycarbonyl isocyanate in dry acetone followed by treatment with sodium ethanolate in ethanol at 65°C and finally attaching onto the Merrifield resin. Using such a method, the solid-phase attachment was used to both purify the 3*H*-pyrazolo[1,5-*a*][1,3,5]triazin-4-ones via a scavenging process and also introduce a

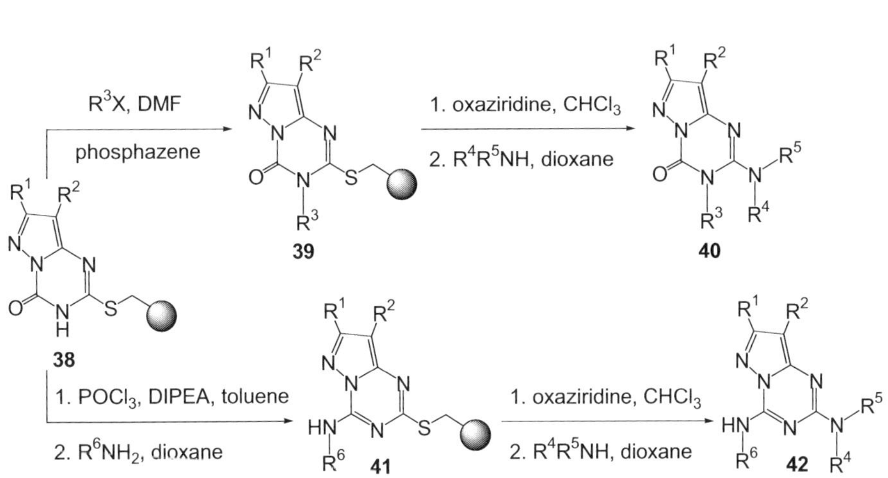

Scheme 12.10. Synthesis of pyrazolo[1,5-*a*][1,3,5]triazines and pyrozolo[1,5-*a*][1,3,5]triazin-4-ones.

new point of diversity at the sulfur-based safety-catch linkage. Treatment of **38** with alkyl halides in the presence of phosphazene followed by oxidation with *N*-(phenylsulfonyl)-3-phenyloxaziridine and subsequent cleavage with various amines gave **40** in good yields and high purities. In addition, treatment of **38** with phosphorous oxychloride in the presence of DIPEA afforded the corresponding 4-chloro intermediate that underwent nucleophilic substitution with amines to furnish resin **41**. Oxidation with *N*-(phenylsulfonyl)-3-phenyloxaziridine followed by cleavage with amines gave **42** in good yields and high purities.

CYCLATIVE CLEAVAGE. Cyclative cleavage refers to the intramolecular cyclization of the resin-bound substrate, which results in the bond that attaches the substrate to the solid support being cleaved simultaneously. The advantage of cyclative cleavage is the ability to generate the final cyclized product in high purity since any uncyclized side products would remain on the solid support. However, this kind of cleavage is restricted to substrates that contain the structural requirements for ring closure. Examples of cyclative cleavage are shown next.

Pyrazoline, Isoxazoline, and Benzo[b][1,4]diazepine Derivatives. The sulfone linker **44** derived from **43** is a versatile and robust tether and many new applications of this linker have been explored. To demonstrate the use of **44** as a versatile and robust tether that offers various on-resin functionalization or cleavage with additional changes, Lam and coworkers have highlighted the preparation of the five-membered pyrazoline/isoxazoline **48**[23a] as well as the fused benzodiazepine **47**[23b] (Scheme 12.11). The polymer-bound sodium sulfinate resin **43** was treated with benzyl bromide using NBu_4I/KI/DMF to yield *S*-alkylated sulfone

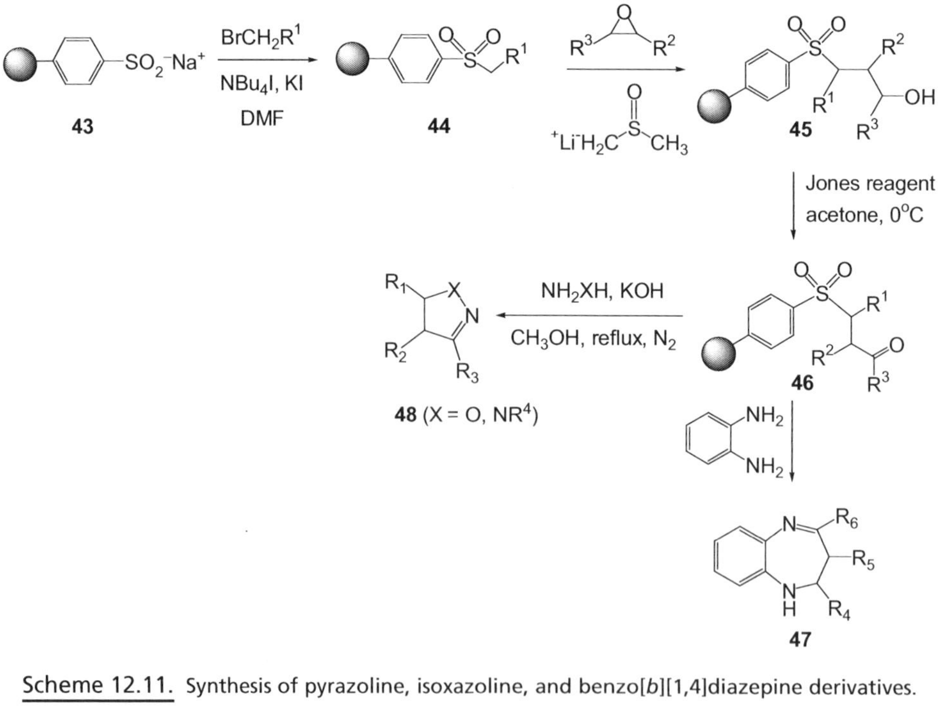

Scheme 12.11. Synthesis of pyrazoline, isoxazoline, and benzo[*b*][1,4]diazepine derivatives.

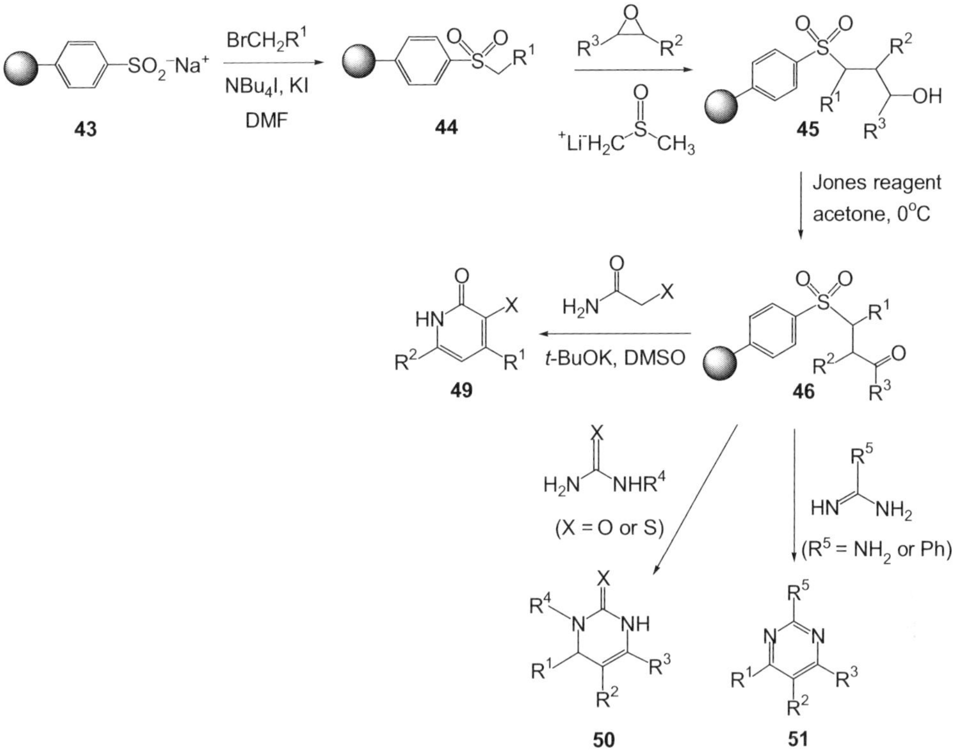

Scheme 12.12. Synthesis of pyrimidine derivatives.

resin **44**. Subsequent sulfone anion alkylation with epoxide afforded the γ-hydroxyl sulfone **45**. Oxidation of **45** using Jones reagent yielded γ-ketosulfone resin **46**. The traceless product was released by a one-pot elimination–cyclization process using *ortho*-phenylene diamine and phenyl hydrazine as cyclization reagent to afford fused benzodiazepine **47** and five-membered pyrazoline/isoxazoline **48**.

Pyrimidine Derivatives. Lam and coworkers have also published the preparation of 3,4,6-trisubstituted-2-pyridones **49**, pyridine-2-thiones, and pyrimidine-2-ones **50** and pyrimidines **51**[23b],[23c] using sodium benzenesulfinate resin **43** to form the key intermediate γ-ketosulfone resin **46**. The traceless products **49**, **50**, and **51** were release via a one-pot elimination–cyclization process using alkyl amides, guanidine/thioureas, and benzamidine, respectively (Scheme 12.12).

Sodium benzenesulfinate resin **43** can also be used to prepare a traceless solid-phase synthesis for 3,4-dihydropyrimidine-2-ones **54** and **55** (Scheme 12.13).[23d] This strategy highlighted the sulfinate acidification to yield resin-bound benzenesulfinic acid **52**, followed by the condensation of urea or thiourea with aldehydes and sulfinic acid.[23e] A one-pot cyclization–dehydration process with 1,3-dicarbonyl compounds or *β*-ketoesters (generated *in situ* by treating the latter reagents with KOH/EtOH) afforded **54**, while cyclization with a mixture of pyrrolidine and *β*-ketoacid in ethanol followed by the addition of $TsOH{\cdot}H_2O$ gave the ester form of **55**. When THF was used as a solvent, the free carboxylic acid form of **55** was obtained in comparable yields.

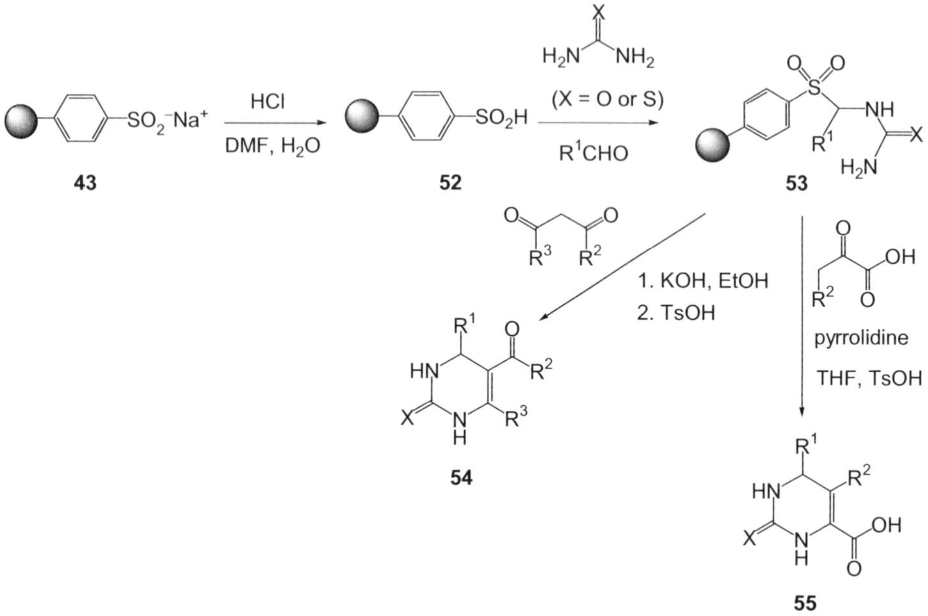

Scheme 12.13. Synthesis of 3,4-dihydropyrimidine-2-ones.

Oxazoles, Imidazoles, and Thiazoles. Resin-bound benzenesulfinic acid **52** obtained from sodium benzenesulfinate resin **43** (Scheme 12.13) could be treated with excess TEA and aldehyde in the presence of thiazolium catalyst to provide α-ketoamide *in situ*. Further reactions of the α-ketoamide with PPh_3/I, Lawesson's reagent, and EtOH/amine generated substituted oxazoles **56**, imidazoles **57**, and thiazoles **58** (Scheme 12.14).[23e]

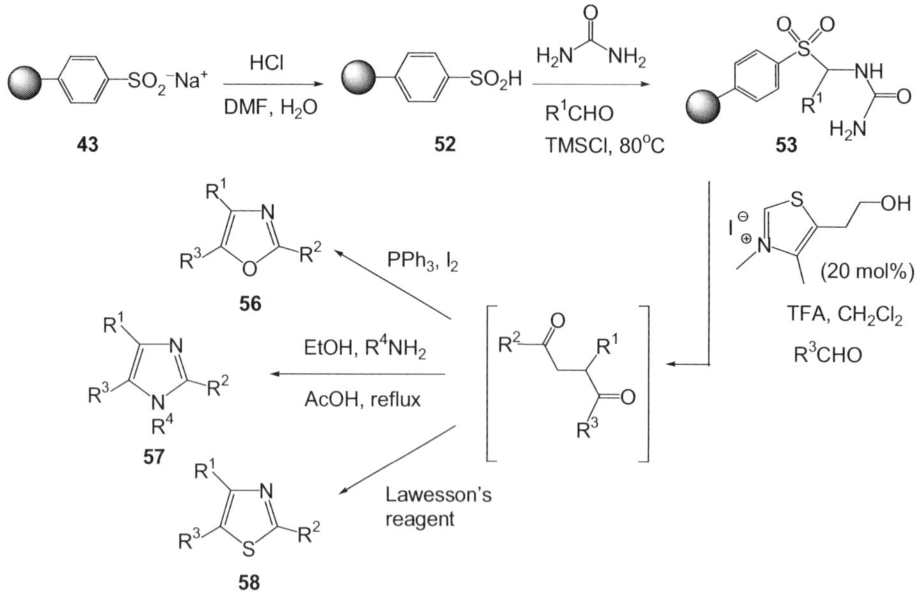

Scheme 12.14. Synthesis of oxazoles, imidazoles, and thiazoles.

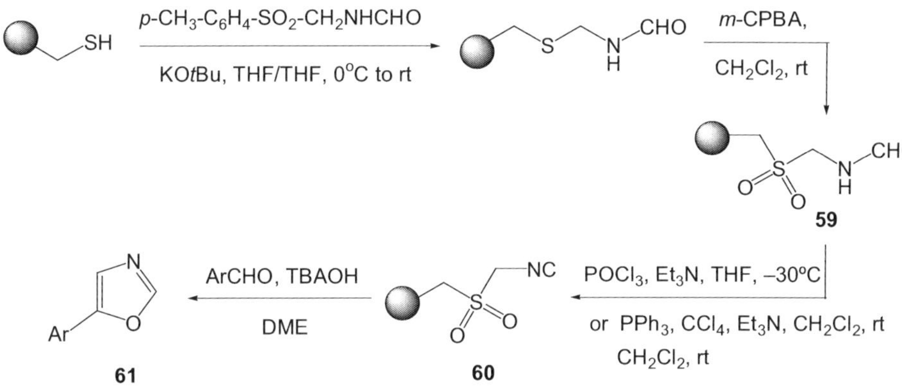

Scheme 12.15. Synthesis of oxazoles.

Kulkarni and Ganesen[24] have also demonstrated the synthesis of oxazoles by adapting the procedure reported by Leusen and coworkers.[25] A solid-phase version of *p*-tolylsulfonylmethyl isocyanide (TosMIC) **60** was prepared from TentaGel-SH (Scheme 12.15). Condensation of the **60** with TBAOH in DME produced aromatic-substituted oxazoles **61** in 25–50% yield.[24]

Furans. Timm and coworkers have devised a strategy to synthesize 2,5-disubstituted furans using sulfone carbanion-mediated solid support as a traceless linker (Scheme 12.16).[26] They demonstrated the usefulness of the highly reactive basic carbanionic species of the

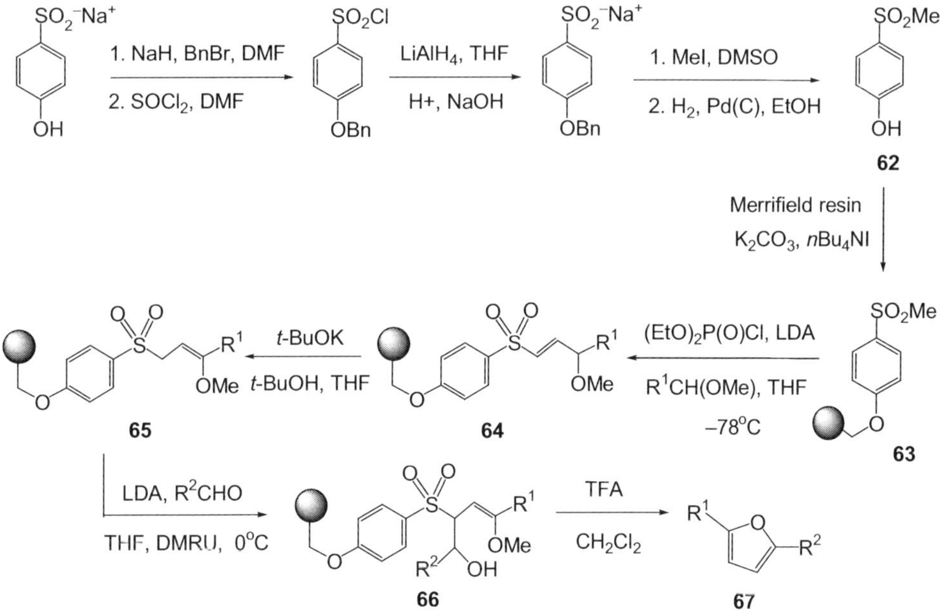

Scheme 12.16. Synthesis of 2,5-disubstituted furans.

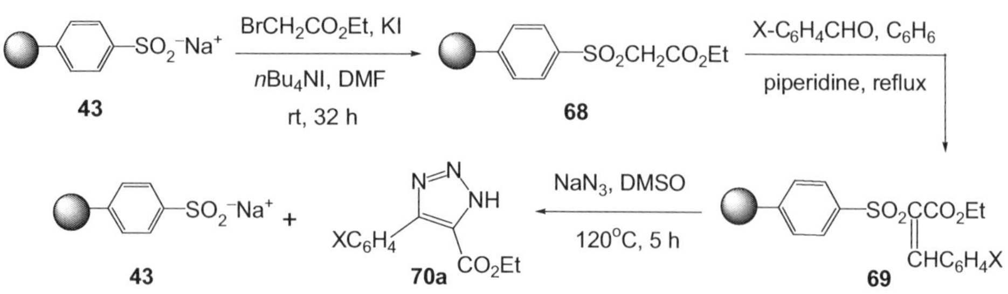

Scheme 12.17. Synthesis of 4,5-disubstituted 1,2,3-triazoles.

sulfone moiety. The aryl methyl sulfone **62** was attached directly to the Merrifield resin to afford resin-bound (arylsulfonyl)methane **63**. Subsequent deprotonation followed by addition of diethyl chlorophosphate resulted in a color change, indicating the formation of a carbanionic species stabilized by the sulfonyl and phosphonyl groups. Addition of 2-methoxyaldehydes afforded the resin-bound **64**. Conversion of the geometry in **64** using potassium *tert*-butoxide yielded the aryl sulfone-supported resins **65**. Optimum conversion of **65** to **66** was observed using DMPU in THF. Finally, treatment of resin **66** with trifluoroacetic acid cleaved the resin and afforded 2,5-disubstituted furans **67** in 13–32% overall yield.

Triazole Derivatives. The synthesis of 4,5-disubstituted 1,2,3-triazoles using sodium polystyrylsulfinate resin **43** was first reported by Huang and coworkers (Scheme 12.17).[27] The synthesis involved the reaction between arylpropiolates and sodium azide followed by oxidation. Resin-bound ethyl polystyrylsulfonylacetate **68** was prepared using ethyl 2-bromoacetate in DMF with potassium iodide and tetrabutylammonium iodide as a phase transfer catalyst. Subsequent reaction of resin **68** with benzaldehyde using piperidine as a catalyst afforded arylidene polystyrylsulfonyl acetate **69**. Cyclization of **69** with sodium azide in DMSO at 120°C for 5 h afforded the 1,2,3-triazoles **70a**.

In 2006, Lam and Gao reported a regioselective procedure for the synthesis of 1,2,3-triazoles via a [3 + 2] cycloaddition of resin-bound vinyl sulfone and sodium azide in the presence of an alkyl halide.[28] The resin-bound vinyl sulfone **71** was effectively generated by the Knoevenagel condensation of resin **44** (Scheme 12.18). Resin **71** then acted as the vinyl sulfone dipolarophile and the electronegative sulfone that was eliminated in the reaction directed the regiochemistry of the cycloaddition.

This synthetic strategy was also applied to the synthesis of monosubstituted 1,2,3-triazoles. Treatment of β-iodosulfone **72** with TEA yielded the vinyl sulfone **73** (Scheme 12.18). Final cycloaddition carried out using sodium azide in DMF cleaved the sulfone linker and afforded monosubstituted 1,2,3-triazoles **70c**. Using microwave irradiation for the cyclization and cleavage step reduced the reaction time dramatically from over 1 day to 1 h. This enabled a variety of reagent to be used in each step of the reaction, leading to efficient library generation.

Lam and Raghavendra have also highlighted the synthesis of 1,2,3-triazoles through an intramolecular cyclization of a diazo intermediate that was generated from the Bamford–Stevens reaction between a tosyl hydrazone and a base.[29] Polystyrene sulfonyl hydrazide **74** was treated with 1,1-dichloroacetone in THF to yield polymer-bound α-dichloro carbonyl sulfonylhydrazone **75** (Scheme 12.19). Alternative reaction of **74** with 1,1,1-trichloroacetaldehyde and pyridine derivatives yielded resin **76** and **77**. When treated with excess amine,

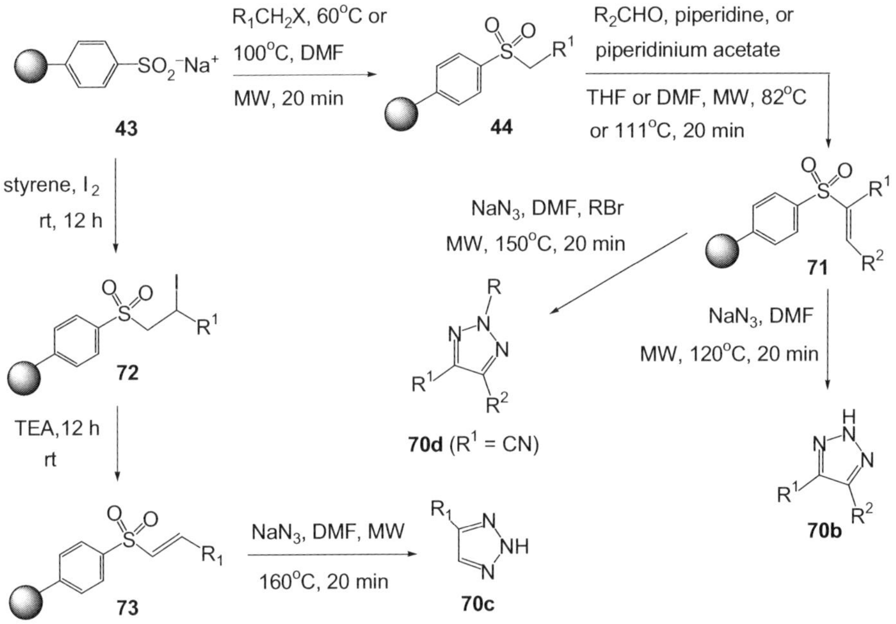

Scheme 12.18. Regioselective synthesis of 1,2,3-triazoles.

the diazo intermediate resins **75**, **76**, and **77** cyclized into the respective 1,2,3-triazole products **78a**, **78b**, and **78c**.

Pyrrole Derivatives. The 1,3-dipolar addition reactions of nitrile oxides to dienes to form substituted pyrroles were demonstrated by Kurth et al. (Scheme 12.20).[30] Polymer

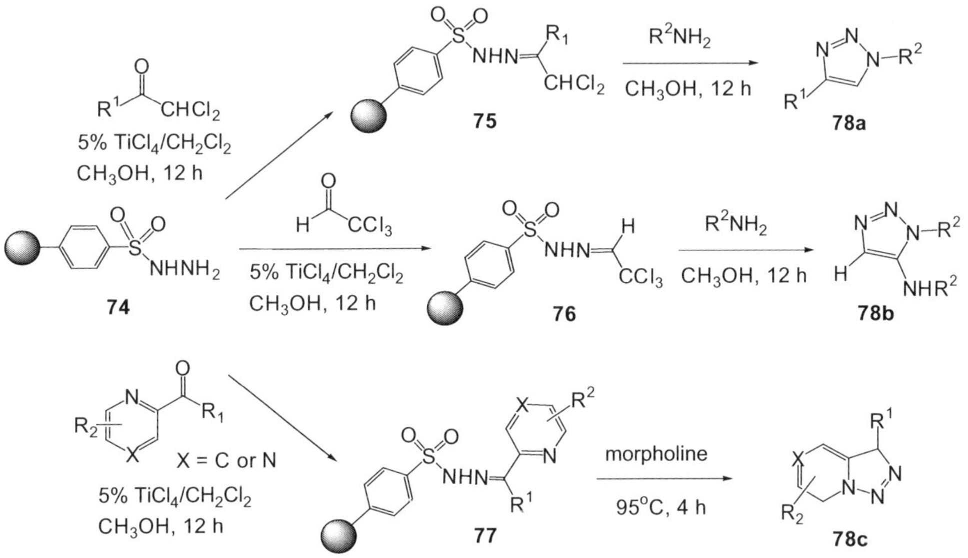

Scheme 12.19. Synthesis of 1,2,3-triazoles via cyclization of a diazo intermediate.

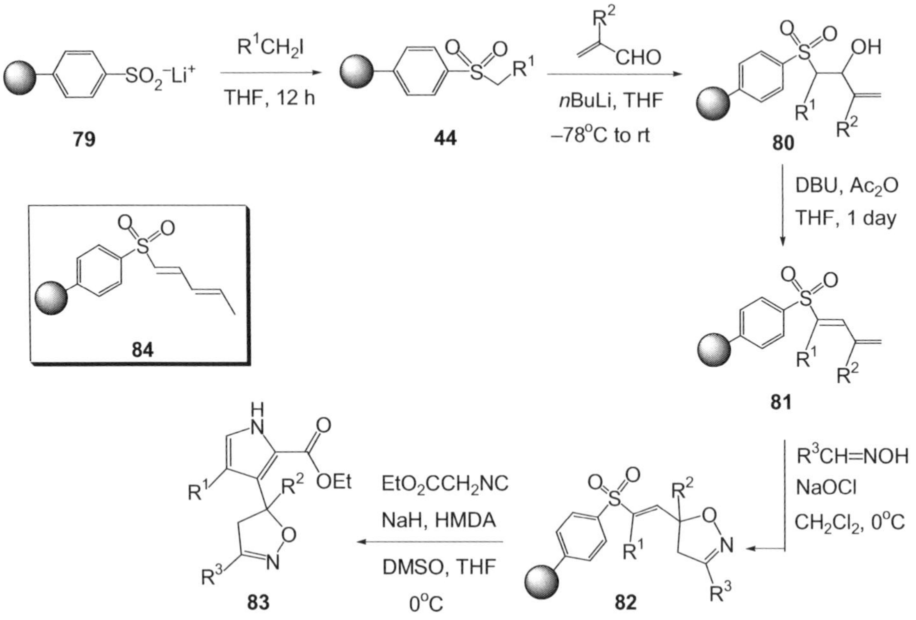

Scheme 12.20. Synthesis of pyrroles.

support 1,3-butadienes **81** were used as dipolarophiles in this methodology. Treatment of lithium benzenesulfinate resin **79** with iodoalkane in THF yielded phenylalkyl sulfone **44**. Reaction of **44** with *n*BuLi generated the α-anion that underwent a condensation reaction with α,*β*-unsaturated aldehydes to yield the allylic alcohol resin **80**. Acetylation followed by *in situ* elimination with DBU yielded resin **81**. Regioselective construction of the isoxazolino derivatives **82** was achieved by the 1,3-dipolar cycloaddition of nitrile oxides (generated via Huisgen method). The author demonstrated that with diene **81a** ($R^1 = H$), the cycloaddition occurred exclusively at the terminal double bond to give **82a** ($R^1 = H$). However, diene **84**, which has only internal double bonds, failed to react with nitrile oxides under the same reaction conditions. Pyrrole annulations of resin **82** with sodium hydride-derived enolate of ethyl isocyanoacetate afforded the desired substituted pyrrole derivatives **83**.

Kurth and coworkers have also adopted the traceless solid-phase sulfone linker strategy for the synthesis of 4,5,6,7-tetrahydroisoindole pyrrole derivatives **88** (Scheme 12.21).[31] *S*-alkylation of lithium benzenesulfinate resin **79** with *trans*-3,4-dibromosulfolane in the presence of pyridine at 80°C afforded 3-(phenylsulfonyl)-3-sulfolene **85**. Upon heating, **85** released SO_2 and the butadiene intermediate **86** that was generated *in situ* underwent cycloaddition with *N*-phenylmaleimide to furnish vinyl sulfone resin **87**. Subsequent Micheal addition of TosMIC in the presence of *t*-BuOK resulted in the cyclization, elimination of the sulfone group, and tautomerization to give the pyrrole derivative **88**. Reaction of sulfolene resin **85** with alternative dienophiles such as ethyl fumarate and diisopropyl fumarate followed by cleavage using TosMIC in the presence of *t*-BuOK led to cyclized product **92** (12–15% overall yield). Treatment with cyclohexadiene or cyclopentadiene in the presence of xylene at 145°C gave sulfone-bound resin **89**. Cleavage using TosMIC in the presence of *t*-BuOK led to the formation of the regioselective pyrrole product **90**.

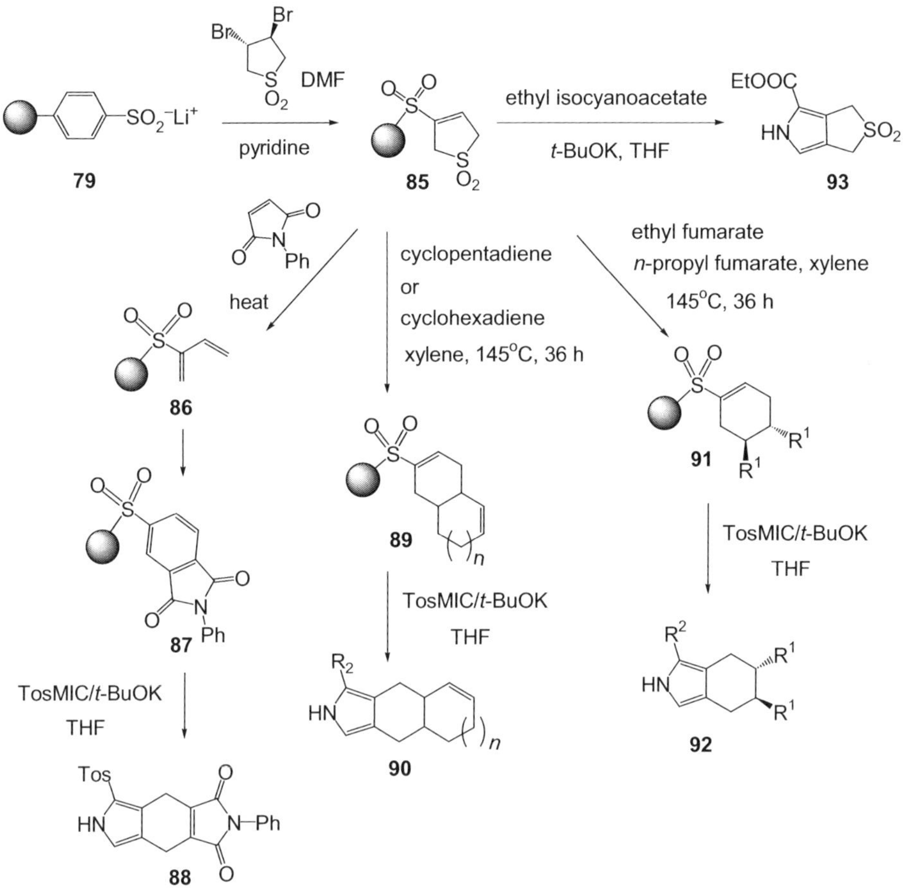

Scheme 12.21. Synthesis of 4,5,6,7-tetrahydroisoindole pyrrole derivatives.

1,3-Oxazolidin-2-ones. 1,3-Oxazolidin-2-ones **98** (Scheme 12.22) was prepared via solid-phase sulfone chemistry by first attaching 1,2-diols onto the resin-bound sulfonyl chloride **95** to yield resin **96**.[32] Resin **95** was in turn prepared from sulfonic acid resin **94**. Treatment of resin **96** with *p*-toluenesulfonyl isocyanate followed by cycloelimination with concurrent detachment from the resin gave **98** as the major product and, at times, cyclic

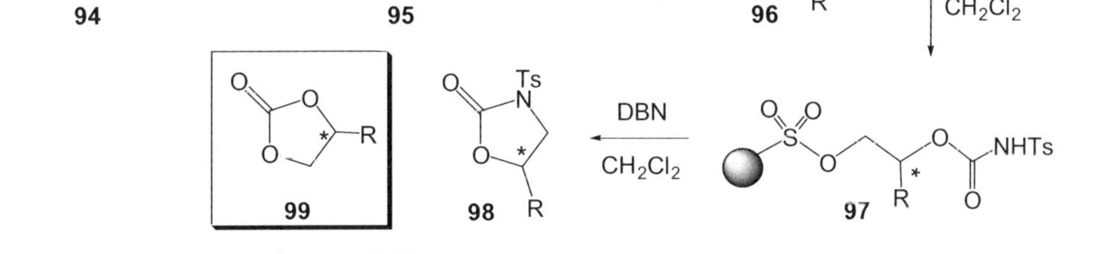

Scheme 12.22. Synthesis of 1,3-oxazolidin-2-ones.

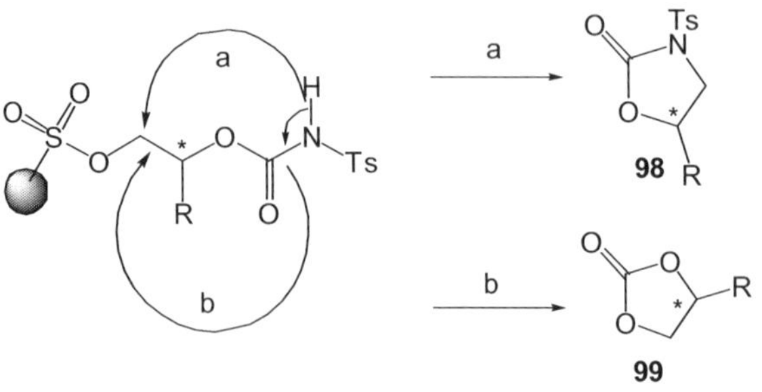

Scheme 12.23. Cycloelimination of resin **97**.

carbonate **99** as minor product. Compound **99** was formed by ring closure via the oxygen atom of **97** (due to steric hindrance) and subsequent hydrolysis as shown in Scheme 12.23. Using this method, oxazolidinones of high enantiopurity were obtained when enantiopure 1,2-diols were employed.

Urea Derivatives. Sun et al. have developed a strategy to synthesize substituted ureas **102** from sodium benzenesulfinate resin **43** (Scheme 12.24).[33] 2-Chloroethanol was first coupled with the resin **43** to afford resin **100**, which can easily react with carboxylic acid to form ester linkages that can be cleaved under basic or acidic conditions. Treatment of resin **100** with Boc-protected glycine using DCC/DMAP afforded resin **101** whose sulfone linkage was cleaved using HCl in dioxane to yield the cyclized urea derivatives **102**.

Pyridazines. The solid-phase synthesis of 3,6-disubstituted pyridazine derivatives using resin-bound sodium benzenesulfinate **43** was described by Lam et al. (Scheme 21.25).[34] The reaction of resin **103** with α-bromoketone substrates in the presence of K_2CO_3/DMF afforded a higher yield of **104** compared to using NaOMe/THF as base. Subsequent reaction with hydrazine led to the condensation and cleavage of the sulfone linker to give the corresponding 3,6-disubstituted pyridazines **105**.

Pavine and Isopavine. A chiral *tert*-butanesulfinamide resin **107** (Scheme 12.26) has been developed from tertiary alcohol **106** by Ellman and coworkers.[35] Condensation of resin **107** with 3,4-dimethoxyphenyl acetaldehyde with excess $Ti(OEt)_4$ followed by addition of 3-methoxyphenylmagnesium bromide afforded the desired amine **108**. The sulfinyl

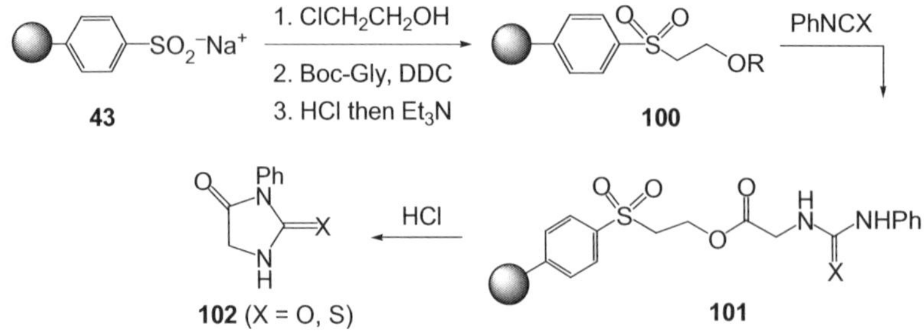

Scheme 12.24. Synthesis of urea derivatives.

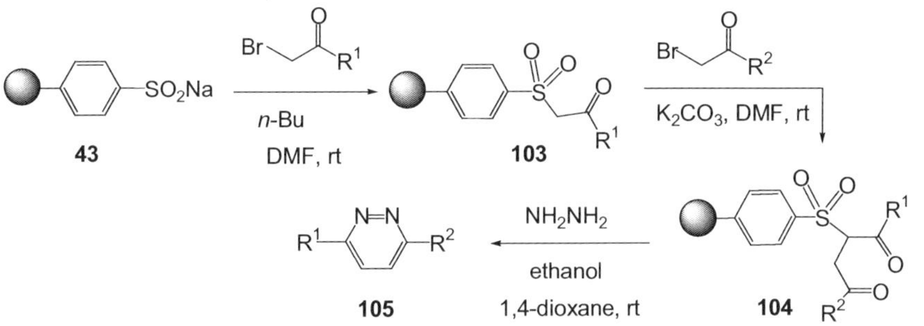

Scheme 12.25. Synthesis of 3,6-disubstituted pyridazines.

nitrogen was then allylated by deprotonation using KO^{t}Bu, followed by addition of allyl bromide. Oxidation of the sulfinyl group using *m*CPBA afforded a sulfone linker resin that is stable under acidic conditions. Catalytic osmylation with excess *N*-methylmorpholine *N*-oxide in THF followed by cleavage of the glycol using $Pb(OAc)_4$ in CH_2Cl_2/acetic acid yielded the aldehyde cyclization precursor **109**. Cyclization of the aldehyde-bound resin in the presence of dilute HCl in CH_2Cl_2 afforded the pavine derivatives **110** with complete selectivity. Similarly, the isopavine alkaloid **111** was obtained with complete selectivity through cyclization using formic acid and cleavage with diluted triflic acid. The sulfone linker **109** was applicable for the multistep asymmetric synthesis of natural products such as pavine and isopavine derivatives.

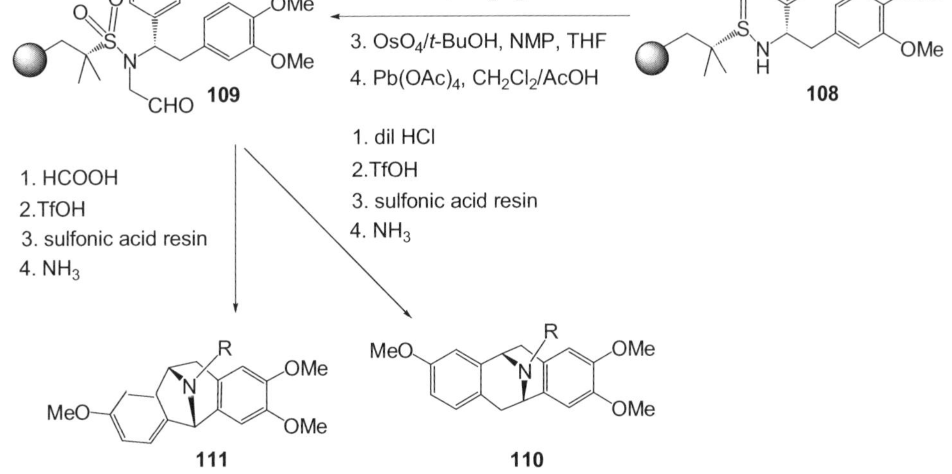

Scheme 12.26. Synthesis of pavine and isopavine.

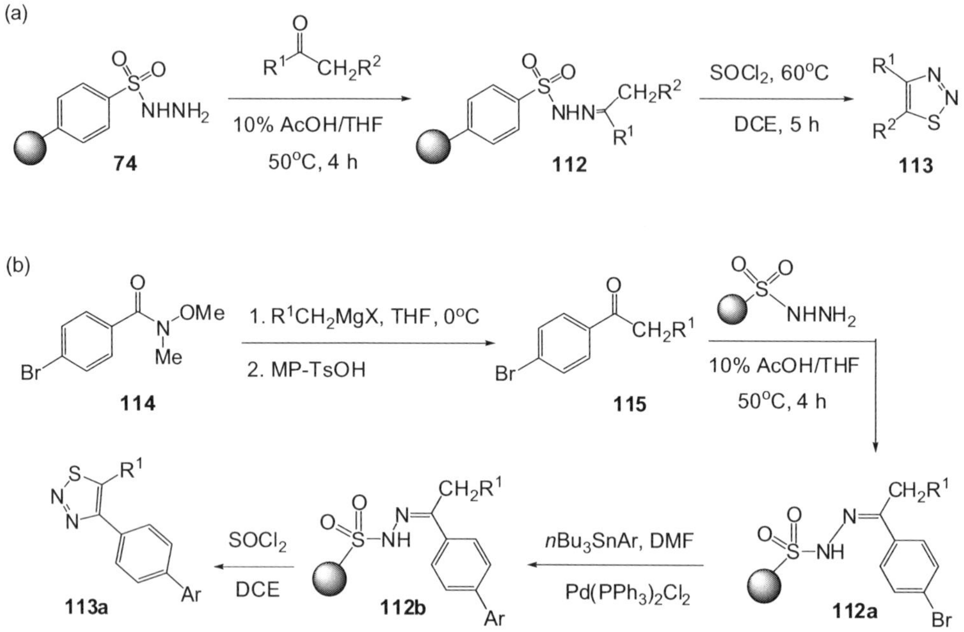

Scheme 12.27. Synthesis of 1,2,3-thiadiazoles.

1,2,3-Thiadiazoles. Sulfonyl hydrazide resin **74** could also serve as a sulfone linker in SPS. Porco and coworkers have demonstrated the use of resin **74** for the preparation of sulfonylhydrazone resin **112**, which was subsequently applied to the synthesis of 1,2,3-thiadiazoles **113** (Scheme 12.27a).[36] Resin **74** has also been efficiently employed in the "resin capture" of ketones in a hybrid solution/solid-phase synthesis of 1,2,3-thiadiazoles (Scheme 12.27b). The resin-bound sulfonylhydrazones with aryl halide substituents **112a** could be subjected to Stille reaction for the generation of more diverse thiadiazole structures **113a**.

Direct Cleavage

Synthesis of Substituted Indoles via Pd-Catalyzed Cyclization. Application of sulfone linkers to generate substituted indoles proceeds through a palladium-catalyzed cyclization reaction.[37] This approach to 2-substituted indoles is based on the coupling reaction between 2-iodoanilines and terminal alkynes (Scheme 12.28a). By introducing a strong electron-withdrawing sulfone group on the amine, the sp^2–sp coupling and indole cyclization occur in a one-pot reaction under relatively mild reaction conditions. The cleavage of the sulfone linker was also achieved using mild conditions. 2-Iodoanilines were first loaded onto commercially available sulfonyl chloride polystyrene resin **95**. The sulfone-bound iodoanilines **116** then underwent palladium-mediated heteroannulation with terminal alkynes to give indoles **117**. Cleavage of the sulfone linker was then carried out using TBAF to release the substituted indoles **118** in 85–100% yield and high purity (85–100%).

Variation at 3-position of indole can be carried out by direct mercuration of **117** to yield **119**. Reaction of **117** with mercury(II) acetate and a catalytic amount of $HClO_4$ afforded the versatile resin-bound organomercurial **119**. This 3-indolylmercury resin **119** can be converted to other organometallics or coupled with a substrate to introduce diversity in

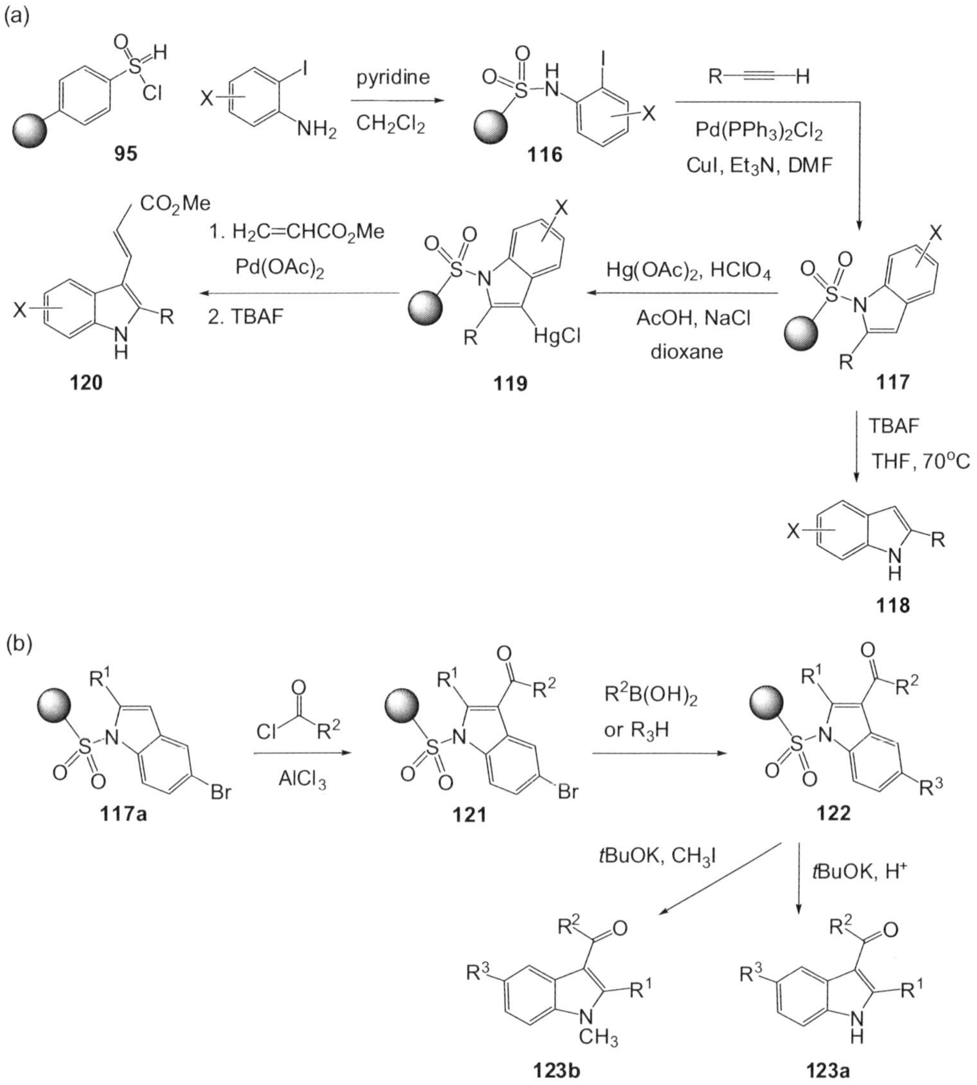

Scheme 12.28. Synthesis of substituted indoles.

the indole template. When reacted with methyl acrylate in $Pd(OAc)_2$ followed by cleavage with TBAF, 2,3-disubstituted indoles **120** were obtained.

Using the method adapted from Zhang et al.,[38] treatment of 4-bromo-2-iodoanilino sulphonamide resin **117a** with acyl chloride in the presence of $AlCl_3$ afforded the resin-bound intermediate **121** (Scheme 12.28b). This intermediate underwent Sonogashira coupling reaction to yield the alkynylated indole or Suzuki coupling to introduce an aryl group into the template (resin **122**). Cleavage of the resin can be achieved by saponification using *t*-BuOK at room temperature to yield **123a**, which could be methylated at N1 using MeI to yield **123b**.

Tin-Mediated Nitro Reduction to Form Benzylimidazoles. Tumelty et al. have developed a traceless solid-phase approach to a diverse group of substituted benzimidazoles

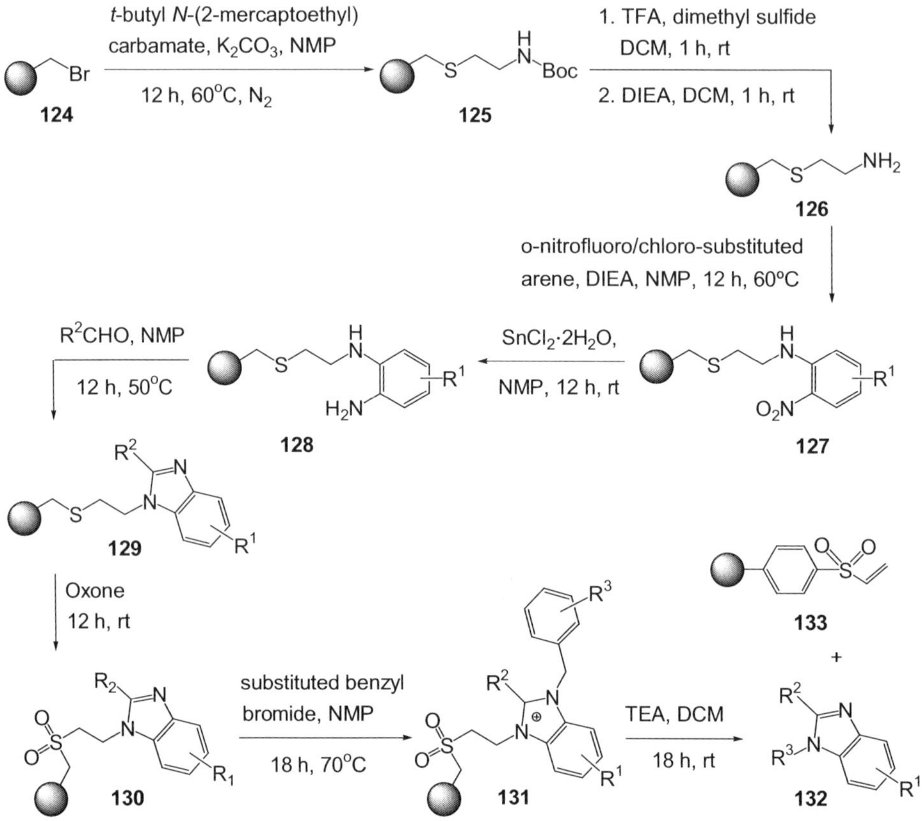

Scheme 12.29. Synthesis of benzylimidazoles.

using tin-mediated nitro reduction strategy (Scheme 12.29).[39] TentaGel-Br resin **124** when treated with *tert*-butyl *N*-(2-mercaptoethyl)carbamate afforded *t*-Boc-protected resin **125**. After deprotection under acidic conditions, resin **126** was subjected to aromatic displacement reactions to yield **127**. Reduction of nitro group on **127** with tin(II) chloride afforded the secondary phenylene diamine resin **128**. The cyclization reaction of resin-bound **128** with a variety of aromatic and aliphatic aldehydes occurred with no exogenous oxidants to yield benzimidazole resin **129**. Subsequent oxidation of the sulfide into sulfone linker **130** using Oxone was crucial for the activation of the resin for base cleavage. The resin-bound intermediate **130** was then converted into a quaternary salt **131** using benzyl bromide. Cleavage from the resin using TEA in DCM afforded benzylimidazoles **132** and the vinyl sulfone resin **133**.

Synthesis of Oxindoles Using Samarium(II) Iodide for Sulfone Cleavage. α-Sulfanyl-*N*-aryl acetamide attached to resin via the sulfur atom **134** undergoes efficient Pummerer cyclization upon activation of sulfur group to yield oxindole derivatives **135** (Scheme 12.30). Oxidation of **135** to sulfone **136** using Oxone followed by alkylation yielded the alkylated sulfone **137**, which was then cleaved from the resin in a traceless manner using samarium(II) iodide to yield the heterocyclic product **138**.

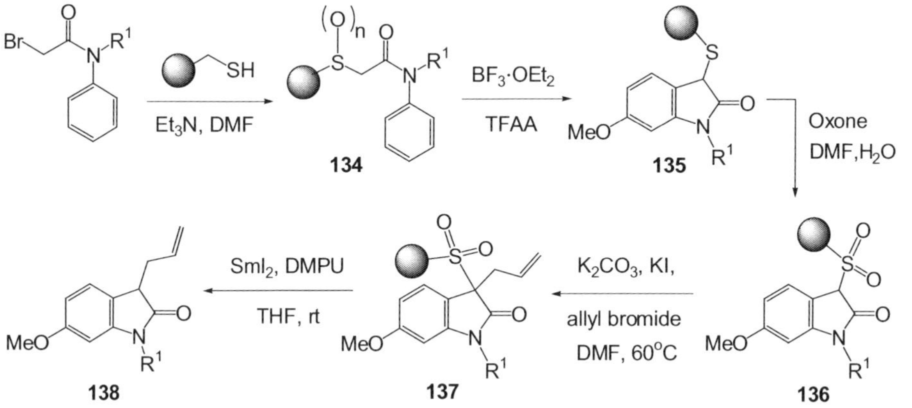

Scheme 12.30. Synthesis of oxindoles.

Synthesis of Butenolides/Butyroactones Using Mg/HgCl₂ for Sulfone Cleavage. Sheng et al. have reported the synthesis of butenolides and butyroactones using the sulfone-bound resin **140** (Scheme 12.31).[40] The procedure for the synthesis of butenolides/butyroactones involved the *S*-alkylation of lithiophenylsulfinate resin **79**, followed by sulfonyl anion alkylation with an epoxide to yield **45**. Subsequent acylation

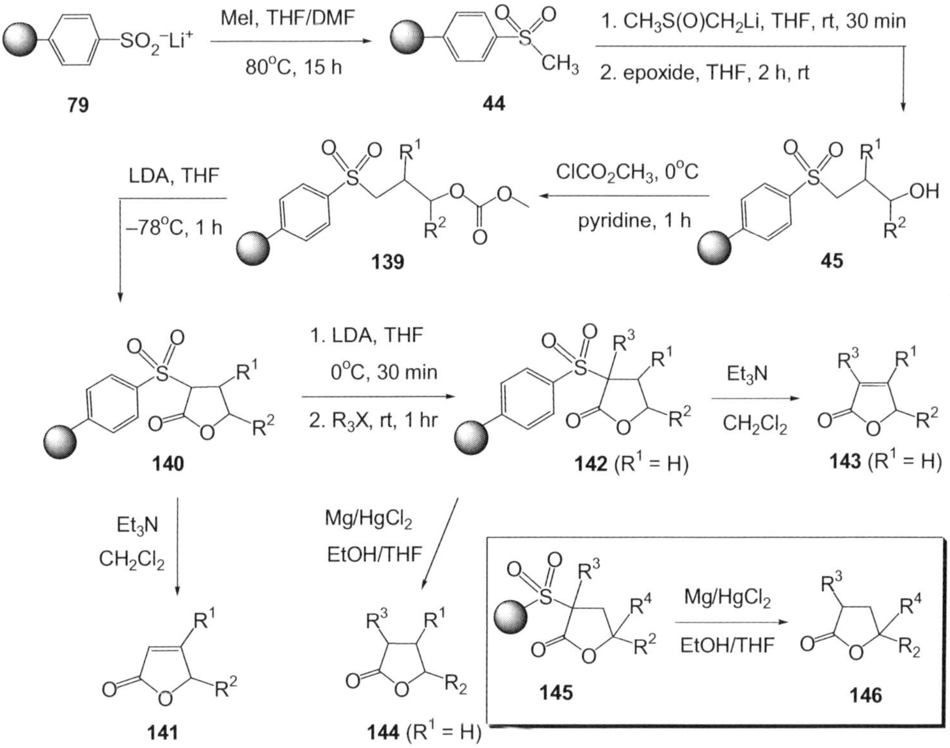

Scheme 12.31. Synthesis of butenolides and butyroactones.

of the γ-hydroxyl sulfone resin **45** afforded **139**. Intramolecular acylation–cyclization of the α-sulfonylcarbanion using LDA in THF at −78°C for 1 h provided the substituted dihyrdofuran-2(3*H*)-ones on solid support **140**, which upon treatment with excess triethylamine in dichloromethane yielded the corresponding butenolides **141**. Selected resin **140** ($R^1 = H$) could be further α-alkylated to give resin **142** and subsequent elimination under the same conditions described above gave the respective 3,5-disubstituted butenolide **143**. Resin **142** could also be cleaved using reducing agent $Mg/HgCl_2$ in EtOH/THF. To demonstrate the versatility of this method, substituted butyrolactones **146** were also obtained from the polymeric sulfone **145** using $Mg/HgCl_2$ in EtOH/THF.

Synthesis of Isoxazolocyclobutanones and Isoxazolinacyclobutenones Using Swern Oxidation for Sulfone Cleavage. Resin-bound lithium benzenesulfinate **79** has also been applied in the synthesis of isoxazolocyclobutanones and isoxazolinacyclobutenones (Scheme 12.32).[41] Resin **79** was first converted to the polymer-supported 3-vinylcyclobutanol

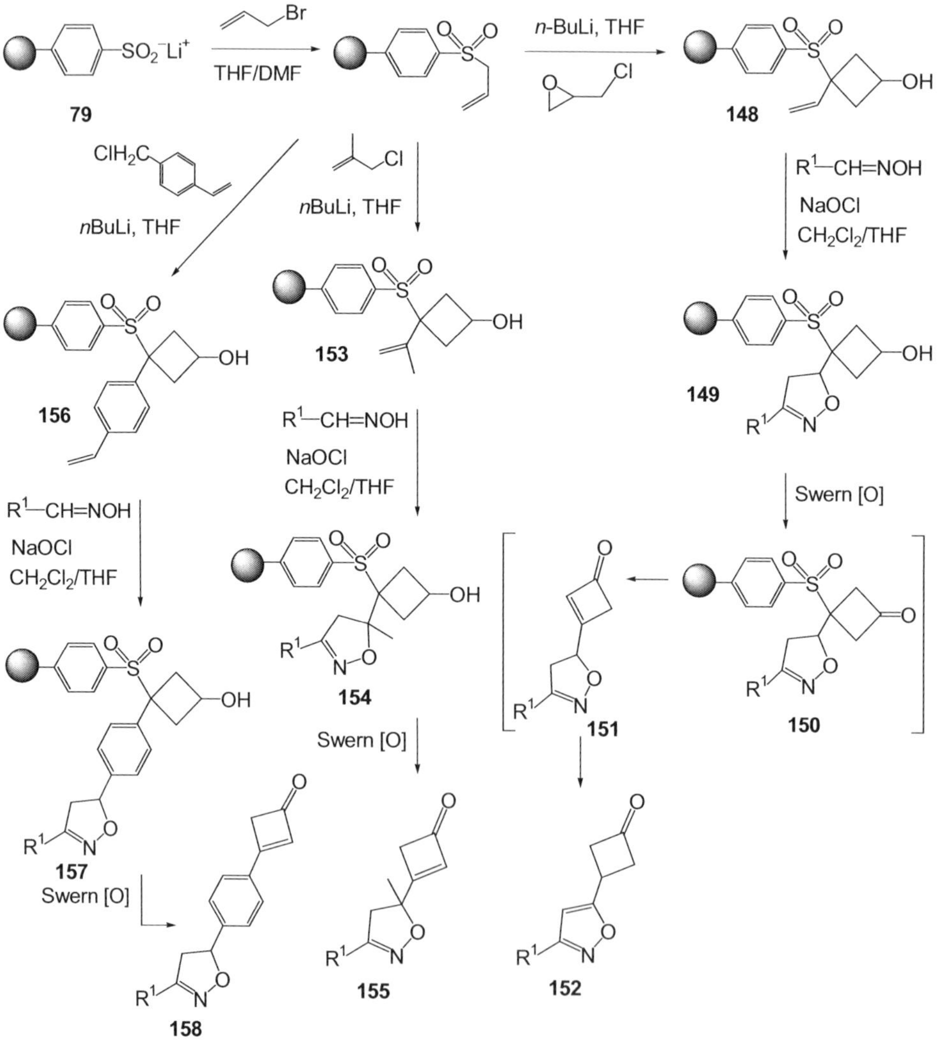

Scheme 12.32. Synthesis of isoxazolocyclobutanones and isoxazolinacyclobutenones.

148 via *S*-alkylation using allyl bromide and α,α-dialkylation with epichorohydrin. Treatment of **148** with substituted benzaldeyhde oxime in the presence of NaOCl in CH_2Cl_2/THF (the Huisgen method[42]) afforded the regioselective isoxazolino resin **149** via 1,3-dipolar cycloaddition. Using Swern oxidation, resin **149** was oxidized to **150**, accompanied by concomitant sulfinate elimination to give **151**. Owing to the high energies involved in placing the double bond in the cyclobutyl ring, **151** isomerized into isoxazolocyclobutanone **152**. To demonstrate the versatility of this chemistry, Kurth and coworkers[40] also introduced a "spacer" between the isoxazolino and cyclobutenone moieties using 3-chloro-2-methylpropene and 4-vinylbenzyl as alkylating agents. The *S*-alkylated resin **153** and **156** were then α,α-dialkylated with epichlorohydrin to yield resins **154** and **157** whose sulfone linker was then cleaved via Swern oxidation to afford **155** and **158**, respectively.

Synthesis of Pyrimidine Derivatives. In 2002, Lam and coworkers reported the use of polystyrene/1% divinylbenzene sodium sulfinate **43** as solid-phase support for the synthesis of pyrimidine derivatives (Scheme 12.33).[43] Resin **43** in NBu_4I/KI/DMF was allowed to react with 1,3-dichloropropane-2-one at room temperature to yield the polymer-supported α-haloketones **159**. Treatment of resin **159** with substituted aminopyridines in anhydrous DME yielded the imidazo[1,2-*a*]pyridine resin **160**. Alkylation of resin **160** occurs selectively with *cis*-epoxide and not the *trans* isomer to give resin **161**, which could be

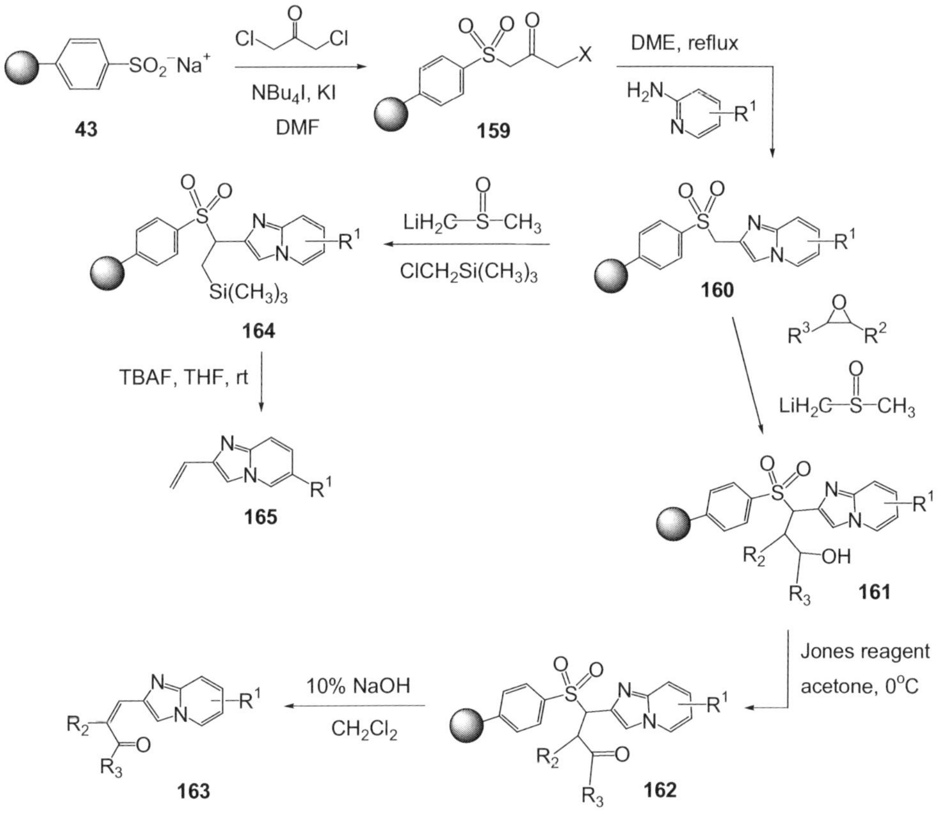

Scheme 12.33. Synthesis of pyrimidine derivatives.

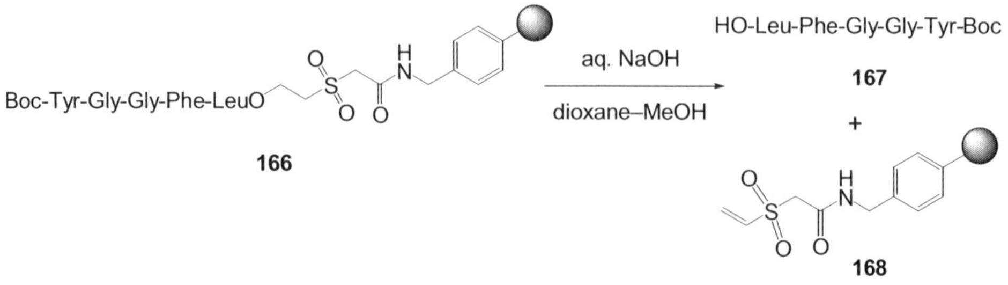

Scheme 12.34. Synthesis of peptide **167**.

oxidized via Jones oxidation to give resin **162**. Cleavage of resin **162** with 10% NaOH in DCM afforded imidazo[1,2-*a*]pyridine **163**. Alternatively, reaction of resin **160** with dimesyl anion followed by chloromethyltrimethylsilane in THF afforded the *β*-silyl sulfone resin **164**. Cleavage of sulfone linker using TBAF/THF generated imidazo[1,2-*a*]pyridines **165**. The use of sodium benzenesulfinate resin **43** in the reaction demonstrates the stability of the sulfone linker, which benefits the synthetic route by making isolation of the products easier and allows greater diversity to be introduced to the imidazo[1,2-*a*] pyridines template.

Other Cleavage Methods

Synthesis of 2-Substituted-4-piperidones Involving β-Elimination of Sulfone. *β*-Elimination of sulfone linkage was first described by Schwyzer et al. using a 2-(4-carboxyphenylsulfonyl)ethanol linker for the synthesis of peptides and oligonucleotides.[11] An application of *β*-elimination of sulfone linker for peptide synthesis was also demonstrated by Katti et al.[44] C-terminal peptide **167** was generated after cleavage from the sulfone linker **166** (Scheme 12.34).

β-Elimination of the sulfone linker has also been used in the formation of aryl sulfonamides, dehydroalanine, and heterocyclic 2-substituted-4-piperidones.[45,8c] Pendant phosphonium sulfone intermediate **169** was subjected to Wittig reagent to form sulfone-bound vinyl ketones **170** (Scheme 12.35). Benzylamine as nucleophile and base was then

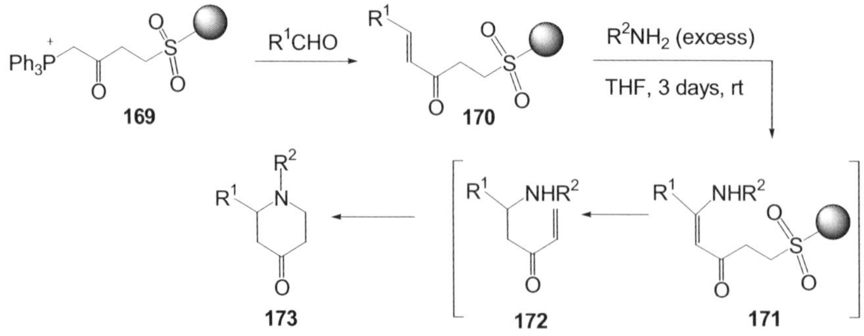

Scheme 12.35. Synthesis of 2-substituted-4-piperidones.

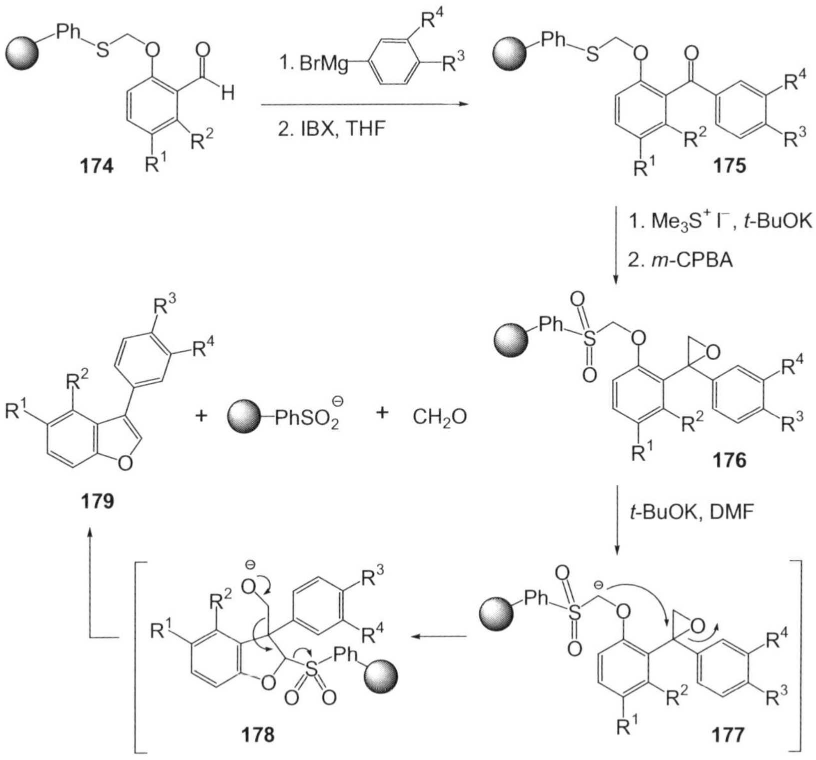

Scheme 12.36. Synthesis of 3-arylbenzofurans.

used to promote a Michael addition in resin **170**, which induces elimination of the sulfone group.[46] A second Michael addition on **172** yields 2-substituted-4-piperidones **173** in 50–60% yield.

Synthesis of 3-Arylbenzofurans via 1,4-Conjugate Elimination of Sulfone. Nicolaou et al. have reported a methodology to synthesize 3-arylbenzofurans by the cyclofragmentation–release pathway (Scheme 12.36).[47] 3-Arylbenzofuran **179** was obtained in a traceless manner when the alkoxide intermediate **178** obtained from a 5-*exo*-trig cyclization collapsed into **179** with the concomitant expulsion of formaldehyde and the phenylsulfinate resin. All products cleaved were extremely pure, but the main limitation for this strategy lies in the necessity for the generation of an epoxide **176** and both aryl groups are required for the regioselective epoxide opening.

12.2.2.3 Sulfone Linkers in Oligosaccharide Synthesis. The nucleophilic displacement of alkylsulfonated groups such as mesylates and tosylates are classical reactions in organic synthesis and this concept has been adapted by Roush and coworkers in their synthesis of oligosaccharides.[48a] Sulfonyl chloride resin **182** was prepared by treating Merrifield resin with the lithium anion of isopropyl methanesulfonate **180** to first obtain the sulfonic acid resin that was in turn treated with thionyl chloride to yield resin **182** (Scheme 12.37a). In the oligosaccharide synthesis, resin **182** was treated with glucose derivatives to yield monosaccharide resin-bound intermediate **183**. Subsequent reaction with iodo acetate **184** or galactosyl trichloroacetimidate **185**

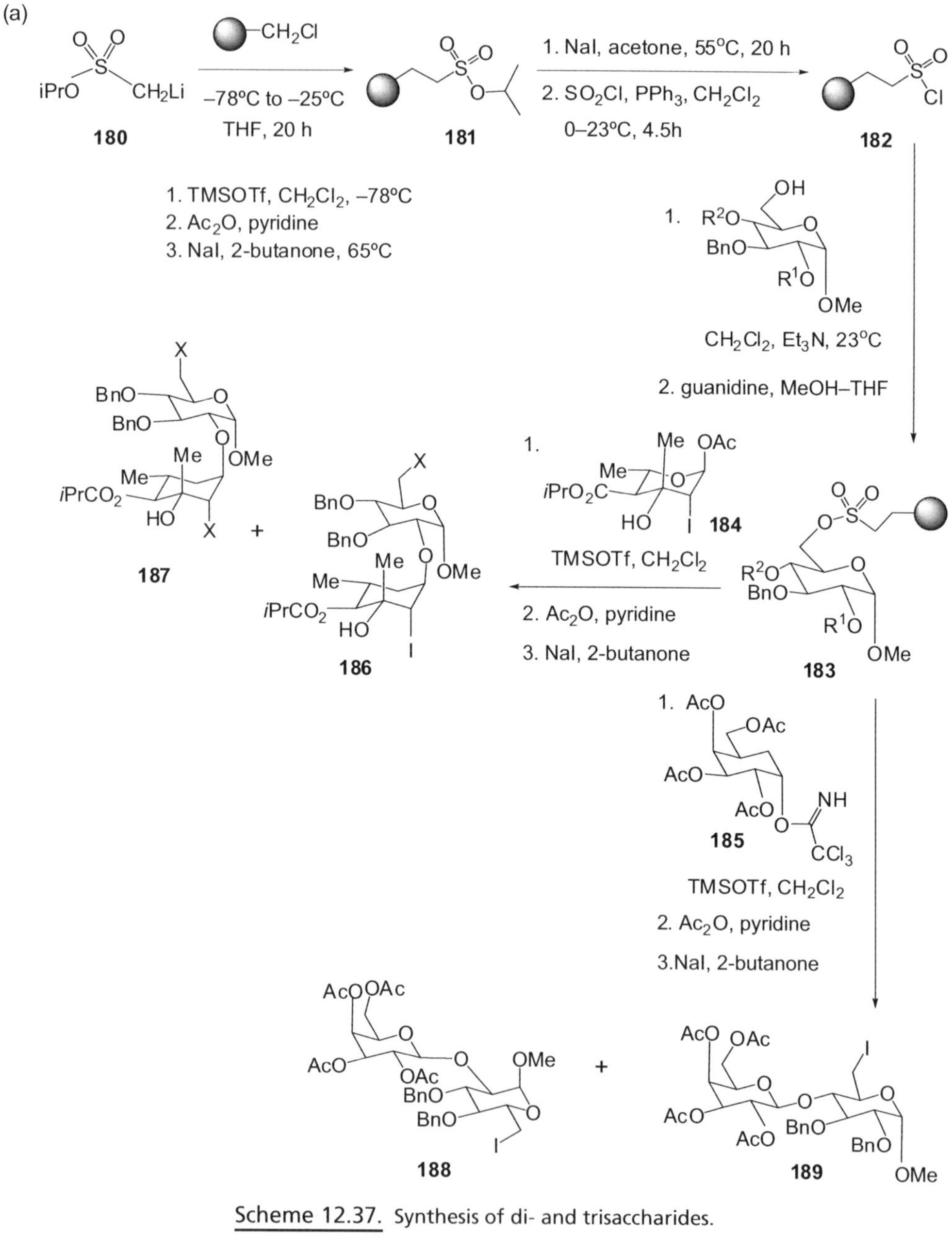

Scheme 12.37. Synthesis of di- and trisaccharides.

afforded the respective diastereometrically pure disaccharides **186**, **187**, **188**, and **189**. Displacement of disaccharides from the resin was achieved using iodide ion followed by reduction. Other nucleophiles such as AcO^- and N_3^- were also suitable for cleaving the oligosaccharides from the resin. As a more demanding test of this methodology, Roush and Sebesta have synthesized trisaccharide **188**, a functionalized precursor of a trisaccharide unit in olivomycin A (Scheme 12.37b).[48b]

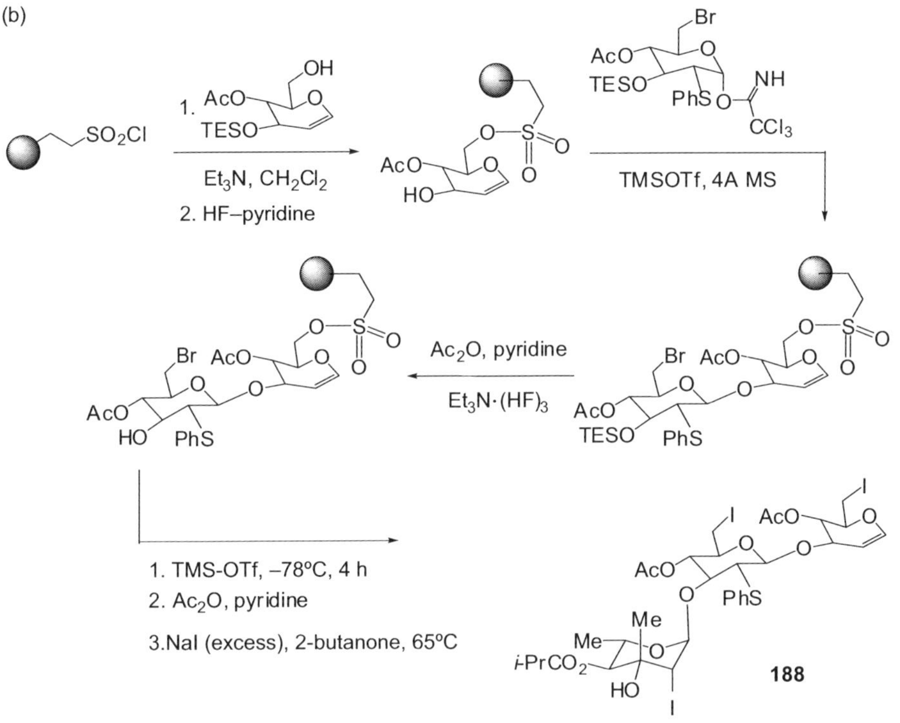

Scheme 12.37b (*continued*).

12.3 CONCLUSIONS

Besides heterocyclic synthesis, the sulfone linker unit has also been applied to the preparation of a variety of other substrates.[8c,14,15,44,49] The diverse cleavage strategies that can be applied to cleave substrates from sulfone linkers make them a versatile multifunctional linker. This together with the simple purification process, compared to solution-phase synthesis, adds to the potential of sulfone chemistry being applied to solid-phase synthesis.

REFERENCES

1. R. B. Merrifield, *J. Am. Chem. Soc.* **1963**, *85*, 2149.
2. (a) R. R. Schmidt, S. Jonke, K. Liu, *ACS Symp. Ser.* **2007**, *960*, 209; (b) M. Grathwohl, N. Drinnan, M. Broadhurst, M. L. West, W. Meutermans, *Methods Enzymol.* **2003**, *369*, 248; (c) W. Christ, L. Krock, O. J. Plante, B. Castagner, P. H. Seeberger, PCT Int. Appl. WO 2010011828, **2010**; (d) W.-C. Haase, P. H. Seeberger, in Seeberger, P. H. (Ed.), *Solid Support Oligosaccharide Synthesis and Combinatorial Carbohydrate Libraries*, Wiley, **2001**, pp. 1–14.
3. (a) F. Morvan J.-J. Vasseur, *Collect. Symp. Ser.* **2008**, *10*, 54; (b) E. Pedroso, N. Escaja, M. Frieden, A. Grandas, in Herdewijn, P. (Ed.), *Oligonucleotide Synthesis: Methods and Applications*, Humana Press Inc. , **2005**, pp. 101–126.
4. (a) P. Seneci, *Solid-Phase Synthesis and Combinatorial Technologies*, Wiley, **2002**, pp. 91–135; (b) J. Buchele, H. A. Mayer, *Chem. Commun.* **1999**, 2165; (c) N. Ljungdahl, K. Bromfield, N. Kann, *Top. Curr. Chem.* **2007**, *278*, 89.

5. P. J. H. Scott, in Scott, P. J. H. (Ed.), *Linker Strategies in Solid-Phase Organic Synthesis*, Wiley, **2009**, 591–656.
6. (a) S. Patai, Z. Rappoport, C. Stirling, *The Chemistry of Sulfones and Sulfoxides*, Wiley, Chichester, **1988**;(b) N. S. Simpkins, *Sulfone in Organic Synthesis*, Pergamon Press, Oxford, **1993**;.(c) S. Patai, Z. Rappoport, *The Synthesis of Sulfones, Sulfoxides and Cyclic Sulfides*, Wiley, Chichester, **1994**.
7. C. Najera, M. Yus, *Tetrahedron*, **1999**, *55*, 10547.
8. (a) T. Masquelin, D. Sprenger, R. Baer, F. Gerber, Y. Mercadal, *Helv. Chim. Acta.* **1998**, *81*, 646; (b) F. E. K. Kroll, R. Morphy, D. Rees, D. Gani, *Tetrahedron Lett.* **1997**, *38*, 8573; (c) M. Yamada, T. Miyajima, H. Horikawa, *Tetrahedron Lett.* **1998**, *39*, 289; (d) E. Flanigan, G. R. Marshal, *Tetrahedron Lett.* **1970**, *11*, 2403; (e) L. M. Gayo, M. J. Suto, *Tetrahedron Lett.* **1997**, *38*, 211.
9. D. L. Marshall, I. E. Liener, *J. Org. Chem.* **1970**, *35*, 867.
10. L. A. McAllister, S. Brand, de Gentile, R. D. J. Procter, *Chem. Commun.* **2003**, 2380.
11. R. Schwyzer, E. Felder, P. Failli, *Helv. Chim. Acta.* **1984**, *67*, 1316.
12. Q. S. Ren, W. Q. Huang, B. L. He, *Youji Huaxue* **1989**, *9*, 463.
13. A. J. Hagen, M. J. Farrall, J. M. J. Frechet, *Polym. Bull.* **1981**, *5*, 111.
14. T. M. Fyles, C. C. Leznoff, *Can. J. Chem.* **1976**, *54*, 935.
15. C. Halm, J. Evarts, M. J. Kurth, *Tetrahedron Lett.* **1997**, *38*, 7709.
16. P. Heinonen, H. Lönnberg, *Tetrahedron Lett.* **1997**, *38*, 8569.
17. D. Obrecht, C. Abrecht, A. Grieder, J. M. Villalgordo, *Helv. Chim. Acta.* **1997**, *80*, 65.
18. S. K. Srivastava, W. Haq, P. M. S. Chauhan, *Comb. Chem. High Throughput Screening* **1999**, *2*, 33.
19. (a) C. L. Gibson, S. La Rosa, C. J. Suckling, *Org. Biomol. Chem.* **2003**, *1*, 1909; (b) C. L. Gibson, S. La Rosa, C. J. Suckling, *Tetrahedron Lett.* **2003**, *44*, 1267.
20. (a) S. Ding, N. S. Gray, Q. Ding, X. Wu, P. G. Schultz, *J. Comb. Chem.* **2002**, *4*, 183; (b) V. Brun, M. Legraverend, D. S. Grierson, *Tetrahedron* **2002**, *58*, 7911.
21. (a) S. M. Khersansky, Y. T. Chang, *J. Comb. Chem.* **2004**, *6*, 474; (b) T. A. Neubert, Y. T. Chang, *J. Am. Chem. Soc.* **2003**, *125*, 11804.
22. T. Masquelin, D. Sprenger, R. Baer, F. Gerber, Y. Mercadal, *Helv. Chim. Acta.* **1998**, *81*, 646.
23. (a) Y. Chen, Y.L. Lam, Y. H. Lai, *Org. Lett.* **2003**, *5*, 1067; (b) K. H. Kong, Y. Chen, X. Ma, W. K. Chui, Y. L. Lam, *J. Comb. Chem.* **2004**, *6*, 928; (c) W. W. Li, Y. Chen, Y. L. Lam, *Tetrahedron Lett.* **2004**, *45*, 6545; (d) W. W. Li, Y. L. Lam, *J. Comb. Chem.* **2005**, *7*, 721; (e) W. W. Li, Y. L. Lam, *J. Comb. Chem.* **2005**, *7*, 644.
24. B. A. Kulkarni, A. Ganesan, *Tetrahedron Lett.* **1999**, *40*, 5633.
25. B. E. Hoogenboom, O. H. Oldenziel, A. M. van Leusen, *Org. Synth.* **1977**, *57*, 102.
26. E. A. Arvanitis, D. Craig, A. P. Timm, *ARKIVOC* **2002**, *9*, 19.
27. C. X. Li, W. Q. Huang, Y. Lu, B. L. He, *Chin. Chem. Lett.* **1991**, *2*, 773.
28. Y. Gao, Y. L. Lam, *Org. Lett.* **2006**, *8*, 3283.
29. M. S. Raghavendra, Y. L. Lam, *Tetrahedron Lett.* **2004**, *45*, 6129.
30. (a) S. H. Hwang M. J. Kurth *J. Org. Chem.* **2002**, *67*, 6564; (b) S. H. Hwang, M. J. Kurth, *Tetrahedron Lett.* **2002**, *43*, 53.
31. W. C. Cheng, M. M. Olmstead, M. J. Kurth, *J. Org. Chem.* **2001**, *66*, 5528.
32. P. Ten Holte, L. Thijs, B. Zwanenburg, *Tetrahedron Lett.* **1998**, *39*, 7407.
33. W. Sun, S. Cheng, W. Huang, *Tetrahedron Lett.* **2001**, *42*, 1973.
34. Y. Chen, Y. L. Lam, S. Y. Lee, *Chem. Lett.* **2001**, 274.
35. D. R. Dragoli, M. T. Burdett, J. A. Ellman, *J. Am. Chem. Soc.* **2001**, *123*, 10127.
36. Y. Hu, S. Baudart, J. A. Porco, Jr., *J. Org. Chem.* **1999**, *64*, 1049.

37. (a) T. H. Wu, S. Ding, N. S. Gray, P. G. Schultz, *Org. Lett.* **2001**, *31*, 3827; (b) T. Okauchi, M. Itonaga, T. Minami, T. Owa, K. Kitoh, H. Yoshino, *Org. Lett.* **2000**, *2*, 1485.
38. (a) H. C. Zhang, H. Ye, A. F. Meretto, K. K. Brumfield, B. E. Maryanoff, *Org. Lett.* **2000**, *2*, 89; (b) H. C. Zhang, H. Ye, K. B. White, B. E. Maryanoff, *Tetrahedron Lett.* **2001**, *42*, 4751.
39. D. Tumelty, K. Cao, C. P. Holmes, *Org. Lett.* **2001**, *3*, 83.
40. (a) S. R. Sheng, L. Xu, X. L. Zhang, *J. Comb. Chem.* **2006**, *8*, 805; (b) S. R. Sheng, P. G. Huang, W. Zhou, H. R. Luo, S. Y. Lin, X. L. Liu, *Synlett* **2004**, *14*, 2603.
41. W. C. Cheng, M. Wong, M. M. Olmstead, M. J. Kurth, *Org. Lett.* **2002**, *4*, 741.
42. R. Huisgen, *Angew Chem., Int. Ed. Engl.* **1963**, *2*, 565.
43. Y. Chen, Y. L. Lam, Y. H. Lai, *Org. Lett.* **2002**, *4*, 3935.
44. S. B. Katti, P. K. Misra, W. Haq, K. B. Mathur, *J. Chem. Soc., Chem. Commun.* **1992**, 843.
45. C. G. Echeverria, *Tetrahedron Lett.* **1997**, *38*, 8933.
46. A. Bacro, S. Benetti, C. De Risi, P. Marchetti, G. P. Pollini, V. Zanirato, *Tetrahedron Lett.* **1998**, *39*, 7591.
47. K. C. Nicolaou, S. A. Snyder, A. Bigot, J. A. Pfefferkorn, *Angew Chem., Intl. Ed.* **2000**, *39*, 1093.
48. (a) J. A. Hunt, W. R. Roush, *J. Am. Chem. Soc.* **1996**, *118, 9998;* (b) D. P. Sebesta, W. R. Roush, *J. Org. Chem.* **1992**, *57*, 4799.
49. (a) X.-Y. Zhao, K. W. Jung, K. D. Janda, *Tetrahedron Lett.* **1997**, *38, 977;* (b) X.-Y. Zhao, W. A. Metz, F. Sieber, K. D. Janda, *Tetrahedron Lett.* **1998**, *39, 6433;* (c) X.-Y. Zhao, K. D. Janda, *Tetrahedron Lett.* **1997**, *38, 5437;* (d) X.-Y. Zhao, K. D. Janda, *Bioorg. Med. Chem. Lett.* **1998**, *8, 2439;* (e) W. C. Cheng, C. Halm, J. B. Evarts, M. M. Olmstead, M. J. Kurth, *J. Org. Chem.* **1999**, *64, 8557;* (f) W. Huang, S. Cheng, W. Sun, *Tetrahedron Lett.* **2001**, *42, 1973;* (g) R. Schwyzer, E. Felder, P. Failli, *Helv. Chim. Acta.* **1984**, *67*, 1316.

13

SOLID-PHASE ORGANIC RADIOSYNTHESIS

Raphaël Hoareau and Peter J. H. Scott

13.1 INTRODUCTION

Positron emission tomography (PET) is a noninvasive technique for functional molecular imaging in living human subjects, using bioactive molecules tagged with positron-emitting radionuclides (radiopharmaceuticals), that was recently reviewed by Ametamey et al.[1] PET imaging has undergone consistent development following the initial clinical success of $[^{18}F]$ fluorodeoxyglucose ($[^{18}F]$FDG), a radiolabeled analogue of glucose approved for human use by the U.S. Food and Drug Administration (FDA). As demand for radiopharmaceuticals continues to grow, many standard organic chemistry reactions have been adapted for radiochemical syntheses,[2–5] which allow preparation of different types of radiopharmaceuticals that find application in areas such as oncology,[6–12] cardiology,[13–17] and neurology.[18–29]

Owing to the short half-lives of the radioactive isotopes involved (e.g., $t_{1/2}$ of carbon-11 = 20 min, $t_{1/2}$ of fluorine-18 = 110 min), radiopharmaceuticals for PET (and related SPECT) imaging are typically produced on-site daily or at facilities in close proximity to the PET imaging center. Radiopharmaceutical production typically requires trained radiochemists, with specific and expensive facilities. Moreover, the short-lived radionuclides involved means that radiochemical reactions must be fast, efficient, and amenable to automation, and new techniques that facilitate radiopharmaceutical production and the ability to meet clinical demand for PET tracers are in demand. Solid-phase synthesis is an attractive synthetic technique to the radiochemist as it bears many of these desirable characteristics. Despite this, solid-phase organic radiosynthesis (SPOR) is a relatively new

Solid-Phase Organic Synthesis: Concepts, Strategies, and Applications, First Edition.
Edited by Patrick H. Toy and Yulin Lam.

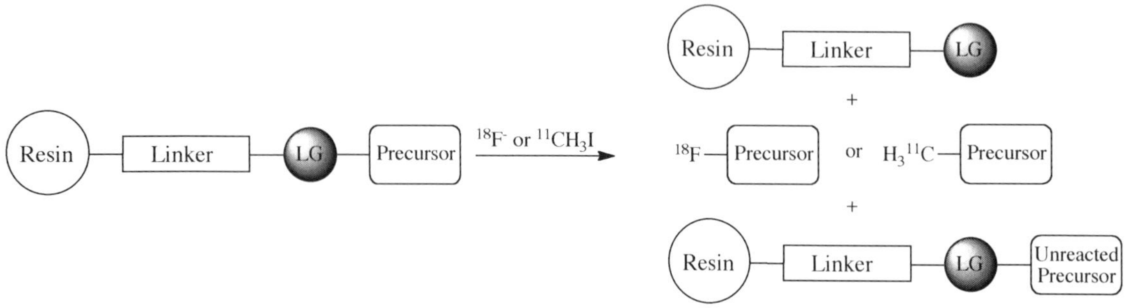

Scheme 13.1. General approach of SPOR.

area of research and only a few peer-reviewed articles have been published to date. In fact, the majority of the SPOR reports can be found in the patent literature, albeit frequently with limited experimental details. Nevertheless, this chapter introduces the main strategies for SPOR reported to date and reviews the current state of the art.

The general concept of solid-phase radiosynthesis, described in Scheme 13.1, is to attach the precursor onto the resin by a reactive substituent (LG, leaving group). There are plenty of methods for conjugating the radiopharmaceutical precursor moiety to a polymer resin. According to Scheme 13.1, the precursor can be attached directly to the commercially available polystyrene resin, with or without the use of a linker unit to separate the reactive center from the polymer network. Linker strategies for solid-phase radiochemistry have been reviewed,[30] and literature suggests that the linker strategy is more successful for producing radiopharmaceuticals. SPOR with fluorine-18 usually needs no additional cleavage as, typically, the fluorination releases the labeled product in solution. In contrast, a resin-dependent cleavage is needed in SPOR with, for example, carbon-11. SPOR with fluorine-18 frequently demonstrates proof-of-concept through the radiosynthesis of [^{18}F]FDG, even if its SPOR is still a challenge compared to the Hamacher method.[31] The main idea behind SPOR is to provide alternative radiopharmaceutical production techniques to the molecular imaging community that can simplify and facilitate production, while also contributing to cost reduction measures in healthcare in, for example, the United States. Moreover, the option of transitioning SPOR techniques into single use kits for automated radiopharmaceutical synthesis modules could play a significant role in development of current good manufacturing practice (cGMP) compliant radiopharmaceutical syntheses recently mandated by the U.S. FDA. The following sections present successful strategies for SPOR, classified by the isotope used for radiolabeling, that have been reported to date.

13.2 SOLID-PHASE ORGANIC RADIOSYNTHESIS WITH FLUORINE-18

13.2.1 Radiolabeled Peptides with Fluorine-18

Peptide radiolabeling has been investigated by Sutcliffe et al. using conjugation of a prosthetic group, such as 4-[^{18}F]fluorobenzoic acid (FBA, **3**) and 2-[^{18}F]fluoropropionic acid (FPA, **5**), with a resin-supported peptide as shown in Scheme 13.2.[32–35] Originally, the strategy was straightforward with prosthetic group **3**.[32] The two-step conjugation produced

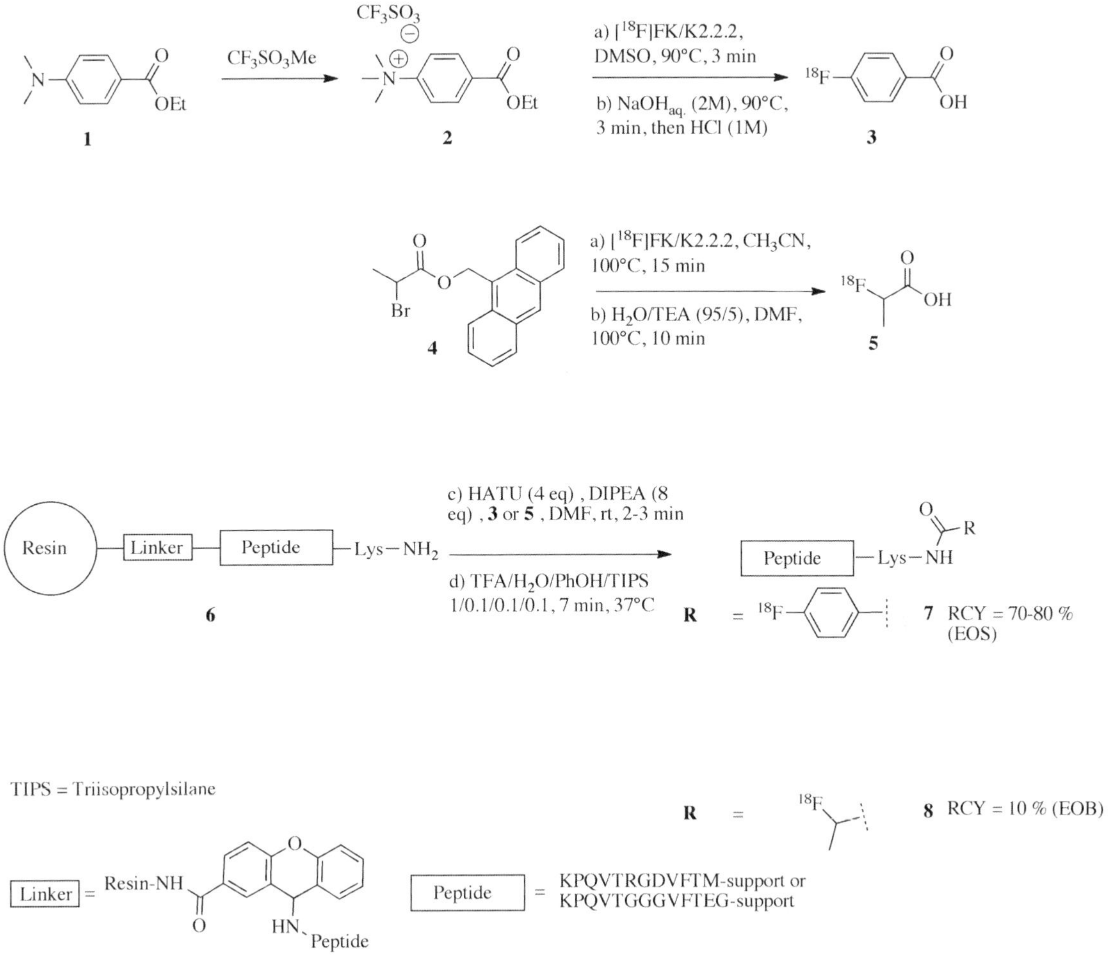

Scheme 13.2. Radiolabeling peptides according to Sutcliffe et al.

radiolabeled peptide **7** from 70% to 80% radiochemical yield (RCY) at the end of synthesis (EOS). Unfortunately, this good result was not reproducible with FPA **5**, **3**, or other radiolabeled peptides,[33,34] and this multistep peptide radiolabeling usually occurs with around 10% RCY from end of bombardment (EOB).

13.2.2 Solid-Phase Organic Radiosynthesis of [^{18}F]FDG

The synthesis reported by Hamacher et al. has been routinely used to prepare [^{18}F]FDG in PET production centers around the world. The synthesis requires no HPLC purification, and the described RCY was excellent (>90%, EOB).[31] As noted above, [^{18}F]FDG is the most widely used fluorine-18-labeled radiopharmaceutical to date, and so the synthesis has also been adapted for solid phase. SPOR strategies for preparation of [^{18}F]FDG, shown in Scheme 13.3, have been reported by Wadsworth et al.[36,37] and Brady et al.[38–41] and illustrate proof-of-concept. Notably, these authors used a concept based on perfluorosulfonate linker units to attach the precursor to the resin. The cleavage step used standard radiofluorination

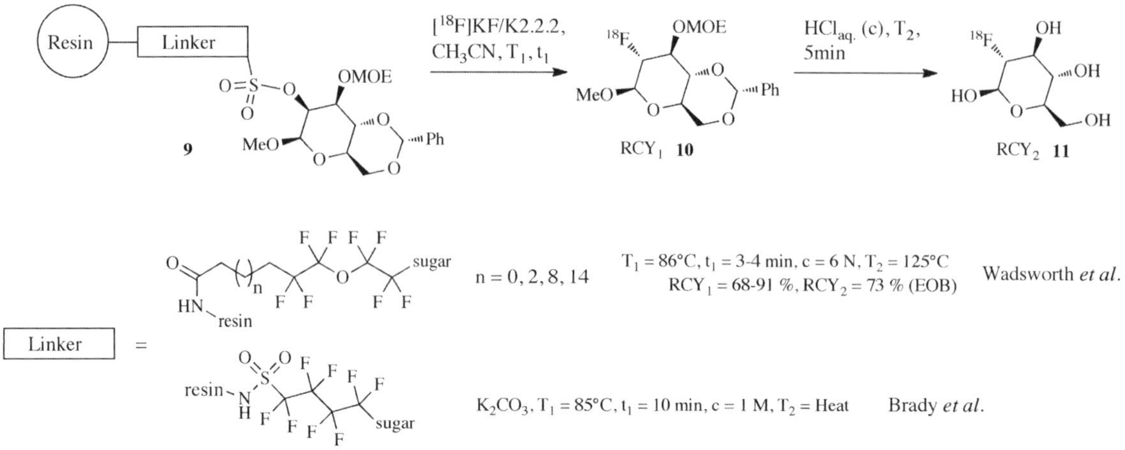

Scheme 13.3. Solid-phase organic radiosynthesis of [^{18}F]FDG according to Wadsworth et al. and Brady et al.

conditions employing potassium carbonate and Kryptofix-2.2.2 (K-2.2.2) as a phase transfer catalyst and furnished the labeled fully protected sugar **10**. Subsequent acidic deprotection provided [^{18}F]FDG (**11**) in 73% RCY (EOB) according to Wadsworth's strategy. Brady et al. reported no yield.

13.2.3 Fluorine-18 Displacement of Supported Aryliodonium

Aryliodoniums are alternative precursors to fluorinate inactivated aryl groups with electron-donating or no substituent.[42,43] Brady et al. patented the concept of aryliodonium-supported resins for use in fluorine-18 solid-phase radiochemistry,[38–41] and Scheme 13.4 illustrates this proof-of-concept with the SPOR of uracil **15** with fluorine-18.

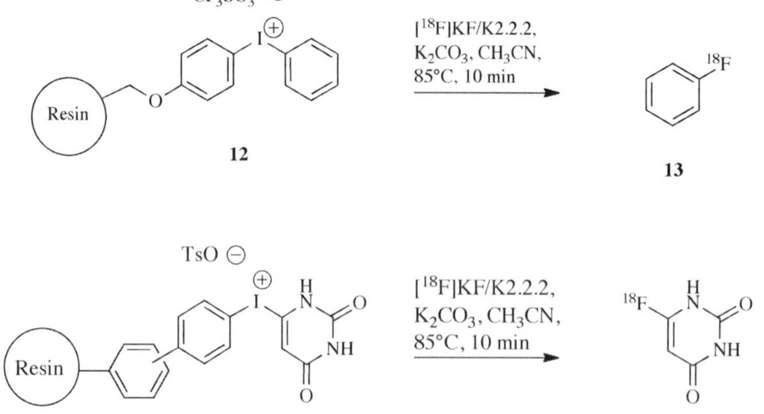

Scheme 13.4. Nucleophilic aromatic substitution of supported aryliodonium salts with fluorine-18 according to Brady et al.

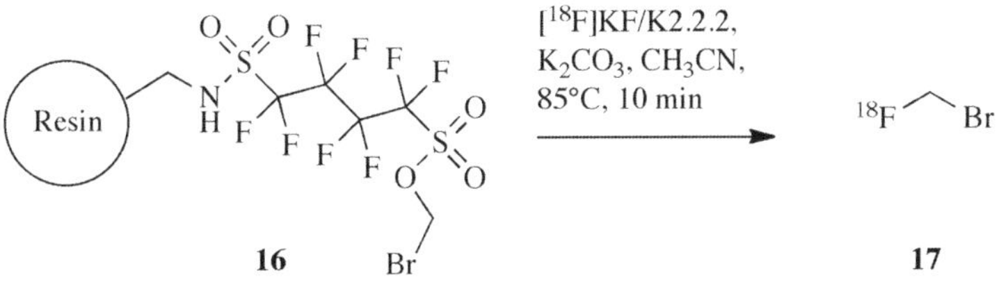

16 17

Scheme 13.5. Solid-phase organic radiosynthesis of $^{18}FCH_2Br$ according to Brady et al.

13.2.4 Solid-Phase Organic Radiosynthesis of $^{18}FCH_2Br$

Brady et al. also patented a SPOR method for preparing the volatile [^{18}F]fluorobromomethane **17**. This strategy also employs a perfluorosulfonyl linker unit and is illustrated in Scheme 13.5.[38–41,44] The radiosynthetic cleavage occurred under classical fluorination conditions, although radiochemical yields were not reported in the patent.

13.2.5 Solid-Phase Organic Radiosynthesis of [^{18}F]FluoroDOPA

The SPOR of [^{18}F]fluoroDOPA **20**, described in Scheme 13.6, used the same radiochemistry as employed in the corresponding solution-phase radiosynthesis. [^{18}F]FluoroDOPA **20** is synthesized by electrophilic fluorine. The fully protected organometallic precursor is attached to the support *via* a triorganotin moiety in which one of the methyl substituents is replaced by a propyl linker. The cleavage step occurs upon fluorine incorporation, and subsequent acidic treatment furnishes [^{18}F]fluoroDOPA **20**.

13.2.6 Solid-Phase Organic Radiosynthesis of β-Amyloid PET Tracers

Brady et al. patented the SPOR of two putative Alzheimer's disease PET tracers.[38–41] Scheme 13.7 illustrates the radiosynthesis, by nucleophilic cleavage with fluorine-18, of two fluorous analogues of β-amyloid PET tracers such as PIB (Pittsburgh compound B),[45,46] and DDNP (1,1-dicyano-2-[6-(dimethylamino)-2-naphthalenyl]propene).[47,48] Precursors are attached to the resin by their sulfonyl moiety, bound to an aryl or alkyl group that could

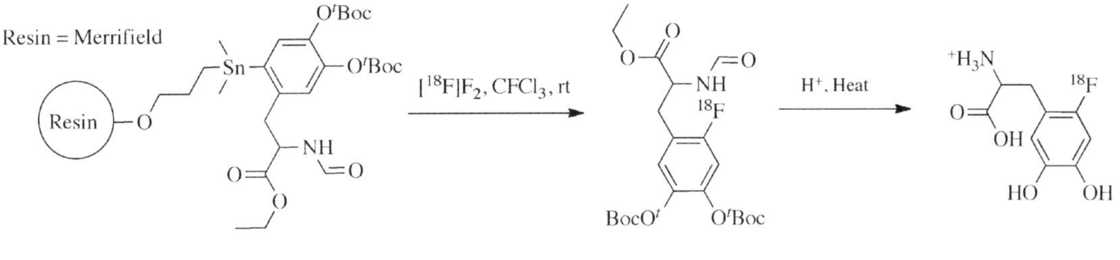

18 19 20

Scheme 13.6. Solid-phase organic radiosynthesis of [^{18}F]fluoroDOPA according to Wadsworth et al.

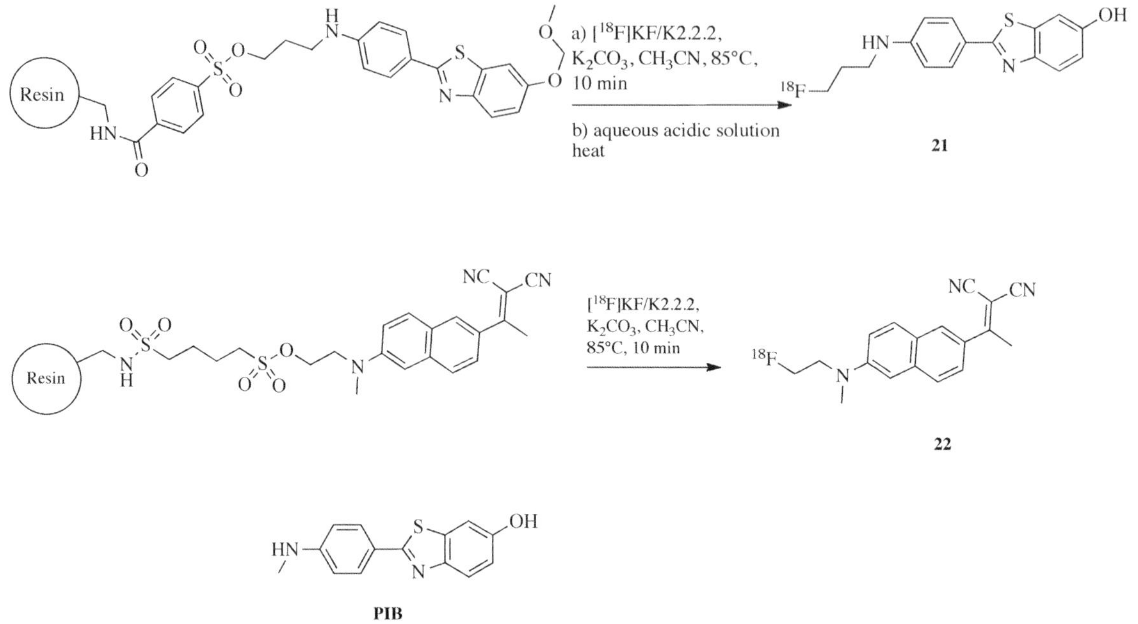

Scheme 13.7. Solid-phase organic radiosynthesis of [^{18}F]-labeled analogues of PIB (**21**) and DDNP (**22**) according to Brady et al.

be considered as linkers improving the resin reactivity. The authors did not report yields for these radiochemical syntheses.

13.2.7 Solid-Phase Organic Radiosynthesis of Oncological PET Tracers

The SPOR of two fluorous analogues of a tumor PET tracer, aminocyclobutane carboxylic acid (ACBC, **24** and **26**) is described in Scheme 13.8.[49] The FACBC ([^{18}F]fluoro-aminocyclobutane carboxylic acid) precursor was fully protected and attached to the resin by a perfluoroalkyl ether linker. FACBC **24** was obtained in two steps: cleavage by a fluorination step, followed by deprotection under acidic conditions. The corresponding methylene-FACBC analogue (**25**) was obtained by the same strategy. Notably, no linker was

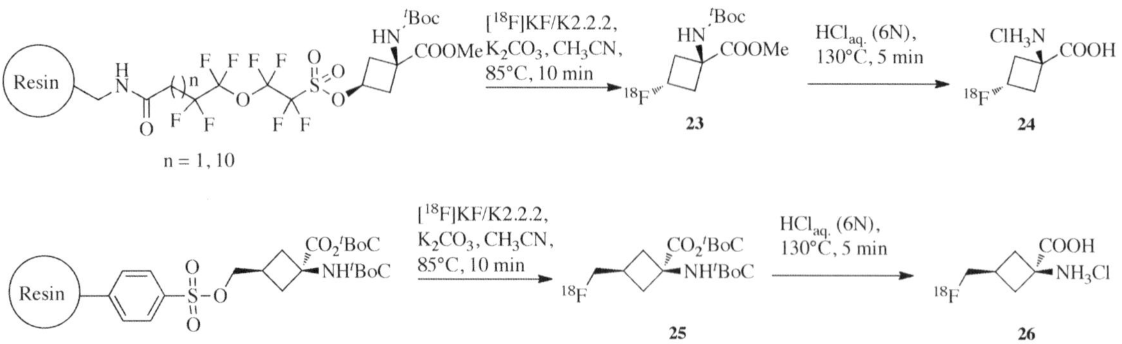

Scheme 13.8. Solid-phase organic radiosynthesis of FACBC derivatives according to Brown et al.

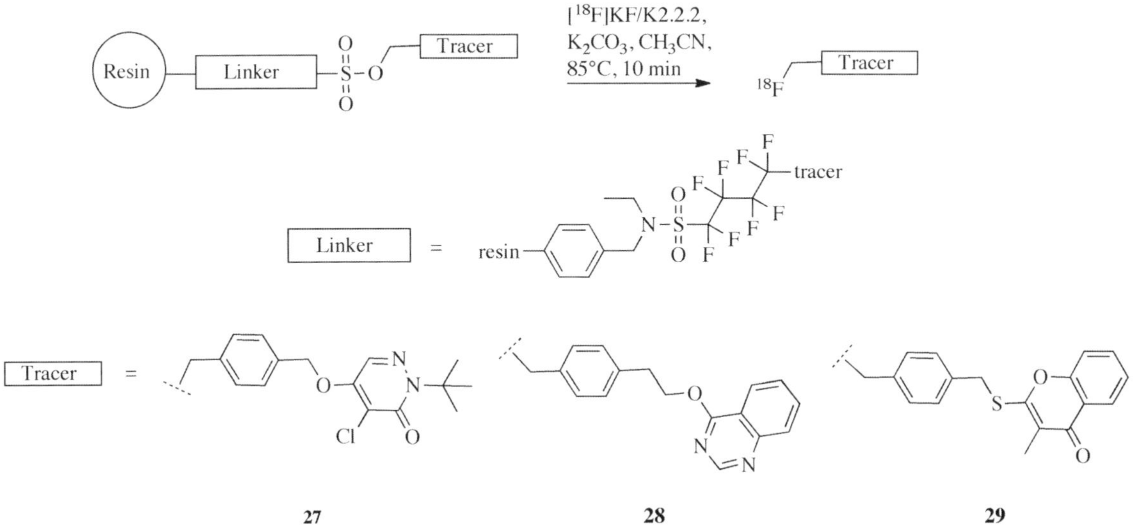

Scheme 13.9. Solid-phase organic radiosynthesis of NADH:ubiquinone oxidoreductase inhibitors as PET tracers for brain tumors according to Cheesman and coworkers.

employed to attach the precursor of **25**, suggesting that the methylene is a sufficient spacer for the fluorination. The authors did not report any yields for these syntheses.

Scheme 13.9 describes the SPOR of three brain tumor PET tracers known as NADH: ubiquinone oxidoreductase inhibitors. As in the case of previous examples, a perfluoroalkylsulfonyl linker was used to attach the precursors to the resin and fluorination, with concomitant cleavage, could be achieved using standard radiofluorination conditions.[50]

13.3 SOLID-PHASE ORGANIC RADIOSYNTHESIS WITH CARBON-11

The literature contains only a few examples of SPOR with carbon-11. For example, Maclean et al. published one elegant method (Scheme 13.10) to obtain a library of [^{11}C]-labeled tracers by reacting a base-cleavable-supported arylpiperazine resin with [^{11}C] methyl iodide to form the supported methylpiperazinium.[51,52] This quaternary ammonium is an excellent leaving group and an elimination reaction under basic conditions furnished the [^{11}C]methylarylpiperazine. The RCY (EOB) were relatively low (5–10%) but the radiochemical purities (RCP) were greater than 95%, making them suitable for direct *in vivo* injection after formulation. Then, tracer properties can be screened and evaluated *in vivo*. Unfortunately, this approach is limited to methylpiperazine-containing compounds, or all other structures containing a reactive moiety toward methyl iodide under neutral conditions.

Robins and Turton patented a SPOR method to produce [^{11}C]carbonyl-containing compounds.[53] Scheme 13.11 describes the chloride exchange between the acyl chloride-supported resin and the carboxylate salt obtained by quenching [^{11}C]carbon dioxide with a Grignard reagent. Any [^{11}C]acyl chloride remaining in solution was quenched with an alcohol or an amine to furnish, respectively, the ester or amide [^{11}C]-labeled compound. The later could be reduced to the corresponding [^{11}C]-labeled primary amine as desired. This method has the advantage of not employing complex technology, using high pressure

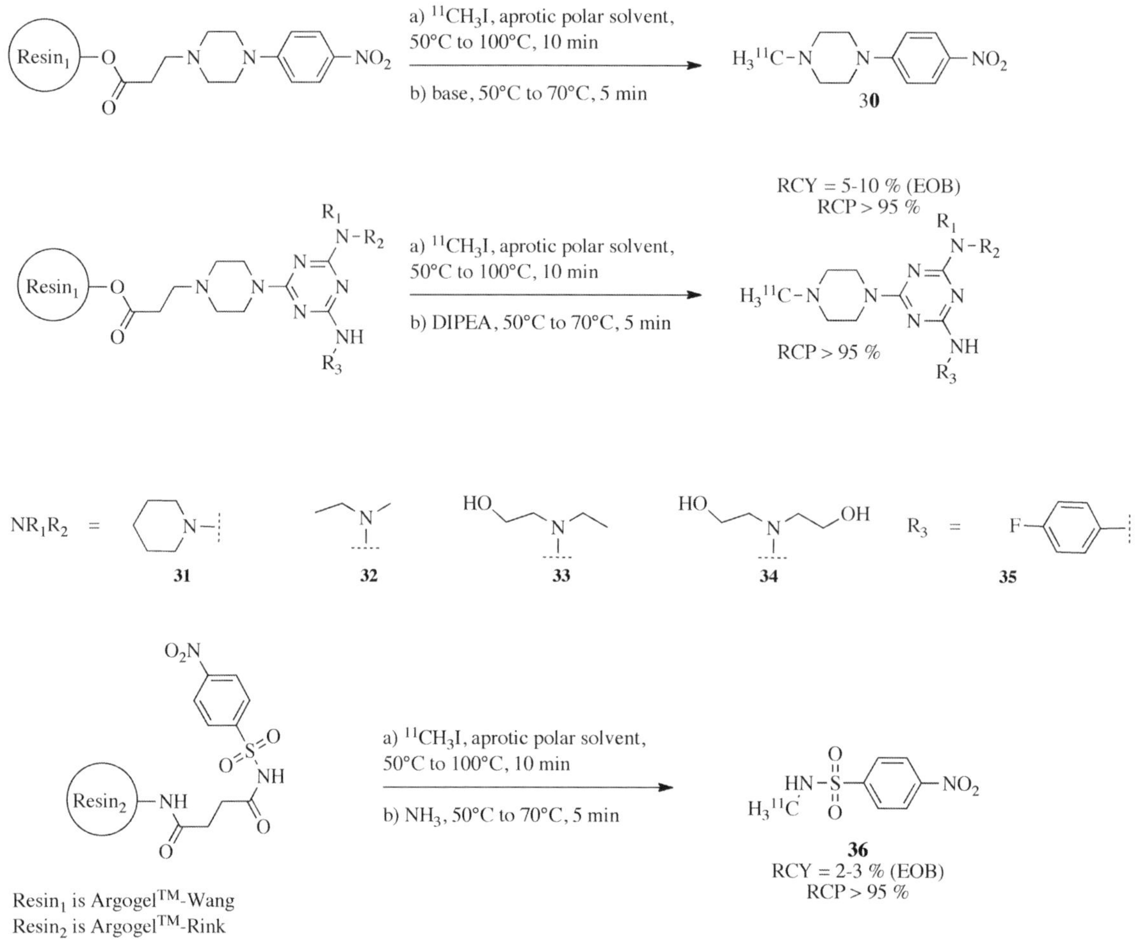

Scheme 13.10. Solid-phase organic radiosynthesis of a carbon-11-based PET tracers library.

and palladium catalysts, to obtain [^{11}C]-labeled carbonyl compounds and their derivatives.[54]

13.4 SOLID-PHASE ORGANIC RADIOSYNTHESIS WITH OTHER RADIOISOTOPES

13.4.1 Solid-Phase Purification of Copper-64 Metalloradiopharmaceuticals

Betts et al. developed a new methodology to purify copper-64 radiopharmaceuticals.[55] Scheme 13.12 illustrates the purification technique based on a commercial DMAP-supported resin. In the approach A, the resin was able to selectively purify a pre-formed solution mixture of copper-64 and zinc-containing metalloradiopharmaceuticals. By the Jahn–Teller effect, the DMAP ligand selectively bound to the zinc-containing complex and the [^{64}Cu]-labeled metalloradiopharmaceuticals remained in solution after filtration of

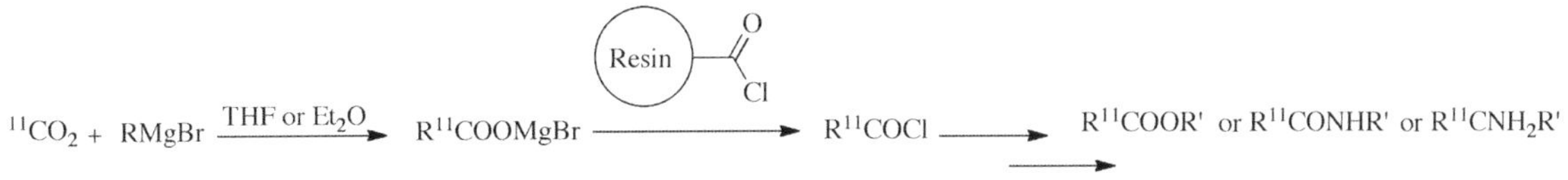

Scheme 13.11. Solid-phase organic radiosynthesis of carbon-11-labeled acyl chlorides according to Robins and Turton.

the resin. The alternative approach B showed that a copper-64 complex could be obtained by transmetallation from a precomplexed zinc radiopharmaceutical trapped on the DMAP-supported resin. After this metal exchange, the ^{64}Cu complex was released from the resin by the Jahn–Teller effect. This approach is interesting when the synthesis of the metalloradiopharmaceutical is slow. The idea is to precomplex any chelate with Zn^{II}, attach this complex to the resin, and then proceed to the fast copper-64 exchange. The purified copper-64 metalloradiopharmaceutical remains in solution after filtration of

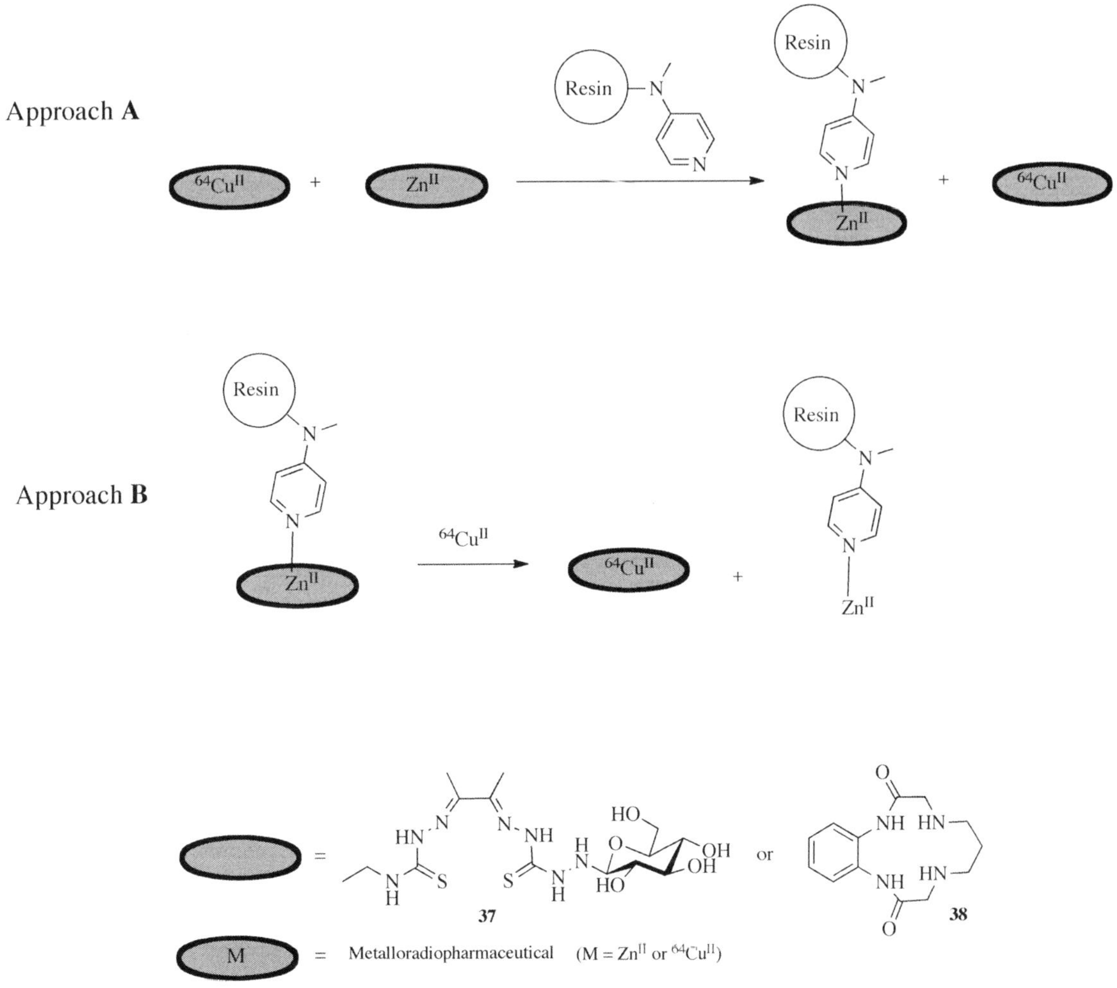

Scheme 13.12. Solid-phase purification of copper-64 complexes according to Betts et al.

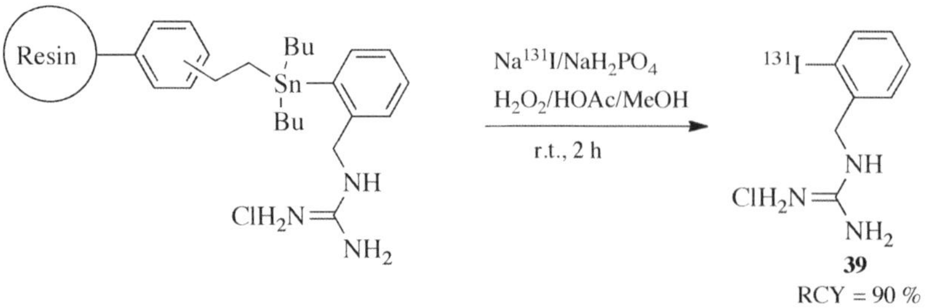

Scheme 13.13. SPOR of [^{131}I]MIBG according to Hunter and Zhu.

the resin. The authors pointed out the high radiochemical purity (>97%), while yields were not reported.

13.4.2 Solid-Phase Radiosynthesis of [^{131}I]MIBG

Finally, Hunter and Zhu described the iodine-131 *meta*-iodobenzylguanidine ([^{131}I]MIBG) SPOR illustrated in Scheme 13.13.[56] The authors investigated the electrophilic aromatic substitution of a polymer-supported precursor linked by the triorganotin moiety to the resin. The same strategy was used to obtain [^{18}F]DOPA by SPOR, as previously described (Scheme 13.6). The electrophilic iodide-131 was generated *in situ* by oxidation of Na^{131}I with H_2O_2/AcOH in methanol. Subsequently, the electrophilic aromatic substitution occurred at the C–Sn bond, with concomitant cleavage of the radiolabeled product [^{131}I]MIBG (**39**). The excellent radiochemical yield (90%), the absence of released tin (<10 ppb), and the claimed high chemical purity make this method adaptable for kit production of [^{131}I]MIBG. This successful iodide-131 SPOR should stimulate development of the corresponding electrophilic fluorine-18 SPOR. One possible way of success may be to obtain softer fluoronium than [^{18}F]F_2.

13.5 CONCLUSIONS

SPOR is a new research area, oriented to easily produce PET tracers. Patents constitute most of the literature and while experimental details are often sparse, they do contain valuable proof-of-concept studies involving isotope incorporation as the cleavage step. Such cleavage reactions typically require a linker between the resin and the precursor, and, to date, perfluoroalkylsulfonyl linkers and their derivatives seem promising for SPOR. Other notable examples are the labeling of peptide by SPOR and the use of DMAP-supported resin to purify metalloradiopharmaceuticals. Such examples suggest that SPOR, like solid-phase organic synthesis (SPOS) before it, could play an important role in radiopharmaceutical production and medicinal chemistry. As this research area develops further, there is also scope for incorporation of SPOR concepts into single use kits suitable for production of radiopharmaceuticals in compliance with cGMP.

REFERENCES

1. S. M. Ametamey, M. Honer, P. A. Schubiger, *Chem. Rev.* **2008**, *108*, 1501.
2. R. Schirrmacher, C. Wängler, E. Schirrmacher, *Mini Rev. Org. Chem.* **2007**, *4*, 317.

3. P. W. Miller, N. J. Long, R. Vilar, A. D. Gee, *Angew. Chem., Int. Ed.* **2008**, *47*, 8998.
4. L. Cai, S. Lu, V. Pike, *Eur. J. Org. Chem.* **2008**, 2843.
5. F. Wuest, M. Berndt, T. Kniess, *Ernst Schering Res. Found. Workshop* **2007**, *62*, 183.
6. J. Mercer, *J. Pharm. Pharm. Sci.* **2007**, *10*, 180.
7. N. Orichi, T. Higuchi, T. Ishikita, M. Miyakubo, H. Hanaoka, Y. Iida, K. Endo, *Cancer Sci.* **2006**, *97*, 1291.
8. L. Didier, *J. Fluorine Chem.* **2006**, *127*, 1488.
9. H. Fukuda, *Int. Congr. Ser. 1264* **2004**, 158.
10. A. M. Scott, in Bailey, D. L., Townsend, D. W., Valk, P. E., Maisey, M. N. (Eds.), *Positron Emission Tomography: Basic Sciences*, Springer, 2005, 311.
11. K. Kubota, *Ann. Nucl. Med.* **2001**, *15*, 471.
12. S. Gambhir, *Nat. Rev. Cancer* **2002**, *2*, 683.
13. M. Schwaiger, S. Ziegler, S. G. Nekolla, *J. Nucl. Med.* **2005**, *46*, 1664.
14. M. Maisey, *Nucl. Med. Commun.* **2000**, *21*, 234.
15. T. Amol, M. Ayse, A. Abass, A. Luis, *Radiol. Clin. North Am.* **2005**, *43*, 107.
16. L. Giovanni, *Eur. J. Nucl. Med. Mol. Imaging* **2006**, *33*, 621.
17. J. Knuuit, F. Bengal, *Heart* **2008**, *94*, 360.
18. R. Hilker, A. V. Thomas, J. C. Klein, S. Weisenbach, E. Kalbe, L. Burghaus, A. H. Jacobs, K. Herholz, W. D. Heiss, *Neurology* **2007**, *65*, **1716**.
19. C. Wu, V. Pike, Y. Wang, *Curr. Top. Dev. Biol.* **2005**, *70*, 171.
20. M. T. Toczek, R. E. Carson, L. Lang, Y. Ma, M. V. Spanski, M. G. Der, S. Fazilat, L. Kopylev, P. Herscovitch, W. C. Eckelman, W. H. Theodore, *Neurology* **2003**, *60*, 749.
21. W. Zhang, M. Kung, S. Oya, C. Hou, H. Kung, *Nucl. Med. Biol.* **2007**, *34*, 89.
22. A. Nordberg, *Lancet Neurol.* **2004**, *3*, 519.
23. A. Troiano, J. Stoessl,in Broderick, P. Rahni, D. Kolodny, E. (Eds), *Bioimaging in Neurodegeneration*, Humana Press Inc. , Totowa, NJ, **2005**.
24. R. Sanchez-Pernaute, A.-L. Brownell, B. Jenkins, O. Isacson, *Toxicol. Appl. Pharmacol.* **2005**, *207*, S251.
25. D. Brooks, *Mol. Imag. Biol.* **2007**, *9*, 217.
26. H. Kung, *Int. Congr. Ser. 1264* **2004**, 3.
27. N. Seneca, L. Cai, J.-S. Liow, S. Zoghbi, R. Gladding, J. Hong, V. Pike, R. Innis, *Nucl. Med. Biol.* **2007**, *34*, 681.
28. J. Natsume, Y. Kumakura, N. Bernasconi, J.-P. Soucy, A. Nakai, P. Rosa, M. Fedi, F. Dubeau, F. Andermann, R. Lisbona, A. Bernasconi, M. Diksic, *Neurology* **2003**, *60*, 756.
29. D. Burn, J. O'Brien, *Mov. Disord.* **2003**, *18*, S88.
30. B. G. Hockley, P. J. H. Scott, M. R. Kilbourn,in Scott, P. J. H. (Ed.), *Linker Strategies in Solid-Phase Organic Synthesis*, Wiley, **2009**, 577.
31. K. Hamacher, G. Blessing, B. Nebeling, *Int. J. Radiat. Appl. Instrum. A* **1990**, *41*, 49.
32. J. L. Sutcliffe-Goulden, M. J. O'Doherty, S. S. Bansal, *Bioorg. Med. Chem. Lett.* **2000**, *10*, 1501.
33. J. L. Sutcliffe-Goulden, M. J. O'Doherty, P. K. Marsden, I. R. Hart, J. F. Marshall, S. S. Bansal, *Eur. J. Nucl. Med.* **2002**, *29*, 754.
34. J. Marik, S. H. Hausner, L. A. Fix, K. J. Gagnon, J. L. Sutcliffe, *Bioconjugate Chem.* **2006**, *17*, 1017.
35. S. H. Hausner, D. DiCara, J. Marik, J. F. Marshall, J. L. Sutcliffe, *Cancer Res.* **2007**, *67*, 7833.
36. S. K. Luthra, F. Brady, H. J. Wadsworth,WO Patent, 03/002489A2, 2003.
37. L. J. Brown, D. R. Bouvet, S. Champion, A. M. Gibson, Y. Hu, A. Jackson, I. Khan, N. Ma, N. Millot, H. J. Wadsworth, R. C. D. Brown, *Angew. Chem., Int. Ed.* **2007**, *46*, 941.

38. S. K. Luthra, F. Brady, H. J. Wadsworth, A. M. Gibson, M. E. Glaser,WO Patent, 03/002157A1, 2003.
39. F. Brady, S. K. Luthra, Y. Zhao,WO Patent, 2004/056726A1, 2004.
40. F. Brady, S. K. Luthra, E. G. Robins,WO Patent, 2004/056400A1, 2004.
41. F. Brady, S. K. Luthra, A. M. Gibson,WO Patent, 2004/056399A2, 2004.
42. V. W. Pike, I. A. Franklin, *J. Chem. Soc., Chem. Commun.* **1995**, 2215.
43. T. L. Ross, J. Ermert, C. Hocke, H. H. Coenen, *J. Am. Chem. Soc.* **2007**, *129*, 8018.
44. A. P. Kudchadker, S. A. Kudchadker, R. P. Shukla, P. R. Patnaik, *J. Phys. Chem. Ref. Data* **1979**, *8*, 500.
45. W. E. Klunk, C. A. Mathis,WO Patent, 2006/014382A1, 2006.
46. W. E. Klunk, C. A. Mathis, Y. Wang,WO Patent, 2004/083195A1, 2004.
47. A. Noda, Y. Murakami, S. Nishiyama, D. Fukumoto, S. Miyoshi, H. Tsukada, S. Nishimura, *Synapse* **2008**, *62*, 472.
48. E. D. Agdeppa, V. Kepe, J. Liu, S. Flores-Torres, N. Satyamurthy, A. Petric, G. M. Cole, G. W. Small, S. C. Huang, J. R. Barrio, *J. Neurosci.* **2001**, *21*, RC189.
49. A. M. Gibson, L. J. Brown, R. C. D. Brown,WO Patent, 2004/056725A1, 2004.
50. D. S. Casebier, R. R. Cesati, III, E. H. Cheesman,US Patent, 2007/0036716A1, 2007.
51. D. Maclean, J. M. Star-Lack,WO Patent, 02/102250A1, 2002.
52. D. Maclean, J. Zhu, M. Chen, R. Hale, N. Satymurthy, J. R. Barrio, *J. Am. Chem. Soc.* **2003**, *125*, 10168.
53. D. R. Turton, E. Robins,WO Patent, 2005/090266A1, 2005.
54. O. Rahman, T. Kihlberg, B. Långström, *J. Org. Chem.* **2003**, *68*, 3558.
55. H. M. Betts, P. J. Barnard, S. R. Bayly, J. R. Dilworth, A. D. Gee, J. P. Holland, *Angew. Chem., Int. Ed.* **2008**, *47*, 8416.
56. D. H. Hunter, X. Zhu, *J. Label. Compd. Radiopharm.* **1999**, *42*, 653.

14

SOLID-PHASE SYNTHESIS OF DYES AND THEIR APPLICATION AS SENSORS AND BIOIMAGING PROBES

Marc Vendrell, Hyung-Ho Ha, Sung Chan Lee, and Young-Tae Chang

14.1 INTRODUCTION

Solid-phase methodologies have been widely used in the context of sensor development. Initial applications came after the expansion of solid-phase peptide synthesis and primarily made use of the immobilization process to facilitate the screening of potential peptide-based sensors in a high-throughput manner. Their subsequent adjustment to nonpeptidic biomolecules considerably spread out their scope from the preparation of resins to label bioactive molecules to combinatorial approaches to accelerate the derivatization of fluorescent scaffolds. Once combinatorial chemists succeeded in adapting fluorescent heterocycles to solid-phase synthesis, small molecule fluorescent dyes started to be increasingly implemented in sensor discovery programs as powerful bioimaging probes (Figure 14.1). Diversity-oriented fluorescent libraries (DOFLs) have been effectively prepared by solid-phase synthesis and demonstrated that subtle structural modifications within a fluorescent core can be distinguishable by large macromolecules and hence lead to selective fluorescent probes. This chapter will review the progress of solid-phase approaches for the preparation of sensors over the last years and outlook the role that these methodologies will play in probe development in the near future.

Solid-Phase Organic Synthesis: Concepts, Strategies, and Applications, First Edition.
Edited by Patrick H. Toy and Yulin Lam.

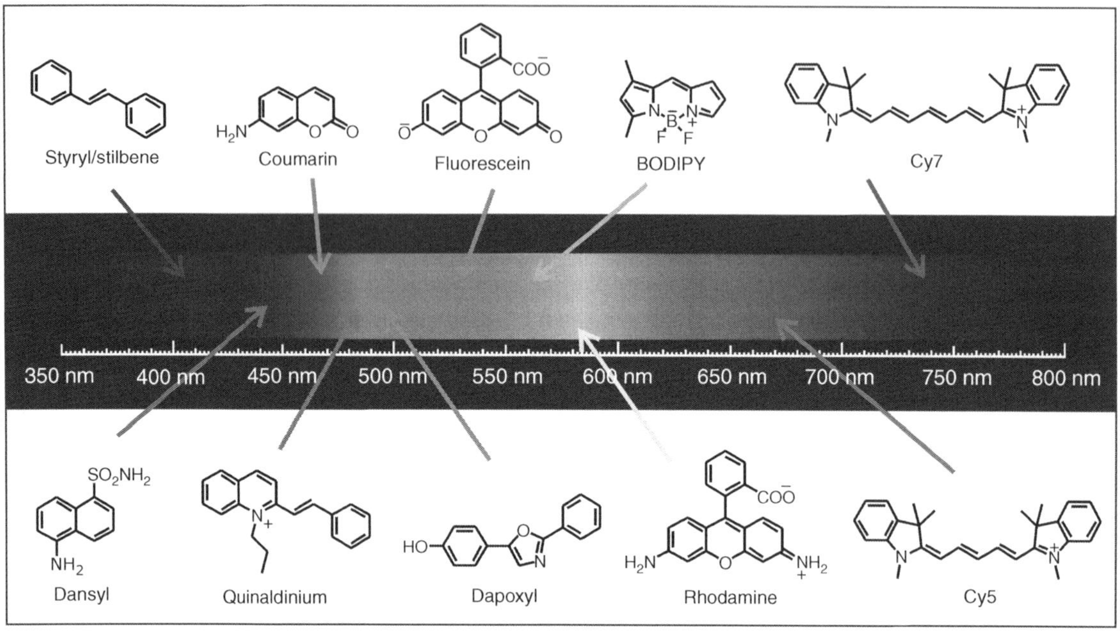

Figure 14.1. Structures of dyes synthesized by solid-phase methodologies and their approximate fluorescence emission spectral properties. (See the color version of this figure in Color Plates section.)

14.2 ON-BEAD SENSORS

The first steps of solid-phase approaches in sensor development were linked to the emergence of solid-phase peptide synthesis. Split-and-mix strategies[1–3] rendered "one-bead one-peptide" libraries that could be screened against any analyte of interest in a solid-phase assay. Colorimetric or fluorescence changes were generally used to report the interaction and enabled the discovery of sensors for a number of biomolecules (e.g., tripeptides[4] and ATP[5]).

Although it was clear that the immobilization of the sensors could overcome the purification challenges derived from solution-phase chemistry and facilitate the screening process, random combinatorial peptide libraries were gradually replaced by rational design-based approaches. Advantages of on-bead sensors include lower reagent consumption, continuous readout, and the possibility of using solvents in which the sensors may show low solubility. These properties were exploited to construct on-bead sensors that conjugated well-known recognition motifs (e.g., polyamines, crown ethers, boronic acids) to suitable fluorescent reporters. Most of these sensors depended on the photoinduced electron transfer (PET) mechanism, in which the direct interaction between the analyte and a nitrogen lone pair of the fluorophore was transduced into a signal change (Figure 14.2a).[6] Because the solid support might drastically change the sensing properties of the probes, spacers were usually placed between the resin and the main core structure to minimize such interference. Several examples following the pioneering contribution of Singh et al. with nanomolar aqueous copper sensors[7] have been reported for the detection of alkali and heavy metal ions[8,9] as well as sugars.[10] Furthermore, since PET-based probes entailed the incorporation of binding domains containing anilinic or benzylic nitrogen atoms, on-bead sensors without implementing a known recognition motif in their architecture were also described:

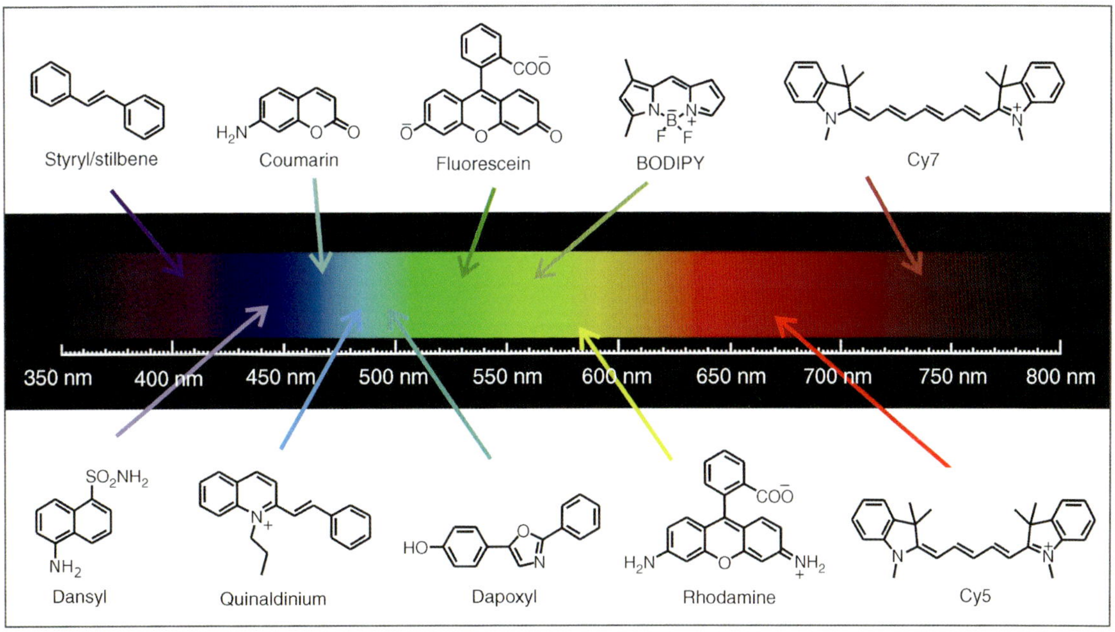

Figure 14.1 Structures of dyes synthesized by solid-phase methodologies and their approximate fluorescence emission spectral properties.

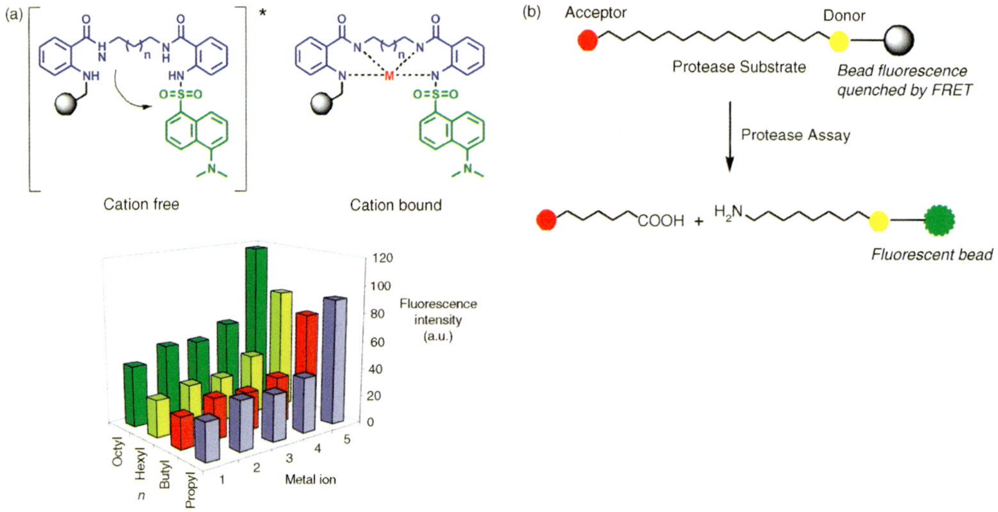

Figure 14.2 On-bead sensors. (a) Metal sensors using a dansyl fluorophore following a PET mechanism. (b) Schematic representation of a peptide-based FRET sensor to monitor protease activity. Adapted with permission from Refs 8 (left) and 15 (right). Copyright 2000–2009 American Chemical Society.

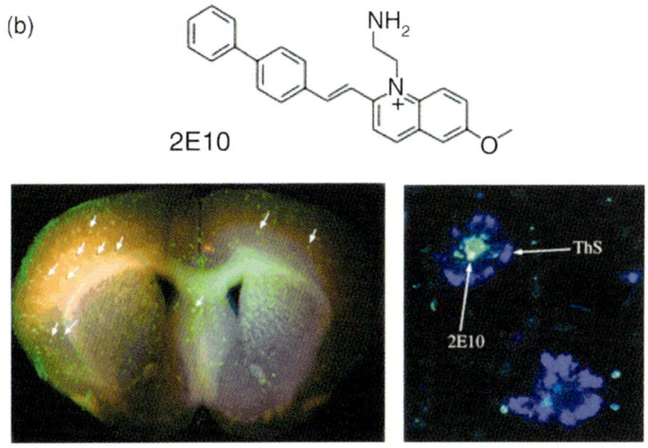

Figure 14.9b Diversity-oriented fluorescent library based on the styryl scaffold. (b) (left) labeling amyloid deposits (white arrows) in mouse brain tissue with 2E10; (right) confocal microscope images of amyloid deposits after incubation with 2E10 (green) and thioflavin S (ThS, blue). Copyright Wiley-VCH Verlag GmbH & Co. KGaA. Reproduced with permission from Ref. 46.

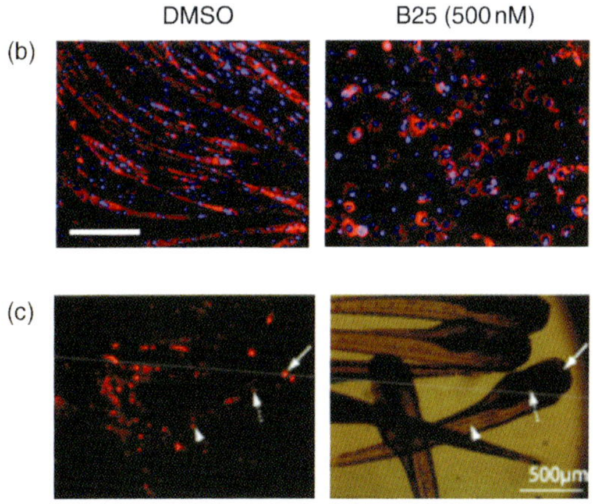

Figure 14.12b and c Rosamine compounds as fluorescent probes. (b and c) diverse applications of rosamine sensors derived from image-based live cell and organism screenings. Reproduced with permission from Refs 53, 54, and 56. Copyright 2007–2010 American Chemical Society; Reproduced by permission of The Royal Society of Chemistry. Ref 57.

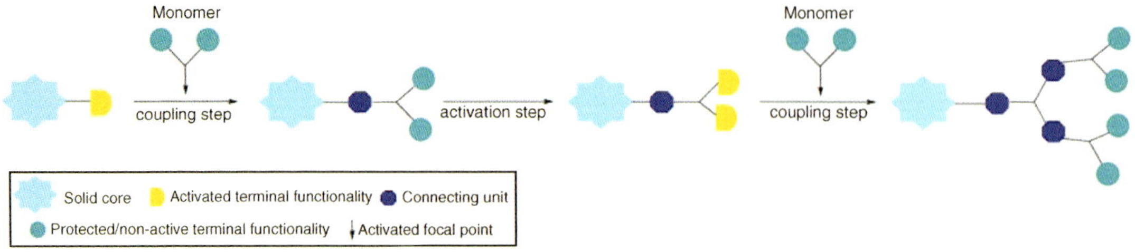

Scheme 15.1 The divergent approach to dendron assembly on solid support.

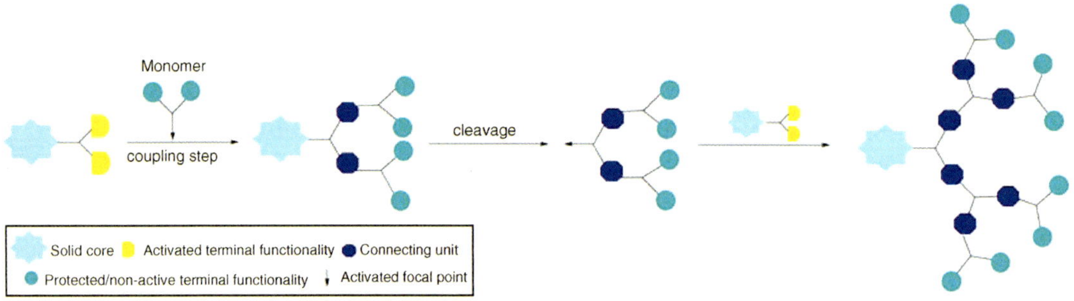

Scheme 15.2 The convergent approach to dendron assembly on solid support.

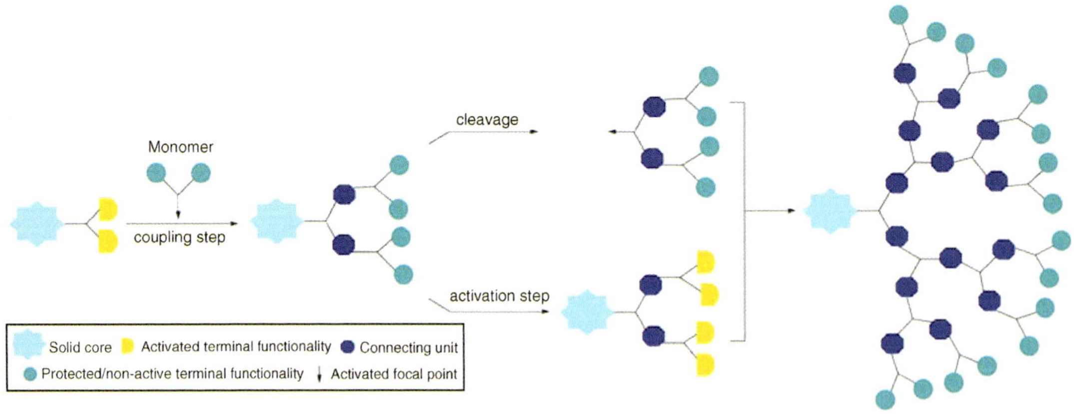

Scheme 15.3 The divergent/convergent approach to dendron assembly on solid support.

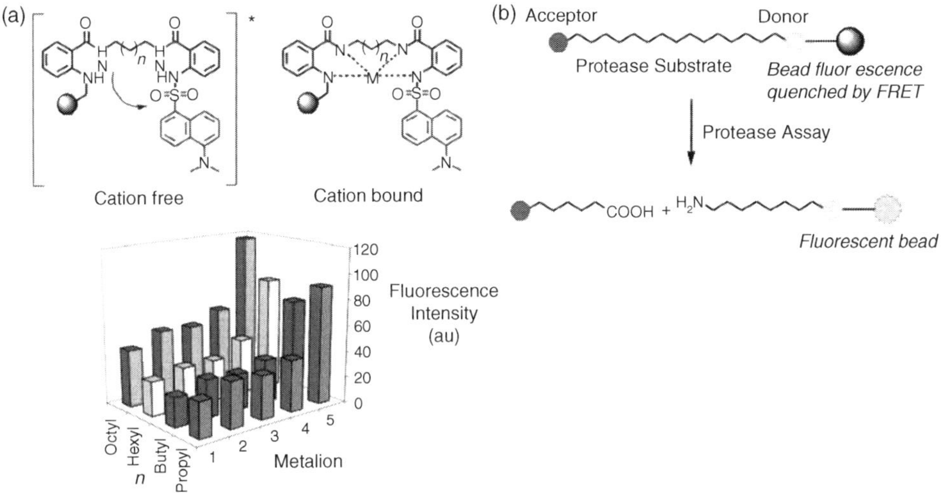

Figure 14.2. On-bead sensors. (a) Metal sensors using a dansyl fluorophore following a PET mechanism. (b) Schematic representation of a peptide-based FRET sensor to monitor protease activity. Adapted with permission from Refs 8 (left) and 15 (right). Copyright 2000–2009 American Chemical Society. (See the color version of this figure in Color Plates section.)

biarylpyridine mercury sensors[11] and aza-BODIPY probes to detect gaseous HCl[12] are some representative examples.

On-bead sensors have also been developed on the basis of the fluorescence energy transfer (FRET) mechanism (Figure 14.2b). FRET signaling is less stringent than PET in terms of binding domain structure but requires large changes in the distance between donor and acceptor, which results in their main application within peptide or nucleic acid structures. On-bead FRET sensors constructed by split-and-mix techniques were initially described in the characterization of proteolytic enzymes[13] and later applied to the discovery of protease substrates and inhibitors.[14,15]

14.3 SOLID-PHASE APPROACHES IN FLUORESCENT LABELING

Many bioimaging techniques, especially those involving live cells, are not compatible with the use of immobilized fluorescent sensors. For such cases, solid-phase methods to efficiently label small biomolecules, peptides, or nucleic acids with fluorescent tags have been designed. Fluorescent peptides can be prepared by incorporating unnatural amino acids including a reporter (e.g., fluorogenic, solvatochromic) that is compatible with solid-phase peptide synthesis protocols.[16–18] These approaches impart a minor change in the peptide structure, but require some prior knowledge of the residue within the sequence that can be modified without altering the properties of the peptide. More versatile strategies have been devised by loading a fluorophore (e.g., coumarin, fluorescein) as the C-terminal amino acid and building the rest of the peptide afterward. Although Rink amide resins are the most popular supports for this purpose, many resins can be used and the main limitation arises from the need of orthogonally protected fluorescent reporters.[19–21] An interesting example of this procedure was described by Rademan and coworkers, who loaded an aminofluorescein derivative as the C-terminal residue via the phenol group to a tritylpolystyrene resin.

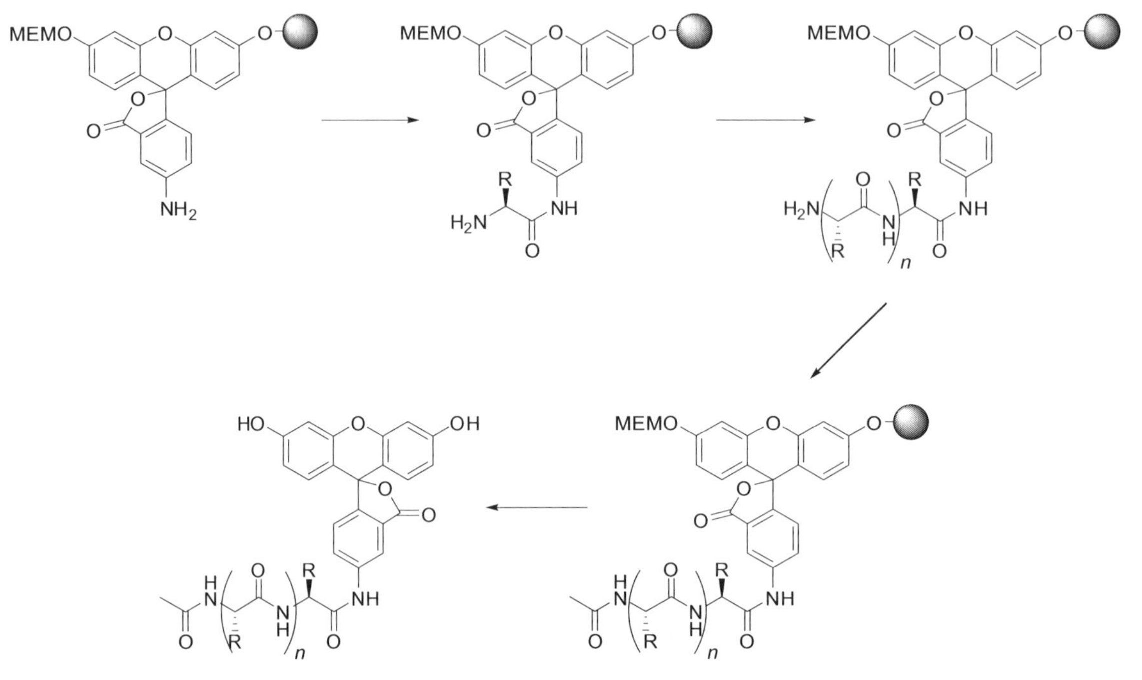

Figure 14.3. Solid-phase synthesis of C-terminal fluorescein-labeled peptides. Copyright Wiley-VCH Verlag GmbH & Co. KGaA. Reproduced with permission from Ref. 22.

Further construction of a small peptide library led to high-affinity polarization probes for polyproline-specific GYF domains (Figure 14.3).[22]

Alternative labeling methods are "solid-phase organic tagging" resins, which incorporate a reactive group within the bioactive molecule to be further derivatized with a reporter tag.[23] This strategy is particularly useful when orthogonally protected fluorophores are not available or incompatible with the reaction conditions employed in solid-phase synthesis. Finally, other protocols have also been established for specific applications, such as labeling small amounts of antibodies with fluorescent molecules.[24]

14.4 SOLID-PHASE DERIVATIZATION OF FLUORESCENT SCAFFOLDS

The need for understanding recognition events in chemistry and biology has directed great efforts toward the development of sensors based on small molecule fluorescent probes. However, predicting the properties of organic fluorophores, even when they are similar to known structures, is difficult, and transferring such information into the design of new fluorescent sensors remains a big challenge in probe development. Combinatorial chemistry approaches, which were initially envisaged for drug discovery, appeared in the last years as a promising tool to circumvent our limited abilities in this field.[25] The first obstacle that combinatorial chemists faced in this endeavor was the adaptation of fluorescent heterocycles to solid-phase synthesis. One of the first examples was reported for 3-oxo-3*H*-benzopyrans, commonly designated as coumarins. With several reports of the coumarin chemistry in solution,[26,27] solid-phase approaches were sequentially designed. Watson and Christiansen first described the synthesis of coumarin-3-carboxylates obtained by

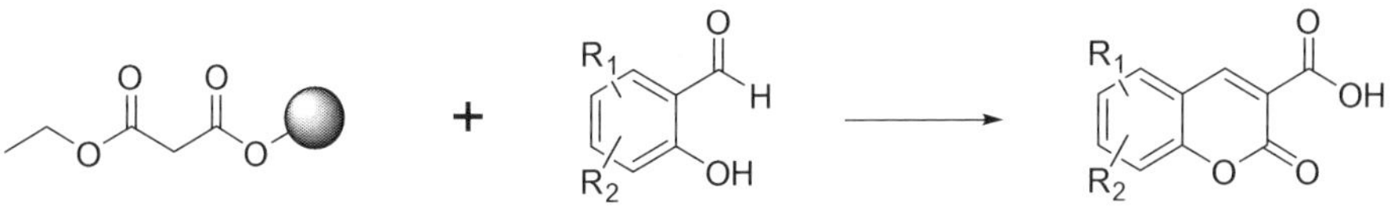

Figure 14.4. Solid-phase synthesis of substituted coumarin-3-carboxylates based on Knoevenagel condensation. Reprinted from Ref. 28 with permission from Elsevier.

Knoevenagel condensation reaction between a Wang resin-supported ethylmalonate and hydroxyaryl aldehydes (Figure 14.4).[28]

This procedure facilitated the modification and purification of coumarin derivatives, but it was restricted to one-dimensional chemical diversity. Lam and coworkers overcame this limitation and improved the overall synthetic efficiency using 7-fluoro-4-methyl-6-nitrocoumarin-3-carboxylic acid as a scaffold with three potential derivatization points (Figure 14.5). After loading the carboxylic acid onto a Rink amide resin, the aryl fluoride underwent aromatic nucleophilic substitution and the nitro group was reduced to render an additional heterocyclic ring that could incorporate further diversity. This versatile synthetic route yielded a wide spectrum of coumarin dyes, such as imidazocoumarins, lactam coumarins, and thioimidazocoumarins.[29,30]

Solid-phase procedures for the synthesis of fluorescent compounds do not significantly differ from those used in medicinal chemistry or drug discovery programs. Therefore, well-known solid-phase strategies can be easily applied to the derivatization of fluorescent scaffolds. For instance, traceless linkers were developed to release loaded compounds without leaving any trace of their linkage to the solid phase.[31] Park and colleagues recently described the synthesis of unsymmetrical *trans*-stilbenes using a sulfonate-based traceless linker strategy.[32] Unsymmetrical stilbenes are typically prepared by Wittig and Horner–Wadsworth–Emmons reactions of substituted benzaldehydes with benzyl phosphonium ylides or phosphonates. In this solid-phase approach, the authors prepared a collection of stilbenes via a nickel(0)-catalyzed cleavage/cross-coupling reaction of the supported stilbenesulfonates with different aryl Grignard reagents and evaluated the resulting structure–fluorescence relationships (SFR) (Figure 14.6). Traceless linkers strategies have been successfully used in the diversification of other fluorescent structures, such as thiazole compounds.[33,34]

Catch-and-release strategies can also be accommodated to the derivatization of fluorescent scaffolds. Mason and Balasubramanian reported a catch-and-release approach

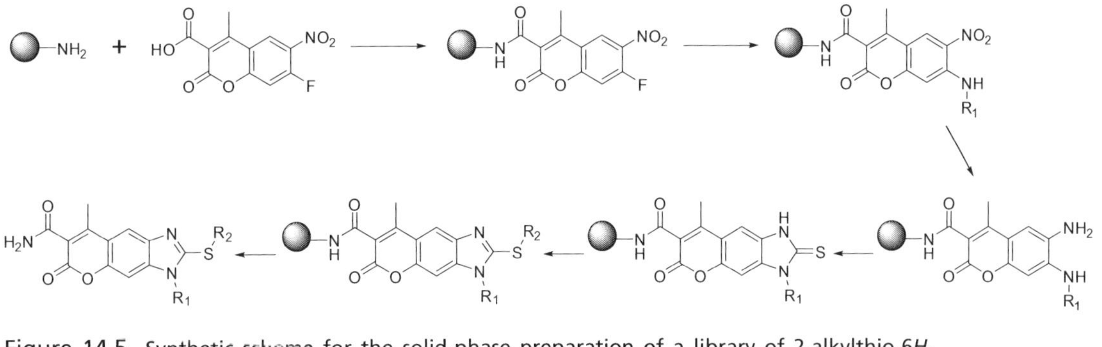

Figure 14.5. Synthetic scheme for the solid-phase preparation of a library of 2-alkylthio-6*H*-pyrano[2,3-*f*]benzimidazole-6-ones. Adapted with permission from Ref. 29. Copyright 2004 American Chemical Society.

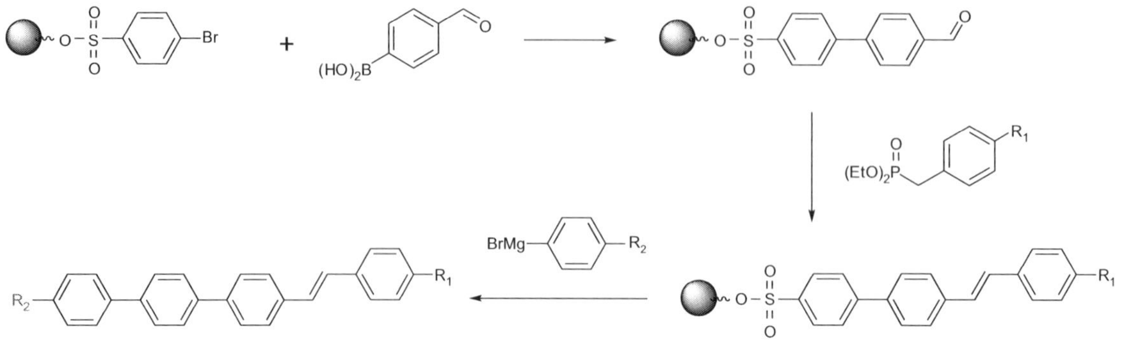

Figure 14.6. Traceless linker strategy for the solid-phase synthesis of unsymmetrical *trans*-stilbenes. Adapted with permission from Ref. 32. Copyright 2010 American Chemical Society.

to synthesize unsymmetrical trimethine cyanine dyes with a sulfonyl chloride resin: hemicyanine intermediates were loaded onto the solid support and cleaved with an heterocyclic carbon nucleophile to render the corresponding cyanines in good purities (Figure 14.7).[35] Three years later, the authors extended such strategy to pentamethine and water-soluble cyanine derivatives.[36]

The broad application of cyanine dyes (e.g., biomolecular labeling,[37] *in vivo* imaging,[38,39] and proteomics[40]) has attracted a general interest in their solid-phase synthesis. Isacsson and Westman reported the first synthesis of asymmetric cyanine dyes,[41] and Erdem et al. described the preparation of phthalocyanines with asymmetrical substitution using a polyethylene glycol-based support with a Wang-type linker[42]. In the latter, the authors surmounted the purification problems derived from the phthalocyanine mixtures by using a hydrophilic support that allowed the complete removal of the undesired symmetrical phthalocyanine product. A remarkable example of the spectral versatility of cyanine dyes was recently published by Bradley and colleagues, who applied a catch-and-release strategy

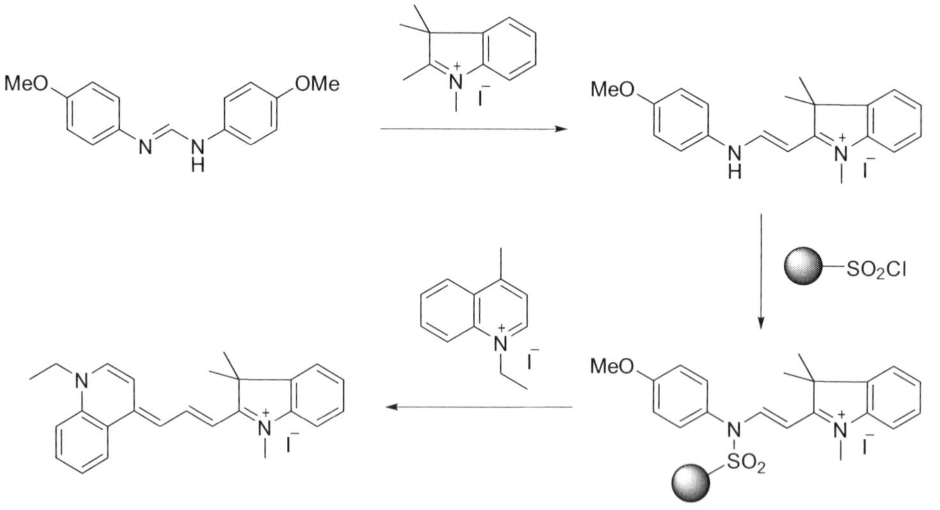

Figure 14.7. Catch-and-release strategy for the synthesis of asymmetrical cyanine dyes. Reproduced with permission from Ref. 35. Copyright 2002 American Chemical Society.

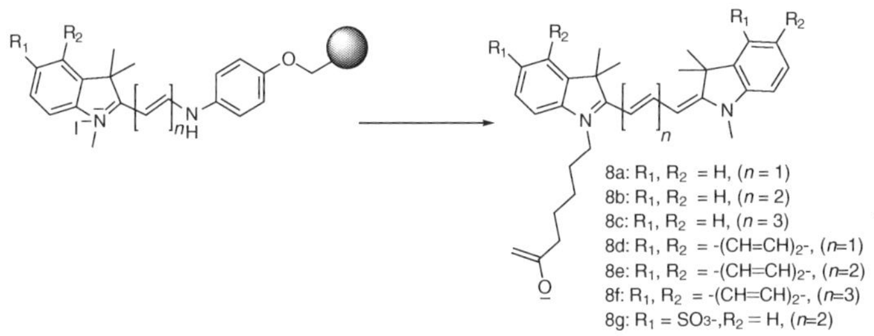

Dye	$\lambda_{max, abs}$ (nm)	$\lambda_{max, em}$ (nm)	Yied
8a	547	561	68%
8b	640	660	84%
8c	743	765	86%
8d	587	603	49%
8e	678	704	92%
8f	781	808	51%
8g	647	667	94%

Figure 14.8. Solid-phase synthesis of unsymmetrical functionalized cyanine dyes with diverse fluorescent properties. Reproduced by permission of The Royal Society of Chemistry. Ref 43.

to obtain cyanine compounds covering an extensive range of fluorescent properties, with emission maxima ranging from 561 to 808 nm (Figure 14.8).[43]

14.5 DIVERSITY-ORIENTED FLUORESCENT LIBRARIES

In view of the utility of solid-phase chemistry for the derivatization of fluorescent scaffolds, Chang and coworkers pioneered the use of this methodology to construct DOFLs for sensor development.[44,45] They envisioned that the inherent characteristics of fluorescent small molecules (e.g. intrinsic fluorescence, chemical stability, solubility in aqueous media, and cell permeability) could accelerate the discovery of novel sensors for biological applications. With DOFLs tested in a high-throughput manner, selective probes could be quickly identified and, at the same time, the best applications for families of fluorescent compounds could be determined. The discovery of new probes would follow an evolutive mechanism (diversity generation followed by selection and further modification) so that every evolution step, together with the obtained SFR, would improve the sensing properties of every generation of compounds. Solid-phase synthesis plays a key role in probe development using DOFLs, since large collections of fluorescent compounds need to be synthesized in a short period in order to maximize both the efficiency of sensor discovery and the compilation of meaningful SFR information.

Li et al. reported the first solid-phase synthesis of a DOFL based on the styryl structure.[46] Styryl compounds are simple fluorescent molecules that can be prepared by condensation of aldehydes and pyridinium salts. After testing a series of resins and reaction routes, the authors employed a 2-chlorotrityl polystyrene resin to load two amino alcohols with different length chains. The alcohol groups were then mesylated and subsequently treated with four picoline and three quinoline derivatives to render solid-supported pyridinium salts that were then condensed with 64 chemically diverse aldehydes under microwave irradiation (Figure 14.9a). Final cleavage with TFA–DCM (1:99) yielded a collection of 320 styryl dyes with very diverse spectral properties, and they were evaluated as amyloid probes without further purification.

The identification of selective styryl dyes as amyloid plaque-staining agents (Figure 14.9b) proved that (1) solid-phase synthesis could be used to derivatize positively charged scaffolds and prepare libraries of fluorescent molecules with enhanced properties (i.e. water solubility, cell permeability) as biological sensors, (2) subtle structural modifications in fluorescent small molecules were distinguishable by large macromolecules, contradicting the traditional belief that highly conjugated hydrophobic dyes may

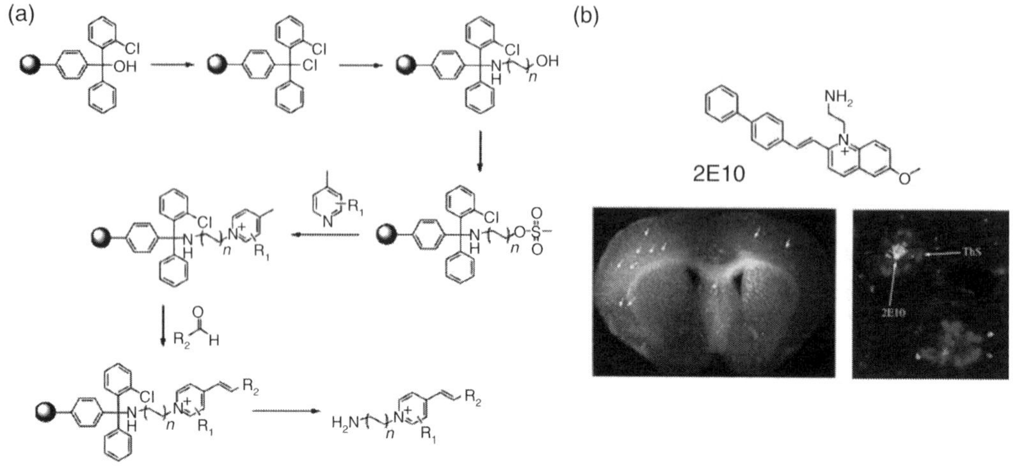

Figure 14.9. Diversity-oriented fluorescent library based on the styryl scaffold. (a) Synthetic scheme; (b) (left) labeling amyloid deposits (white arrows) in mouse brain tissue with 2E10; (right) confocal microscope images of amyloid deposits after incubation with 2E10 (green) and thioflavin S (ThS, blue). Copyright Wiley-VCH Verlag GmbH & Co. KGaA. Reproduced with permission from Ref. 46. (See the color version of this figure in Color Plates section.)

lack specificity, and (3) a conveniently designed primary screening successfully narrowed down a large DOFL and minimized costs and efforts in the search of probes for complex biological samples. Furthermore, with the SFR data compiled, the authors designed a second-generation library with blood–brain barrier (BBB) permeability that enabled the application of styryl amyloid plaque-staining agents for *in vivo* imaging.[47]

DOFLs incorporating chemical diversity by condensation of aldehyde moieties have been developed for other positively charged fluorescent structures, such as benzimidazole and quinaldine. Wang and Chang constructed a library of benzimidazolium dyes and exploited the potential electrostatic interactions to evaluate their fluorescence response against negatively charged biomolecules. An imidazolium compound with a 2-phenylindole motif was identified as a selective turn-on sensor of GTP, showing an 80-fold fluorescence increase and an outstanding selectivity against all other nucleosides and nucleotides.[48] Benzimidazolium compounds have been later reported as fluorescent probes for other anionic molecules, such as heparin.[49] Besides, Chang and coworkers described the construction of a quinaldinium library and the discovery of novel fluorescent chymotrypsin sensors.[50] In this case, the authors optimized a solid-phase approach where the quinaldine scaffold was loaded as a quaternary salt to reduce the potential side reactions and improve the overall yields and purities of subsequent quinaldinium dyes. In both benzimidazolium and quinaldinium libraries, an amine-derivatized tritylpolystyrene resin was used as the solid support (Figure 14.10). Derivatization by Knoevenagel condensation followed by a cleavage using weak acidic conditions (generally TFA–DCM (5:95)) yielded the final derivatives in good purities. Interestingly, the Knoevenagel condensation not only provided a suitable platform to introduce a broad chemical diversity but also rendered fluorescent compounds with a wide range of emission properties in both libraries (benzimidazolium dyes: 450–600 nm; quinaldinium dyes: 440–640 nm).

Environmentally-sensitive fluorescent structures are interesting scaffolds for probe development. On account of the high sensitivity of the dapoxyl core (2-pyridyl-5-aryl-

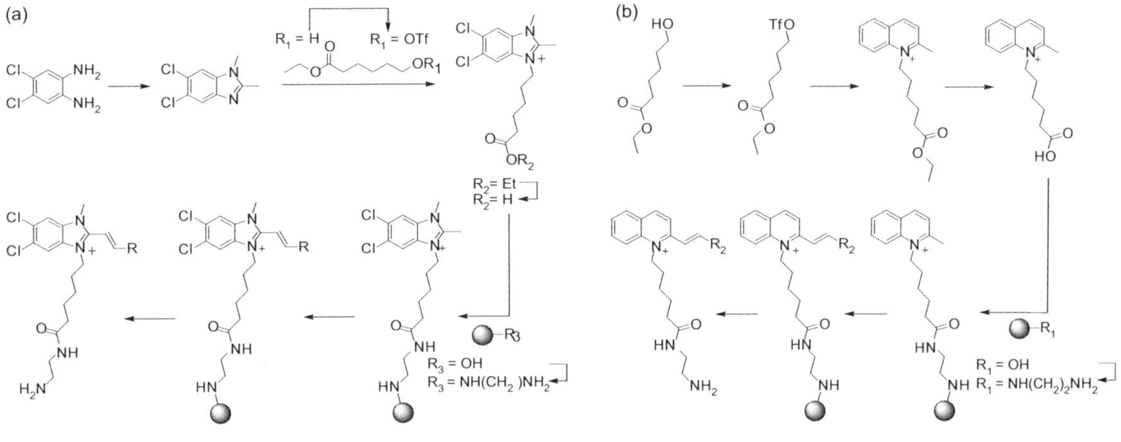

Figure 14.10. Solid-phase synthesis of fluorescent libraries based on (a) benzimidazolium and (b) quinaldinium scaffolds. Reproduced with permission from Refs 48, 50. Copyright 2006–2008 American Chemical Society.

oxazole) to polarity changes and its capacity to track different protein conformations,[51] Min et al. constructed a dapoxyl-based library and evaluated its potential in protein sensing.[52] As the first step, one phenol-containing 2-aminoacetophenone was loaded onto an acid labile solid support and modified with different benzoic acid derivatives to obtain a series of supported 2-acylaminoacetophenones (Figure 14.11). The solid-phase synthesis of dapoxyl dyes required the adjustment of the typically harsh cyclodehydration conditions (i.e. neat sulfuric acid employed in solution-phase approaches) to milder reagents that were compatible with the lability of the resin. An optimized procedure using Ph_3PCl_2 in thc presence of triethylamine followed by the final acidic cleavage yielded a set of dapoxyl dyes, from which one compound (**A41-S**) showed a 55-fold fluorescence intensity increase upon binding to human serum albumin (HSA). Further competition assays with reported HSA ligands identified the interaction of **A41-S** at the binding site I of the protein. This specificity among the different HSA binding pockets manifested the exceptional discriminative power of environmentally-sensitive fluorophores.

Although the rapid synthesis, broad chemical diversity, and adaptability to high-throughput screenings of solid-phase approaches had clearly benefited the development

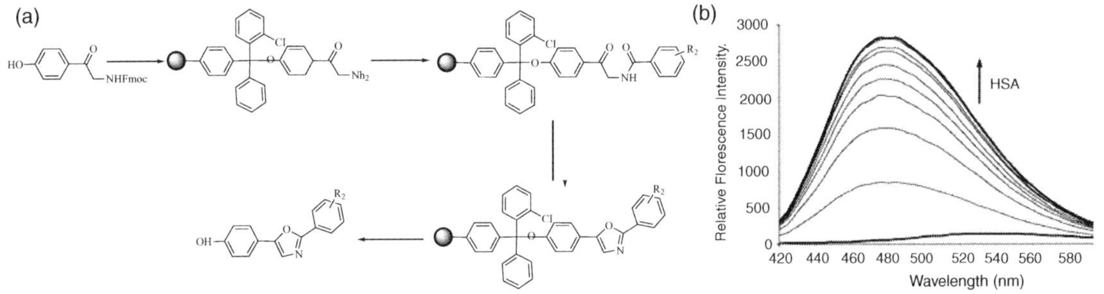

Figure 14.11. Solid-phase synthesis of dapoxyl dyes. (a) Library generation; (b) fluorescence response of **A41-S** to increasing concentrations of HSA. Reproduced with permission from Ref. 52. Copyright 2007 American Chemical Society.

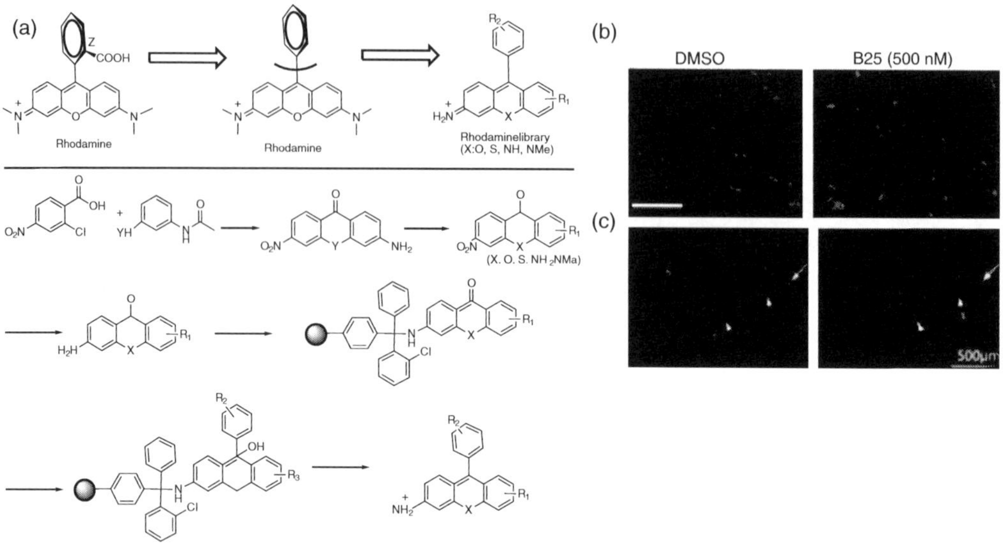

Figure 14.12. Rosamine compounds as fluorescent probes. (a) Solid-phase library synthesis; (b and c) diverse applications of rosamine sensors derived from image-based live cell and organism screenings. Reproduced with permission from Refs 53, 54, and 56. Copyright 2007–2010 American Chemical Society; Reproduced by permission of The Royal Society of Chemistry. Ref 57. (See the color version of this figure in Color Plates section.)

of DOFLs for sensor discovery, it was unclear whether solid-phase syntheses could be used to optimize the photophysical properties of the libraries and improve their characteristics as fluorescent probes. Compounds with moderate quantum yields are likely to exhibit fluorescence intensity fluctuations and thus have a higher potential as sensors. A good example of this adaptation was reported by Ahn et al., with the solid-phase synthesis of rosamine compounds.[53] The authors designed a rosamine library on basis of the highly fluorescent rhodamine core. Since the rosamine structure lacked the 2′-carboxylic acid of rhodamine, its enhanced molecular flexibility resulted in a decrease in the fluorescence quantum yield, while maintaining high extinction coefficients and photostabilities (Figure 14.12a). The solid-phase approach facilitated the synthesis of a 240-member rosamine library by circumventing acidic reflux conditions and tedious purification steps. The combination of 12 unsymmetrical xanthones (containing oxygen, sulfur, and nitrogen bridges) and 33 Grignard reagents plus a final acidic cleavage with concomitant dehydration yielded the final compounds with diverse quantum yields (0.0003–0.89) and wide absorption (480–545 nm) and emission (530–605 nm) ranges (Figure 14.12a). Rosamine dyes are extremely versatile fluorescent probes. In addition to their environmental sensitivity that led to the discovery of protein sensors,[54] their cell permeability makes them easily adaptable to image-based high-throughput screenings. The emerging interest in understanding complex biological processes (e.g. stem cells self-renewal and differentiation) together with the major technical improvements implemented in high-content screening methodologies have drastically expanded the potential applications of fluorescent probes in bioimaging. The recent discoveries of rosamine-based cell state sensors (Figures 14.12b)[55,56] and zebrafish neural tracers, from an organism-based screening (Figure 14.12c),[57] validate the significant role that DOFLs will continue to play in the development of high-performance fluorescent sensors for contemporary scientific problems.

14.6 CONCLUSIONS

Solid supports have significantly contributed to sensor development, both as screening platform and synthetic tool. Regarding the latter, solid-phase approaches facilitated the diversification of potential sensors, especially for those scaffolds that encompass tedious purification steps in solution-phase chemistry. The compatibility of the chemical structures with the lability of the resins and the design of combinatorial modifications that render broad chemical and spectral diversities are key points in such methodologies. Having validated the power of fluorescent small molecules as bioimaging probes, the adaptation of solid-phase chemistry to fluorescent structures with enhanced photophysical properties (e.g. highly photostable cores such as BODIPY or near-infrared dyes for *in vivo* applications) will certainly improve the current optical imaging toolbox in the near future.

14.7 ACKNOWLEDGMENTS

The authors gratefully thank the financial support from the National University of Singapore (NUS) (Young Investigator Award: R-143-000-353-123) and from the intramural funding of A*STAR (Agency for Science, Technology and Research, Singapore) Biomedical Research Council.

REFERENCES

1. R. A. Houghten, C. Pinilla, S. E. Blondelle, J. R. Appel, C. T. Dooley, J. H. Cuervo, Generation and use of synthetic peptide combinatorial libraries for basic research and drug discovery. *Nature* **1991**, *354*, 84–86.
2. K. S. Lam, S. E. Salmon, E. M. Hersh, V. J. Hruby, W. M. Kazmierski, R. J. Knapp, A new type of synthetic peptide library for identifying ligand-binding activity. *Nature* **1991**, *354*, 82–84.
3. A. Furka, F. Sebestyen, M. Asgedom, G. Dibo, General method for rapid synthesis of multicomponent peptide mixtures. *Int. J. Pept. Protein Res.* **1991**, *37*, 487–493.
4. R. Xu, G. Greiveldinger, L. E. Marenus, A. Cooper, J. A. Ellman, Combinatorial library approach for the identification of synthetic receptors targeting vancomycin-resistant bacteria. *J. Am. Chem. Soc.* **1999**, *121*, 4898–4899.
5. S. E. Schneider, S. N. O'Neil, E. V. Anslyn, Coupling rational design with libraries leads to the production of an ATP selective chemosensor. *J. Am. Chem. Soc.* **2000**, *122*, 542–543.
6. A. G. Fang J. V. Mello, N. S. Finney, Structural studies of biarylpyridines fluorophores lead to the identification of promising long wavelength emitters for use in fluorescent chemosensors. *Tetrahedron* **2004**, *60*, 11075–11087.
7. A. Singh, Q. Yao, L. Tong, W. C. Still, D. Sames, Combinatorial approach to the development of fluorescent sensors for nanomolar aqueous copper. *Tetrahedron Lett.* **2000**, *41*, 9601–9605.
8. G. Pina-Luis, A. Ochoa-Teran, I. A. Rivero, Solid phase synthesis of *N*-alkyl-bis-*o*-aminobenzamides for metal ion sensing based on a fluorescent dansyl platform. *J. Comb. Chem.* **2009**, *11*, 83–90.
9. I. A. Rivero, T. Gonzalez, G. Pina-Luis, M. E. Diaz-Garcia, Library preparation of derivatives of 1,4,10,13-tetraoxa-7, 16-diaza-cycloctadecane and their fluorescence behavior for signaling purposes. *J. Comb. Chem.* **2005**, *7*, 46–53.
10. D. Stones, S. Manku, X. Lu, D. G. Hall, Modular solid-phase synthetic approach to optimize structural and electronic properties of oligoboronic acid receptors and sensors for the aqueous recognition of oligosaccharides. *Chemistry* **2004**, *10*, 92–100.

11. J. V. Mello, N. S. Finney, Reversing the discovery paradigm: a new approach to the combinatorial discovery of fluorescent chemosensors. *J. Am. Chem. Soc.* **2005**, *127*, 10124–10125.
12. A. Palma, M. Tasior, D. O. Frimannsson, T. T. Vu, R. Meallet-Renault, D. F. O'Shea, New on-bead near-infrared fluorophores and fluorescent sensor constructs. *Org. Lett.* **2009**, *11*, 3638–3641.
13. M. Meldal, I. Svendsen, K. Breddam, F. I. Auzanneau, Portion-mixing peptide libraries of quenched fluorogenic substrates for complete subsite mapping of endoprotease specificity. *Proc. Natl. Acad. Sci. USA* **1994**, *91*, 3314–3318.
14. C. W. Tornoe, S. J. Sanderson, J. C. Mottram, G. H. Coombs, M. Meldal, Combinatorial library of peptidotriazoles: identification of [1,2,3]-triazole inhibitors against a recombinant Leishmania mexicana cysteine protease. *J. Comb. Chem.* **2004**, *6*, 312–324.
15. G. Rosse, E. Kueng, M. G. Page, V. Schauer-Vukasinovic, T. Giller, H. W. Lahm, P. Hunziker, D. Schlatter, Rapid identification of substrates for novel proteases using a combinatorial peptide library. *J. Comb. Chem.* **2000**, *2*, 461–466.
16. G. Loving, B. Imperiali, A versatile amino acid analogue of the solvatochromic fluorophore 4-*N*, *N*-dimethylamino-1,8-naphthalimide: a powerful tool for the study of dynamic protein interactions. *J. Am. Chem. Soc.* **2008**, *130*, 13630–13638.
17. N. Jotterand, D. A. Pearce, B. Imperiali, Asymmetric synthesis of a new 8-hydroxyquinoline-derived alpha-amino acid and its incorporation in a peptidylsensor for divalent zinc. *J. Org. Chem.* **2001**, *66*, 3224–3228.
18. A. R Katritzky, M. Yoshioka, T. Narindoshvili, A. Chung, J. V. Johnson, Fluorescent labeling of peptides on solid phase. *Org. Biomol. Chem.* **2008**, *6*, 4582–4586.
19. D. J. Maly, F. Leonetti, B. J. Backes, D. S. Dauber, J. L. Harris, C. S. Craik, J. A. Ellman, Expedient solid-phase synthesis of fluorogenic protease substrates using the 7-amino-4-carbamoylmethylcoumarin (ACC) fluorophore. *J. Org. Chem.* **2002**, *67*, 910–915.
20. R. Fischer, O. Mader, G. Jung, R. Brock, Extending the applicability of carboxyfluorescein in solid-phase synthesis. *Bioconjug. Chem.* **2003**, *14*, 653–660.
21. Q. Zhu, D. B. Li, M. Uttamchandani, S. Q. Yao, Facile synthesis of 7-amino-4-carbamoylmethylcoumarin (ACC)-containing solid supports and their corresponding fluorogenic protease substrates. *Bioorg. Med. Chem. Lett.* **2003**, *13*, 1033–1036.
22. V. Uryga-Polowy, D. Kosslick, C. Freund, J. Rademann, Resin-bound aminofluorescein for C-terminal labeling of peptides: high-affinity polarization probes binding to polyproline-specific GYF domains. *Chembiochem* **2008**, *9*, 2452–2462.
23. D. Bonnet, S. Riche, S. Loison, R. Dagher, M. C. Frantz, L. Boudier, R. Rahmeh, B. Mouillac, J. Haiech, M. Hibert, Solid-phase organic tagging resins for labeling biomolecules by 1,3-dipolar cycloaddition: application to the synthesis of a fluorescent non-peptidic vasopressin receptor ligand. *Chemistry* **2008**, *14*, 6247–6254.
24. E. Lundberg, M. Sundberg, T. Graslund, M. Uhlen, H. A. Svahn, A novel method for reproducible fluorescent labeling of small amounts of antibodies on solid phase. *J. Immunol. Methods* **2007**, *322*, 40–49.
25. N. S. Finney, Combinatorial discovery of fluorophores and fluorescent probes. *Curr. Opin. Chem. Biol.* **2006**, *10*, 238–245.
26. M. S. Schiedel, C. A. Briehn, P. Bauerle, Single-compound libraries of organic materials: parallel synthesis and screening of fluorescent dyes. *Angew. Chem., Int. Ed. Engl.* **2001**, *40*, 4677–4680.
27. K. Sivakumar, F. Xie, B. M. Cash, S. Long, H. N. Barnhill, Q. Wang, A fluorogenic 1,3-dipolar cycloaddition reaction of 3-azidocoumarins and acetylenes. *Org. Lett.* **2004**, *6*, 4603–4606.
28. B. T. Watson, G. E. Christiansen, Solid phase synthesis of substituted coumarin-3-carboxylic acids via the Knoevenagel condensation. *Tetrahedron Lett.* **1998**, *39*, 6087–6090.

29. A. Song, J. Zhang, C. B. Lebrilla, K. S. Lam, Solid-phase synthesis and spectral properties of 2-alkylthio-6*H*-pyrano[2,3-f]benzimidazole-6-ones: a combinatorial approach for 2-alkylthioimidazocoumarins. *J. Comb. Chem.* **2004**, *6*, 604–610.

30. A. Song, J. Zhang, K. S. Lam, Synthesis and reactions of 7-fluoro-4-methyl-6-nitro-2-oxo-2*H*-1-benzopyran-3-carboxylic acid: a novel scaffold for combinatorial synthesis of coumarins. *J. Comb. Chem.* **2004**, *6*, 112–120.

31. J. Tulla-Puche, F. Albericio (Eds), *The Power of Functional Resins in Organic Synthesis*. Wiley-VCH, Weinheim, **2008**, pp. 450–455.

32. C. H. Cho, C. B. Kim, K. Park, Solid-phase synthesis of unsymmetrical *trans*-stilbenes. *J. Comb. Chem.* **2010**, *12*, 45–50.

33. T. Lee, D. Lee, I. Y. Lee, Y. D. Gong, Solid-phase synthesis of thiazolo[4,5-b]pyridine derivatives using Friedlander reaction. *J. Comb. Chem.* **2010**, *12*, 95–99.

34. X. Fei, S. Yang, B. Zhang, Z. Liu, Y. Gu, Solid-phase synthesis and modification of thiazole orange and its derivatives and their spectral properties. *J. Comb. Chem.* **2007**, *9*, 943–950.

35. S. J. Mason, S. Balasubramanian, Solid-phase catch, activate, and release synthesis of cyanine dyes. *Org. Lett.* **2002**, *4*, 4261–4264.

36. S. J. Mason, J. L. Hake, J. Nairne, W. J. Cummins, S. Balasubramanian, Solid-phase methods for the synthesis of cyanine dyes. *J. Org. Chem.* **2005**, *70*, 2939–2949.

37. R. B. Mujumdar, L. A. Ernst, S. R. Mujumdar, A. S. Waggoner, Cyanine dye labeling reagents containing isothiocyanate groups. *Cytometry* **1989**, *10*, 11–19.

38. R. Trehin, J. L. Figueiredo, M. J. Pittet, R. Weissleder, L. Josephson, U. Mahmood, Fluorescent nanoparticle uptake for brain tumor visualization. *Neoplasia* **2006**, *8*, 302–311.

39. R. Weissleder, V. Ntziachristos, Shedding light onto live molecular targets. *Nat. Med.* **2003**, *9*, 123–128.

40. R. Tonge, J. Shaw, B. Middleton, R. Rowlinson, S. Rayner, J. Young, F. Pognan, E. Hawkins, I. Currie, M. Davison, Validation and development of fluorescence two-dimensional differential gel electrophoresis proteomics technology. *Proteomics* **2001**, *1*, 377–396.

41. J. Isacsson, G. Westman, Solid-phase synthesis of asymmetric cyanine dyes. *Tetrahedron Lett.* **2001**, *42*, 3207–3210.

42. S. S. Erdem, I. V. Nesterova, S. A. Soper, R. P. Hammer, Solid-phase synthesis of asymmetrically substituted "AB3-type" phthalocyanines. *J. Org. Chem.* **2008**, *73*, 5003–5007.

43. M. Lopalco, E. N. Koini, J. K. Cho, M. Bradley, Catch and release microwave mediated synthesis of cyanine dyes. *Org. Biomol. Chem.* **2009**, *7*, 856–859.

44. J. S. Lee, Y. K. Kim, M. Vendrell, Y. T. Chang, Diversity-oriented fluorescence library approach for the discovery of sensors and probes. *Mol. Biosyst.* **2009**, *5*, 411–421.

45. M. Vendrell, J. S. Lee, Y. T. Chang, Diversity-oriented fluorescence library approaches for probe discovery and development. *Curr. Opin. Chem. Biol.* **2010**, *14*, 383–389.

46. Q. A. Li, J. S. Lee, C. Ha, C. B. Park, G. Yang, W. B. Gan, Y. T. Chang, Solid-phase synthesis of styryl dyes and their application as amyloid sensors. *Angew. Chem., Int. Ed.* **2004**, *43*, 6331–6335.

47. Q. Li, J. K. Min, Y. H. Ahn, J. H. Namm, E. M. Kim, R. Lui, H. Y. Kim, Y. Ji, H. Z. Wu, T. Wisniewski, Y. T. Chang, Styryl-based compounds as potential *in vivo* imaging agents for beta-amyloid plaques. *Chembiochem* **2007**, *8*, 1679–1687.

48. S. L. Wang, Y. T. Chang, Combinatorial synthesis of benzimidazolium dyes and its diversity directed application toward GTP-selective fluorescent chemosensors. *J. Am. Chem. Soc.* **2006**, *128*, 10380–10381.

49. S. L. Wang, Y. T. Chang, Discovery of heparin chemosensors through diversity oriented fluorescence library approach. *Chem. Commun.* **2008**, 1173–1175.

50. S. L. Wang, Y.K. Kim, Y.T. Chang, Diversity-oriented fluorescence library approach (DOFLA) to the discovery of chymotrypsin sensor. *J. Comb. Chem.* **2008**, *10*, 460–465.

51. Q. Wang, D. S. Lawrence, Phosphorylation-driven protein–protein interactions: a protein kinase sensing system. *J. Am. Chem. Soc.* **2005**, *127*, 7684–7685.
52. J. Min, J. W. Lee, Y. H. Ahn, Y. T. Chang, Combinatorial dapoxyl dye library and its application to site selective probe for human serum albumin. *J. Comb. Chem.* **2007**, *9*, 1079–1083.
53. Y. H. Ahn, J. S. Lee, Y. T. Chang, Combinatorial rosamine library and application to *in vivo* glutathione probe. *J. Am. Chem. Soc.* **2007**, *129*, 4510–4511.
54. Y. H. Ahn, J. S. Lee, Y. T. Chang, Selective human serum albumin sensor from the screening of a fluorescent rosamine library. *J. Comb. Chem.* **2008**, *10*, 376–380.
55. B. K. Wagner, H. A. Carrinski, Y. H. Ahn, Y. K. Kim, T. J. Gilbert, D. A. Fomina, S. L. Schreiber, Y. T. Chang, P. A. Clemons, Small-molecule fluorophores to detect cell-state switching in the context of high-throughput screening. *J. Am. Chem. Soc.* **2008**, *130*, 4208–4209.
56. Y. K. Kim, H. H. Ha, J. S. Lee, X. Bi, Y. H. Ahn, S. Hajar, J. J. Lee, Y. T. Chang, Control of muscle differentiation by a mitochondria-targeted fluorophore. *J. Am. Chem. Soc.* **2010**, *132*, 576–579.
57. J. Li, H. H. Ha, L. Guo, D. Coomber, Y. T. Chang, Discovery of novel zebrafish neural tracers by organism-based screening of a rosamine library. *Chem. Commun.* **2010**, 2932–2934.

15

DENDRITIC MOLECULES ON SOLID SUPPORT: SOLID-PHASE SYNTHESIS AND APPLICATIONS

Kerem Goren and Moshe Portnoy

15.1 INTRODUCTION

The unique properties of dendritic molecules include high degree of structural control, well-defined shape, molecular weight and size, multivalency, high loading and proximity of the peripheral groups, and isolated microenvironment of the interior sites.[1]

While the majority of the materials based on the dendritic architecture are prepared and used in solution, the amount of dendritic compounds prepared on or postsynthetically conjugated to solid insoluble support is constantly increasing. Although the dendritic components of such composite materials usually constitute only a minor part of the total volume, they can significantly alter the support properties, imparting entirely new features and functions on the solid support.

There are a number of modes for conjugation of a dendritic moiety to a support. Dendritic wedges can be coupled to the support through the focal point or through the peripheral functionalities, whereas dendrimers are usually immobilized through the peripheral functional groups (Figure 15.1).[2,3] Since this book deals with solid-phase synthesis, we choose to focus on the dendritic structures branching outward from the support (i.e., hybrid materials in which the dendrons are connected to the support via their focal point, Figure 15.1a) that are prepared stepwise on the support (rather than by a single postsynthetic immobilization step).

In this chapter, we will first discuss the synthetic approaches and sequences that lead to the supported dendrons and then review the applications of these materials. A number of

Solid-Phase Organic Synthesis: Concepts, Strategies, and Applications, First Edition.
Edited by Patrick H. Toy and Yulin Lam.

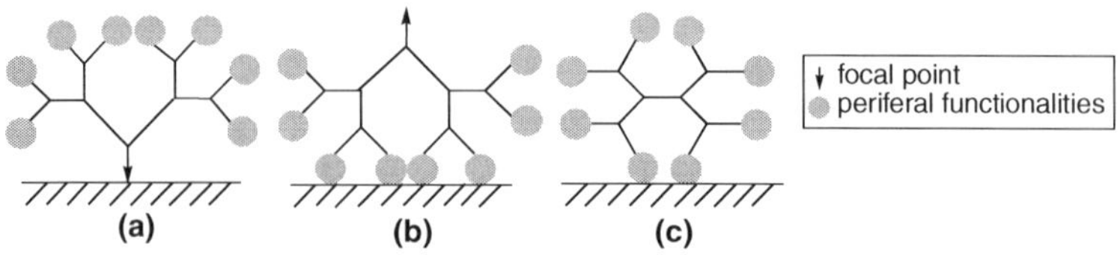

Figure 15.1. Possible modes of coupling dendritic fragments to support. Dendrons coupled through the focal point (a) or the peripheral groups (b). Dendrimers coupled through the peripheral groups (c).

review articles surveyed some of the topics that are brought together and summarized in this chapter.[4]

15.2 SYNTHESIS

15.2.1 General Synthetic Schemes

In addition to the aforementioned postsynthetic immobilization,[5] the materials of the chosen architecture can be assembled via stepwise growth of the dendron on the support, applying the technique of solid-phase synthesis. This approach is used in the majority of the cases and is the leading route to most dendritic motifs used for the dendronization of insoluble supports.

Solid-phase synthesis of dendrons is mostly carried out on polystyrene, polyacrylamide, or silica supports. While in solution dendrons and dendrimers are assembled via both divergent and convergent routes, the former is much more suitable for the synthesis of polymer-supported dendrons. On solid support such systems are, indeed, mostly assembled via the divergent approach (Scheme 15.1).[1,6] In solution, the divergent approach suffers, particularly for higher generations, from incomplete conversion at the growth step and the difficulty in purifying "perfect" dendrimers from similar defective structures. The technique of synthesis on solid support has a number of inherent advantages over synthesis in solution.[7] Two of the advantages—the possibility to drive the reaction to completion using a large excess of soluble reagents and the ease of separation/purification—are used to eliminate or significantly reduce the aforementioned difficulties in divergent solid-phase dendron synthesis. Many of the reported synthetic schemes for the solid-phase dendron synthesis are characterized by very high conversions at each step and high uniformity of the dendritic grafts.

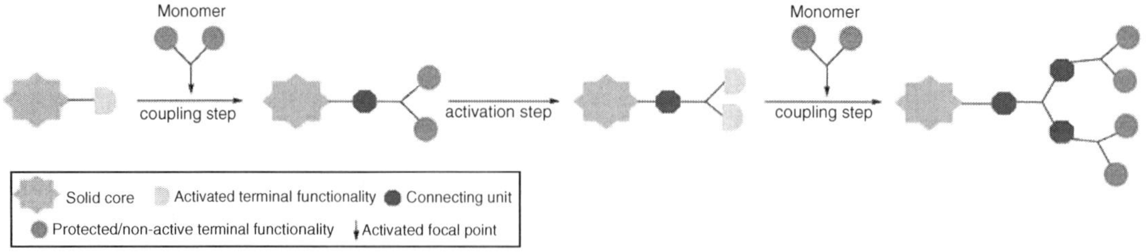

Scheme 15.1. The divergent approach to dendron assembly on solid support. (See the color version of this figure in Color Plates section.)

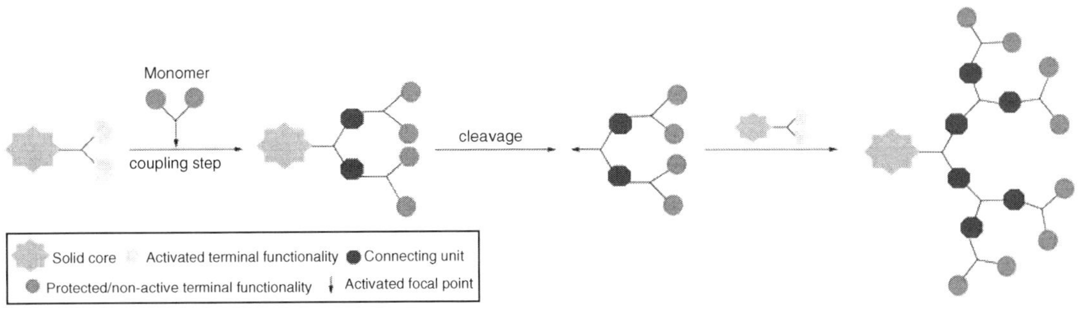

Scheme 15.2. The convergent approach to dendron assembly on solid support. (See the color version of this figure in Color Plates section.)

In a single report, one of the first dealing with solid-phase dendron synthesis, Moore and coworkers used a convergent approach.[8] According to this strategy, the assembly of each dendritic generation on solid support is followed by the dendron cleavage and coupling of the cleaved dendritic wedge to the support-bound branching unit (first-generation dendron), thus forming the next dendritic generation (Scheme 15.2).

Few years later, again in a single report, a combined divergent/convergent strategy was introduced by Wang and coworkers.[9] According to this approach, each dendritic generation assembly is followed by the division of the resin into two portions. Dendrons of the nth generation are cleaved from one portion (as in the convergent approach) and their focal point is covalently attached to the activated termini of the supported dendrons of the second portion, thus rapidly forming the generation $2n$ (Scheme 15.3). Although conceptually elegant, the divergent/convergent approach, as well as the previously mentioned convergent strategy, suffers from one of the characteristic features of solid-phase synthesis—the need to use a significant excess of soluble reagents for each synthetic step. Accordingly, the cleaved dendritic wedges, designated for reassembly on the resins, must be produced in large amounts. This demands an efficient, high purity synthesis, the use of relatively high loading on the support, and the use of large quantities of the support in the initial steps of the synthesis.

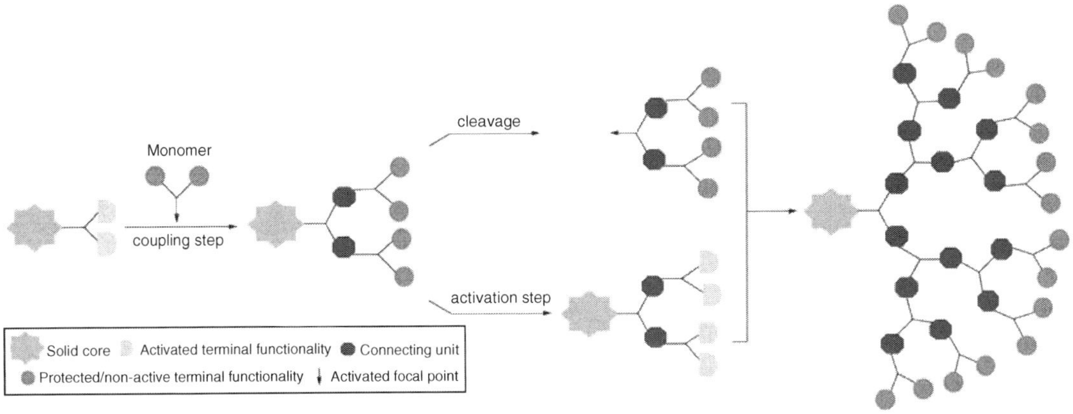

Scheme 15.3. The divergent/convergent approach to dendron assembly on solid support. (See the color version of this figure in Color Plates section.)

15.2.2 Preparation of Polyamide Dendrons

Many dendrons immobilized on insoluble support are, in fact, peptides or related polyamide structures. A number of reviews extensively surveyed peptide dendrimers, both soluble and support bound.[10]

15.2.2.1 Polylysine Dendrons. The most prominent of the support-bound peptidic dendrons is polylysine. The synthesis of up to tenth-generation dendritic polylysine was patented in 1981 by Denkewalter et al.,[11] but this structure remained largely unexplored until the pioneering reports of Tam published in 1988.[12] The dendrons were prepared on phenylacetamidomethyl (PAM) polystyrene with a β-Ala spacer by a divergent approach through a conventional solid-phase peptide synthesis (SPPS). Although based on a nonsymmetrical monomer, these structures are true dendrons and were prepared to the third (and later the fourth) generation. The major emphasis in these reports was on the application; the novelty of the synthetic concept and the dendritic nature of the polylysine were only recognized later.[13]

Polymer-bound polylysine dendritic cores were widely used as synthetic intermediates, and, accordingly, a number of schemes for their synthesis have been reported. Initially, the dendrons were prepared via the Boc SPPS method using a Boc-Lys(Boc)-OH building block (Scheme 15.4a).[12,14] For assembly of each layer, two coupling cycles were carried out, first with the preformed symmetrical anhydride and second with carbodiimide coupling agent.[12,14a,b] Later simpler coupling protocols based on DCC or HBTU/HOBt reagents were reported.[15] Subsequently, a number of groups prepared the same dendrons via the

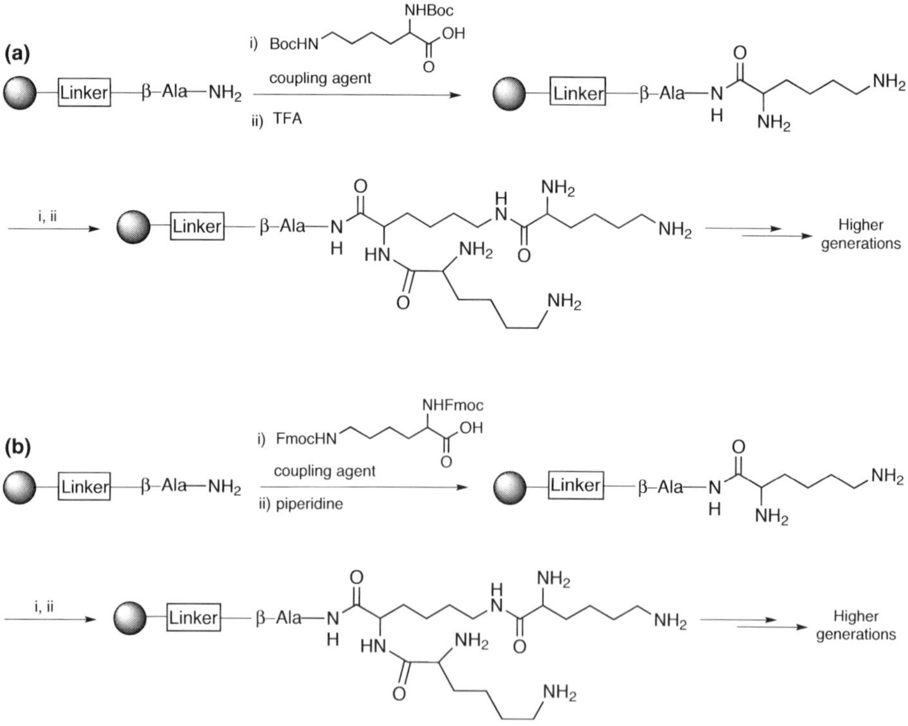

Scheme 15.4. Synthesis of polylysine dendrons on solid support.

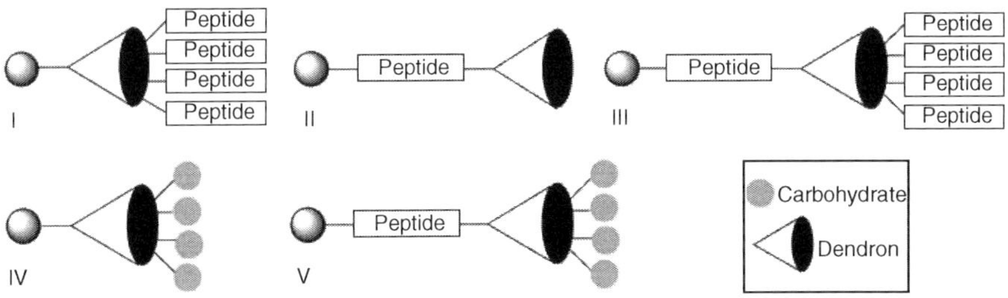

Figure 15.2. Various types of conjugates generated on solid support with polylysine dendrons.

Fmoc SPPS method, using, for instance, a Fmoc-Lys(Fmoc)-OC_6F_5 building block (Scheme 15.4b).[16a,b] Alternatively, DCC/HOBt, HBTU/HOBt, BOP, or PyBOP coupling protocols using the free acid were applied.[16c–e] In order to distinguish between the α- and ε-terminal amine groups of the dendron, the Boc strategy with a Boc-Lys(Fmoc)-OH building block in the first or last coupling was applied.[15a] TentaGel™ (polyethylene glycol (PEG)-grafted polystyrene) and polyethylene glycol polyacrylamide copolymer (PEGA) resins, equipped with a range of linkers, have also been adopted as supports for polylysine synthesis.[16e,17]

A variety of conjugates, incorporating polylysine dendrons, have been generated on these supports. The first structures synthesized were those with peptides decorating the dendron periphery (Figure 15.2, I).[12] Later peptides were also introduced between the support and the dendritic focal point (Figure 15.2, conjugates of types II and III).[15b,17,18] In subsequent works, carbohydrates were introduced as peripheral functionalities on the dendrons (leading to conjugates of types IV and V, Figure 15.2).[16e,17a,19] Hybrid dendrons incorporating PEG linear segments between lysine branching units were recently reported.[20]

15.2.2.2 Dendrons Combining Natural and Artificial Amino Acids. Of the naturally occurring amino acids, only lysine and glutamic acid have been used as branching units in the solid support-bound dendrons.[21] However, a variety of other natural amino acids have been used in combination with synthetic branching units for the construction of supported dendrons. In a series of publications, Reymond and coworkers reported the preparation of peptide dendrimers via the straightforward divergent synthesis using the Fmoc SPPS technique on polystyrene or TentaGel supports. The first series of the peptide dendrons (dubbed "1.1.1" dendrons) were assembled on supports equipped with the Rink amide linker using 2,3-diaminopropionic acid (DAP), (1,3-diaminoisopropyloxy)acetic acid, bis(aminomethyl)acetic acid, or 3,5-diaminobenzoic acid as branching units.[22] The naturally occurring amino acids (mainly histidine, serine, and aspartic acid, and also alanine, phenylalanine, threonine, and others) connected the branching diamino acids, whereas cysteine was located near the focal point (Figure 15.3a).

Mostly BOP or TBTU coupling protocols were applied for the assembly of dendrons. However, in the case of the coupling of amino acids to the polymer-bound 3,5-diaminobenzoic acid branching unit, only monoacylation occurred under these conditions and the dendritic growth was restricted.[22c] To achieve coupling to both amino groups of the branching unit, the symmetrical anhydrides of the *N*-Fmoc amino acids (prepared using DIC) were used. The revealed chemoselectivity of the BOP-induced coupling enabled preparation of dendrons with differentially substituted branches (Figure 15.3a-2).[22c] Alternatively, dendrons with differentially substituted branches were prepared using the

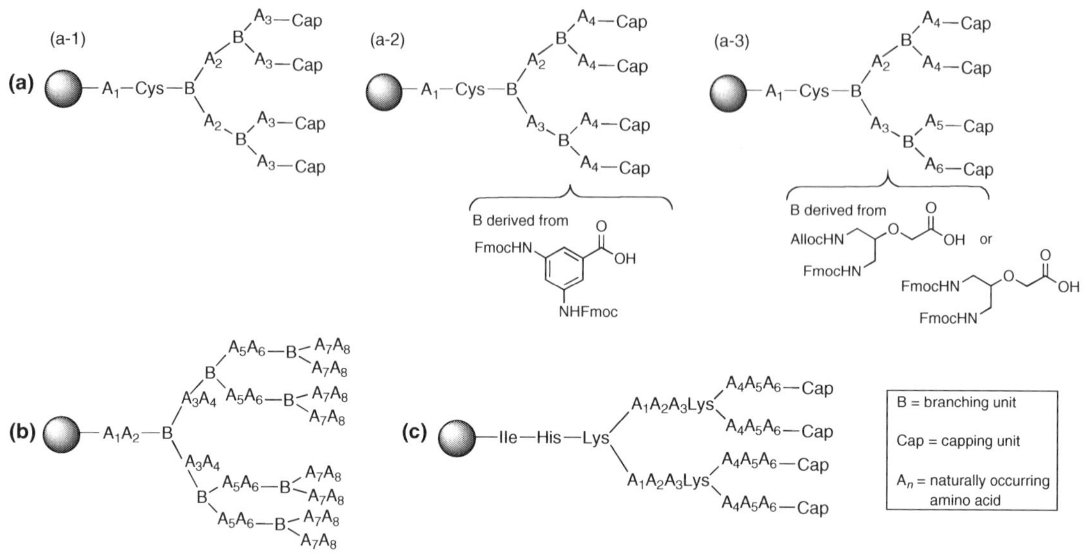

Figure 15.3. Peptide dendrons prepared by Reymond et al. on solid support.

orthogonally protected (1,3-diaminoisopropyloxy)acetic acid branching unit (Fmoc and Alloc protecting groups, Figure 15.3a-3).[22d]

The second series of Reymond's peptide dendrons (dubbed as "2.2.2.2" dendrons) with dipeptides (mostly His-Ser dyad, but also other sequences) connecting the artificial branching units (DAP diamino acid) was assembled on Rink amide TentaGel resin using BOP- or PyBOP-based coupling protocols.[23] Up to fourth-generation dendrons were prepared based on this design (Figure 15.3b). These dendrons were also prepared in a combinatorial format yielding a 65,536-membered library of third-generation dendrons via split-and-mix approach.[23a,b]

Recently, a third type of dendron based on tripeptides connecting the lysine branching units was described (Figure 15.3c).[24] A library of second-generation dendrons and, subsequently, selected dendrons of the third generation, based on this design, were prepared on TentaGel resin using Fmoc-protected building blocks and BOP/PyBOP coupling protocols. Although these dendrons are based on a proteinogenic branching amino acid and formally belong to the polylysine type of dendritic molecules, this work evolves from the other projects carried out by Reymond's group and therefore is described in this section.

A similar design, with artificial amino acids as branching units, separated by linear peptide segments, was applied in the synthesis of dendritic peptides on polystyrene by Albericio, Giralt, Royo, and coworkers (e.g., the synthesis in Scheme 15.5).[25] Mostly *cis*-4-amino-L-proline (L-Amp) and imidazolidine-2-carboxylic acids were used as the branching units. Linear pentaproline segments connected the branching units into the dendritic structure. The dendrons were assembled by a Boc SPPS technique. The synthesis followed, in general, the divergent approach, though parts of the dendrons were prepared separately on support, cleaved, and reassembled on support into larger structures, thus imparting some characteristics of the convergent approach onto the synthesis. After the initial screening of various synthetic approaches, the method based on solid-phase synthesis of the branched module Fmoc-Pro_5-L-Amp(Fmoc-Pro_5)-OH from Fmoc-L-Amp(Boc)-OH using DIC, BOP, or PyAOP coupling agents was selected. Following its synthesis on support, the branched module was cleaved and reimmobilized on the PAM resin-bound hexaproline linear

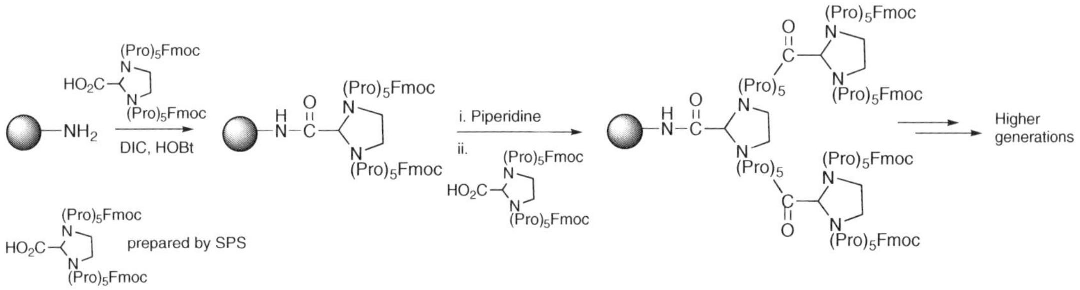

Scheme 15.5. Synthesis of polyproline dendrons on solid support.

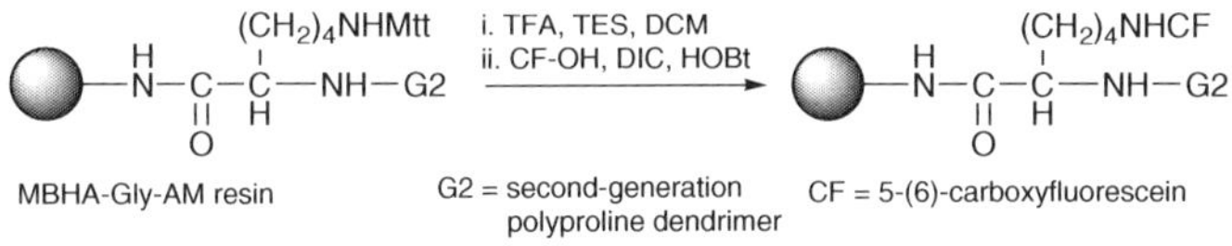

Scheme 15.6. Synthesis of polyproline dendrons fluorescently labeled near the focal point.

segment H-Pro$_6$-PAM resin (to form the first generation) or on the first-generation dendron (after the removal of the terminal Fmoc protecting groups) to form the second-generation polyproline structure (Scheme 15.5). While the replacement of L-Amp by imidazolidine-2-carboxylic acid did not lead to a significant improvement in the yield or purity, introduction of the flexible Gly residue at the N-terminal position of the monomeric building block (i.e., replacement of the (Pro)$_5$ spacer connecting the branching amino acids by Gly(Pro)$_5$) resulted in a more efficient synthesis. The higher efficiency was reflected in a higher yield and a higher level of purity, enabling construction of dendrons up to the fourth generation.[25d] Introduction of an ε-Mtt-protected lysine residue between the polymer core and the dendron enabled postsynthetic tagging near the focal point (e.g., fluorescent tag, Scheme 15.6).[25e] While preparing the polyproline dendrons, Albericio and Royo observed fragility of the beads, suggested that the bead breakdown resulted from the "saturation of solid support," and proposed calling the phenomenon "stress of the bead".[26]

Starting from 1999, Alper, Arya, and coworkers prepared a variety of support-bound dendrons. Among other systems, polyamide dendrons, based on 3,5-diaminobenzoic acid branching units, were prepared to the third generation on a polystyrene support (Rink amide MBHA resin) via the divergent Fmoc chemistry-based approach with DIC/HOBt coupling agents.[27] In the first of the reported dendron types, the branching amino acid moieties were connected by the Phe-Gly dipeptide spacer (Figure 15.4a).[27a] In the more advanced version of these polymer-bound dendrons, the repeating motif was Lys-3,5-diaminobenzoic acid-Gly; thus, an additional branching point was added and each generation provided an AB$_4$-type unit (Figure 15.4b).[27c]

15.2.2.3 Dendrons Made of Artificial Amino Acids. Dendrons based on the already mentioned 3,5-diaminobenzoic acid were among the first dendron types prepared on solid support. In 1991, Fréchet and coworkers reported the preparation of polyamido dendrons on Merrifield polystyrene using Boc-protected 3,5-diaminobenzoic acid or its bis-β-alanine conjugate as monomers (Scheme 15.7).[28]

Because of the low efficiency of coupling, the synthesis of the dendron based on the shorter unit stalled at the third generation (19% coupling yield only). With the second

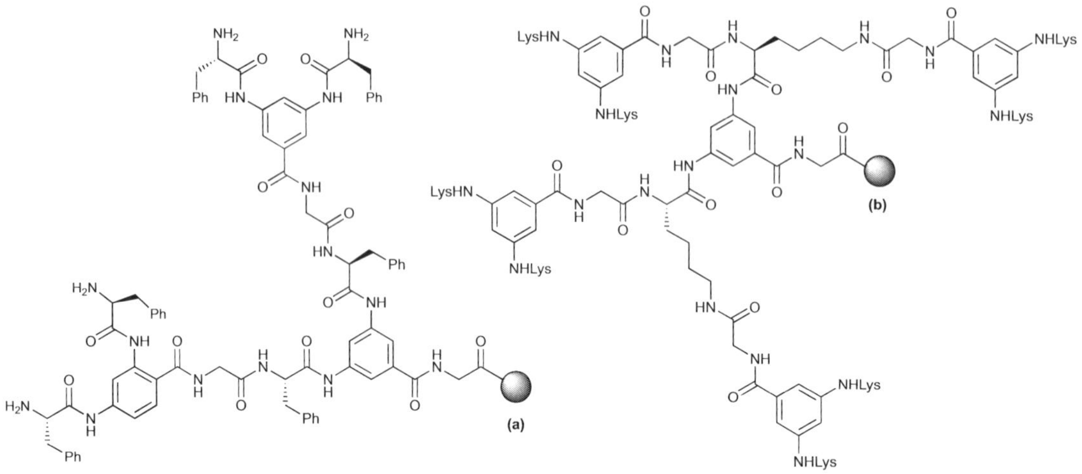

Figure 15.4. Polyamide dendrons prepared by Alper and coworkers on polystyrene support.

monomer, possessing elongated branches and somewhat more reactive terminal amine functionalities, the efficiency of the synthesis was improved and it proceeded with modest 70–80% coupling yields up to the fourth generation. However, it was not possible to proceed further. The cleavage of the dendritic molecules, followed by size exclusion chromatography (SEC), revealed a broad molecular weight distribution and confirmed the nonuniform nature of the dendrons of higher generations. According to the authors, the incomplete couplings may be a result of the difficult coupling chemistry chosen for the propagation step, the restricted access of the reagents to the propagation sites on the polymer, or the steric crowding of the peripheral units frequently observed in the dendrimer chemistry.

A number of groups have reported preparation of support-bound dendrons based on trifurcated branching units. The dendrons prepared divergently by Lee and coworkers on TentaGel and PS-*co*-PEG-NH_2 resins were based on symmetric, TRIS-derived triamino acid monomer and were constructed up to the formal fifth and seventh generations, respectively, using the BOP/HOBt coupling method (Scheme 15.8).[29] Since only the spectrophotometric determination of Fmoc release from the molecules, attached to the dendritic termini, was used to deduce the efficiency of the dendron growth, the characterization of the dendrons was rather limited and no indication as to the dendron homogeneity was provided. Chan and coworkers used a nonsymmetric triamino acid, 4-azalysine, to prepare (via divergent growth) first- and second-generation peptide-like

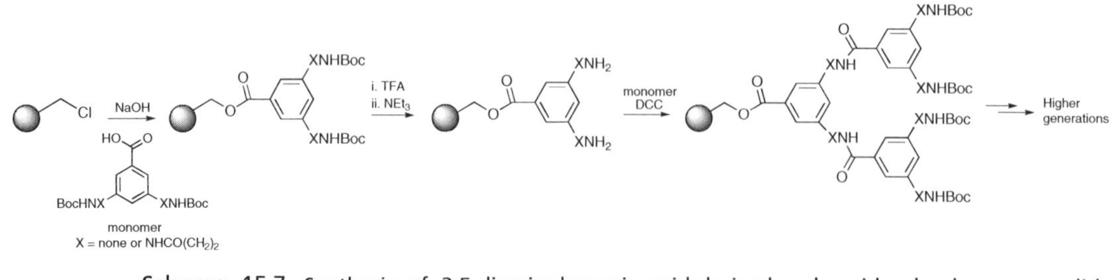

Scheme 15.7. Synthesis of 3,5-diaminobenzoic acid-derived polyamide dendrons on solid support.

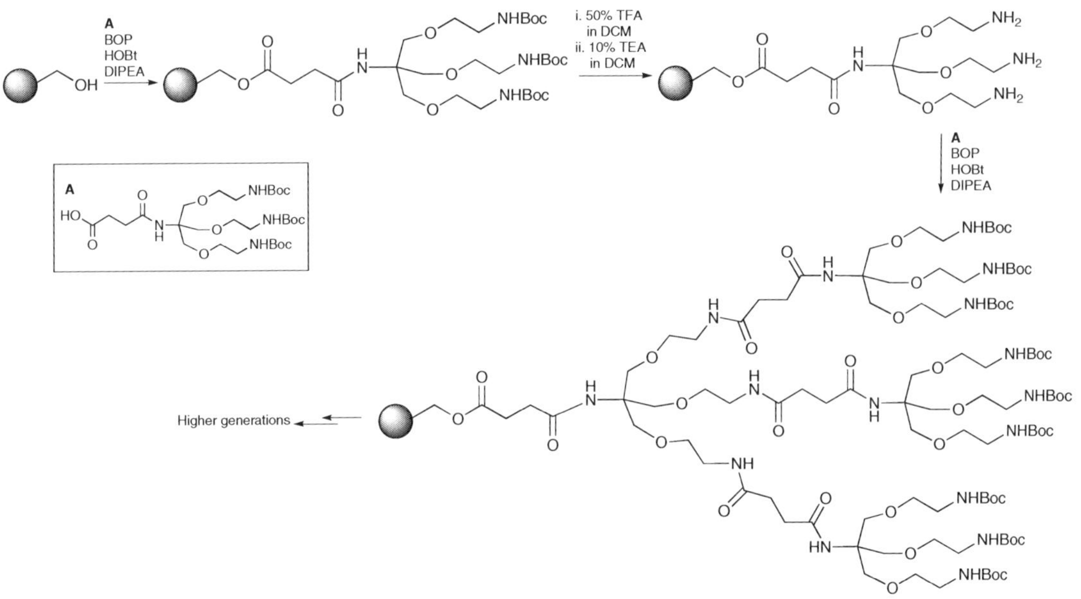

Scheme 15.8. Synthesis of supported polyamide dendrons prepared by Lee et al.

dendrons on polystyrene using Boc-protected monomer and DIC/HOAt (first generation) and HATU (second generation) coupling protocols (Figure 15.5).[30]

In a similar manner, Bradley and coworkers used a lysine-imitating *N*-Fmoc-*N*-(6-*N′*-Fmoc-aminohexyl)glycine monomer and HOBT/DIC coupling protocol for synthesis of up to third-generation peptoid dendron on Rink-amide resin (Figure 15.6).[31] The synthesis was facilitated by microwave irradiation.

The DAP, which was used by Reymond et al. in the synthesis of peptide dendrimers (see above), was adopted by Wickstrom and coworkers as a single building block for assembly of polyamide peptide-like dendrons on a PNA–peptide chimera presynthesized on a TentaGel Sieber resin (Scheme 15.9).[32] Aminoethoxyethoxyacetate spacers were used to separate the various parts of this hybrid structure (peptide, PNA, and dendron) and HATU coupling

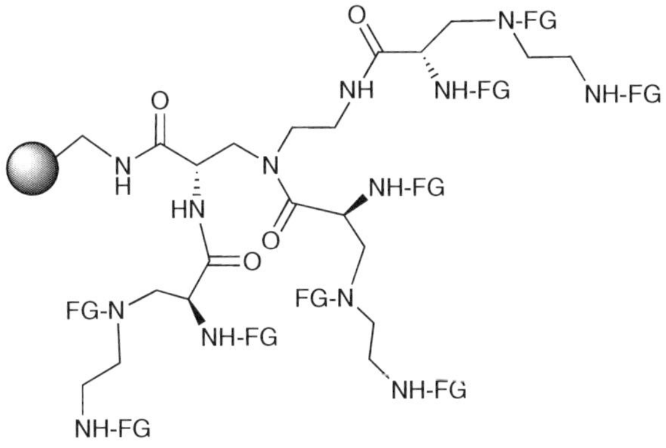

Figure 15.5. Supported polyamide dendrons derived from 4-azalysine.

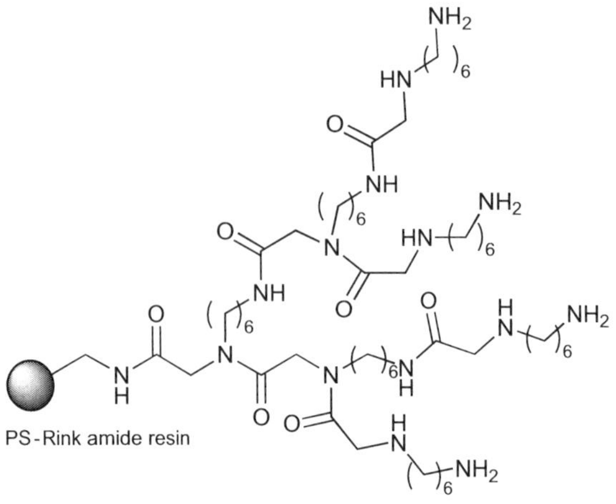

Figure 15.6. Supported peptoid dendrons prepared by Bradley et al.

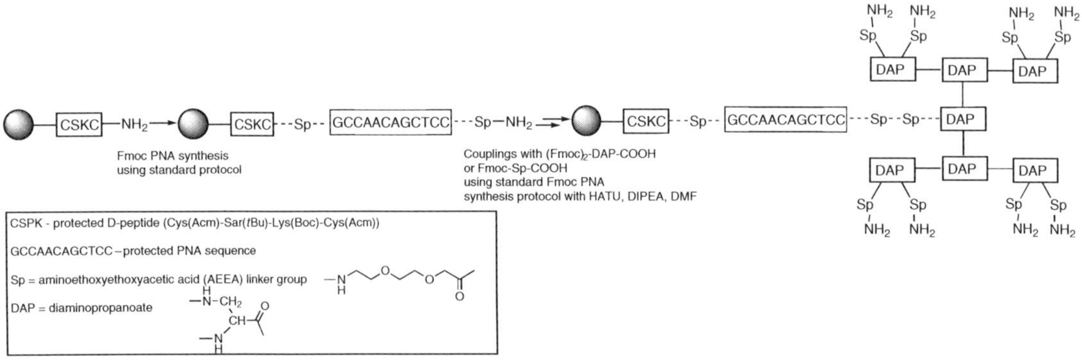

Scheme 15.9. Synthesis of DAP-based dendrons on the polymer-supported PNA–peptide grafts.

reagent was applied for the synthesis of the dendrons that were prepared up to fourth generation.

Publications from a few groups have described preparation of other amide bond-linked dendrons on polystyrene, glass, and polypropylene support.[33] These were based on diamino-, triamino-, or pentaamino carboxylic acid monomeric units, either prepared in solution or assembled from simpler building blocks on solid support.

15.2.2.4 Polyamide Dendrons from Alternative Building Blocks.

In 2006, Bai and coworkers prepared silica-bound polyamide dendrons via an alternative approach. Instead of using amino acid heterofunctional building blocks, the dendritic growth was envisioned via an alternating use of homofunctional 1,3,5-benzenetricarbonyl trichloride and (+)-1,2-diphenylethylenediamine (Scheme 15.10).[34] The terminal amine group was capped with phenyl isocyanate. In this way, the authors attempted to prepare dendrons of up to third generation. The elemental analysis demonstrated imperfect growth of the dendrons. This can be attributed to the incomplete reactions or, more likely, to the inter- and intramolecular cross-linking (cyclization in the intramolecular case) during the amidation reactions. Such unwanted side reactions are inherent to this approach and probably occur in spite of the excess of the soluble reagents used in the synthesis.

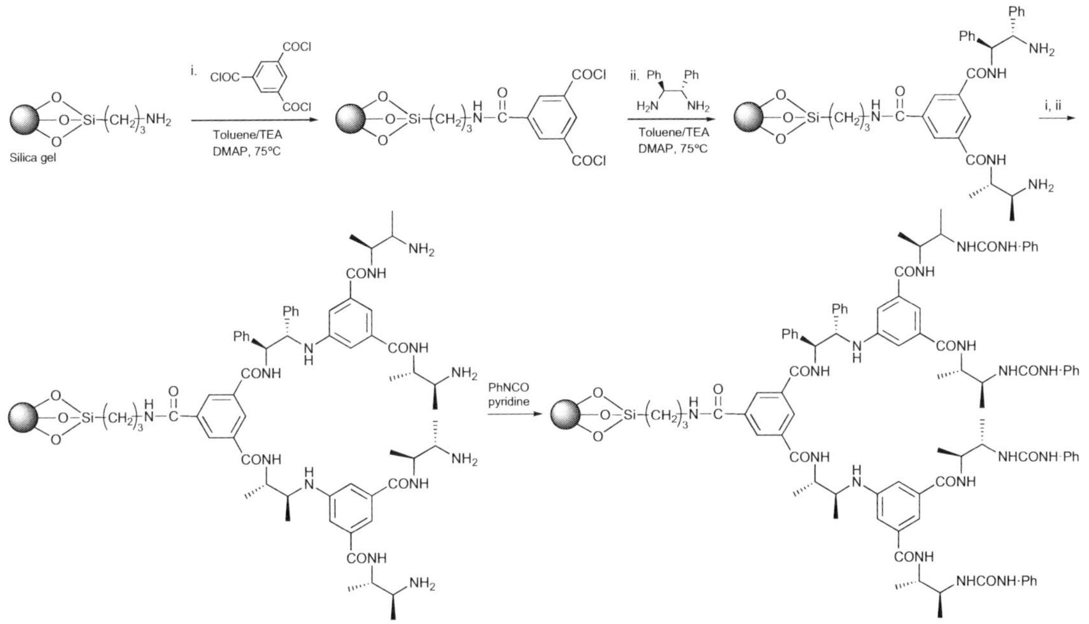

Scheme 15.10. Synthesis of polyamide dendrons from two homofunctional building blocks.

15.2.3 Preparation of Polyamidoamine Dendrons

Thus far, all the dendritic structures presented in this chapter were constructed via the amide bond forming reactions. A somewhat different synthetic strategy led to the synthesis of the polyamidoamine (PAMAM) dendrons on insoluble supports. In 1997, Bradley and coworkers introduced the solid-phase synthesis of PAMAM dendrons on organic polymer support. Initially, these dendrons were prepared on TentaGel up to the fourth generation (although the third generation was considered by the authors to be the optimal for synthetic purposes) (Scheme 15.11).[35] The assembly of the dendrons followed the divergent synthetic strategy developed by Tomalia et al. for PAMAM dendrimers in solution and was based on alternating steps of double Michael addition of terminal primary amines to methyl acrylate and aminolysis of the formed terminal esters with α,ω-alkanediamines (e.g., 1,3-propanediamine).[36] Later, the approach was extended to PAMAM dendrons on polystyrene beads equipped with a short PEG spacer.[37] Basso and Bradley investigated the issue of unwanted site–site interactions on third-generation PAMAM dendronized TentaGel during solid-phase

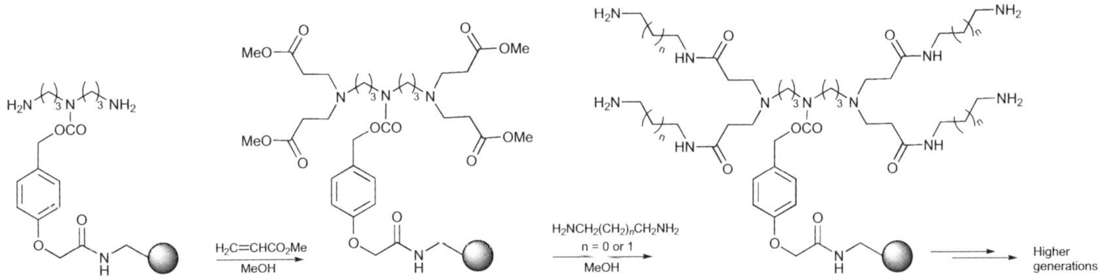

Scheme 15.11. Synthesis of PAMAM dendrons on solid support.

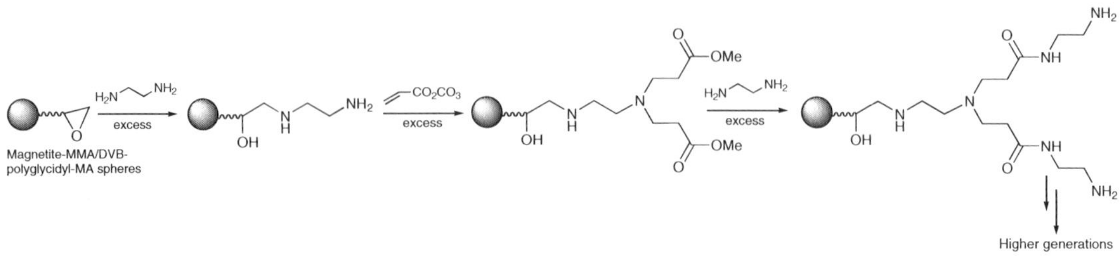

Scheme 15.12. Synthesis of PAMAM dendrons on multilayer microspheres.

synthesis.[38] It was demonstrated that for reaction at ambient temperature, under reagent excess conditions standard for solid-phase synthesis, the site–site interaction problem is negligible, although the G3 dendronized support proved somewhat more susceptible than the parent TentaGel to this type of side reactions.

In recent years, PAMAM on polystyrene was also synthesized by other groups.[39,40] A similar methodology was used by Wang and coworkers to assemble PAMAM dendrons on multilayer microspheres consisting of a magnetite core covered by an intermediate methyl methacrylate–divinylbenzene polymer shell and polyglycidyl methacrylate outer layer (Scheme 15.12).[41]

PAMAM dendrons were also prepared on silica, as early as 1998, by Tsubokawa et al. and by many other groups, such as Arya and Alper, Kunitake, Rhee, Bu, Kawi, Qiu, Imae, and Yang.[42–50] Usually, 3-aminopropyl-functionalized silica was employed as the parent support. In most studies, ethylenediamine-based PAMAM was constructed via the approach developed for solution by Tomalia and pioneered on solid support by Bradley.[35–37]

Using ultrafine silica (Tsubokawa),[42a] and later regular amorphous silica (Rhee and Alper),[43c,d,45] the researchers examined the obtained content of the amino group (or other terminal group) per gram of silica (or total mass) and compared it with theoretical values. They found that the observed values were much lower than the theoretical ones and that the ratio of the two values declined rapidly as the generation increases. Thus, when second-generation dendrons were constructed on particles with an initial amine content of 0.9 mmol/g SiO_2, the observed value was only 45% of the theoretical one, whereas for the fourth generation, it decreased to about 16%.[42a,45] Remarkably, when dendrons were constructed on silica with a lower initial loading of the amino groups, the deviation of the observed values from the theoretical ones was smaller and the decline in the ratio of the two values as a function of generation was less steep.[42a,45,46a]

The observed deviation most likely resulted from incomplete propagation reactions caused by the increased steric hindrance in the pores and on the surface of the silica, accompanying the dendrimer growth. IR studies have supplied evidence of an incomplete amidation step.[42a] Bu et al. investigated the reaction kinetics of propagation of PAMAM dendrons on solid spherical silica gel.[46b,c] Their findings point to steric hindrance during the Michael addition step and a site–site cross-linking reaction, occurring during the amidation step, as the mechanistic reasons for structural defects of the dendrons. As in the previously described studies, data from this study demonstrated that the preparation of dendrons on a support with a reduced loading of the initiator amino groups proceeds with a remarkably enhanced efficiency and produces more perfect dendritic structures than synthesis on support with a higher initial loading. Alper and Zweni demonstrated that, indeed, the efficiency of the PAMAM dendron synthesis on silica decreases with the increase in the length of the α,ω-diamine involved in the synthesis.[43d] This phenomenon can be explained by a higher percentage of the cross-linking side reaction accompanying lengthening of the diamine. Overall, the characterization results suggest that the PAMAM preparation on silica produced, at least in the earlier studies, a hyperbranched polymer rather than a perfect dendritic structure.[42a,43c]

A detailed analysis of the PAMAM-on-silica synthesis and properties published subsequently by Kunitake and coworkers pointed out that not only the generation and initial loading of the amino sites, but also the propagation step parameters, such as the reaction time and temperature, define the degree of branching and structural uniformity of the dendrons.[44b] The structures with the highest degree of perfection seem to be those prepared at low temperatures with long reaction times and low loading of amine initiation sites.

An additional important factor, affecting the efficiency and uniformity of the PAMAM dendron growth, is the nature of the support. In the earlier studies described above, amorphous silica with broad pore size distribution and average pore size of 6 nm was employed. A series of articles by Alper, Sayari, and coworkers described preparation of dendrons on a periodic MCM-41 silica (pore size 6.5 nm), a pore-expanded MCM-41 silica (pore size 10.6 nm), and large-pore Davisil silica (pore size 18 nm).[51] In-depth characterization of the dendronized silica, which included solid-state NMR, FTIR, nitrogen adsorption studies, elemental analysis, and thermogravimetry, showed that the dendrons are grown inside the silica pores, completely filling them, as the third or fourth generation is assembled. The physical limit on the dendron size, dictated by the pore dimension and shape, limits the propagation of the dendritic oligomers. Yet, for the pore-expanded silica, an efficient synthesis of the fourth-generation dendrons with the overall yield of 71% could be achieved.[51b] Improvement in the PAMAM preparation process through quaternary ammonium salt catalysis of the aza-Michael step was later demonstrated by Qiu and coworkers on mesoporous silica.[48] Shortening of the reaction times by microwave irradiation during the PAMAM synthesis was demonstrated by Yang and coworkers.[50]

The PAMAM grown on silica can be used as it is,[42a,44,52–54] or with small organic or organometallic moieties attached to the periphery of the dendrons.[42b,43,45,46a,47,49,51,55,56] Alternatively, a polymer (e.g., vinyl polymer) can be grafted on the dendron termini, thus forming core–shell particles with the dendritic interface between the inorganic core and the organic outer shell.[42c]

In a number of cases, PAMAM dendrons were also prepared on other polymer particle cores, such as alumina, chitosan, cellulose, zirconia–urea-formaldehyde resin, silica-coated magnetite, and carbon nanotubes.[57] Like the silica support (see above), at least in the case of the chitosan, the theoretically predicted propagation of the dendrons was not achieved, presumably due to the steric hindrance in the support pores.

15.2.4 Preparation of Polyurea Dendrons

Triply branched polyamidourea dendrons were developed by Bradey and Fromont in 2000 (Scheme 15.13).[58] The TRIS-derived monomer, an AB_3-type symmetrical isocyanatotriester, was prepared in three steps in solution. The dendrimer growth involved the reaction of the isocyanate functionality of the monomer with a resin-bound primary amine (forming urea linkage), followed by the reaction of the terminal ester groups with propane-1,3-diamine, forming an amide and restoring a primary amine terminus. The synthesis was monitored by both on-resin and cleavage analysis, thus ensuring clean and uniform growth of the dendritic wedges. Although the dendrons were constructed on aminomethyl polystyrene or polystyrene, bearing diamine linker, only to the second generation, their triply branched nature and the large size of the starting resin beads resulted in dendronized

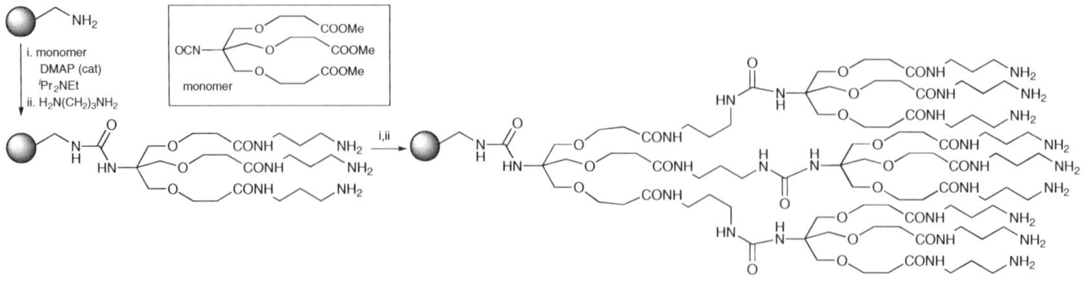

Scheme 15.13. Synthesis of polyamidourea dendrons on solid support.

beads with a very high loading of the terminal groups (up to 230 nmol/bead). Recently, facilitation of this synthetic scheme using microwave irradiation was demonstrated.[59]

To avoid possible macrocyclization during the alkoxide displacement by a diamine (a step also common to PAMAM synthesis), alternative monomers for preparing urea-bridged triply branched dendrons were prepared and used.[60] The dendrons were assembled on solid support, up to the third generation, from Boc-protected monomeric triaminoisocyanates via a urea-connector formation, followed by deprotection of the terminal amines (Scheme 15.14). The assembly took place on polystyrene as well as TentaGel resins with up to 19-fold increase in the loading of the terminal groups.

The study of the properties of these dendronized beads revealed that dendronization increases the bead fragility. The attempts to assemble third-generation dendrons on large polystyrene beads (250–300 μm) resulted in breakage of the beads, as a result of isobutylene and CO_2 gas bubble evolution during the deprotection step. Increasing the length of the monomeric arms improved the situation only marginally. The authors assumed that electronic repulsion between the dendrimers (positively charged under these conditions) resulted in the rupture of the polystyrene matrix. These results correlate with the previously mentioned "stress of the bead" observed by Albericio and coworkers.[26] Replacement of the Boc protecting group by the Dde group enabled deprotection under milder conditions, without any gas evolution, thus improving the quality of the formed dendronized beads.[60c]

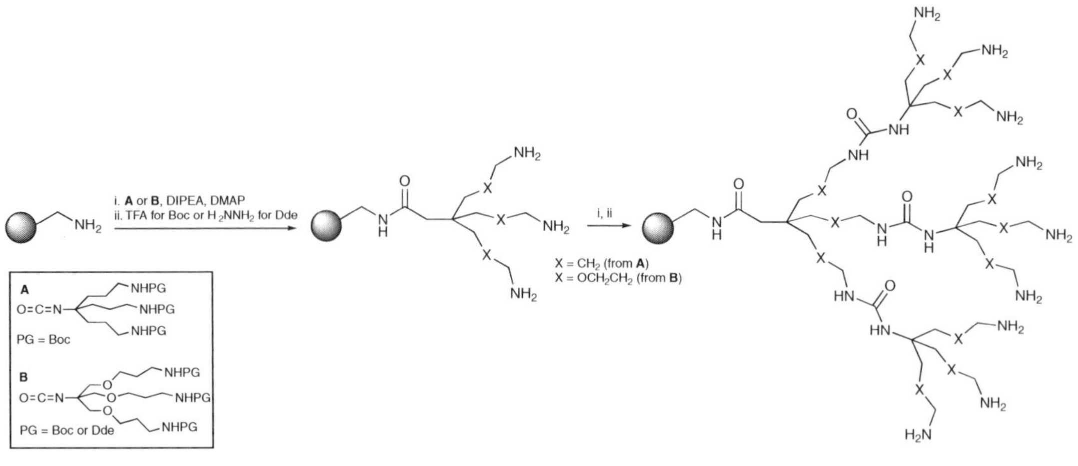

Scheme 15.14. Synthesis of polyurea dendrons on solid support.

Subsequently, this synthetic methodology was extended to synthesis of polyurea dendrons on SynPhase™ crowns and lanterns, radiation-grafted polyethylene.[61] This polymer support displays greater mechanical strength than the polystyrene beads and consequently the evolution of isobutylene and CO_2 during the deprotection step did not rupture the polymer matrix (although during the deprotection of the second-generation dendrons on crowns, roughening of the surface texture was observed).

On these supports, the first-generation dendrons could be prepared quantitatively, but the second-generation dendrons could be assembled only on crowns. This limitation was attributed to the very high density of the terminal amino groups on the unmodified lanterns and, as a result, to severe steric crowding after the assembly of the first-generation dendrons. This steric crowding is probably also responsible for the inhibited reaction kinetics (as was demonstrated for Fmoc release reaction and amino acid coupling) on dendronized crowns and lanterns. It is likely that due to the same reason, mixed anhydrides formed with EEDQ from Fmoc-protected amino acids showed better results than DCC-activated acids in the amidation of the terminal amino groups of the dendrons.

15.2.5 Preparation of Polyester Dendrons

Fréchet and coworkers explored the synthesis of polyester dendrons on solid support.[62] These dendrons, based on 2,2-bis(hydroxymethyl)propanoic acid monomers, were assembled up to the fourth generation on poly(2-hydroxyethyl methacrylate-*co*-ethylene dimethacrylate) resin and decorated with chiral proline derivatives (Scheme 15.15). Two methods for preparation of these materials were reported. The solid-phase divergent approach yielded polymer with higher loading but a less precise architecture than the second approach. The second approach, the divergent synthesis of the dendrons in solution, followed by the focal point deprotection and then attachment to the solid support, led to well-defined structures on the polymer, but relatively low loading.

Synthesis of dendrigrafts of citric acid on acrylonitrile/acrylic acid copolymer solid films via repetitive DCC-induced esterification reaction were reported by Kish et al.[63] Since after each coupling cycle the unreacted sites cannot be distinguished from the carboxylic ends of the units that had just been attached, it is likely that a range of dendrons of different generations (rather than monodispersed dendrons of a certain generation) populate the surface of the film.

15.2.6 Preparation of Polyether Dendrons

The supported dendritic molecules, described thus far, were all based on carbonyl-containing connecting groups (amides, ureas, esters). These groups are vulnerable to a variety of reaction conditions (reductive, hydrolytic, nucleophilic) and, therefore, are unsuitable for a wide range of synthetic and catalytic applications. Moreover, the coordinating nature of the

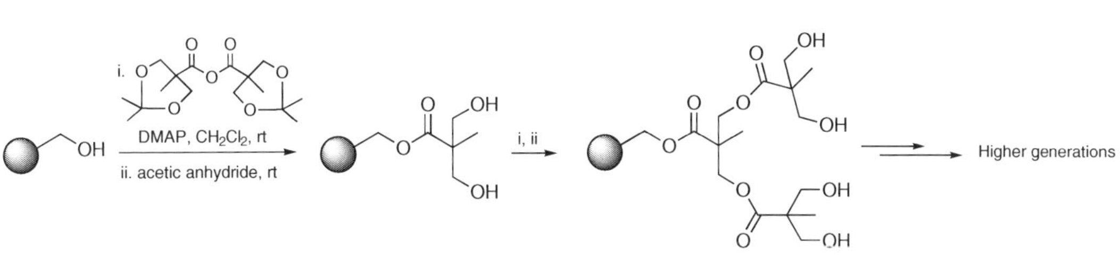

Scheme 15.15. Synthesis of polyester dendrons on solid support.

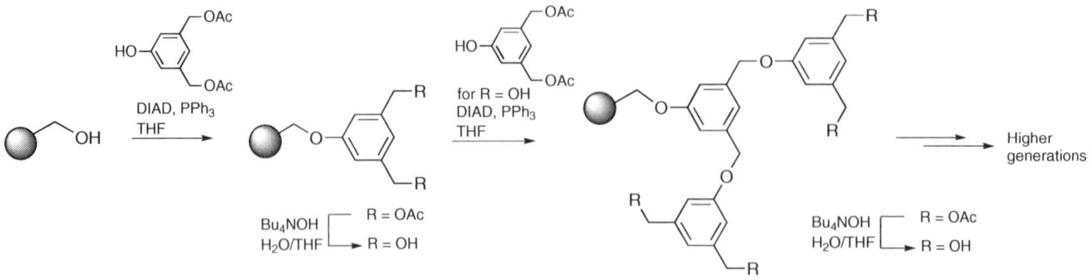

Scheme 15.16. Synthesis of polyether dendrons on solid support by Bradley et al.

amido group can affect the functioning of the constructs in some applications (e.g., metal coordination and catalysis). For these reasons, alternative dendritic motifs were considered.

Ethers are chemically and coordinatively more inert than amides. Indeed, Fréchet-type polyether dendrimers prepared in solution demonstrate a broad spectrum of chemical and coordinative compatibilities.[64] Accordingly, polyether dendron synthesis on solid support was developed, initially by Bradley and coworkers and later by Portnoy and Dahan[65,66,68] Instead of the 3,5-dihydroxy benzyl alcohol derivatives used in the soluble Fréchet-type dendrimers, a more SPOS suitable 3,5-bis(hydroxymethyl)phenol unit was conceived as a building block for the dendron construction.

The first reported synthesis of the polystyrene-supported aryl benzyl ether dendrons was based on a 3,5-bis(acetoxymethyl)phenol building block, which was prepared in solution from the commercially available dimethyl-5-hydroxyisophthalate in four steps. Using this building block, Bradley and coworkers constructed the dendrons up to the fourth generation through two repetitive steps (Mitsunobu coupling and ester hydrolysis, Scheme 15.16).[65]

Adopting an alternative synthetic strategy, Portnoy and Dahan prepared the dendron, again via two repetitive steps (Mitsunobu coupling and ester reduction), using the dimethyl-5-hydroxyisophthalate itself without any need for building block preparation in solution (Scheme 15.17a).[66] Although the coupling reagent inducing the Mitsunobu reaction (the Castro reagent) imposes very high conversion and purity at the coupling step of this synthetic scheme, it has to be prepared from costly starting materials and has a limited shelf life.[67] To avoid this obstacle, Portnoy and coworkers changed the synthetic scheme (Scheme 15.17b), replacing the Mitsunobu coupling by chlorodehydroxylation–nucleophilic substitution sequence, thus adding a third synthetic step during each generation assembly but eliminating the need for the Mitsunobu special reagent.[68]

The same group introduced a tetrafurcated building block that can be incorporated into the dendritic structures.[69] Tetra-, octa-, and hexadecaol dendritic grafts were prepared, starting from Wang resin or the aforementioned first- and second-generation polyether dendronized resins (Scheme 15.18). The unit is built on hydroxyl-terminated resins in four highly efficient steps: attachment of the aldehyde (or oxidation of the terminal alcohol into aldehyde), Knoevenagel condensation with a malonate diester, Michael addition of the same diester, and reduction that restores the hydroxyl terminal groups. Combining this reaction sequence with the aforementioned polyether dendron synthesis, the loading of the terminal hydroxy groups can be increased up to 10-fold compared to the parent Wang resin.

Recently, assembly of first-generation dendrons, bearing from two to four homo- or heteroprotected amines, via Mitsunobu coupling of resin-bound 3,5-dihydroxybenzoate with protected amino- or diaminoalcohols was described by Gellerman et al. (Scheme 15.19).[70] It was envisioned that in the case of the heteroprotected diamine arms,

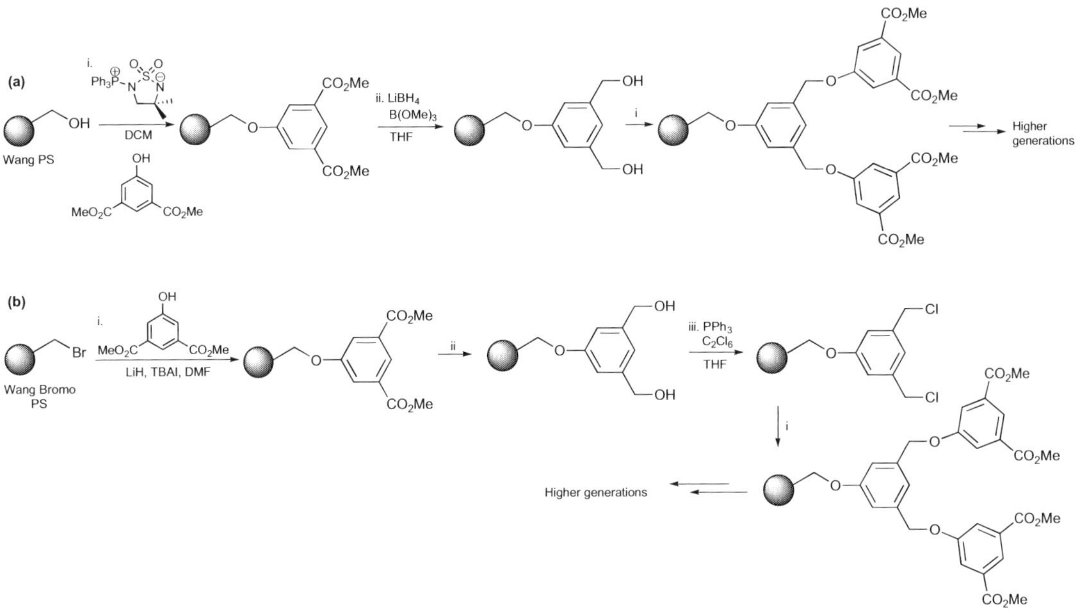

Scheme 15.17. Syntheses of polyether dendrons on solid support by Portnoy et al.

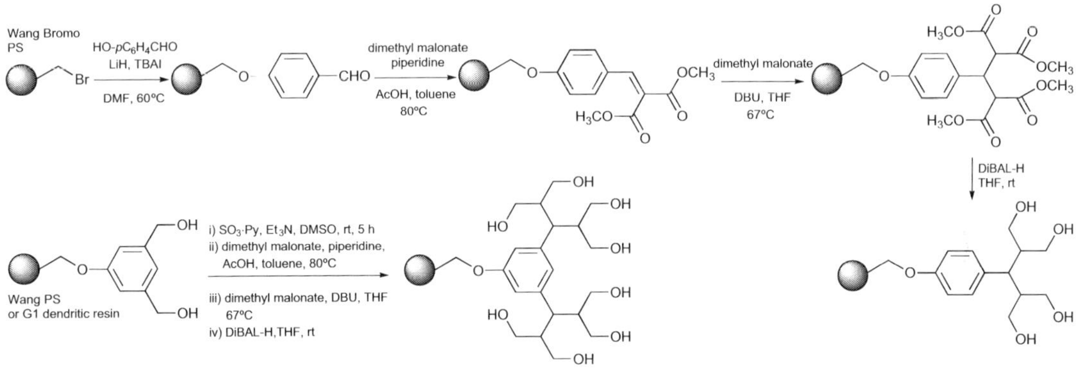

Scheme 15.18. Synthesis of tetrafurcated polyether dendrons on solid support.

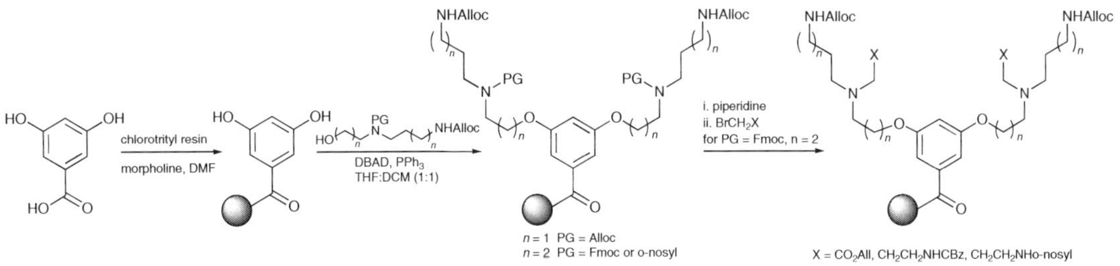

Scheme 15.19. Synthesis of amine carrying ether dendrons by Gellerman et al.

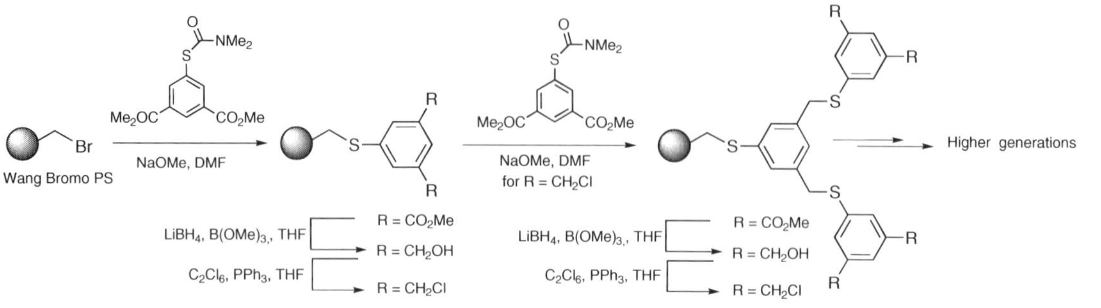

Scheme 15.20. Synthesis of polythioether dendrons on solid support.

sequential deprotection and functionalization of the amino groups will lead to multifunctional dendrons with predetermined end group arrangement.

15.2.7 Preparation of Polythioether Dendrons

Polythioether dendrons, the sulfur analogues of the poly(aryl benzyl ether) dendrons, were assembled by Portnoy and coworkers to the fourth generation. These dendrons were prepared using the protected thiophenol monomer building block, which can be efficiently synthesized from the hydroxyisophthalate diester in multigram quantities.[71] The deprotection of the thiophenolate performed *in situ* is followed by three steps (similar to those previously described for the polyether dendrons in Scheme 15.17b) for each generation assembly (Scheme 15.20).[72]

15.2.8 Preparation of Polyamine Dendrons

Portnoy and Dahan also prepared polyamine dendrons, analogous to the aryl benzyl ether and thioether dendrons.[73] Although the 5-amino isophthalate diester is commercially available, its nucleophilicity is too weak to enable dendron assembly through a route analogous to those in Schemes 15.17b and 15.20. Following the attachment of the first module, the growth of the dendron is achieved via a repetitive reduction, oxidation, and reductive amination sequence (Scheme 15.21). These poly(aryl benzyl amine) dendrons were assembled divergently to the third generation.

The feasibility of a synthetic approach for solid-phase assembly of aliphatic polyamine dendritic structures via tandem hydroformylation/reductive amination of olefins and phthalimide protecting group removal was demonstrated by Eilbracht and Koç.[74] However,

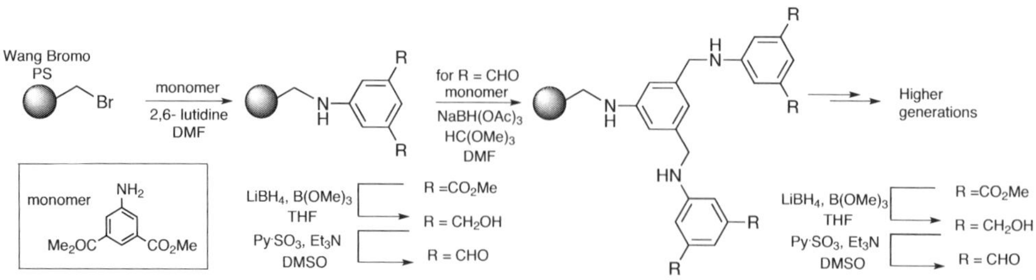

Scheme 15.21. Synthesis of polyamine dendrons on solid support.

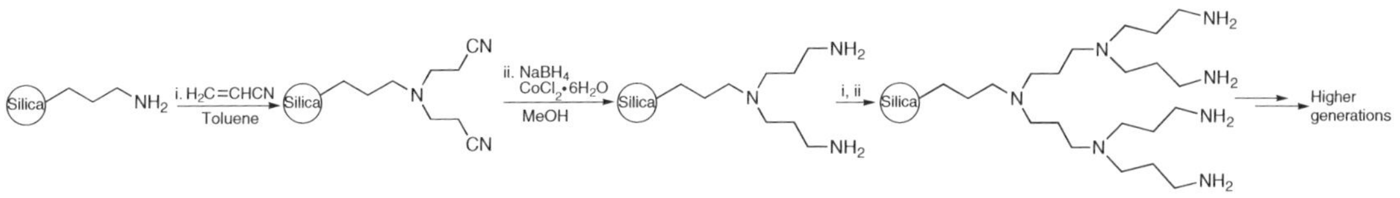

Scheme 15.22. Synthesis of poly(propylenimine) dendrons on solid support.

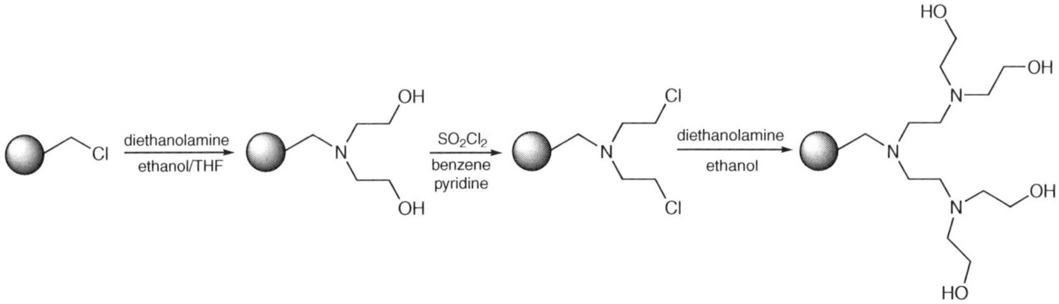

Scheme 15.23. Synthesis of polyamine dendrons by Qu et al.

only the simple protected first-generation structure was prepared on Wang polystyrene through this approach.

The "classic" poly(propylenimine) dendrons were prepared on silica gel by Liu et al. via a divergent approach that was based on a repeating sequence of amine Michael addition to acrylonitrile and nitrile reduction (Scheme 15.22),[75] a route pioneered in solution many years ago by Vögtle and Meijer.[76] Synthesis of fourth-generation dendrons was reported, though the publication lacked the characterization details to support the formation of truly dendritic architecture.

Hydroxyl-terminated polyamine dendrons of the first and second generations were obtained on macroporous polystyrene by Qu and coworkers.[77] The authors used diethanolamine monomer for the divergent synthesis of the dendrons. This was immobilized by nucleophilic substitution, while the terminal hydroxy groups were activated by the chlorodehydroxylation reaction (Scheme 15.23). The elemental analysis, used for quantification of the dendrimer loading, revealed 80–90% yield for each generation assembly.

15.2.9 Preparation of Dendrons Based on 1,3,5-Triazines

A number of research groups exploited use of triazine branching units, and particularly the stepwise selective substitution pattern of triazine chlorides (e.g., cyanuric chloride), to prepare a range of polymer-supported dendrons.[78] Marsh et al. prepared first- and second-generation dendrons, assembled divergently from cyanuric chloride and bis(3-hydroxypropyl)amine on Wang polystyrene using nucleophilic substitution of heterocyclic chlorides by alcohols and amines (Scheme 15.24).[79] In the first-generation shell, twofold branching originating from the oxy-diaminotriazine units was doubled by the branching of the aliphatic aminodiol spacer.

Subsequently, this synthetic approach was first expanded by Su and coworkers, who prepared melamine (2,4,6-triamino-1,3,5-triazine) dendrons on silica, replacing the aminodiol spacer by a simple 1,6-hexanediamine.[80] These dendrons were assembled via a

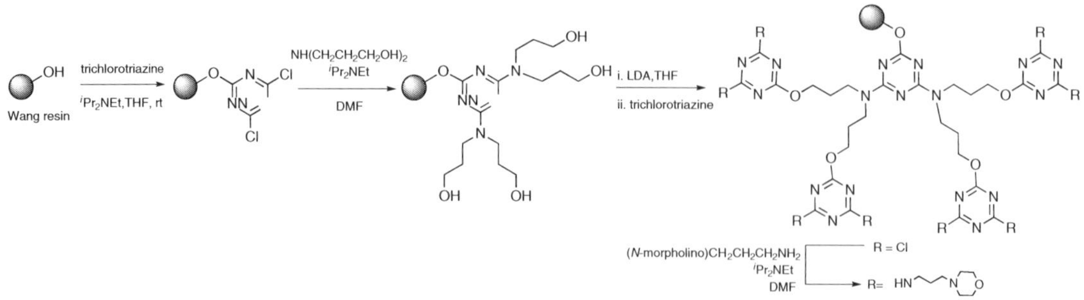

Scheme 15.24. Synthesis of 1,3,5-triazine-based dendrons on polystyrene support.

simple solid-phase divergent approach up to the fourth generation. Similarly, Marsh and coworkers themselves used cyanuric chloride and ethylenediamine to prepare up to the third-generation melamine-based dendrons on polystyrene, PEGA, and SynPhase resins, as well as silica gel.[81] In the same work, it was demonstrated that replacement of a diamine spacer by diethylenetriamine bifurcated building block doubled the degree of branching for each generation, thus forming second-generation dendrons with eight terminal groups on PEGA (Scheme 15.25). The new strategy, which replaced the amino alcohol spacers by di- or triamines, enables more efficient and high-yielding synthesis to be carried out under less restrictive reaction conditions.

Somewhat more rigid melamine-based dendrons were synthesized by Simanek and coworkers, who used piperazine or 4-aminomethyl piperidine, instead of aliphatic linear chain diamines, to connect the triazine units. Divergent synthesis on SBA-15 mesoporous silica afforded up to fourth-generation dendrons, mostly in the silica pores, forming organic–inorganic hybrids that were extensively characterized (Scheme 15.26a).[82] The characterization included thermogravimetric, elemental, and mass spectrometric analysis, IR spectroscopy, nitrogen porosimetry, and transmission electron microscopy (TEM). As in the case of PAMAM dendrons,[51] this characterization demonstrated gradual filling of the pores, with the fourth-generation dendrons reaching the "critical" size, as dictated by the pore diameter, beyond which it was not possible to further expand the dendritic structure.

In another study,[83] two approaches for synthesis of the polymelamine dendrons on silica gel were experimentally compared: the divergent solid-phase synthesis (Scheme 15.26b)

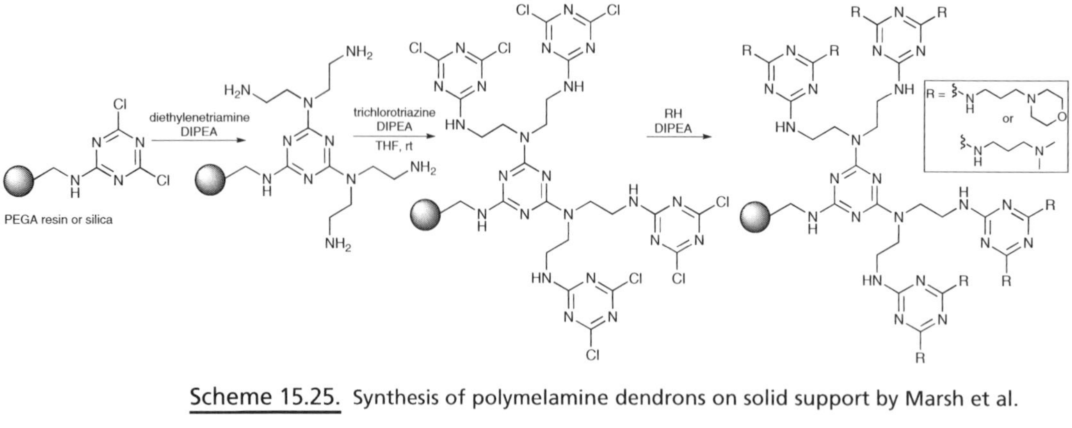

Scheme 15.25. Synthesis of polymelamine dendrons on solid support by Marsh et al.

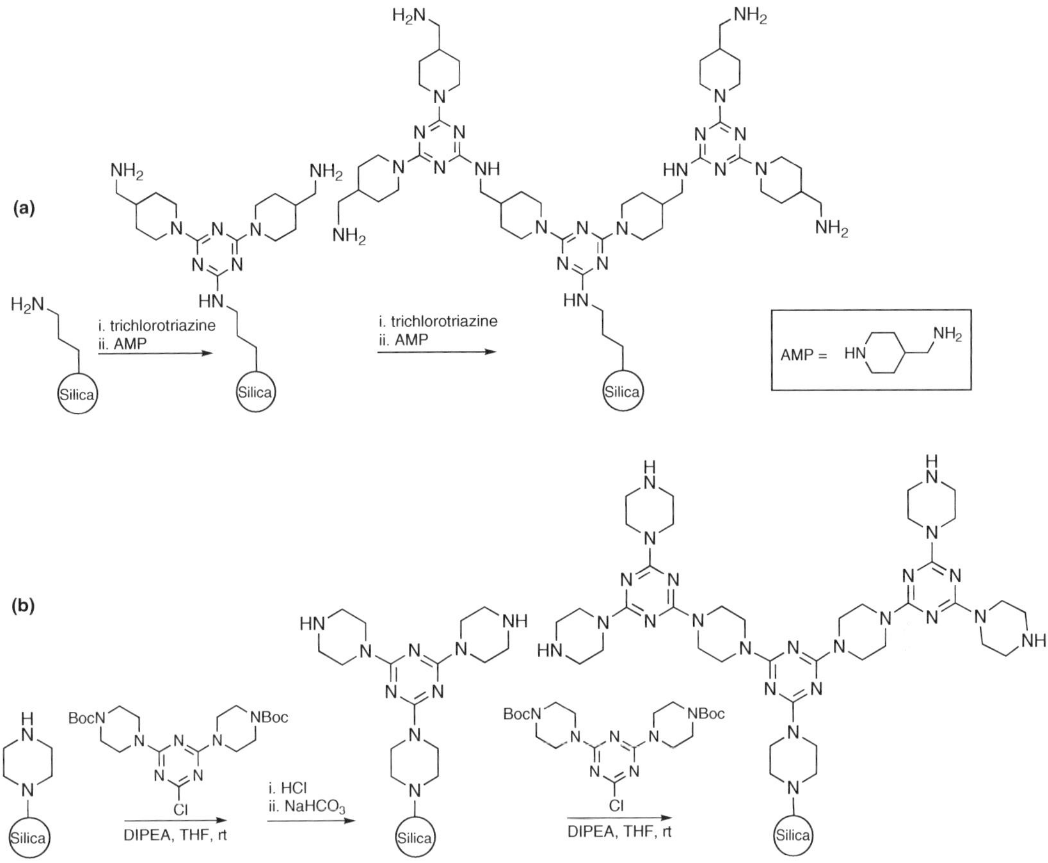

Scheme 15.26. Synthesis of polymelamine dendrons on silica gel by Simanek and coworkers.

and the convergent assembly in solution, followed by focal point attachment to the support. The dendron preparation was monitored and the products were extensively characterized. Conclusions of the study were very similar to those reported earlier by Fréchet and coworkers.[62] The postsynthetic attachment of the dendrons yields highly uniform, architecturally perfect structures, but is limited by the demanding solution synthesis and relatively low loading of the dendron on the support. On the other hand, the divergent strategy benefits from the ease and technical simplicity of the solid-phase synthesis and yields higher dendron loading, but suffers from relatively high degree of the structural defects—hyperbranched polymers are formed on the surface along with the perfect dendrons.

15.2.10 Preparation of Poly(arylacetylene) Dendrons

The purely solid-phase synthesis-based convergent and divergent/convergent approaches were used in the mid-1990s by Moore and coworkers and later by Wang and coworkers to prepare poly(phenylacetyene) dendrons on polystyrene.[8,9] Both syntheses used the Sonogashira coupling to assemble the next dendritic shell and MeI-induced cleavage of aryl triazenes for cleavage of the dendritic building blocks from the support (Schemes 15.27 and 15.28). Owing to the divergent element in Wang's synthesis, the deprotection of the terminal acetylene groups from TMS was required to activate the polymer-bound dendrons

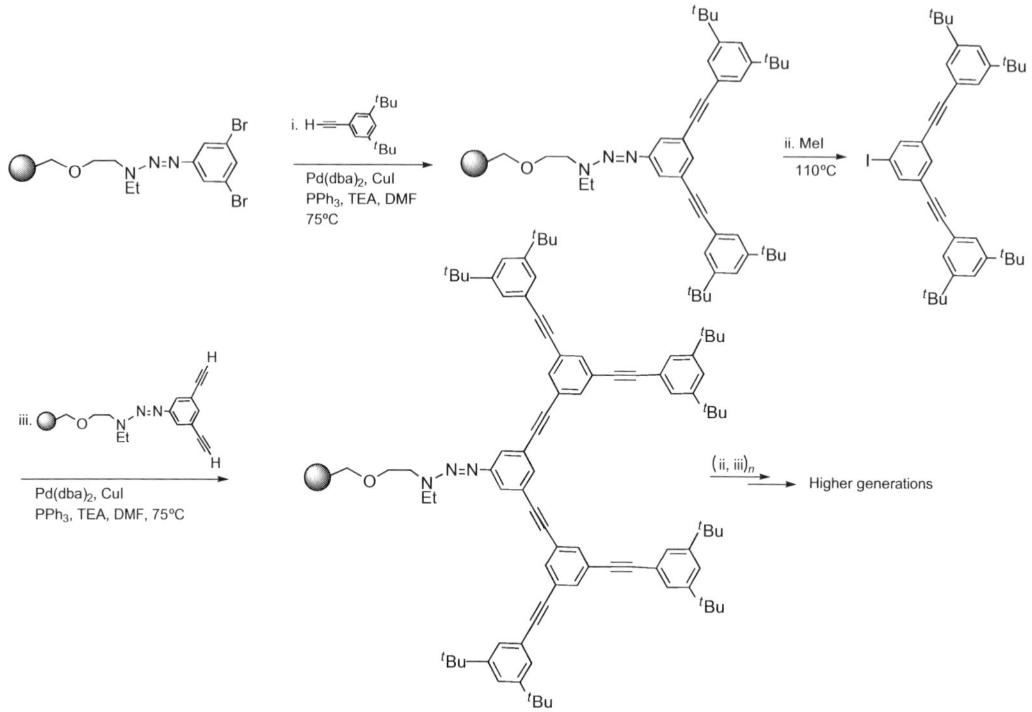

Scheme 15.27. Synthesis of poly(arylacetylene) dendrons on solid support via the convergent approach.

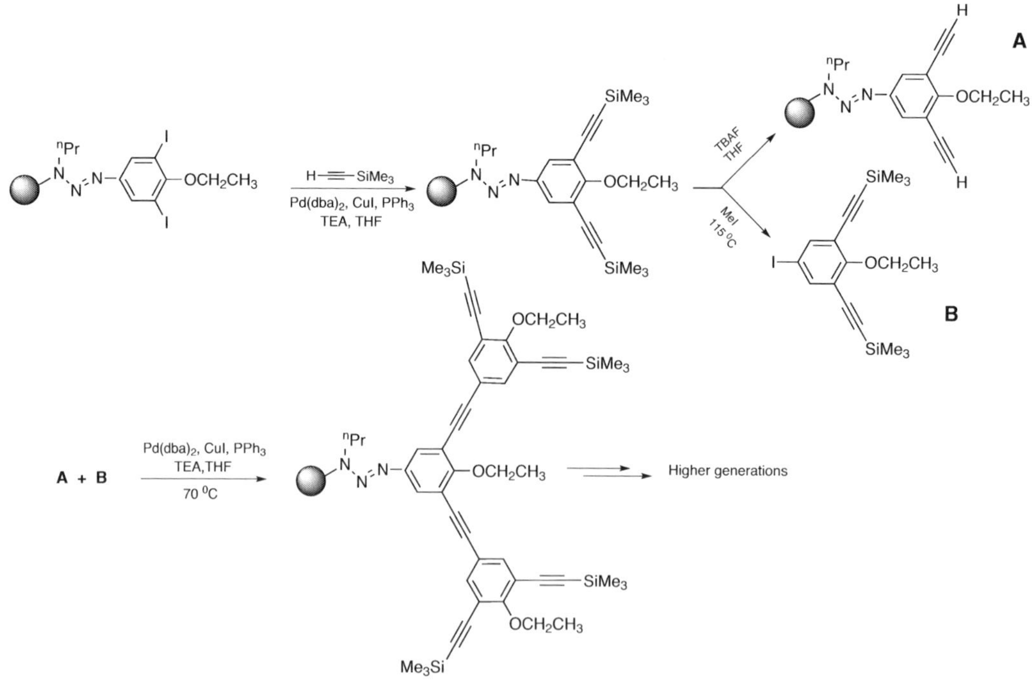

Scheme 15.28. Synthesis of poly(arylacetylene) dendrons on solid support via the divergent/convergent approach.

for the next coupling. On the other hand, Moore's convergent approach was based on permanent "capping" of the dendritic termini with 3,5-di-*tert*-butylphenyl groups. The dendrons in both cases were prepared up to fourth generation (third according to Moore, who used a different generation numbering). In the convergent case, efficient synthesis was reported, demonstrating the feasibility of large-scale solid-phase synthesis of the low-generation dendrons. Attempts to proceed beyond the fourth-generation dendrons succeeded only on the resin, with a low loading of the resin-bound branching monomer, and were fruitless beyond the fifth generation. Though in the divergent/convergent approach the estimation of the success of each new layer assembly is straightforward (because of the cleavage step included in the synthesis), yields have not been reported.

15.2.11 Coordination-Linked Dendrons

A divergent growth of coordination dendrons on gold surfaces was reported by Rubinstein and coworkers.[84] Starting from disulfide bishydroxamate, inserted into the thiol SAM on gold, the dendron was repeatedly reacted with Zr^{4+} ions (from $Zr(acac)_4$ precursor) and trimesic acid-derived building block, bearing three bishydroxamate branches (Scheme 15.29).

Elipsometric, contact angle, and atomic force microscopy (AFM) studies demonstrated that although high dilution of the anchoring units in the SAM ensures lateral separation between the dendrons of lower generation, at higher generation coalescence between

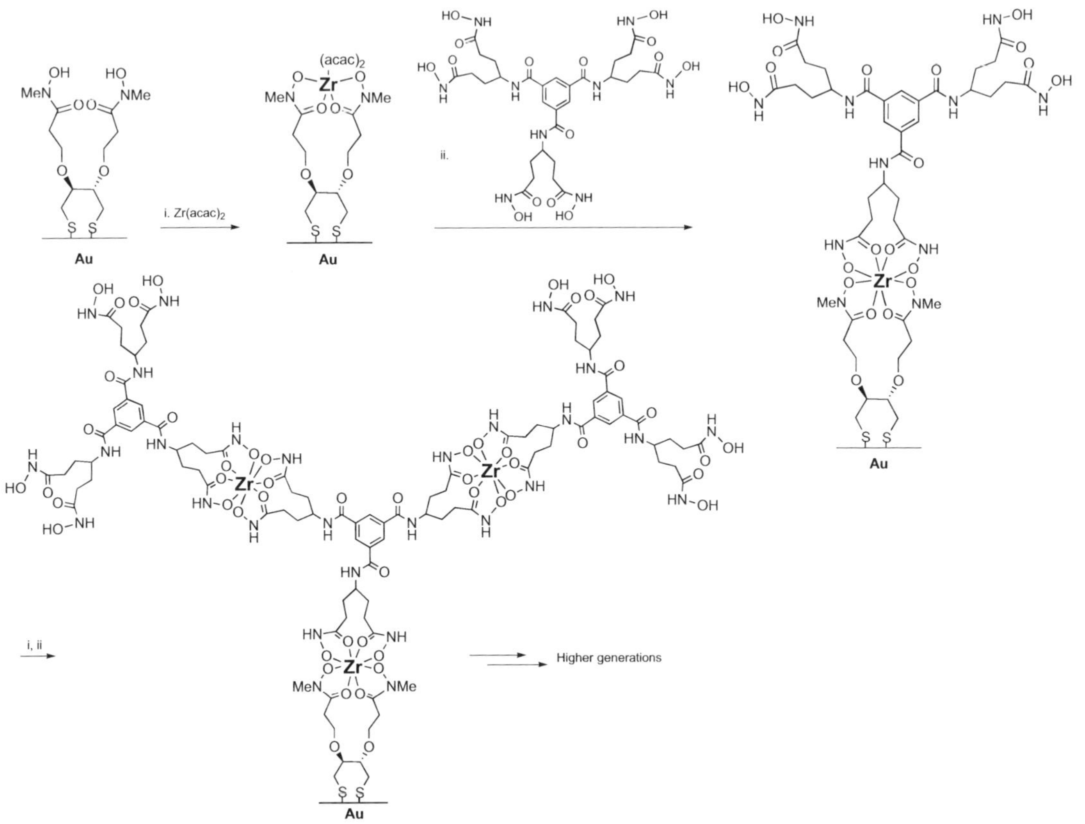

Scheme 15.29. Preparation of coordination-linked dendrons on solid support via the divergent approach.

individual dendrons occurs, and a growth mode, similar to that of multilayers prepared from SAMs of the anchoring units only, is observed. Though, in the study, the dendron assembly cycles were repeated up to the formal thirteenth generation and there was a good correlation between the generation and the dendron height (as demonstrated by AFM), it appears that for higher generations an interlinked network of hyperbranched structures, rather than a landscape of isolated perfect dendrons, was obtained.

15.3 APPLICATIONS OF DENDRONIZED SUPPORTS

15.3.1 Dendronized Supports as Synthetic Intermediates

Although part of the dendron-support hybrids are prepared for the subsequent use as new materials, in other cases these constructs are only intermediates *en route* to production of soluble dendrons. In these cases, the solid-phase synthesis technique was chosen as a synthetic approach to the preparation of dendrons, designated for application in solution.

Some of these dendrons are cleaved immediately after formation; others are used as supported templates for preparation of soluble dendritic conjugates, which are initially constructed on support, but thereafter are cleaved together with the dendritic component. In some comparative studies, this approach showed better results than the preparation of the dendritic conjugate in solution.[85]

For instance, the polylysine dendrons on solid support were used mostly for preparation of MAPs (multiple antigenic peptides, see Section 15.2.2.1). The applications of MAPs were extensively reviewed in the past.[10] Following cleavage from the support, peptide-polylysine dendron conjugates are used in a variety of biomedical applications, particularly as immunogens for the elicitation of cytotoxic immune response and antibody production. MAPs bearing pathogenic antigens have been used as synthetic vaccines. In these applications, the dendron plays the role of an alternative hapten carrier, imitating a larger protein structure.

The advantages of MAPs as immunogens, as summarized by Sadler and Tam, include simplicity in design and synthesis, versatility for investigating various immune responses, reliability of generating site-specific antibodies, and ease of generation in the laboratory.[10a] Antiparasite, antiviral, antimicrobial, antifungal, antiprion, and antitumor immune system responses were induced by MAPs.[86,87] Some research groups produced mannose-conjugated, antigen-bearing polylysine dendrons (type V in Figure 15.2) to enhance the immunogenicity of the constructs.[16e,17a] Among the related applications of MAPs, produced from support-dendron hybrids, are immunoassay and serodiagnosis reagents, as well as inhibitors of viral action.[86] MAPs themselves were used as antimicrobial agents.[16d]

MAP-like systems can also be used as biomineralization templates,[88] metal ion complexants,[89] protein mimics,[90] and molecular probes for enzymatic activity.[20,91] Arginine-bearing polylysine dendrons demonstrated strong antiangiogenic activity.[92] Similar arginine-decorated MAP-like polylysines, as well as loligomers, MAP-like dendrons decorated with cytoplasmic translocation signal, and nuclear localization signal peptides, were studied as cell and nucleus membrane translocation agents and vehicles for delivery of exogenous molecules into cells and particular cell compartments.[93,94] Cationic lipidic polylysine dendrons (e.g., Figure 15.7), prepared by Florence and coworkers, self-assembled into higher order structures, dendrisomes, which can be used as drug or oligonucleotide delivery vehicles,[18b,95] as well as for the formation/modification of nanoparticles.[18c,96]

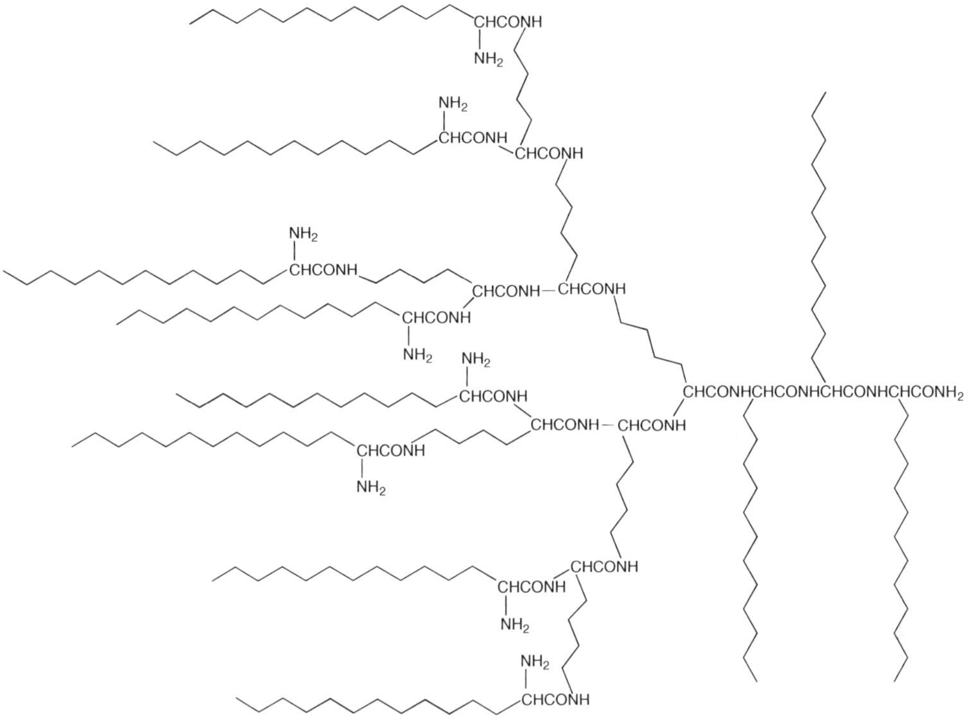

Figure 15.7. Cationic lipidic polylysine dendrons prepared by Florence et al.

Another two types of supported dendrons used for generation of dendritic gene delivery devices are the polystyrene-bound polyurea and peptoid dendrons developed by Bradley and coworkers.[58–60,31] They demonstrated that these dendrons, once cleaved from the support, complex efficiently plasmid DNA and transfect it successfully into cells.[97] The highest efficiency (comparable or better than that displayed by the commercial transfection agent Superfect) was observed for bis-G3 dendrons capped with lysine.[59] In general, the efficiency increased with the dendron generation (positive effect).

Bradley and coworkers reported two techniques that make use of dendron-peptide conjugates equipped with fluorescent reporters for investigation of proteolytic activity of proteases.[98] The polyurea dendron-based constructs were prepared on support and cleaved for use in solution. The use of the dendritic core causes high local concentration of the fluorophore and leads to internal self-quenching of fluorescence in the intact proteolytic activity probe, without any need for a separate quenching moiety (e.g., Scheme 15.30).[98a] This quenching between the fluorophores, decorating the dendron–multipeptide constructs, ceases upon enzymatic cleavage of the peptide tethering the fluorophore. The specific proteolysis results in both termination of the self-quenching and amplification of the fluorescent signal due to the release of multiple fluorophore-containing fragments.

The second technique was based on efficient quenching of the fluorescence of the fluorophore-terminated dendron by the quencher tethered to the focal point by the peptidic substrate of the protease (Scheme 15.31).[98b] Upon cleavage, strong amplification of the fluorescence enables detection of the specific proteolytic activity. The normalized amplification observed with the first-generation dendritic conjugate surpassed that of the linear analogue.

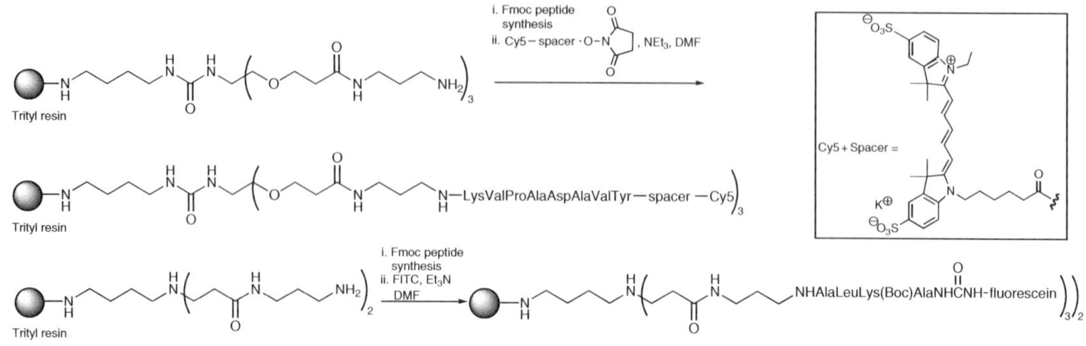

Scheme 15.30. Preparation of dendritic probes based on internal self-quenching of fluorophores for detecting specific proteolytic activity.

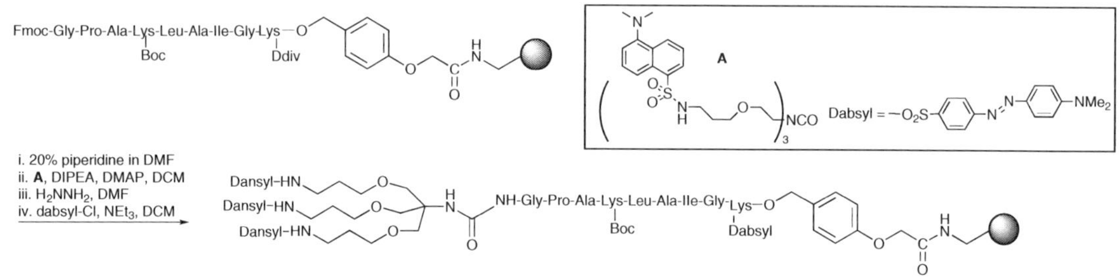

Scheme 15.31. Preparation of a dendritic probe based on fluorophore–quencher pair for detecting specific proteolytic activity.

The secondary structure of the polyproline dendrons (see Section 15.2.2.2), cleaved after their assembly on solid support, was studied and found to be notably influenced by the dendritic architecture.[25b,d] Internalization of these dendrons by rat kidney cell and complexation of drugs by the dendrons suggest the possibility of their application as drug delivery agents.[25b] Additional demonstration of the cell membrane crossing, gene delivery, and therapeutic potential of the polyproline dendrons was described for the second-generation dendrons, labeled with a fluorescent tag at the focal point and modified with positively charged residues at the periphery, while still attached to the support (see Scheme 15.6).[25e] Following their cleavage, these dendritic conjugates underwent toxicity tests as well as study of their internalization and distribution within cells and gene delivery patterns.

The DAP polyamide dendrons conjugated to PNA–peptide chimera and decorated with Gd(III) complexes of DO3A ligand were tested as plausible candidates for genetic molecular imaging using MR modality (Scheme 15.32).[32b] The T_1 relaxivity was increased with the dendron generation.

The peptide dendrons prepared by Reymond and coworkers via solid-phase synthesis and cleavage (see Figure 15.3) were examined in aqueous solution and in a combinatorial "off-bead" on silica assay as artificial esterases (see also Section 15.3.4).[22a–d,23,99] These histidine-containing dendritic structures displayed enzyme-like esterolytic activity, cooperativity of the histidine residues, and a strong positive dendritic effect. The modular design enabled synthesis of structures based on His-Ser-Asp triad and His-Ser dyad in the library

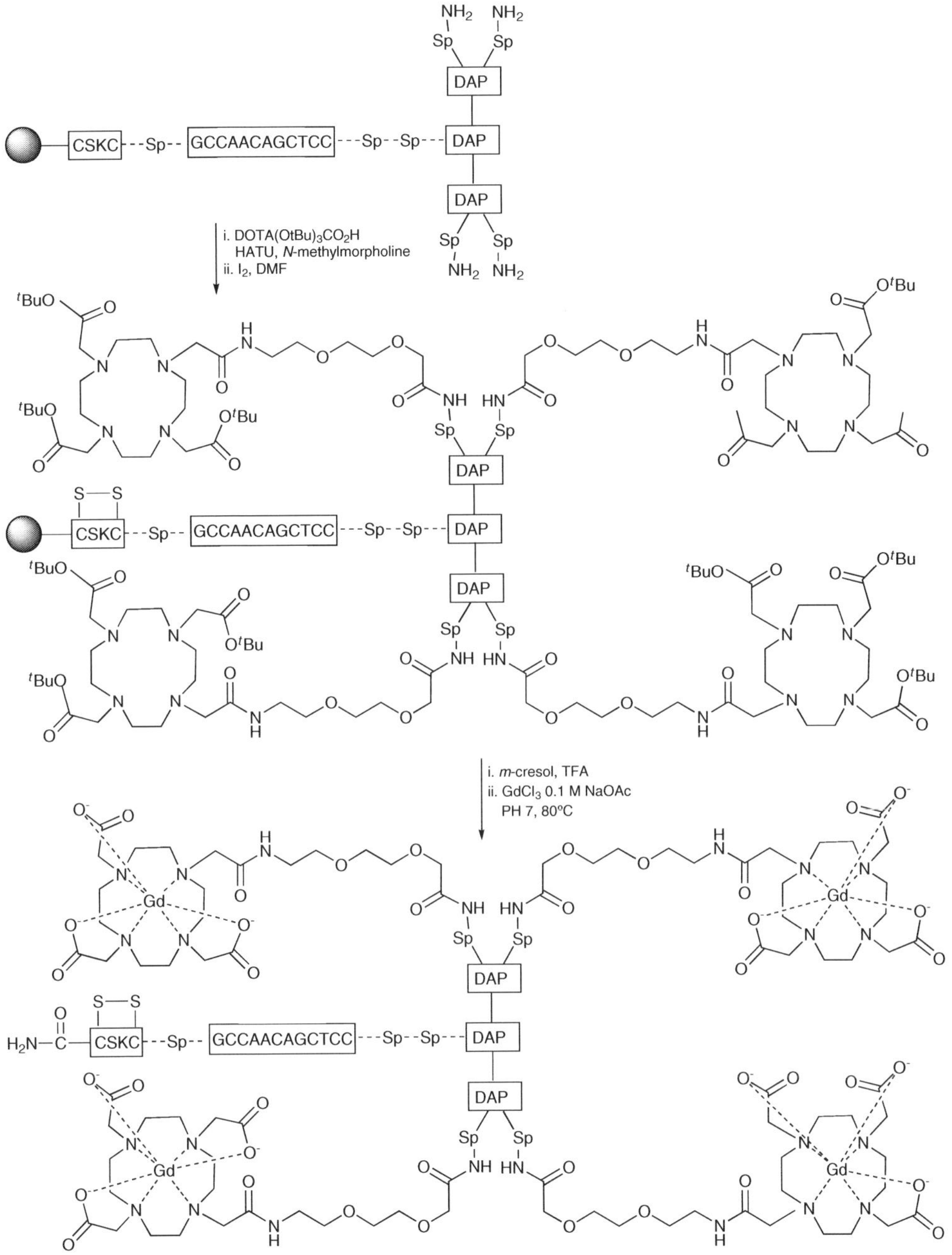

Scheme 15.32. Synthesis of peptide-PNA-DAP dendron constructs decorated with DO3A-gadolinium complexes.

format. "Mutant" structures, incorporating other amino acids, were prepared and examined, leading to His-Thr-based dendrons, which displayed superior catalytic properties.[23d] Peptide dendrons, incorporating Lys and Pro amino acids, were also catalytically active in asymmetric aldol addition.[23e] The dendrons of the "1.1.1" series were functionalized, subsequent to their SPPS synthesis and cleavage, with glycoside moieties on the periphery and colchicine alkaloid at the near-focal cysteine side chain (see Figure 15.3a).[22e] These glycopeptide-colchicine conjugates were studied as drug delivery vehicles.

15.3.2 High-Loading Dendronized Supports for Solid-Phase Synthesis

In contrast to the structures discussed in the preceding section, which were used as synthetic intermediates and distinguished by the cleavage of the support-dendron bond after their synthesis on support, some of the dendronized insoluble materials are designed as high-loading supports for subsequent solid-phase synthesis. The cleavage that usually follows such synthesis will break the bond between the dendritic termini and substrate and leave the support-dendron bond intact.

Preparation of such high-loading supports was extensively explored and promoted by Bradley and coworkers. The PAMAM, polyamidourea, and polyurea dendrons, they prepared (see Sections 15.2.3 and 15.2.4), were designed to increase the per-bead loading of the reactive sites (rather than per weight loading). Thus, the PAMAM dendrons on TentaGel increased the loading of the NH_2 groups on resin more than ninefold, with the loading per bead increasing up to 9.6 nmol/bead in the fourth generation.[35] The high per-bead loading of the support enabled the cleavage and analysis of the products from a single bead, using NMR, MS, and HPLC techniques.[37b]

With the polyamidourea dendrons the increase in loading was even more dramatic (up to 230 nmol/bead).[58] The high per-bead loading of the terminal functional groups enabled the analysis of compounds cleaved from a single bead for these dendrons (e.g., biaryls prepared on the resin were analyzed by HPLC and NMR).

The high-loading dendronized supports were used by Bradley and coworkers to prepare peptides and biaryls, as well as amidine, aryl ether, and hydantoin libraries (see, for instance, Scheme 15.33a and b).[35,58,60,61,100] Polyurea dendronized polystyrene was used as a support for preparation of a library of antibacterial triazine-based compounds (Scheme 15.33c).[101] The high loading of the beads allowed screening of the antibacterial activity of compound released from a single bead. Similarly, a library of the (poly)arginine-lipid conjugates, prepared on high-loading polyurea dendronized support, was screened by a "single bead release" technique for DNA transfection activity.[102]

The utility and versatility of other polyamide dendron-modified high-loading supports were demonstrated by Chan and coworkers in SPOS of sulfonamides, carboxamides, and other small molecules,[30] as well as by Lee and coworkers in SPOS of pentapeptides.[29]

15.3.3 Dendronized Supports for Multivalent Molecular Recognition

Lysine-based peptide dendrons prepared by Reymond and coworkers in a library format were decorated with fucosylacetic acid and the formed multivalent glycopeptide conjugates were screened on solid support for the lectin LecB binding (Scheme 15.34).[24] The chosen candidates were resynthesized and subjected to binding tests in solution. These tests demonstrated that the enhanced binding of these glycopeptides originated from their multivalency and that the architecture of the dendron, combined with the nature of its building blocks, modulated the binding strength.

Glycoside-decorated amino-ether dendrons of up to third generation were prepared in solution and subsequently covalently immobilized on alumina chip by Pieters, Liskamp, and coworkers.[103,104] The study of the binding of fluorescence-tagged lectins to the glycoclusters on the chip revealed the multivalency enhancement of the binding.

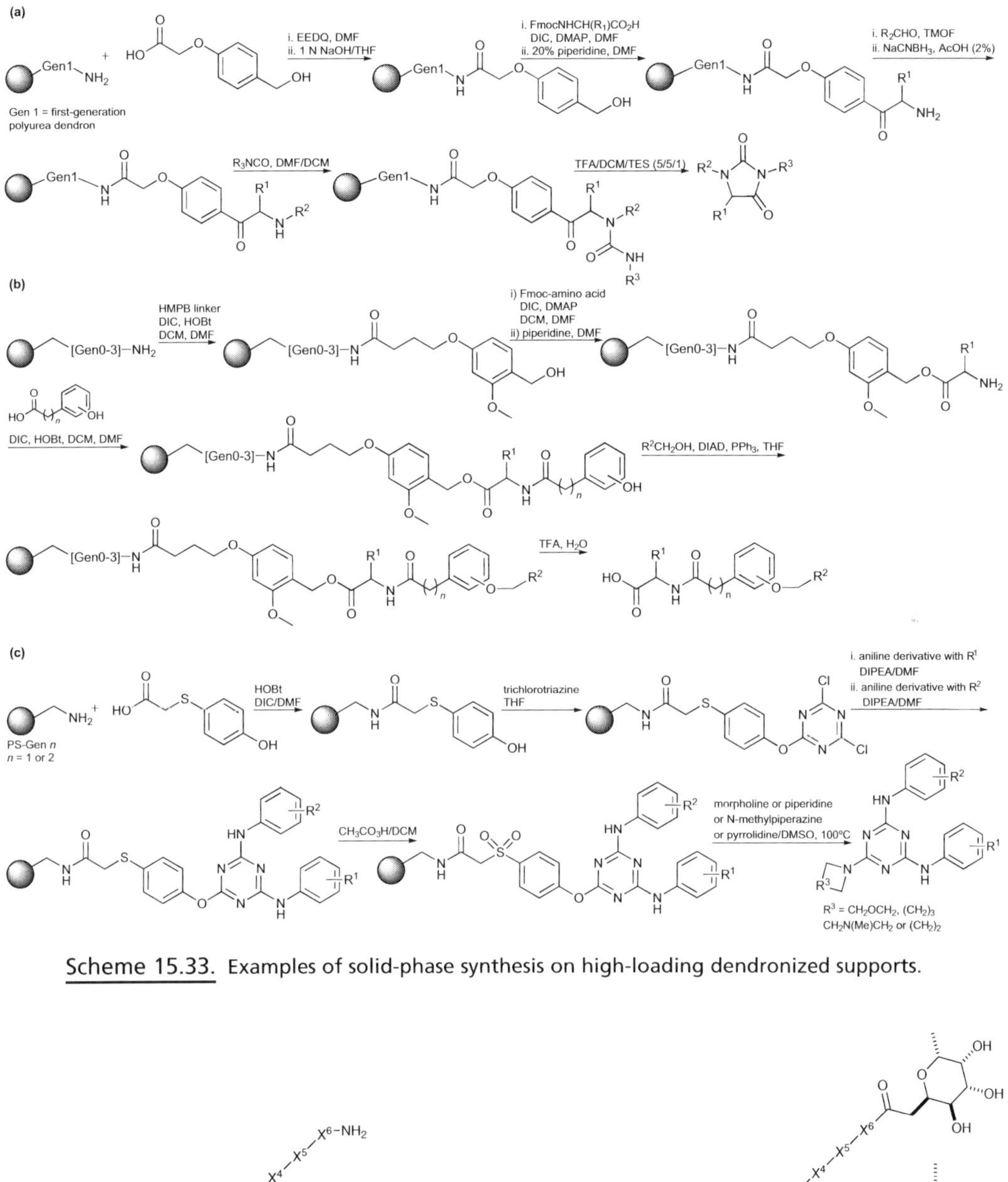

Scheme 15.33. Examples of solid-phase synthesis on high-loading dendronized supports.

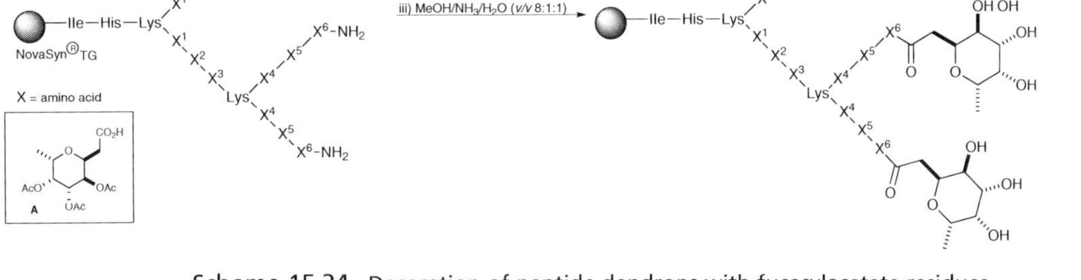

Scheme 15.34. Decoration of peptide dendrons with fucosylacetate residues.

Differences in the extent of this enhancement for different lectins were attributed to the variations in the distance between the binding sites on the lectins.

15.3.4 Supported Dendritic Catalysts

One of the most important applications of the dendronized supports is the preparation of support-bound dendritic catalysts. Such systems, based on outward branching dendrons, were prepared and explored by the groups of Alper and Arya, Reymond, Portnoy, Kawi, Rhee, and Sreekumar.[105] A number of heterogeneous dendritic catalysts based on a different design (e.g., polymerized dendrimers bearing catalytic units) are also known, but are beyond the scope of this book.[106]

The majority of supported dendritic catalysts are based on phosphine ligands coordinated to a transition metal fragment. Dahan and Portnoy decorated the previously described polyether dendritic templates with monodentate phosphine (thus forming formally polydentate ligands) via an esterification reaction of diphenylphosphinobenzoic acid (Scheme 15.35). These ligands were complexed to Pd and Co precursors to form catalytic systems for Heck, Suzuki and Pauson–Khand reactions.[73,107] The use of the dendritic template significantly improved the activity and selectivity of the catalytic systems (a positive dendritic effect). Analogous Pd-containing catalysts were also prepared on polymer-bound thioether and amine dendrons. Although less active than the parent polyether catalyst, these systems also displayed a positive dendritic influence. The lower activity of the thioether- and amine-based dendritic Pd catalysts is probably due to the coordinative interference of the dendritic backbone with the active metal fragments and emphasizes the important impact of the dendritic template architecture on the catalytic outcome.

Bidentate ligands were used by Alper and coworkers. The bis(diphenylphospinomethyl)amine ligands were prepared on primary amine-terminated PAMAM dendrons on silica as well as polyamido dendrons on polystyrene via the double Mannich-like reaction with formaldehyde and diphenylphosphine (Scheme 15.36a).[27,43,51,57e,108] Subsequently, Alper and coworkers subjected the dendronized ligand-decorated supports to complexation with rhodium and palladium precursors in order to prepare active catalysts for a number of important chemical transformations (Scheme 15.36b). Initially, the dendronized rhodium catalysts were tested in the hydroformylation reaction and carbonylative ring expansion of

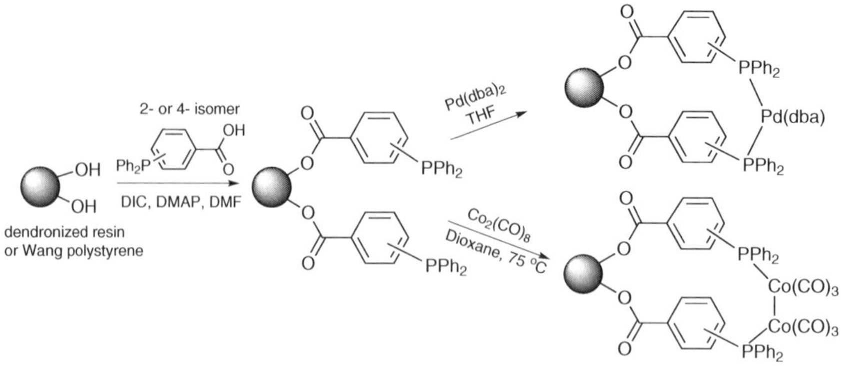

Scheme 15.35. Preparation of dendronized resins decorated with monodentate phosphines and their complexes.

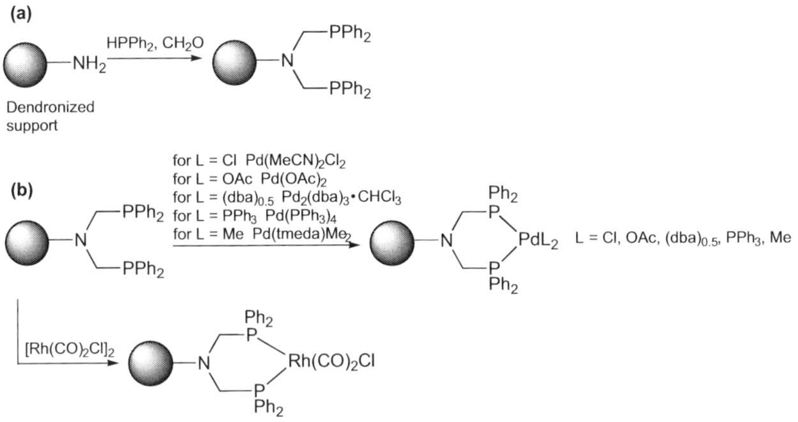

Scheme 15.36. Preparation of dendronized resins decorated with bidentate phosphines and their complexes.

aziridines.[27,43a,b,51a,c,108a] Subsequently, palladium-based complexes were used for carbonylation of aryl halides,[108b–d] hydroesterification and hydroamidation of olefins,[43c,108e] hydroamidation of alkynes,[108f] Heck reaction,[108g] and oxidation and hydrogenation of olefins.[43d,108h]

The authors observed severe steric problems, restricting the formation of the higher generation catalysts (third and above) due to the limited size of the pores. The problem was especially pronounced for silica with its fixed-size pores. A number of approaches were taken in order to overcome this obstacle. Thus, in the hydroformylation reaction, the improvement for the third- and fourth-generation catalysts was achieved through elongation of the diamine fragment of the branching module of the PAMAM dendron.[43b] The extension of the branch length, obtained by substitution of the 1,2-diaminoethane by 1,4-diaminobutane, 1,6-diaminohexane, or 1,12-diaminododecane in the PAMAM synthesis, relieved steric crowding, led to an increase in the metal catalyst loading for the higher dendron generations, and resulted in an improvement in the activity and recyclability of the catalysts. Similar trend was observed for Pd-containing catalysts.[43c,d]

Performance of the catalyst is influenced by not only the loading of the active metal, but also the morphology of the support, as demonstrated by Alper, Sayari, and coworkers.[51a,c] PAMAM dendrons synthesized on a mesoporous or large-pore silica and decorated with diphosphine-Rh complexes demonstrate a notably higher activity in the hydroformylation reaction, compared to those prepared on amorphous silica. On MCM-41 silica, only zeroth- and first-generation catalysts were active (the second-generation catalyst was nonactive and the third generation could not be prepared), whereas on Davisil, large-pore silica zeroth to third-generation catalysts were prepared and examined.

The observed dendritic effect on the effectiveness of the catalyst preparation and on the normalized activity of the systems is explained by the gradual filling of the silica pores for the higher generations and the increasing difficulty in accessing the reactive sites. Surprisingly, the recycling is possible only with MCM-41 and amorphous silica-supported catalysts, but could not be achieved for the Davisil-based systems.

In the case of polystyrene-supported catalysts, an additional study explored the influence of the isolation of the catalyst environment on the outcome of the hydroformylation reaction. For this purpose, first- and second-generation dendrons (very similar to those described above) were constructed and the biphosphine-Rh complex was assembled only on

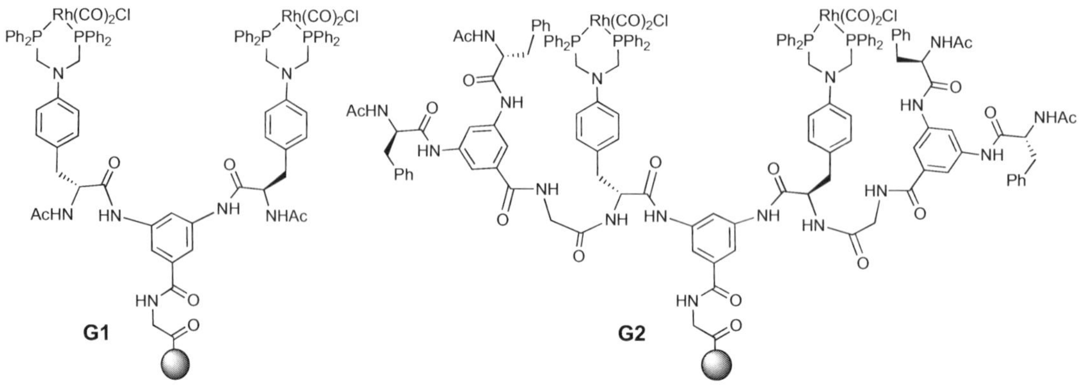

Figure 15.8. Supported, internally functionalized dendritic catalyst for hydroformylation.

the first-generation module (Figure 15.8).[27b] In the second-generation dendrons, the outer layer modules did not carry metal and were used to isolate the catalytic site from the environment. Although the metal loading of the second-generation catalyst was lower than that of the first, the reactivity and recyclability of the second-generation-derived catalyst were notably better with some substrates (e.g., vinyl benzoate). This interesting dendritic effect again emphasizes the influence of the dendritic template architecture on the catalytic outcome.

In a subsequent study,[27c] the polystyrene-bound lysine incorporated dendrons with four propagation sites in each monomer (see Figure 15.4b) were converted into first- and second-generation catalysts bearing 4 and 16 Rh complexes, respectively, on each of the dendrons. These new catalytic systems showed even higher reactivity (enabling room temperature transformation), excellent regioselectivity (higher than the previous systems, probably as a result of the lower reaction temperature), and outstanding recyclability in the hydroformylation reaction. Moreover, these systems showed excellent activity and selectivity in the carbonylative ring expansion of aziridines to β-lactams.[108a]

Portnoy and coworkers synthesized bidentate phosphine ligands on a polyether dendronized polystyrene using an alternative approach (Scheme 15.37).[68] Serinol was immobilized on the chloromethyl-terminated dendrons through its amino group. Subsequently, the chlorodehydroxylation followed by the substitution of chlorides by diphenylphosphide anion led to the formation of chelating bisphosphine ligands on the polyether dendronized polystyrene. Complexes of Pd formed with these polymer-supported ligands catalyzed the Heck and amidocarbonylation reactions. Noteworthy, while a positive dendritic effect was again observed for the amidocarbonylation of bromobenzene, the catalysis of the Heck reaction of this substrate with methyl acrylate was negatively affected by the dendronization of the support.

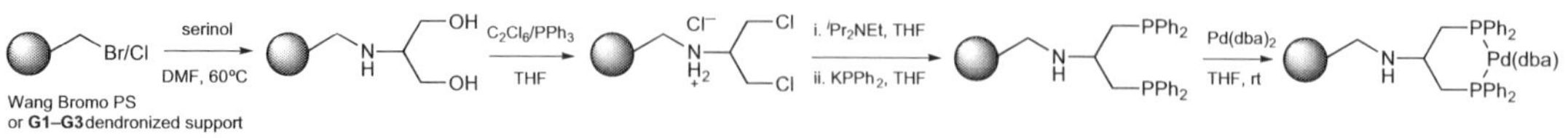

Scheme 15.37. Synthesis of bidentate phosphines and their Pd complexes on dendronized supports by Portnoy and coworkers.

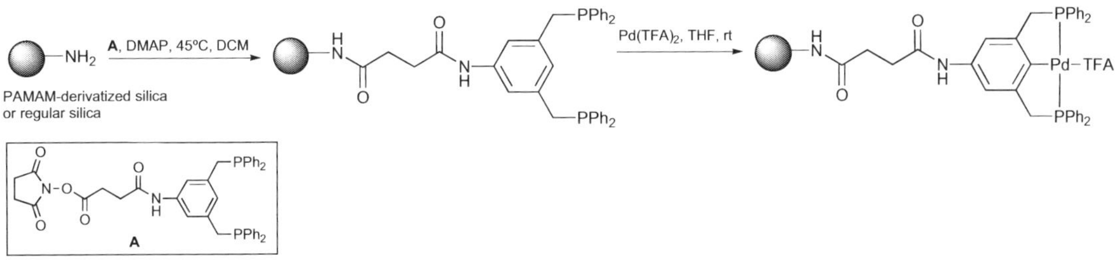

Scheme 15.38. Decoration of dendronized support with pincer complexes based on a tridentate PCP ligand.

Only one case of a tridentate phosphine-containing ligand on dendronized support is known. A tridentate PCP ligand with a succinate spacer was immobilized on PAMAM-covered silica via the acylation of the terminal amino groups with an activated NHS ester of the spacer (Scheme 15.38).[109] This ligand underwent efficient coordination and cyclometallation with Pd(II) precursor to form the very stable pincer PCP-Pd(II) complexes, which were tested in the intramolecular hydroesterification of olefins. The dendritic catalysts showed higher activity than their nondendritic analogue (G0), with the second-generation catalyst being the most active.

In a few examples, metal complexes with chelating N,O-ligands were prepared on dendronized supports. Kawi and coworkers prepared a PAMAM-on-silica template with amino terminal groups converted into salicylimines (Scheme 15.39).[46a] These salen-imitating ligands were complexed with Mn(II) and used as catalysts in olefin epoxidation. Remarkably, the catalytic activity per Mn equivalent increased dramatically with the dendron generation. Weck and coworkers used a supported dendritic catalyst based on a true salen ligand. They immobilized a protected first-generation Newkome-type amino-tricarboxylate dendron on polystyrene support.[110] Following their deprotection, the carboxylic dendritic termini were functionalized with a hydroxymethyl-bearing chiral salen ligands, which were subsequently complexed with a Co precursor (Scheme 15.40). The catalytic system exhibited excellent activity and enantioselectivity in hydrolytic kinetic resolution of epoxides.

Iminophenols immobilized on PAMAM dendronized polystyrene and cellulose were used by Lei and coworkers to tether Sn complexes, forming active catalysts for the Baeyer–Villiger oxidation of ketones (Scheme 15.41).[39,57c] Probably due to the already mentioned difficulty in building fully branched PAMAM dendrons on solid support, the catalytic activity decreased with the generations. This tendency was diminished by preparing the dendritic catalysts on macroporous polystyrene, particularly that with a high percentage of cross-linking.[39b]

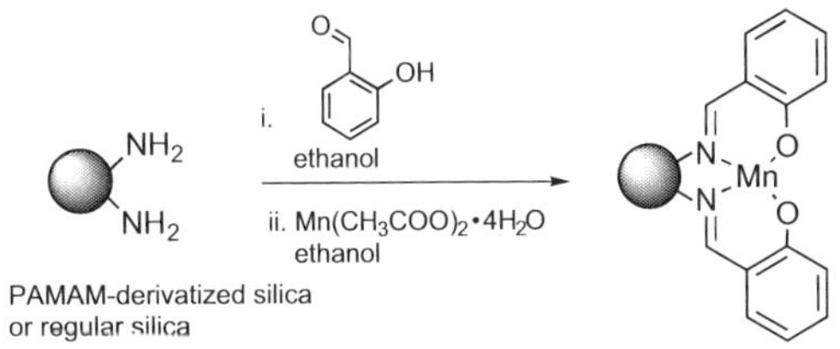

Scheme 15.39. Decoration of dendronized silica gel with salicylimine ligands and their Mn complexes.

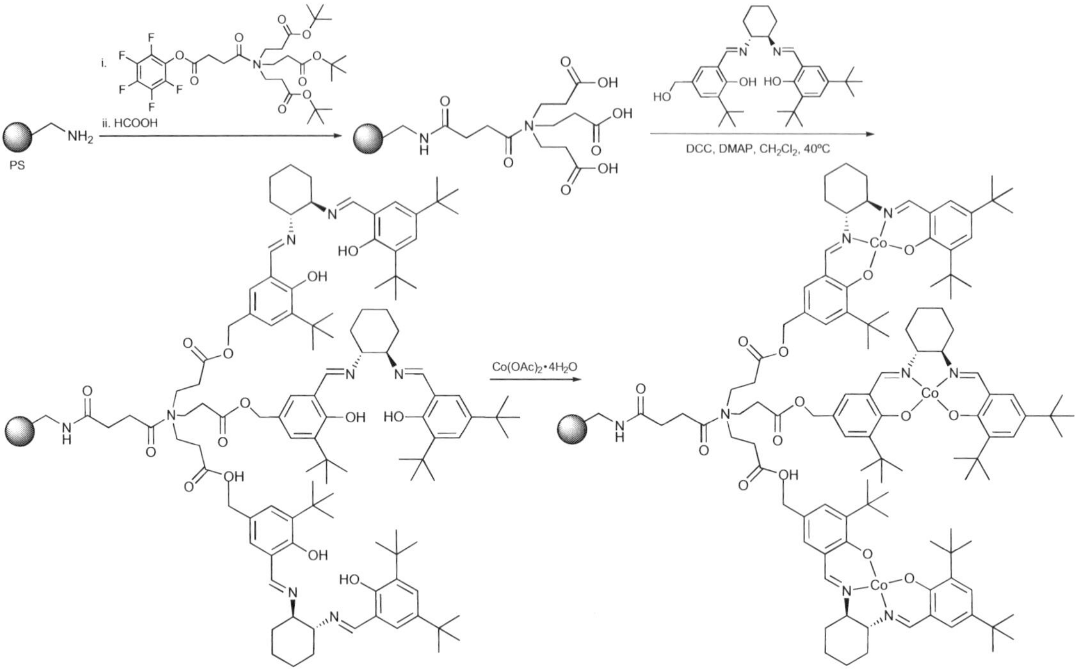

Scheme 15.40. Immobilization of salen ligands on dendronized support and their complexation with Co.

PAMAM itself was used as a multivalent macromolecular ligand, probably due to its multiple amino groups, in order to complex and immobilize metal ions, complexes, and nanoparticles with catalytic capabilities. Thus, Kawi and coworkers used PAMAM-on-silica and PAMAM-on-alumina templates to immobilize Rh(I) complexes as hydroformylation catalysts.[47,57a] Passivation of the silica OH sites outside the pores of SBA-15 silica resulted in a tighter binding of rhodium complexes inside the pores and led to a series of catalysts that displayed a positive dendritic effect up to the second PAMAM generation.[47] Sreekumar and Krishnan used PAMAM on polystyrene to complex Mn(II) precursors and catalyze the oxidation of secondary alcohols.[40c]

Gao and Jiang used this "ligandless" approach to prepare supported dendritic hydrogenation catalyst.[52,57f] Pd(0) nanoparticles were prepared and stabilized in the interior of the PAMAM dendrons grown on SBA-15 silica or on silica-covered magnetite particles (Scheme 15.42). According to the approach pioneered in solution by Crooks and et al.,[111] the dendron served both as a template for the assembly and stabilization of the metal particle and as a nanoreactor for controlling its reactivity. On silica, the second-generation heterogeneous catalyst was the most active and achieved the highest TOF in hydrogenation

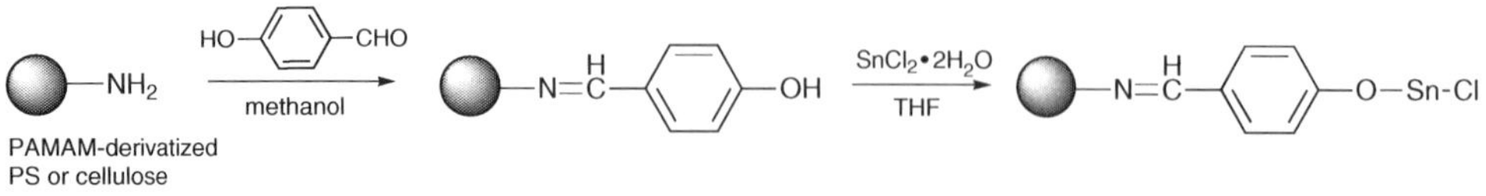

Scheme 15.41. Decoration of dendronized support with phenol ligands and their Sn complexes.

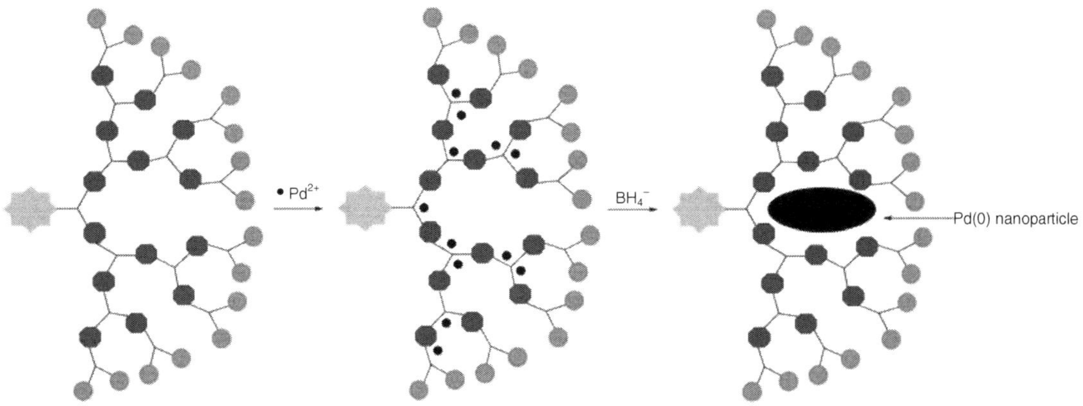

Scheme 15.42. The approach for preparation of dendron-stabilized metal nanoparticles immobilized on solid support.

of allyl alcohols, surpassing the activity of the homogeneous dendrimer/Pd nanoparticle analogue, whereas the third-generation catalyst exhibited the best chemoselectivity toward the monohydrogenation product.

A number of metal-free (organocatalytic) moieties decorating supported dendrons also promoted catalytic processes. Dendritic peptides, prepared by Reymond and coworkers (see Sections 15.2.2.2 and 15.3.1), exhibited esterolytic activity not only in solution, following their cleavage from the support, but also while still bound to the TentaGel resin.[23a,b] The dendrons of the library of the "2.2.2.2" series were subjected to the high-throughput screening for the ability to catalyze the hydrolysis of 8-butyryloxypyrene-1,3,6-trisulfonate. Soaking the library beads with this fluorogenic substrate and their subsequent spreading on a dry surface allowed direct selection of beads that carried the most active dendron structures. The consensus sequences of the most catalytically active dendrons included the His-Ser dyad in the outermost layer and hydrophobic amino acids (Ile and Val) in the innermost part of the dendron.

Similar peptide dendrons from the "2.2.2.2" series prepared in a library format were screened, while still on the beads, for potential aldolase activity.[23e] Probes forming colored enaminones or releasing fluorescent fragments via enolization or retro-aldol reactions were used for the screening. Lysine- and proline-rich sequences were identified with the enaminone-forming and enolization-sensitive probes.

PAMAM dendrons on insoluble support can themselves serve as basic catalysts due to their multiple tertiary and primary amine sites, as demonstrated by Kapoor et al. on silica and by Sreekumar and coworkers on polystyrene.[40a,b,54] PAMAM promoted the Knoevenagel condensation and ring opening of epoxides by aniline. The catalytic systems on polystyrene demonstrated a positive dendritic effect, possibly as a result of generating a special isolated environment near the catalytic sites.

Portnoy and Goren immobilized *N*-alkylated imidazoles on polystyrene-bound polyether dendrons via three different synthetic routes (Scheme 15.43).[112] All these systems catalyzed the Baylis–Hillman addition of methyl vinyl ketones to aromatic aldehydes, displaying a very strong positive dendritic effect and a significant enhancement effect of water as a cosolvent. Remarkably, substrates that did not undergo the reaction with the nondendritic immobilized or soluble *N*-alkyl imidazole catalysts underwent a smooth reaction with one of the catalytic second-generation dendrons.

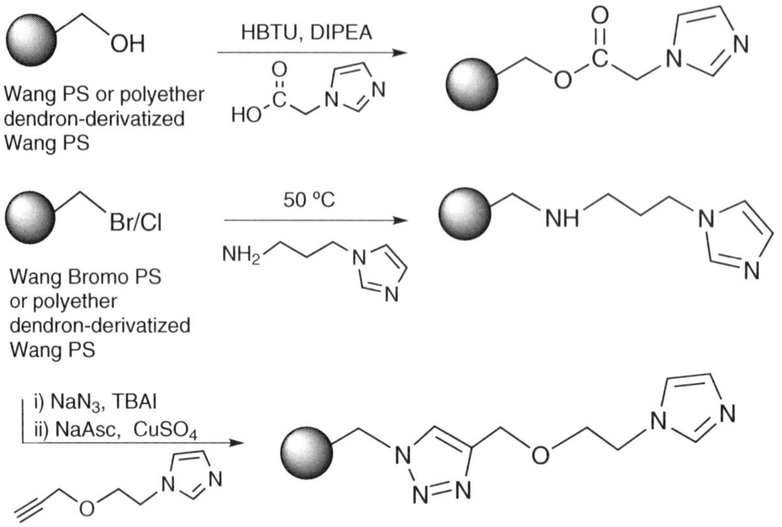

Scheme 15.43. Decoration of dendronized polystyrene with *N*-alkylated imidazole catalytic units.

The first published example of an enantioselective reaction carried out with a catalyst immobilized on dendronized supports was described by Chung and Rhee.[45] The PAMAM-on-silica template was functionalized with a chiral aminoalcohol ephedrine, and the obtained systems were used to catalyze the enantioselective addition of diethylzinc to benzaldehyde (Scheme 15.44). The enantioselective addition of dialkylzinc reagents to aldehydes is one of the main approaches to the synthesis of chiral secondary alcohols and a yardstick for the chiral induction capability of amino alcohols and other related catalytic systems.[113] The test of the catalysts of first to fifth generation revealed a remarkable dependence on the generation and general loading of the dendrons. With dendrons derived from silica with a lower initial loading of the amino groups (0.24 mmol/g), a strong positive dendritic effect on the conversion, chemoselectivity, and enantioselectivity of the catalytic reaction was observed up to the fourth generation. Moreover, the fourth-generation catalyst could be recycled three times without a loss of activity, chemoselectivity, or enantioselectivity. On the other hand, when the analogous catalysts were prepared on silica with a higher initial loading of 0.9 mmol/g, the opposite trend (a negative effect) was encountered, with the fifth-generation catalyst promoting formation of a practically racemic product. These dramatic differences might be in part a reflection of the differences in the perfectness of dendron growth, as extensively discussed for the PAMAM-on-silica in Section 15.2.3.

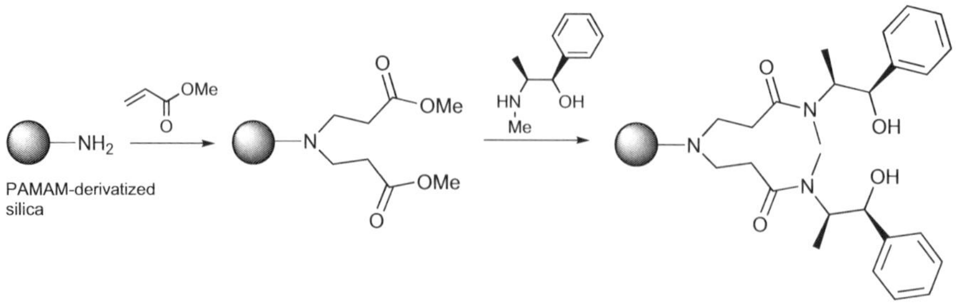

Scheme 15.44. Decoration of PAMAM-derivatized silica with a chiral aminoalcohol catalyst.

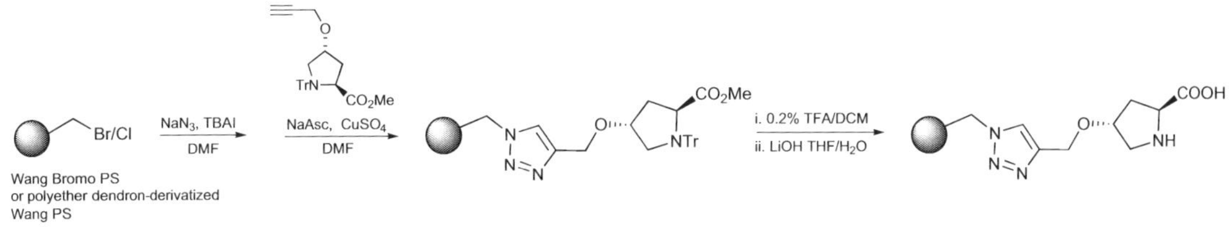

Scheme 15.45. Decoration of dendronized polystyrene with chiral hydroxyproline-derived catalytic units.

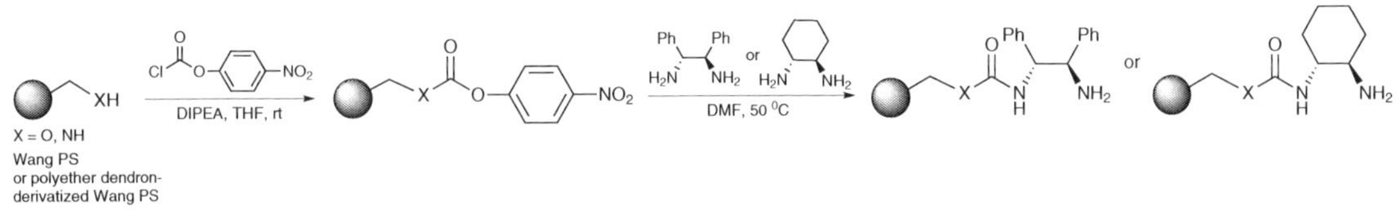

Scheme 15.46. Preparation of chiral aminourea and aminocarbamate catalysts on dendronized polystyrene.

Portnoy and coworkers immobilized chiral hydroxyproline derivatives on polystyrene support functionalized with polyether dendrons (Scheme 15.45).[114] These catalysts promoted the aldol addition of acetone to aromatic aldehydes with excellent enantioselectivities, significantly superior to those achieved in the same reaction with analogous catalyst lacking the dendritic interface. The same group prepared polymer-supported chiral bifunctional aminocarbamate and aminourea catalysts for nitro-Michael reaction (Scheme 15.46).[115] However, in this case the dendritic catalysts were inferior to their simpler dendron-lacking analogues.

15.3.5 Dendronized Supports in Separation Processes

The potential of dendronized supports to serve as stationary phases for various separation techniques has been explored by a number of research groups. One of the most important applications in this field is the use of dendron-modified support as chiral stationary phase (CSP) for separation of enantiomers. An unsuccessful attempt to do so was reported by Mitchell and coworkers, who used poly-L-glutamate dendrons postsynthetically immobilized on silica.[5b,116]

The use of dendronized polymers as CSPs was successfully established by Fréchet and coworkers.[62] The polymer-bound polyester dendrons (not chiral themselves, see Section 15.2.5 and Scheme 15.15) were decorated with L-proline indananilide chiral selectors (Scheme 15.47). It was found that these materials enable effective separation of enantiomers of the tested protected amino acids with a very high separation factors. In general, the separation factor α increased with the selector loading. However, the structure and morphology of the dendritic interface also play a role, since there were some exceptions from this rule. It was demonstrated that the selectivity increases with the distance of the selector from the support core. The specific selectivity of the CSP (defined as the separation factor normalized by the selector loading) was notably higher for CSPs built from perfect dendrons generated in solution, compared to CSPs prepared by solid-phase divergent dendron synthesis. It was also enhanced for higher generations. The authors conclude that this validates the positive synergistic effect of multiple selectors located in well-defined positions within the dendritic CSP.

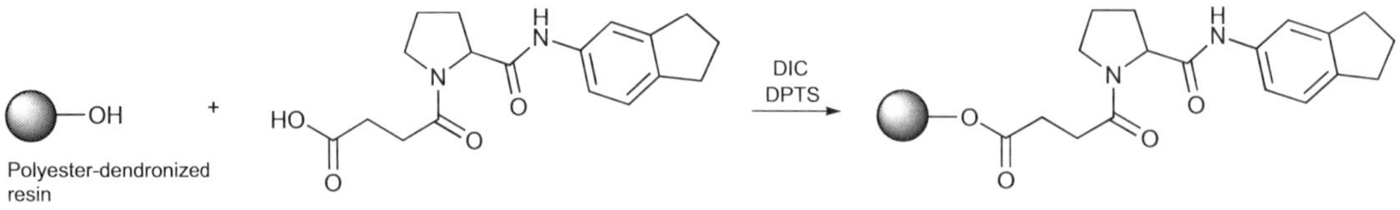

Scheme 15.47. Preparation of CSPs with resin-supported dendrons.

Later, Bai and coworkers used both, dendrons built from chiral monomers and achiral dendritic backbone decorated with chiral selectors, to prepare CSPs. First, they tested the polyamido dendrons prepared from benzenetricarbonyl trichloride and enantiopure stilbene diamine capped with phenylurea function (see Section 15.2.2.4 and Scheme 15.10).[34] The separation ability decreased with the generation, possibly as a result of imperfect dendron architecture. Subsequently, the authors decorated silica-bound PAMAM with chiral selectors: *N*-tosyl-L-phenylalanine and a chiral unit derived from 2-amino-1,2-diphenylethanol (Scheme 15.48).[55] CSPs derived from third- and second-generation dendrons showed the highest separating ability. The second selector was inferior to the first, in spite of possessing an additional chiral center.

In another study, the behavior of PAMAM dendronized silica (see Section 15.2.3) as a stationary phase for site-exclusion chromatography demonstrated that its performance is governed by the local microstructure of the dendrons, which could be manipulated by the dendron growth conditions.[44b] It was suggested that modification of the dendrons by suitable ligands will generate column packing for multiseparation applications.

The use of dendronized silica as a stationary phase in capillary gas chromatography was pioneered in 2001 by Newkome et al.,[117a] who used the triply branched dendrons prepared in solution and immobilized on fused silica inside the capillary column. The dendritic architecture provides unique selectivity in separation of a Grob test mixture of compounds. The same dendron-silica hybrids were used by the authors for solventless capillary microextraction (CME) and preconcentration in chemical analysis.[117b]

Su and coworkers used silica gel with immobilized polymelamine, poly(amidoamine), or poly(propylenimine) dendrons as the microcolumn packing for the flow injection preconcentration and separation of traces of Pt(IV) ions for flame atomic absorption spectroscopy determination.[75,80,118] Polyamine dendron-functionalized polystyrene was

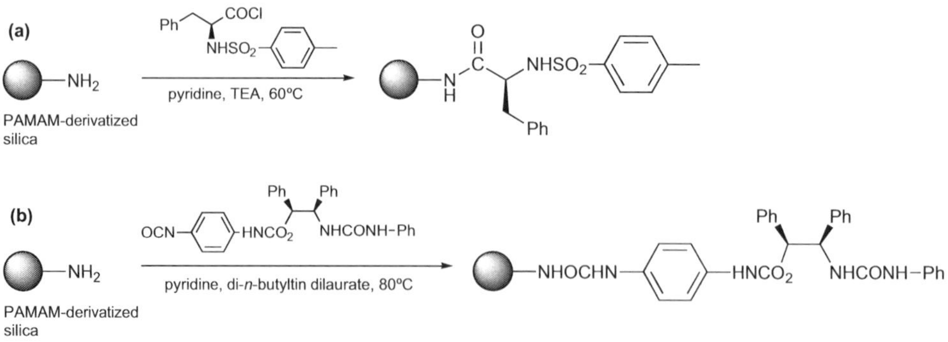

Scheme 15.48. Preparation of CSPs by Bai and coworkers.

used for sequestering metal ions (Ag^{+}, Cu^{2+}, and Hg^{2+}) from aqueous solutions.[77] The resins demonstrated good adsorption capacities, particularly for Cu^{2+}, due to the chelating nature of the polyamine dendrons. Such dendronized resins could potentially be applied for removal of metals from industrial wastewater.

Imae and coworkers demonstrated the applicative potential of ampiphilic dendrons, prepared by functionalization of the end groups of PAMAM-on-silica templates with long aliphatic residues, for adsorption of anionic targets from water and water purification.[49a] The same system was very recently proposed as stationary phase for ultrahigh-pressure liquid chromatography.[49b]

Removal of another pollutant, atrazine herbicide, by polymelamine dendrons on silica was demonstrated by Simanek and coworkers.[83] The authors revealed two trends. First, the dendrons of higher generation increased the sequestering ability of the hybrid material relative to that of the support modified with the lower generation dendrons. Second, this ability was higher for the dendrons divergently grown on silica, compared to the materials obtained by the focal-point immobilization of the dendrons presynthesized in solution (see Section 15.2.9).

Polystyrene supports modified with the similar triazine-based dendrons were used by Marsh et al. as insoluble scavengers for removal of excess nucleophilic or electrophilic reagents from the reaction mixtures.[79] Dendrons, decorated with tertiary amines, effectively removed electrophiles (protons), whereas the similar resin with chlorotriazine terminal groups sequestered the excess nucleophiles (amines).

15.3.6 Dendronized Surfaces for Immobilization of Biomacromolecules

Poly(amidoamine) dendrons immobilized on gold surface and cross-linked with poly (propylenimine) dendrimers have been used as bioreactive surfaces for conjugation of biological macromolecules *en route* to fabrication of DNA microarrays, protein microchips, and other biosensors.[119] Polyamide pseudodendrons on glass and polypropylene were designed for similar purposes.[33c] With a similar goal in mind, aldehyde-terminated first- and second-generation Fréchet-type polyether dendrons immobilized postsynthetically on silica were used by Lemcoff and coworkers for protein binding and for demonstration of a positive dendritic effect on the binding.[120]

15.3.7 Other Applications

Capitalizing on the PAMAM-on-silica grafting methodology that they pioneered several years ago,[42a] the group of Tsubokawa recently reported immobilization of polybromoaromatics to prepare silica particles with flame retardant activity (Scheme 15.49a).[56a] Such activity of epoxy resins mixed with the formed dendronized silica hybrids was improved compared to the use of simple polybromoaromatics dispersed in the resins. The same silica-bound PAMAM constructs were used by Tsubokawa and coworkers for immobilization of capsaicin, the principal pungent constituent of hot peppers, in order to prepare silica particles with biorepellent activity (Scheme 15.49b).[56b]

Peptide Reymond-type dendrons with cysteine or cystine near the focal point or at the core were immobilized on gold surface to directly visualize single macromolecules using STM.[121] For a similar goal of studying dendron–dendron interaction by AFM-based single-molecule force spectroscopy, Zhang and coworkers immobilized polyether dendrons on gold plate and PEG-fitted polyether dendron on the AFM tip (Figure 15.9).[122]

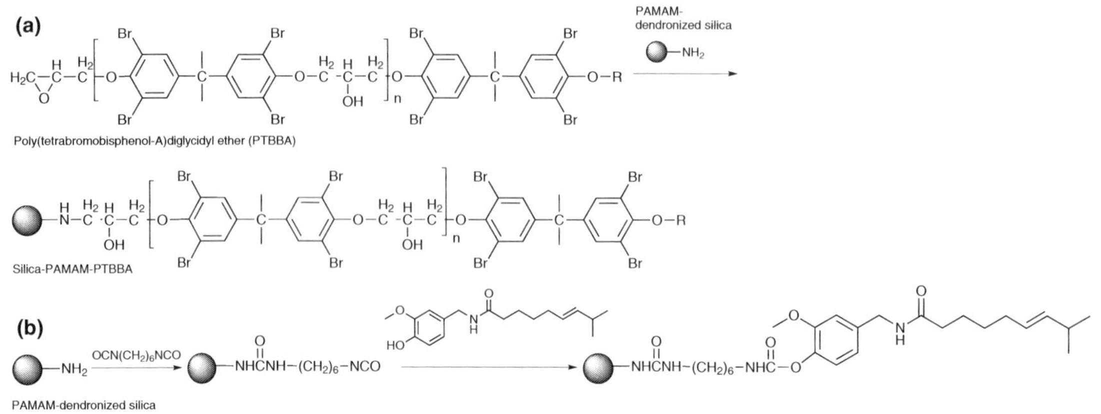

Scheme 15.49. Derivatization of PAMAM-on-silica support with flame-retardant (a) and biorepellent (b) terminal moieties.

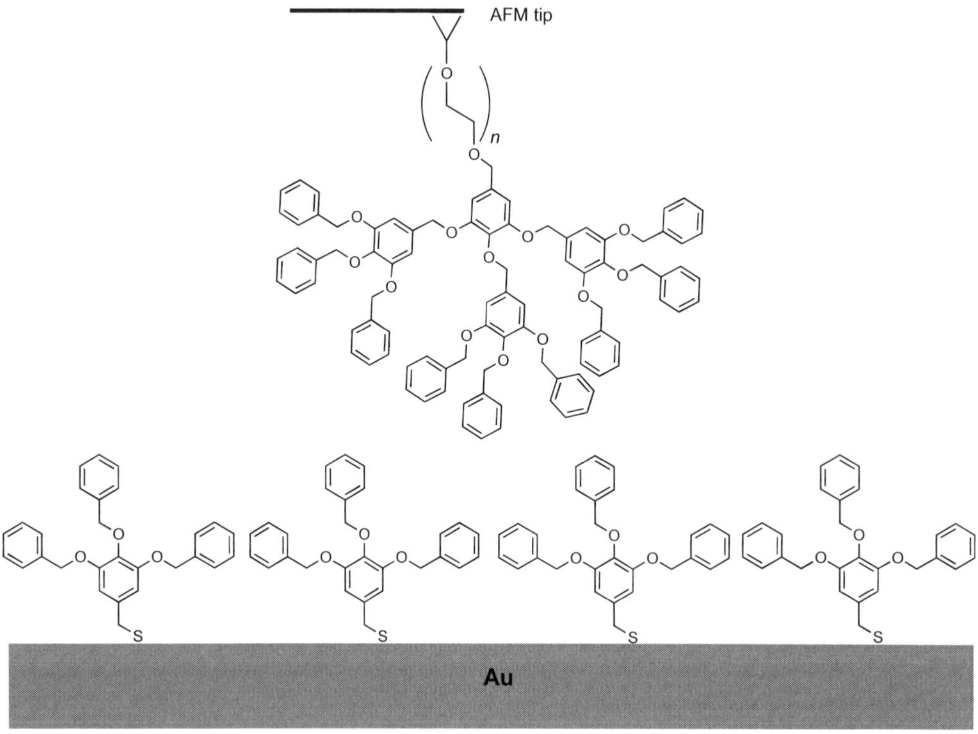

Figure 15.9. The use of dendrons by Zhang et al. for study of intermolecular interaction by AFM-based technique.

15.4 CONCLUSIONS

The dendrons on solid support, reviewed in this chapter, display a substantial structural and applicative diversity. Although these hybrids are usually prepared on polystyrene or silica cores, additional organic (PEGA, hydroxyethyl methacrylate) and inorganic (carbon nanotubes, glass) supports have also been dendronized. The dendritic motifs, originally limited to polylysine and PAMAM, have been expanded to polyurea, polyester, polyether, polyamine, polymelamine, and other dendritic structures. The applications of these hybrid materials range far beyond synthetic intermediates of immunogenic polypeptides and now encompass drug delivery vehicles, catalysts, high-loading supports for synthesis, probes for enzymatic activity, stationary chromatographic phases, and a number of other applications. The uniqueness of the dendritic architecture, leading to dramatically altered performance of the hybrid materials compared to the parent supports, was prominent in a number of fields, particularly catalysis and separation.

Regarding the synthetic aspects of the field, we may expect continued activity. As aforementioned, the range of the dendritic motifs readily accessible on insoluble supports was significantly expanded in the past decade. While additional types of dendrons can certainly be designed and prepared in the future, the arsenal of "available" dendrons is broad and diverse enough to suit most applications. On the other hand, one may expect continuous improvement of the existing or development of new synthetic routes to dendrons analogous or closely resembling those described in the literature and reviewed in this chapter. Though many synthetic schemes, discussed above, were brought to a very high efficiency, others are still far from optimal. It is possible that achieving perfect dendron growth in these cases requires alternative synthetic schemes (e.g., "click chemistry"-type reactions), different supports (e.g., macroporous resins), or a better match between the type of the support and the dendron or the chemistry used in its synthesis. The solid-phase synthesis approach in its current, mature state provides a multitude of synthetic tools (supports, linkers, reactions, and techniques) that can eventually enable the accomplishment of this goal.

While one may expect that additional efforts will be devoted to the preparation of dendronized supports with higher loading and lower level of structural defects, recent studies of hyperbranched polymers grafted on solid support (including silica-supported PAMAM, which is in some cases better described as a hyperbranched polymer rather than dendron) demonstrated their potential as comparable substitutes of perfectly dendronized supports. Taking into account that some of these hyperbranched grafts are generated in one-pot polymerization and possess high degrees of branching, it can be predicted that the routes to such composite materials and the studies of their applications will be substantially expanded in the future.

As to the applications of the dendrons on solid support, it seems that to date only the "top layer" of their full potential has been revealed. Uses of MAPs, particularly as immunogenic conjugates, are well established and may have reached maturity, but some of the other applications, discussed in this chapter, are likely to be expanded. The positive dendritic effects on activity, selectivity, and recyclability observed in catalysis, and the broad range of the reactions tested with the supported dendritic catalysts, are encouraging, as they point to the possibility of tailor-made suitable dendritic catalysts for improving a specific catalytic property for a given transformation. One may expect that the dendritic architecture will also open the way to heterogeneous bifunctional catalysts, which are underrepresented today in comparison to their soluble analogues. Dendritic architecture may add additional diversity elements to combinatorial studies of the supported catalysts. Moreover, supported dendron-incorporating constructs may emerge as important tools in discovering novel catalytic activities because they enable preorganized immobilization of potential catalysts and multiple substrate units.

The use of multivalency or preorganization of terminal groups has proven beneficial for various uses of the hybrid materials in separation studies. We may expect that the ability of chiral dendrons (internally chiral or bearing chiral terminal groups) to resolve enantiomer mixtures will be further exploited. One may also anticipate that the increased awareness of environmental issues will lead to expanded exploration of the use of dendronized solids for removal of pollutants from water (drinking or waste water).

In addition to the already broadly studied applications, there are a number of those that have barely been proposed, sometimes supported by a limited feasibility study, for example, the use of supported dendrons as synthetic intermediates for the preparation of drug or gene delivery vehicles, for the construction of nanoparticles, and for the assembly of probes for the study of biochemical processes. The use of dendronized supports as tools for multivalent molecular recognition has also hardly been explored. New applications may emerge, though naturally they are more difficult to foresee. For instance, support-bound probes for investigation of enzymatic activity, which will exploit the multivalency and the proximity of the dendritic arms, may follow the examples of the detachable dendritic probes.[98]

Today, dendronized supports are well-established tools in the arsenal of synthetic, catalytic, and polymer chemists and their uses are extending to the areas of bio- and materials sciences. One may hope that new and fascinating synthetic schemes and applications will follow.

REFERENCES

1. (a) G. R. Newkome, C. N. Moorefield, F. Vögtle, *Dendrimers and Dendrons: Concepts, Synthesis, Applications*, 2nd edn., Wiley-VCH, Weinheim, **2001**; (b) J. M. J. Fréchet, D. A. Tomalia, *Dendrimers and Other Dendritic Polymers*, 2nd edn., Wiley, Chichester, **2002**.
2. For dendritic wedges coupled to the support through the periphery, see (a) M. Cardona, S. H. Jannach, H. Huang, Y. Itojima, R. M. Leblanc, R. E. Gawley, G. A. Baker, E. B. Brauns, *Helv. Chim. Acta* **2002**, *85*, 3532; (b) Y. S. Choi, C. W. Yoon, H. D. Lee, M. Park, J. W. Park, *Chem. Commun.* **2004**, 1316; (c) S. J. Oh, J. Ju, B. C. Kim, E. Ko, B. J. Hong, J. G. Park, J. W. Park, K. Y. Choi, *Nucleic Acids Res.* **2005**, *33*, e90.
3. For dendrimers coupled to the support, see. (a) B. Miksa, S. Slomkowski, M. M. Chehimi, M. Delamar, J. P. Majoral, A. M. Caminade, *Colloid. Polym. Sci.* **1999**, *277*, 58; (b) R. Benters, C. M. Niemeyer, D. Wöhrle, *Chembiochem* **2001**, *2*, 686; (c) S. S. Mark, N. Sandhyarani, C. Zhu, C. Campagnolo, C. A. Batt, *Langmuir* **2004**, *20*, 6808; (d) C. Grüttner, V. Böhmer, A. Casnati, J. F. Dozol, D. N. Reinhoudt, M. M. Reinoso-Garcia, S. Ruderhausen, J. Teller, R. Ungaro, W. Verboom, P. Wang, *J. Magn. Mater.* **2005**, *293*, 559; (e) M. J. Archer, B. Lin, Z. Wang, D. A. Stenger, *Anal. Biochem.* **2006**, *355*, 285.
4. (a) T. Kehat, K. Goren, M. Portnoy, *New J. Chem.* **2007**, *31*, 1218; (b) A. Dahan, M. Portnoy, *J. Polym. Sci. A* **2005**, *43*, 235; (c) S. Lebreton, S. Monaghan, M. Bradley, *Aldrichim. Acta* **2001**, *34*, 75; (d) R. Haag, *Chem. Eur. J.* **2001**, *7*, 327; (e) R. J. M. Klein Gebbink, C. A. Kruithof, G. P. M. van Klink, G. van Koten, *Rev. Mol. Biotechnol.* **2002**, *90*, 183; (f) A. S. H. King, L. J. Twyman, *J. Chem. Soc., Perkin Trans. 1* **2002**, 2209.
5. For postsynthetic immobilization of dendrons, (a) see M. Driffield, D. M. Goodall, A. S. Klute, D. K. Smith, K. Wilson, *Langmuir* **2002**, *18*, 8660; (b) B. T. Mathews, A. E. Beezer, M. J. Snowden, M. J. Hardy, J. C. Mitchell, *New J. Chem.* **2001**, *25*, 807; (c) G. R. Newkome, K. S. Yoo, C. N. Moorefield, *Des. Monomers Polym.* **2002**, *5*, 67.
6. S. M. Grayson, J. M. J. Fréchet, *Chem. Rev.* **2001**, *101*, 3819.
7. J. S. Früchtel and G. Jung, *Angew. Chem., Int. Ed.* **1996**, *35*, 17.
8. P. Bharathi, U. Patel, T. Kawaguchi, D. J. Pesak, J. S. Moore, *Macromolecules* **1995**, *28*, 5955.

9. C. Chi, J. Wu, X. Wang, X. Zhao, J. Li, F. Wang, *Tetrahedron Lett.* **2001**, *42*, 2181.
10. (a) K. Sadler, J. P. Tam, *Rev. Mol. Biotechnol.* **2002**, *90*, 195; (b) L. Crespo, G. Sanclimens, M. Pons, E. Giralt, M. Royo, F. Albericio, *Chem. Rev.* **2005**, *105*, 1663; (c) P. Niederhafner, J. Sebestík, J. Jezek, *J. Pept. Sci.* **2005**, *11*, 757.
11. R. G. Denkewalter, J. Kole, W. J. Lukasavage,US Patent 4,410,688, **1983**.
12. (a) D. N. Posnett, H. McGrath, J. P. Tam, *J. Biol. Chem.* **1988**, *263*, 1719; (b) J. P. Tam, *Proc. Natl. Acad. Sci. USA* **1988**, 85, 5409.
13. The words *dendron, dendrimer,* or *dendritic* did not appear in the early reports.
14. (a) J. P. Tam and F. Zavala, *J. Immunol. Methods* **1989**, *124*, 53; (b) M. J. Francis, G. Z. Hastings, F. Brown, J. McDermed, Y. A. Lu, J. P. Tam, *Immunology* **1991**, *73*, 249; (c) C. Rao, J. P. Tam, *J. Am. Chem. Soc.* **1994**, *116*, 6975.
15. (a)J. P. Tam, Y. A. Lu, Proc. Natl. Acad. Sci. USA **1989**, 86, 9084;(b) T. Sakthivel, I. Toth, A. Florence, *Pharm. Res.* **1998**, *15*, 776; (c) V. Cavallaro, P. Thompson, M. Hearn, *J. Pept. Sci.* **2001**, *7*, 262.
16. (a) J. W. Drijfhout, W. Bloemhoff, *Int. J. Pept. Protein Res.* **1991**, *37*, 27; (b) G. W. McLean, A. M. Owsianka, J. H. Subak-Sharpe, H. S. Mardsen, *J. Immunol. Methods* **1991**, *137*, 149; (c) B. Nardelli, Y. A. Lu, D. R. Shiu, C. Delpierre-Defoort, A. T. Profy, J. P. Tam, *J. Immunol.* **1992**, *148*, 914; (d) J. P. Tam, Y. A. Lu, J. L. Yang, *Eur. J. Biochem.* **2002**, *269*, 923; (e) E. A. B. Kantchev, C. C. Chang, D. K. Chang, *Biopolymers* **2006**, *84*, 232.
17. For polylysine dendron on Tentagel and PEGA resins, see (a) C. Grandjean, C. Rommens, H. Gras-Masse, O. Melnyk, *Angew. Chem., Int. Ed.* **2000**, *39*, 1068; (b) O. Melnyk, J. S. Fruchart, C. Gandjean, H. Gras-Masse, *J. Org. Chem.* **2001**, *66*, 4153.
18. (a) G. Purohit, T. Sakthivel, A. T. Florence, *Int. J. Pharm.* **2001**, *214*, 71; (b) K. T. Al-Jamal, T. Sakthivel, A. T. Florence, *J. Pharm. Sci.* **2005**, *94*, 102; (c) B. Singh, A. T. Florence, *Int. J. Pharm.* **2005**, *298*, 348.
19. T. Becker, A. Kaiser, H. Kunz, *Synthesis* **2009**, 1113.
20. D. Kozak, P. Surawski, K. M. Thoren, C. Y. Lu, L. Marcon, M. Trau, *Biomacromolecules* **2009**, *10*, 360.
21. For polyglutamic acid dendrons, see Ref. 5b.
22. (a) A. Esposito, E. Delort, D. Lagnoux, F. Djojo, J. L. Reymond, *Angew. Chem., Int. Ed.* **2003**, *42*, 1381; (b) D. Lagnoux, E. Delort, C. Douat-Casassus, A. Esposito, J. L. Reymond, *Chem. Eur. J.* **2004**, *10*, 1215; (c) C. Douat-Casassus, T. Darbre, J. L. Reymond, *J. Am. Chem. Soc.* **2004**, *126*, 7817; (d) A. Clouet, T. Darbre, J. L. Reymond, *Adv. Synth. Catal.* **2004**, *346*, 1195; (e) D. Lagnoux, T. Darbre, M. L. Schmitz, J. L. Reymond, *Chem. Eur. J.* **2005**, *11*, 3941.
23. (a) A. Clouet, T. Darbre, J. L. Reymond, *Angew. Chem., Int. Ed.* **2004**, *43*, 4612; (b) A. Clouet, T. Darbre, J. L. Reymond, *Biopolymers* **2006**, *84*, 114; (c) E. Delort, T. Darbre, J. L. Reymond, *J. Am. Chem. Soc.* **2004**, *126*, 15642; (d) E. Delort, N. Q. Nguyen-Trung, T. Darbre, J. L. Reymond, *J. Org. Chem.* **2006**, *71*, 4468; (e) J. Kofoed, T. Darbre, J. L. Reymond, *Org. Biomol. Chem.* **2006**, *4*, 3268.
24. E. Kolomiets, M. A. Swiderska, R. U. Kadam, E. M. V. Johansson, K. E. Jaeger, T. Darbre, J. L. Reymond, *ChemMedChem* **2009**, *4*, 562.
25. (a) L. Crespo, G. Sanclimens, M. Royo, E. Giralt, F. Albericio, *Eur. J. Org. Chem.* **2002**, 1756; (b) L. Crespo, G. Sanclimens, B. Montaner, R. Pérez-Tomás, M. Royo, M. Pons, F. Albericio, E. Giralt, *J. Am. Chem. Soc.* **2002**, *124*, 8876; (c) G. Sanclimens, L. Crespo, E. Giralt, M. Royo, F. Albericio, *Biopolymers* **2004**, *76*, 283; (d) G. Sanclimens, L. Crespo, E. Giralt, F. Albericio, M. Royo, *J. Org. Chem.* **2005**, *70*, 6274; (e) G. Sanclimens, H. Shen, E. Giralt, F. Albericio, M. W. Saltzman, M. Royo, *Biopolymers* **2005**, *80*, 800.
26. G. Sanclimens, L. Crespo, M. Pons, E. Giralt, F. Albericio, M. Royo, *Tetrahedron Lett.* **2003**, *44*, 1751.

27. (a) P. Arya, N. R. Rao, J. Singkhonrat, H. Alper, S. C. Bourque, L. E. Manzer, *J. Org. Chem.* **2000**, *65*, 1881; (b) H. Alper, P. Arya, S. C. Bourque, G. R. Jefferson, L. E. Manzer, *J. Am. Chem. Soc.* **2001**, *123*, 2889; (c) S. M. Lu, H. Alper, *J. Am. Chem. Soc.* **2003**, *125*, 13126.
28. K. E. Uhrich, S. Boegeman, J. M. J. Fréchet, S. R. Turner, *Polym. Bull.* **1991**, *25*, 551.
29. J. K. Cho, D. W. Kim, J. Namgung, Y. S. Lee, *Tetrahedron Lett.* **2001**, *42*, 7443.
30. (a) A. Mahajan, S. R. Chhabra, W. C. Chan, *Tetrahedron Lett.* **1999**, *40*, 4909; (b) S. R. Chhabra, A. Mahajan, W. C. Chan, *J. Org. Chem.* **2002**, *67*, 4017.
31. J. J. Diaz-Mochon, M. A. Fara, R. M. Sanchez-Martin, M. Bradley, *Tetrahedron Lett.* **2008**, *49*, 923.
32. (a) N. V. Amirkhanov, E. Wickstrom, *Nucleosides Nucleotides Nucleic Acids* **2005**, *24*, 423; (b) N. V. Amirkhanov, I. Dimitrov, A. W. Opitz, K. Zhang, J. P. Lackey, C. A. Cardi, S. Lai, N. J. Wagner, M. L. Thakur, E. Wickstrom, *Biopolymers* **2008**, *89*, 1061.
33. (a) J. Katajisto, T. Karskela, P. Heinonen, H. Lönnberg, *J. Org. Chem.* **2002**, *67*, 7995; (b) H. Neubert, A. T. Kicman, D. A. Cowan, S. S. Bansal, *Tetrahedron Lett.* **2002**, *43*, 6723; (c) M. Beier, J. D. Hoheisel, *Nucleic Acids Res.* **1999**, *27*, 1970.
34. S. H. Huang, S. R. Li, Z. W. Bai, Z. Q. Pan, C. Q. Yin, *Chromatographia* **2006**, *64*, 641.
35. V. Swali, N. J. Wells, G. J. Langley, M. Bradley, *J. Org. Chem.* **1997**, *62*, 4902.
36. D. A. Tomalia, H. Baker, J. Dewald, M. Hall, G. Kallos, S. Martin, J. Roeck, J. Ryder, P. Smith, *Polym. J.* **1985**, *17*, 117.
37. (a) N. J. Wells, A. Basso, M. Bradley, *Biopolymers* **1998**, *47*, 381; (b) N. J. Wells, M. Davies, M. Bradley, *J. Org. Chem.* **1998**, *63*, 6430.
38. A. Basso and M. Bradley, *Tetrahedron Lett.* **2003**, *44*, 2699.
39. (a) C. L. Li, Z. W. Yang, S. Wu, Z. Q. Lei, *React. Funct. Polym.* **2007**, *67*, 53; (b) J. Wang, R. Wang, Z. Zhang, C. Li, Z. W. Yang, Z. Q. Lei, *J. Macromol. Sci. A* **2008**, *45*, 672.
40. (a) G. R. Krishnan, K. Sreekumar, *Eur. J. Org. Chem.* **2008**, 4763; (b) G. R. Krishnan, K. Sreekumar, *Polymer* **2008**, *49*, 5233; (c) G. R. Krishnan, K. Sreekumar, *Appl. Catal. A* **2009**, *353*, 80.
41. H. Liu, J. Guo, L. Jin, W. Yang, C. Wang, *J. Phys. Chem. B* **2008**, *112*, 3315.
42. (a) N. Tsubokawa, H. Ichioka, T. Satoh, S. Hayashi, K. Fujiki, *React. Funct. Polym.* **1998**, *37*, 75; (b) N. Tsubokawa, K. Kotama, H. Saitoh, T. Nishikubo, *Compos. Interfaces* **2003**, *10*, 609; (c) Y. Taniguchi, K. Shirai, H. Saitoh, T. Yamauchi, N. Tsubokawa, *Polymer* **2005**, *46*, 2541.
43. (a) S. C. Bourque, F. Maltais, W. J. Xiao, O. Tardif, H. Alper, P. Arya, L. E. Manzer, *J. Am. Chem. Soc.* **1999**, *121*, 3035; (b) S. C. Bourque, H. Alper, L. E. Manzer, P. Arya, *J. Am. Chem. Soc.* **2000**, *122*, 956; (c) J. P. K. Reynhardt, H. Alper, *J. Org. Chem.* **2003**, *68*, 8353; (d) P P. Zweni, H. Alper, *Adv. Synth. Catal.* **2004**, *346*, 849.
44. (a) K. Sakai, T. C. Teng, A. Katada, T. Harada, S. Uemura, Y. Asami, M. Sakata, M. Kunitake, C. Hirayama, *Chem. Lett.* **2001**, 510;(b) K. Sakai, T. C. Teng, A. Katada, T. Harada, K. Yoshida, K. Yamanaka, Y. Asami, M. Sakata, C. Hirayama, M. Kunitake, *Chem. Mater.* **2003**, *15*, 4091.
45. (a) Y. M. Chung, H. K. Rhee, *Chem. Commun.* **2002**, 238;(b) Y. M. Chung, H. K. Rhee, *C. R. Chim.* **2003**, *6*, 695.
46. (a) J. Bu, Z. M. A. Judeh, C. B. Ching, S. Kawi, *Catal. Lett.* **2003**, *85*, 183; (b) J. Bu, R. Li, C. W. Quah, K. J. Carpenter, *Macromolecules* **2004**, *37*, 6687; (c) R. Li, J. Bu, *Korean J. Chem. Eng.* **2004**, *21*, 98.
47. (a) P. Li, S. Kawi, *Catal. Today* **2008**, *131*, 61; (b) P. Li, S. Kawi, *J. Catal.* **2008**, *257*, 23.
48. C. Wang, G. Zhu, J. Li, X. Cai, Y. Wei, D. Zhang, S. Qiu, *Chem. Eur. J.* **2005**, *11*, 4975.
49. (a) C. C. Chu, N. Ueno, T. Imae, *Chem. Mater.* **2008**, *20*, 2669; (b) C. C. Chu, N. Ueno, T. Imae, *J. Nanosci. Nanotechnol.* **2010**, *10*, 5324.
50. C. Zhang, P. Su, M. U. Farooq, Y. Yang, X. Gao, E. Hongjun, *React. Funct. Polym.* **2010**, *70*, 129.

51. (a) J. P. K. Reynhardt, Y. Yang, A. Sayari, H. Alper, *Chem. Mater.* **2004**, *16*, 4095; (b) J. P. K. Reynhardt, Y. Yang, A. Sayari, and H. Alper, *Adv. Funct. Mater.* **2005**, *15*, 1641; (c) J. P. K. Reynhardt, Y. Yang, A. Sayari, H. Alper, *Adv. Synth. Catal.* **2005**, *347*, 1379.
52. Y. Jiang, Q. Gao, *J. Am. Chem. Soc.* **2006**, *128*, 716.
53. R. Qu, Y. Niu, C. Sun, C. Ji, C. Wang, G. Cheng, *Microporous Mesoporous Mater.* **2006**, *97*, 58.
54. M. P. Kapoor, Y. Kasama, T. Yokoyama, M. Yanagi, S. Inagaki, H. Nanbu, L. R. Juneja, *J. Mater. Chem.* **2006**, *16*, 4714.
55. (a) C. Q. Yin, B. J. He, S. H. Huang, J. Y. Zhang, Z. W. Bai, Z. Y. Li, *Chirality* **2008**, *20*, *846;*(b) B. J. He, C. Q. Yin, S. R. Li, Z. W. Bai, *Chirality* **2010**, *22*, 69.
56. (a) T. Yamauchi, A. Yuuki, G. Wei, K. Shirai, K. Fujiki, N. Tsubokawa, *J. Polym. Sci. A* **2009**, *47*, 6145; (b) T. Yamauchi, T. Saitoh, K. Shirai, K. Fujiki, N. Tsubokawa, *J. Polym. Sci. A* **2010**, *48*, 1800.
57. (a) P. Li, W. Thitsartarn, S. Kawi, *Ind. Eng. Chem. Res.* **2009**, *48*, 1824; (b) N. Tsubokawa, T. Takayama, *React. Funct. Polym.* **2000**, *43*, 341; (c) C. Li, J. Wang, Z. Yang, Z. Hu, Z. Lei, *Catal. Commun.* **2007**, *8*, 1202; (d) S. Lei, S. Yu, C. Zhao, *J. Chromatogr. Sci.* **2001**, *39*, 280; (e) R. Abu-Reziq, H. Alper, D. Wang, M. L. Post, *J. Am. Chem. Soc.* **2006**, *128*, 5279; (f) Y. Jang, J. Jiang, Q. Gao, M. Ruan, H. Yu, L. Qi, *Nanotechnology* **2008**, *19*, 075714; (g) B. Pan, D. Cui, F. Gao, R. He, *Nanotechnology* **2006**, *17*, 2483; (h) L. Tao, G. Chen, G. Mantovani, S. York, D. M. Hddleton, *Chem. Commun.* **2006**, 4949.
58. C. Fromont, M. Bradley, *Chem. Commun.* **2000**, 283.
59. S. E. How, A. Unciti-Broceta, R. M. Sanchez-Martin, M. Bradley, *Org. Biomol. Chem.* **2008**, *6*, 2266.
60. (a) S. Lebreton, N. Newcombe, M. Bradley, *Tetrahedron Lett.* **2002**, *43*, 2475; (b) S. Lebreton, N. Newcombe, M. Bradley, *Tetrahedron Lett.* **2002**, *43*, 2479; (c) S. Lebreton, S. E. How, M. Buchholz, B. Yingyongnarongkul, M. Bradley, *Tetrahedron* **2003**, *59*, 3945.
61. S. Lebreton, N. Newcombe, M. Bradley, *Mol. Diversity* **2003**, *6*, 19.
62. F. H. Ling, V. Lu, F. Svec, J. M. J. Fréchet, *J. Org. Chem.* **2002**, *67*, 1993.
63. S. Akbari, M. H. Kish, A. A. Entezami, *Polym. Int.* **2008**, *57*, 846.
64. H. F. Chow, C. F. Leung, G. X. Wang, J. Zhang, *Top. Curr. Chem.* **2001**, *217*, 1.
65. A. Basso, B. Evans, N. Pegg, M. Bradley, *Chem. Commun.* **2001**, 697.
66. A. Dahan, M. Portnoy, *Macromolecules* **2003**, *36*, 1034.
67. (a) J. L. Castro, R. Baker, A. R. Guiblin, S. C. Hobbs, M. R. Jenkins, M. G. N. Russell, M. S. Beer, J. A. Stanton, K. Scholey, R. J. Hargreaves, M. I. Graham, V. G. Matassa, *J. Med. Chem.* **1994**, *37*, 3023; (b) J. L. Castro, V. G. Matassa, *J. Org. Chem.* **1994**, *59*, 2289.
68. A. Mansour, T. Kehat, M. Portnoy, *Org. Biomol. Chem.* **2008**, *6*, 3382.
69. A. Dahan, H. Dimant, M. Portnoy, *J. Comb. Chem.* **2004**, *6*, 305.
70. G. Gellerman, S. Shitrit, T. Shalit, O. Granot, A. Albeck, *Tetrahedron* **2010**, *66*, 878.
71. For a specific synthesis, see H. Hioki, W. C. Still, *J. Org. Chem.* **1998**, *63*, 904. For a general procedure, see M. S. Newman, H. A. Karnes, *J. Org. Chem.* 1966, *31*, 3980.
72. A. Dahan, A. Weissberg, M. Portnoy, *Chem. Commun.* **2003**, 1206.
73. A. Dahan, M. Portnoy, *J. Am. Chem. Soc.* **2007**, *129*, 5860.
74. F. Koç, P. Eilbracht, *Tetrahedron* **2004**, *60*, 8465.
75. P. Liu, X. Wu, Q. Pu, Z. Su, *Annal. Chim.* **2005**, *95*, 695.
76. (a) E. Buchleier, W. Wehner, F. Vögtle, *Synthesis* **1978**, 155; (b) E. M. M. de Brabander-van den Berg, E. W. Meijer, *Angew. Chem., Int. Ed.* **1993**, *32*, 1309.
77. C. Sun, R. Qu, C. Ji, C. Wang, Y. Sun, Z. Yue, G. Cheng, *Talanta* **2006**, *70*, 14.
78. For a review on triazine-based dendrimers, see M. B. Steffensen, E. Hollink, F. Kuschel, M. Bauer, E. E. Simanek, *J. Polym. Sci. Part A* **2006**, *44*, 3411.

79. A. Marsh, S. J. Carlisle, S. C. Smith, *Tetrahedron Lett.* **2001**, *42*, 493.
80. X. Wu, P. Liu, Q. Pu, Z. Su, *Anal. Lett.* **2003**, *36*, 2229.
81. S. J. Dilly, S. J. Carlisle, A. J. Clark, A. R. Shepherd, S. C. Smith, P. C. Taylor, A. Marsh, *J. Polym. Sci. A* **2006**, *44*, 2248.
82. E. J. Acosta, C. S. Carr, E. E. Simanek, and D. F. Shantz, *Adv. Mater.* **2004**, *16*, 985.
83. E. J. Acosta, S. O. Gonzalez, E. E. Simanek, *J. Polym. Chem. A* **2005**, *43*, 168.
84. M. Wanunu, A. Vaskevich, A. Shanzer, I. Rubinstein, *J. Am. Chem. Soc.* **2006**, *128*, 8341.
85. G. P. Vlasov, I. I. Tarasenko, G. A. Pankova, I. E. Il'ina, *Polym. Sci. B* **2009**, *51*, 483.
86. See the comprehensive lists in Ref. 10a and c.
87. (a) For recent reports, see. R. M. Aguilar, F. J. Talamantes, J. J. Bustamante, J. Munoz, L. R. Trevino, A. O. Martinez, L. S. Haro, *J. Pept. Sci.* **2009**, *15*, 78; (b) W. Kowalczyk, B. G. de la Torre, D. Andreu, *Bioconjugate Chem.* **2010**, *21*, 102.
88. J. Ziegler, R. T. Chang, D. W. Wright, *J. Am. Chem. Soc.* **1999**, *121*, 2395.
89. A. Kolozsi, I. Vosekalna, T. Martinek, E. Larsen, B. Gyurcsik, *Dalton Trans.* **2009**, 5647.
90. E. K. Liebler, U. Diederichsen, *Org. Lett.* **2004**, *6*, 2893.
91. A. K. Galande, S. A. Hilderbrand, R. Weissleder, C. H. Tung, *J. Med. Chem.* **2006**, *49*, 4715.
92. S. Kasai, H. Nagasawa, M. Shimamura, Y. Uto, H. Hori, *Bioorg. Med. Chem. Lett.* **2002**, *12*, 951.
93. (a) S. Futaki, I. Nakase, T. Suzuki, Z. Youjun, and Y. Sugiura, *Biochemistry* **2002**, *41*, 7925; (b) S. Futaki, *Int. J. Pharm.* **2002**, *245*, 1.
94. (a)K. Sheldon, D. Liu, J. Ferguson, J. Gariépy, Proc. Natl. Acad. Sci. USA **1995**, *92*, 2056; (b) D. Singh, R. Kiarash, K. Kawamura, E. C. LaCasse, J. Gariépy, *Biochemistry* **1998**, *37*, 5798; (c) D. Singh, S. K. Bisland, K. Kawamura, J. Gariépy, *Bioconjugate Chem.* **1999**, *10*, 745; (d) S. K. Bisland, D. Singh, J. Gariépy, *Bioconjugate Chem.* **1999**, *10*, 982; (e) J. Gariépy and K. Kawamura, *Trends Biotechnol.* **2001**, *19*, 21; (f) R. D. Brokx, S. K. Bisland, J. Gariépy, *J. Controlled Release* **2002**, *78*, 115.
95. (a) K. T. Al-Jamal, T. Sakthivel, A. T. Florence, *Int. J. Pharm.* **2003**, *254*, 33; (b) K. T.Al-Jamal, C. Ramaswamy, B. Singh, A. T. Florence, *J. Drug Delivery Sci. Technol.* **2005**, *15*, 11; (c) K. T. Al-Jamal, T. Sakthivel, A. T. Florence, *Colloids Surf. A* **2005**, *268*, 52.
96. T. Sakthivel and A. T. Florence, *Int. J. Pharm.* **2003**, *254*, 23.
97. S. E. How, B. Yingyongnarongkul, M. A. Fara, J. J. Diaz-Mochón, S. Mittoo, M. Bradley, *Comb. Chem. High Throughput Screening* **2004**, *7*, 423.
98. (a) J. M. Ellard, T. Zollitsch, W. J. Cummins, A. L. Hamilton, M. Bradley, *Angew. Chem., Int. Ed.* **2002**, *41*, 3233; (b) M. Ternon, M. Bradley, *Chem. Commun.* **2003**, 2402; (c) M. Ternon, J. J. Diaz-Monchón, A. Belsom, M. Bradley, *Tetrahedron* **2004**, *60*, 8721.
99. N. Maillard, T. Darbre, J. L. Reymond, *J. Comb. Chem.* **2009**, *11*, 667.
100. (a) A. Basso, N. Pegg, B. Evans, M. Bradley, *Eur. J. Org. Chem.* **2000**, 3887; (b) A. Basso, B. Evans, N. Pegg, M. Bradley, *Tetrahedron Lett.* **2000**, *41*, 3763.
101. S. Lebreton, N. Newcombe, M. Bradley, *Tetrahedron* **2003**, *59*, 10213.
102. B. Yingyongnarongkul, M. Howarth, T. Elliott, M. Bradley, *J. Comb. Chem.* **2004**, *6*, 753.
103. H. M. Branderhorst, R. Ruitjenbeek, R. M. J. Liskamp, R. J. Pieters, *ChemBioChem* **2008**, *9*, 1836.
104. A similar design of a glycocluster prepared via a postsynthetic attachment of dendrons to resin-bound peptides was also reported by another group: H. A. Shaikh, F. D. Sonnichsen, T. K. Lindhorst, *Carbohydr. Res.* **2008**, *343*, 1665.
105. See Ref. 4a and b (reviews) and references therein; for recently published examples, see Refs 40, 47, 57a.
106. (a) H. Sellner, P. B. Rheiner, D. Seebach, *Helv. Chim. Acta* **2002**, *85*, 352; (b) P. B. Rheiner, D. Seebach, *Polym. Mater. Sci. Eng.* **1997**, *77*, 130; (c) H. Sellner, D. Seebach, *Angew. Chem.,*

Int. Ed., **1999**, *38*, 1918; (d) H. Sellner, C. Faber, P. B. Rheiner, D. Seebach, *Chem. Eur. J.* **2000**, *6*, 3692; (e) M. T. Reetz, D. Giebel, *Angew. Chem., Int. Ed.* **2000**, *39*, 2498; (f) J. W. Kriesel, T. D. Tilley, *Polym. Prepr.* **2000**, *41*, 566; (g) J. W. Kriesel, T. D. Tilley, *Chem. Mater.* **2000**, *12*, 1171; (h) P. E. Berget, J. M. Teixeira, J. L. Jacobsen, N. E. Schore, *Tetrahedron Lett.* **2007**, *48*, 8101.

107. (a) A. Dahan, M. Portnoy, *Org. Lett.* **2003**, *5*, 1197; (b) A. Dahan, M. Portnoy, *Chem. Commun.* **2002**, 2700.
108. (a) S. M. Lu, H. Alper, *J. Org. Chem.* **2004**, *69*, 3558; (b) S. Antebi, P. Arya, L. E. Manzer, H. Alper, *J. Org. Chem.* **2002**, *67*, 6623; (c) S. M. Lu, H. Alper, *J. Am. Chem. Soc.* **2005**, *127*, 14776; (d) S. M. Lu, H. Alper, *Chem. Eur. J.* **2007**, *13*, 5908; (e) R. Touzani, H. Alper, *J. Mol. Catal. A* **2005**, *227*, 197; (f) S. M. Lu, H. Alper, *J. Am. Chem. Soc.* **2008**, *130*, 6451; (g) H. Alper, P. Arya, S. C. Bourque, G. R. Jefferson, L. E. Manzer, *Can. J. Chem.* **2000**, *78*, 920; (h) P. P. Zweni, H. Alper, *Adv. Synth. Catal.* **2006**, *348*, 725.
109. R. Chanthateyanonth, H. Alper, *Adv. Synth. Catal.* **2004**, *346*, 1375.
110. P. Goyal, X. Zheng, M. Weck, *Adv. Synth. Catal.* **2008**, *350*, 1816.
111. R. W. J. Scott, O. M. Wilson, R. M. Crooks, *J. Phys. Chem. B* **2005**, *109*, 692.
112. K. Goren, M. Portnoy, *Chem. Commun.* **2010**, *46*, 1965.
113. L. Pu, H. B. Yu, *Chem. Rev.* **2001**, *101*, 757.
114. (a) T. Kehat, M. Portnoy, *Chem. Commun.* **2007**, 2823; (b) K. Goren, T. Kehat, M. Portnoy, *Adv. Synth. Catal.* **2009**, *351*, 59.
115. L. Tuchman-Shukron, M. Portnoy, *Adv. Synth. Catal.* **2009**, *351*, 541.
116. B. T. Mathews, A. E. Beezer, M. J. Snowden, M. J. Hardy, J. C. Mitchell, *Chromatographia* **2001**, *53*, 147.
117. (a) G. R. Newkome, K. S. Yoo, A. Kabir, A. Malik, *Tetrahedron Lett.* **2001**, *42*, 7537; (b) A. Kabir, C. Hamlet, K. S. Yoo, G. R. Newkome, A. Malik, *J. Chromatogr. A* **2004**, *1034*, 1.
118. X. Z. Wu, P. Liu, Q. S. Pu, Q. Y. Sun, Z. X. Su, *Talanta* **2004**, *62*, 918.
119. M. Yang, E. M. W. Tsang, Y. A. Wang, X. Peng, H. Z. Yu, *Langmuir* **2005**, *21*, 1858.
120. O. Iliashevsky, L. Amir, R. Glaser, R. S. Marks, N. G. Lemcoff, *J. Mater. Chem.* **2009**, *19*, 6616.
121. E. Delort, E. Szocs, R. Widmer, H. Siegenthaler, J. L. Reymond, *Macromol. Biosci.* **2007**, *7*, 1024.
122. W. Shi, Y. Zhang, C. Liu, Z. Wang, X. Zhang, *Langmuir* **2008**, *24*, 1318.

16

OLIGOSACCHARIDE SYNTHESIS ON SOLID, SOLUBLE POLYMER, AND TAG SUPPORTS

Katsunori Tanaka and Koichi Fukase

16.1 INTRODUCTION

Synthesis of oligosaccharides and glycoconjugates has played important roles in elucidation of their biological functions. However, substantial time and efforts are often needed, since it requires multiple step transformations involving iterative deprotection–glycosylation procedures. Therefore, extensive efforts have been made to establish the efficient synthesis of oligosaccharides by means of solid-phase method and polymer-supported solution method. In solid-phase synthesis, the starting material binds to the polymer support, which allows the product to be isolated by filtration after each reaction. Finally, the product is cleaved from the resin. The operation is simple and rapid, and hence solid-phase synthesis enables rapid preparation of a large number of compounds and thus increases the speed of development of a lot of functional molecules including new drugs and new materials. However, there have been several critical issues for solid-phase oligosaccharide synthesis; for example, the synthetic intermediate cannot be purified. Thus, each glycosylation and deprotection therefore must proceed in a high yield. Recent progress in glycosylation and protective group manipulation solved these problems to make solid-phase synthesis of oligosaccharides practical.[1] In this chapter, the authors have mainly focused on the recent 10-year achievements of the solid-supported techniques for the oligosaccharide synthesis and acknowledged the pioneering and earlier work in this field.[1] In addition, they have also described not only solid-phase synthesis but also polymer-supported solution synthesis as well as tag-supported synthesis, which have proved to be useful for rapid preparation of oligosaccharides.

Solid-Phase Organic Synthesis: Concepts, Strategies, and Applications, First Edition.
Edited by Patrick H. Toy and Yulin Lam.

16.2 SOLID-PHASE METHODS FOR SYNTHESIS OF OLIGOSACCHARIDES

16.2.1 New Linkers and Protection Groups for Solid-Phase Synthesis of Oligosaccharides

The choice of linkers is critically important for both solid-phase and polymer-supported solution synthesis. Selective cleavage of the linkers, leaving other protecting groups intact, and then purification and final deprotection is a rational strategy for oligosaccharide synthesis. This is distinctly different from the solid-phase synthesis of peptides, in which all the side-chain protecting groups and linkers are removed simultaneously at the end of the synthesis on the solid supports. In the case of oligosaccharides, the purification in protected forms is generally more preferable owing to the hydrophilic properties of completely deprotected oligosaccharides. Traditionally, the linkers have been directly attached to the anomeric positions as thioglycoside, 1-*O*-acyl, photolabile *o*-nitrobenzyl glycoside, *p*-alkoxybenzyl glycoside, base labile 9-hydroxymethylfluorene, or silylethyl glycoside.[1] On the other hand, the ester-type, the silyl ether-type, or the *p*-acylaminobenzyl-type linkers have also been used to obtain the anomerically pure oligosaccharides after cleavage from the resins. More recently, the unique linkers that are bifunctional, temporary, and cleavable under the completely chemoselective manners by the recently appeared transition metal-catalyzed reactions have been developed and successfully applied to the solid-phase synthesis of oligosaccharides. The development of such linkers enabled the quite challenging oligosaccharide assembly on the solid supports, such as *N*-glycans or glycosylphosphatidylinositol of complex and branched structures, even by the automated systems, which will be described in Section 16.2.3. Schmidt and coworkers have developed the *o*-nitro-phenoxyacetate linker, which can be readily and quantitatively cleaved by treatment with methylamine under mild conditions (Scheme 16.1).[2] By repeating the sequences of glycosylation with glucosyl trichloroacetimidate in the presence of TMSOTf (trimethylsilyl triflate), and Et_3N-induced deprotection of Fmoc (9-fluorenylmethoxycarbonyl) group, an efficient solid-phase synthesis of di-, tri-, and tetrasaccharides **1** containing glucose β(1-4) and β(1-6) linkages has been achieved.

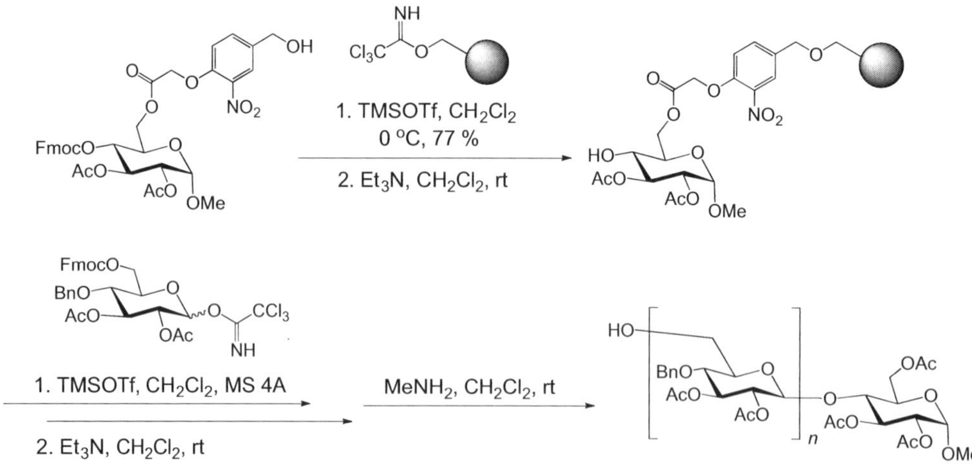

1 (*n* = 1–3)

Scheme 16.1. Synthesis using the o-nitro-phenoxyacetate linker.

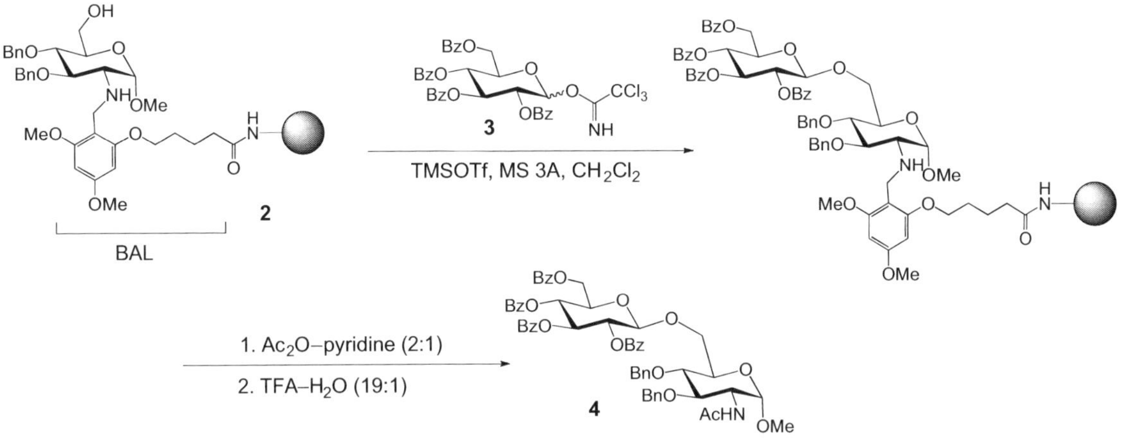

Scheme 16.2. Synthesis using the tris(alkoxy)benzylamine linker.

Jensen and coworkers have developed the tris(alkoxy)benzylamine linker (BAL) on the polystyrene support, which is introduced to the amino sugars by the reductive amination of the corresponding aldehyde in the presence of $NaBH_3CN$ in DMF (dimethylformamide)–AcOH (Scheme 16.2).[3,4] This linker proved to be stable to concentrated TFA (trifluoroacetic acid) and other Lewis acids, but readily cleaved by TFA–H_2O (19 : 1) after *N*-acetylation. Thus, polystyrene-bound glucosamine derivative **2** was glycosylated with glucosyl trichloroacetimidate **3** using TMSOTf as an activator. After the acetylation of the C2 amino group by Ac_2O in pyridine, β(1,6)-linked disaccharide **4** was liberated from the solid support in 82% by TFA treatment. The high stability of the linker under the strongly acidic conditions allows the use of excess Lewis acid for glycosylations on the solid phase.

N-1-(4,4-Dimethyl-2,6-dioxocyclohexylidene)ethyl linker has also been introduced between the C2 amino group of the glucosamine and a solid support.[5] This linker was stable during the glycosylation with the thioglycoside in the presence of the methyl trifluoromethanesulfonate, but readily cleaved by hydrazine, primary amines, or even ammonia.

Boons and coworkers have used the polystyrylboronic acid as a temporary polymeric support for oligosaccharide synthesis (Scheme 16.3).[6,7] Polystyrylboronic acid **5**, easily prepared from the corresponding bromide by treatment with trimethyl borate and *n*-BuLi in high loading capacity, was introduced on the 4,6-*O* positions of the acceptor **6** by heating in pyridine. The immobilized boronate acceptor **7** was glycosylated with thioglycoside donor

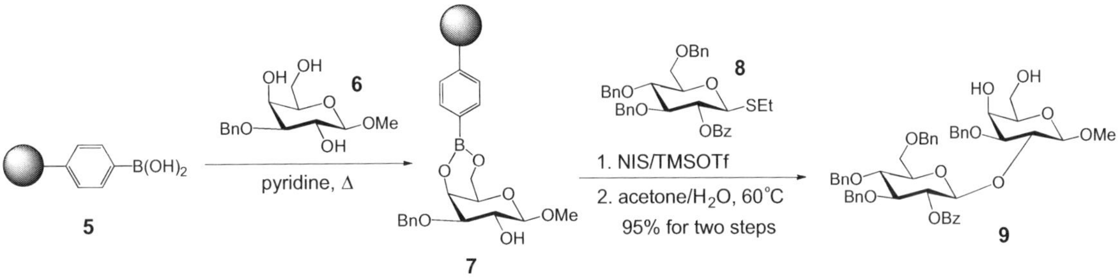

Scheme 16.3. Synthesis using the polystyrylboronic acid linker.

8 by using NIS (*N*-iodosuccinimide)/TMSOTf as the promoter to give the β(1-2)-linked disaccharide **9** in 95%, after *in situ* cleavage from the solid support by heating in acetone/water. The polystyrylboronic acid was recovered and reused for subsequent loading to the acceptors and glycosylation.

The recent exciting progress in the development of efficient and stable metathesis catalysts and their use in the metathesis reactions both in solution phase and on solid phase have prompted several groups to investigate the metathesis-based linker for solid-supported synthesis. The olefin-based linkers have attracted much attentions for solid-phase oligosaccharide synthesis since the isolated olefins are inert under wide range of the acid-catalyzed glycosylation and deprotection conditions, and at the end of the synthesis, these linkers can be cleaved under mild and olefin-selective conditions by both intramolecular and intermolecular metathesis. Seeberger and coworkers have applied the octanediol linker, which is readily prepared from the cycloocta-1,5-diene, to the solid-phase oligosaccharide synthesis, especially to their programmed automated synthesis (see Section 16.2.3). The octanediol linker is cleaved from the resin by cross (intermolecular) metathesis with ethylene gas in the presence of the Grubbs' first-generation catalyst to give the *n*-pentenyl glycosides (Scheme 16.4).[8] The *n*-pentenyl moiety thus generated offers much opportunity as the glycosylating donor or an anchor to connect the synthesized oligosaccharides to proteins, labels, or surface of the solid support in preparing the oligosaccharide microarrays.[9,10]

Van der Marel and coworkers have designed a self-cleavable linker via tandem ring-closing metathesis (Scheme 16.5).[11] Their triene linker on Merrifield resin **10** was glycosylated twice with mannosyl trichloroacetimidate **11** in the presence of TMSOTf in CH_2Cl_2 at $-20°C$ to provide the α-linked disaccharide **12**, which was treated with the Grubbs' second-generation ruthenium catalyst to liberate the cyclopent-2-enyl mannnoside **13** in 67% yields from **10** (average 90% for each step). A cyclopent-2-enyl moiety was then isomerized to the corresponding vinyl ether glycoside by treatment with $(PPh_3)_3RhCl$ and DBU in EtOH, which was hydrolyzed by I_2 in THF–H_2O. The self-cleavable linker **10**, which provides the thermodynamically stable cyclopentene products both on the solid supports and on the products, prevents the catalyst deactivation due to covalent attachment

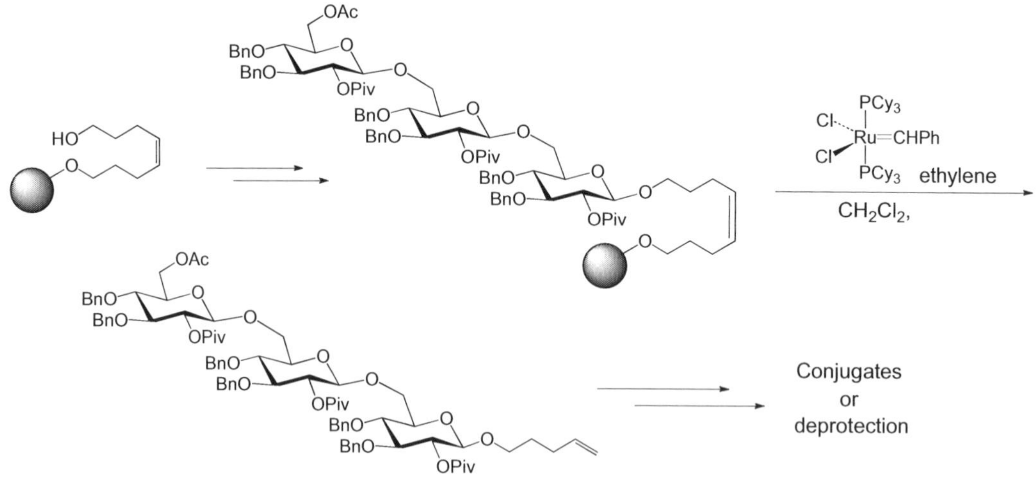

Scheme 16.4. Release from resin by cross-metathesis.

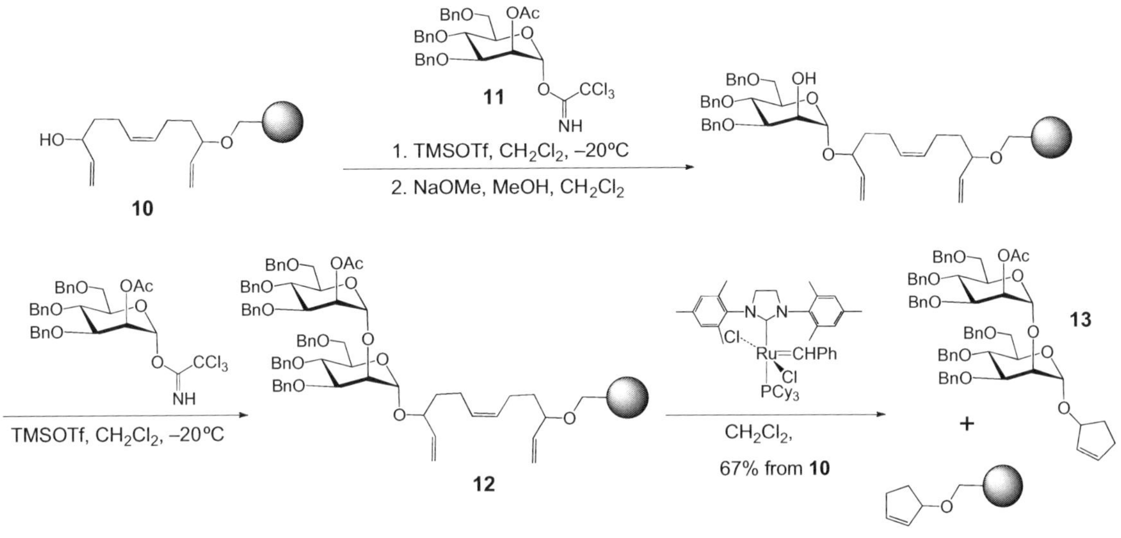

Scheme 16.5. A self-cleavable diene linker via tandem ring-closing metathesis.

of the ruthenium catalyst to solid supports or the products *via* C=C double bonds during metathesis reaction.

Schmidt and Knerr have also independently developed the self-cleavable diene linker **14**, and solid-phase synthesis of the branched hexasaccharide **16** has been achieved (Scheme 16.6).[12] The ring-closing metathesis of **15** provided the allyl glycoside **16**, of which allyl protecting group was readily removed by Ir-mediated isomerization followed by hydrolysis.

Fukase and coworkers have developed two types of alkyne linkers (Scheme 16.7).[13] A propargyl glycoside-type linker between a sugar residue and a solid support was formed by

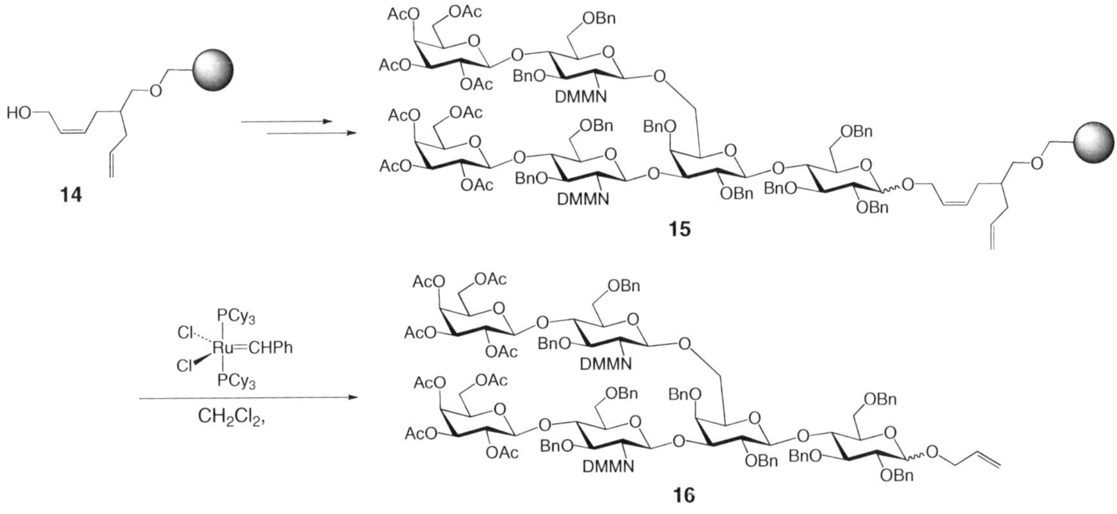

Scheme 16.6. Synthesis of a branched hexasaccharide using a self-cleavable diene linker.

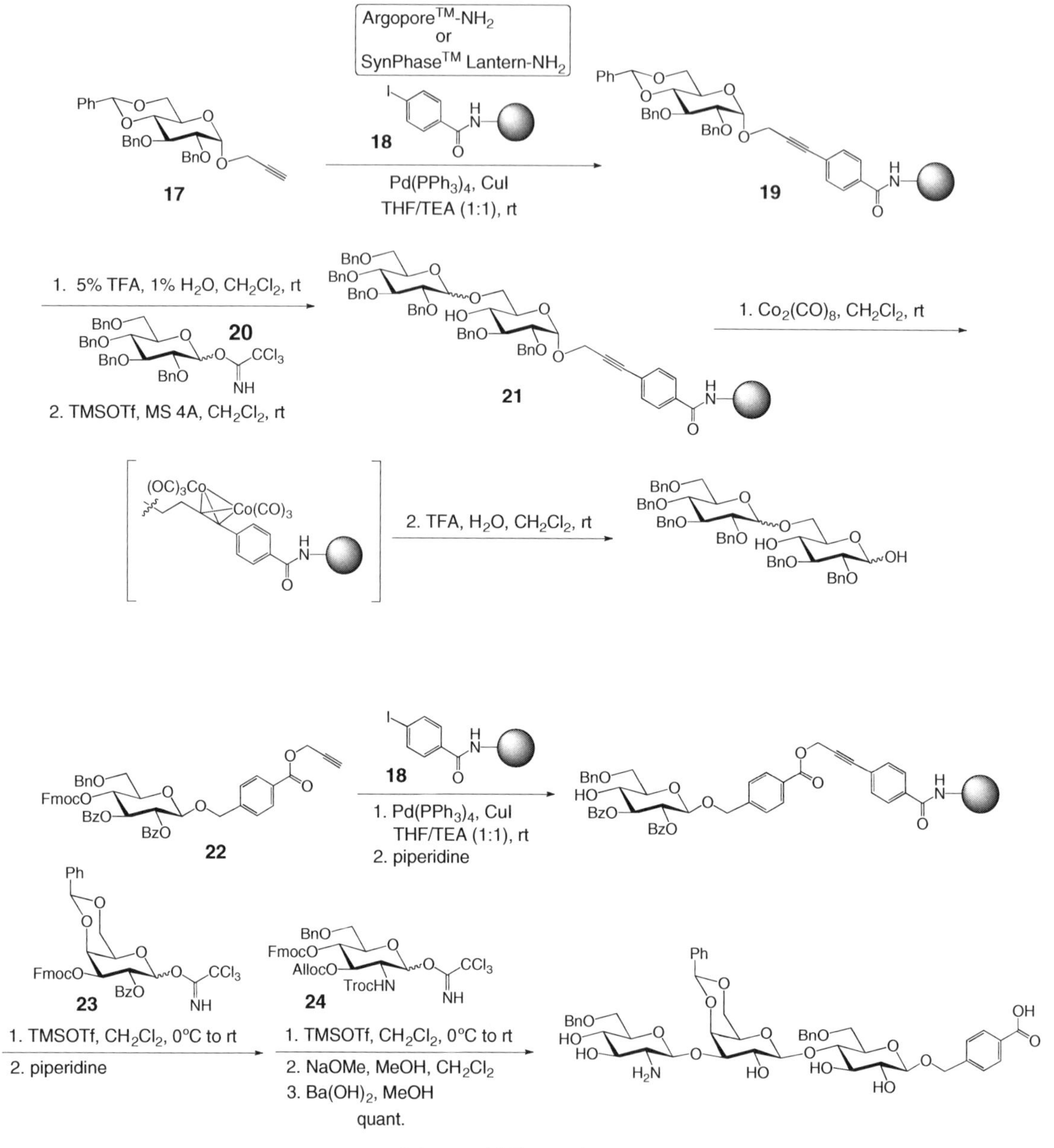

Scheme 16.7. Synthesis using alkyne linkers.

the Sonogashira reaction; the propargyl glycoside **17** was reacted with polymer-supported iodobenzene **18** in the presence of $Pd(PPh_3)_4$ and CuI in THF/Et_3N to provide **19**. After 4,6-*O*-benzylidene group was removed by treatment with TFA, the liberated C6 hydroxyl was glycosylated with glucose trichloroacetimidate **20** using TMSOTf as a promoter to give the disaccharide **21**. The alkyne-based linker is thus stable during the glycosylation or the TFA treatment but can be readily cleaved with TFA after the conversion to the corresponding alkyne–cobalt complex by reacting with $Co_2(CO)_8$. Alternatively, a propargyl ester-type linker was also generated by the same Sonogashira reaction of a 4-(propargyloxycarbonyl)benzyl (Bn)

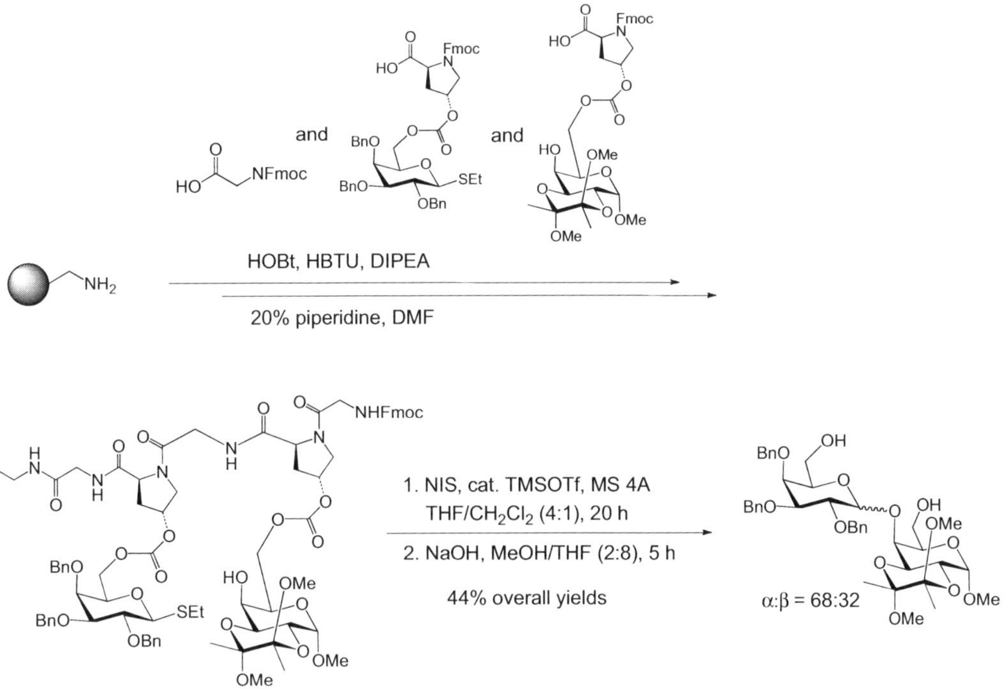

Scheme 16.8. Synthesis using a peptide-based linker.

glycoside **22** with the polymer-supported iodobenzene **18**. After sequences of the glycosylation reactions with imidates **23** and **24**, the trisaccharide was released from the resin by treatment with TFA of the corresponding alkyne–cobalt complex. The latter ester-type linker is generally advantageous since the carboxybenzyl glycoside, liberated from the resin under the mildly acidic conditions, is readily purified and the linker can also be cleaved under the nucleophilic conditions (NaOMe in $MeOH/CH_2Cl_2$). Noteworthy is that the Sonogashira reaction was found to proceed only at spatially reactive sites on the solid support where the reagents access readily, so that the subsequent reactions including glycosylation and Fmoc deprotection on the solid phase proceeded smoothly and quantitatively, resulting in high total yields of the desired oligosaccharides.

The peptide-based linker has also been investigated by Warriner and coworkers (Scheme 16.8).[14] Since the solid-phase synthesis of peptide is well established, both glycosyl acceptors and donors were initially conjugated to the 3-hydroxyproline template and introduced on the aminomethyl Novagel™ resin using standard Fmoc peptide coupling protocol (a sequence of amide formation by using HOBt (1-hydroxybenzotriazole), HBTU (2-(1*H*-benzotriazole-1-yl)-1,1,3,3-tetramethyluronium hexafluorophosphate), and DIPE, followed by Fmoc deprotection by piperidine). They examined the stereoselectivities of glycosylation between acceptors and donors that were loaded to a peptide template, and 12 members of disaccharide library was constructed.

Ellman's sulfonamide linker has also been used as the safety-catch-type linker for the synthesis of the glycodendrimer to enhance the interaction with the glycan receptors and lectins by the glycan multivalency effects (Scheme 16.9).[15] Thus, the hexavalent α-mannose introduced on the solid-supported dendrimer **25** was activated for the nucleophilic

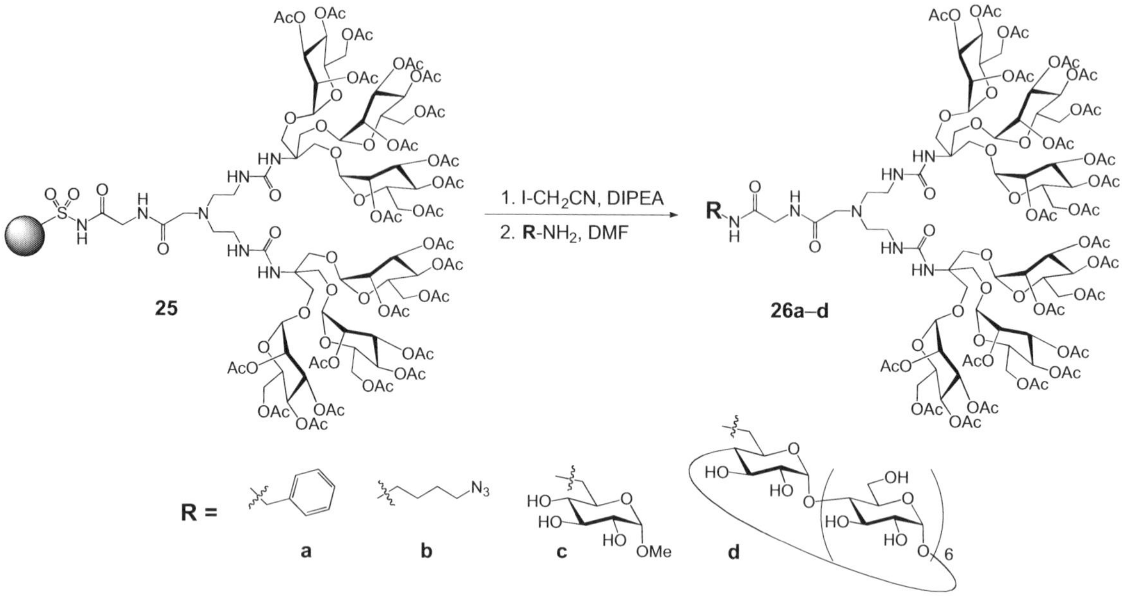

Scheme 16.9. Glycodendrimer synthesis using a safety-catch-type linker.

attack by the reaction with the iodoacetonitrile in DMF, giving rise to a variety of the substituted dendriomers **26a–d** in more than 70% yields. Even the βCD conjugate of mannose dendrimer **26d** could be synthesized in 68% by combining the mannose-specific lectin-binding properties and inclusion capabilities. The similar safety-catch approach using the sulfonamide linker has also been applied to the solid-supported synthesis of the hyaluronic acid oligosaccharides.[16]

On library-directed approach, Takahashi and coworkers have developed a synthetic strategy for the combinatorial solid-phase synthesis of the tree-type carbohydrate clusters using three types of orthogonally cleavable linkers, namely, photocleavable 2-nitrobenzyl-type linker (orthogonal cleavage site 1), base-cleavable ester linker (orthogonal cleavage site 2), and cobalt complexation/acid-cleavable alkyne linker (orthogonal cleavage site 3) (Figure 16.1).[17] The combined use of three types of spacers, two types of carbohydrates, and two types of orthogonally cleavable linkers provided 12 members of carbohydrate cluster library (six dimers and six tetramers).

In addition to the importance of linkers described above, the precise choice of protecting groups is also an important factor for the oligosaccharide assembly to be successful on the solid supports. The utilization of the protecting groups that are stable under the glycosylation conditions but are readily removed under the mild conditions and compatible to other protecting groups and linkers are highly desirable for combinatorial and high-throughput oligosaccharide synthesis. The excellent orchestration of the protecting groups manipulation, leading to the construction of the complex oligosaccharide structures, will be discussed in Section 16.2.3. It is also preferable if the protecting group serves as a tag to both identify the compounds and indicate their protection states.

Kasumi and coworkers have reported unique protecting group manipulations by using (N-α-1-ethylpropylglycine)$_n$ ((EPG)$_n$) for both solution- and solid-phase synthesis of oligosaccharides (Scheme 16.10).[18] The hydroxyl groups of oligosaccharides were

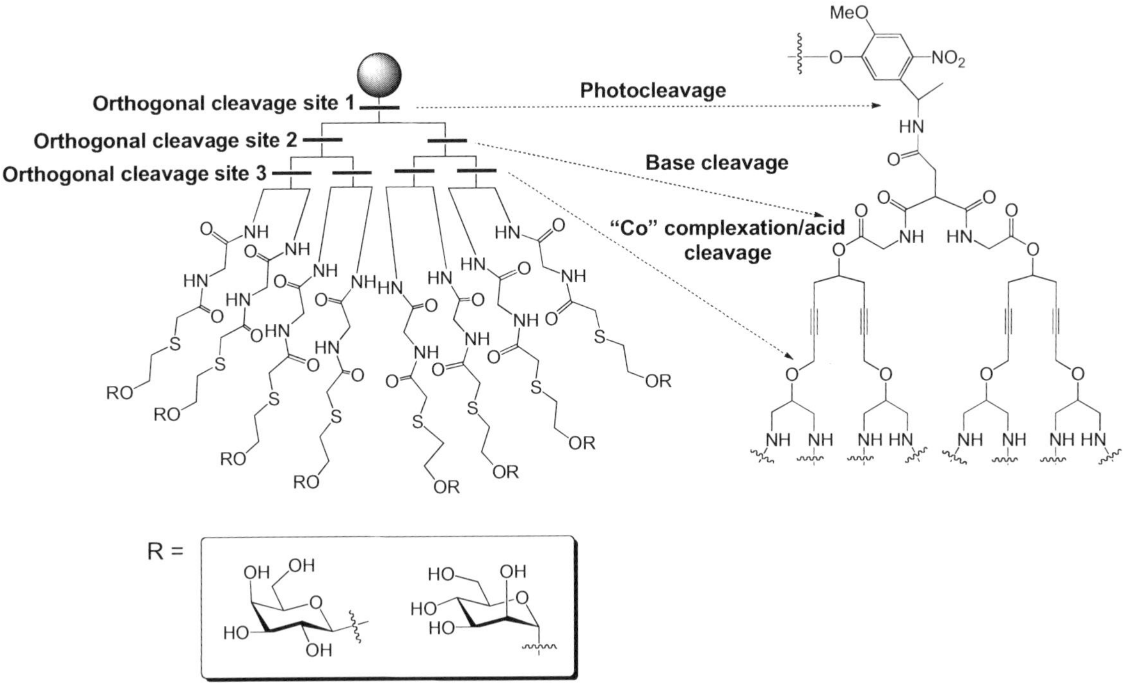

Figure 16.1. Synthesis of tree-type carbohydrate clusters using orthogonally cleavable linkers.

protected by $(EPG)_n$ group with a different degree of polymerization that can be uniquely identified. The $(EPG)_n$ groups were then deprotected successively from the highest to the lowest degree of polymerization by repeating the Edman degradation cycle. Thus, the galactose derivative **27** with $(EPG)_3$ at the C6 hydroxyl and $(EPG)_1$ at the C4 hydroxyl was

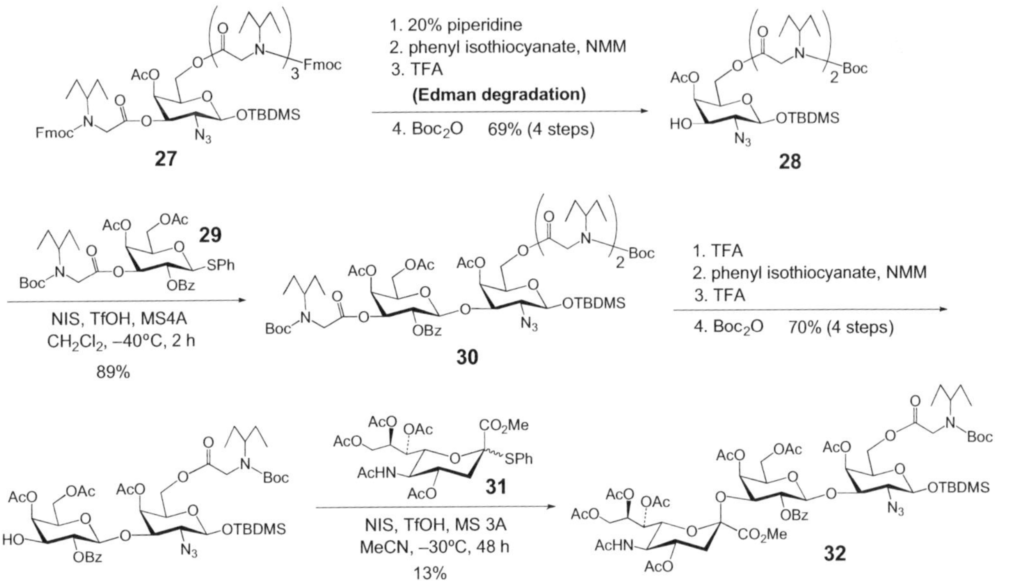

Scheme 16.10. Oligosaccharide synthesis using an iterative Edman degradation strategy.

subjected to the Edman degradation, that is, by (i) Fmoc deprotection by 20% piperidine, (ii) phenyl isothiocyanate treatment, (iii) degradation by TFA, and (iv) Boc protection of newly appearing amino group to provide the acceptor **28** having $(EPG)_2$ at the C6 hydroxyl in 69% yield for four steps. This acceptor was glycosylated with thiogalactose **29** bearing one EPG at the C3-position by using NIS and TfOH as the activators to give **30** in 89% yield. Repetition of Edman degradation cycle liberated the C3′ hydroxyl in 70% yields, which was further sialylated with thioglycoside **31** to give **32** in 13% yield. They have applied this idea of the protecting group manipulation to the solid-phase oligosaccharide synthesis and even automated synthesis of a tribranched pentasaccharide, a PSGL-1 derivative[19]; the four different monosaccharide donors of $GalN_3$, Gal, GlcNHTCA, Fuc were successfully glycosylated on the 2-deamino-2-azide-glucosamine acceptor on the solid supports using an aappAAPPTec automated solid-phase synthesizer (Vantage).

16.2.2 Application of Unique Glycosylation Methods in Solution to Solid-Phase Synthesis of Oligosaccharides

A variety of recently developed glycoslylation methods in solution have been applied to the solid-phase synthesis of oligosaccharides. A few unique examples are described in this section. Demchenko and coworkers have applied their stable donors possessing the thioimidoyl leaving groups, that is, *S*-benzoxazolyl (SBox) or *S*-thiazolinyl (STaz), to the solid-phase oligosaccharide synthesis (Scheme 16.11).[20] Taking advantage that these leaving groups are smoothly activated by TMSOTf, MeOTf, or AgOTf under the mild conditions and that these are selectively activated in response to the protection group patterns based on the armed and the disarmed concept, they succeeded in preparing a number of the disaccharides on solid supports; these donors are suitable for both glycosyl acceptor-bound and glycosyl donor-bound strategies. Scheme 16.11 shows the synthesis of the trisaccharide. In this example, ethylthio group was also applied as an orthogonal leaving group that can be selectively activated by NIS/TfOH but not simply by TMSOTf for the first glycosylation with SBox donor on the solid supports.

Random glycosylation strategy is a promising method for obtaining a certain oligosaccharide library in a single procedure. According to the pioneering Hindsgaul protocol,[21,22] for instance, the lactosyl or the glucosamyl acceptors with all free hydroxyls were

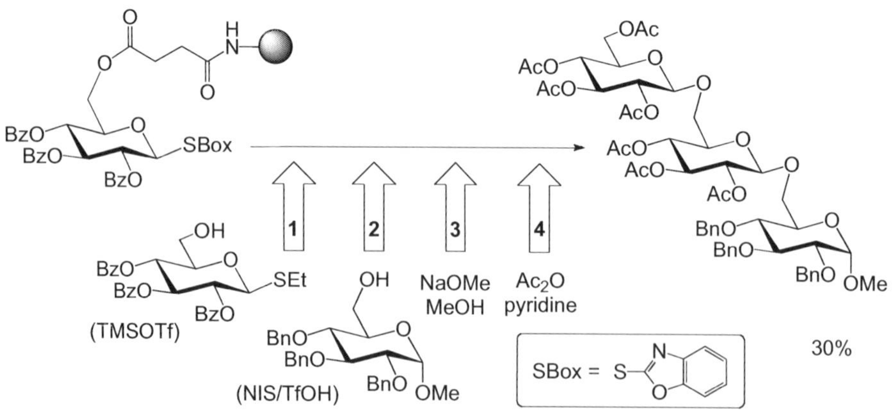

Scheme 16.11. Synthesis using thioimidoyl leaving groups.

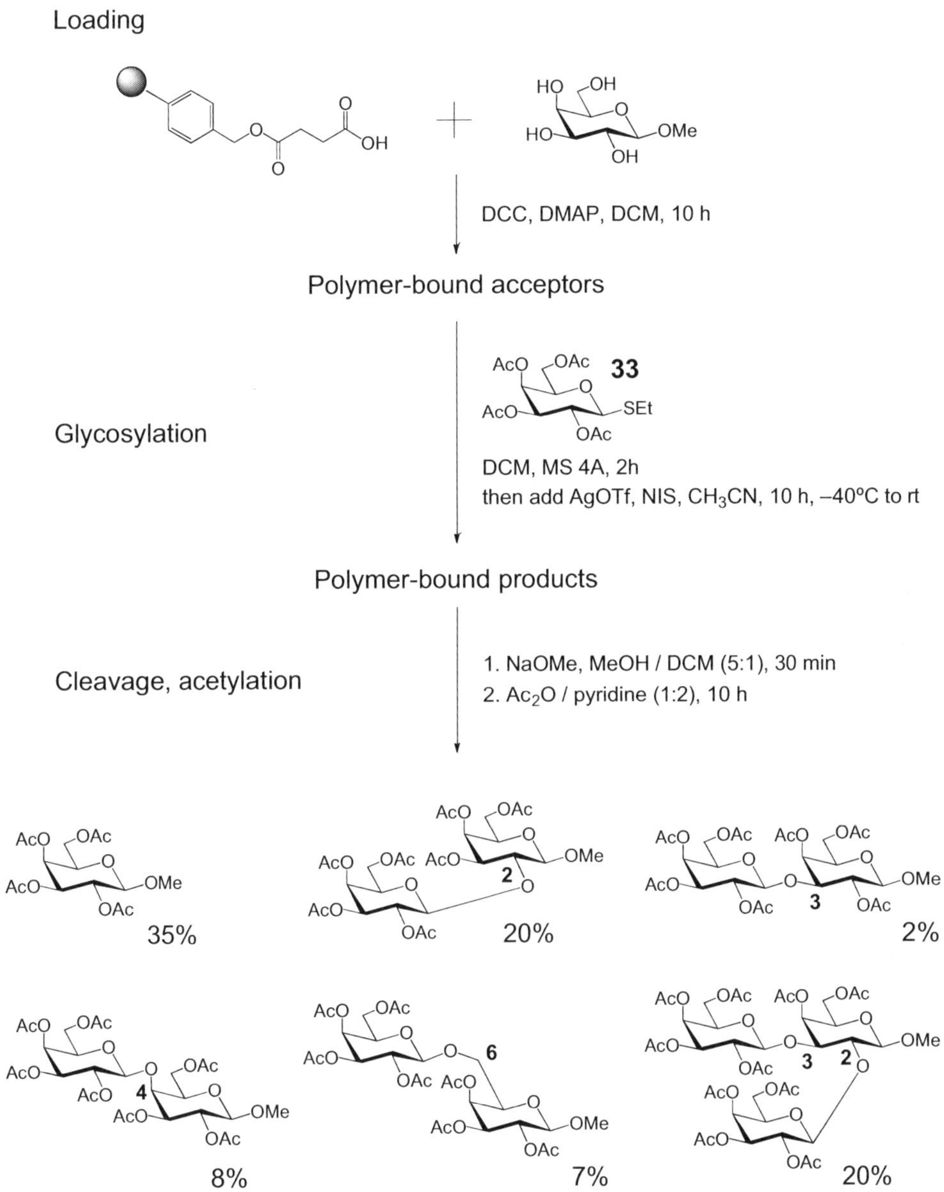

Scheme 16.12. Oligosaccharide synthesis by a random glycosylation strategy.

glycosylated "randomly" with the appropriate donors, such as fucosyl trichloroacetimide, to give glycoside isomers with all possible anomeric configurations and linkage positions. This random strategy was also applied to the solid-supported synthesis by Thiem and coworkers (Scheme 16.12).[23] They have loaded the galactose acceptor with four free hydroxyls on the Wang resin through a succinic acid linker by acylation of one of the hydroxyls, and the polymer loaded acceptor was then randomly glycosylated with the thiogalactosyl donor **33** in the presence of NIS/AgOTf as an activator in CH_2Cl_2/MeCN at room temperature. After cleavage from the resin by treatment with NaOMe and peracetylation with Ac_2O in pyridine, a variety of randomly glycosylated disaccharides were obtained. They have also examined the random glycosylation by loading the thiogalactosyl donor on the solid supports, but the former procedure gave the superior results in terms of the overall yields and versatility of the products. Comparison of this "random" protocol with the conventional protection-based approach in solution or on solid supports, reported by the same authors in the same year, is interesting to be noted.[24]

Although the 1,2-*trans*-glycosylation, where the neighboring participation effects could be used, has now been widely applied to the solid-supported glycosylation and even to the automated synthesis of the complex oligosaccharides (see Section 16.2.3), the formation the 1,2-*cis*-glycosides is still quite challenging; when one might think to apply the solid-supported synthesis, the "perfect" *cis*-selectivity is necessary, since even the formation of a trace amount of the *trans*-stereoisomer cannot be removed on solid supports and the intractable mixtures of products are eventually obtained after the cleavage from the solid supports when several such *cis*-glycosides need to be installed. Boons and coworkers elegantly achieved the synthesis of the biologically well-defined branched 1,2-*cis*-linked pentaglucoside, α-glucans, containing the α(1-6)-linked glucosyl backbone branched with α(1-3)-linked glucoside moieties, based on the chiral auxiliary-mediated 1,2-*cis*-glycosylations developed in solution earlier in their laboratory (Scheme 16.13).[25] The method uses the (*S*)-(phenylthiomethyl)benzyl group as the chiral auxiliary at the C2 hydroxyl of the glucosyl imidate donors (Figure 16.2); upon activation by TMSOTf, the neighboring group participation by the C2 auxiliary results in the formation of the thermodynamically stable "six-membered" anomeric sulfonium ion, which is displaced by the acceptor hydroxyl stereoselectively, yielding the *cis*-1,2-glycosides.

For the synthesis of the α-glucans **44** on the solid supports, they prepared the imidates **36**, **38**, **40**, and **41** as the building blocks (Scheme 16.13). Thus, the carboxypolystyrene resin (Advanced ChemTech, 2.0 mmol/g) was initially functionalized by the *p*-xylylene glycol ester linker at the loading amount of ~0.15 mmol/g, which they found to be optimal for the subsequent glycosylation. After the first glycosylation with the imidate **34**, typically using 2.0 equiv throughout the synthesis, in CH_2Cl_2 at −40°C in the presence of TMSOTf as an activator, the Fmoc at C6 hydroxyl was removed by treatment with piperidine in DMF (1/9 v/v). For the critical *cis*-1,2-glycosylation, the imidate **36** with auxiliary was preactivated in a separate flask with a stoichiometric amount of TMSOTf at −40°C to produce the intermediary sulfonium ion. The solution containing the sulfonium intermediate was then added to a cooled (at −40°C) suspension of the solid-supported acceptor **35** and 2,6-di-*tert*-4-methylpyridine (DTBMP) in CH_2Cl_2. The reaction was gradually warmed to rt over 5 h and then shaken for an additional 11 h. Repeating the chiral auxiliary-assisted 1,2-*cis*-glycosylation procedure on the solid supports, in combination with the selective (orthogonal) deprotection of Fmoc and branching C3-Alloc (by $Pd(PPh_3)$ in THF/AcOH), provided the pentasaccharide **42** on solid supports. They checked each glycosylation step by cleaving the glycosides from the small portion of the resin (~5.0 mg), and glycosylation was repeated for the complete consumption of the starting acceptor on the solid supports, for example, double glycosylation. The glycan structures and "perfect" stereoselectivity in each glycosylation steps were also analyzed by comparing the authentic samples separately prepared in solution. For glycosylation at C3 hydroxyl of the acceptor **37**, namely, α(1-3)-linked glucoside formation (process from **37** to **39**), the (*S*)-(phenylthiomethyl)benzyl auxiliary group at C2 hydroxyl of **37** inhibited the glycosylation, and hence this was in advance converted into the acetyl group by the reaction with acetic anhydride in the presence of $BF_3{\cdot}OEt_2$. Finally, deprotection of auxiliary and Fmoc of the solid-supported **42**, cleavage

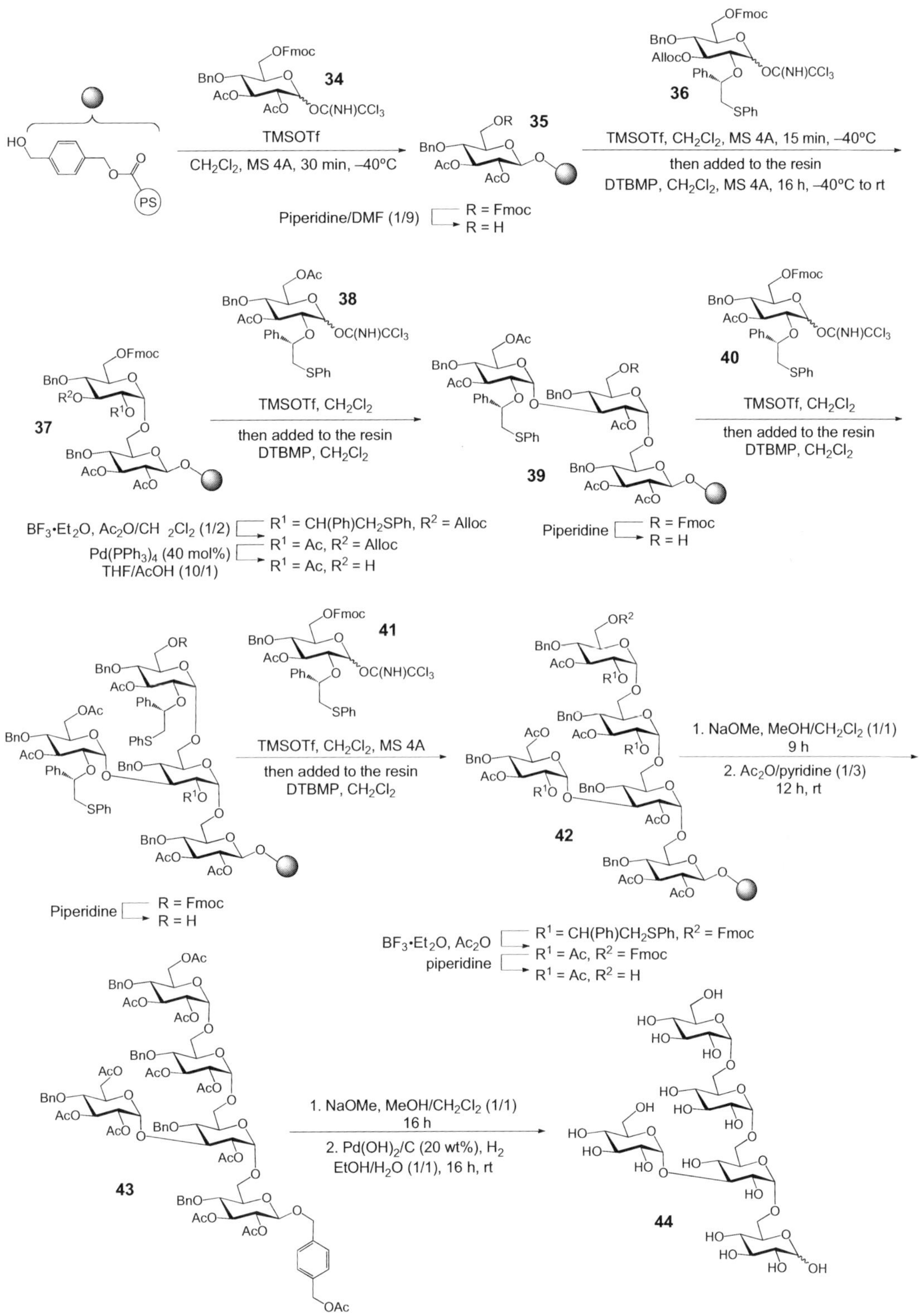

Scheme 16.13. Solid-supported synthesis of an α-glucan.

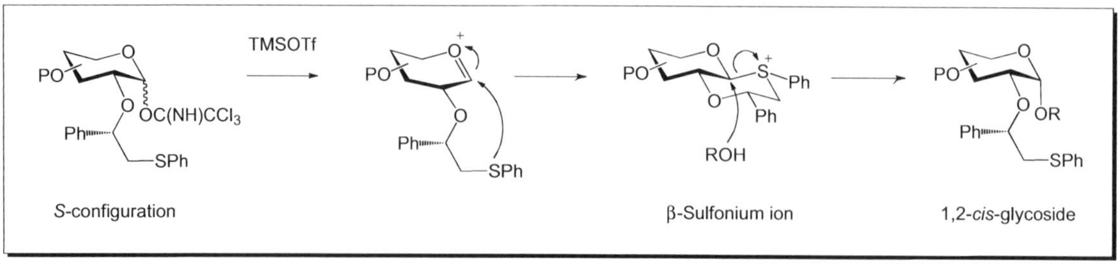

Figure 16.2. 1,2-*cis*-Glycosylation using a (*S*)-(phenylthiomethyl)benzyl auxiliary.

from the resin by treatment with sodium methoxide, and peracetylation yielded **43**, which was purified by the size-exclusion chromatography, LH-20, and HPLC; **43** was obtained in 25% overall yield, per step of 90%. The final deprotection of **43** by methanolysis and hydrogenolysis by $Pd(OH)_2/C$ (20 wt%) in the hydrogen atmosphere successfully provided the pentasaccharide **44**. The loading and glycosylation efficiency and rapid and convenient reaction monitoring still need to be improved; the successful synthesis of α-glucans has opened up new possibilities for the synthesis of the hitherto difficult complex glycan structures by automated systems. The other trials on the formation of *cis*-1,2-glycosyl bond linkages will be discussed in Section 16.2.3.

Acceleration effects of sonication on the solid-supported glycosylation has also been reported by Chang and coworkers (Scheme 16.14).[26] Trimannose derivative **46** could thus be obtained in 50% overall yields from the solid-supported monosaccharide **45** through the sequence of the Ac deprotection, NIS/TMSOTf-mediated glycosylation under the sonication, and nucleophilic displacement of the benzenesulfonyl "traceless" linker by the sodium azide.

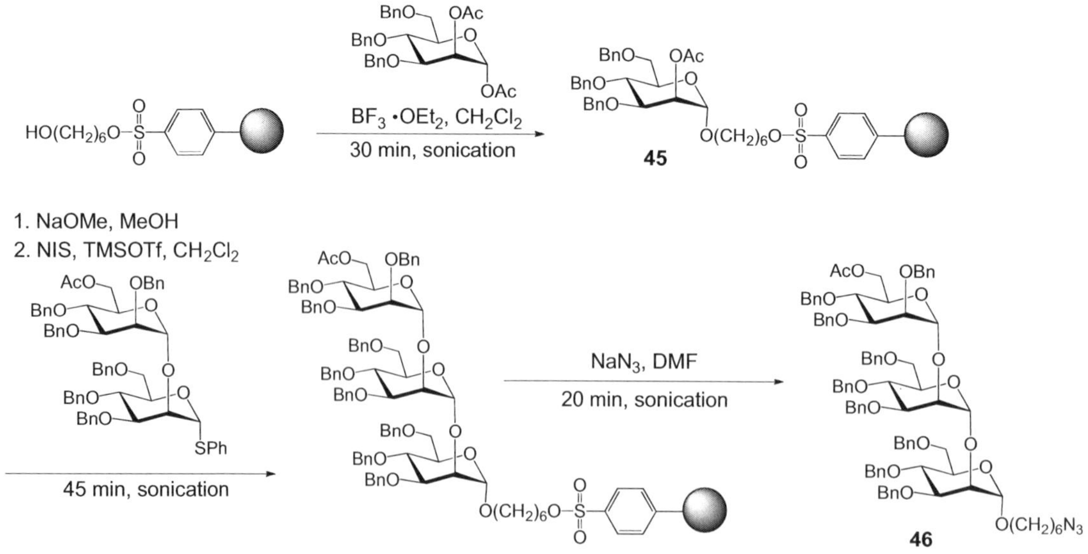

Scheme 16.14. Sonication-accelerating effects on solid-supported glycosylation.

16.2.3 Solid-Phase Synthesis of Complex Oligosaccharides

New methods on the solid supports, together with the accumulation of information on the acceptor/donor reactivity in the solution phase, led to the synthesis of complex oligosaccharides on the solid supports.[27–31] Schmidt and coworkers have achieved solid-phase synthesis of a small library of *N*-glycan (Scheme 16.15).[32–34] They used the hydroxymethylbenzyl benzoate linker attached to the Merrifield resin. Glycosylation on the resin was successfully performed using the glycosyl trichloroacetimidates,[35] such as glucosamine **47**, mannosylβ(1-4)glucosamine **48**, mannose **49**, and galactose **50**, which allow chain extension, branching, and chain termination for the synthesis of heptasaccharide *N*-glycan **51**; the tedious β-mannosyl linkage in **48** was constructed in advance in solution phase. *O*-Benzyl and *O*-benzoyl (Bz) groups were used as permanent protecting groups, while *O*-Fmoc and *O*-phenoxyacetyl groups were used as temporary protecting groups for the sugar extension on the solid phase. The *N*-dimethylmaleoyl (DMM) group was selected for the protection of glucosamine **47**, since the neighboring effects ensure the β-selective glycosylation, and also this protecting group confers the high reactivity on the glucosamine **47** as an acceptor on the polystyrene.[36] Their solid-phase synthesis involves (i) glycosylation under the TMSOTf catalyst, (ii) selective removal of the temporary protecting groups, Fmoc by Et_3N and phenoxyacetyl group by 0.5 equiv of NaOMe in $MeOH/CH_2Cl_2$, and (iii) products cleavage from the resin by 4.0 equiv of NaOMe in $MeOH/CH_2Cl_2$ and then *O*-acetylation for convenient product isolation. The precise design of linker, protecting group manipulations, and stereoselective glycosylation on the resin enabled the successful preparation of a small library of 17 *N*-glycan structures.

Later on, they have also investigated the orthogonality of the ester protecting groups at the vicinal hydroxyls on solid supports, that is, Fmoc, phenoxyacetyl (PA), allyloxycarbonyl (Alloc), and levulinoyl groups (Lev), at the C3 and C4 diols of the glucosamine as the model case.[37] They found that Fmoc and Lev or Alloc in the Fmoc/Lev and Fmoc/Alloc pairs could be selectively removed under their respective conditions, namely, Et_3N for Fmoc, hydrazine acetate for Lev, and Pd(0) for Alloc. These basic experimental data would further expand the applicability of Schmidt's protocol to a variety of linear and branched *N*-glycan structures.

Using the ideas similar to Schmidt's procedure, Fukase and coworkers recently achieved the first solid-supported synthesis of a complex-type *N*-glycan **52** with a nonreducing end sialic acid (Scheme 16.16).[38] Two challenging glycosyl bond formations (β-mannosylation and α-sialylation) were constructed in advance in solution-phase or under microfluidic conditions; the synthesis of the fragment **c**, Manβ(1-4)GlcNTroc, was achieved by their highly β-selective mannosylation method, which utilized the trimethylsilyl tetrakis (pentafluorophenyl)borate, $TMSB(C_6F_5)_4$, as a Lewis acid/cation trap activator,[39] and by using microfluidic conditions.[40] On the other hand, the quantitative and α-selective sialylation ($\alpha : \beta > 20 : 1$) was achieved for the large-scale preparation of the NeuNAcα (2-6)Gal disaccharide (fragment **d**) by using C5-substitued phthalimide or azide imidates through the mechanism of fixed-dipole moment effects under microfluidic conditions (see Section 16.3.4).[41,42] Then, the suitably protected mono- and disaccharide donors (fragments **a**–**d**), as the longer storable *N*-phenyltrifluoroacetimidates, could be stereoselectively glycosylated on JandaJel™, which was found to be the optimal solid support for glycosylation, with the aid of neighboring group participation. An *O*-Fmoc group was used as a temporary protecting group for the sugar extension on the solid phase, while an azidochlorobenzyl (AzClBn) group was applied at the C3 hydroxy group of the branching mannose. These temporary protecting groups allowed *N*-glycans to be prepared with diverse structures at the C3′ and C6′ hydroxy groups. A trichloroethoxycarbonyl (Troc) group was

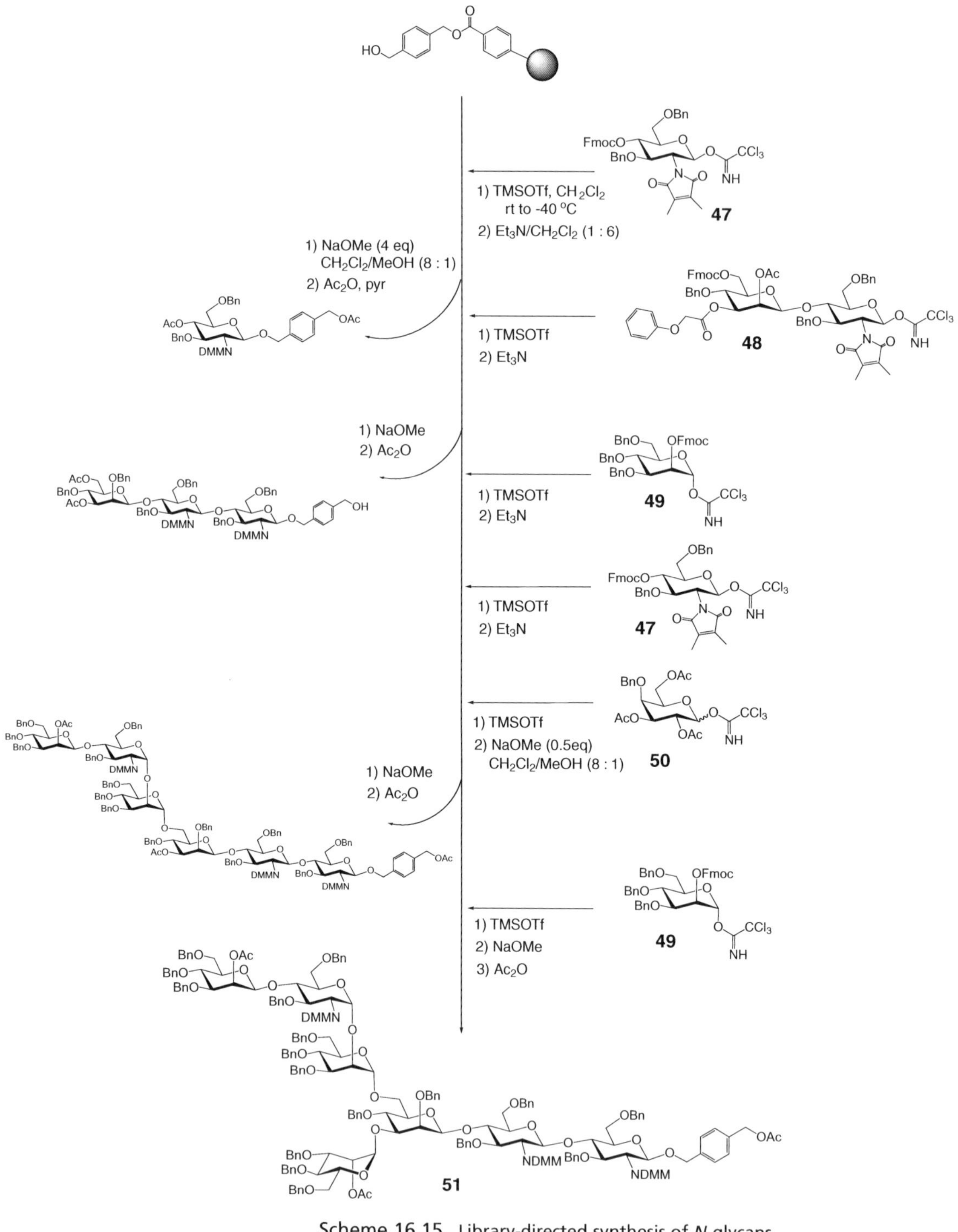

Scheme 16.15. Library-directed synthesis of *N*-glycans.

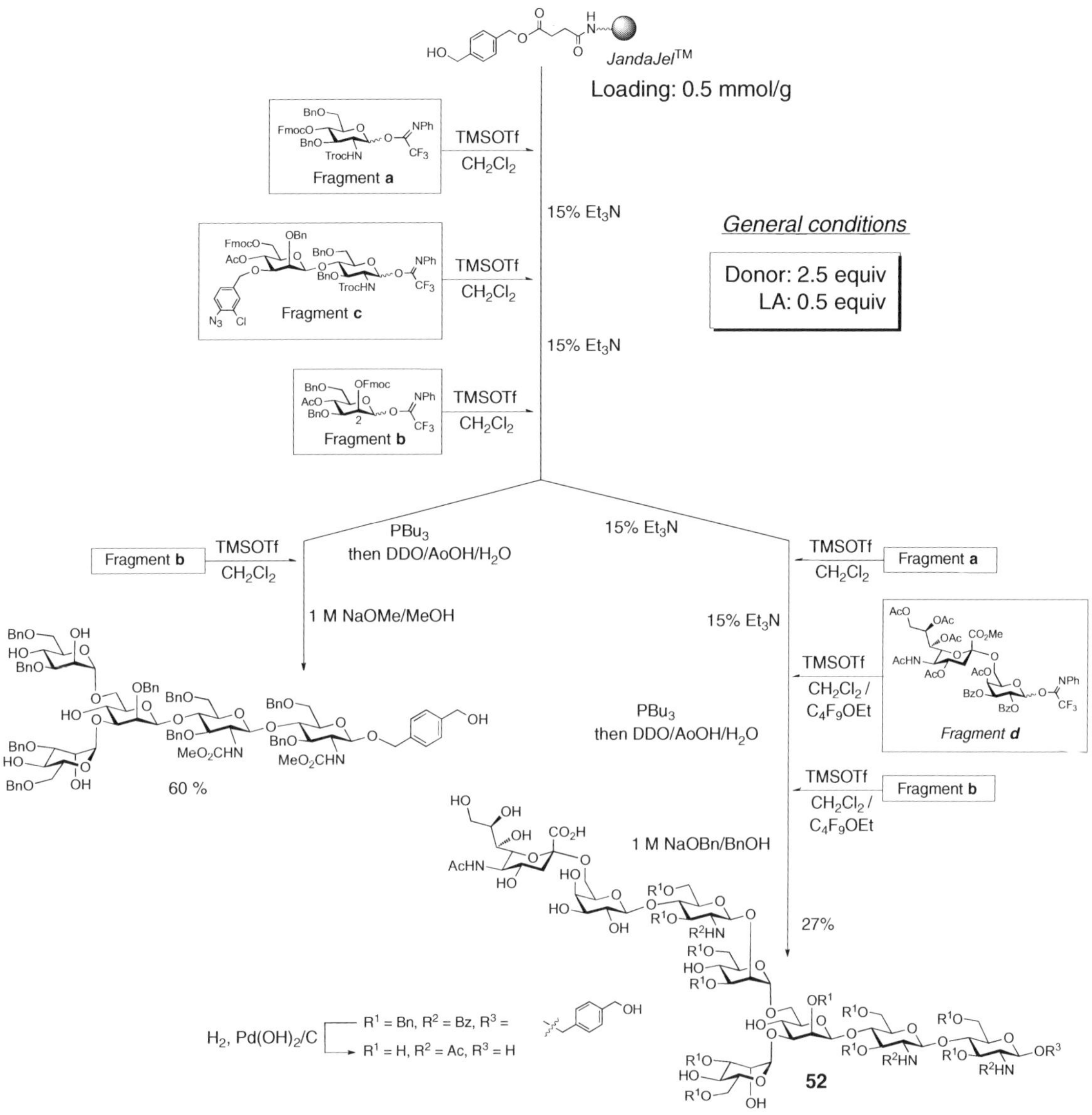

Scheme 16.16. Solid-supported synthesis of sialic acid containing *N*-glycan.

applied to protect the nitrogen atom on glucosamine because this protecting group not only ensured glycosylation selectivity by the neighboring participation, but also conferred high reactivity as the glycosyl acceptor. After an efficient sugar extension on solid supports, by sometimes using fluorous solvents to enhance the glycosylation reactivity on the extended oligosaccharide structure on the solid supports (reagent concentration effects on the solid supports), alcoholysis of the *p*-xylylene glycol ester linkage by NaOBn provided the octasaccharide protected with hydrogenolysis-sensitive groups in 27% overall yield; hydrogenolysis and *O*-acetylation successfully provided the desired *N*-glycan with non-reducing sialic acid. Because a variety of natural and unnatural *N*-glycans could be easily

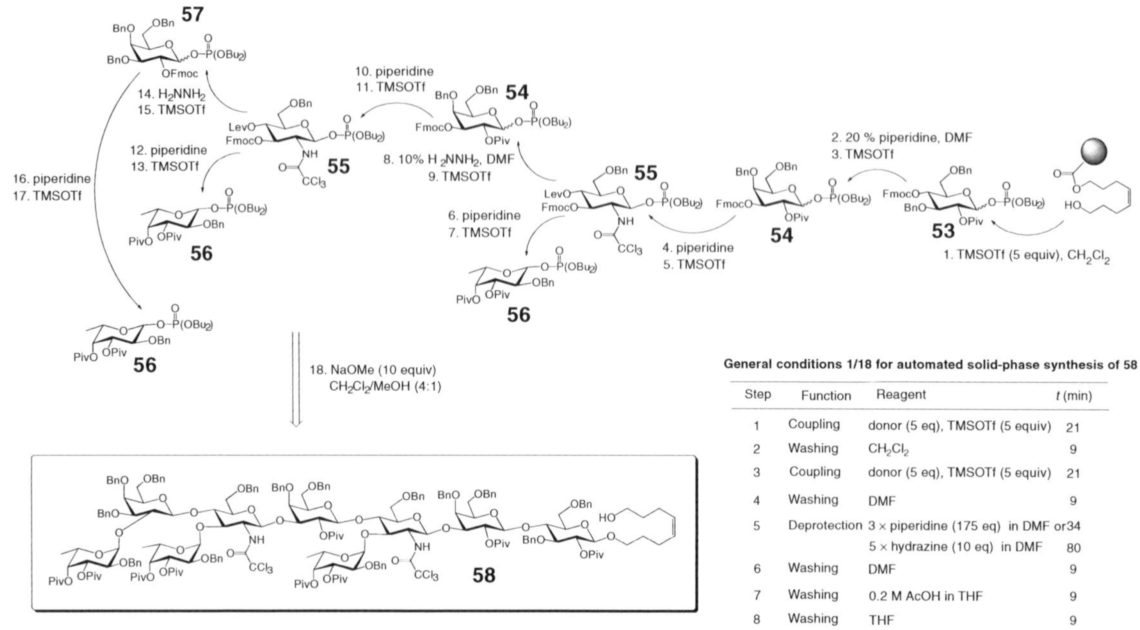

General conditions 1/18 for automated solid-phase synthesis of 58

Step	Function	Reagent	t (min)
1	Coupling	donor (5 eq), TMSOTf (5 equiv)	21
2	Washing	CH_2Cl_2	9
3	Coupling	donor (5 eq), TMSOTf (5 equiv)	21
4	Washing	DMF	9
5	Deprotection	3 × piperidine (175 eq) in DMF or	34
		5 × hydrazine (10 eq) in DMF	80
6	Washing	DMF	9
7	Washing	0.2 M AcOH in THF	9
8	Washing	THF	9
9	Washing	CH_2Cl_2	9

Scheme 16.17. Automated synthesis of Lewis blood group oligosaccharides.

prepared by glycosylating the imidate donors or their slightly structural variants (such as fucosylated congener of the glucosamine donor), this protocol might be applicable to a general *N*-glycan synthesis, even in the library-directed automated synthesis of these diverse structures of *N*-sialoglycans.

An assembly of the oligosaccharides on the solid supports gave a quite exciting and promising opportunity for automation. Seeberger and coworkers have designed the first automated oligosaccharide synthesizer on solid phase (Scheme 16.17).[43–53] A peptide synthesizer was reengineered to function as the automated oligosaccharide synthesizer and provide an easy access to structures as large as dodecasaccharides about 20 times faster than the previously reported methods. Even the oligosaccharides with branched structures are now accessible and a series of oligosaccharides of biological relevance, such as glycosylphosphatidylinositol (GPI),[54] *N*-glycan core pentasaccharide,[55] Lewis blood group oligosaccharides,[56] or tumor-associated antigens Gb3 and Globo-H,[57] have been prepared. Their solid-phase synthesizer uses the glycosyl phosphate building blocks and the octanediol linker attached to the resin that allow a simple automated coupling/deprotection cycle for the oligosaccharide assembly. The released oligosaccharides from solid supports were applied for biological purposes, that is, bioconjugation or microarray.[9,10] An example of the automated synthesis of Lewis blood group oligosaccharide **58** is shown in Scheme 16.17.[56] Five monomer units **53–57**, which allow the stereoselective construction of five different glycosidic bonds in **58**, were designed for the automated synthesis of **58**. Especially, fucosyl phosphate **56** was selected based on the solution study; this donor showed the excellent selectivity in constructing the difficult α(1-2)-*cis* and α(1-3)-*cis* fucose linkages common to Lewis antigens. The octanediol linker was attached to the resin by ester linkage that is cleaved by the nucleophilic bases such as NaOMe. Fmoc was used as the temporary protecting group for hydroxyls not only because it is quite stable under the glycosylation conditions and easily cleaved by the nonnucleophilic amines, but also it facilitates the qualitative evaluation on the efficiency of each glycosylation and deprotection cycle by monitoring the UV/vis of the dibenzofurubene derivative after deprotection; this monitoring method is valuable for the automated synthesis. For the branching linkage of **55**, levulinoyl ester, being cleaved by hydrazine treatment, was chosen as another temporary protection group for elongation of sugar chains. General automated cycles involve (i) glycosylation by activating the phosphate donors with TMSOTf in CH_2Cl_2 at $-15°C$ and (ii) selective removal of temporary protecting groups, Fmoc by 20% piperidine in DMF, and levulinoyl group by 10% hydrazine in DMF, as shown in Scheme 16.17. If the glycosylation was not completed by the dibenzofurubene UV analysis, another cycle of the glycosylation was performed, that is, double glycosylation. More important, a series of washing steps was performed after each deprotection to swell the resin and purge any remaining basic materials for the next glycosylation cycles. At the end of the synthesis, the product was cleaved from the resin by treatment with 10 equiv of NaOMe in MeOH/CH_2Cl_2, and by this automated method, **58** was produced in 6.5% overall yields in as short as 23 h.

Notably, they have recently investigated the construction of the hitherto difficult *cis*-1,2-glycosidic bond linkages, such as β-mannosidic[58] and *cis*-galactosidic linkages to the axial C4 hydroxyl of galactose[57] on the automated systems, and realized the moderate to high yield and selectivity. Especially for the β-mannosylation, they used the Kim's carboxybenzyl glycoside donors to readily be activated by Tf_2O and DTBMP to access anomeric α-triflates.[59] Boons and coworkers have also succeeded in synthesizing the α(1-6) and α(1-3) branched α-glycan derivatives by completely controlling the few *cis*-1,2-glycosidic bonds formation, by applying their chiral auxiliary-mediated reaction on solid supports, as described in Section 16.2.2 (Scheme 16.13). Further improvements in the automated synthesis might allow everyone, even nonchemists, to prepare complex and biologically relevant oligosaccharides, similar to peptides and nucleotide synthesizers, which will speed up the elucidation of their biological functions as well as clinical applications of the oligosaccharide-based vaccines.

16.2.4 Solid-Phase Methods for Purification of Synthesized Oligosaccharides

The isolation of products is sometimes a tedious and time-consuming procedure. Especially in the area of combinatorial synthesis of oligosaccharides in solution or on solid phase, a simple and effective isolation method is necessary because many compounds are handled at the same time. For instance, chromatography is a quite popular method for isolation of products of organic synthesis. However, it is somewhat tedious and consumes a considerable amount of solvents, and optimization of the separation conditions sometimes requires much time. Therefore, techniques for polymer-supported purification, such as "catch-and-release" purification methods, polymer-bound reagents, and scavenger resins, have been developed. Fukase and coworkers have developed the 4-azido-3-chlorobenzyl (ClAzb) group not only for safety-catch-type protection group of hydroxyls but also for a tag for catch-and-release purification (Scheme 16.18).[60,61] A desired compound possessing the ClAzb group can be selectively caught by a solid-supported phosphine by the specific reaction between the azido function and the phosphine, and the product on the polymer support was separated from other compounds simply by rinsing. The desired compound was then released by treatment with DDQ (Scheme 16.18). A glucose acceptor bound to a macroporous polystyrene (ArgoPore^{TM}-NH_2) **59** was glycosylated with thioglycoside **60** using NBS and $Sn(OTf)_2$ in the presence of MS 4A to give a mixture of disaccharide **61** and unreacted **59**. The resin was treated with NaOMe in MeOH to cleave the linker, and a

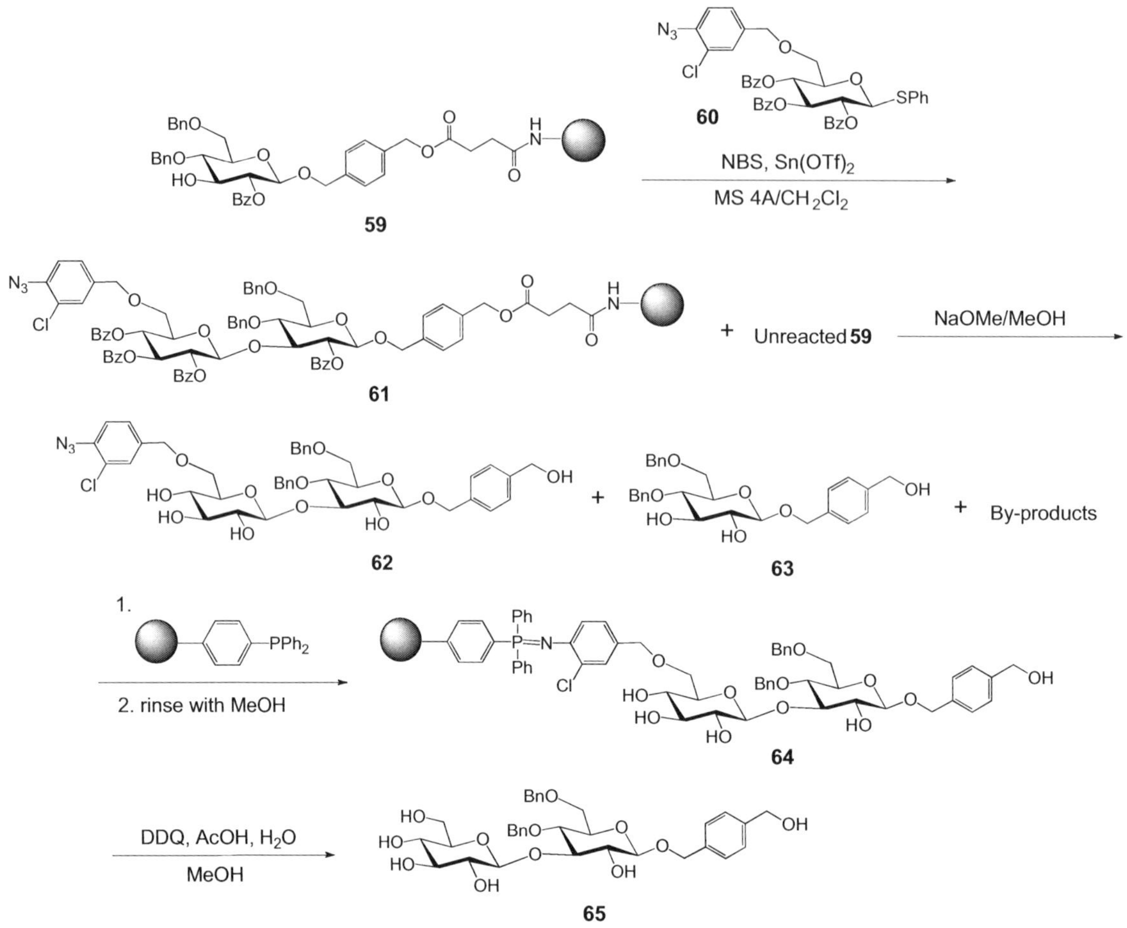

Scheme 16.18. 4-Azido-3-chlorobenzyl for catch-and-release purification.

mixture of disaccharide **62**, monosaccharide **63**, and other by-products was obtained. The mixture was treated with triphenylphosphine–polyethylene glycol–polystyrene copolymer resin in MeOH and the resulting solid-supported disaccharide **64** was treated with DDQ. By this procedure, a highly pure disaccharide **65** was obtained in 38% yield from **59**.

Similarly, the Levulinyl group (Lev), which is widely used for the hydroxyl protecting group for the oligosaccharide synthesis, was used for the catch-and-release purpose by Izumi et al.[62]; the ketone carbonyl of the Lev group in the crude trisaccharide **68**, which was produced by double glycocylation of glucose acceptor and donor, was smoothly reacted with aminooxy resin (hydroxylamine Wang resin), and thus captured trisaccharide was rinsed by CH_2Cl_2 to remove non-Lev containing impurities. Solid-supported trisaccharide was then released by treatment with NaOH in MeOH/CH_2Cl_2 to afford the pure **69** in 88% overall yield after passing through the ion exchange resin Dowex 50W-X4 (Scheme 16.19).

Ito and coworkers have combined the soluble polymer-supported synthesis with the catch-and-release protocol using the chloroacetyl (ClAc) tag (Scheme 16.20).[63–65] Glucosamine acceptor bound to a low molecular weight polyethylene glycol (PEG) **70** was glycosylated with a thiogalactoside donor **71** in the presence of dimethyl(methylthio) sulfonium triflate (DMTST) in CH_2Cl_2 to provide the disaccharide **72** in 94% yield.

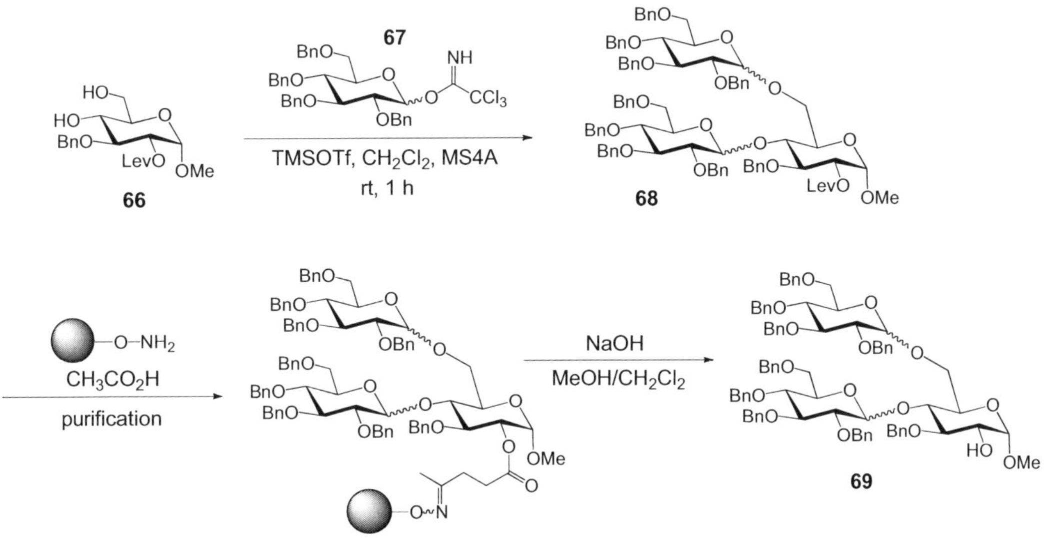

Scheme 16.19. Levulinyl for catch-and-release purification.

Compound **72** was then successfully captured by the Fmoc-cystein-loaded Wang resin at the ClAc tag in the presence of diisopropylethylamine (DIPEA). After washing the resin, **73** was released from the resin by Fmoc deprotection with 4-(aminomethyl)piperidine followed by the spontaneous intramolecular cyclization to afford **74** with excellent purity

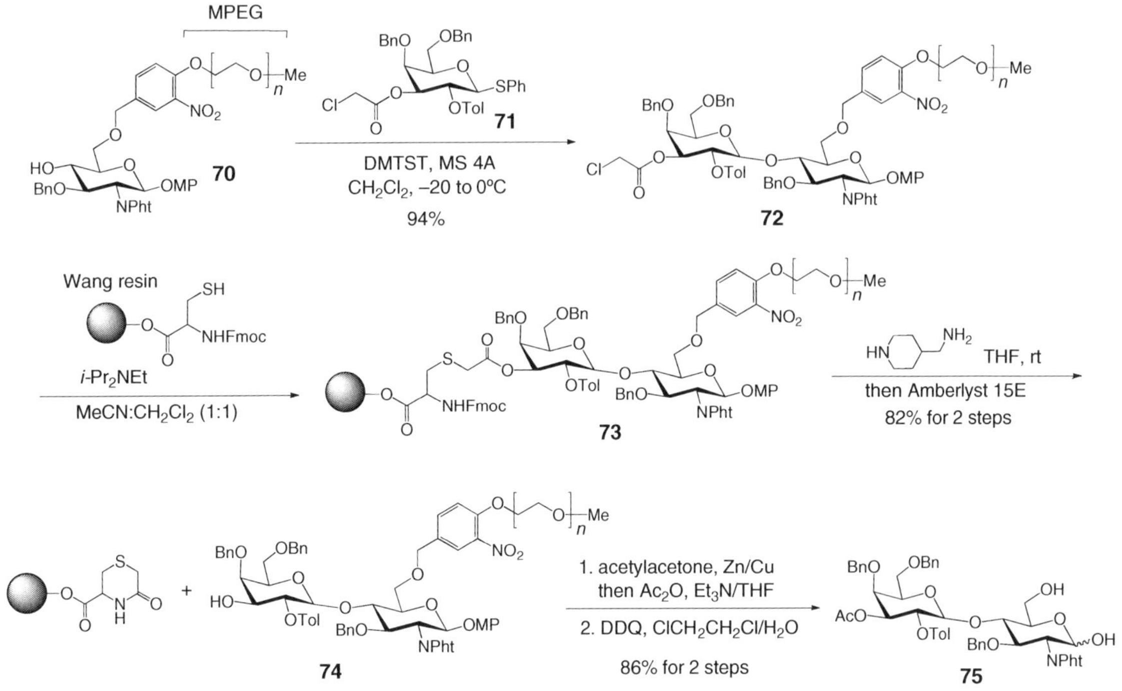

Scheme 16.20. Catch-and-release strategy using a chloroacetyl tag.

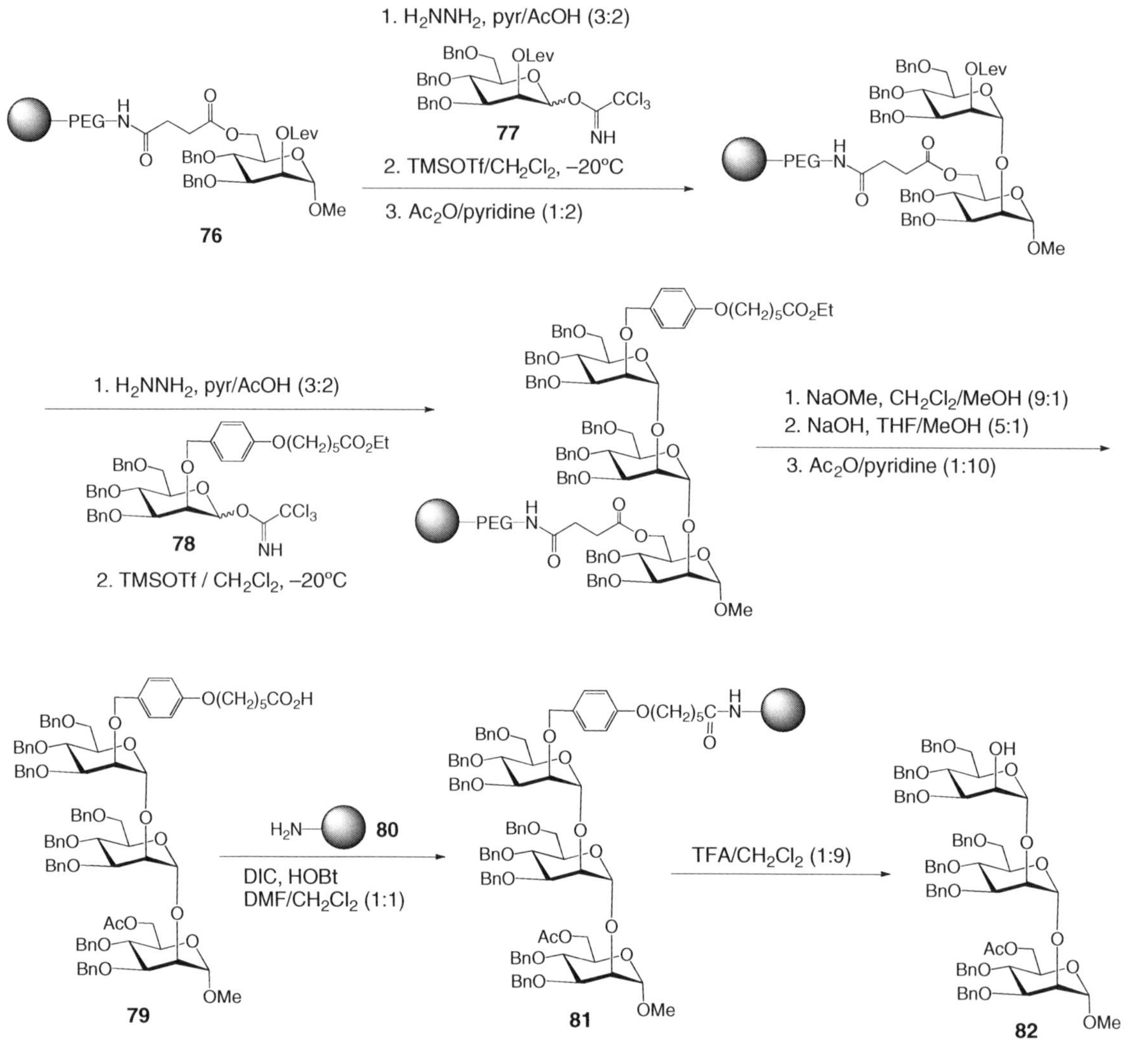

Scheme 16.21. Synthesis of α-linked trimannose using a catch-and-release strategy.

(82% for two steps). The reduction of the nitro group of PEG linker to amine by the reaction with acetylacetone and Zn/Cu, followed by acetylation and DDQ oxidation, gave the disaccharide **75** in 86% yield for two steps.

The idea of catch-and-release was applied to the solid-phase synthesis of α-linked trimannose **82** by Guo and Wu (Scheme 16.21).[66] The mannose acceptor **76**, loaded on the PEG-grafted polystyrene *via* amide linkage, was glycosylated twice with the mannosyl trichloroacetimidate donors by using TMSOTf as an activator in CH_2Cl_2, that is, first with **77** and second with **78** bearing the ester (precursor of carboxylate) group as a tag. After cleavage from the resin and hydrolysis of the ester tag under basic conditions (NaOMe, CH_2Cl_2/MeOH, and then NaOH, THF/MeOH), a trimannose derivative **78** was efficiently caught out from mono-, di-, and other by-products by the amino-functionalized "fishing" resin **80** by acylation. The pure compound **82** was then released from the solid support by treatment of **81** with TFA.

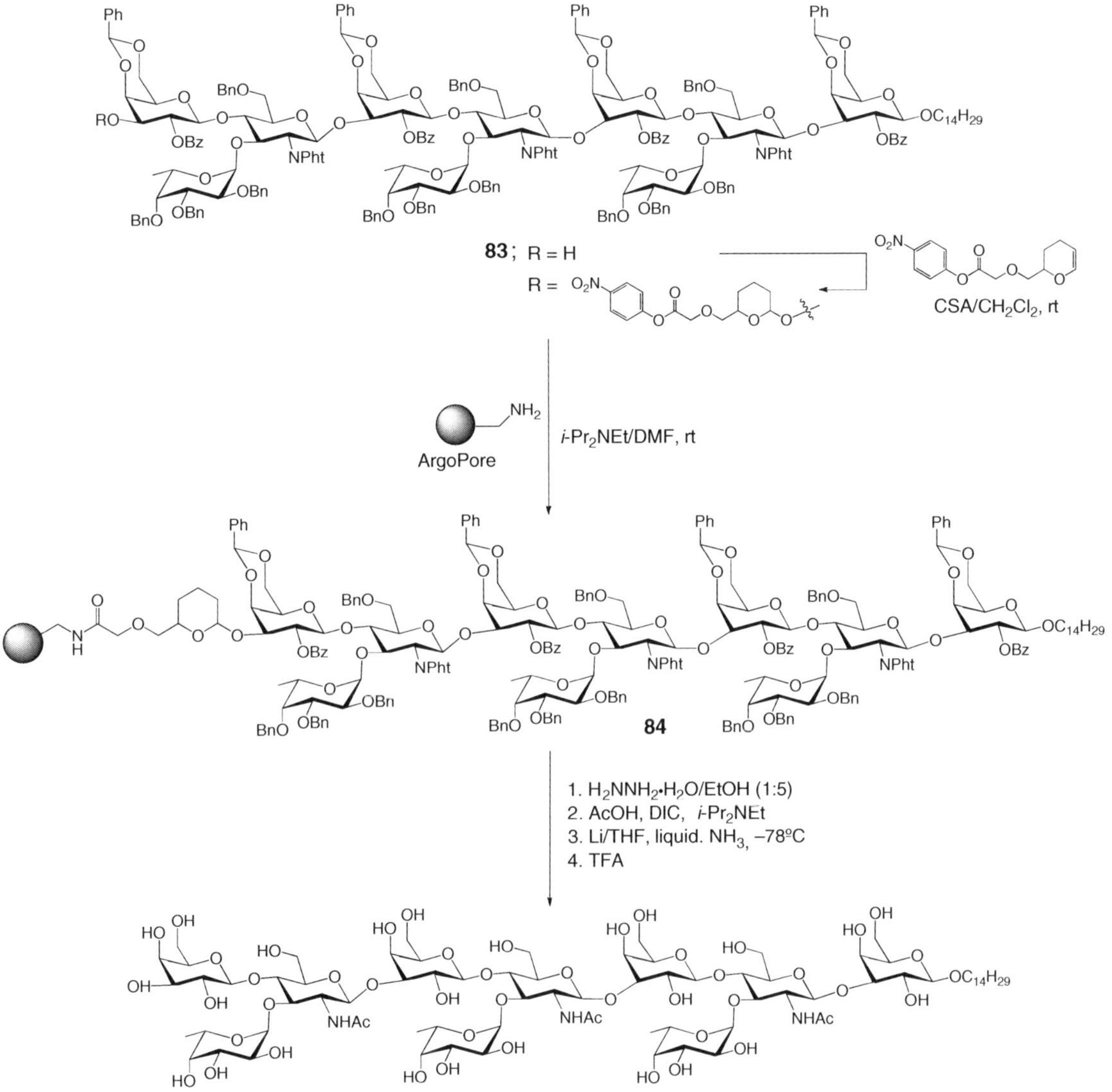

Scheme 16.22. Solid-supported deprotection of oligosaccharides.

On the other hand, an efficient polymer-assisted method for deprotection of the protected oligosaccharides has been demonstrated by Takahashi and coworkers (Scheme 16.22).[67] The deprotection of synthesized oligosaccharides, including the cleavage of various *O*-protecting groups and the replacement of *N*-protecting groups with *N*-acetyl group, is sometimes troublesome under standard conditions. Furthermore, the complete deprotection of the synthesized oligosaccharides frequently requires careful selection of the reaction solvents to prevent the partially deprotected intermediates from participating. The authors first loaded the synthesized oligosaccharide **83** to the aminomethyl ArgoPore through a tetrahydropyranyl (THP) linker. The protected oligosaccharide on polymer-supported **84** thus obtained was subjected to (i) deprotection of *N*-phthalimide groups by hydrazine, (ii) acetylation of the resulting free amino groups, (iii) removal of the benzyl groups by Birch reduction, and (iv) deprotection of 4,6-*O*-benzylidene groups and simultaneous cleavage from the THP linker by TFA to provide the fully deprotected trimeric Lewis X epitope **85** in 58% overall yields from **83**. Although loading and release of the synthesized oligosaccharides might be accompanied by slight loss of materials, this method allows the practical use for the library synthesis of complex oligosaccharides.

The polymer-supported methods for removal of unreacted acceptors during the glycosylation process also provide a rapid access to the automated synthesis and the parallel library synthesis of oligosaccharides. Ley and coworkers have applied the polymer-supported tosyl chloride (PS-TsCl) to remove the hydroxyl-containing contaminants, including the acceptors.[68] The unreacted acceptor was also sequestered and recovered by Dondoni et al. based on catch-and-release protocols.[69] By this method, the hydroxyls of the unreacted acceptors were reacted with the trichloroacetyl isocyanate under neutral conditions, and the resulting trichloroacetyl carbamates (urethanes) were extracted by the highly basic and nonnucleophilic polymer-supported BEMP (2-*tert*-butylimino-2-diethylamino-1,3-dimethylperhydro-1,3,2-diazaphosphorine on polystyrene) by making use of strong ionic interactions. The pure acceptor was released from the resin by KOH treatment and again used for glycosylation.

Seeberger and coworkers have successfully used two capping methods to remove the unreacted acceptors during their automated synthesis of oligosaccharides (Scheme 16.23 and see section 16.2.3).[70,71] Both 2-azido-2-methylpropionic acid anhydride **86** (A-tag) and (heptadecafluorodecyl)diisopropylsilyl triflate **87** (F-tag) were applied to cap the hydroxyls of unreacted glycosyl acceptors during each incomplete glycosylation step. At the end of the synthesis, the tagged compounds were efficiently removed after the cleavage from the resin by cross-metathesis with ethylene. For A-tag, the azide moiety was reduced by tributyl phosphine, and the resulting amino derivative was fished out by an isocyanate silica gel scavenging resin. On the other hand, the F-tagged compounds were removed by column chromatography packed with tridecafluoro($Si(CH_2CH_2C_6F_{13})_3$)-functionalized silica gel. By this method, they successfully simplified the purification process for the automated synthesis of β-linked (1-6)-triglucose and α-linked (1-2)-mannose.

16.2.5 Monitoring of Solid-Phase Reactions

Real-time monitoring of the polymer-supported reaction is nearly impossible by simple and routine methods such as TLC and HPLC. In the case of solid-phase peptide synthesis, the end of the reaction can be readily checked by the ninhydrin test. In the solid-phase synthesis of oligosaccharides, a convenient method for checking the hydroxyl group on the polymer support has not been developed yet. In most cases, the reaction is checked after small-scale cleavage of products. If the Fmoc protecting group is used, the efficiency in glycosylation and deprotection on the resins can be evaluated by UV absorption of dibenzofurubene derivatives when it is deprotected by mild bases, such as piperidine or Et_3N.

Recent advances in NMR have proven to be effective for direct and nondestructive monitoring of the progress in oligosaccharide synthesis. Wong et al. used the inverse gated decoupling technique of ^{13}C NMR to monitor the glycosylations on the TentaGel resin, in which two ^{13}C-enriched carbonyl tags were introduced both on the glycine linker as an internal standard and on the acetyl protecting group of the introducing glycosyl donors.[72] Sialyl Lewis X branched tetrasaccharide was synthesized by this monitoring method.

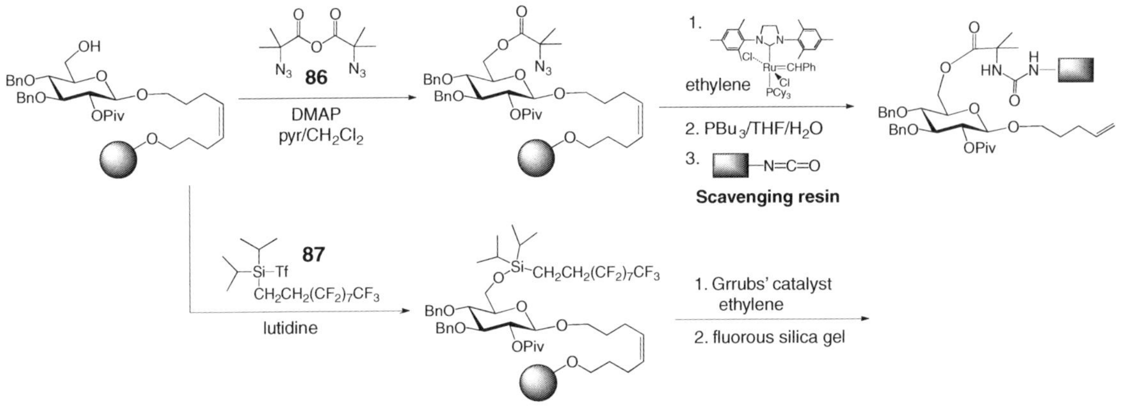

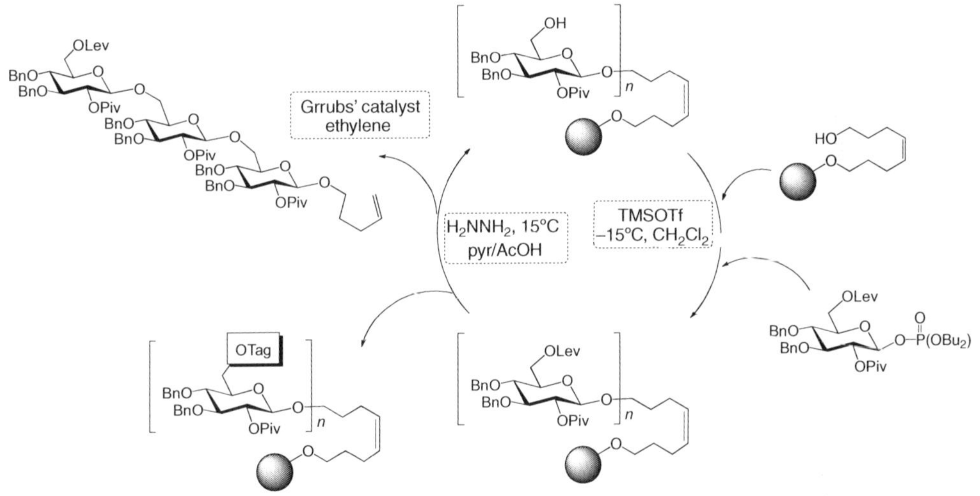

Scheme 16.23. Capping methods for automated oligosaccharide synthesis.

In a similar way, Kihlberg and coworkers have applied the ^{19}F-tag both on the linker and the protecting groups, that is, *p*-fluorinated benzyl ethers, benzoates, and benzylidene acetals of the glycosyl reagents (Scheme 16.24).[73–75] The gel-phase ^{19}F NMR allowed highly sensitive, simple, and quantitative analysis of the oligosaccharide assembly *via* continuous glycosylation and deprotection process on the resin. ArgoGel resin with 3-fluoro-4-hydroxymethyloxybenzamide linker was esterified with glucosyl derivative **88**, which was protected with two *o*-fluorobenzoates and 4,6-*O*-*p*-fluorobenzylidene acetal. The loading yield was easily evaluated based on the integration of resonances from the fluorine atoms in the linker ($\delta = -134.4$ ppm), the 4,6-*O*-*p*-F-benzylidene group (-113.2 ppm), and two *o*-F-benzoyl groups (-109.0 and -110.1 ppm). Followed by the reductive opening of the 4,6-*O*-*p*-F-benzylidene group by $NaCNBH_3$ in the presence of HCl gas, the acceptor **89** was glycosylated with thiogalactoside **90** and subsequently with **91**, each bearing the fluorine tag in different positions of 4,6-*O*-benzylidene groups, that is, ortho- ($\delta = -120.9$ ppm) or meta-positions ($\delta = -113.9$ ppm) of aromatic groups. The sequence

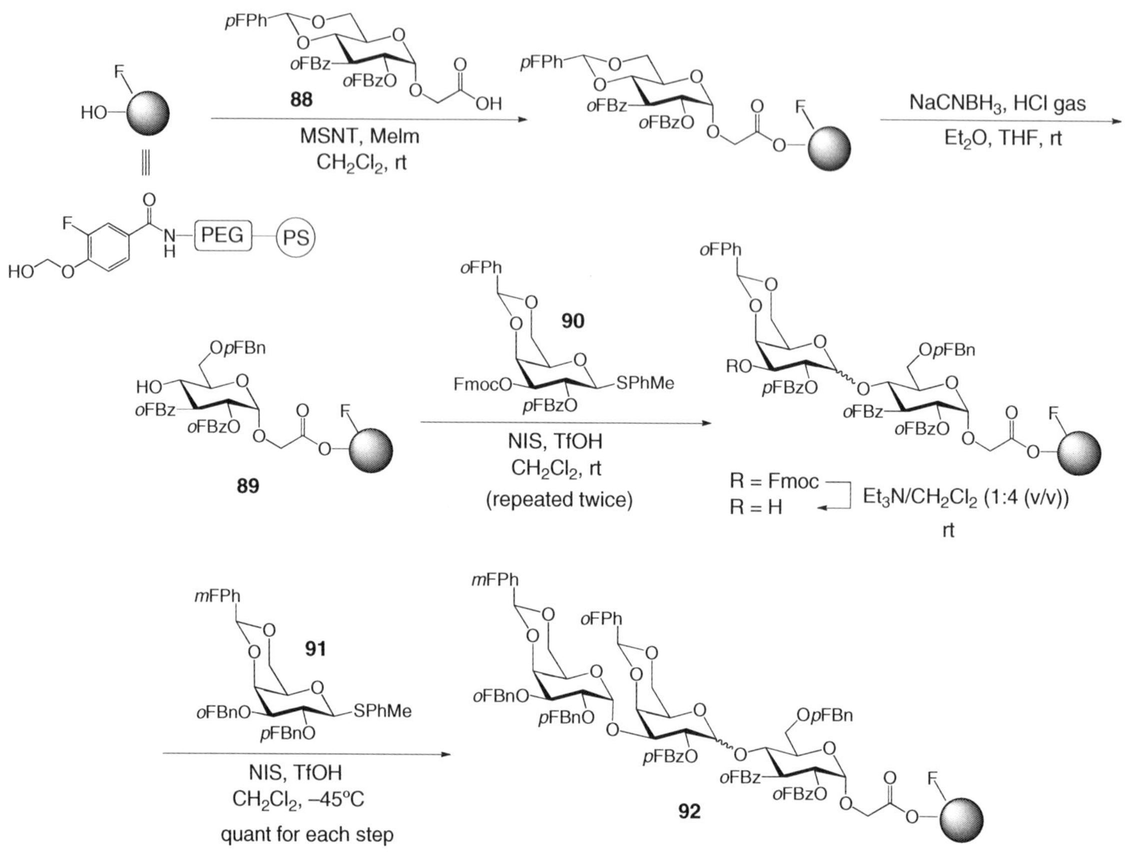

Scheme 16.24. Solid-supported oligosaccharide synthesis by ^{19}F NMR monitoring.

of the reactions provided the trisaccharide **92**, α-Gal epitope (Galα(1-3)Galβ(1-4)Glc), which is responsible for hyperacute rejection in xenotransplantation of porcine organs. Each step was evaluated to proceed quantitatively based on this ^{19}F NMR analysis.

Ito and Manabe have developed the practical and convenient method for the real-time monitoring of solid-phase oligosaccharide synthesis based on "on-resin color tests," which enable side-by-side detection of glycoside bond formation and acceptor consumption (Scheme 16.25).[76] They have used two color test systems: the modified Taddei's test with the cyanuric chloride–Disperse Red conjugate **98** was applied to detect the presence of nucleophilic free hydroxyl and amino groups on the resin, whereas the chloroacetyl group was used as a "temporary" hydroxyl protecting group, which can be detected by a red color generated by the action of *p*-nitrobenzylpyridine (PNBP) **99** under basic conditions. Based on these monitoring methods, the repeating units of the immunoactive oligosaccharide schizophyllan were synthesized (Scheme 16.25). The TentaGel-loaded acceptor **93**, which is negative for PNBP test and positive for Disperse Red test, was glycosylated with 3- and 6-*O*-ClAc-protected trichloroacetimidate **94** in the presence of $BF_3{\cdot}OEt_2$ in CH_2Cl_2 to provide β(1-3)-linked disaccharide **95**. The completion of the reaction was easily checked by the positive response for PNBP test and the negative response for Disperse Red test. After the selective deprotection of ClAc group by hydrazinedithiocarbonate treatment, the free hydroxyls at 3′- and 6′-positions were subjected to further glycosylation with glucosyl

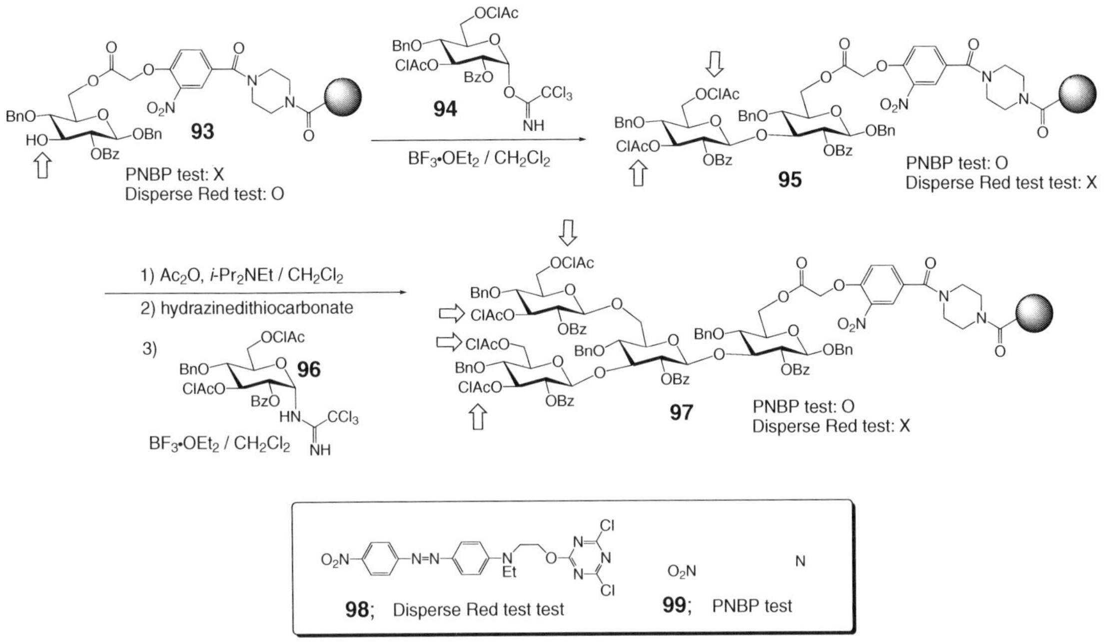

Scheme 16.25. "On-resin color tests" for solid-supported synthesis.

trichloroacetimidate **96** to give resin-bound tetrasaccharide **97**, which was easily detected by deep red color in the PNBP test.

Nitrophthalimidobutyric acid (NPB) was recently applied to the protecting group of hydroxyls for colorimetric monitoring of fluorous-phase and solid-phase synthesis of oligosaccharides (Scheme 16.26).[77] The hydroxyl-containing polymer support grafted with polystyrene and C_8H_{17} fluorous tag was glycosylated with galactosaminne imidate **100** with the NPB at the C6 hydroxyl in the presence of TMSOTf. After the NPB group was removed by the hydrazine acetate, the glycosylation yield was calculated to be 98% by monitoring the orange-colored nitrophthalhydrazide (accompanied by the butyrolactone

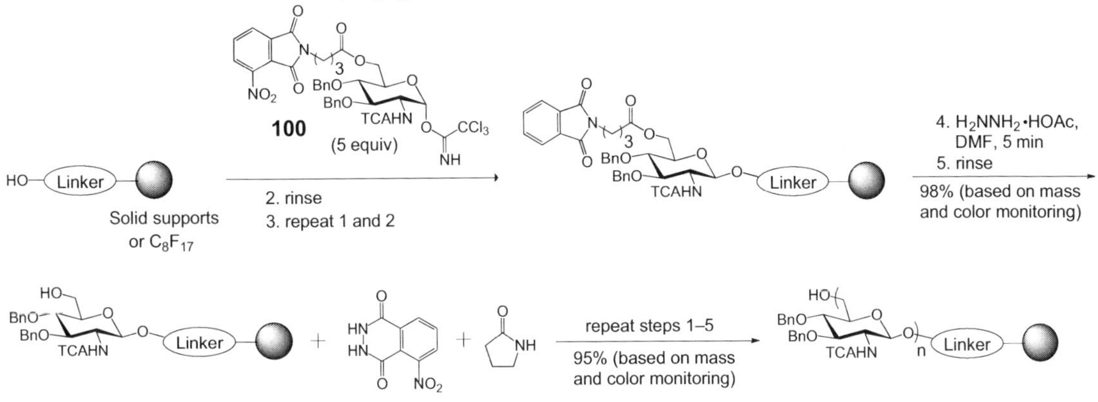

Scheme 16.26. Nitrophthalimidobutyric acid test for colorimetric monitoring.

formation). The sequence of the glycosylation and the colorimetric quantification of the reaction led to an efficient oligosaccharide synthesis.

16.3 POLYMER-SUPPORTED AND TAG-ASSISTED OLIGOSACCHARIDE SYNTHESIS IN SOLUTION

A phase tag strategy has been developed for high-throughput synthesis as a hybrid system that combines the merits of solid-phase synthesis (easy separation) and solution-phase synthesis (homogeneous reaction conditions). In this strategy, a compound having a tag group is easily separated from untagged molecules. Various phase tags such as soluble polymers, fluorous, hydrophobic, and basic tags have been reported.

16.3.1 Polymer-Supported Synthesis of Oligosaccharides

Polymer-supported solution synthesis has several advantages over solid-phase synthesis. The reaction can be carried out in solution and the solubility of the reactants allows reaction kinetics and anomeric stereocontrol similar to those observed in solution chemistry. Furthermore, in some cases, the process of the reaction can be easily checked by the conventional TLC or MALDI-TOF-MS (matrix-assisted laser desorption ionization mass spectrometry). Out of a large number of the polymer-supported synthesis so far reported, the polyethylene glycol monomethyl ether (MPEG) has been widely used for oligosaccharide synthesis.[78,79] Polyethylene glycol is soluble in a variety of solvents such as CH_2Cl_2 and the protic solvents but insoluble in diethyl ether. The polymer-bound products are therefore isolated by precipitating the polyethylene glycol supports using ether. Martin-Lomas and coworkers have succeeded in synthesizing heparin-like oligosaccharide **107** using both MPEG as a soluble polymer support and Merrifield-type resin (polyethylene glycol-grafted polystyrene) as a solid support used for scavenging the unreacted acceptors (Scheme 16.27).[80,81] The disaccharide acceptor **101** bound to MPEG through a succinic ester linkage at the uronic acid carboxylate was glycosylated with the trichloroacetimidate acceptor **102** in the presence of TMSOTf to provide the α(1-4)-linked tetrasaccharide **104** with the complete stereoselectivity. This glycosylation step was followed by capping with the Merrifield-type resin **103** functionalized with the carboxylic acid to esterify the unreacted C4 hydroxyl group of the acceptor **104**. The outcome of the α-stereoselectivity and the glycosylation could directly be monitored by NMR. Removal of the benzylidene group of **104** by EtSH in the presence of *p*-TsOH, benzoylation of the C6 hydroxyl, second glycosylation with **105**, and capping of the unreacted acceptor **104** with scavenger resin **103** provided the key intermediate **106**, which was successfully transformed to hexasaccharide **107**.

Ito and coworkers have reported the MPEG-supported synthesis of the trisaccharide **111**, the outer branching fragment of the tetraantennary complex-type glycan chain, by applying the self-cleavable Wang resin-type linker (Scheme 16.28).[82] The mannosyl acceptor **108** loaded on the MPEG was glycosylated with the lactosamine-derived trichloroacetimidate **109** by using TMSOTf, providing β-glycoside **110** in 77% yield. MPEG-bound trisaccharide **110** was then treated with $Sn(SPh)_2$ in the presence of thiophenol to release **111** from the soluble support *via* selective reduction of the nitro group followed by the spontaneous intramolecular cyclization.

A simple formylacetal (CH_2) has also been used as a linker for oligosaccharide synthesis on MPEG. This small linker allows the hindered hydroxyls, such as 4-hydroxy

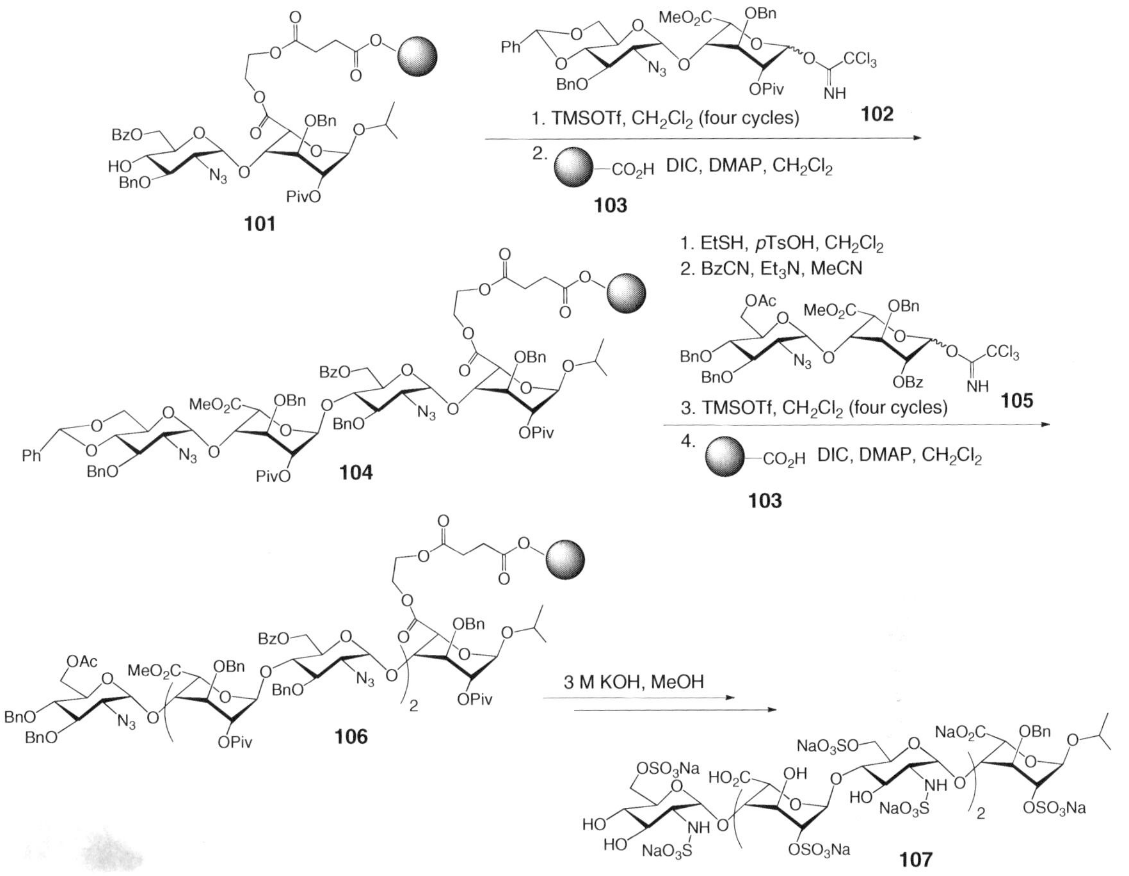

Scheme 16.27. Synthesis of a heparin-like oligosaccharide using soluble and solid supports.

group of glucose, to be installed on the MPEG.[83] The formylacetal linker is stable under acidic conditions mostly used for glycosylation, but readily cleaved by Lewis acids (TMSI or $Ce(OTf)_x$) or trifluoroacetic acid. The simplified workup and purification procedure was also reported by use of the solid acid catalyst.

16.3.2 Tag-Assisted Synthesis of Oligosaccharides

Tag-assisted methods in solution phase, other than MPEG supports, have also emerged as an attractive strategy for combinatorial and high-throughput synthesis of oligosaccharides. A highly fluorinated compound is soluble in fluorous solvents, such as perfluorohexane, and is readily separated from nonfluorinated compounds through simple fluorous–organic solvent partitioning. Therefore, the fluorous synthesis is an attractive alternative to solid-phase synthesis, directed toward combinatorial chemistry and parallel synthesis. Although several fluorous protecting groups are reported for this purpose, their application is limited to the synthesis of simple small molecules because the fluorine content in each fluorous protecting group is low.[84] Inazu and coworkers have developed the fluorous tags with high fluorine contents and applied to the oligosaccharide synthesis (Scheme 16.29).[85,86] They initially investigated the Bfp-OH tag (bisfluorous chain-type propanoyl), which was introduced at

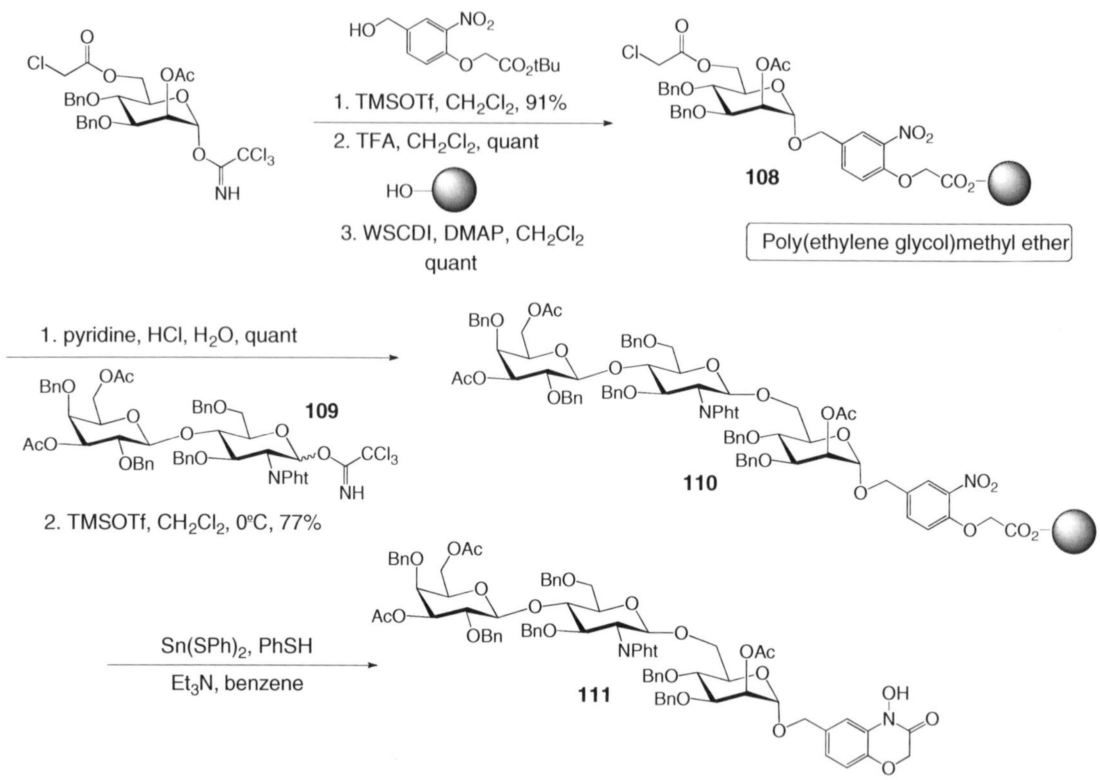

Scheme 16.28. MPEG-supported synthesis of tetraantennary complex-type *N*-glycans.

three of four hydroxyls on the glycosyl acceptors for the synthesis of a simple tetrasaccharide as well as the Gb3 oligosaccharide. Later, Hfb-OH tag (hexakisfluorous chain-type butanoyl) with more fluorous chains was designed to enhance the efficiency of fluorous extraction; a rapid synthesis of trisaccharide **118** was achieved by introducing only one fluorous tag (Scheme 16.29). Hfb tag was first attached to the anomeric hydroxyl of the glucose derivative **112** by using PyBOP (1*H*-benzotriazole-1-yloxy)tripyrrolidinophosphonium hexafluorophosphate) and DMAP, and the deprotection of TBDPS group by HF–pyridine in THF provided the glucose acceptor **113**. Compound **113** was then glycosylated with glucosyl trichloroacetimidate **114** by using TMSOTf in a mixed solvent of $EtOC_4F_9$ and Et_2O to provide disaccharide **115**. The fluorous tag-containing **115** was extracted with the fluorous solvent FC-72 by partitioning the product mixtures between FC-72 and an organic solvent. No additional purification, such as silica gel chromatography, was carried out. After removing the TBDPS group of **115** by HF–pyridine treatment in THF, the second glycosylation was performed with **116** to provide trisaccharide **117** after the partitioning by the fluorous/organic solvents. Finally, the Hfb group of **117** was removed by the treatment with NaOMe in $EtOC_4F_9$/MeOH to afford the crude **118**, which was extracted into MeOH layer from the partitioning mixture of CF-72 and MeOH. By this method, pure **118** was obtained in 67% yield from **112** after silica gel chromatography, while the methyl ester derivative of Hfb-OH tag, which was extracted into the CF-72 layer during the partitioning process, was collected and reused after hydrolysis with NaOH.

Huang and coworkers have also used the fluorous extraction to efficiently purify desired oligosaccharides from the mixtures of their one-pot products (Scheme 16.30).[87] Thus, after

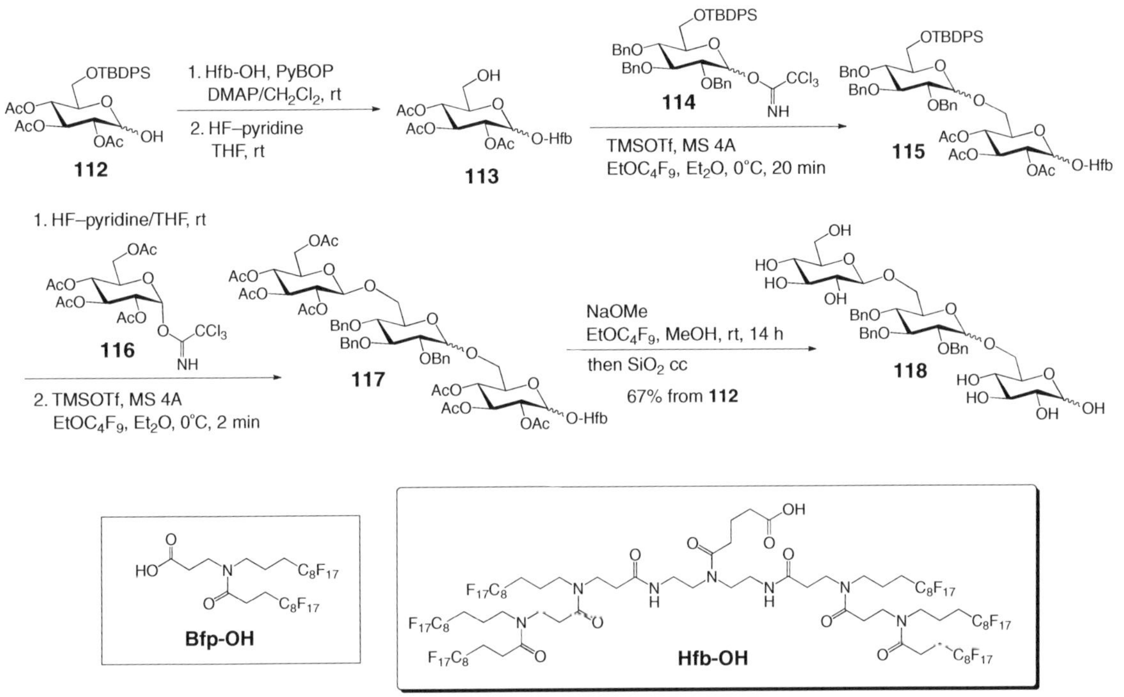

Scheme 16.29. Synthesis using fluorous tags.

synthesizing the Lexis X trisaccharide **119** in one pot by using the thioglycosyl donors and *p*TolSCl/AgOTf as the activators, C_8F_{17} fluorous hydrazide was reacted with the reducing end ketone of the desired trisaccharide. Separation from the nonfluorous impurities by the fluorous solid-phase extraction, cleavage of the hydrazone by treatment with 0.5% TFA in acetone, and finally the fluorous extraction gave the desired Lewis X in 62%, the overall yield including the one-pot glycan assembly and the purification process. Owing to the

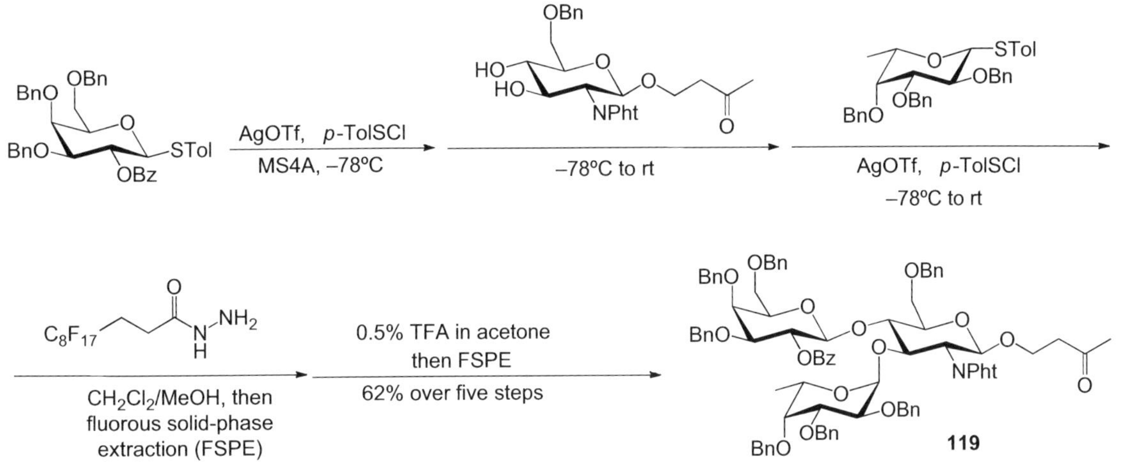

Scheme 16.30. Fluorous extraction in one-pot oligosaccharide synthesis.

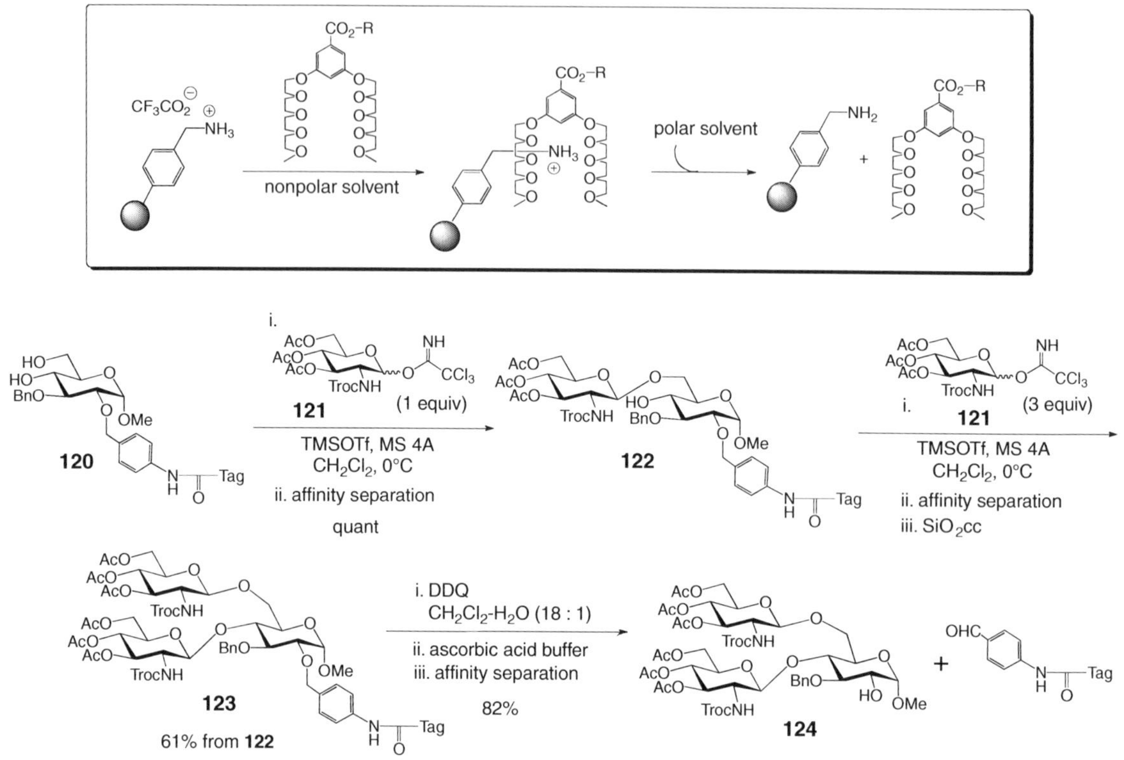

Scheme 16.31. Synthesis based on affinity separation for oligosaccharide synthesis.

simplified purification procedure, both the linear and branched oligosaccharides have been synthesized in just a few hours. The difference in the fluorous extraction efficiency of the glycosamine with "mono"-C_8F_{17} fluorous tag and with "di"- C_8F_{17} fluorous tags has also been examined to apply the fluorous extraction strategy to the iterative and more preferably the automated oligosaccharide synthesis.[88]

Fukase and coworkers have developed a concept termed synthesis based on affinity separation (SAS) in which the desired tagged compound is separated from the reaction mixture by solid-phase extraction using specific molecular recognition (Scheme 16.31).[89–92] They first employed the interaction between a crown ether (32-crown-10) as a tag on the desired compounds and the ammonium ion for SAS. After each reaction cycle, the reaction mixture was applied to the column loaded with aminomethylated polystyrene resin (TFA form). By using nonpolar eluents such as CH_2Cl_2 and toluene, the tagged compound was selectively adsorbed on the column, whereas other untagged impurities were washed off. Subsequent desorption by CH_2Cl_2–Et_3N or CH_2Cl_2–MeOH (1 : 1) afforded the desired compound with high purity. However, since the crown ether tag needed to be prepared by the tedious synthesis, the commercially available short-chain PEG was used as a second-generation tag. Since Triton X-100 is a detergent having a PEG chain and a hydrophobic moiety, the tagged compounds show good solubility in many organic solvents. However, the chain length of Triton X-100 is heterogeneous and Triton X-100 with the shorter chain length showed only weak binding. A podand-type ether with a pseudobenzo-31-crown-10 structure, introduced as third-generation tag, was much easier to synthesize and found to show high affinity to the ammonium ion on the solid support.[92] The present SAS method has been successfully applied to the oligosaccharide synthesis (Scheme 16.31). The glycosylation between acceptor **120**, having the tag moiety *via* an acylaminobenzyl linker, and *N*-Troc glucosamine trichloroacetimidate **121** was affected by using TMSOTf. After the affinity separation, the resulting disaccharide **122** was further subjected to glycosylation with an excess amount of the donor **121**. Although the affinity separation could not remove a small amount of the unreacted disaccharide **122** from the desired trisaccharide **123** (since both bear the tag moieties), subsequent separation by silica gel column chromatography afforded pure **123** in 61% yields from **122**. Cleavage of the acylaminobenzyl linker in **123** was achieved by treatment with excess DDQ followed by the affinity column separation to provide the pure trisaccharide **124** in 82% yield.

Alternatively, Rademann and coworkers have combined solid-phase and solution-phase reactions for the synthesis of oligosaccharides by using the hydrophobic tag, a concept called hydrophobically assisted switching synthesis (HASP) (Scheme 16.32).[93,94] They applied a sufficiently long, double C18 hydrocarbon chain **125** as a fully reversible tag for reactions both in solution and on the solid support. The hydrophobic tag was introduced to the anomeric position of L-rhamnose, and α(1-2)-selective glycosylation was first conducted in solution (CH_2Cl_2) with rhamnosyl trichloroacetimidate **126** in the presence of TMSOTf as a Lewis acid (quantitative yield). The C18 silica support was then added to the reaction mixture, solvents were evaporated, and all untagged materials were easily removed by washing with $MeOH/H_2O$. After 2-*O*-phenoxyacetyl group was deprotected on the solid support by the reaction with NaOMe in $MeOH/H_2O$ (92%), followed by the washing step, a sufficiently pure compound was released from the solid support by addition of $MgSO_4$ and eluting by CH_2Cl_2. After HASP cycle was repeated for a few times, the

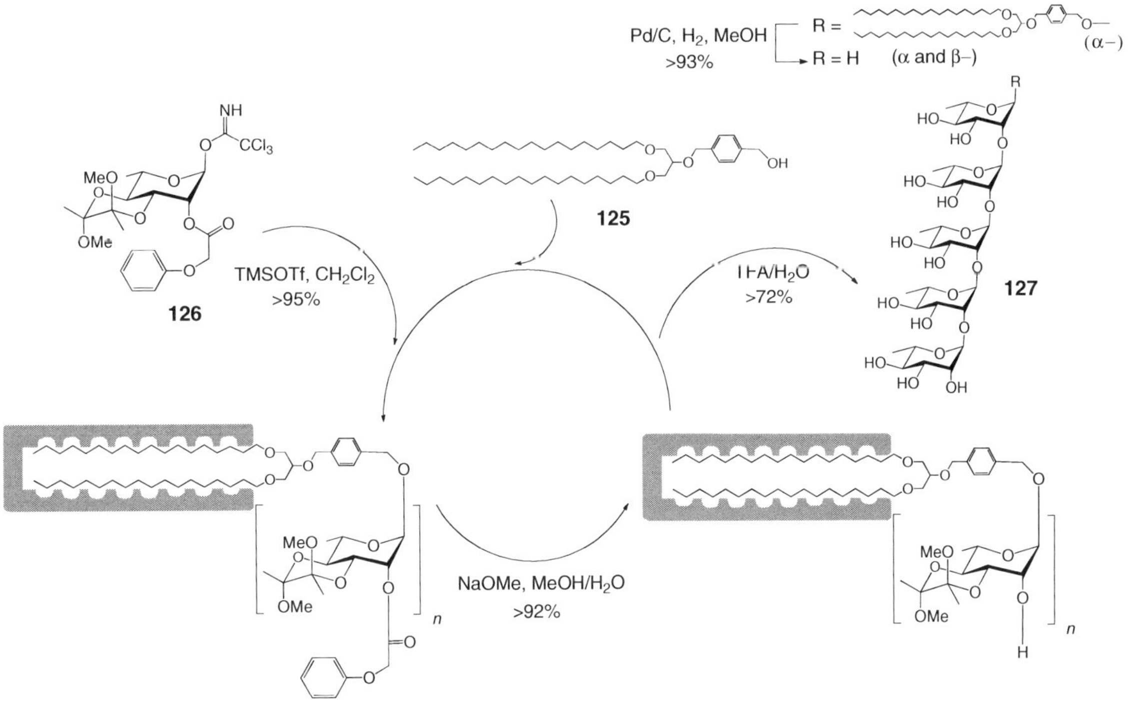

Scheme 16.32. Hydrophobically assisted switching synthesis.

3,4-*O*-butane-2,3-diacetal protecting group was removed by TFA. Finally, hydrogenation provided α(1-2)-liked oligorhamnose **127**, with an average yield of 94% per step. Each reaction could be directly followed by TLC and MS (molecular sieves) for this HASP method.

16.3.3 Polymer-Supported Enzymatic Synthesis of Oligosaccharides

Since the chemical synthesis of oligosaccharides requires many synthetic steps, including protection and deprotection procedures, the enzymatic approach has attracted much attention for the rapid synthesis of oligosaccharides. In addition, the perfect regio- and stereoselectivities of enzymatic methods with glycosyltransferases are quite attractive. Several transferases such as β(1,4)-galactosyltransferase, α(1,3)-fucosyltransferase, and α-sialyltransferase have been used for polymer-supported enzymatic synthesis.[95–100] The selection of the polymer support is very important for the polymer-supported enzymatic synthesis of oligosaccharides.

Nishimura and coworkers have developed the catch-and-release strategy between solid-phase and water-soluble polymer supports, the so-called polymer blotting method, which allows the rapid and efficient synthesis of glycopeptides.[101] The method involves (i) conventional solid-phase synthesis of glycopeptides that contain mono-, di-, and trisaccharides on Thr or Ser residues and Blase (glutamic acid-specific protease)-sensitive amino acid sequence, (ii) removal of the protecting groups and release from the resin, (iii) attachment of the released glycopeptides to the water-soluble polymers *via* oxime anchor, (iv) sugar elongations using glycosyltransferases, and (v) cleavage of the desired glycopeptides from the soluble supports by Blase-mediated hydrolysis. They have applied three glycosyltransferases, namely, β(1,4)-galactosyltransferase, α(2,3)-(*O*)-sialyltransferase, and α(2,3)-(*N*)-sialyltransferase, and successfully prepared 6 kinds of the parallel and 36 kinds of combinatorial libraries of MUC1 mucin glycopeptides. The applications of the method to other members of glycopeptides have also been reported by the same group.

Wong and coworkers have investigated the enzymatic oligosaccharide synthesis on a thermoresponsive polymer support (Scheme 16.33).[102] The copolymers of *N*-*i*-propylacrylamide (NIPAm) and functionalized monomers are thermoresponsive and exhibit inverse temperature-dependent solubility in water; they are soluble in cold water but become

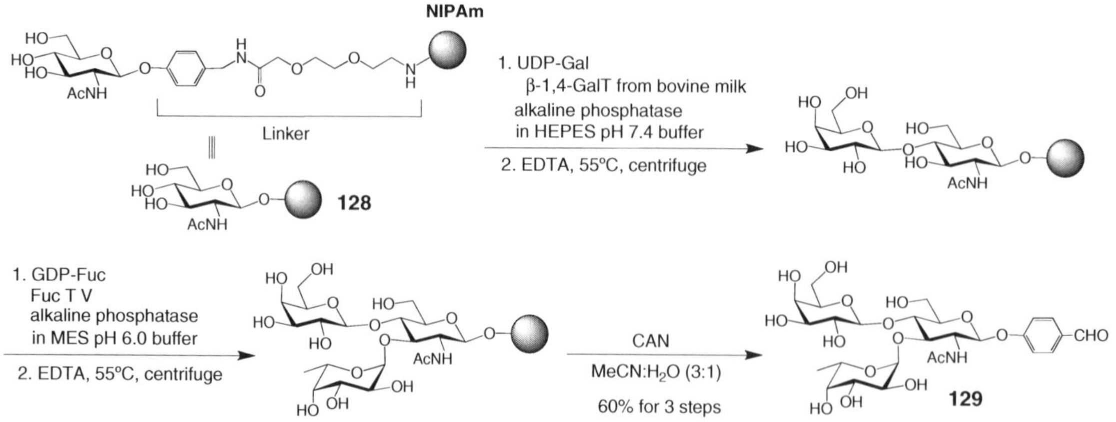

Scheme 16.33. Enzymatic synthesis on a thermoresponsive polymer support.

insoluble and precipitate from the solution when the temperature is higher than their lower critical solution temperature. The NIPAm polymer support greatly simplified the purification process for oligosaccharide synthesis in water; the carbohydrate acceptors attached to the NIPAm support were used for enzymatic glycosylation, and the glycosylation products can be isolated simply by thermal precipitation and centrifugation. The availability of direct NMR analysis also makes the method attractive. Galactosylation of the immobilized GlcNAc **128** was performed by using 100 mU bovine milk β(1,4)-galactosyltransferase with 1 equiv of UDP (uridine diphosphate)-galactose donor in *N*-(2-hydroxyethyl)piperazine-*N'*-(2-ethanesulfonic acid) (HEPES) buffer at pH 7.4. Followed by the precipitation of the product by heating the solution to 55°C, it was further fucosylated using the bovine α(1,3)-fucosyl transferase and GDP-fucose donor. After cleavage from the NIPAm support by CAN treatment, the trisaccharide Le^X**129** was obtained in 60% yield for three steps without chromatographic separation of the intermediates. They also have immobilized several enzymes on the same NIPAm polymers, such as subtilisin BPN′, which were found to show comparable activities as their soluble forms. For example, after the NIPAm-subtilisin conjugate was used for proteolysis of a modified form of the glycoprotein ribonuclease (RNase B), the immobilized enzyme can be easily recovered for reuse by gentle heating and precipitation. The methods are expected to be applied to glycosyltransferases and this thermoresponsive polymer-supported procedure will greatly facilitate enzyme-catalyzed organic synthesis.

16.3.4 Microfluidic Methods for Oligosaccharide Synthesis

Continuous flow synthesis using a microfluidic system has emerged as an innovative technology in organic synthesis, that is, from small-scale optimization of the reaction conditions, high-throughput synthesis, to large-scale production. This method has various advantages in terms of reaction control, such as fast mixing, effective temperature control, and precise residence time control. Since the reaction can be performed under the flow process, the method is readily applicable to the library construction by preparing the stock solutions of glycosyl acceptors and donors and flowing each through syringe or HPLC pumps. On the other hand, once the reaction conditions are optimized for small-scale operation, the same conditions are directly applied to large-scale synthesis, that is, for process synthesis on industrial scale, by using several microreactors under the flow process. Seeberger and coworkers have reported the first application of microfluidic system to optimize the glycosylation on analytical scale.[103] The reaction performed in microreactor was monitored by HPLC, which is directly connected to the microreactor. The composition of products was analyzed by changing the concentration, reaction time, and temperature; the optimal conditions for α-mannosylation and the different reactivity due to the protection patterns of acceptors have been found.

Fukase et al. have combined a microreactor and their SAS protocol (synthesis based on affinity separation, Section 16.3.2) in pursuit of the high-throughput oligosaccharide synthesis (Figure 16.3).[92] They used a combination of an IMM micromixer and a stainless tube reactor, equipped with a stainless column as an affinity separation unit. The glycosylation was carried out in the micromixer by combining the CH_2Cl_2 solution of acceptor **130** and donor **131** with the solution of TMSOTf in CH_2Cl_2 under continuous flow. The mixture was directly introduced to the stainless steel affinity column and the flow channel was switched to wash the column with CH_2Cl_2. During the time, the desired disaccharide **132** was trapped in the column, but then eluted with CH_2Cl_2/MeOH (1 : 1); the product **132** was obtained in 91% yield.

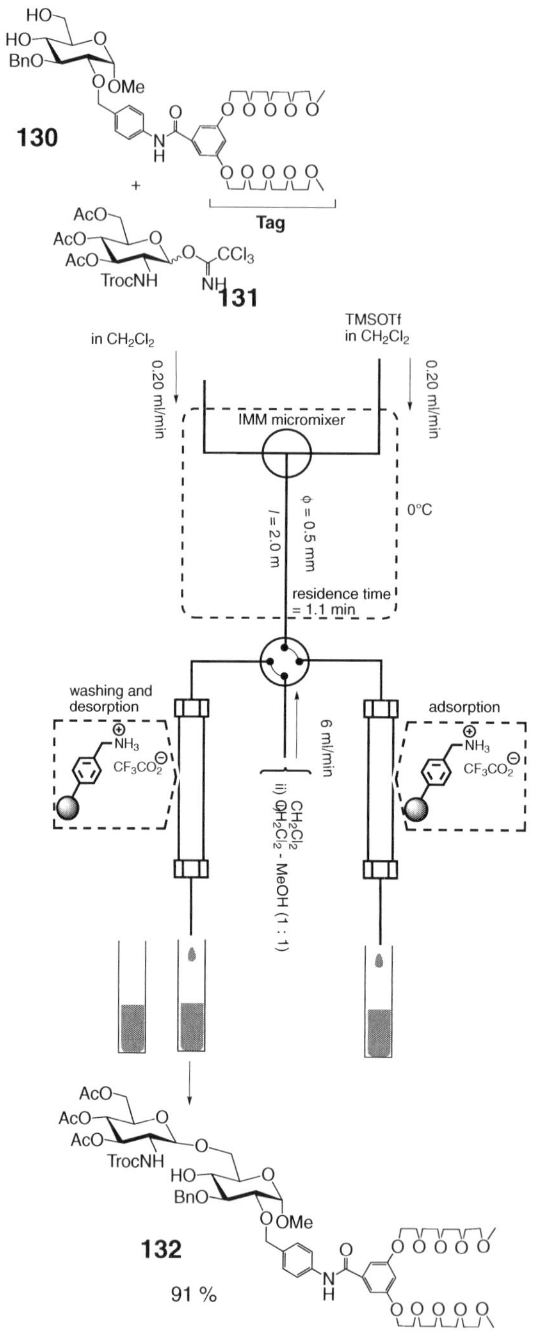

Figure 16.3. An integrated microfluidic and SAS strategy.

A successful α-sialylation under the microfluidic conditions has also been reported by the same group, of which reactivity is totally different from that observed in solution.[42] The highly reactive sialyl donor developed in this laboratory, bearing *N*5-phthalyl (Pht) or azide groups and the *N*-phenyl trifluoroacetimidate leaving group,[41] could not be used for the

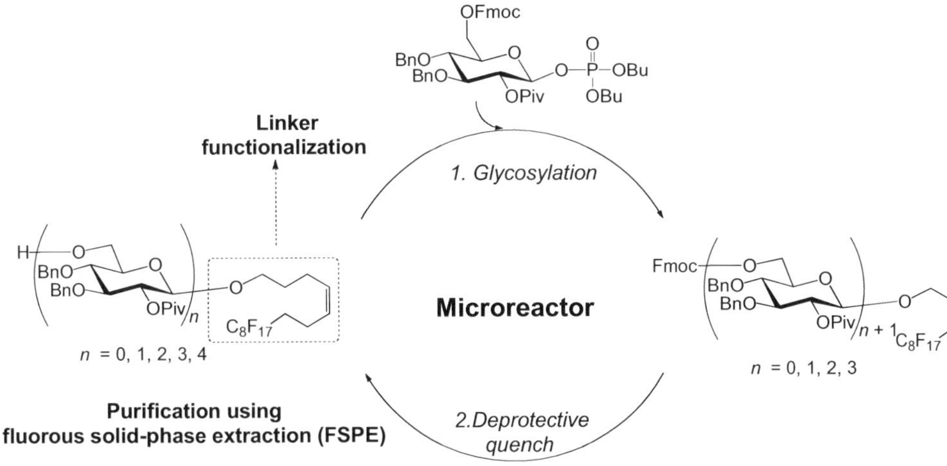

Scheme 16.34. Microfluidic and fluorous extraction techniques for oligosaccharide synthesis.

large-scale α-sialylation with galactose acceptor in a batch apparatus. Owing to the inadequate mixing and inefficient heat transfer under batch conditions, the exposure of reactive sialyl donor to Lewis acid mainly leads to the glycal formation. They have circumvented the problems by using the microfluidic systems and realized the quantitative α-sialylation with perfect α-selectivity, even in a large scale. Significantly, improved glycosylation could be achieved not only for α-sialylation but also for β-mannosylation[40] and *N*-glycosylation,[104] and the combination of these microfluidic glycosylation[105–107] and the solid-supported glycoslylation led to the successful synthesis of complex-type *N*-glycan with sialic acid as described in Section 16.2.3 (Scheme 16.16).

The combined use of microfluidic technique and fluorous solid-phase extraction for the oligosaccharide and monosaccharide building block synthesis has also been reported. Seeberger and coworkers have performed the iterative glycosylation in a silicon-based microreactor using the Fmoc-protected glycosyl phosphate and TMSOTf as an activator, of which protocol was applied to their automated oligosaccharide synthesis (Section 16.2.3)[108]; the microfluidic glycosylation was performed both in small scale for optimizing the conditions and in productive scale. After the oligosaccharides were produced by each cycle of glycosylation, it was efficiently purified by fluorous extraction by making use of the C_8F_{17} tag, purposely introduced at the reducing end of the initial glycosyl acceptor (Scheme 16.34).

On the other hand, Mizuno and coworkers have prepared the monosaccharide acceptor with the dense fluorous-containing tag using the two Y-shaped micromixers (inner diameter of 500 μm); after the sugar substrate containing fluorous tag and reagent were mixed by the first micromixer and further reacted in a Teflon tube under fluidic conditions, the mixture was partitioned by a fluorous ($MeOC_4F_9$–FC-72) and organic solvents by the second micromixer (Scheme 16.35).[109] The fluorous solution separated by the separating funnel was then directly subjected to the next reaction without further purification. By this protocol, fluorous tag-containing glucose acceptor **133** was obtained in 55% overall yield. Although the application to the glycosylation using this donor has not been reported, the "online" fluorous separation under microfluidic conditions is promising for the automated and combinatorial synthesis of oligosaccharides.

These leading examples will not only open a new way to high-throughput and combinatorial library synthesis, favorably by combination with the programmed

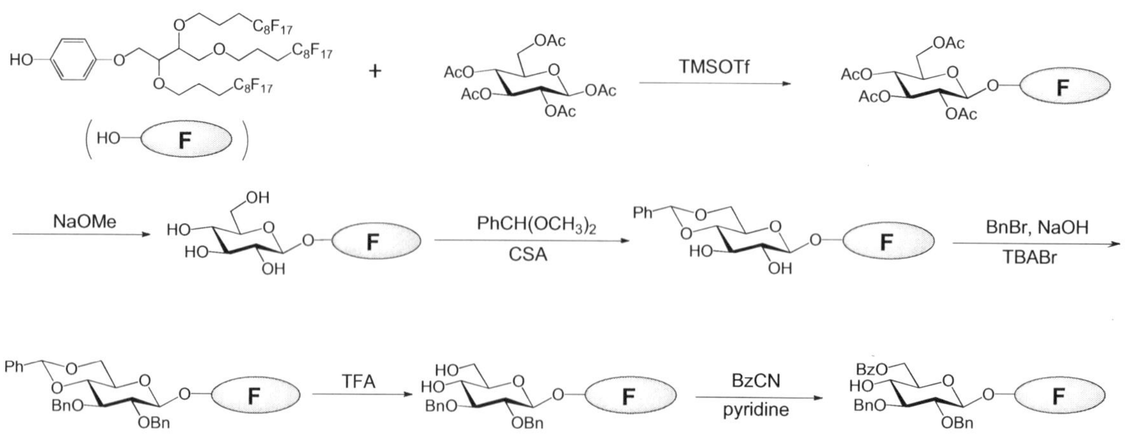

133

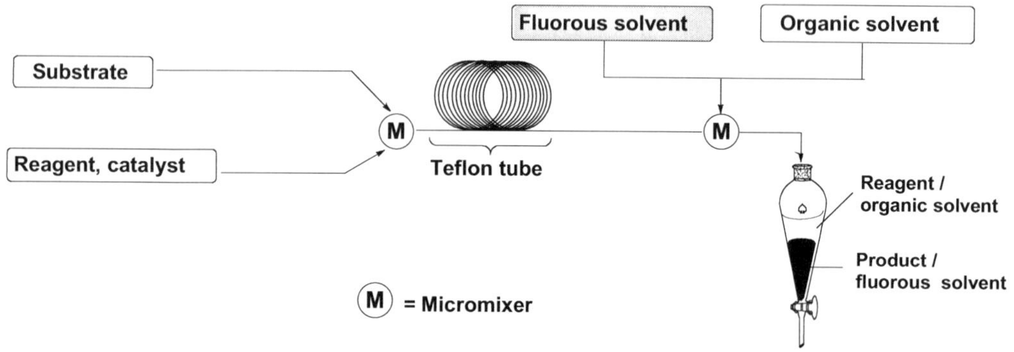

Scheme 16.35. Integration of microfluidic and "online" fluorous separation.

automation of glycosylation analysis sequences, but also provide a new insight into the reactivity of glycosylation. Once an oligosaccharide of biological relevance has been found from the microfluidic reactions, it will be provided on an industrial scale by the same apparatus.[106,107]

16.4 CONCLUSIONS

Synthetic oligosaccharides and glycoconjugates often play a decisive role in elucidation of their biofunctions, since chemical synthesis can provide homogeneous preparations without contamination of other bioactive compounds. For the clinical application, the use of synthetic specimens is also important to avoid side effects due to possible contamination. Although oligosaccharide synthesis on solid, soluble polymer, and tag supports has been dramatically developed as described, a high level of technical expertise is still required for the oligosaccharide synthesis. As the direct construction and assay techniques of the oligosaccharides library on the array,[110] as well as the utilization of the new surface platform for the iterative glycan synthesis,[111] have rapidly been emerging, further improvements in the oligosaccharide synthesis on the solid supports is expected to establish the general and efficient methods, which will speed up the elucidation of the biological functions of oligosaccharides as well as clinical applications of oligosaccharide-based drugs.

16.5 ACKNOWLEDGMENTS

We thank Yosuke Uchinashi, Takuya Miyagawa, and Dr. Guang-ming Bao for searching some references and drawing a part of the schemes. A part of the work described in this chapter was supported by Grants-in-Aid for Scientific Research No. 19681024 and 19651095 from the Japan Society for the Promotion of Science, Collaborative Development of Innovative Seeds from Japan Science and Technology Agency (JST), New Energy and Industrial Technology Development Organization (NEDO, project ID: 07A01014a), research grants from Yamada Science Foundation, as well as Molecular Imaging Research Program, Grants-in-Aid for Scientific Research from the Ministry of Education, Culture, Sports, Science and Technology (MEXT) of Japan.

REFERENCES

1. K. Fukase, Combinatorial and solid-phase methods in oligosaccharide synthesis, in Fraser-Reid, B. Tatsuta, K. Thiem, J. (Eds), *Glycoscience II*, Springer, Berlin, 2001, p. 1621.
2. X. Wu, M. Grathwohl, R. R. Schmidt, *Org. Lett.* **2001**, *3*, 747–750.
3. J. F. Tolborg, K. J. Jensen, *Chem. Commun.* **2000**, 147–148.
4. L. Petersen, K. J. Jensen, *J. Chem. Soc., Perkin Trans. 1* **2001**, 2175–2182.
5. N. Drinnan, M. L. West, M. Broadhurst, B. Kellam, I. Toth, *Tetrahedron. Lett.* **2001**, *42*, 1159–1162.
6. G. Belogi, T. Zhu, G. Boons, *Tetrahedron. Lett.* **2000**, *41*, 6965–6968.
7. G. Belogi, T. Zhu, G.-J. Boons, *Tetrahedron. Lett.* **2000**, *41*, 6969–6972.
8. L. G. Melean, W.-C. Haase, P. H. Seeberger, *Tetrahedron. Lett.* **2000**, *41*, 4329–4333.
9. J. L. de Paz, C. Noti, P. H. Seeberger, *J. Am. Chem. Soc.* **2006**, *128*, 2766–2767.
10. J. L. de Paz, P. H. Seeberger, *QSAR Comb. Sci.* **2006**, *25*, 1027–1032.
11. M. S. M. Timmer, J. D. C. Codee, H. S. Overkleeft, J. H. van Boom, G. A. van der Marel, *Synlett* **2004**, 2155–2158.
12. L. Knerr, R. R. Schmidt, *Eur. J. Org. Chem.* **2000**, 2803–2808.
13. M. Izumi, K. Fukase, S. Kusumoto, *Synlett* **2002**, 1409–1416.
14. I. Burt, T. Dean, S. Warriner, *Chem. Commun.* **2004**, 454–455.
15. A. Diaz-Moscoso, J. M. Benito, C. O. Mellet, J. M. G. Fernandez, *J. Comb. Chem.* **2007**, *9*, 339–342.
16. J. L. de Paz, M. Mar Kayser, G. Macchione, P. M. Nieto, *Carbohydr. Res.* **2010**, *345*, 565–571.
17. T. Amaya, H. Tanaka, T. Takahashi, *Synlett* **2004**, 497–502.
18. S. Komba, M. Kitaoka, T. Kasumi, *Eur. J. Org. Chem.* **2005**, 5313–5329.
19. S. Komba, T. Terauchi, S. Machida, *J. Appl. Glycosci.* **2009**, *56*, 193–206.
20. M. C. Parlato, M. N. Kamat, H. Wang, K. J. Stine, A. V. Demchenko, *J. Org. Chem.* **2008**, *73*, 1716–1725.
21. O. Kanie, F. Barresi, Y. Ding, J. Labbe, A. Otter, L. S. Forsberg, B. Ernst, O. Hindsgaul, *Angew. Chem., Int. Ed. Engl.* **1995**, *34*, 2720–2722.
22. Y. Ding, J. Labbe, O. Kanie, O. Hindsgaul, *Bioorg. Med. Chem.* **1996**, *4*, 683–692.
23. K. Agoston, L. Kroger, G. Dekany, J. Thiem, *J. Comb. Chem.* **2009**, *11*, 813–819.
24. K. Agoston, L. Kroger, A. Agoston, G. Dekany, J. Thiem, *Carbohydr. Res.* **2009**, *344*, 1428–1433.
25. T. J. Boltje, J.-H. Kim, J. Park, G.-J. Boons, *Nat. Chem.* **2010**, *2*, 552–557.

26. C. T. Tanifum, J. Zhang, C.-W. T. Chang, *Tetrahedron Lett.* **2010**, *51*, 4323–4327.
27. T. Takahashi, A. Okano, T. Amaya, H. Tanaka, T. Doi, *Synlett* **2002**, 911–914.
28. J. Lopez-Prados, C. Felix, N.-C. Reichardt, J.-L. de Paz, E. Q. Morales, J.-L. Martin-Lamos, *Org. Biomol. Chem.* **2005**, *3*, 764–786.
29. O. Kanie, I. Ohtsuka, T. Ako, S. Daikoku, Y. Kanie, R. Kato, *Angew. Chem., Int. Ed.* **2006**, *45*, 3851–3854.
30. F. Roussel, L. Knerr, R. R. Schmidt, *Eur. J. Org. Chem.* **2001**, 2067–2073.
31. F. Roussel, M. Takhi, R. R. Schmidt, *J. Org. Chem.* **2001**, *66*, 8540–8548.
32. S. Jonke, K.-G. Liu, R. R. Schmidt, *Chem. Eur. J.* **2006**, *12*, 1274–1290.
33. X. Wu, M. Grathwohl, R. R. Schmidt, *Angew. Chem., Int. Ed.* **2002**, *41*, 4489–4493.
34. X. Wu, R. R. Schmidt, *J. Org. Chem.* **2004**, *69*, 1853–1857.
35. F. Roussel, L. Knerr, M. Grathwohl, R. R. Schmidt, *Org. Lett.* **2000**, *2*, 3043–3046.
36. X. Wu, R. R. Schmidt, *Eur. J. Org. Chem.* **2004**, 2826–2832.
37. S. D. Markad, R. R. Schmidt, *Eur. J. Org. Chem.* **2009**, 5002–5011.
38. K. Tanaka, Y. Fujii, H. Tokimoto, Y. Mori, S. Tanaka, G.-M. Bao, E. R. O. Siwu, A. Nakayabu, K. Fukase, *Chem. Asian J.* **2009**, *4*, 574–580.
39. S. Tanaka, M. Takashina, H. Tokimoto, Y. Fujimoto, K. Tanaka, K. Fukase, *Synlett* **2005**, 2325–2328.
40. K. Tanaka, Y. Mori, K. Fukase, *J. Carbohydr. Chem.* **2009**, *28*, 1–11.
41. K. Tanaka, T. Goi, K. Fukase, *Synlett* **2005**, 2958–2962.
42. S. Tanaka, T. Goi, K. Tanaka, K. Fukase, *J. Carbohydr. Chem.* **2007**, *26*, 369–394.
43. M. C. Hewitt, P. H. Seeberger, *J. Org. Chem.* **2001**, *66*, 4233–4243.
44. K. R. Love, R. B. Andrade, P. H. Seeberger, *J. Org. Chem.* **2001**, *66*, 8165–8176.
45. O. J. Plante, E. R. Palmacci, R. B. Andrade, P. H. Seeberger, *J. Am. Chem. Soc.* **2001**, *123*, 9545–9554.
46. O. J. Plante, E. R. Palmacci, P. H. Seeberger, *Science* **2001**, *291*, 1523–1527.
47. A. Bartolozzi, P. H. Seeberger, *Curr. Opin. Struct. Biol.* **2001**, *11*, 587–592.
48. M. C. Hewitt, P. H. Seeberger, *Org. Lett.* **2001**, *3*, 3699–3702.
49. L. A. Marcaurelle, P. H. Seeberger, *Curr. Opin. Struct. Biol.* **2002**, *6*, 289–296.
50. E. R. Palmacci, O. J. Plante, M. C. Hewitt, P. H. Seeberger, *Helv. Chim. Acta* **2003**, *86*, 3975–3990.
51. P. H. Seeberger, *Chem. Commun.* **2003**, 1115–1121.
52. P. H. Seeberger, D. B. Werz, *Nature* **2005**, *4*, 751–763.
53. P. H. Seeberger, *Chem. Soc. Rev.* **2008**, *37*, 19–28.
54. M. C. Hewitt, D. A. Snyder, P. H. Seeberger, *J. Am. Chem. Soc.* **2002**, *124*, 13434–13436.
55. D. M. Ratner, E. R. Swanson, P. H. Seeberger, *Org. Lett.* **2003**, *5*, 4713–4720.
56. K. R. Love, P. H. Seeberger, *Angew. Chem., Int. Ed* **2004**, *43*, 602–603.
57. D. B. Werz, B. Castagner, P. H. Seeberger, *J. Am. Chem. Soc* **2007**, *129*, 2770–2771.
58. J. D. C. Codee, L. Krock, B. Castagner, P. H. Seeberger, *Chem. Eur. J.* **2008**, *14*, 3987–3994.
59. K. S. Kim, J. H. Kim, Y. J. Lee, Y. J. Lee, J. Park, *J. Am. Chem. Sci.* **2001**, *123*, 8477–8481.
60. K. Egusa, K. Fukase, Y. Nakai, S. Kusumoto, *Synlett* **2000**, 27–32.
61. K. Egusa, S. Kusumoto, K. Fukase, *Synlett* **2001**, 777–780.
62. M. Izumi, R. Okamoto, M. Sato, S. Nakajima, N. Baba, K. Fukase, *Chem. Lett.* **2008**, *37*, 1030–1031.
63. H. Ando, S. Manabe, Y. Nakahara, Y. Ito, *Angew. Chem., Int. Ed.* **2001**, *40*, 4725–4728.
64. S. Manabe, Y. Ito, *Chem. Pharm. Bull.* **2001**, *49*, 1234–1235.

65. S. Hanashima, S. Manabe, Y. Ito, *Synlett* **2003**, 979–982.

66. J. Wu, Z. Guo, *J. Org. Chem.* **2006**, *71*, 7067–7070.

67. H. Tanaka, T. Ishida, N. Matoba, H. Tsukamoto, H. Yamada, T. Takahashi, *Angew. Chem., Int. Ed.* **2006**, *45*, 6349–6352.

68. R. N. MacCoss, P. E. Brennan, S. V. Ley, *Org. Biomol. Chem.* **2003**, *1*, 2029–2031.

69. A. Dondoni, A. Marra, A. Massi, *Angew. Chem., Int. Ed.* **2005**, *44*, 1672–1676.

70. E. R. Palmacci, M. C. Hewitt, P. H. Seeberger, *Angew. Chem., Int. Ed.* **2001**, *40*, 4533–4437.

71. F. R. Carrel, P. H. Seeberger, *J. Org. Chem.* **2008**, *73*, 2058–2065.

72. T. Kanemitsu, C.-H. Wong, O. Kanie, *J. Am. Chem. Soc.* **2002**, *124*, 3591–3599.

73. M. Mogemark, M. Elofsson, J. Kihlberg, *Org. Lett.* **2001**, *3*, 1463–1466.

74. M. Mogemark, M. Elofsson, J. Kihlberg, *ChemBioChem* **2002**, *12*, 1266–1269.

75. M. Mogemark, M. Elofsson, J. Kihlberg, *J. Org. Chem.* **2003**, *68*, 7281–7288.

76. S. Manabe, Y. Ito, *J. Am. Chem. Soc.* **2002**, *124*, 12638–12639.

77. K.-S. Ko, G. Park, Y. Yu, N. L. Pohl, *Org. Lett.* **2008**, *10*, 5381–5384.

78. R. Geurtsen, G.-J. Boons, *Eur. J. Org. Chem.* **2002**, 1473–1477.

79. B. Quiclet-Sire, A. Wilczewska, S. Z. Zard, *Tetrahedron Lett.* **2000**, *41*, 5673–5677.

80. R. Ojeda, J.-L. de Paz, M. Martin-Lomas, *Chem. Commun.* **2003**, 2486–2487.

81. R. Ojeda, O. Terenti, J.-L. de Paz, M. Martin-Lomas, *Glycoconjugate J.* **2004**, *21*, 179–195.

82. S. Manabe, Y. Nakahara, Y. Ito, *Synlett* **2000**, 1241–1244.

83. M. Oikawa, T. Tanaka, S. Kusumoto, M. Sasaki, *Tetrahedron Lett.* **2004**, *45*, 787–790.

84. L. Manzoni, *Chem. Commun.* **2003**, 2930–2931.

85. T. Miura, K. Goto, D. Hosaka, T. Inazu, *Angew. Chem., Int. Ed.* **2003**, *42*, 2047–2051.

86. T. Miura, T. Inazu, *Tetrahedron Lett.* **2003**, *44*, 1819–1821.

87. B. Yang, Y. Jing, X. Huang, *Eur. J. Org. Chem.* **2010**, 1290–1298.

88. G. Park, K.-S. Ko, A. Zakharova, N. L. Pohl, *J. Fluor. Chem.* **2008**, *129*, 978–982.

89. S.-Q. Zhang, K. Fukase, S. Kusumoto, *Tetrahedron Lett.* **1999**, *40*, 7479.

90. S.-Q. Zhang, K. Fukase, M. Izumi, Y. Fukase, S. Kusumoto, *Synlett* **2001**, 590.

91. Y. Fukase, S.-Q. Zhang, K. Iseki, M. Oikawa, K. Fukase, S. Kusumoto, *Synlett* **2001**, 1693.

92. K. Fukase, M. Takashina, Y. Hori, D. Tanaka, K. Tanaka, S. Kusumoto, *Synlett* **2005**, 2342–2346.

93. J. Bauer, J. Rademann, *J. Am. Chem. Soc.* **2005**, *127*, 7296–7297.

94. J. Bauer, K. Brandenburg, U. Zahringer, J. Rademann, *Chem. Eur. J.* **2006**, *12*, 7116–7124.

95. S. L. Flitsch, *Curr. Opin. Chem. Biol.* **2000**, *4*, 619–625.

96. S. Nishiguchi, K. Yamada, Y. Fuji, S. Shibatani, A. Toda, S. Nishimura, *Chem. Commun.* **2001**, 1944 1945.

97. N. Bezay, G. Dudziak, A. Liese, H. Kunz, *Angew. Chem., Int. Ed.* **2001**, *40*, 2292–2295.

98. K. Haneda, T. Inazu, M. Mizuno, R. Iguchi, H. Tanabe, K. Fujimori, K. Yamamoto, H. Kumagai, K. Tsumori, E. Munekata, *Biochim. Biophys. Acta* **2001**, *1526*, 242–248.

99. J. F. Tolborg, L. Peterson, K. J. Jensen, C. Mayer, D. L. Jakeman, R. A. J. Warren, S. G. Withers, *J. Org. Chem.* **2002**, *67*, 4143–4149.

100. K. Naruchi, T. Hamamoto, M. Kurogochi, H. Hinou, H. Shimizu, T. Matsushita, N. Fujitani, H. Kondo, S.-I. Nishimura, *J. Org. Chem.* **2006**, *71*, 9609–9621.

101. M. Fumoto, H. Hinou, T. Ohta, T. Ito, K. Yamada, A. Takimoto, H. Kondo, H. Shimizu, T. Inazu, Y. Nakahara, S.-I. Nishimura, *J. Am. Chem. Soc.* **2005**, *127*, 11804–11818.

102. X. Huang, K. L. Witte, D. E. Bergbreiter, C.-H. Wong, *Adv. Synth. Catal.* **2001**, *343*, 675–681.

103. D. M. Ratner, E. R. Murphy, M. Jhunjhhunwala, D. A. Snyder, K. F. Jensen, P. H. Seeberger, *Chem. Commun.* **2005**, 578–580.

104. K. Tanaka, T. Miyagawa, K. Fukase, *Synlett* **2009**, 1571–1574.

105. K. Tanaka, K. Fukase, *Synlett* **2007**, 164–166.

106. K. Tanaka, K. Fukase, *Org. Process Res. Dev.* **2009**, *13*, 983–990.

107. K. Tanaka, K. Fukase, *Beilstein J. Org. Chem.* **2009**, *5*, No. 40.

108. F. R. Carrel, K. Geyer, J. D. C. Codee, P. H. Seeberger, *Org. Lett.* **2007**, *9*, 2285–2288.

109. H. Kawakami, K. Goto, M. Mizuno, *Chem. Lett.* **2009**, *9*, 906–907.

110. L. Ban, M. Mrksich, *Angew. Chem., Int. Ed* **2008**, *47*, 3396–3399.

111. P. Pornsuriyasak, S. C. Ranade, A. Li, M. C. Parlato, C. R. Sims, O. V. Shulga, K. J. Stine, A. V. Demchenko, *Chem. Commun.* **2009**, 1834–1836.

INDEX

Solid-Phase Organic Synthesis: Concepts, Strategies, and Applications, First Edition.
Edited by Patrick H. Toy and Yulin Lam.